culture

Working
with

REFER

THE NATURAL SELECTION OF THE CHEMICAL ELEMENTS

The Natural Selection of the Chemical Elements

The Environment and Life's Chemistry

R. J. P. WILLIAMS
Emeritus Research Professor,
University of Oxford

and

J. J. R. FRAÚSTO da SILVA
Professor of Analytical Chemistry, Instituto Superior Técnico,
Universidade Técnica de Lisboa

CLARENDON PRESS · OXFORD

Oxford University Press, Great Clarendon Street, Oxford OX2 6DP

Oxford New York
Athens Auckland Bangkok Bogota Bombay Buenos Aires
Calcutta Cape Town Dar es Salaam Delhi Florence Hong Kong
Istanbul Karachi Kuala Lumpur Madras Madrid Melbourne
Mexico City Nairobi Paris Singapore Taipei Tokyo Toronto
and associated companies in
Berlin Ibadan

Oxford is a trade mark of Oxford University Press

Published in the United States
by Oxford University Press Inc., New York

© R. J. P. Williams and J. J. R. Fraústo da Silva, 1996

First printed 1996
First published in paperback 1997

A catalogue record for this book is available from the British Library

Library of Congress Cataloging-in-Publication Data

Williams, R. J. P. (Robert Joseph Paton)
The natural selection of the chemical elements: the environment and
life's chemistry/R. J. P. Williams and J. J. R. Fraústo da Silva.
Includes bibliographical references and index.
1. Chemistry.   I. Silva, J. J. R. Fraústo da.   II. Title.
QD31.2.W57   1995   540–dc20   95-24776
ISBN 0 19 855843 0 (Hbk)
*   0 19 855842 2 (Pbk)*

Printed in Great Britain by
Bath Press Ltd
Bath

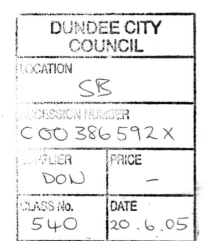

In my hunt for the secret of life, I started my research in histology. Unsatisfied by the information that cellular morphology could give me about life, I turned to physiology. Finding physiology too complex, I took up pharmacology. Still finding the situation too complicated, I turned to bacteriology. But bacteria were even too complex, so, I descended to the molecular level, studying chemistry and physical chemistry. After twenty years' work, I was led to conclude that to understand life we have to descend to the electronic level and to the world of wave mechanics. But electrons are just electrons and have no life at all. Evidently on the way I lost life; it had run out between my fingers.

ALBERT SZENT-GYÖRGYI
Personal Reminiscences

Preface

The main objective of this book is to show the relationship of every kind of material around us, living and non-living, to the properties of the chemical elements of the periodic table. The book must bring together, therefore, all the relevant subject matters that are based on chemistry, including not only inorganic and organic chemistry (and their principles, usually separated under the heading physical chemistry), but also geochemistry, environmental chemistry and the chemistry of living organisms, biological chemistry. We shall demonstrate that, through the elements, all these last three chemistries, which in themselves are specific and selected (see below), interact with man's industrial inorganic and organic chemistry. To develop this discussion we first stress the ways in which chemical elements bind together by describing the underlying principles of chemical affinity related to the electronic configuration of their atoms. This is one mode of natural selection. (An earlier mode of selection was the formation of the nuclei of the elements themselves, which also obeys 'natural' laws.) We stress next the way in which a physical chemist describes the appearance of order and disorder in the three main states of matter—solid, liquid and gas. Within these broad physical divisions the different types of non-biological chemical combinations of elements are placed in a selective arrangement forced upon them by binding and based on two determining influences. The first is thermodynamic stability, generating equilibrium, a principle much used in laboratory and industrial inorganic chemistry and to some extent in geochemistry. The second is kinetic stability, by which combinations of elements are held (trapped) in thermodynamically unstable conditions for long periods; the obvious examples are organic chemicals, but we again find relevant cases in geochemistry. There is a tendency among chemists to treat the combinations of elements by using bond theories, which often forces consideration of stoichiometric association in molecules or crystals, rather than by discussing phases first in which stoichiometry is far from generally observed. A point of departure in this book is that the physicists' approach to materials through *phase structure* is strongly stressed so that the physical states of matter are analysed using the phase rule in both its simplest form and in more complicated ways in which the sizes of phases and the fields of forces to which they are exposed are taken into account together with the binding tendencies for formation of chemically selected, *co-operative* condensed states.

A common feature of the above considerations of selectivity is that the compounds formed are treated as being in time-independent stationary states that do not require a constant supply of energy or material to maintain them, that is they can be looked upon as being permanently in the shape of molecules and/or crystals.

There is a third mode of element selection that uses both of the above-mentioned driving principles to some degree only, namely the selection of chemical elements for functional value to an organisation such as a living system. This selection mode is not so obviously restricted by factors such as strength of element bonding, molecular structure or physical phase equilibria, although all are involved, nor is it based on kinetically trapped stability alone, even though kinetics is important for organised systems of chemicals. The new feature, so apparent in living systems, which are organised rather than ordered, is the drive for *survival* of the system based on continuous turnover or flow of both chemicals and energy. This is a steady state, not a stationary condition. A living cell is of this kind, as is a whole plant or a whole animal. Such living organisation is analogous to other non-living organisations, for example the flow of water in geophysical features such as rivers, the movement of planets around the sun or the progress of a continuous process in industry. This third mode of element selection is *natural* in the same way as the two previous modes are natural, that is, they all arose spontaneously from underlying laws of nature based upon the properties of atoms. It will be shown to be related to the Darwinian selection of species, although the reasons for survival may differ. By giving this book the title *The natural selection of the chemical elements* we wish, therefore, to encompass all the existing activities in the classes of chemical systems we have now mentioned.

To present the nature of substances around us, including those in organisations, within the context of conventional chemistry we need to outline the principles of all the factors that come into play. We give, therefore, what is at first an elementary, perhaps oversimplified, treatment not only of structure but also of thermodynamics and kinetics in the opening chapters of the book. These chapters concern structure and the balance of order and disorder of elements in physical states and in chemical compounds, including the energetics of chemical systems and the kinetics of change. Much of the presentation of these principles, which constitutes Part I of the book, is orthodox, but we stress observables and less traditional aspects and approaches to the discussion of the materials around us while we avoid the description of the theories and the tools by which the facts have been discovered. In particular, we have avoided any lengthy description of binding theory, of thermodynamics and of the principles of techniques such as spectroscopy and magnetism. This allows us to describe the properties of matter with a minimal use of mathematics. The approach we use also differs from standard texts in that we wish to include the full range of observed materials. For example, when discussing structure we have to consider polymer molecules as well as small molecules and atoms since the polymers pack in a different way in non-repeating lattices. Although much of the treatment of kinetics is orthodox, we have again considered it convenient to make some departures from traditional treatments. One is to describe flow and physical barriers to flow, and the other is to describe steady states of both physical flow and chemical reactions in terms of feedback circuits. In these considerations sources of energy are important and are given due consideration.

Given the outline of principles in Part I, Part II tackles the material world, illustrating the selective development of chemical elements in compounds and in particular physical states in three different ways. The first is the description of the evolution of Earth, largely restricted to the selective evolution of inorganic chemicals before there was life. This description is, in essence, a treatment of competition between elements for partners in conditions of equilibrium, but we show that progressive cooling in the universe and then on Earth itself has determined that some inorganic chemicals remained out-of-equilibrium in kinetic traps. The second is an analysis of the evolution of organic compounds using examples from abiotic or bioorganic chemistry to discuss the early organic chemicals (on Earth) found in living systems (but not their organisation) and the ways in which they are made. Here selecting of stationary states only arises and we note the distribution of such chemicals in different phases, namely, gas, liquid and solid. The third complex evolutionary pattern is that of organised biological chemistry. In its description we touch very briefly on the possible origin of life, but our real purpose is to show how and why life has evolved into such complicated organisms, separated into species. The foremost feature to have in mind is that the selection of the elements in any organism is now a 'holistic' compromise so as to optimise the continuously operating whole. Consequently, no chemical is to be treated except within the concept of the total flow in a given steady state, and energy capture too must be knitted into all processing. Thus, selection is for *fitness* and chemical selection becomes a part of Darwinian selection of the fittest, but it can only operate within the limitations imposed by the electronic structures of the elements. Obviously, the fitness concept concerns not just chemical properties but the physical properties of states of matter (microphases) and their dynamic organisation (steady states).

Of course, steady states of flowing matter, which also require and dissipate energy, have an output that accumulates and therefore backinteracts with the existing activities. The changes that eventually occur in different occupied niches force the evolution of different living organisms due to the stresses imposed on them by the environment (which includes the chemical outputs of the organisms themselves). In effect, while it is now clear that organisms may drive in part the environment (and did so strongly in the past), it is also clear that environmental changes, no matter how they came about, will also drive biological evolution. This is a major theme of the book, especially as today man's industrial activity is beginning to affect the environment.

There is, then, a fourth stage of 'natural' selection which is that of man's industry and its development for his functional needs and survival. Here we observe that, even though life's evolution was based on change of ambient reducing conditions to more oxidising conditions at fixed low temperature and using a fixed set of some 15–20 elements, industry depends on using practically all of the elements of the periodic table and more reducing conditions at ever higher temperature to generate a new range of chemicals. It follows that, since these two processes of making chemicals, biological and industrial, are substantially different, we must ask how they will affect one another when they become of comparable magnitude. Man is not on his own in this concern as we shall show. In effect, if we want to preserve the present network of life rather than to assist the development of some other, then it is quite possible that we shall have to abandon careless exploitation not just of natural fuels but of many mineral resources. In turn man's life will have to be seen in the near future, say, in the next century, in terms of a more settled rather than as a progressive materialistic activity. Such a change of attitude will force a change of style. The centre pin of education in order to evaluate whether such a change of attitude is necessary must be a proper appreciation of the chemistry of the world in which we live, for it is very much a chemical world, and of the chemistry of living systems. This involves a respect not only for living forms but also for the whole of our material ('dead') world. Hence, just as we accept today a duty to look after 'living' things so we may have to accept a duty to look after 'dead' material.

In the course of this book we attempt to explain this view, which differs radically from the view of nature held by man for many thousands of years, and even today, which gave him a higher level of priority and justified his 'right' to exploit endlessly. It is appropriate to start the book, therefore, with earlier formulations of ideas concerning the material world and then to describe how they have evolved. In the last chapter we attempt an appraisal of the limitations on man's exploitation of chemistry that compatibility with biological chemistry may demand, and come again to the earlier formulations of ideas contrasting them with today's formulations so as to show how the present knowledge of chemistry forces a continuous re-evaluation of man in nature. We cannot be sure that today's paradigms will be valid tomorrow.

In this book it is our intention to give science graduates and especially teachers an opportunity to appreciate the involvement of chemistry within everything that is around us—the world we live in, the immense universe still inaccessible to our understanding, life itself—in a unified approach. It is obviously complementary reading, not an undergraduate text, although we would like it to contribute to a reconsideration of the content of general introductory chemistry courses in which fundamental aspects and ideas are still conspicuously absent. In a period when the impact of man's activities on the environment is giving reasons for concern it is imperative to make society aware of the importance that a better understanding of chemistry has for the future of mankind. This must begin and be consolidated in the educational institutions. If this book contributes in any way to such awareness then it will have served its purpose.

Oxford and Lisbon   RJPW
October 1995   JJRFdS

Acknowledgements

Our special thanks go to Mrs S. Compton who typed the manuscript and to Oxford University Press for their skilful assistance in its final preparation.

The work reviewed here reflects in large part the efforts of a very distinguished group of research scientists from all over the world, who have exchanged views with us. We wish to thank them for all that they brought to our knowledge. We trust that the integrated view in this book will be seen as the result of an international effort to examine a further part of the wonders of chemistry within geology, biology and man's industry.

J. J. R. Fraústo da Silva acknowledges a grant for the preparation of this book from the Junta Nacional de Investigação Cientifica e Tecnologica (JNICT)—Portugal. Acknowledgements are also due to Dr João de Deus Ramos for guidance to the literature of ancient Chinese philosophy and to Mrs Teresa Maria Carreiras da Silva and Mrs Maria Idalisa Figueiredo dos Santos for invaluable secretarial help.

R. J. P. Williams acknowledges the generous support of Wadham College, Oxford, of Oxford University, of two British Research Councils (the Medical Research Council and the Science and Engineering Research Council), and of The Royal Society over some forty years.

We have both benefited from our association with the Inorganic Chemistry Laboratory at Oxford over very many years.

Throughout the book every effort has been made to acknowledge sources of illustrations and permissions are given in the figure caption text. In addition to the material given in the main text, the publishers would like to add the following information:

Fig. 5.21	Reprinted with permission of John Wiley & Sons, Inc.
Fig. 7.1	From *The shadows of creation* by Riordan and Schramm. © 1991 Michael Riordan and David N. Schramm. Used with permission of W. H. Freeman and Company.
Figs 7.23, 7.24, 7.29, and 13.15	From *Biology*, Second edition by Claude A. Villee. © 1989 by Saunders College Publishing. Reprinted by permission of the publisher.
Fig. 11.31	From *Five kingdoms*, Second edition by Margulis and Schwartz. © 1988 by W. H. Freeman and Company. Used with permission.
Figs 12.2, 15.3, 15.4, and 15.5	From *Earth*, Fourth edition by Press and Siever. © 1986 by W. H. Freeman and Company. Used with permission.
Fig. 12.25	From *Invertebrate zoology*, Fifth edition by Robert D. Barnes. © 1987 by Saunders College Publishing. Reproduced by permission of the publisher.
Fig. 14.9	From *Understanding Earth* by Press and Siever. © 1994 by W. H. Freeman and Company. Used with permission.
Fig. 15.7	From *Inventeurs et Decouvertes* by Albert Bettex. Librairie Hachette, Paris.

Every effort has been made to identify and contact copyright holders and obtain their permission to reproduce illustrations. If any have been overlooked inadvertently the publishers will be happy to rectify any omissions in subsequent editions of this work.

Contents

Part II · The observed natural selection of chemical elements in both abiotic and biotic systems during their evolution

Units of Energy and Work and the Values of Some Physical Constants

The joule, SI unit of energy

$$1J = 1 \text{ kg m}^2 \text{ s}^{-2}$$
$$= 1 \text{ N m (newton meter)}$$
$$= 1 \text{ W s (watt second)}$$
$$= 1 \text{ C V (coulomb volt)}$$
$$= 0.24 \text{ cal (thermochemical calorie)}$$
$$= 6.242 \times 10^{18} \text{ eV (electron-volt)}$$

Thermochemical calorie

$$1 \text{ cal} = 4.184 \text{ J}$$

Large calorie

$$1 \text{ Cal} = 1 \text{ kcal} = 4.184 \text{ kJ}$$

Electron-volt

$$1 \text{ eV} = 1.602 \times 10^{-19} \text{ J (joule)}$$

Work required to raise 1 kg 1 m on earth

$$(\text{at sea level}) = 9.807 \text{ J}$$

Free energy of hydrolysis of 1 mol of ATP

at pH 7, millimolar concentrations $= -12.48 \text{ kcal} = -52.2 \text{ kJ}$

Work required to concentrate 1 mole of a substance

1000-fold, e.g., from 10^{-6} to 10^{-3} M
$$= 4.09 \text{ kcal} = 17.1 \text{ kJ}$$

Avogadro's number, the number of particles in a mole

$$N = 6.0220 \times 10^{23}$$

Faraday $1F = 96{,}485 \text{ C mol}^{-1} \text{ (coulombs per mole)}$

Coulomb $1C = 1 \text{ A s (ampere second)}$
$$= 6.241 \times 10^{18} \text{ electronic charges}$$

Electronic charge $1e = 1.602 \times 10^{-19} \text{ C (coulomb)}$

The Planck constant $h = 6.626 \times 10^{-34} \text{ J s (Joule second)}$

The Boltzmann constant

$$k_B = 1.3807 \times 10^{-23} \text{ J deg}^{-1}$$

The gas constant, $R = N \, k_B$

$$R = 8.3144 \text{ J deg}^{-1} \text{ mol}^{-1}$$
$$= 1.9872 \text{ cal deg}^{-1} \text{ mol}^{-1}$$
$$= 0.08206 \text{ 1 atm deg}^{-1} \text{ mol}^{-1}$$

and at $25°$ $RT = 2.479 \text{ kJ mol}^{-1}$

The unit of temperature is °K (or simply K); $0°C = 273.16°K$

Speed of light in vacuum

$$c = 3.0 \times 10^8 \text{ m s}^{-1} \text{ (metres per second)}$$

Mass of the electron

$$m_e = 9.10956 \times 10^{-31} \text{ kg}$$
$$= 0.511 \text{ MeV (mega electron-volt)}$$

Mass of the proton

$$m_p = 1.67261 \times 10^{-27} \text{ kg}$$
$$= 939.5 \text{ MeV (mega electron-volt)}$$

Mass of the neutron

$$m_n = 1.675 \times 10^{-27} \text{ kg}$$
$$= 939.5 \text{ MeV (mega electron-volt)}$$

Wave number unit, cm^{-1} (waves per centimetre)

$$1 \ cm^{-1} = 1.9862 \times 10^{-23} \ J \ molecule^{-1} \ (\text{joules per molecule})$$

$$= 2.8593 \ cal \ mol^{-1} \ (\text{thermochemical calories per mole})$$

Frequency unit, s^{-1} (cycles per second)

Part I

The principles of the natural selection of chemical elements into physical states and chemical combinations

1

The development of man's ideas concerning nature

The Intellect: 'Apparently there is colour, apparently sweetness, apparently bitterness, actually there are only atoms and the void'.
The Senses: 'Poor Intellect, do you hope to defeat us, while from us you borrow your very evidence?'

Democritus (*c.* 420 BC)

1.1 The early views

Ever since man became self-conscious he has tried to build up knowledge about himself and his relationship with his surroundings, even with the universe. One approach through chemistry and the associated physics of chemicals, which we will use in this book, has existed for thousands of years in developing form. It is based on the concept that all that we observe can be broken down into basic primary components, of which there are a limited number. The

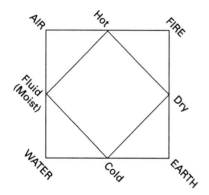

Fig. 1.1 The four primary manifestations of underlying 'form', as four entities or 'elements' (in upper case letters) connected by qualities.

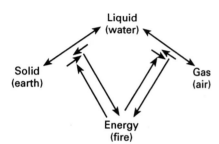

Fig. 1.2 A modern view of the four 'elements'.

primary components, as we shall see, were not thought to be material atoms† in early views, such as those of the most prominent Greek philosophers, for example, Aristotle (384–322 BC), but often incorporated vague concepts such as 'earth' or 'fire' (Fig. 1.1). One of the early underlying difficulties then was the attempt to include all observables in one system of very few separate concepts, only four, an approach that lasted in various guises for nearly 2000 years and that is worth describing briefly not only so as to compare it with the approach we all adopt today but because it reveals some of the fundamental difficulties facing our description of the natural world to which we shall return in the last chapter.

According to Aristotle, the basis of the material world was a primitive matter that had, however, only potential existence until it was imposed upon by 'form'. By 'form' he did not mean shape only, but all that conferred upon a body its specific properties. (Today we might say that there exists a set of elementary particles that define all matter and space and are held together by 'fields of force'.) Thus, underlying this thinking is the concept of some deeper holistic system. In its simple manifestations, form gave rise to the four primary entities (or 'elements'), distinguished by their pairs of 'qualities'. The examination of the world led Empedocles and then Aristotle and his followers to suppose that these primary entities (or 'elements') were air, fire, water and earth, each with some pair of general 'qualities', for example hot and dry (FIRE) or cold and fluid (moist) (WATER), one of which was predominant over the other: in earth, dryness; in water, coldness; in air, moisture; in fire, heat (see Fig. 1.1). None of the four entities was considered to be unchangeable; they might pass into one another through the medium of the 'quality' that they possess in common. To account for the substance and the brightness of the bodies 'floating' in the sky Aristotle admitted a fifth element, pure, eternal and incorruptible, which Plato had called ether,‡ a subtler kind of air. The history of these ideas is readily accessible (see references 1–4 in 'Further reading') and the reasoning behind them, as well as the choices of primary entities, is clear enough and certainly not naïve given the state of knowledge at the time.

We stress now only one point. The description in these terms contains the three physical states (classes of phases) of matter—gas, liquid and solid, that is air, water and earth, that can be changed into one another—and what we recognise now as energy (fire). It is worth noting immediately that this division of the observable features of the environment is stated in terms of physical rather than today's chemical terminology. We can rewrite the diagram of the Greek philosophers as in Fig. 1.2. Written in this way it is an extremely attractive division even today (see Chapter 4). It so happens that, if we still believed that (under the influence of form) matter is continuously adjustable within the limitation of three physical states of matter, solid/liquid/gas, and not further separable into inviolate chemical components, then the above division could be said to be a correct description of material and its changes in

† As a matter of fact 'atoms' were suggested as the ultimate indivisible pieces of all matter by other early philosophers, such as Democritus (*c.* 470–*c.* 380 BC), but their ideas did not gain wide acceptance although they were revived later by Epicurus (*c.* 342–*c.* 270 BC) and his followers and much praised and elaborated by the poet Lucretius (*c.* 95–*c.* 55 BC) in his long poem *De rerum natura*.

‡ From this fifth element derives the expression 'quinta essentia', referring to the highest possible quality or purity.

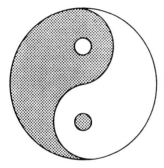

Fig. 1.3 The *Ta-chi* symbol of balance between opposite tendencies. Consider for example order and disorder (Section 1.8 and Chapter 3) and pairs of waves of opposed phase (Section 1.5).

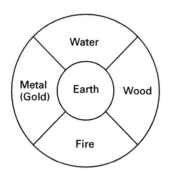

Fig. 1.4 The five 'elements', 'agents' or 'movers' (*wu-hsing*), which in Chinese philosophy are not material but more like five fundamental stages of any process in space–time, and which have been associated with spatial directions (north, south, east and west—with the earth in the centre), seasons of the year, musical notes, organs of the body and even with five human senses or attitudes, as well as with 'chemical' materials.

physical if not in chemical terms. In effect, today we express the physical relationships in a simple phase diagram of a one-component system (no chemical distinctions), as shown in Fig. 4.1. What is more, the phases, of water for example, can be changed into one another by a supply of energy, which, in a general sense, is to be likened to 'fire', (Chapter 3).

The Greeks had no chemistry, but it would be quite wrong to push the idea of a purely philosophical (and physical) attitude in the Greek world too far—they were looking for cosmological explanations but also for explanations of the 'chemical' behaviour of materials on Earth and their possible applications (see footnote † on p. 4). Of course, they were not alone in this search—thoughtful man everywhere has always puzzled in this way about his environment. For example, a parallel set of ideas developed in China, but here there is a greater difficulty in being sure of one's ground. The basic concepts of the Chinese approach to the nature of the universe and the material world are contained in the *I Ching* (book of changes) ascribed to Wen Wang (*c.* 1200 BC) and in the *Shu-Ching* (book of stories) of the Chou dynasty (722–211 BC). The first of these introduces the *yin* and *yang* principles (combined in the *Ta-chi* symbol (Fig. 1.3)), regarded as representing opposed but complementary cosmic forces which originated all things that exist. (Attraction and repulsion would be one such pair.) This is a holistic concept. The second book refers to the 'five things' or 'five movers' (notice the idea of flow in 'movers' and the 'magic number' five) of which everything is composed—water, fire, wood, metal and earth—which might change into one another in a continuous and permanent cyclic manner (see Fig. 1.4). There is here an analytical approach parallel to that of the Greeks, but a product from a living system, wood, and one from minerals, metal, are included. (It so happens that the separation of wood from metal is a *chemical* separation within solids or 'earth'.) These concepts were integrated into the Taoist philosophy, which originated in the writings of Lao-Tzu (sixth century BC), dealing mainly with ethics and social and political reform, and were later elaborated and extended by Chuang-Tzu (fourth century BC), and who was more concerned with the Universe and the material world of observation or experience. While Lao-Tzu's emphasis is on 'permanence', 'spontaneity' and 'eternity', although the idea of constant 'flow' is present, Chuang-Tzu's emphasis is on *cyclic* 'change' within unity or oneness. (Note that flow and change introduce *time* as of the essence of substance, being absolutely interlocked with matter.) It is possible that the major contributor to the ideas expressed in the *Shu-Ching* (book of stories) was Tsou Yen (fourth century BC), who was possibly the real founder of all Chinese scientific thought and the first notable Chinese alchemist. He divided the material world into the above-mentioned five elements, but at the same time he believed that there was a sequence of change amongst them all from which the designation of 'five movers' derived. (The ideas were incorporated even into dynastic relationships, that is, periods of society dominated by the spirit of fire (*yang* characteristic) would be followed by a period under the influence of water (*yin* characteristic).) The sense of constant cyclic change, frequently embedded in Chinese thought and expressed in the paradigmatic *Ta-chi* symbol (Fig. 1.3), is connected to the ideas of the flowing of material (see Fig. 1.5) from one shape to another and from one place to another, always in balance and always returning on itself. We shall have occasion to return to these kinetic concepts from Chapter 7 onwards. Note also references 1–7 in 'Further reading'.

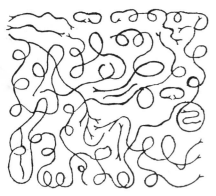

Fig. 1.5 Taoist diagram of 'change', representing the permanent flow and transformation inherent to the physical world (reproduced from *Tao Tsang*, a Taoist canon). The representation is of turbulence or *chaotic* motion as an unavoidable state of being. Note the sense of direction implied by arrows (accidentally?). NB. Turbulence in the universe preceded organisation.

Fig. 1.6 Simplified coat of arms of Niels Bohr, in *Niels Bohr*, edited by S. Rozental, North Holland Publishing Co, Amsterdam, 1967.

In fact, implicit in both these early analyses, Greek and Chinese, is the fundamental idea that a reductionist approach must have an underlying holistic system associated with the notion of cyclic change. This ancient set of ideas is not so different from the search by physicists today for a unifying theory of forces despite the obvious separation from human experience based on our senses.† Put in another way, if we want to understand the material world around us, its diversity but yet its oneness, we have to explain everything from first principles. Hence we must start from the 'big bang', which generated a time–space linked set of observables, and we must follow change to the present day. A considerable part of this description of what is around us concerns the changing selection of chemical elements with time observed in the universe, on Earth and in life, which is to be discussed in this book. We shall not hesitate to look at this chemistry from a holistic as well as a reductionist point of view.

This brief and necessarily incomplete description of the Greek and Chinese approaches to the interpretation of the universe may hopefully have shown their similarities and differences—the first, which influenced occidental culture, is more rational, abstract, intellect-based; the second, which impregnated oriental culture and the way of living, is more naturalistic, concrete, experience-based. Curiously, the first put the emphasis on the static composition of matter and the other on the dynamic character of its transformations, hence on flow, though clearly both features are present in some writings within either culture. Thus, it should be added that about the time (sixth century BC) that the Chinese developed their approach to nature, elaborating their concepts of permanence, spontaneity and flow, a Greek philosopher, Heraclitus of Ephesus, proposed 'fire' as a primordial 'element' and advocated a similar idea of a changing world, of eternal becoming, well expressed in his sentence 'everything flows'. For him too, all changes in the world derived from the dynamic and cyclic conjugation of opposed pairs, which, however, formed a unity containing and transcending them—the *Logos*. How curious it is then to observe that the most prominent Greek philosophers rejected two major ideas (of today) that other less prominent colleagues—Democritus and Heraclitus—had advanced: atoms and flow. Perhaps even more curious is to observe that, odd as they seem now to us, the concepts of Aristotle and his followers lasted for at least 2000 years without significant philosophical improvement. (We must always be aware that we may have fallen into a similar paradigm trap.)

The reason for this was that, all through this period, the role of man in the material world was regarded as a distinctly superior one, in which he bore, through his self-conscious existence, a special relationship to the underlying 'forces' (or form) in the universe. Some, in the tradition of Plato, believed in a rational world and in the power of rational thinking which, without recourse to experience (despised by many of the Greek philosophers and their cultural inheritors to this day), could uncover the fundamental principles and patterns of Nature; others sought life's meaning in religions, which allowed man contact with God, the omniscient and omnipotent Prime Mover, renouncing any search for other necessarily 'blasphemous' explanations. Both of these attitudes inhibited development of an empirical scientific appreciation of

† It is revealing to notice that Niels Bohr used the *Ta-chi* symbol in his coat of arms (see Fig. 1.6) together with the motto '*Contraria sunt complementa*', possibly referring to the wave–particle dualism.

nature, no matter what other 'truth' they may contain. We shall mention these early views now and then in the rest of the book, but we only really return to them in the last chapter.

A radical change in attitude caused by the development of the experimental approach, which became of every-increasing practical importance, put an end to simple faith in intellect-based rational or mystical arguments. In the experimental approach no final system is *a priori* supposed to lie beneath the world observable through our *senses*, while models, which by their nature are reductionist, can be proposed provided they are consistent with observation. Observation now includes study with the help of powerful instruments external to man. However, we must never forget that behind this approach there remains the ever-present wish and struggle to describe the objects and activities we observe within a holistic view that we shall see has to contain a description of the very factors the ancient thinkers uncovered.

1.2 The development of modern views

Some elements and their symbols

Element	Symbol
aluminium	Al
argon	Ar
carbon	C
chlorine	Cl
copper	Cu (from cuprum) *
gold	Au (from aurum) *
helium	He
hydrogen	H
iodine	I
iron	Fe (from ferrum) *
lead	Pb (from plumbum) *
magnesium	Mg
mercury	Hg (from hydrargyrum) *
neon	Ne
nitrogen	N
oxygen	O
silicon	Si
silver	Ag (from argentum) *
sodium	Na (from natrium) *
sulphur	S
tin	Sn (from stannum) *
tungsten	W (from wolfram) *
uranium	U
zinc	Zn

* The symbols for these elements indicate their early discovery.

It has become clear very recently, that is, in the last 200 years, that early attempts, based on the human senses, especially touch and sight, to give an impression of the whole universe in simple systematic terms, such as those used by the Greeks and the Chinese, are inadequate. The difficulty is partly due to the size (scale) of the building units of the objects around us, which are not open to man's senses, and partly to the misguided early wish to establish just a small number of useful concepts, Fig. 1.1, (in the sense of valuable for further development of understanding) to associate with observable phenomena. The universe is now known to be exceedingly large and complicated, and yet explanations of it depend on the discussion of extremely small physical 'objects', much smaller than atoms, namely electrons, protons and neutrons (see Table 1.1) as well as other more fundamental particles that we hardly need to consider here. Our sight and touch do not really help us in this endeavour. Unfortunately, both the very, very large and the very, very small can only be dealt with by using theoretical physical and mathematical treatments that are outside the reach of the comprehension of most of us and can only be studied by using instruments with sensitivities greatly outside the range of our senses. In large part, however, neither the very, very large on the astronomical scale nor the very, very small on the particle physics scale need concern us greatly in this book, especially as we shall limit ourselves mainly to the context of Earth and especially life on the Earth. (We shall state unequivocally that, to the best of our knowledge, life is where the natural selection of the chemical elements, such as many of those in the marginal illustration, has evolved its greatest degree of sophistication and must command our greatest interest (see Chapters 13 and 16).) We live in a world that is dominated by a somewhat less uncomfortable chemical atomic scale (where visual models are still extremely useful) than is required to describe the universe. (It is worth noting that the development of the instruments of physics between the fifteenth and eighteenth centuries removed the idea that the Earth is the centre of a small universe, and that here again a few early thinkers had thought that this idea was incorrect.)

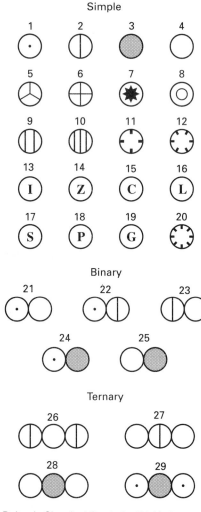

Dalton's Chemical Symbols. (1) Hydrogen, (2) Azote, (3) Carbon, (4) Oxygen, (5) Phosphorus, (6) Sulphur, (7) Magnesia, (8) Lime, (9) Soda, (10) Potash, (11) Strontia, (12) Baryta, (13) Iron, (14) Zinc, (15) Copper, (16) Lead, (17) Silver, (18) Platinum, (19) Gold, (20) Mercury, (21) Water, (22) Ammonia, (23) 'Nitrous Gas', (24) Ethylene, (25) Carbon Monoxide, (26) Nitrous Oxide, (27) Nitric Acid, (28) Carbon Dioxide, (29) Methane.

Detailed examination of the materials available around us, through chemical analysis and preparation over many centuries and using more and more elaborate equipment, gradually showed that the composition of matter was far more complex than had been thought. The search for understanding was (and still is) not just driven by scientific curiosity, of course, but by the use to which the knowledge might be put. But, whatever the reasons, the universality of this empirical analytical approach linked to chemistry was of major importance since it began to reveal a lengthy list of common substances to which all matter could be reduced. Progress was slow because one weakness in the early analyses, say, until the end of the eighteenth century, was that they were at best qualitative and thus did not allow clear-cut quantitative knowledge of the elementary composition of any of the materials. Most scientists were also hindered in their interpretation by mistaken viewpoints, for example, that matter was indefinitely continuous.† Furthermore, it was thought that living organisms might contain a special extra feature, a 'vital force', that was also treated as open to discovery by analysis. It was not easily realised either that an initial separation in thinking about matter and energy (contrast Fig. 1.1) would be helpful even if there would have to be a synthesis of ideas later.

The change in thinking about the substances around us, which can be traced back many centuries as it gathered acceptance, depended on increasing sophistication in the experimental study of materials, mainly more careful quantitative chemical analyses, over about 200 years, say, from 1650 to 1850. Using experiments such as the breaking down of solids by heating, or their reaction with acids, or by dissolving them in water while following the weights and volumes of the reactants and the substances produced, it was finally realised that all of the materials around us (on earth, in air and water) were made from a limited number of 'chemical elements' of fixed weights, which were later represented by symbols—first, circles with dots, lines and letters (Dalton, 1808, see marginal illustration) and then just by a letter code (Berzelius, 1813).

As a result, we can now write down on an atomic basis and in this symbolic language anything chemical that we study around us. Using the techniques available in early days, around 1800–50, the chemical elements could not be broken down further; hence, they had to be made of the same kind of indivisible pieces, which were called atoms of the (chemical) elements by analogy with the ultimate material pieces of Democritus. We stress that it has turned out that everything we see, smell, hear or feel on this Earth can be related to these elements, without further analysis. (This is due to the fact (see later) that even all biological systems including ourselves are made up in this atomic fashion.)

Thus, by the year 1860, it was known that all of one part, the material part (we come to fire (energy) later), of the observable world could be broken down into such chemical indivisible pieces—atoms—and at that time (by 1860) there were discovered some 60 elements, made up of distinct atoms that, if we remain on this (atomic) scale, could not be dissected further. From them effectively all things by then investigated could be made. All the material world

† It was Boyle, who was born in 1627 and was later called the 'father of chemistry', who argued strongly against this idea and in favour of a corpuscular nature of matter, thus reviving the ideas of Democritus some 2000 years earlier.

became therefore open to us in a language of combinations of letter symbols,

$$mA + nB \longrightarrow A_mB_n,$$

atoms   combination of atoms.

This 'alphabet' has subsequently been found to require about 100 symbols, equivalent to atomic elements, most of which must have two letters, of course, for example Cu (copper) or Cl (chlorine) but we do not need a more complex code for all of the chemical atomic description of living or dead matter. (Note that there are only about 100 elements for reasons related to the stability of bare nuclei, not atoms, which is a subject into which we shall not enquire deeply.) The remarkable nature of this discovery must never be under-estimated.

Obvious questions arise immediately. Will any A join up with any B? Will n and m be simple? Will $A_mB_n \dots$ resist heat, water, etc.? The answers to many of these questions have been obtained, but the variety of ways of putting together these approximately 100 elements, $A + B + C$ etc., and n and m, etc., that is in any ratio one wishes at a time, is so immense that man is still searching for new properties and possible functional values. However, just as we now know that there are no more than approximately 100 types of chemical elements (of which only 90 occur naturally on Earth (see later)), so we know the major rules for the combining ratios of some (note not all) of them, for example Proust's law of constant combination (1797) and Dalton's law of multiple proportions (1830). We shall see later, that all such combinations are connected to loss or gain of energy (fire). We have also discovered (quite probably) most if not all of the properties and major uses to which the combinations can be put!

A second question is: why do atoms stick together in special combinations? This question relates to the early discussion concerning 'form' and is answered today by proposing forces—electrical, magnetic and gravitational, for example (see Table 1.2). To understand today's answer we must look at the characteristics of the naturally occurring 90 elements.

1.3 The periodic table and the electronic structure of atoms

The pattern of the simple ways in which elements combine was used during the second half of the last century to form a table—the periodic table (Fig. 1.7)—which shows similarity in potential for combination of the elements in given ratios relative to, say H (hydrogen) down a group, while within a group atomic weight increases. At the same time, the chemical and physical properties of the elements and their compounds undergo systematic change along periods.

After hydrogen, which is the lightest and is taken as the reference element, and helium, which is a noble (inert) gas, there is a period of eight elements, from lithium to neon. Across this period the combining numbers with hydrogen are in the order 1; 2; 3; 4; 3; 2; 1; zero, for the eight elements from number 3 to 18 (see Fig. 1.8). We represent the corresponding compounds as $LiH, BeH_2, BH_3, CH_4, NH_3, H_2O, HF$, (Ne). Thus we have first a sequence of two

Group number

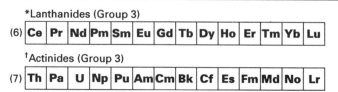

	1	2	3	4	5	6	7	8	9	10	11	12	13	14	15	16	17	18
1	H																	He
2	Li	Be											B	C	N	O	F	Ne
3	Na	Mg											Al	Si	P	S	Cl	Ar
4	K	Ca	Sc	Ti	V	Cr	Mn	Fe	Co	Ni	Cu	Zn	Ga	Ge	As	Se	Br	Kr
5	Rb	Sr	Y	Zr	Nb	Mo	Tc	Ru	Rh	Pd	Ag	Cd	In	Sn	Sb	Te	I	Xe
6	Cs	Ba	La*	Hf	Ta	W	Re	Os	Ir	Pt	Au	Hg	Tl	Pb	Bi	Po	At	Rn
7	Fr	Ra	Ac†	Db	Jl	Rf	Bh	Hn	Mt	110	111							

*Lanthanides (Group 3)

(6)	Ce	Pr	Nd	Pm	Sm	Eu	Gd	Tb	Dy	Ho	Er	Tm	Yb	Lu

†Actinides (Group 3)

(7)	Th	Pa	U	Np	Pu	Am	Cm	Bk	Cf	Es	Fm	Md	No	Lr

Fig. 1.7 The periodic table of elements, displayed in the modern 'long' form. Each element is denoted by its atomic number or symbol. The groups have been numbered according to the recent International Union of Pure and Applied Chemistry (IUPAC) recommendation.

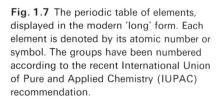

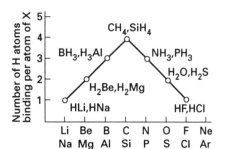

Fig. 1.8 Number of H atoms binding per atom of element X in the two eight-element periods, comprising atomic numbers 3–10 and 11–18.

elements, H and He, and then a sequence of eight elements. Both finish with gaseous monatomic elements, He and Ne, which are very reluctant to react and have been referred to as 'inert gases' (presently called 'noble gases').

Now we can develop this table of weights and combining ratios for the 90 elements. If we do this we find that, after the periods of two, H and He, and then eight elements, Li to Ne, there is a repeat of eight, then a longer stretch of 18 followed by a second repeat of 18, then of 32 and a new repeat of 32 (incomplete). Thus, we obtain the periodic table in a numerical order, which is called the table of the *atomic number* of the elements with a form that gives 18 groups† in parallel with chemical properties. (The atomic *weight* is dominant in nuclear structure (and weight), but the atomic *number* relates to the combinatorial chemistry of atoms, see below.) Furthermore, the physical properties of the elements, such as apparent atomic volume, follow the same periodic trend (see Fig. 2.2). The periodic table is the centre of chemistry and has to be the major key to the understanding of the properties of materials of every kind, mineral and biological, in our surrounds. It must contain an essential clue to the material world on Earth (not in the stars), which man has longed to appreciate ever since he became self-conscious and started to observe.

The periodic table has an even more detailed internal numerical content, a shell-like structure that is built in a curious way, that is,

$$(1 \times 2) = 2; (1 \times 2) + (3 \times 2) = 8; (1 \times 2) + (3 \times 2) = 8;$$
$$(1 \times 2) + (5 \times 2) + (3 \times 2) = 18;$$
$$(1 \times 2) + (5 \times 2) + (3 \times 2) = 18;$$
$$(1 \times 2) + (7 \times 2) + (5 \times 2) + (3 \times 2) = 32;$$
$$(1 \times 2) + (7 \times 2) + (5 \times 2) + (3 \times 2) = 32.$$

[In the last shell only 25 elements are known at present; after the first six of these, the elements do not occur naturally on Earth or elsewhere—they are

† Although there are 32 elements in the longest series, 15 of them, the lanthanides, are so similar that they are treated collectively as a member of group 3 (Table 1.7).

not stable. The last naturally occurring element (92) of any stability is uranium, and we all know the implications of its instability—atomic fission energy. (Scientists have actually gone on to synthesise several new unstable heavier elements (93–111), some in extremely minute amounts, as they had already done to obtain elements 43 (technetium) and 61 (prometium), which have not been found in the Earth presumably because they disintegrate as fast as they are formed.) This is one feature of matter with which we have to come to terms—atomic fission (and synthesis)—which is a property not of atoms but of the nuclei of atoms. This feature is not a chemical property and will concern us later in only two particular aspects—element abundance (section 8.3) and sources of energy (Chapter 3).) The lack of nuclear stability at high atomic weights limits the number of elements and the stability of a given nucleus helps to control its abundance.] As far as the chemistry of the elements is concerned, we believe that the Greeks would have loved the mathematical simplicity of the above numerical rules that build the periodic table had they but been aware of the atomic composition of matter and the structure of atoms. But, of course, these only became known more than 2000 years after the breakdown of their civilisation.

Having discovered the periodic table, experimental scientists, physicists, looked for the nature of the atom they had been forced to postulate, but which they could not see or measure directly, so as to explain the periodicity of properties, that is, the peculiarities of the above shell-like structure. Only after a series of discoveries using more and more sophisticated equipment, was it possible to establish a succession of models of the 'structure' of atoms, the ultimate constituents of all the chemical elements. [X-rays were discovered by Plucker in 1859; electrons, which were found to carry the unit of negative charge, were discovered by J. J. Thomson in 1897; radioactivity by Becquerel in 1896; quanta of energy by Planck in 1901; the central nucleus by Rutherford in 1911; the relation between nuclear charge (positive charge, see Table 1.1, found to be in units and equal to the number of protons) and atomic number by Moseley in 1913–14, and the behaviour of electrons both as particles and as waves by Davisson and Germer in 1927, amongst several others]. In fact, the *atomic number* turned out to be the number of electrons (and hence of protons in the nucleus) associated with each atom. It was then necessary to explain the pattern of these numbers in the periodic table. This involves us in a description of the electrical forces that hold protons (positively charged) and electrons (negatively charged) together and returns us to the discussion of fields of force (form) and their related energies.

1.4 Fields of force

The material world can be described in terms of massive objects such as planets and stars or, at the other extreme, in terms of the 'fundamental' particles which constitute atoms. The particles of major relevance for us in this book are the electron, the proton and the neutron, Table 1.1.

(Although both the proton and the neutron are now known to be composites of *quarks*, the quarks are never found free. There are still other particles, some stable and some unstable, among which are those that,

Table 1.1 Characteristics of the main chemically important subatomic particles*

Particle	Symbol	Mass (kg)	Electrical charge (coulomb)
Proton	p	1.67265×10^{-27}	$+1.602189 \times 10^{-19}$ $(+e)$
Neutron	n	1.67495×10^{-27}	Zero
Electron	e	0.910953×10^{-30}	$-1.602189 \times 10^{-19}$ $(-e)$

* *Note.* Giving objects masses and charges is a way of expressing quantitatively their interactions which are observables.

according to modern theory, 'mediate' the four fundamental forces of nature: gravitation, electromagnetism and the 'strong' and 'weak' nuclear forces, (Table 1.2). Moreover, each particle also has a corresponding antiparticle. All of these must be considered in a more comprehensive discussion of the physical world (see references 15 and 16 in 'Further reading'). In the present book, fortunately, we do not need to deal with the complexity underlying the existence of atoms. We shall, therefore, consider the interactions between both large and small parts of matter with and without charges assuming the presence of protons, neutrons and electrons only.)

The way in which uncharged masses, such as the Earth, mass m_1, and the sun, mass m_2, interact is through what is called the *gravitational* force, about which we know little although it is readily measurable. The energy of interaction is found to be proportional to $(m_1 m_2)/r$ where r is the distance from the Earth to the sun, but the expression can be applied to any two bodies of masses m_1 and m_2 at a distance r. Through this force the Earth is held in a fixed orbit by its motion round the sun. Gravitational force on Earth has a major influence in chemical and biological systems causing separation on a vertical axis (see Section 6.3.5).

We describe the interactions in the nuclei through what are termed *strong* forces, which act on protons and neutrons at very short distances. These strong forces give rise at particular atomic numbers to a range of stable nuclei of *atomic weights* given by the summed mass of protons and neutrons. It was found that, while a given element has a unique *atomic number*, it can have several atomic weights associated with it, giving various *isotopes* for each element. Thus, closely related numbers of neutrons are found associated with a single number of protons, for example, hydrogen (1p) exists with a little

Table 1.2 The four forces of nature

Force	Relative strength	Particles acted upon	Range of action	Example
Strong	1	Protons and neutrons (quarks)	10^{-15} m	Holds nuclei together
Electromagnetic	1/137	Charged particles	Infinite	Holds atoms together
Weak	1/10 000	Protons and neutrons (quarks), electrons	10^{-16} m	Radioactive decay
Gravitational	6×10^{-39}	Everything	Infinite	Holds the solar system together

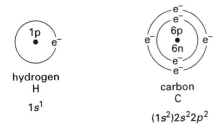

An early representation of the outer and inner electrons (e^-) and the nucleus (including protons (p) and neutrons (n)).

deuterium ($1p + 1n$) but they hardly differ in chemistry. We do not need in this book to describe the strong forces involved. The same is true for the so-called *weak* forces, which are also operative between particles in the nuclei of atoms and account for phenomena such as radioactive decay.

The most interesting interaction for the understanding of chemical properties turns out to be that between electrically charged particles, the *electromagnetic* forces between electrons and protons (nuclei). The fact that the atoms are electrically neutral tells us that protons and electrons are present in equal amounts. Unbalanced charge is not stable and attracts opposite charge so that nuclei of n protons pull to themselves n electrons. The general law of this attraction for static charges was discovered early in the nineteenth century to give an energy proportional to $(z_1 z_2)/r$ where z_1 is the charge of one sign on a site at a distance r from a charge of the opposite sign, z_2, on a second site. This allows us to make an electrostatic model of electrons held by protons, in which the small nuclei hold a central position (see marginal illustration) while the electrons, apparently equally small, *together* fill a large space at a distance from the nuclei due to their motions around the centre. This description does not explain why they do not collapse together or why there is shell structure, 2, 8, 18, 32 in the periodic table. The atomic number is then both the number of protons and the number of electrons and it is this number that governs chemistry through the electrons on the atom surface.

Now we saw earlier that the electron shells are built of blocks—$1 \times 2 = 2$, $3 \times 2 = 6$, $5 \times 2 = 10$ and $7 \times 2 = 14$—which are usually referred to by chosen letters as s, p, d and f (see Fig. 1.9).

$$\frac{s}{1 \times 2} \qquad \frac{p}{3 \times 2} \qquad \frac{d}{5 \times 2} \qquad \frac{f}{7 \times 2}$$

It has been discovered that the nature of this internal form of the shells, which reflects relative electrostatic energies, can be appreciated only from controlling limitations, restrictions,† on the permitted strengths of the electromagnetic interactions, (see below). The sequence of electron shells, 2, 8, 18, 32, shows

† A restriction on an energy of interaction is counter-intuitive in that it implies that an electron can only be held in special spatial relationships with a nucleus, that is, at particular average distances. The explanation given is that electrons are also waves and that waves can only be trapped in space with given wavelengths (see text). This mathematical treatment and description is forced upon us by experiment, and is quite at odds with our sensory experience.

Fig. 1.9 The structure of the periodic table showing the subshell character.

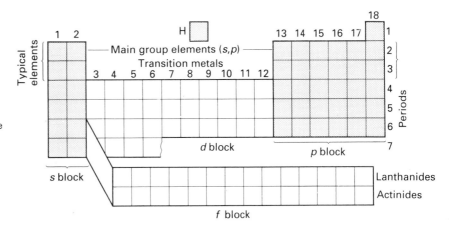

that they are *combinations* of s, p, d and f blocks, that is, s, s+p, s+p+d, s+p+d+f. The successive major shells can now be designated by an integer, n=1, 2, 3, ..., called the *principal quantum number*, which we must relate to a *major* restriction on electrostatic interactions (the energy of the electrons) while in addition there must be further restrictions in each of the subshells (blocks) of each shell. Thus, for n=1 there is only one s block and two elements; for n=2 there is one block s and one block p and hence eight elements; for n=3 there is one block s, one block p and one block d, hence 18 elements; etc. For each n there is then a secondary restriction. In each complete block there are 2, 6, 10 and 14 electrons, respectively, and electrons are seen to be associated in pairs. We can refer here to Pauli's exclusion principle, which will be seen to be a third restriction on permitted electrostatic interaction; this principle says that each permitted condition for an electron occurs in two possible ways that have the same energy but opposite 'spin'.

To be satisfying (for the scientist), the restrictions had to be given more detailed and appropriate constructs within a model. They can be viewed, and were so viewed at first, as restrictions on motions of electrons round nuclei and even later as restrictions on motion 'through' nuclei, for which only certain 'angular and radial momenta' are allowed, that is, the momentum of electrons is *quantised*.† *It is the restriction on types of radial and angular momenta that then gives the restrictive set of numbers and values of the electrostatic energy.* This is a counter-intuitive model, but it fit the data and gave rise to the idea that electrons travelled held by nuclei in special orbits, a description related to the movements of planets around the sun (Fig. 1.10; compare Fig. 1.20). The restricted motions, today expressed in terms of 'orbitals' so as to keep the anology with classical models, and the stability of electrons in pairs are now described more effectively by oscillating standing waves, *wave functions*, in three-dimensional space. The oscillations are again associated with quantised momenta, introducing more and more spherical and planar nodes in the waves. The idea derives from a theory (wave mechanics) in which the electron, like other very small particles, is admitted to have a particle–wave dual character (see Fig. 1.6). It is not too difficult to see how a wave-like model of an electron in three-dimensional space gives rise to restricted solutions of one of a kind (s), three of a kind (p), five of a kind (d) and so on, related to re-restrictions on radial and angular momentum. Waves trapped in any 'box' always are as

---

† The angular momentum mvr is a 'conserved' quantity related to energy, which can be represented by a vector perpendicular to the plane of the 'orbit' of a moving mass m with a speed v round a centre at a distance r (see Fig. 1.10). The permitted motions of electrons were later related to the fitting of waves in, or on the surface of, a sphere (see Fig. 1.11).

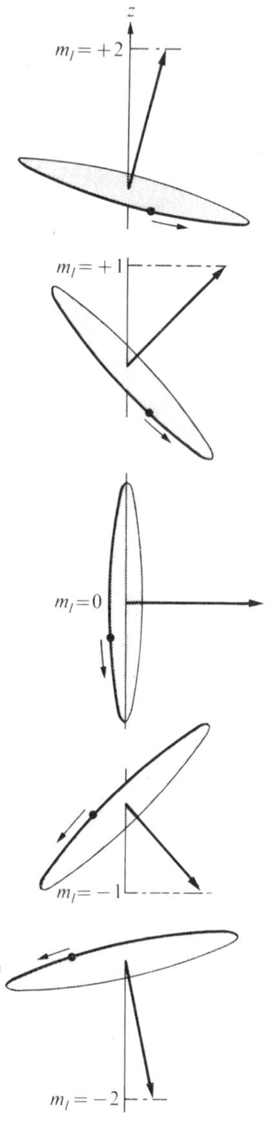

Fig. 1.10 In order to describe the observed energy states of electrons in atoms it was necessary to have a model which restricted motion of the electron around the nucleus as well as its distance from the nucleus, dependent on n. It was observed that restrictions on angular momentum could be devised which gave the correct number of states, m_l, and their energies. In essence the restrictions were that the angular momentum about an axis, z, could only have integral values up to maximum of ±1 or ±2 or ±3, etc. for a given subshell p, d, f. The resulting picture of electron motion is shown for a total angular momentum of ±2 (3d orbital). It is easily seen that this quantisation gives subshells p, d, f, etc. contaning 3, 5, 7, etc. states which differ in angular momentum and so give a satisfactory account of the periodic table. Since a circulating electric charge gives a magnetic field the states were given 'magnetic', m_l, quantum numbers.

in musical instruments (see Fig. 1.11(a) and (b)).† In this sense the electron, as a wave, is trapped in a spherical box by the central nucleus. The waves extend radially to infinity (Fig. 1.11(c)(i)). This mathematical modelling of electrons in atoms, (see references 10–14 in 'Further reading'), is given here to show that, of necessity, that is to explain the periodic table, the description has had to become abstract. We shall hardly use this approach in this book but it lies behind a full appreciation of chemistry.

It is immediately clear that there exists a direct relationship between the way in which the last orbitals, *outermost* type of shells, are filled with electrons and the chemical properties of the corresponding elements in the periodic table since the table was originally built on this principle—elements with analogous properties (families) are in the same column (Fig. 1.8). Furthermore, it is immediately permissible to think that there must exist a relationship between the chemical properties of the elements and the energy with which electrons are held in those shells, specified firstly by the quantum number n and then, within each level, by the type of orbital, that is by the $n(s, p, d, \text{or } f)$ combinations. These energies can be obtained experimentally by spectroscopic techniques. (Notice that not only have we lost the idea of continuity of weight of matter by accepting atomic conditions, but we have also abandoned continuity of energy by accepting quantised conditions.)

The energy necessary to remove an electron from an atom in its most stable or *ground* state is called the *ionisation energy*, usually expressed in electron-volts, eV, but frequently replaced by the numerically equivalent *ionisation potential* (*IP*) expressed in volts, V. For hydrogen, or any other element, the energy released on capture of the electron by the neutral atom is called *electron affinity energy*, expressed in the same units as the ionisation energy. In effect, the electron affinity energy of any element X equals the ionisation energy of the mononegative ion, X^-, of that element. Obviously the electron affinity, expected to be zero on simple electrostatic theory, is much lower than the ionisation energy. (It is the peculiarities of the 'motion' of the electron that give the electron affinity for a neutral atom a non-zero value.)

Naturally, for any atom, there are successive ionisation potentials, but the second, third, etc. refer to the removal of an electron from progressively a more positively charged ion, not from the neutral atom, and hence they are of higher binding strength. The ionization potentials were found to vary periodically with the atomic number of the elements (see Fig. 1.12 for IP_1) and their successive values increase, sometimes rapidly, until an electron configuration of particular stability is reached, for example, ns^2 or ns^2np^6 corresponding to the outer shells of the noble gases helium ($1s^2$) and then of the sequence neon to radon, but note also the stability of $(n-1)d^{10}$ and $(n-2)f^{14}$. The ionisation potentials then fall, only to rise again in the sequence of shells and subshells. It was found too that the order of energy of electrons in orbitals follows that of the principal quantum number for s orbitals, that is $1s < 2s < 3s$, etc., and for the lighter elements, which have fewer electrons, one also finds the order $s < p < d < f$ for each value of n. (We shall show later that this leads to an

† Note that, just as there are two possible wave forms of opposite phase and equal frequency in any box, there are two 'spin' states for each electron in each orbital. In a curious way this picture is related to the 'Yin/Yang' sign (see Figs 1.3 and 1.6) as noted by Bohr. Wave mechanics gives a mathematical 'explanation' of the properties of matter but must not be considered to be more than a modelling of behaviour.

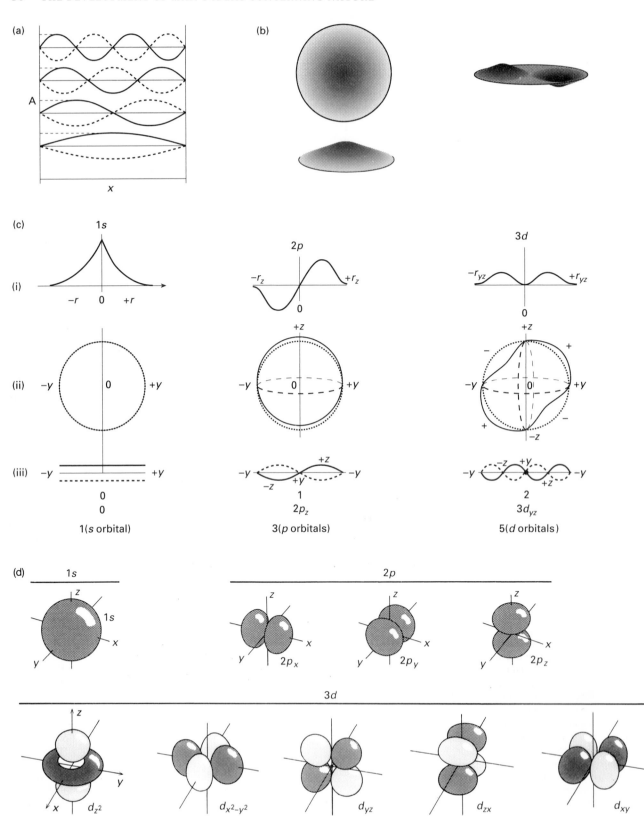

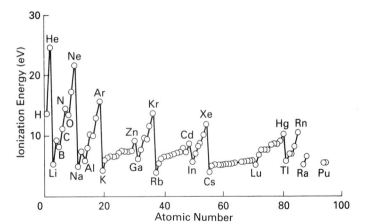

Fig. 1.12 First ionisation energies of the elements as a function of atomic number.

explanation of the periodic trends of chemical and physical properties.) For heavier elements the situation becomes more complicated and inversions of the order of subshell energies are observed, for example, $4s < 3d$ or $6s < 4f$ (Fig. 1.13). This is why the elements of the d blocks are called transition elements (three series) and those of the f blocks are called internal transition elements (two series, lanthanides and actinides). The first correspond to the filling of $(n-1)d$ orbitals after the ns and before the np orbitals and the second to the total filling of internal $(n-2)f$ orbitals before the $(n-1)d$ orbitals. These changes are deduced from the study of the atomic spectra of the elements concerned and are reflected in several properties, especially, for our purpose, in the chemical properties that decide their compounds and their functional value (see Chapters 2 to 7). Today we can calculate and appreciate why the energies are in this order using wave mechanics although a really fundamental understanding of the duality of electron behaviour is not open to us (see references 10–12 in 'Further reading'). (Clearly, in the equation $(ze)/r$ for the interaction energy of an electron with a nucleus, the charge, z, 'seen' by the electron cannot be the charge on the atom in an ionised state; if this had been so, all ionisation energies would have been related to r for a given charged state. We consider that the electron in its 'orbital' sees an *effective nuclear charge*, z_{eff}, much higher than the ionic charge. The charge seen is particular to a given element since nuclear charge is dependent on the element and on the

Fig. 1.11 (a) Examples of allowed standing waves on a line or string of length L. The extremes of the oscillations are shown by solid and dashed lines. All the allowed waves conform to the condition $n\lambda/2 = L$, where n is an integer and λ is the wavelength, with nodes at their ends. There are two possible waves of each wavelength where the dashed lines have the opposite phase from the full lines (compare Fig. 1.3). A is the amplitude. (b) Representation of waves on a plane surface, a drum, of diameter L. There are now two kinds of waves: (i) radial, obtained by beating the drum centrally and with wavelengths $L = \lambda/2, 3\lambda/2, 5\lambda/2 \ldots$ and (ii) angular waves created by beating the drum off centre and which run round the centre with radial wavelengths proportional to $c = \lambda, 2\lambda, 3\lambda \ldots$ (c = circumference). There are two orthogonal angular waves for each wavelength and four possible waves taking phase into account. (c) Representations of waves for electrons in an atom, i.e. waves trapped in spherical space: (i) gives an impression of the *radial* nature, r, of $1s$, $2p$ and $3d$ 'waves' (see panel (b)). Higher n states are introduced via interior radial nodes. (ii) gives the *angular* nature of waves in a sphere which are opened out circumferentially in (iii) to give the picture in the form of panel (a). The nucleus is at 0 and there are two waves of opposite phase always. (d) Cloud density representation of electrons in the s, p, and d orbitals. Notice the 3-degenerated and the 5-degenerated sets of p and d orbitals, which, when all filled, give a spherical distribution of charge like the s orbitals. The impression of an edge to an atom is given by taking 90 per cent of the cloud while in fact the cloud dies out at infinity. The density comes from the square of the wave amplitude in (c). Note that the impression of a particle, an electron, based on sense experience is lost entirely.

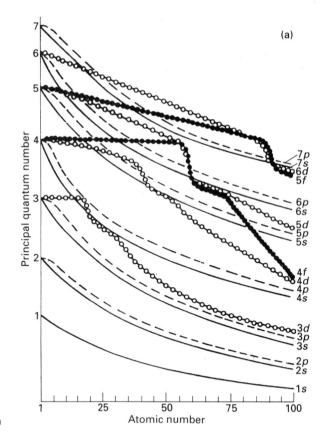

Fig. 1.13 (a) Energy dependence of atomic orbitals as a function of atomic number. (b) A simpler illustration of the predicted sequence of orbital energies for electrons in atoms: *s* levels can hold two electrons each; *p*, *d* and *f* levels, respectively, 6, 10 and 14. The diagram indicates the number of electrons that can be accommodated in each successive electron shell (row in the periodic table), ending with the filled-shell noble gas elements in group 18, and finishing at uranium (92).

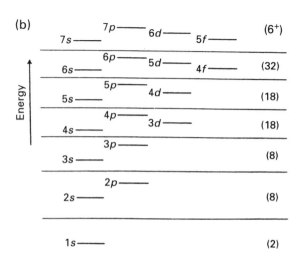

particular orbital motion of the electron that allows it to 'penetrate' the inner electron core to some degree, hence the electrons of the core do not shield the charge of the nucleus completely).

Absorption of energy of appropriate frequency, v, can cause transitions of electrons from lower, E_1, to higher energy levels, E_2, that is, into different orbitals, according to the equation $E_2 - E_1 = hv$ where h is Planck's constant (see Section 1.7). The electrons in energy level E_2 can return to E_1 by loss of energy (photons) (Section 1.7). (In both cases only certain transitions are allowed, again a consequence of the 'wave-like' character of matter and radiation (see 'Further reading'). It is this interaction that allows matter to be energised by radiation from the sun.)

1.5 Secondary consequences of (orbital) motion restrictions: shapes of atomic combinations

A consequence of the treatment of the wave-like motion of the electrons in proximity to the nucleus is that the electrons are not just characterised in terms of *radial* shell density and energy but also in terms of *angular-dependent* density (Fig. 1.11). While in an individual atom the coordinates of angular motion, x, y, z, are of equal energy and not identifiable, they are no longer so in a non-spherical electrostatic field (Fig. 1.14). The non-spherical fields of major chemical concern in this book will be those due to neighbouring atoms (nuclei and electrons) and will be shown to give rise to the peculiarities of *observed* non-isotropic shape in atomic associations, molecules and condensed systems (Fig. 1.15) due to the fact that some arrangements of atoms around a central one are more stable than others in ways that are not obvious from simple considerations; for example, not all triatomic molecules are linear. Thus we are faced with the strange finding that another obvious and ordinary feature of

Description of shape	Shape	Examples
Linear	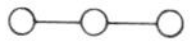 O—O—O	HCN, CO_2
Angular		H_2O, O_3, NO_2^-
Trigonal planar	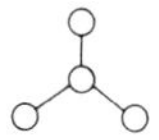	BF_3, SO_3, NO_3^-, CO_3^{2-}
Trigonal pyramidal	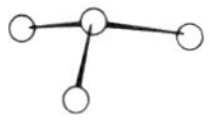	NH_3, SO_3^{2-}
Tetrahedral	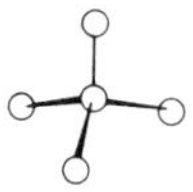	CH_4, SO_4^{2-}, NSF_3
Square planar tetragonal	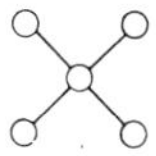	XeF_4
Square pyramidal	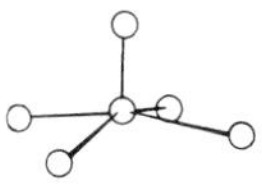	$Sb(Ph)_5$
Trigonal bipyramidal	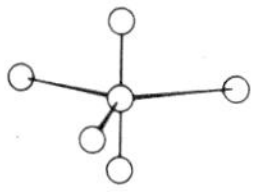	$PCl_5(g)$, SOF_4
Octahedral	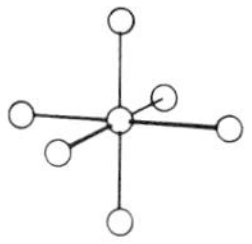	SF_6, PCl_6^-, $IO(OH)_5$

Fig. 1.15 The description of some molecular shapes (see Section 2.2.4).

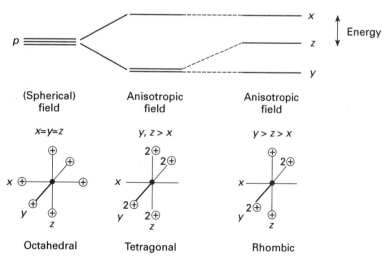

Fig. 1.14 The splitting of the energy states of the three p orbitals which are degenerate in the absence of anisotropy of fields.

things around us—shape (as well as energy states)—has a quality associated with it in atomic (molecular) systems due to electron 'motion' and that this 'motion' is in turn 'explained' in terms of the wave-like character of very small masses, something that lies beyond our capacity to understand the 'real' nature of matter.

1.6 Fire (energy), temperature and pressure

While our understanding of matter, by which we mean material ultimately composed of atoms that themselves consist of protons, neutrons and electrons, has increased and deepened, a similar development has taken place in our ideas concerning not only the origin of matter but also the connected character of two other central thoughts in ancient times—flow and fire. As mentioned above, we have replaced fire by energy, and later we have similarly replaced flow by *anisotropic momentum*. Energy is defined for a particle in motion by $\frac{1}{2}mv^2$, kinetic energy, where m is the mass and v is the velocity of the particle.

Once it has been agreed that material is best described by atoms, then clearly the atoms of a gas must be far from one another and free to move relative to one another in what is of necessity a chaotic motion (see Fig. 1.5). Since it is necessary to apply the heat (energy), of say a fire, to generate the gas state, it follows that a scale of the heat content of the gas, the so-called temperature, T, must be related to an averaged function of the motion of the particles.† The system pressure, p, at fixed volume, V, also increases with temperature, such that, for one mole,

$$pV = RT$$

where R is the molar gas constant. This is the expression of the so-called *perfect gas law*, which was found by experiment and can be derived theoretically. Thus, the kinetic theory of gases gives a model for this relationship in terms of the atoms in a homogeneous gas and their motions: temperature is a measure of the average random translational kinetic energy of the 'molecules', and pressure arises from the averaged force resulting from repeated impacts of molecules on containing walls.

Generalising these concepts to any particles, we may say that pressure, equal in all directions, is associated with the average momentum of the number of randomly moving particles in unit volume and it is the relative *random* energy of the motion in that volume that generates the temperature. Pressure is different from the *directional motion*, flow, of a rest mass due to an inhomogeneity of forces acting on particles and generating a velocity downfield—the corresponding energies here are not related to temperature. Obviously, a flowing stream is not different in temperature from its banks. We must consider later the difference between isotropic velocity, related to

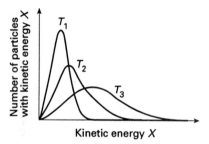

Fig. 1.16 The distribution of kinetic energy of particles with change of temperature, $T_3 > T_2 > T_1$.

† The detailed theory is more explicit in that it describes the distribution of kinetic energies of the different units in the gas (Fig. 1.16). This distribution parallels that of black-body radiation at a given temperature (Section 1.9). These distributions become important when we describe disorder in systems in Chapter 3.

pressure and temperature, and anisotropic velocity, flow. Before we turn to this problem there is a second aspect of 'fire', energy, that we must tackle—the nature of radiant energy including light.

1.7 Light and radiant energy

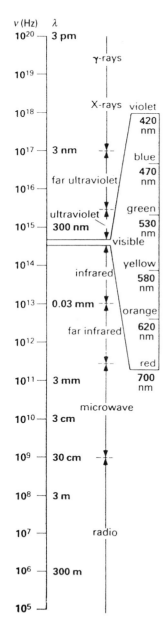

ν (Hz)   λ

10^{20} — 3 pm

γ-rays

10^{19} —

X-rays   violet

10^{18} —

420 nm

10^{17} — 3 nm

blue

470 nm

far ultraviolet

10^{16} —

green

ultraviolet

10^{15} — 300 nm

530 nm

visible

10^{14} —

yellow

infrared

580 nm

10^{13} — 0.03 mm

orange

far infrared

620 nm

10^{12} —

red

10^{11} — 3 mm

700 nm

microwave

10^{10} — 3 cm

10^9 — 30 cm

10^8 — 3 m

10^7 —

radio

10^6 — 300 m

10^5 —

Fig. 1.17 The electromagnetic spectrum: $E = h\nu$. The frequency is in Hertz, cycles per sec.

One of the features of our surroundings is the presence of radiant energy, light and heat, the nature of which remained a puzzle for thousands of years. *Radiant energy* was found by experiment to be separate from *rest mass*. It was discovered by Planck that radiation had to be described as a sequence of independent energy packets, or quanta, *photons*, with both particulate and wave-like character, but that these can only be assigned a *relativistic mass, m,* related to frequency υ, and given by Einstein's equation $m = E/c^2$, where c is the speed of light and $E = h\upsilon$ is the photon energy.† All forms of radiant energy can be related to such *photons*, for which relativistic mass and energy are interconvertible. A scale of radiant energy frequency (or wavelength λ where $\lambda = c/v$) has been established (Fig. 1.17). The photons extend from those of very large wavelength, radio waves of the lowest energy, all the way up to those of very short wavelength, γ-rays and beyond. They all travel with the speed of light, c. It is very clear that this packet or quantum description is entirely counter to the experience of light by our senses which is continuous, but it gives a huge and *complete* range for the electromagnetic spectrum observable by instruments.

Now, any body with rest mass at a fixed temperature is found to radiate energy; the distribution of this radiant energy (photons) over the entire range of wavelengths (frequencies) is known as *black-body* radiation and at equilibrium is related to the temperature of the body concerned (see Fig. 1.18). These photons, which, as stated, have relativistic mass only, must be kept separate in one's mind from particles with *rest mass*, especially since particles such as the electron, the proton and the neutron have a variable energy associated with their particulate random motions and which are related to temperature and pressure as described above. The temperature of particles is then characterisable by either its balanced incoming and outgoing radiant energy or its random kinetic energy, which are strictly related in a system at equilibrium, as indeed are the distributions of Figs. 1.16 and 1.18.‡

Temperature is, therefore, a very important variable from the point of view of radiant energy and rest mass random motion. Our knowledge of its range and methods of measurement are of critical importance. Scientists, such as Celsius in 1742, devised a scale of measurement of temperature based on the degree of agitation of liquid water from the melting of ice (conventionally 0°C) to the vaporisation of the water (conventionally 100°C) through, for instance,

† The term relativistic mass carries the implication only that the light photon behaves with mass-like character much as an electron behaves with wave-like character. Such words carried over from sense experiences must not be taken literally, however. There are new approaches to these problems almost every 20 years.

‡ Note that a body at a fixed temperature in an environment at the same temperature radiates photons and transfers kinetic energy to the environment but receives exactly the same energy from the environment.

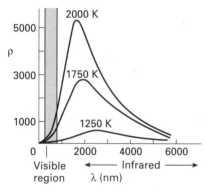

Fig. 1.18 The energy per unit volume per unit wavelength (λ) range, ρ, in a black-body cavity at several temperatures. Note how the energy density increases in the visible region as the temperature is raised, and how the peak shifts to shorter wavelengths. The total energy density (the area under the curve) increases as the temperature is increased (as T^4). N.B. Figure 1.16 is plotted with the opposite sense on the energy axis.

Table 1.3 Temperature and visible appearance of a solid*

Temperature (K)	Visible appearance
1000	Dull red
1200	Cherry red
1500	Orange white
1800	White heat

* An experienced observer estimates correctly the temperature of a glowing body to within ± 100 K.

the expansion of mercury in a thermometer. If we extend this scale upwards we can ultimately reach the temperature of millions of degrees found in the stars where the atoms of the Earth were formed at the periphery and where molecules are not stable and fall apart to basic atoms. As the agitation of the atoms is reduced, that is, as the elementary atoms cool, they come together in particular ways according to the amounts of each element available and their affinity for one another (see Chapters 2 and 8).

The point on one scale where water freezes defines 0°C and in this solid state, ice, molecules have lost 'translational' motion although the atoms can still vibrate in the lattice. Other liquids do not freeze at 0°C and some alcohols and oils (for example, antifreezes) freeze only at, say, -30°C. Some gases, for example methane, in contrast to water vapour, do not become liquids, never mind solids, except at temperatures lower than -200°C, and we now know that the molecules and atoms that stay in the gaseous state to the lowest temperature, namely hydrogen and helium, only freeze at temperatures lower than -250°C. Cooling still further effectively stops all motion in all materials, even atomic vibrations, and at -273°C (called 0 K, where K stands for Kelvin) there is effectively no motion (kinetic energy) in the system (but see later for a paradox). As Francis Bacon said, 'heat itself, its essence and quiddity, is motion, nothing else' (F. Bacon, *Novum Organum*, 1620). Apart from material itself, we and the ancient scientists are indeed correct when we say that *random motion* in material together with radiation ('fire') is another 'essence' represented by the temperature, for example, ~ 300 K or 27°C at which we live. The relationship between this form of energy and the 'heat' of fire is obvious, but it is clear to us all that the kind of energy we commonly (and incorrectly) call heat is largely created from another form of energy, radiant energy, such as light. For instance, when coloured material absorbs light it becomes hotter, and hot materials radiate higher energy light at higher temperatures, (Table 1.3). That light is also given off by 'fire' is quite obvious to the senses. Thus, random motion and light are forms of energy, both of which are related to the ancient concept of 'fire'.

Experimental work indicates that there is no upper limit to temperature as far as we know (although we must be very careful since there may be a problem we have not foreseen). However, we do know that if we cool things down we come to a limit of -273°C, which is not a long way from where we are compared to the upper values that have already been obtained.

We have described temperature as related to kinetic energy and we see no reason to suppose that kinetic energy should not obey the same rules for large and small systems. But here we hit the paradox mentioned above: if we look at matter (for example, electrons) as waves and take into account the fact that there is a relationship between frequency or wavelength and energy ($E = h\upsilon = hc/\lambda$) then, since an atom or electron has a wavelength, it must have an associated motion and kinetic energy *even at absolute zero of temperature*— that is, it has a 'zero point' energy. This is observed to be true, which means that certain 'motions' cannot be removed from material by cooling it even down to absolute zero. [A simple calculation in terms of quantum mechanics shows that the value of the zero-point energy is $E_0 = \frac{1}{2}h\upsilon$ and its value depends on the associated wavelength (or frequency) of the particle considered (for example, for one electron in a 10-Å box $E_0 \sim 3 \times 10^{-20}$ J, about 4.3 kcal mol^{-1}). Once again we meet with incomprehensible observations.]

1.8 Fields and flow

It is generally accepted that gravitational and electrical fields have been generated in the universe by the big bang expansion of rest mass in simultaneously created space and their distribution has been affected by its subsequent non-homogeneous development. It is rest mass that gives rise to gravity and it is rest mass that carries charge. The fact that when masses were formed certain of them were oppositely 'charged' (Table 1.1) meant that they moved to one another to form elements, atoms, by electrical attraction.

Due to the energy of the big bang all rest masses flowed outwards from a centre with directed radial momentum while the isotropic gravitational field acted in the opposed sense but more weakly. This *outward motion* gives a sense of 'time' not directly present in considerations of energy, for example, temperature. The system broke up due to fluctuations and subsequent turbulences that were amplified over time and so there were formed systems of stars and planets. Again when electrons were captured by nuclei they too took on trapped radial and angular motion, flow (see Section 1.4).

It was also turbulence after the big bang that generated local zones of different temperature and density, so that, imposed on the outward *flowing system* of rest masses, there were created new local gravitational fields. These fields induced anisotropic momenta within the universe and hence patterns of flowing mass, for example, spiral nebulae (Fig. 1.19). Our planetary system is a further example in which local further turbulence affected the matter associated with the sun, throwing it in an extended cloud with angular momentum around that star (Fig. 1.20). The local gravitational field plus the induced momentum forced the Earth and other planets to proceed in an 'organised' pattern of orbits.

Meanwhile, the local differential cooling allowed atoms to attract one another through electrical (atomic) fields and so form molecules. The molecules being locally energised also entered into patterns of flow, which we see in the air and waters around us, but flow also extended to the deep Earth even causing the drift of continents. We see that, due to motion in a system of

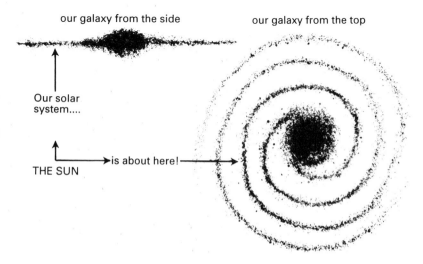

our galaxy from the side   our galaxy from the top

Our solar system....

THE SUN is about here!

Fig. 1.19 An impression or model of a galaxy taken using light, but many objects are missing. The size is unbelievable for the senses but found to be true by machine measurement.

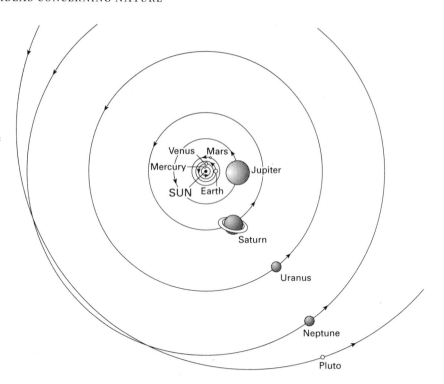

Fig. 1.20 Schematic diagram of the solar system. The diameters of the planets are shown in approximately the correct proportion to each other, as are the sizes of the orbits. However, the sizes of the orbits are *not* in correct proportion to the diameters of the planets. (If the orbits had been drawn on the same scale as the sizes of the planets, this page would have had to be more than 1 kilometre wide!) The diameter of the sun is about 10 times that of Jupiter, so the sun could not be shown at the centre of this figure; on the scale we are using for the planets, it would have covered up most of the figure!

fields, there is a tendency for many patterns of *flow* of matter to appear that may be stable for long periods of time. We (like the Greeks and Chinese before us) are constantly aware of the flows of air and water, even of molten rocks, on and in the Earth, which form shaped features, for example, the Gulf Stream, the high jet-winds and even the continents themselves. These flows are made possible by gravitational separation based on density. However, as we saw above, it is microscopic (atomic) electrical fields of flowing electrons (held by protons) that generate the *ordered* chemicals we see around us. To understand the macroscopic *flow* of electronic or atomic charge we must postulate electric fields due to uncompensated charges on surfaces creating bulk electrostatic gradients as first noted in the seventeenth century. These are responsible for the currents we call electricity where electrons move in wires, but they are also responsible for the *flow* of charged particles, ionised atoms or groups of atoms in biology. All these motions are clearly *organised*, in some cases by *geological structures* and in others by *living structures* and, most recently, by *man's constructs*.

To allow a deeper understanding of these aspects of ordered flow we have to examine the three fundamental concepts of *order*, *disorder* and *organisation*.†

1.9 Order, disorder and organisation of rest masses

Almost all the description of matter we have given so far is related to the ordering of particles in atoms or condensed systems. Any inspection of our

† We shall come to see in Chapter 2 that to discuss order by itself we need only refer to energy but to discuss organisation we shall have to include flow and disorder.

environment shows, however, that much of it is disordered, for example, in liquids and gases. (We can extend this description to the appearance of the stars in the sky.) Taking any one substance, such as water on Earth, we see that, in fact, part of it is present as ice, part as liquid water and part as water vapour; all are in balance at a particular finite temperature. In the presence of forces there is then a balance or equilibrium of ordered and disordered arrangements in given conditions of temperature and pressure, such that

$$\text{Ordered} \rightleftharpoons \text{Disordered}.$$

Without further explanation we state that order and disorder increase together as an isolated system evolves by expanding and cooling. (We show this in more detail in Chapter 3.) The appearance of local order is then assisted by forces (see above) and opposed by random motion, for example, temperature.

As we shall see in subsequent chapters, to measure the degree of disorder of a system we use a statistically meaningful thermodynamic quantity called the entropy, S. The entropy is related to the number of distinguishable microstates compatible with a particular macroscopic state of the system or, in other words, related to the probability of a macroscopic state which again depends on the number and type of moving particles per unit volume and on the temperature and pressure.

We shall reserve the word *organisation* for *a structure in which particular motion is maintained*, that is, the motion is not random and produces a pattern that is more or less permanent. A temporary appearance of a pattern, a transitory condition, can occur in any system where there is turbulence, but this is not called an organisation. It follows that in an organised system the directed motion itself must play a role in maintaining the structure. Thus, it must be that organisation is related to *directed* anisotropic momentum (that is, to flow). Where does such momentum come from? How can it maintain structure or pattern?

There is then a question as to how *organisation* appeared and the forms it will take, a question that we leave until Chapter 7. Here we must state simply that it is due to the evolution of anisotropic flow locally, which requires gases or liquids in fields, that is, structured motion.

A simplified example may help. The planets were probably formed around the sun through an accidental fluctuation in the universe when a gas cloud was sucked from within our galaxy and sent whirling in a particular plane. The gas cloud condensed to give the sun and the planets, which have retained the directional momentum, relative to the sun, that became the central mass. The planets are trapped in their orbits by: (1) the gravitational field of the sun; and (2) the component of their velocity, their angular momentum, which is at any time effectively a field, centrifugal force, in the opposite direction, that is, equivalent to a force acting away from the sun. The two 'forces' are in balance, so that the planetary motions may be said to be 'organised' in a dynamic pattern. To maintain a pattern flows are necessarily under balanced forces. In this sense all kinds of 'permanent' cycles are organised and have dynamic structural features, which we shall discuss in Chapter 15.

A second example is the pattern of water flow in a river system. The flow is maintained by two opposing fields—the radiation field of the sun and the gravity field of the Earth—but it is guided by the preset structure of the Earth's surface. The structure arose through fluctuations of the Earth's core but is changed slowly by the flow. This allows us to suggest that organisation can continue to develop within developing structure (even self-assembling structure) so long as energy is supplied to material continuously from fields. We do not wish to extend this description further here except to point out that, if it is useful, it has to describe all biological activities including those of man's industrial society (Chapters 10–14).

We shall be concerned throughout this book with investigating the role of chemical elements in the ordered structures and organised flows that are clearly of the essence of much of what we observe including life. Here again we sense the shadow of Chinese and some Greek thinkers—there are fundamental problems of long standing to be tackled.

1.10 The evolving universe

We wish to put the whole of the above discussion of matter, energy and momenta in the context of the evolving universe, and ask why ordered and organised structures were and are being formed in such a system while much remains disordered, that is, without apparent order or organisation.

As shown in Table 1.4, many ordered systems and structures were progressively formed in the expanding universe, while pressure and temperature were falling down to the present background level of about 3 K (actually, 2.736 K according to the latest measurements made by the COBE). We ascribed this ordering to the existence of nuclear, electromagnetic and gravitational forces that act on the various particles and bodies, from nucleons to planetesimals, forming different types of aggregates. It is clear that, in the absence of such forces, disorder would prevail. This is because disordered states have a higher degree of probability—there are more ways of arranging particles in a disordered (random) way than in an ordered way and, if the number of particles is very large (or non-identical particles are involved), the

Table 1.4 The formation of our universe

Time	Average T (K)	Situation
0	(Infinite?)	Creation
10^{-43} s	10^{32}	Inflation
10^{-34} s	10^{27}	Hot big bang
10^{-3} s	10^{12}	Quarks and exotic particles fill universe
1 s	10^{10}	Neutrons and protons
3 min	10^{9}	D, He, (Li)
10^{5}–10^{6} years	3×10^{3}	Decoupling matter/radiation
10^{6}–10^{10} years	$<10^{3}$	Atoms and galaxies, compounds
1.05×10^{10} years	$<10^{3}$	Earth
1.15×10^{10} years		Life on Earth
1.5×10^{10} years*	3	Today's universe

* Recent data (1994) from the Hubble satellite telescope suggest that the universe is not quite so old, but this is still an open problem.
NB. The temperature is very different locally.

probability of disordered states increases many fold (Sections 3.2.2 and 3.3.1). Now detailed examination shows (see Chapter 3) that, in fact, the number of 'particles' in the universe increases as structures form due to the transformation of high-energy radiation photons (for example, light) to low-energy radiation (for example, heat). Thus, although scientists often state that disorder is increasing in the universe, in fact, while it increases more rapidly than order, order also is increasing.

There is a more particular question within this analysis. Why are there so many diverse types of material around us, some ordered, some dynamic and organised as in life? We need to understand this variability of chemical combination and its limitations. What we see around us are the many combinations that have survived, yet there could be millions of other possible combinations. After all there are 90 elements and, in principle, we can combine any number at a time in any amounts. However, only some combinations are found, implying that there are restrictions on what is permitted, just as there are restrictions on the numbers and types of atoms. Thus we need to appreciate *survival* of compounds both in the mineral and the living world. It will be shown that there are two general reasons for survival— first, stability in unchanging and effectively unalterable combinations, which we shall describe through *thermodynamic* analysis; second, controlled rates of change in fixed (cyclic) sequences, which we shall describe by *controlled kinetics*. The first leads to order with no time-dependent features, the second to organisation, that is, flowing activity, but neither is infinitely variable in a surviving form.

1.11 The limitations to understanding

The finding that electromagnetic radiation can be treated as waves means that we know today the full range of radiant energy from infinitely long to infinitesimally short wavelengths (Fig. 1.17). There is no more to be known, except the link to gravity. Since the other form of energy, of masses, is kinetic energy, and its limits, from zero to the speed of light, are also known, we can state that, just as we know the only *atomic* forms of matter that can occur (Fig. 1.7), so we know the only forms that energy can take. Furthermore, there are restrictions on the way in which energy and matter can be combined. These observations once again are counter-intuitive, do not derive from sense experience but are profoundly valuable with or without deep understanding.

When we return to the descriptions of objects around us by the Greeks (Fig. 1.1) and the Chinese (Fig. 1.4) we see that the puzzles for them in the macroworld were due to limitations on their capacity for measurement. We have come a long way since then and improved enormously our knowledge of many different properties of matter, but, strangely indeed, after more than 2000 years, including 300–400 years of intensive exploration using instruments of extreme sophistication, we have come full circle discovering that the mystery of what is around us is still inaccessible to our comprehension of material objects *based upon our senses* and can only be approached in part through extremely elaborate mathematical language of which the implications are not understood and in a strange way not understandable. As Democritus asked when facing the possibility that matter was composed of unseeable atoms, 'Is this a victory or a defeat of the intellect?' The next obvious problem for us to face is: does a parallel limitation apply to insight into life, that is, is there some hidden factor not discovered as yet, or is life just a complicated chemical machine where the problems of the very, very large (the universe) and the very, very small (much smaller than atoms) are of trivial concern but the complexity of chemical atomic organisation is very, very great? We may fail to understand life through fundamental difficulties with conceptions based on reductionist approaches as has happened in physics and chemistry, but also through the inability of our senses to understand holistic complexity when, for example, we might be forced to accept the solution produced by a computer but have no deep confidence that it is the genuine answer. In this book, and in agreement with the view presented in Fig. 1.21, a simplified form of Fig. 16.4,

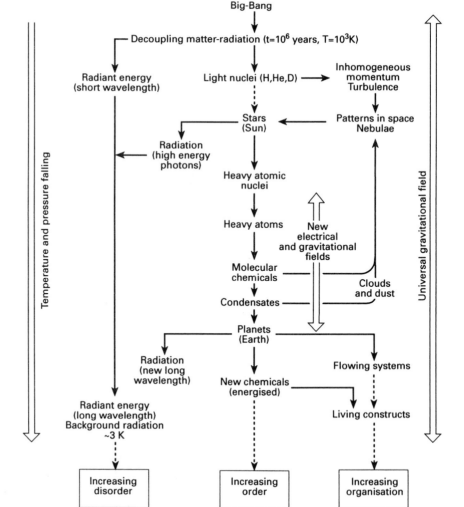

Fig. 1.21 An outline scheme given as an indication of the development of the objects we have detected around us, and some of their underlying interactions and energy, since the beginning of time.

we shall assume that it is the complexity of life's organisation that we have failed to understand so far. This complexity, we shall maintain, is not of the kind usually described in biochemical texts, but is based on a fundamental use of many chemical elements in integrated feedback systems—life is, indeed, we shall state, related to flow, of energy as well as of materials, that is, to organisation. This book will therefore attempt to describe life that is around us in a historical evolution of dynamic interactions between atoms.

Before we can follow the above development of ideas we must describe in outline what is known concerning the way atoms combine to form molecules and then condense in isolated solid and liquid phases, atomic ordering. This is the subject of Chapter 2. In Chapter 3 we re-introduce the idea of *balanced* systems between order and disorder, that is between solids, liquids and gases, and connect this with energy. Chapter 4 illustrates such balances for bulk phases, while Chapter 5 looks at the way balanced order $\rightleftarrows$ disorder conditions appear for minor components within a liquid. In Chapter 6 we turn to the problem of describing equilibria for limited amounts of materials, held in small compartments and by fields, before discussing, in Chapter 7, the rate at which balance is approached or inhibited by barriers to change of both compartments and chemical combinations, that allows us to introduce the ideas of time and flow. This concludes Part I, which deals with all the basic principles we believe we need. To approach an understanding, that is, a chemical understanding, of the present condition of the universe and especially of Earth and ourselves in terms of the concepts of Part I we must go back to the formation of the chemical elements in the universe at the beginning of Part II, Chapter 8. As we then proceed forward we shall see that the chemical elements are selected by underlying 'laws of force'—first at the level of the stability of nuclei, then of atoms in combination, then by physical properties of the compounds formed that separated them in zones, that is, mineral phases. From this selection there resulted, in particular, our Earth. In it, the very composition of the zones, with the peculiarities of temperature, pressure and availability at the surface, made new chemistry possible. New processes of element selection, physical and chemical, then generated abiotic organic chemistry (Chapter 9).

Further processes of element selection, physical and chemical, based on flowing systems, evolved. Through some fluctuation (of unknown probability)—we have no other explanation—life started and then moved forward by evolution (Chapters 10–13), that is, the very composition of the crust, the sea and the atmosphere, with the peculiarities of their temperature, pressure and element availability at the surface, made life possible in organised flowing systems. Elements were selected to assist the survival value of chemical systems, but the environment changed slowly and so allowed life's chemistry to evolve. Finally, with self-conscious man, selection of elements for functional value has become deliberate (Chapter 14). All of this development is called here the natural selection of the chemical elements, but it leads to a question. Are man's deliberate selection procedures compatible with those of biological systems? To put this question in context we look at the cycles of elements on Earth in Chapter 15. Finally, in Chapter 16 we attempt a summary of the evolving natural selection of the elements and the problems that exist for our understanding and actions in the near and distant future.

Further reading

There are several books, at different levels, on the history of chemistry. The following provide sufficient information to complement the contents of this chapter.

1. Asimov, I. (1956). *A short history of chemistry*. Anchor Books, New York. Simple and concise.
2. Partington, J. R. (1960). *A short history of chemistry* (3rd edn), Harper Torchbooks. Harper and Brothers, New York. A compact version of the monumental treatise on the history of chemistry by the same author.
3. Brock, W. H. (1992). *The Fontana history of chemistry*. Fontana Press, London.
4. Holmyard, E. J. (1957). *Alchemy*. Penguin Books, Ltd, Harmondsworth, Middlesex. A very readable and informative history of 'Egyptian art'.
5. Fung Yu-Lan (1983). *A history of Chinese philosophy*, Vols 1 and 2. Princeton University Press, Princeton, New Jersey. A source of inspiration for many more recent authors.
6. Needham, J. (1956). *Science and civilization in China*. Cambridge University Press, Cambridge. Vol. 2, particularly Chapter 13. A very good reference to the Chinese ideas on Nature.
7. Schwartz, B. I. (1985). *The world of thought in ancient China*. Belknap Press, Harvard University, Cambridge, Massachusetts. A more recent guide to Chinese thought, which rectifies some overly enthusiastic interpretations of the ancient philosphical texts.

More information on Chinese philosophy can be found, for example, in the *Encyclopedia Britanica*, under 'Taoism' (and see also given references).

The following books are three pleasant and (as far as possible) clear presentations of the strange world of the quantum and of wave mechanics for the layman.

8. Davies, P. (1980). *Other worlds*. Dent, London.
9. Gribbin, J. (1984). *In search of Schrödingers's cat*. Bantam books, Corsi, London.
10. Polkinghorne, J. C. (1988). *The quantum world*. Pelican Books, London.

Current experiments trying to confirm predictions of quantum theory are described in the following article.

11. Horgan, J. (1992). *Scientific American* **267**, 72–80.

Textbooks at undergraduate level, dealing with the principles of atomic physics and wave mechanics include the following.

12. Kerwin, L. (1963). *Atomic physics—an introduction*. Holt, Rinehart and & Winston, New York. Chapters 1 and 2. An excellent simple introduction to atoms and atomic properties.
13. Atkins, P. W., Clugston, M. J., Frazer, M. J. and Jones, R. A. Y. (1988). *Chemistry—principles and applications*. Longman Group, London. Chapter 1.
14. Day Jr, M. C. and Selbin, J. (1969). *Theoretical inorganic chemistry* (2nd edn). Reinhold Book Corporation, New York. Chapters 1–3. At a slightly more advanced level than the previous references.

The principles of cosmology of the early universe and models for its creation are described in many books; the following texts are particularly helpful and commendable.

15. Kaufmann, J. III. (1991). *Universe* (3rd edn). W. H. Freeman and Co, New York. Chapters 5, 28 and 29. An authoritative textbook of astronomy, informative, clear and extremely well illustrated.

16. Pagels, R. (1992). *Perfect symmetry*. Simon and Schuster, New York (1985) or Penguin Group (1992). A very clear, elegant and accurate presentation of cosmology for the layman.

17. Gribbin, J. (1994). *In the beginning—the birth of the living universe*. Penguin Books, London. A wide-ranging over-view of cosmology and evolution for the layman.

See also

18. Rouvray, D. H. (1995). John Dalton: the world's first stereochemist. *Endeavour*, (*New series*) **19**, 52–7.

19. Lawrence, C., Rodger, A., and Compton, R. (1996). *Foundations of physical chemistry*. Oxford University Press, Oxford.

2

Order in chemical systems: elements and their combinations

The same letters variously selected and combined
Signify heaven, earth, sea, rivers, sun
Most having some letters in common.
But the different subjects are distinguished
By the arrangement of letters to form the words.
So likewise in the things themselves,
When the intervals, passages, connections, weights
Impulses, collisions, movement, order,
And position of the atoms interchange,
So also must the things formed from them change.

Lucretius, *De rerum natura* (c. 57 BC)

A Elements and stoichiometric combination of elements

2.1 Introduction to chemical binding

In Chapter 1 we indicated that through careful analysis it has become clear that there are only some 90 elements *on Earth*† and that, putting to one side man's synthetic capability of making heavier nuclei, there are no more to be found. This conclusion is based on the firm knowledge of the unitary nature of nuclear structures, protons plus neutrons, which become generally unstable above atomic number 92 (Fig. 2.1), and on the fact that the electronic shells are completed in an understood systematic numerical manner, namely, 2, 8, 8, 18, 18, 32, (+6) electrons. There are no gaps for new elements. It was necessary to describe these findings since it is against the background of this sure knowledge that we can begin to see how these 90 or so elements have been selected naturally during the development of the universe.

By this natural selection we imply that there are reasons for: (1) the limitation of the number of elements to 90; (2) the appearance of some of these elements in larger amounts than others; and (3) the selective coming together of one or more of the atomic elements in gases, liquids and solids with molecular or continuous structures. There are yet further reasons for: (4) the incorporation of some elements and not others into biological structures and organisation; and (5) their use in man's constructions. The scene is not yet set for us to pursue this sequence of events historically since we need to describe the chemical and physical principles that have led from the production of atoms to that of chemicals, planets, life and man's artefacts. We shall use as little of modern theory as possible in describing these principles since this theory of forces acting between atoms is conceptually difficult and not indispensable for the understanding of the actual relevant phenomena. Throughout this chapter we shall be examining separated and isolated, chemically *ordered* species in particular phases and not real chemicals that equilibrate in systems between phases and chemical species (see Chapter 3). It is the *ordering* of atoms in space and the properties of the associated atoms that we wish to appreciate first.

2.2 The atomic and physical properties of the elements

Before we describe the combination of different elements in any detail it is convenient to look at trends in the atomic and physical properties of the elements themselves in a variety of physical states at room temperature. We notice first that, of the elements occurring naturally on Earth, that is 90, most (67) form metals and all but one of these (Hg) are solid, while only 11 are atomic or molecular gases and the remaining 12 form solid or liquid (one, Br) non-metals (Fig. 2.1). Notice how unusual the liquid state is at 300 K, that is 27°C. However, the properties of many elements, especially borderline cases, are not independent of the forms in which they appear. The example of carbon,

† As mentioned in Chapter 1, the elements of atomic number 43 and 61, technetium and promethium, respectively, are not found on Earth presumably because they have disintegrated completely. They can, however, be synthesised in nuclear reactors, and the atomic spectrum of technetium has been observed in some stars.

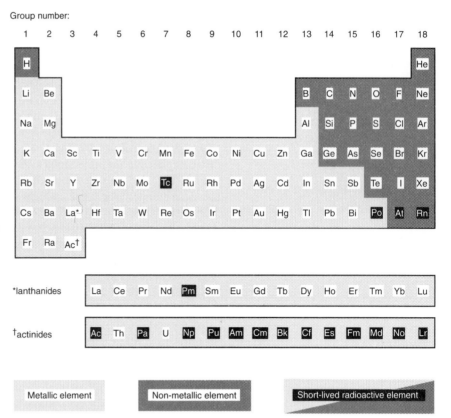

Fig. 2.1 The division between metallic and non-metallic elements in the periodic table. (From Cox, P. A. (1995) *The elements on Earth.* Oxford University Press, Oxford.)

which may occur in an amorphous state, as conducting graphite, as molecular fullerene, C_{60}, or as insulating diamond, clearly illustrates this point (see Fig. 2.7). At first, therefore, we shall consider the most stable form only of each element while making only occasional reference to one or two exceptional less stable forms that are observed now and then.

The physical state in which chemically simple substances (or indeed compounds) occur reflects, of course, the energy of binding, 'bonding', the holding together of their structural units (atoms or molecules) relative to the energy of thermal motion that tends to keep them apart (see Chapter 3). We have to consider the nature of these bonds to understand why some elementary substances are solids, some are liquids and some are gases at a given temperature, and then why, within each of these divisions of state, many different physical and chemical properties are found.

As explained in Section 1.4, bonding must depend on the nature of combinations of electrons from the shells of the atoms. Electrons effectively fill space around atomic nuclei, and the nuclei themselves, which are so small (Table 1.1), take no part in the binding together of atoms. Moreover, it is only at the very surface of the atoms that electrons can interact favourably since the inner space is occupied by fully filled inner electron shells that exclude one another (Section 1.3). One of the features of atomic combination will therefore be the extent to which the outermost, partly filled shells of one atom can be used to stabilise electrons by sharing them with neighbouring atoms. A rough impression is given if one thinks of the outer negatively charged electrons on an atom A as being able to feel the attractive force of the positively charged nucleus not only of A itself but of neighbouring atoms of the same kind as A, where A is any element with an incomplete shell of electrons, for example carbon. We have already stated that the packing together of atoms of a particular element results in different properties as we cross the rows of the periodic table from metals at the left-hand side to the final member of each row where one finds a noble monatomic gas, that is He, Ne, Ar, Kr, Xe, Rn. The very finding of the noble monatomic gases at the completion of the electron shells tells us that it is indeed the degree of completeness of an outer shell that largely decides the combining properties of an element. It also tells us that completed shells of electrons tend to keep atoms apart.

The simplest observable properties of the elements will now be taken in turn against the background of the periodic table. We ask first what is the spread of electrons away from nuclei, that is, what is the size of the atoms, and then what is the energy with which shared electrons are held by nuclei in combinations of atoms. It is intuitively obvious that these two properties will affect the strength with which the outermost electrons can cause atoms to interact through overlapping electron 'clouds' (Section 1.4) and the way they fill space, therefore determining the properties of matter.

2.2.1 Atomic sizes and structures of the elements

Using a variety of experimental data we know now that atoms in combination apparently have radii of around 10^{-8} cm (1 angstrom (Å) or 0.1 nanometer (nm)). To illustrate just one (indirect) approach to the occupation of space by atoms and its variation in the periodic table, Lothar Meyer's atomic volume curve (Fig. 2.2) gives a good impression. The trends follow the shell structures of atoms. (In passing, note that many other physical properties of elementary substances have been found to follow regular trends analogous to this, for

Fig. 2.2 The variation of atomic volume with atomic number.

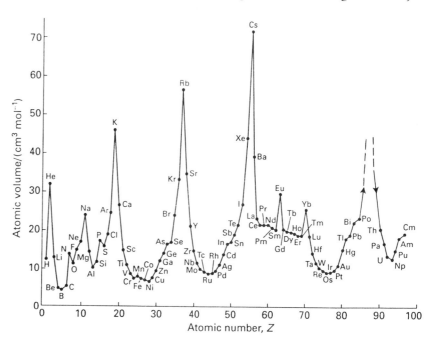

Table 2.1 Trends in physical properties of the elements

Substance	Melting point (°C)	Boiling point (°C)	Density (g cm^{-3})	Heat of vaporisation (kcal mol^{-1})
Li	180.5	1347	0.534	35.3
Na	97.8	883	0.971	23.7
K	63.2	774	0.862	18.9
Rb	39.0	688	1.532	18.1
Cs	28.6	679	1.873	15.9
B	2300	3658	2.34	120.7
Al	660.3	2467	2.70	69.6
Ga	29.8	2403	5.91	64.7
In	156.2	2080	7.31	55.5
Tl	303.5	1463	11.85	39.7
F$_2$	−219.7	−188	1.52 (liq.)	0.78
Cl$_2$	−101.0	−34	2.03 (liq.)	4.9
Br$_2$	−7.3	59	4.05 (liq.)	7.3
I$_2$	+113.5	184	4.93 (solid)	10.0
He	−272.7	−269	0.125 (liq.)	0.02
Ne	−248.7	−246	1.44 (liq.)	0.4
Ar	−189.4	−186	1.66 (liq.)	1.6
Kr	−156.6	−152	2.28 (liq.)	2.2
Xe	−119.9	−107	3.54 (liq.)	3.0

Data from Emsley, J. (1991). *The elements.* Clarendon Press, Oxford.

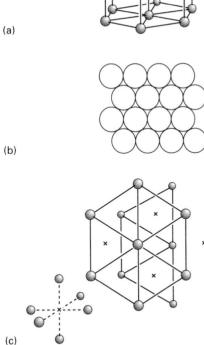

(a)

(b)

(c)

Fig. 2.3 Packing of metal atoms in a close-packed hexagonal (c.p.h.) lattice. (a) Stacking of close-packed planes in the c.p.h. structure. (b) A fragment of a plane of close-packed equal spheres showing space filling. (c) Octahedral holes (x) in the c.p.h. structure.

example, density, melting points, boiling points, heats of vaporisation, etc. Table 2.1 gives some examples.) It is obvious that from atomic volumes we can obtain *apparent* atomic radii. The atomic volumes themselves can be obtained simply by dividing atomic weights by the densities of the elementary substances, but a major problem does arise in that densities depend also on the physical state of an element; hence anomalous results are likely to be obtained due to atomic packing differences between elements. (In some cases, as mentioned above, there is even more than one state of an element; for instance, the density of *white* tin is 7.3 and that of *grey* tin is 5.75; hence this procedure leads to two different values for the atomic radius of tin!) We must, therefore, turn to the way in which the atoms are packed.

Through the use of X-ray diffraction the structures of all the elements in their crystalline states have been found. Metals are observed to form close-packed arrays of atoms, 12 or 8 near-neighbours, (Figs 2.3 and 2.4), while condensed diatomic gases have only one close neighbour (for example, H$_2$ and N$_2$) and non-conductors have anything from two to four near-neighbours (for example S and C (see Figs 2.7 and 2.8). (In the two forms of tin there are 12 (white metallic form) and 4 (grey non-metal form) near-neighbours, of any one tin atom.) Obviously, simple occupation of space is misleading concerning atomic size. We would think intuitively, however, that, given the building-up process of the periodic table (Fig. 1.9) and the energies of removing electrons (Fig. 1.12), the intrinsic size of atoms would be continuously changing, falling along each period before suddenly rising at the beginning of a new period, and indeed theoretical treatments indicate that this is so, but now we meet the conceptual difficulty of considering atoms in isolation.

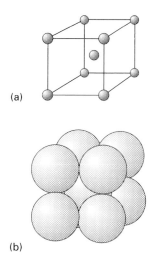

(a)

(b)

Fig. 2.4 (a) Packing of metal atoms in the body-centred cubic (b.c.c.) lattice. (b) Packing of equal shperes in a b.c.c. lattice but to illustrate space filling properly all the balls should touch.

In effect, there is a fundamental problem concerning the physical meaning of the atomic radius since, according to the orbital theory (see Chapter 1), the electrons are not strictly held close to nuclei but die away exponentially with distance from them in a probabilistic distribution 'cloud' to infinity; consequently, the isolated atoms have no defined boundaries. In principle, we could postulate that the radius coincides with the distance to maximum electron density (so-called orbital radius), but, even if it were possible to measure this distance in isolated individual atoms, the results would be questionable and of limited practical interest since we would have to relate this to sizes in structures. To make a long story short, rather than assigning definite radii to atoms, to compare with theoretical atomic radii, we determine from experimental data several 'operational' types of radii. These radii are derived, in fact, from the ways in which atoms are packed together and are therefore dependent on packing. It is found that as the packing changes from that in metals (Figs 2.3 and 2.4) with 8 or 12 near-neighbours, to that in monatomic gases, the radii need selected specification, for example, *metallic* radii for 8 or 12 coordination numbers (number of nearest neighbours), and diatomic, tetrahedral or octahedral *covalent* radii, etc. within molecular structures. Thus, before we give radii, we have to look at atom packing in more detail. To do this we go back to the periodic table and ask the reason for the trend

metals—non-metallic liquids or solids—molecular gases—atomic gases

across each period (Fig. 2.1), which is certainly correlated with atomic volumes in all the different periods (Fig. 2.2) and with the way in which the atoms pack. These physical properties turn out to give us the clue to why we have difficulty in discussing atomic size. The metals, as we have seen, occupy two-thirds of the periodic table—at the left, in the middle and toward the bottom. They exhibit a series of specific properties, particularly that of conducting electricity in a manner that is little altered by the increase of temperature (slight decrease). This indicates that some electrons in metallic solids are effectively free from individual atoms; hence we can think of metallic structure as a close packing of positive atomic ions strongly held together by their 'mobile glue' of negative electrons. Here binding has a *collective* lattice image rather than an atom-to-atom linkage, and close packing of spheres is just the most economical use of space (Figs 2.3 and 2.4). Atomic volume, density and apparent covalent (metallic) radii (Table 2.2) are therefore linked

Table 2.2 Metallic radii (in Å)*

Li	Be											Al			
1.57	1.12														
Na	Mg											Al			
1.91	1.60											1.43			
K	Ca	Sc	Ti	V	Cr	Mn	Fe	Co	Ni	Cu	Zn	Ga			
2.35	1.97	1.64	1.47	1.35	1.29	1.27	1.26	1.25	1.25	1.28	1.37	1.53			
Rb	Sr	Y	Zr	Nb	Mo	Tc	Ru	Rh	Pd	Ag	Cd	In	Sn		
2.50	2.15	1.82	1.60	1.47	1.40	1.35	1.34	1.34	1.37	1.44	1.52	1.67	1.58		
Cs	Ba	Lu	Hf	Ta	W	Re	Os	Ir	Pt	Au	Hg	Tl	Pb	Bi	
2.72	2.24	1.72	1.59	1.47	1.41	1.37	1.35	1.36	1.39	1.44	1.55	1.71	1.75	1.82	

* The values refer to co-ordination number 12. (See Section 2.2): Wells, A. F. (1984). *Structural inorganic chemistry* (5th edn). Clarendon Press, Oxford.

for metals as a group. We need to understand why their collective close-packed structures appear, as opposed to the structures adopted by non-metals; the latter structures are clearly indicative of a very local atom-to-atom linkage, that is, they are not a lattice property (for example, the structure of H_2).

2.2.2 Ionisation potential and electron affinity

That the trend corresponding to the progressive decrease of metallic character from left to right in the periodic table is related to the trend of decreasing tendency *of the atoms* to give up electrons *per se* and not only to a lattice is shown by the decreasing ease of the ionisation process,

$$M(g) \longrightarrow M^+(g) + e$$

where g refers to the gas state and e is the electron. This tendency is measured by the ionisation potential (I.P.$_1$) of the atom M, which does in general increase from left to right across each period in the table (Figs 2.5 and 1.12). From naïve theory of electrostatics the energy of interaction between a negative charge, the electron, and a positive charge, the nucleus, is $z_{eff}e/r$ where z_{eff} is the *effective nuclear charge*, that is, the charge *seen* or *felt* by the electron, e is the charge of an electron and r is the distance between the electron and the nucleus. The effective nuclear charge is used since even the most loosely held electron sees more than the unit charge on the nucleus because it penetrates the inner electron core (see Section 1.4 and reference 6 in the 'Further reading'). The 'size' of the atomic cores of electrons is a determinant factor, but the 'effective' nuclear charge acting on the electron and the shell from which the electron is removed are clearly equally important (Section 1.4) in determining the ionisation potential. Note, for example, that there are 15 elements from La to Lu, all metals, where the ionisation properties of outer 6s

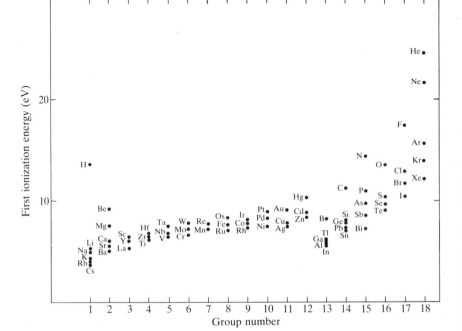

Fig. 2.5 First ionization energies of the elements (that is, the energy required to remove one electron from a neutral atom) plotted against their position (group number) in the periodic table. (After Cox, P. A. (1989). *The elements*. Oxford University Press, Oxford.)

electrons hardly change at all while the $4f$ shell is filled and three series where the metals go through maxima or minima in ionisation properties while the d shells ($3d$, $4d$, $5d$) are filled. The obvious suggestion is that the $4f$ electrons and orbitals are hidden (from bonding) and are, in effect, part of the core, and that the d electrons and orbitals though exposed at first are hidden later. Theoretical arguments indicate that this reasoning is correct (see references 2–6 in 'Further reading'). These peculiarities in the filling of d and f shells limit the availability of some orbitals and electrons for binding (Fig. 1.13) and correlate with the variations in physical properties (Fig. 2.2). We note that, as well as low ionisation energies, the metallic elements have very incomplete outer shells so that repulsion between outer electrons is low. Starting at the left hand side of any row of the table the metal bonding becomes stronger as the number of electrons available from each atom increases, for example from Na to Al and from Cs to W in the later rows and then falls back toward Group 12.

As we approach the right side of the periodic table, we find solid or liquid, non-conducting substances and then gases. As stated in Section 1.3, among these, at the extreme right, there is a group of six that are composed of atoms with very stable, complete, external shells of electrons, $1s^2$ or $ns^2 np^6$, for $n = 2$ to 6; these are the monatomic, noble gases He, Ne, Ar, Kr, Xe and Rn. They have closed non-interacting electron shells and hardly bind to other atoms of the same kind. When they do condense they form low-density close-packed structures due to van der Waals forces (Section 2.6).

The right-hand top part of the periodic table is formed from atoms with high ionisation potentials, small cores and nearly completed shells that are reluctant to lose electrons and, in fact, have a tendency not only to hold on to their own electrons but to pull as many electrons as possible from neighbouring atoms, up to the eight of the fully filled shell, as close to their nuclei as possible, as shown by their electron affinity (Table 2.3). This causes the elements to form molecules or continuous more open lattices of low co-ordination number in which the structures show that bonding electrons must be held very largely locally in pairs (see the Pauli exclusion principle) in the space between *pairs of nuclei*. The electrons not used in bonding, the majority, now repel all electrons of other atoms thus giving the impression of the

Table 2.3 Electron affinities of the main group elements (in eV)*

H							He
0.754							−0.5
Li	Be	B	C	N	O	F	Ne
0.618	−0.5	0.277	1.263	−0.07	1.461	3.399	−1.2
					−8.75		
Na	Mg	Al	Si	P	S	Cl	Ar
0.548	−0.4	0.441	1.385	0.747	2.077	3.617	−1.0
					−5.51		
K	Ca	Ga	Ge	As	Se	Br	Kr
0.502	−0.3	0.30	1.2	0.81	2.021	3.365	−1.0
Rb	Sr	In	Sn	Sb	Te	I	Xe
0.486	−0.3	0.3	1.2	1.07	1.971	3.059	−0.8

* To convert to kJ mol^{-1}, multiply by 96.485. The first values refer to the formation of the ion X$^-$ from the neutral atom X; the second values to the formation of X^{2-} from X$^-$.
Source: Hotop, H. and Lineberger, W. C. (1985). *J. Phys. Chem. Ref. Data* **14**, 731. Positive values correspond to energy released and negative values to energy absorbed.

electrons in bonds being localised. As a consequence, elements next to noble gases are composed of diatomic molecules X_2, namely hydrogen, H_2, and the halogens, F_2, Cl_2 and so on. All the other non-metal liquid or solid elements are either composed of discrete molecules, for example Br_2, I_2, P_4, S_8 (see Fig. 2.6), or have their atoms bonded in continuous low co-ordination number networks, for example graphite or diamond (see Fig. 2.7). Naturally, these

Fig. 2.6 Some examples of discrete molecules (Br_2, I_2, P_4 and S_8). Alternative forms for P and S are shown, but there are many other forms for sulphur including linear polymers, and a third form for phosphorus (black phosphorus).

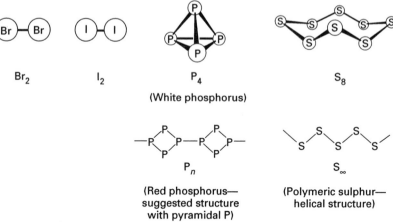

Br_2

I_2

P_4
(White phosphorus)

S_8

P_n
(Red phosphorus—
suggested structure
with pyramidal P)

S_∞
(Polymeric sulphur—
helical structure)

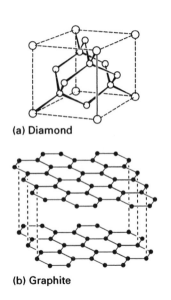

(a) Diamond

(b) Graphite

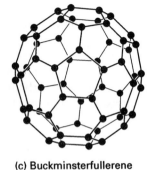

(c) Buckminsterfullerene

$+ C≡C–C≡C–C≡C +$

(d) Polyacetylene

Fig. 2.7 Different forms of carbon.

latter substances, having continuous lattices, are solid, whereas the physical state of the molecular substances depends on the energy of secondary *intermolecular* forces (see later). (All physical states depend on temperature, of course.) It is clear, too, that alternative arrangements, for example in rings or chains and in tetrahedra or branching chains, satisfy the co-ordination rules equally and, in fact, different forms exist for S and P (Figs 2.6 and 2.7), a finding that touches on the problem of shape (Section 2.2.4). (NB. The different forms are not equally stable.)

All this indicates that, unless the atom has a completed outer ns^2np^6 electron shell, that is, unless it is a noble gas, then it has either: (1) a low affinity for electrons, a very incomplete shell structure, and a tendency to give up electrons to go back to the noble gas structure of lower atomic number while generating a collective electron gas around the nuclei in a lattice, that is, it is a metal; or (2) a higher affinity for electrons (and a nearly filled shell) which is used to achieve the magic number structure of the next stable electron shell of a noble gas of higher atomic number by sharing electrons with other atoms and forming 'bonds', that is, it is a non-metal. Just as the formation of metals depends on low ionisation energy and electron affinity (and a large number of empty orbitals), the formation of discrete molecules or networks depends on the atom's high ionisation energy and high electron affinity (and a large number of filled orbitals). Since in elementary substances there is only one kind of atom, unequal electron transfer is out of question; hence equal sharing of electrons (so called homonuclear *covalent bonding*) is the only possibility toward the end of each period.

There is, then, a strong tendency for the later elements of a row to form the molecules with the smallest number of atoms using each unpaired electron to form one bond, but, somewhat strangely, molecules of O (O_2) and N (N_2) are diatomic gases while S (S_8) and P (P_n where $n = 4$ or is very large) are solids. To be consistent, in the cases of O_2 and N_2 the building toward the noble gas shell must be represented by double O=O or triple N≡N bonds rather than single bonds as, for example, in H—H, F—F, Br—Br or I—I. This rule extends in part to carbon, which can form —C≡C— in polyacetylene, $\diagup_{C=C}\diagdown$ in graphite and fullerene as well as $\diagup_{C}\diagdown$ in diamond (Fig. 2.7). Silicon, by way of contrast, almost always gives single bond structures $\diagup_{Si}\diagdown$. The three atoms C, N, O are a very special group in that they readily form rather strong multiple bonds as well as single bonds, a fact that contributes to making organic chemistry a distinct branch of chemistry and that is connected to the chemistry of life's polymers. The formation of multiple bonds is possible everywhere in the periodic table, but they are generally weaker other than with the elements C, N and O.

This description still remains somewhat problematic, however, since, when we consider elements down a group, that is, when heavier atoms come together, the homonuclear 'covalent bond' description for light non-metals merges with the lattice collective description where both descriptions are derived apparently from the same set of atomic orbitals, and the elements become metals. For instance, as stated before, we know that tin, Sn, exists in both forms, metal and non-metal, while silicon is a non-metal and lead a metal. All may be considered to use 'hybrid' sp^3 orbitals (see references). We wish to avoid the sophisticated mathematics necessary to analyse these situations although they are now understood in essence. We can give no more than a qualitative impression, which is that, when atoms are very small and have a relatively high electron affinity and ionisation potential, the two tendencies, that is, to hold electrons close to their own nuclei and to reach a noble gas electron structure, can be better satisfied in a short diatom structure than in longer-range interactions and, in turn, than in metallic structures. For example, the formation of multiple bonds O=O rather than single-bond chains —O—O—O— can be thought of in terms of optimalising bonding locally while avoiding electron/electron repulsion in other directions between the *small* completed shells. Multiple bonding is therefore a correlate of very small electron cores and resultant very short bond lengths. (Throughout the periodic table the most important double bonds of heavier elements remain those formed with oxygen.) As we go to heavier and *larger* atoms the interaction becomes more complicated for three reasons: (1) the electron–electron repulsion is less and electron affinity is less so that packing as in metals is more likely; (2) atom cores become larger; and (3) there are new possible orbital combinations. When one reaches argon, for example, the $3s$ and $3p$ orbitals are filled, $3s^2 3p^6$, but the $3d$ orbitals remain unoccupied and the $4s$ orbitals of potassium and calcium are filled next. As we proceed across the row from sodium to argon the atoms are larger than in the row lithium to neon so that short-range repulsion is less and the binding between atoms is no longer simply related to the orbitals and electrons of one period. In forming —S—S—S— chains rather than S=S molecules, for example, there is some help from the $3d$ as well as from $3s$ and $3p$ orbitals. Electron attraction to nuclei is as important as electron–electron repulsion in deciding the units formed. Thus, going down any group there is a switch from very localised binding, often in molecules, to more delocalised binding of electrons in the solid state.

2.2.3 Empirical covalent and van der Waals radii of atoms

Against the background of these difficulties, which are intrinsic to the nature of electrons in atoms, we can give a set of empirical radii, deduced by making allowance for the complexity of the structure from which they were obtained (see references 5, 6 and 9 in 'Further reading') and which are useful in thinking about the way in which space can be filled along the direction to the nearest neighbour(s). These are so-called *covalent* radii. The values in Table 2.4 (compare metallic radii in Table 2.2) follow expectation in that they fall across a whole short period and increase down a group (Fig. 2.8). In long periods the rises and falls of the radii can again be explained by the lack of full availability of *d* and *f* orbitals and electrons. The way space is filled in other directions can also be defined by the closest distance of approach of *non-bonded* atoms. These values are given in Table 2.5 and are called *van der Waals* radii (see Section 2.6). Together, the two sets allow us to describe the degree to which space in any material is 'unoccupied'. This is very useful in the understanding of small molecules, crystals and large folded structures such as proteins. It is generally true that very little space is left empty (see Figs. 2.3, 2.4

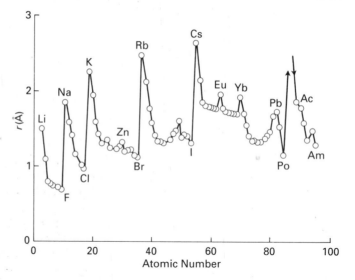

Fig. 2.8 The variation of atomic radius, *r* (metallic and covalent) through the periodic table. Note the contraction of radii following the lanthanides in period 6. We have included non-metallic elements in the figure using radii derived by taking half the distance between like atoms in a molecule. (After Shriver *et al.* (1994), reference 3 in 'Further reading'.)

Table 2.4 Covalent radii for some elements (Å)*

B	0.88	Al	1.26	Ga	1.26	In	1.44
C—	0.77	Si	1.17	Ge	1.22	Sn	1.40
C=	0.67						
C≡	0.60						
N—	0.70	P—	1.10	As—	1.21	Sb—	1.41
N=	0.63	P=	1.00	As=	1.11	Sb=	1.31
N≡	0.55						
O—	0.66	S—	1.04	Se—	1.17	Te—	1.37
O=	0.62	S=	0.94	Se=	1.07	Te=	1.27
F	0.64	Cl	0.90	Br	1.11	I	1.28
Ne	0.71	Ar	0.98	K	1.12	Xe	1.31

Data from Ball, M. C. and Norbury, A. H. (1974). *Physical data for inorganic chemists.* Longman Group, London.
* The radii are those in common geometries mostly octahedral or tetrahedral.

Table 2.5 van der Waals radii (Å)

				H	He
				1.20	1.80
C	N	O		F	Ne
1.70	1.5	1.40		1.35	1.60
	P	S		Cl	Ar
	1.9	1.85		1.80	1.90
	As	Se		Br	Kr
	2.0	2.00		1.95	2.00
	Sb	Te		I	Xe
	2.2	2.20		2.15	2.20

Data from Ball, M. C. and Norbury, A. H. (1974). *Physical data for inorganic chemists.* Longman Group, London.

and 2.7). It is space that allows motion, of course, so that liquids generally have more space available. We are not, however, in a position to consider motion in condensed systems yet, and we must be aware of electron motion in metallic substances as well as of atomic or molecular motion in liquids and gases.

2.2.4 Shapes of molecules formed from single elements

In Section 1.5 we mentioned that the shape as well as the size of atoms in molecules arises from the nature of the electrons and their motion around atomic nuclei. We can now elaborate with examples taken from the elements. While metals have close-packed infinite structures in which we can consider that each atom is a sphere and hence that molecular shape does not arise, and, of course, diatomic molecules have a simple linear structure, non-metal elements that form more than one bond to each atom give rise to shapes such as those shown in Figs 1.15 and 2.6 and 2.7. The shapes of molecules (not only of elementary substances) are due to the available bonding orbitals of atoms and their character, and to the number of electrons in each atom. In Fig. 2.6 it is seen that bonding pairs of electrons and non-bonding pairs of electrons occupy space roughly symmetrically, for example, the bond angles in P_4 and S_8 are tetrahedral. In structures with double bonds, for example, graphite, the angles become $120°$, again for symmetry reasons. The ability to form different molecular shapes is all part of the ability of atoms to bind with different kinds of symmetry and affinity.

The macroscopic shape of all materials, for example, of crystals of metallic elements, is a quite different property and is related to the surfaces that they form. Here we deal with different *phases*, and phases are defined by boundaries, not by local internal bonding (see Chapters 4 and 6). Local molecular shape and crystal morphology are not closely related. The way in which molecules pack together in solids has not been described yet since it is more convenient to tackle this subject later (Section 2.6). Surfaces will be described in Section 6.2.3.

2.2.5 The binding energies of the elements

The physical properties of the elements reflect the strength as well as the continuity of binding or structure (Table 2.1). For a continuous metallic solid we expect the energy of binding to reflect the number of electrons available and the ability of nuclei to hold electrons (see Chapter 3.5 of reference 5, in 'Further reading'). Down groups 1 and 2 we see that the number of available electrons is fixed, but, as the ability to retain electrons gets weaker (see Fig. 2.5), bonding in metals weakens. Since there is a charge separation of electrons (minus charge) and effective atomic ions (plus charge) in metals, and therefore there is also a long-range co-operative binding that stabilises the lattice, we find that binding in metals increases along all rows at least for the first few groups. When we go across the first two periods to the non-metals the number of available electrons increases further, as does binding energy per atom, until group 15 when the lattices become discontinuous (N, P) and average bond energy per atom weakens (see Fig. 2.9). As stated earlier, it is

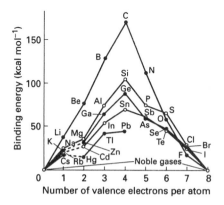

Fig. 2.9 Binding energy of solid non-transition elements relative to their free atoms. (See Chapter 3.5 of Phillips and Williams (1966), reference 5 in 'Further reading'.)

not just the increasing attractive energies of electrons for nuclei that are causing change as we go across periods; there is also the repulsion between increasing numbers of paired electrons in the shell which is being filled. Long-range co-operativity also decreases quickly at and after group 14. It is no accident, then, that the binding strength per atom and the packing change most markedly after group 14 where the number of outer electrons, four, is half the maximum number of electrons in the available orbitals, $2s^2 2p^6$, eight. The net attraction depends approximately on the product $n(8\text{-}n)$, where n is the number of electrons, but in a complicated manner even for the first two rows of the periodic table.

In later periods the available number of electrons increases to group 3 and then, somewhat erratically, to group 9. Instead of an 8-electron 4-orbital rule we have an 18-electron ($s^2 d^{10} p^6$) 9-orbital rule and stability increases with the product $n(18\text{-}n)$ but more irregularly (see Fig. 2.10). After ten electrons fill the d subshell orbitals they do not remain available, becoming part of the core, so that, for example, the binding energy of zinc as a metal resembles that of calcium indicating that it has but two available electrons leaving the $3d^{10}$ electrons unavailable. The weakness of bonding in heavier metal elements requires detailed examination of electron availability, but soon after Zn, Cd and Hg the development of non-metal character gradually asserts itself again, as shown in the structures and weakness of binding in group 15. The parallel with the first two rows is apparent. Later in the periodic table the $4f$ electrons are never available (see Section 2.4.3). All these trends, which explain variations in all other physical properties (see Fig. 2.2, for example), can now be based on a firm theoretical understanding. It would be quite wrong to suppose, however, that there is any *simple* electrostatic explanation for the variation of element properties in the periodic table; in effect, the behaviour of electrons in forming atomic shells is quite contrary to the *simple* expectation of classical electrostatics and can only be understood in terms of a counter-intuitive wave-mechanical model which we are forced to accept as being closer to 'reality' than the ideas that our sense perceptions generate (Chapter 1).

Fig. 2.10 Binding energy of solid elements in the three transition series. The binding energy is relative to the free atom. (See Chapter 3.5 of Phillips and Williams (1966), reference 5 in 'Further reading'.) Note the peculiarity of manganese when the $3d$ shell is half full. The theory for this effect is given elsewhere but stability of a half-filled subshell has great influence, see references 1–3.

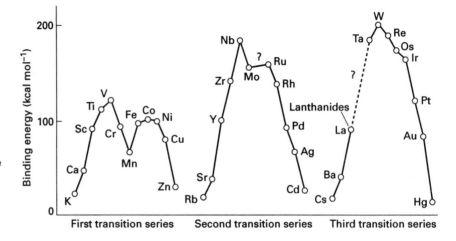

2.2.6 Summary of the properties of the elements and their natural selection

In this chapter we are discussing the nature of the forces between like atoms that cause structure, *order*. We have shown why elementary substances are ordered and also why they are so differently ordered in physical states at or around the temperature of Earth. This is a feature of natural self-selection amongst like atoms. We shall have occasion to refer again to the fact that there are elements in very different self-ordered and self-selected conditions on Earth such as C (coal), N and O (diatomic gases), inert gases, and Au and Cu (metals). These conditions of the stable form of the elements mean that, as far as pure elements are concerned, some are able to flow at 300 K (27°C), for example N_2 and O_2, constituents of air, while most others cannot. Moreover, the elements individually at room temperature provide but two liquids, Br_2 and Hg, of rather uncommon elements. The absence of liquids arises largely because of the dramatic changes in properties after group 14 to 16 from solids to gases. Most elements are solid metals. The solid metals provide strong structures but, perhaps most importantly, they generate a very interesting medium for *electron flow*, not atom flow, which proves to be extremely valuable in man's *organisations* (Chapter 14). These metallic elements have other fortunate properties in that they are easily shaped and take stress without fracture as bulk materials. Meanwhile, the non-metals give rise to idiosyncratic molecular shapes or lattice surfaces. These shapes and surfaces allow constructs with properties quite different from those of metals, but they are of equal functional use as we shall see in proteins and plastics such as carbon fibres. Finally, we must always remember that the close-packed structures especially of heavy elements lead to high density, which is obviously important in a gravitational field such as that of Earth which has a metallic iron core (Chapter 8). Table 2.6 summarises the general features of the various kinds of substances made from one element.

Now, most elements are not found naturally in homo-atomic assemblies but in combination with others, hetero-atomic assemblies, a subject to which we turn next. (As an aside, note that it is largely through man's industry that more or less pure elements, for example copper and iron, occur in objects around us (see Chapter 14).).

Table 2.6 Outstanding general features of elemental substances at 270–370 K

Metals			Semimetals		Non-metals				
Li	Be		B	C (graphite)	C	N	O	F	He
Na	Mg	Al		Si		P	S	Cl	Ar
K	Ca	Sc		(Ge)		As	Se	Br	Ne
Solid			Solid		Solid/gas				
Electronic conductors			Semiconductors to insulators		Insulators				
Malleable			Brittle		Soft or free flowing (gases)				
Hard			Hard						

2.3 The combination of different elements—stoichiometric and non-stoichiometric compounds

The commonly observed materials around us are rarely pure elements but rather combinations of elements, which are often complicated in their composition as we shall see. Most materials were (see Chapter 1) and still are recognised by their phase boundaries, for example gases (air), liquids (water) or minerals (earth) of different kinds. The chemical nature of these materials has been uncovered by analysis of the elements in them and by their synthesis from elements. It was discovered that for combinations of some elements simple rules of proportion occurred, for example, A_xB_y, where x and y are small integers. Such substances are called stoichiometric compounds. These findings have led to useful theories that explain some of the reasons for the observed combination of elements. However, it was also found that in solids (and liquids) just as many non-stoichiometric combinations of different pairs of A with B atoms exist, often over limited, but sometimes over extensive, continuous ranges of composition, requiring theories of condensed phases unrelated to small number ratios. In all these circumstances of A and B combination in solid (or liquid) compounds, a fundamental difference lies in the relative ease with which specific A and B elements hold on to or give up electrons; hence, some form lattices based on close packing, when stoichiometry is often not found, as in alloys, while others form covalently bonded units that are more generally stoichiometric, as in non-conductors. Both types of binding can be related to 'phase diagrams' (see Chapter 4) where, for example, we plot pressure p against temperature T (for fixed composition), composition against pressure or temperature, or composition against composition (different components at constant p and T) indicating over which regions different types of combination are observed (see Fig. 4.1). More generally, therefore, it is the miscibility of A/B mixtures in liquid and solid phases that we need to understand, as well as the nature of the interactions that sometimes limit miscibility to stoichiometric *compound* formation. This fact is often hidden in standard texts on chemistry. To present in an easier form the principles of stoichiometric compound, not yet phase, formation we shall start in an orthodox manner looking for general principles controlling the combinations of two different atoms in isolated phases. The controlling factors that stand out are the differences in character of the combining elements, which we describe under electronegativity and size differences (Section 2.4.1), and differences in combining ability, that is, valence. After we have given a general account of the way in which elements A and B interact (Section 2.4) we shall divide the discussion, first of stoichiometric compounds, into two classes, namely: (1) A and B are like elements, (Sections 2.5–2.8); and (2) A and B are unlike elements (Section 2.9). These combinations produce very different properties in compounds. Second, we turn to non-stoichiometric substances (Chapter 2B) with ranges of composition, and there we can deal with ternary and higher varieties of atom combination most easily. As was the case in the description of the structures and properties of the elements, we are attempting to describe and explain *order* wherever it is found in isolated substances.

2.4 General factors affecting the combination of two different atoms

2.4.1 Electronegativity and size

It is generally considered that the way (and the strength) in which two unlike atoms combine results from their differences in physical atomic properties, that is, size, ionisation energy and electron affinity, while shape is imposed on the combination through orbital symmetry properties. We have already described sizes of atoms (Section 2.2.1) and sizes of ions will be given in Section 2.9. We have also shown that ionisation energies and electron affinity energies depend on essentially the same factors, namely, the nature of atomic orbitals and their energies (see Sections 1.3 and 2.2.2); hence, for practical purposes, it is logical and convenient to combine both the last two energies into just one parameter since both are required in binding. After all, covalent bonding is identified with both giving and taking electrons, pairwise sharing, and metal bonding is identified with giving electrons while the tendency to take electrons is very weak. This combination has been done in various ways. The simplest, proposed by R. S. Mulliken in 1934 is to take an average sum of the electron affinity energy, EA, and the ionisation energy, IP, for each atom, and the resulting new parameter, which measures the tendency of a neutral atom to hold on to shared *pairs* of electrons when combined (in effect the atom partly gives one (IP) and partly takes one (EA) electron simultaneously), is called *electronegativity*. Differences of electronegativity—Δ—between elements then become a criterion for bond type. Large differences correspond to so-called *ionic* bonding, shown as A^+B^-, where one element B takes the electron from A to form ions A^+ and B^-; small differences of Δ correspond to predominant covalent bonding where both elements have high electronegativity, or metallic bonding where both elements have low electronegativity. Covalent bonds are shown as A—B. The difference of electronegativity Δ then influences binding strengths and physical properties in compounds (in some cases through the additional effect of the intensity of intermolecular forces (see later)). The scale of electronegativities given in Table 2.7 has a more empirical

Table 2.7 Pauling electronegativities (H = 2.1)*

Li	Be	B											C	N	O	F
1.0	1.5	2.0											2.5	3.0	3.5	4.0
Na	Mg	Al											Si	P	S	Cl
0.9	1.2	1.5											1.8	2.1	2.5	3.0
K	Ca	Sc	Ti	V	Cr	Mn	Fe	Co	Ni	Cu	Zn	Ga	Ge	As	Se	Br
0.8	1.0	1.3	1.5	1.6	1.6	1.5	1.8	1.8	1.8	1.9	1.6	1.6	1.8	2.0	2.4	2.8
Rb	Sr	Y	Zr	Nb	Mo	Te	Ru	Rh	Pd	Ag	Cd	In	Sn	Sb	Te	I
0.8	1.0	1.2	1.4	1.6	1.8	1.9	2.2	2.2	2.2	1.9	1.7	1.7	1.8	1.9	2.1	2.5
Cs	Ba	La–Lu	Hf	Ta	W	Re	Os	Ir	Pt	Au	Hg	Tl	Pb	Bi	Po	At
0.7	0.9	1.1–1.2	1.3	1.5	1.7	1.9	2.2	2.2	2.2	2.4	1.9	1.8	1.8	1.9	2.0	2.2
Fr	Ra	Ac	Th	Pa	U	Np										
0.7	0.9	1.1	1.3	1.5	1.7	1.3										

* The values given in the table refer to the common oxidation states of the elements. For some elements variation of the electronegativity with oxidation number is observed, for example: Fe(II), 1.8; Fe(III), 1.9; Cu(I), 1.9; Cu(II), 2.0; Sn(II), 1.8; Sn(IV), 1.9. *Source*: L. Pauling, reference 4 in 'Further reading'. See also Allen, L. C. and Huheey, J. E. (1980). *J. inorg. nucl. Chem.* **42**, 1523 and note the relationship to the hard/soft classification of Pearson (see 'Further reading').

origin, but is closely related to that of Mulliken. (It is also related to the scales of so-called 'hardness' and 'softness' (see references 2 and 3 in 'Further reading' and Section 5.8). The variation of electronegativity clearly follows the periodicity of the periodic table. All these scales, derived from atomic properties, take no notice of relative size (see Section 2.2.1) so that we shall need two parameters, electronegativity and size, as well as occupancy and symmetry properties of orbitals to describe the properties of compounds in terms of those of atomic elements. One additional factor, which relates to combining ratios, is described under 'valence'.

2.4.2 Combining ratios—valence

Continuing the general discussion of compounds we note that a most important chemical property of an atomic element is its combining capability, as in A_xB_y. This is classically related to the 'valence' of A and B as shown in the combining ratios in hydrides (Fig. 1.8). Due to the discovery of the complex nature of bonding between atoms, and even when we consider only cases where x and y are whole numbers, it is not always so easy to show a relationship between the observed combining ratio in all the different compounds of an element A and any fixed number to be given for the valence of A. We, therefore, start a description of the observed combining ratios with an operational definition of valence as the number of *electrons* lost or gained by an atom in an ionic compound or half the number shared in a covalent compound. Turning back to Fig. 1.8 for the hydrides, we see that valence as defined increases and then decreases in the first row of the periodic table, that is, it takes the values 1, 2, 3, 4, 3, 2, 1, 0 for Li, Be, B, C, N, O, F, Ne. A similar result is obtained for fluorides or chlorides of these elements. As stated the periodic table was built up in part using the knowledge of such chemical combining ratios—classical valencies.

Valence considerations are not restricted to AB_n units, however, but can be extended to A_mB_n where stoichiometry and valence have a more complicated relationship. Thus we can build from C and H, CH_4 and $CH_3 \cdot (CH_2)_n \cdot CH_3$, as well as branching chains,

$$\begin{array}{cc} -\,CH\,-\,CH_3 \\ | \\ CH_2\,-\,CH_3 \end{array} \quad \text{and} \quad \begin{array}{c} CH_3 \\ \diagdown \\ C\,-\,CH_3 \\ \diagup \\ CH_3 \end{array}.$$

It is clear that the building of larger molecules can use either C—C or C—H bonds and extends all the way from CH_4 to diamond (Fig. 2.7), keeping the same tetrahedral bond angle. In all these cases carbon has a valence of four and hydrogen has a valence of one. We have already observed that carbon itself also forms other types of solids based on 'unsaturated binding' in such substances as graphite, but we can still give carbon a valence of four, using double bonds. The same unsaturated bonding feature arises in carbon compounds with hydrogen; taking again as a starting point the chain $CH_3 \cdot (CH_2—CH_2)_n \cdot CH_3$, then by removing hydrogen we obtain $CH_3 \cdot (CH=CH)_n \cdot CH_3$ and $CH_3 \cdot (C\equiv C)_n \cdot CH_3$. The networks of carbon need not be on a string, and two-dimensional rings such as $(CH_2)_6$, cyclohexane, or $(CH)_8$, cubane, (see marginal illustration) can arise, as can any combination of single, double and triple bonds. Finally, the systems may branch at all carbon

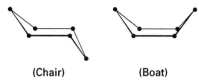

(Chair)   (Boat)

Two forms of cyclohexane, C_6H_{12}; carbon atoms only are shown.

(Cubane)

Cubane, C_8H_8; carbon atoms only are shown.

(a)

H_2

H — H

N_2, CO

:N ≡ N: :C ≡ O:

O_3, SO_2, NO_2^-

NH_3, PCl_3

Molecules that fit noble gas electron structures when bonding is written as shown.

(b)

PCl_5, SO_3^{2-}

PO_4^{3-}, SO_4^{2-}

Molecules that do not fit noble gas electron structure.

atoms into a huge, cross-connected set. In all these cases we consider that carbon still shows a valence of four. This versatility that allows non-metals such as carbon to build structures with hydrogen is shared by other non-metals, but in only a very limited way. In fact, carbon dominates this kind of building of frameworks using C—H bonded units, multiple bonds and fixed valence, and is given a chapter of its own—organic chemistry (see Chapter 9)—for this reason. This building of frameworks also generates selectivity of interaction with other elements, which allowed life to develop.

Silicon builds very different frameworks largely with O, which dominate Earth's surface (Section 2.11). Here the valence of Si is four and that of O is two, for example, in SiO_2. This descriptive procedure is open to a wide extension of valence in many stoichiometric compounds using single and multiple bonds and parallel building of compounds can now be done for combinations between many atoms (see marginal illustrations). Frequently, however, the situation becomes more complicated since, as in the case of carbon, which forms CO and CO_2, several types of oxide of the same element can be formed, for example, M_2O, MO, M_2O_3, etc. Once again we must be aware of the formation of double and triple as well as single bonds, but now another problem is found: some elements show more than one combining ratio even where we are forced to consider all the bonds as single, for example, PCl_3 and PCl_5, or, in the case of ionic metal salts, $M^{2+}(Cl^-)_2$ and $M^{3+}(Cl^-)_3$.

2.4.3 Variable valence in A/B compounds

It has been noted and explained that valence in the hydrides of the first short period increases from one to four through the first four groups of the two short periods of the periodic table and then falls back to zero as electrons are added to the $2s$ and $2p$ shell (Fig. 1.8). For reasons advanced above (Section 2.2.5) and which are explained in advanced texts, elements in the long series of the periodic table do not increase their valence systematically as the electrons are added to the shells. The extreme example is the lanthanide series in which 15 elements in series have extremely similar chemistries and a dominant valence of three, for example, they all form chlorides MCl_3. Thus the change in nuclear charge by 14 and the increase in the number of $4f$ electrons by the same number has little influence on valence or on chemical properties. The theoretical explanation is that these electrons and their empty orbitals are held too close to the nucleus to give rise to combining capability, an argument we have already used to explain the similarity of physical properties of these elements (see Section 2.2.2). In the periodic table there is then the row starting from group 1: Cs (value 1), Ba (value 2), 15 lanthanide elements (value 3), followed by Hf, in group 4, where valence increases again.

In a somewhat parallel fashion, in the three series of 18 elements, that is, the transition metal series caused by filling of d subshells, we observe that the nuclear charge increase of 10 with an increase of 10 electrons from Ca to Zn, from Sr to Cd and from Ba to Hg (overlooking the above lanthanides) gives rise to 10 metals in succession now with variable valence and, at the end of the series, there is more than a passing resemblance in valence between, say, Ca and Zn. These two metals both show a major valence of two, as do Sr and Cd and Ba and Hg, for example, in chlorides MCl_2. Following zinc it is easy to match the valences of the elements Ga, Ge, As, Se and Br with those of Al, Si, P,

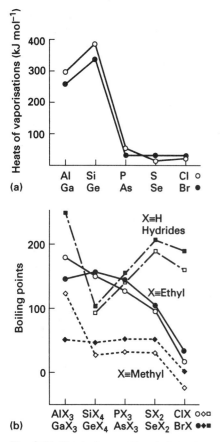

(a)

(b)

Fig. 2.11 Physical properties of elements of groups 13 to 17. (a) Heats of vaporisation; (b) boiling points of their hydrides (K) and alkyls (°C).

Fig. 2.12 Valence patterns for the elements from potassium to zinc, compiled from a survey of aqueous monatomic ions and aqueous oxyions, stoichiometric oxides, halides and other simple salts. NB. Valence and oxidation state (see Chapter 5) are identical for these compounds.

S and Cl in their respective periodic groups, that is, 13, 14, 15, 16 and 17. They also show similar physical property patterns in compounds (see Fig. 2.11(a) and (b) for some examples). Thus we have associated valence of metals, with available electrons to give away, and valence of non-metals with electrons to share. When we turn to the valencies of the 10 elements from group 2 (Ca, Sr, Ba) and to those in group 12 (Zn, Cd, Hg), we find that they show the valence pattern of Fig. 2.12. Each element has more than one valence, differing by single units, and some have several. Noticeably, the number of valences increases from group 3 toward the middle of the series, s^2d^5, as do physical properties, and then decreases. The conclusion is obvious: the number of electrons and empty orbitals available for chemical combination is variable (and depends on the partner in AB compounds) since, as we progress along the series, d electrons become more hidden; at some elements any valence is then almost equally probable, for example, from 2 to 7 for manganese with oxygen. Thus, all these elements show variable valence depending on their chemical partners, and with some partners several valences are of approximately equal stability. The simplest example is iron that forms both ferrous, $FeCl_2$, and ferric, $FeCl_3$, chlorides.

Before we go further, we must add that variable valence is possible for many elements other than transition metals since some electrons are less available than others. Examples are PCl_3 and PCl_5 where, in the first case, two electrons, the $3s^2$ electrons, have been held back from combination, while being used in the second. In effect, variable valence is quite common amongst heavier non-metal compounds (see marginal illustrations on p. 49), but the valence steps are now in units of two and unpaired electrons are rare in contrast to transition metal valence states. In PCl_5 the inert gas state $3s^23p^6$ has been exceeded but, as stated before, there are always present after the first short period empty orbitals, here $3d$, that help to expand the valence possibilities. It is features such as this that make simple theoretical treatments of almost all chemical problems almost impossibly difficult to appreciate fully, so it is preferable to proceed empirically with a background of theoretical appreciation.

In the first two or three rows, somewhat different cases of *apparent* variable valence are clearly known, for example, in compositions such as CO, N_2O, NO, NO_2, N_2O_4, in which the valence of C or N is sometimes apparently *lower* than expected from CH_4 and NH_3, and sometimes higher as in NO_2. The concept of valence as classically defined or as introduced above is not really capable of

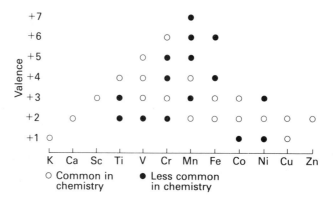

handling these cases. Unfortunately, simple numerical concepts of stoichiometry are once again related to theoretical electronic structures rather than to identifiable bonds, so that pictures (see marginal illustrations on p. 49) are to be treated with caution. This problem is acute since it lies behind the diversity of C, N, O chemistry on which organic chemistry and life are based. As mentioned before, unusual valences with multiple bonds can occur anywhere in the periodic table, but their effects are most striking in the first row. Once again, and in conclusion, we state that the classical concept of 'valence' with its variability in all such compounds is understood, at least in outline, and that it is an essential feature of natural selection of elements for one another, arising from the sizes and subshell electronic structure of atoms.

2.4.4 Available orbitals: structures

Through a long series of studies man has discovered that the ways in which different atoms bind together and the structures, stability and properties of the resultant compounds depend on the same three main factors, as was true for their elements: (1) the size of the atoms; (2) their affinity for electrons, electron affinity, or ease of loss of electrons, ionisation potentials, which are terms dependent on radial interaction with the nucleus and now described by electronegativity and valence; (3) angular binding dependencies, which implies that an atom has a preference for a given shape when combined with other atoms due to the availability of orbitals (see Section 2.2.4). We have already discussed contributions from electronegativity, valence and atom size, and we next turn to effects of availability of orbitals on structure and shape. There are three separate cases to be considered. When the elements are very similar there are: (1) metal combinations (alloys) and (2) non-metal combinations, while the third case (3) is non-metal/metal combination.

1. As with pure metal elements we expect close-packing in alloys although, as sizes become increasingly different, the smaller atoms may pack within the holes left in the larger atom lattices, (Fig. 2.3). Orbitals here are of little consequence, see Section 2.13.

2. For combinations of similar non-metals we expect and find that the orbitals, partly filled with electron pairs of the non-metal and partly filled due to binding, give rise to structures related to the non-metal element structures, Section 2.2.4.

3. In the third case, non-metal/metal combination, packing of *ions* dominates since the non-metal behaves as an 'inert gas' negatively charged ion, for example F^-, Cl^-, O^{2-}, while the metal behaves as an inert gas positively charged ion, for example Na^+, K^+, Mg^{2+}. Packing is largely based on ion size ratios.

There are intermediate combinations, of course, and we have to be aware of the potential polarisation, especially of partially filled cores (Section 2.9.5). Clearly, it is easier to handle these problems under the separate headings of the two cases of like-element binding, Sections 2.5–2.8, and of the one case of unlike-element binding, Section 2.9.

In conclusion to the discussion of general factors affecting A/B combinations, we may say that it is not only structure that is decided by the joint effect of the three factors mentioned above; we shall see that physical and chemical properties, including boiling points, melting points and reactivity, are decided by molecular shape, lattice formation and charge distribution, together with the energies of removal and addition of electrons to the orbitals of atoms involved in these combinations. Since there are many factors operating simultaneously, it is often difficult (or impossible) to say which dominates, just as was the case for the elements themselves. This makes the theory of the properties of combination of different atoms very complicated and we shall again prefer to proceed more or less empirically. By looking first at the structures and properties of assemblies of *very similar atoms* the effects of the three terms—electronegativity, size and symmetry of orbitals—can be seen most clearly, before we look at very different atom combinations in salts. We shall use the same terminology of covalent and metallic bonding in the discussion of compounds as we have used in the description of elements, but we will add partial ionic bonding as well, noting that covalent bonds will always now have some polarity. The polarity generates a stronger *co-operativity* in lattices than we have seen in the condensed states of individual non-metals.

2.5 Stoichiometric combination of two similar non-metal atoms

The coming together of non-metal atoms from closely related positions in the periodic table only leads to some modification of the properties of the elements. Thus, just as the elements themselves, C, N and O, were distinctively associated with both single- and multiple-bond covalent chemistry, so are combinations of C, N and O in the examples of Table 2.8. As stated, extending these combinations to larger numbers of atoms (halogens, sulphur and phosphorus) effectively includes the subject organic chemistry (see Chapter 9). The resulting compounds are stoichiometric, and they have *molecular shapes*. They always have somewhat polar bonds, however, due to the electronegativity differences between A and B and for this reason they may form higher melting solid or higher boiling liquid phases than the constitutive elements. The importance of liquid phases cannot be overstressed since the whole of the solution chemistry on the surface of the Earth and of biology is dependent upon elements that in compounds give liquids. Of course, it will have been noticed that hydrogen has to be added as one of the required non-metal elements to complete (organic) chemistry. This element, in a period of its own, is unique in size, orbital use (*s* only) and both ionisation potential and electron affinity, and is given special discussion (see Section 2.7). Its behaviour is critical in all biological systems and much of organic chemistry since it forms but one bond

Table 2.8 Some examples of A–B combinations of C, N, O

Element	Combinations		
	Single bonds only	**With double bonds**	**With triple bonds**
C	CH_4, CCl_4	CO_2, $CH_2{=}CH_2$	CO, $CH{\equiv}CH$
N	NH_3, N_2H_4, NF_3	NO_2,	NO, $HC{\equiv}N$
O	H_2O, OCl_2, H_2O_2	CO_2, NO_2	CO, NO

and, consequently, it is nearly always found as a terminal atom of a molecular structure, for example, $R—CH_3$, $R—NH_2$, ROH, etc. where R is any radical in such compounds. Within many of these units there is considerable polarity. In addition to small molecules, polymers can be made from these units and these polymers through their shapes and properties dominate biological systems (Chapter 10–13).

Finally, there is the small group of non-metal elements, including B, C, Si, P and S, which when bound together form continuous covalent solids. In recent years special interest has been focused on the properties of elementary substances and compounds that behave like Si, which is a borderline electronic semiconductor. These elementary substances and compounds include C itself (graphite), Si, Ge, BN and extend to GaAs. They have allowed the construction of computer circuits, for example, silicon chips, based on modifiable electronic conduction. Despite the importance of these compounds we shall not need to refer to them again, except in passing, until Chapter 14, since in the natural selection of chemical elements they are very recent additions. Here we shall concentrate attention on the properties of molecular structures.

2.5.1 Small covalent molecules: electronegativity and charge distribution

Within the covalent structures of molecules formed from the same elements there could be only equal distribution of charge, electrons, over atoms. In covalent molecules $A_m B_n$ this can never be the case given the different electronegativities of A and B, so that some asymmetry of charge distribution and hence partial charge of opposite sign in A and B has to be considered, which can be represented as $A^{\delta+} B^{\delta-}$. In other words, the pairs of atoms form dipoles and the molecule is said to be polar. If the molecule is of the form ABA, for example, CO_2, then it has a quadrupole $A^{\delta+} B^{\delta=} A^{\delta+}$ but, if it is linear, it will not now have a dipole. The same will be true for more complicated molecules, for example, $CHCl_3$ and CCl_4 (see Fig. 2.13) where the symmetry of the structure determines properties even where there is polarity of bonds. Clearly, the stronger these polarities the stronger the interaction between molecules and they will tend to give condensed states due to so-called intermolecular forces (see Section 2.6). These forces are directional and therefore depend also on the shapes of molecules, which we will consider next.

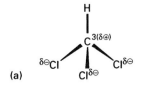

(a)

(b) $H^{\delta\oplus}—Cl^{\delta\ominus}$

(c)

Fig. 2.13 Examples of polar chloroform $CHCl_3$, (a) hydrochloric acid HCl, (b) and of non-polar carbon tetrachloride CCl_4. (c) Bond polarity is indicated by $\delta\oplus$ $\delta\ominus$.

2.5.2 Shapes of small covalent molecules, $A_m B_n$

As already explained in Section 2.4, shapes of covalent molecules formed from different elements arise just as shape arises in the molecules formed from single elements (Section 2.2.4). Shape is obviously important in packing, just as size was important in the packing of atoms. Together with polarity (and polarisability) shape determines the phases that molecules can form and their stability, as well as their ability to dissolve in one another. Such interactions are of immense consequence, so we turn to a more detailed description of them in Section 2.6. Before doing so, however, we will describe in general terms the most likely shapes of individual molecules, not only of organic but of many inorganic covalent molecules. (This topic will be developed further in

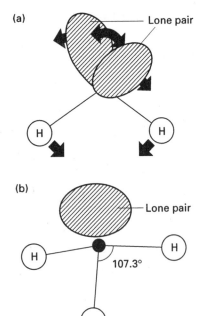

Fig. 2.14 (a) The effect of lone pair repulsions in H_2O. Bonds with pairs of electrons are represented by lines. (b) The structure of ammonia, drawn as for H_2O.

Chapters 8 and 9 on the evolution of inorganic and organic compounds, respectively, and in more complicated ways in later chapters on biological molecules.)

Let us consider the very important case of H_2O. This composition arises because each H can interact with two electrons and each oxygen can interact with eight electrons at a time at most, all electrons going in pairs. Thus we can write the formula of water as $H \overset{x}{\underset{\bullet}{\cdot}} \overset{xx}{\underset{xx}{O}} \overset{x}{\underset{\bullet}{\cdot}} H$ satisfying these electron counting rules. Now, all the eight electrons in four pairs around O repel equally to a first approximation, so we argue that since there are four electron pairs they will form a tetrahedron. This turns out to be nearly correct. If we allow the electrons that see the H nuclei to be pulled away from the O a little then the angle $_H\diagdown^O\diagdown_H$ will close down from 109° (the tetrahedral angle) and we come to the correct answer for the shape of H_2O (Fig. 2.14(a)). The shapes of this kind of molecule are decided by repulsions between bonding and non-bonding pairs of electrons and attractions between electrons and nuclei with a limiting set of four (one s and three p) combined orbitals on oxygen, nitrogen and carbon (which we denote by sp^3) and one s on hydrogen. In this way NH_3, (Fig. 2.14(b)) has a pyramidal shape and CH_4 is a regular tetrahedron. The fact that the numbers of electrons in the s, p, d and f shells are fixed is a fundamental property of the nature of space (and time) as occupied by electrons trapped by nuclei (Section 1.5). Thus we show that combinations of s and p (here) orbitals in which electrons are strongly held decide that molecules have idiosyncratic shapes dependent on composition. These shapes evolved in molecules inevitably on the cooling of the universe. Furthermore, we will see that the shape and charge distribution of molecular H_2O also decided the nature of water as a liquid and ice as a solid (Section 2.6.1).

A set of simple principles based upon the above tendency to minimise the electron repulsion between electron pairs in the valence shells (valence shell electron pair repulsion theory, or VSEPR for short, proposed by Sidgwick and Powell in 1940 and extended by Gillespie and Nyholm in 1957) is all one needs to predict the most likely shapes of most molecules formed from two non-metal atoms. To apply these principles we need to count the number of valence electron pairs that surround a given atom A (or ion $A^{n\pm}$) either shared with other atoms B (bonding pairs, BP) or unshared (lone pairs, LP). Practical rules for counting are given in reference 6 of the 'Further reading'.

The minimising of the electrostatic repulsions between the various pairs of electrons leads to the (symmetrical) shapes indicated in Table 2.9. The real shapes may be somewhat distorted, but we can allow for this taking into account that the order of repelling power must be (LP, LP) > (LP, BP) > (BP, BP). This is due to the fact that lone pairs are more spread in space and bonding pairs are more confined between combining atoms; as a result, bonding pairs tend to move away from lone pairs, as we have seen in the case of H_2O, thereby altering the bond angles to some extent (see Fig. 2.14(a), (b)). According to this theory, the most likely situations for carbon, oxygen and nitrogen are shown in Table 2.10. Naturally, hydrogen does not need consideration since it forms one single covalent bond only. A further important consideration is atom packing within molecules since steric hindrance affects shape as much as attractive interactions.

Table 2.9 Arrangements of electron pairs in valence shells and the shapes of molecules

Electron pairs			Molecular shape* (formula)	Examples
	Number of pairs			
Arrangement	**Bonding (X)**	**Lone (E)**		
Two electron pairs (linear arrangement)				
•—•—•	2	0	Linear (AX_2)	$Ag(NH_3)_2^+$, $(Zn, Cd, Hg)(CH_3)_2$, UO_2^{2+}, CO_2
Three electron pairs (triangular plane)				
	3	0	Triangular plane (AX_3)	$BX_3 (X = F, Cl, Br)$, GaX_3
	2	1	V-shape (AX_2E)	$PbX_2 (X = Cl, Br, I)$, SO_2
Four electron pairs (tetrahedron)				
	4	0	Tetrahedron (AX_4)	BeX_4^{2-}, BX_4^-, CX_4, NH_4^+, GeF_4, Al_2Cl_6
	3	1	Trigonal pyramid (AX_3E)	NX_3, PX_3, $AsX_3 (X = H, F, Cl)$; H_3O^+
	2	2	V-shape (AX_2E_2)	H_2O, SCl_2, SeX_2
Five electron pairs (trigonal bipyramid)				
	5	0	Trigonal bipyramid (AX_5)	PF_5, PCl_5 (gas), PF_3Cl_2, V_2O_5
	4	1	Irregular tetrahedron (AX_4E)	$TeCl_4$, $(S, Se)F_4$
	3	2	T-shape* (AX_3E_2)	ClF_3, BrF_3
	2	3	Linear* (AX_2E_3)	ICl_2^-, I_3^-, XeF_2
Six electron pairs (octahedron)				
	6	0	Octahedron (AX_6)	AlF_6^{3-}, SiF_6^{2-}, PF_6^-, PCl_6^-
	5	1	Square pyramid (AX_5E)	IF_5, BrF_5, ClF_5
	4	2	Square plane* (AX_4E_2)	ICl_4^-, BrF_4^-, XeF_4

* The starred molecular configurations are the ones actually observed, although other molecular geometries are theoretically possible for the particular numbers of electron pairs.

Table 2.10 Common geometries around C, N and O in organic molecules

Element (A)	Geometry around A	Electronic structure	Bond angles (°)*
Carbon	Tetrahedral		109
	Trigonal (planer)		120
	Linear		180
Nitrogen	Tetrahedral		>109 (LP, BP)
	Trigonal		>120 (LP, BP)
	Linear		180
Oxygen	Tetrahedral		>109 (LP, LP)
	Trigonal		>120 (LP, LP)

* LP, Lone pair; BP, bonding pair.

Using these principles, the shapes not only of small molecules formed from non-metal elements but also of most small and large, organic and inorganic, monomeric or polymeric molecules, discrete or in networks, can be reasonably predicted. The examples given in Table 2.9 are clear and can be generalised with but a few exceptions. It is no accident that structures of biological systems, created from these light non-metal elements, have particular shapes, although these shapes also require the consideration of intermolecular forces, as described in Section 2.6. It is these shapes that will generate the peculiarities of biological self-assembly (Chapters 10–13). Note that the molecules fill space very effectively (Fig. 2.15(a)), generating fitting surfaces as in Fig. 2.15(b).

Optical isomers

The absence of symmetry in a mirror image of a molecule (Fig. 2.16) generates optical activity where the molecule rotates the plane of incident polarised light. There are several ways of creating such asymmetry other than in a molecule C(abcd); for example, a screw axis is either left- or right-handed. The importance of this lack of symmetry to reflection is that surfaces of optically active molecules match in very selective ways. This selectivity of shape dominates biological associations of bioorganic molecules (see Section 9.5.1).

Large covalent molecules and their shapes

It is difficult to summarize the importance of the vast number of molecules that can be made by putting together even C and H. As we have seen, together with

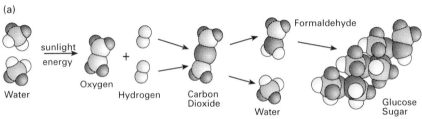

(a)

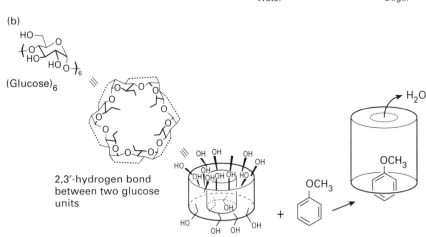

(b)

(Glucose)$_6$

2,3′-hydrogen bond
between two glucose
units

Fig. 2.15 (a) Space-filling models of
some molecules. When bound together
very little free space remains. (Lone pairs
are shown.) (b) Combining molecules of
glucose gives the structure shown as a
space-filling cylinder. Into the cylindrical
hollow other molecules fit, for example,
H_2O which can be displaced by an
aromatic molecule.

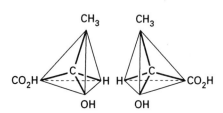

Fig. 2.16 In 1873, J. H. Van't Hoff
explained optical activity for the two forms
of lactic acid in terms of four groups
attached tetrahedrally to a central carbon
atom. These structures, based on
asymmetric carbon, are mirror images of
each other.

O, N and a few other non-metals, they form the subject matter of organic
chemistry. In this section we wish to direct attention to a subject of very deep
concern in the natural selection of elements in combination which is the
variety of precisely stoichiometric molecules of high molecular weight, up to
10^6 (or 10^5 atoms), in proteins or (even larger) in DNA. These polymers are
designed around the local features of H binding to C, N and O, and of N and O
binding to C (see Figs 9.16 and 9.18). The diversity of properties of the
molecules develops from the diversity of atom sequences and virtually all of
them are optically active. While each sequence (linear or branched) is unique,
the properties of individual bonds are not very different from those in simple
molecules. Thus, while individual bonds are fixed, rotation around single
bonds is possible so that many shapes can develop. One such shape is shown in
Fig. 2.15(a) using a space-filling representation. The atomic sequence, given
internal interactions, therefore decides the possible fold of the polymers (see
Fig. 2.15(b) and Section 2.8) so that, from atomic principles, on the scale
10^{-8} cm (ångstrom) there develop shapes at the 10^{-6} cm (micron) size.
Association of polymers then leads to biological cells (microns) and then whole
organisms (up to metres). The feature to recognise is complexity in this *ordered*
arrangement, where the ordered sequences control large-scale spatial
development. It is not just that continuity can be built in particular ways as for
covalent solids AB, but there is a principle of self-assembly more akin to jigsaw
puzzle construction in which there are no repeats and which is only possible
using polymers. (We stress here the *order* in the arrangement only and we are
not yet concerned with motion within the assembly (see Fig. 3.15 and
Chapter 7).) These features arise since co-operative energies although they are
small within the lattices are large enough to cause self-assembly below 100°C,
see Section 6.8.1. Stoichiometry though present may not be apparent.

2.5.3 Bond energy in covalent compounds

For covalent compounds the bonds between two atoms are taken to be localised and can be considered as independent from all the other atoms (a valid approximation in most cases, organic molecules with conjugated multiple bonds clearly excepted), and we can assign to them a *bond energy* value (Table 2.11) such that the sum of bond energies for all bonds in the molecule (in the gas state at $25°C$) gives the energy required to separate it into its component atoms (again in the gas state at $25°C$).†

Table 2.11 Bond energies

Bond	Bond energy (kcal mol⁻¹)	Bond	Bond energy (kcal mol⁻¹)	Bond	Bond energy (kcal mol⁻¹)	Bond	Bond energy (kcal mol⁻¹)
Single bonds							
H—H	104.2	Br—Br	46.1	C—N	69.7	P—Cl	79.1
C—C	83.1	I—I	36.1	C—O	84.0	P—Br	65.4
Si—Si	42.2	C—H	98.8	C—S	62.0	P—I	51.4
Ge—Ge	37.6	Si—H	70.4	C—F	105.4	As—F	111.3
Sn—Sn	34.2	N—H	93.4	C—Cl	78.5	As—Cl	68.9
N—N	38.4	P—H	76.4	C—Br	65.9	As—Br	56.5
P—P	51.3	As—H	58.6	C—I	57.4	As—I	41.6
As—As	32.1	O—H	110.6	Si—O	88.2	O—F	44.2
Sb—Sb	30.2	S—H	81.1	Si—S	54.2	O—Cl	48.5
Bi—Bi	25	Se—H	66.1	Si—F	129.3	S—Cl	59.7
O—O	33.2	Te—H	57.5	Si—Cl	85.7	S—Br	50.7
S—S	50.9	H—F	134.6	Si—Br	69.1	Cl—F	60.6
Se—Se	44.0	H—Cl	103.2	Si—I	50.9	Br—Cl	52.3
Te—Te	33	H—Br	87.5	Ge—Cl	97.5	I—Cl	50.3
F—F	36.6	H—I	71.4	N—F	64.5	I—Br	42.5
Cl—Cl	58.0	C—Si	69.3	N—Cl	47.7		
Multiple bonds							
C=C	146.4	C=N	147	C≡C	194		
N=N	100	C=O	164 to 174	N≡N	226		
O=O	96	C=S	114	C≡N	207 to 213		

Likewise, the *heat* of formation of molecules from their elements in the gas state at $25°C$, approximately equal to the energy of formation (see Chapter 3), is readily related to the sum of the bond energies. These thermodynamic energies show that some combinations of elements are much more stable than others and this could and often does decide which occur naturally (Chapters 3 and 8). One consideration of importance is the competition for partners, say, for any two non-metals like hydrogen and oxygen (see Table 2.12); such competition has dominated purely non-metal geochemistry and biological chemistry. (We turn to parallel considerations involving metals in competition in Section 2.9.2.)

We must next consider what forces cause molecules of low molecular weight formed stoichiometrically from non-metals, such as benzene, to form liquids or even crystals in which the molecules remain clearly identifiable. These

† The values of bond energies, as defined, are actually bond dissociation *enthalpies* (see Chapter 3).

Table 2.12 Heat of formation of non-metal hydrides and oxides

Element X*	XH_n	Heat of formation (kcal mol^{-1})	$XO_{n/2}$	Heat of formation (kcal mol^{-1})
B	BH_3	+3.7	$1/2B_2O_3$	−151
C	CH_4	−18	CO_2	−47
N	NH_3	−11	NO	+21
F	HF	−64	$1/2F_2O$	+21
Si	SiH_4	−15	SiO_2	−205
P	PH_3	+2	$1/2P_2O_3$	−170
S	SH_2	−5	SO_2	−71
Cl	HCl	−22	ClO_2	+24
H	H_2	0†	H_2O	−68.3
O	H_2O	−68.3	O_2	0†

* The only elements with a higher affinity for H than for O are N, O and halogens.
† Convention.

principles also apply to the condensation of single elements in atoms or molecular assemblies, for example, to He, Ar, H_2, N_2 or P_4 and S_8. In part, the same forces generate biological assemblies of large molecules and even their folding. Of course, liquids together with these polymers are of the utmost interest since only they allow life (Chapter 11).

2.6 Assemblies of molecules: intermolecular forces

Just as there are forces that bind atoms in metals and molecules, there are forces between molecules that account for the appearance of molecular substances in liquid and solid states but that are also present in the gaseous state. These forces are rather weak, when compared with interatomic forces (covalent bonds), as may be judged by the generally low energies of vaporisation of molecular substances. Their energies are of the order of 2–10 kcal mol^{-1} (Table 2.13), while bond energies in covalent compounds (see Table 2.11) vary over the range 10–200 kcal mol^{-1}. It is the existence of

Table 2.13 Distribution of van der Waals attraction energy over the three types of intermolecular interaction

	Dipole moment (D)*	Polarizability ($\times 10^{-24}$ cm^3)	Lattice energy (kcal mol^{-1}) Type of interaction			
			Keesom	Debye	London	Total
Ar	0	1.63	0.000	0.000	2.03	2.03
CO	0.12	1.99	0.0001	0.002	2.09	2.09
HI	0.38	5.40	0.006	0.027	6.18	6.21
HBr	0.78	3.58	0.164	0.120	5.24	5.52
HCl	1.03	2.63	0.79	0.24	4.02	5.05
NH_3	1.50	2.21	3.18	0.37	3.52	7.07
H_2O	1.84	1.48	8.69	0.46	2.15	11.30

* D is the Debye unit (change × length) in 10^{-18} e.s.u. × cm.

these forces that also causes the deviations from the perfect gas law accounted for by an additional term, the so-called van der Waals interaction (see references 1–3 in Chapter 3) and for this reason intermolecular forces are frequently referred to as van der Waals forces (see Table 2.13). The three main types of interaction are:

(1) attraction between permanent dipoles in polar molecules (Keesom interactions);

(2) attraction between permanent dipoles in polar molecules and induced dipoles in neighbouring molecules (Debye interactions);

(3) attraction between instantaneous dipoles and induced dipoles in all molecules, atoms or ions (London interactions or dispersion forces).

We will not discuss here the fine details of the theory of these interactions. It is sufficient for our purposes to note that only the Keesom interactions are temperature-dependent (since the motion of molecules affects the relative orientation of the dipoles) and that Debye and London interactions depend strongly on the polarisability of the molecules concerned (and of atoms or ions in the case of London forces). Since the polarisability (deformability) will be higher the larger and less constrained the electron clouds (indeed polarisability has the dimensions of a volume), it is expected that the energy of these interactions will increase with the size of the species considered and the lower the energy with which electrons are retained. The intermolecular distances, r, are also extremely relevant since the interactions under (2) and (3) are inversely proportional to r^6. In Table 2.13 the relative intensity of these forces is compared for some common compounds in the solid state.

This means that all those physical properties that depend on the intensity of the *co-operative* interactions between molecules, for example, melting points, boiling points, heats of fusion and vaporisation, solubility, viscosity, etc. are determined by the energy of van der Waals forces. Some examples are given in Fig. 2.17 where we notice particularly the noble gases and some non-polar molecules in which only London forces are operative. The trends are regular and conform to the theory. The nature of our atmosphere, being largely N_2 and O_2, is dependent on the weakness of interactions between these molecules. More strikingly, the fact that Earth has little or no H_2 or He, abundant elements in the universe, strikingly reflects the fact that these elements form liquids only at extremely low temperatures. Amongst compounds, CO_2 does not condense until about $-80°C$ and is another important component of the atmosphere, most of which has been lost from the original Earth. The series of alkanes C_nH_{2n+2} also provides a good example of the increase of London forces with the increase of size of the molecules (see Table 9.1). The size and shape of these molecules constrain packing and therefore the magnitude of the van der Waals forces. Critically, the formation of oil-like phases, again due to these co-operative forces, created the opportunity for the development of life, since this corresponded to the creation of a new liquid phase on Earth (see Chapters 3, 6 and 7). The formation of liquids, so essential to flow and organisation at temperatures around -50 to $150°C$, is almost entirely limited to those *molecules* that have relatively strong intermolecular forces but with little long-range co-operativity. This allows the assembly of large molecules such as polymers and proteins to be flexible to some degree too when rubber-like features are found.

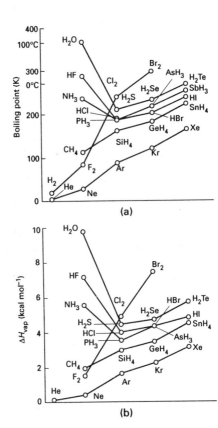

Fig. 2.17 (a) Boiling points and (b) heats of vaporisation ΔH_{vap} of some substances that form molecular crystals. Note that the oxides of C, N, S and several halogens are gases at room temperature unlike H_2O, while most other oxides are solids.

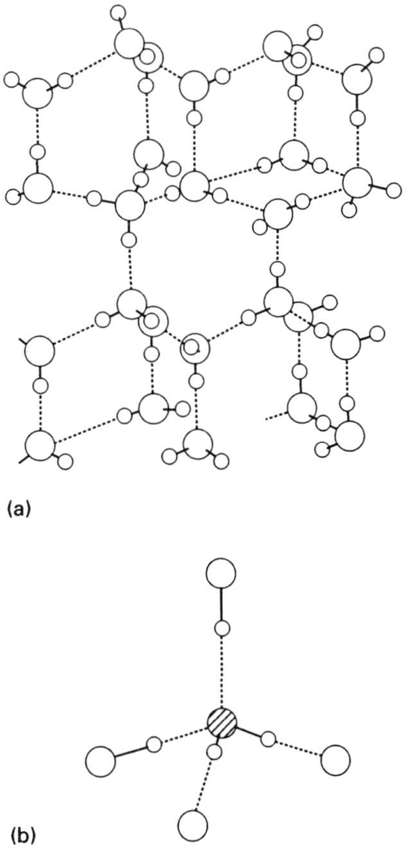

(a)

(b)

Fig. 2.18 (a) The crystal structure of ice in which (b) the H-bonds are linear.

Table 2.14 Estimated energies of hydrogen bonds for selected cases*

Bond	Energy (kcal mol^{-1})
F—H----F	7
N—H----N	6
O—H----O	6
N—H----F	5
O—H----N	4
N—H----O	3

* The values vary somewhat from compound to compound.

Anomalies are apparent when one compares melting points, boiling points or heats of vaporisation of series of compounds, such as,

$$\begin{array}{llll} \text{HF,} & \text{HCl,} & \text{HBr,} & \text{HI} \\ H_2O, & H_2S, & H_2Se, & H_2Te \\ NH_3, & PH_3, & AsH_3, & SbH_3 \end{array}$$

(but not for CH_4, SiH_4, GeH_4, SnH_4),

for which we expect regular trends according to a naïve theory of the effect of polarity. As Fig. 2.17 shows, the reality is different—H_2O, HF and NH_3 exhibit a very striking behaviour. This can only mean that in these compounds there must exist some additional interaction between their molecules that does not exist (or exists to a lesser degree) in the other compounds of the corresponding series.

2.6.1 Hydrogen bonding

If we observe all the above exceptional cases we note they have one characteristic in common: hydrogen atoms are bonded to a very electronegative atom, F, O, or N. In the absence of any other difference from the other compounds in each series this must be the reason for the enhancement of the interactions between the corresponding molecules. The first suggestion to explain this effect was due to Latimer and Rodebush who proposed that hydrogen atoms in such conditions could act as a 'bridge' linking the molecules—the hydrogen bridge or hydrogen bond (as it was called later) theory. One could, therefore, have additional (relative to the normal van de Waals forces) interactions, such as F—H . . . F, O—H . . . O, N—H . . . N or, in heterogeneous systems more generally, O—H . . . N, N—H . . . O, etc. (see Table 2.14), peculiarly due to the nature of the hydrogen atom (see Section 2.7).

Hydrogen bonding is thus a determining factor for the structure of the condensed states of, for example, water, ammonia and hydrogen fluoride, leading to apparent anomalies in their properties. In liquid water the molecules associate forming statistically permanent aggregates and in the solid, ice, the aggregates consolidate and extend into an open crystal structure, shown in Fig. 2.18. The openness of this structure explains why ice is less dense than water (it floats in liquid water) and why molecular inclusion compounds may be formed, for example, $Cl_2(H_2O)_6$ or analogous CO_2, CH_4, etc. containing species in ice.

Clearly, the most critical case of H bonding on Earth is the appearance of *liquid* water. If ice had been more dense than water it would have accumulated at the bottom of the sea and life would have been different. We make a digression on the properties of water while noticing that hydrogen-bond formation will also be shown to be an essential driving parameter in biological association, especially that of polymer matrices in water. In effect, *hydrogen bonding dominates folding of proteins and polynucleotides*, (see Section 6.8.2).

2.6.2 The formation of liquids: water

Of the materials on the surface of the Earth there is a major group of solids that melts only above several hundred degrees centigrade and a minor group of gases that condense only at temperatures below $-50°C$, see Section 3.2.5.

These circumstances arise because, on the one hand, the solids are either metal (alloys), continuous non-metal lattices (coal) or mineral salts in which atom–atom interaction is strong and co-operative or because, on the other hand, the major gases N_2, O_2 and CO_2 have no strong molecular interactions since their molecules are small and not dipolar. Quite exceptional, then, is the molecule H_2O, water, which has a strong dipole, has very strong H bonds, and forms a liquid phase at temperatures ranging from 0 to 100°C, that is, from 273 to 373 K. (Though water as a gas is 'older than the hills', water as a liquid is much younger and is limited to a narrow temperature range.) This liquid is capable of dissolving many compounds, see Chapter 5. As well as many salts that dissolve in water to a greater or lesser degree, we shall see that water dissolves very many polar organic molecules too. For these reasons water became the essential liquid phase for the temperature stability of this planet, but, above all, for life as we know it, which is impossible in its absence. (NB. Solutions in water and then oils were largely a novelty arising as Earth cooled below 100°C.) Solutions in organic solvents, with the exception of solutions in lipids, effectively waited even longer for man's self-conscious activity in their production (see Chapters 9 and 14). Much of the chemistry of low polarity in such non-polar solvents is called organic chemistry, yet the chemistry of life is very largely of polar molecules in water and existed some 4×10^9 years earlier.

2.6.3 The different kinds of liquids

We now know that there are very different kinds of liquid based on non-metal atoms just as there are many kinds of solid based on several other kinds of elements. We distinguish electron-conducting (metals) from non-electron-conducting (insulator) solids, and we must distinguish liquid solvents of high polarity (high dielectric constant) such as water from those of low polarity, for example, petrols and oils. The nature of the highly polar solvents is that, when certain substances AB dissolve in them, they become *conducting* as separated ions A^+ and B^-, whereas in low-polarity solvents they either do not dissolve or do so as AB. The two kinds of common liquids, one due to H bonds (water) and the other due to London forces, do not dissolve in one another. The difference is of immense consequence to the potential function of the solvents in *organised flowing systems* since water (like a metal) is a conducting phase and both solvents allow transfer of material. Notice that there are very few liquids that conduct electrons at low temperature, and they are not abundant, mercury being very rare. We return to the nature of solutions in solvents in Chapters 5, 7 and 9–14.

2.7 Hydrogen in compounds other than water

While we can treat most elements in groups and describe relative properties within periods of the periodic table, hydrogen is in a period and really in a 'group' of its own. It has, as already stated, three characteristics as an atom: (1) very high ionisation energy, *IP*; (2) a very considerable electron affinity, *EA*; and (3) no electron core. The absence of an electron core makes H^+ just a nucleus, a proton, so that it gives rise to strong ionic binding. The corresponding electrostatic energy can overcome the ionisation energy of HX compounds in polar surrounds, for example in water, when $H^+(H_2O)_4$ species are formed with X^-. The ability to ionise HX makes many anions available in water, such as OH^-, NO_3^- and Cl^- from H_2O, HNO_3 and HCl, respectively. However, given its high *IP* and *EA*, hydrogen has a high electronegativity too and can, therefore, form strong covalent bonds. As mentioned above, it forms the outside 'atoms' in a huge variety of organic molecules such as hydrocarbons, fats, proteins, sugars and many solvents (see Chapter 9). Here its electronegativity is closely like that of carbon and the C—H bond is virtually non-polar. Finally, it can pick up an electron to form the rather less stable hydride anion, H^-, with a helium electronic structure. Thus, it sits next to the noble gas helium, at the end of a period, is at the beginning of this period and is in the middle of the same period, thus sharing the properties of $F(F^-)$, $Li (Li^+)$, and covalent C, all at once. This versatility and its vast abundance has made combined hydrogen the centre of life's structural organic chemistry, of the solvent present everywhere, water, and of both one-electron $H^+/H^\bullet/H^-$, covalent, and two-electron H^+/H^- chemistry. This versatility becomes centrally functional in biological chemistry and, indeed, in organic and inorganic chemistry generally, but additionally even in bio-energetics (see Section 10.9). We must never forget that the dominant elements on Earth's *surface* and in life are hydrogen and oxygen, followed by carbon!

2.8 Summary of the outstanding features of non-metal compounds

Before we proceed, certain dominant features of combinations of like non-metal atoms will be surveyed, not only of the kind A_mB_n but including larger numbers of similar elements, ABCD, etc. They are of very great importance in later considerations of the selection of the chemical elements (Chapters 10–14).

The large majority of these combinations give us the common gases, liquids, and soft, low-melting-point solids based on small molecules (Table 2.15). Thus they provide molecules that themselves diffuse readily, sometimes as gases, sometimes in liquids. Many of them are insoluble in water and hence form separate liquid phases. Flow and diffusion within them is relatively free. We shall find that this ability to be *relatively loosely ordered* is of the utmost consequence in biology and for man's industry. Organic synthesis, for example, is totally dependent on the availability of liquids. This property of

Table 2.15 Outstanding features of compounds A_mB_n

Compounds exhibiting		
Ionic properties	**Semi-ionic properties**	**Covalent properties**
Most oxides, sulphides, fluorides, chlorides, etc.; no important hydrides	Some oxides, sulphides and iodides (e.g. Ti_2O_3, Se_xS_y, FeS, CuI, etc.); no important hydrides	Most organic compounds, some oxides (e.g. H_2O), most important hydrides, few important sulphides (e.g. H_2S)

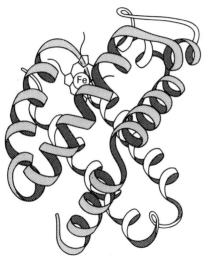

The folded subunit of haemoglobin of over 100 amino-acids in a chain

allowing flow, especially under the influence of a field, becomes the basis of a new possibility, *organisation*, different in kind from that based on metals since it allows flow and rearrangement of non-metal molecules and metal and non-metal ions but not so easily of electrons. At the same time more highly ordered constructs are required and can be built as discussed in Chapters 9 and 13.

Now, as we increase the molecular weight in fair measure but still in a limited way, these combinations generate more *strongly ordered* structures that depend on the sequences of atoms in space for these covalent systems. In any polymer the sequence can be ABCD, ACBD, CDBA, etc. and these polymers, for example, proteins, came to have a wide range of new properties while being individually stoichiometric and shaped (see marginal illustration), including some that are quite rigid and hard, plastics, and some that are softer and elastic. (In fact, we could have drawn attention to these different polymer forms even when describing elements since one of them, covalent sulphur (Fig. 2.6) gives this diversity of forms—gas, liquid, crystal and polymer.) These polymeric substances are easily worked, like metals, and as a result are the basis of one form of building material.

The fact that several of the non-metal compounds are 'liquid crystals' (see later) gives rise to yet other properties. In particular, lipid liquid crystals can be solvents for the free movement of other compounds of comparably low polarity, for example O_2, CO_2, N_2, but not of Na^+Cl^-.

2.9 Stoichiometric combination of very different elements: ionic compounds

We now move to the consideration of a second range of combination of elements that form stoichiometric compounds, namely the combination of elements of very different electronegativity that gives rise to continuous lattices of *ionic* compounds, for example salts, in which, to all intents and purposes, the element of lower electronegativity gives up its electrons to become a smaller positive ion while the element of greater electronegativity takes on more electrons to become a larger negative ion, for example NaCl and MgO. Packing the larger spherical negative ions together gives rise to the same close packing as that for neutral metal atoms and the positive ions can then be placed in the holes to form association with six (octahedral holes) or four (tetrahedral holes) negative neighbours (see Fig. 2.3). The electrostatic forces hold the whole together through *ordered co-operative interaction* over long distances. In this way, though with some elaboration, the solid structures of most of the salts and minerals of the Earth are constructed, including clays, silicate rocks and so on. *Size matching to holes is of the greatest concern here*, and orbital influences are marginal though important with heavy, especially transition metal ions (see ligand field theory in Section 2.9.5 and in references 2, 3 and 5 in 'Further reading').

It becomes important, therefore, to define ion sizes, that is, the apparent sizes in lattices. Here, again, we have some difficulty since we must relate all 'sizes' to one type of packing. The measurements have been made and yield the sizes shown in Table 2.16. There are large changes in radii from Na^+ to K^+ and Mg^{2+} to Ca^{2+}, but small continuous changes across rows of transition metal ions, for example, $Mn^{2+} > Fe^{2+} > Co^{2+} > Ni^{2+} > Cu^{2+} < Zn^{2+}$.

The structure adopted by the ionic (crystalline) compounds depends to a large extent on the relative sizes of the two ions involved. Geometric arguments easily show that on the basis of the 'radius ratio', r^+/r^-, of the positive and negative ions (see Table 2.17) several main structures can be predicted for AB- and AB_2-type compounds (Fig. 2.19). The packing based on

Sodium Chloride (NaCl)

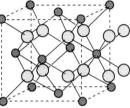

Cesium Chloride (CsCl)

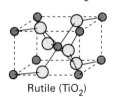

Fluorite (CaF_2)

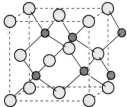

Rutile (TiO_2)

Zinc blende (cubic ZnS)

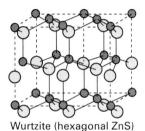

Wurtzite (hexagonal ZnS)

Fig. 2.19 Some common ionic structures. These lattices are infinite in principle.

Table 2.16 Ionic radii (Å) *

		Li⁺	**Be²⁺**						
		0.68	0.30						
O²⁻	**F⁻**	**Na⁺**	**Mg²⁺**		**Al³⁺**				
1.45	1.33	0.98	0.65		0.45				
S²⁻	**Cl⁻**	**K⁺**	**Ca²⁺**	**Ga⁺**	**Ga³⁺**	**Ge²⁺**	**Ge⁴⁺**		
1.90	1.81	1.33	0.94	1.13	0.60	0.93	0.54		
Se²⁻	**Br⁻**	**Rb⁺**	**Sr²⁺**	**In⁺**	**In³⁺**	**Sn²⁺**	**Sn⁴⁺**		
2.02	1.96	1.48	1.10	1.32	0.81	1.12	0.71		
Te²⁻	**I⁻**	**Cs⁺**	**Ba²⁺**	**Tl⁺**	**Tl³⁺**	**Pb²⁺**	**Pb⁴⁺**	**Bi³⁺**	**Bi⁵⁺**
2.22	2.19	1.67	1.29	1.45	0.91	1.21	0.81	1.16	0.74

	Ionic radius (Å) for ionic charge						
	1+	**2+**	**3+**	**4+**	**5+**	**6+**	**7+**
Sc			0.81				
Ti		0.90	0.76	0.60			
V		0.88	0.74	0.60	0.50		
Cr		0.84	0.69	0.56		0.45	
Mn		0.80	0.66	0.54			0.35
Fe		0.76	0.64				
Co		0.74	0.63				
Ni		0.72	0.62				
Cu	0.95	0.69					
Zn		0.70					
Ag	1.13						
Cd		0.92					
Au	1.30						
Hg	1.25	1.05					

* The radii are for 6:6 co-ordination in lattices such as NaCl.

Table 2.17 Radius ratio, co-ordination number and typical lattice structures

Co-ordination number	Geometry	r^+/r^- *	Typical lattice structure
4:4	Tetrahedral	0.414	Zinc blend, ZnS
6:6	Octahedral	0.732	Sodium chloride, NaCl
6:3	Octahedral	0.732	Rutile, TiO_2
8:8	Cubic	1.00	Cesium chloride, CsCl
8:4	Cubic	1.00	Fluorite, CaF_2

* Ideal values for close-packing, that is, hole filling in an anion lattice.

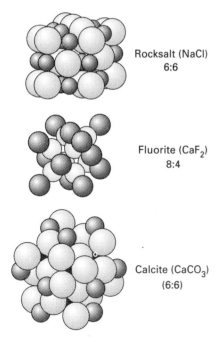

Rocksalt (NaCl)
6:6

Fluorite (CaF_2)
8:4

Calcite ($CaCO_3$)
(6:6)

Fig. 2.20 Some simple crystal structures showing space-filling.

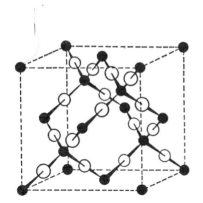

Fig. 2.21 The structure of SiO_2 (β-cristobalite).

radius ratio contributes considerably to the relative stability of large complex anions with large cations, for example $CaCO_3$, and smaller ions with smaller cations, for example $Mg(OH)_2$. As with covalent molecules the packing results in effective space-filling (Fig. 2.20).

There are, of course, other types of crystal structures, not only for AB- and AB_2-types of compounds but for many other stoichiometries, that may be classified in terms of their symmetry elements. We can relate the external appearance of the crystals (morphology) to their classes of symmetry—the external appearance must reflect the regularity of the internal arrangement. Yet, one often finds different morphologies for the same substance since the internal pattern is not very restrictive. For instance, a crystal in the cubic system such as rock salt, NaCl, may show cubic faces but also octahedral, dodecahedral or even other faces depending on the conditions under which the crystal was formed (see Chapter 4).

(Note that these considerations apply also to non-ionic compounds. Substances with three-dimensional networks of covalent bonds form 'covalent crystals', for example diamond and quartz (see Figs 2.7 and 2.21), but even those with discrete molecules may also form 'molecular' crystals in which the structural units are packed due to the effect of van der Waals forces (see Section 2.6) but the crystal phases can have varieties of shape.)

As the elements, *as ions*, become more electronegative (cations) or more electropositive (anions), pure electrostatic interactions give way to partial covalent bonding. As this covalence increases in ionic crystals, the internal site structure begins to depend on occupied and unoccupied outer orbitals and numbers of electrons in unfilled cores. Thus site symmetry, and then shape, is a compromise among many factors. Descriptions are given in several textbooks (see, for example, reference 5 in 'Further reading').

2.9.1 Energetics of ionic compounds

In the case of ionic compounds the concept of bond energy can only be applied to gaseous pairs of ions. We are more frequently interested in the energy of formation of the crystals from their separate ions in the gaseous state at a given reference temperature, for example 25°C, the *lattice energy*,† which can be calculated taking into account all possible electrostatic interactions between the ions (considered as charged spheres) plus the repulsion of electron shells, or derived through adequate thermodynamic cycles from other quantities that can be determined experimentally, for example, heats of dissociation, heats of sublimation, ionisation energies, electroaffinity energies, etc. (see references 2, 5, 6 and 9 in 'Further reading'). The lattice energy is strongly *co-operative* in that the energy of, for example, the NaCl lattice is more than 1.5 times that of the energy of the NaCl molecule per mole of NaCl. This explains why ionic substances are so stable as solids (compare metals, Section 2.2.5). Here we have therefore to adopt, and the same holds in the case of covalent networks, a reference quantity corresponding to a *conventional mole* as if real molecules were formed. (Clearly, in the lattice of an ionic compound, for example NaCl, no particular $M^+ X^-$ (Na^+Cl^-) pair exists.)

Charges, sizes, electronegativity and types of electron shells of the ions involved, as well as their structural arrangements, are the factors determining

† Actually, as defined, it should be called *lattice enthalpy* (see Chapter 3).

the values of the lattice energies per mole. These values (see Table 2.18) are particularly relevant since they are a measure of the internal cohesion of the crystals and are therefore related to many physical properties of the corresponding substances, for example hardness, density, melting and boiling points, solubility, etc. as seen in Table 2.19.

The ability to pack ions together became one of the decisive factors in the natural selection of the chemical elements. In Section 2.5.3 we saw that selection amongst non-metals to give compound formation was based on the matching of energies and sizes of atomic orbitals. In the formation of ionic compounds between metals and non-metals the selection is based on both size and charge. Cations, generally, are smaller and often more highly charged than anions. Thus, the small highly charged cations have a strong preference in a mixture of anions and cations for the smallest anions, for example F^- and O^{2-}. The examples are MgO, Al_2O_3, SiO_2 and Fe_2O_3 and their mixed

Table 2.18 A selection of lattice energies of ionic compounds

Salt	Lattice energy (kcal mol^{-1})	Salt	Lattice energy (kcal mol^{-1})
NaF	217	MgF$_2$	698
NaCl	185	CaF$_2$	631
NaBr	176	BaF$_2$	560
NaI	166	MgCl$_2$	592
KF	193	CaCl$_2$	542
KCl	168	SrCl$_2$	512
KBr	161	BaCl$_2$	489
KI	152	TiO	928
MgO	934	VO	936
CaO	845	MnO	911
SrO	789	FeO	938
BaO	751	CoO	954
MgS	807	NiO	974
CaS	740	CuO	970
SrS	696	ZnO	964
BaS	666	ZnS	864

Data from Ball, M. C. and Norbury, A. H. (1974). *Physical data for inorganic chemists*, Longman, London and Waddington, T. C. *Advances in inorganic chemistry and radiochemistry*, Vol. I. Academic Press, New York (1959).

Table 2.19 Correlation of physical properties of ionic crystals and lattice energies

Salt	Lattice energy (kcal mol^{-1})	Melting point (°C)	Solubility in water (mol l^{-1})	Hardness (Mohs' scale)
NaF	217	992	1.0	
NaCl	185	801	6.0	
NaBr	176	755	9.0	
NaI	166	651	12.3	
BeO	1083	2530±50		9.0
MgO	934	2825±30		6.5
CaO	845	2615±25		4.5
SrO	789	2420		3.5
BaO	751	1920		3.3

compounds, while large cations are found with larger anions, for example NaCl, KCl, $BaSO_4$, $CaCO_3$. We shall discuss the competition between oxide (small) and sulphide (large) in Chapter 8 since it has dominated much of geological and, in considerable part, biological selection. The energy of association is in considerable part due to long-range co-operativity which clearly depends on effective packing. Co-operativity, not just bonding in small units, is increasingly seen to dominate much of chemistry (and biological materials).

2.9.2 Physical properties of ionic solids

Once again it is not out of order to remark immediately upon man's 'natural' selection of elements in compounds that form salts in the preparation of hard, not electron-conducting, solids for buildings, etc., for example, marble, bricks and so on. Notice that, generally speaking, the lattice energies increase with charge, so that materials such as NaF are not as valuable and are more soluble than MgO and AlN. Again MgF_2 is not as useful a solid-state material as SiO_2. Just as in the case of covalent compounds that could give rise to *linear* structures of great value, we note that salts give useful *three-dimensional* substances (compare metals). Underlying the lattice strength is the considerable co-operativity of the lattice due to the long-range forces (depending on $1/r$). The natural selection of elements, once it became biased by usefulness, for example for the survival of a system, living or dead (Chapter 16), is therefore very dependent on the types of compounds the elements form. In effect, what is of interest here is the value of the strength with which order is maintained. Note that for some purposes great strength is required, for example, in static structures, but materials of weak co-operativity have advantages of other kinds, for example elasticity (see Section 2.8 and 6.3.4).

The fact that individual constitutive species are charged in substances such as NaCl could lead one to suppose that they will conduct electricity. This is so, but only in molten salts. However, the salts dissolve in polar solvents, particularly water, and then give rise to conducting solutions. Here there is a parallel with electronic conduction in that the charged particles can flow and so become part of organisations (see also organic charged species in Chapter 9). Once again, usefulness is the dominant requirement that becomes the reason for selection, not binding strength now. For example, hard ionic materials $CaCO_3$ are very different from hard non-metal materials such as diamond.

2.9.3 Variable valence and electronic conductivity in salts

As explained in Section 2.4.3, variable one-electron change in valence is common from Ti to Cu, Zr to Ag, Hf to Bi in the periodic table. This makes it possible to prepare substances of non-integral valence such as Fe_3O_4, really $(Fe^{2+})_1 (Fe^{3+})_2 (O^{2-})_4$. Such a substance has a new important property—it can conduct electrons in the solid state. The character of such 'inorganic' conductors has become potentially very important in man's constructs since superconductors were discovered, for example, in mixed oxides of the kind $Cu(Ln)(O)_n$ (Fig. 2.22) where Ln represents a lanthanide and copper has two valence states. However, it appears to be true that the latter are all non-

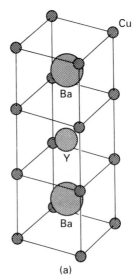

(a)

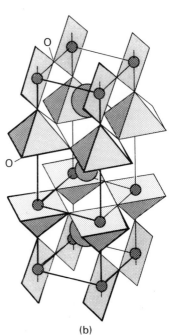

(b)

Fig. 2.22 Structure of the $YBa_2Cu_3O_7$ ('123') superconductor. (a) Metal atom positions. (b) Oxygen polyhedra around the metal ions which are in square-planar and square-pyramidal co-ordination environments (see Table 2.9).

stoichiometric compounds (see Sections 2.10–2.14). Electronic flow is clearly possible in materials other than metals. Biology has taken full advantage of this in an unusual way (Chapters 10–13) and can also use special organic compounds and materials which have the ability to give and to accept electrons to advantage.

2.9.4 Partial covalence in ionic compounds: co-ordination chemistry

The electronic structures of many non-metals and anions, such as Cl^-, O^{2-}, S^{2-}, SO_3^{2-} and CN^-, and molecules, such as H_2O, NH_3 and PR_2, leave them with unshared lone pairs of electrons, while the electronic structures of cations leave them with unoccupied orbitals, some of which have a high electron affinity as measured by the ionization potential, IP, to reach the particular cationic state. There can then be a covalent interaction energy between the non-metal and the metal centres. We can, therefore, compare cations one with another by plotting IP, which here is the dominant term in the electronegativity expression (Section 2.4.1), against the ionic term z/r. This illustrates the tendency at a given size and charge to form more covalent links with the unshared electron pairs of anions or neutral molecules (Fig. 2.23). Anions can be compared on the basis of size and the electron affinity of their atoms. A good electron acceptor metal, a high IP ion, will naturally form a particularly strong interaction with a good electron donor non-metal centre. Orders of binding affinity of cations and anions can then be drawn up (see Chapter 5).

It is quite frequently observed that cations crystallise in salts from solution with such neutral non-metal donors attached as in $[Co(NH_3)_6]Cl_3$. Here the Co^{3+} ion, with a formal valence of three, chloride, with a formal valence of one, and ammonia, in which nitrogen has a formal valence of three, are combined stoichiometrically. In the structure cobalt has formed six extra

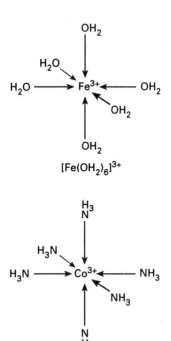

$[Fe(OH_2)_6]^{3+}$

$[Co(NH_3)_6]^{3+}$

Octahedral co-ordination in complex ions.

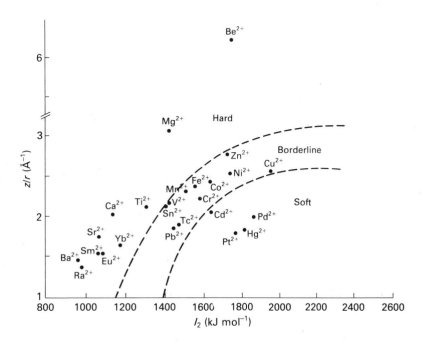

Fig. 2.23 A plot of charge/radius, z/r, against the second ionisation potential, I_2, for divalent ions.

(weak) bonds to ammonia due to the donation of electron pairs of the ammonia ligands. The salts of this kind are called co-ordination compounds, and co-ordinating ligands, unless they give completely, take completely or share electrons *of the central element*, do not contribute to 'valence' as defined in Section 2.4.2.

While co-ordination compounds are of immense importance in solution and in biological chemistry, for example metal–protein binding, they are less significant in the ordered solid state. We therefore describe them in Chapter 5.

2.9.5 Ligand-field effects on structure

When we examine crystal structures more closely we find that the structure around transition metal ions with the same anion partner, for example in oxides, changes in a manner that is not just dependent on ion size. These structural or shape effects are due to the influence of the anionic field upon the *core* electrons and orbitals, that is, the partly hidden d subshells. This is called a ligand-field effect. In Fig. 2.24 we show the d orbitals distributed in relationship to an octahedral distribution of anions. The interaction between the d orbitals and electrons produces a ligand-field effect so that, within the octahedron of anions (donor-ligands), some are attracted by the empty, and to a lesser degree the half-empty orbitals, while some are repelled by the filled orbitals. The full effect on the preferred structure of anions around cations as we progress from d^0 to d^{10} (full subshell) is described in 'Further reading' references 2, 3, 5, 6 and 9. Here we give in Table 2.20 some of the observed preferred stereochemistries around cations (see marginal illustration).

A segment of the spinel (AB_2O_4) unit cell showing the tetrahedral environment of A ions and the octahedral environment of B ions.

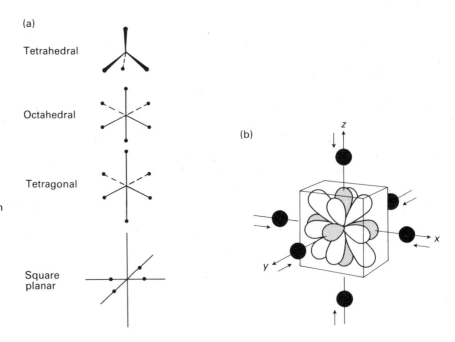

Fig. 2.24 (a) Common co-ordination geometries around a central metal ion. (b) d orbitals in an octahedral field, shaded orbitals, called e_g, point at the ligands and greatest binding is achieved when they are empty. Unshaded orbitals are called t_{2g} and escape repulsion but can make double bonds (see Fig. 2.26).

(a)

Tetrahedral

Octahedral

Tetragonal

Square planar

(b)

Table 2.20 Preferred geometries in simple lattices*

Metal ion	Preferred geometries
Cu^{2+}, Mn^{3+}	Tetragonal > 5-co-ordination > tetrahedral
Ni^{2+}	Octahedral > others
Co^{2+}	Octahedral > tetrahedral > others
Zn^{2+}	Tetrahedral > octahedral > 5-co-ordination
Mn^{2+}	Octahedral > others
Fe^{3+}, Co^{3+}, Cr^{3+}	Octahedral > others

* NB. In this table all metal ions are in high-spin states and liganding atoms are small O, N donors. S-donors favour lower co-ordination numbers. Ligand-field theory, that is polarisation of and binding by the core electrons and orbitals of the metal ion, Table 2.21, can explain the above observations (see inorganic chemistry textbooks cited in the 'Further reading').

2.9.6 Ligand-field energies

The interaction of the *d* core of ions with the surrounding atoms or ligands not only affects structural considerations but also contributes to affinities for partners. The relationship is a slightly difficult one since the ligand-field energy depends on the symmetry of anions disposed around cations but this disposition is also dependent on size factors and covalence. As an example where ligand-field effects can be seen to be important, we can consider the fluorides of the M^{2+} ions, Fig. 2.25. The lattice energies do not follow the ion sizes nor do they follow closely the combined effects of ion size and electronegativity as shown in Fig. 2.23. For the example, the stability of NiF_2 is greater or at least equal to that of CuF_2. The lattice of fluoride ions has octahedral symmetry around ions of this size and it is this ligand field that generates an extra stability for Ni^{2+} ions (Fig. 2.25). Generally speaking, the symmetry-dependent term does not dominate the central field effects of covalence and effects of ion size, so that the order of energy for binding to almost every anion is

$$Mn^{2+} < Fe^{2+} < Co^{2+} < Ni^{2+} < Cu^{2+} > Zn^{2+}.$$

We shall find that this series, the Irving–Williams series, or very closely related ones dominate the chemistry, the biochemistry and the geochemistry of these elements in many compounds.

It is also found that additional interaction energy can be gained by a rearrangement of *d* electrons amongst *d* orbitals so as to increase ligand-field effects—at some internal ion energy cost, of course. The distinction is made between high-spin (the most stable) and low-spin (the least stable) states of free

Fig. 2.25 The experimental lattice energies of the transition-metal fluorides MF_2. The energies, which may be attributed to crystal-field (*ligand-field*) effects, are the differences between the broken and the full line.

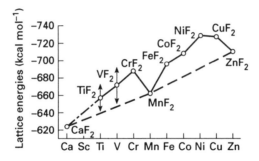

Fig. 2.26 In the absence of ligands the five d orbitals are degenerate, but on the formation of a complex this degeneracy is removed and the orbitals are split into two or more sets. Six d electrons can be arranged in two extreme ways in an *octahedral* complex amongst the orbitals shown in Fig. 2.24. A splitting energy, Δ, describes the separation of the e_g and t_{2g} orbitals. The energy gained by the presence of an octahedral field due to the splitting of the d orbitals depends on the number of d electrons and their orbital packing. Packing of electrons in the low-spin case is opposed by spin-pairing energy in units of P, the energy required to pair electrons. $Fe(H_2O)_6^{2+}$ is high-spin but $Fe^{2+}S_2^{2-}$ (pyrite) is low-spin, see Table 2.21. μ_s = Bohr magnetons.

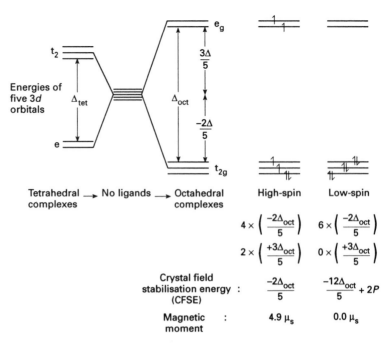

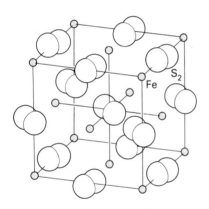

The structure of iron pyrite, FeS_2.

ions (Fig. 2.26). In the low-spin case the $3d$ electrons are paired as far as possible. Spin-pairing increases the availability of d orbitals in the low-spin case and so increases ligand binding energy (Table 2.21.), but it also costs energy. An excellent example is the formation of iron pyrite FeS_2 (see marginal illustration) in which there is low-spin Fe^{2+}, while the stability of most other disulphides MS_2 is much lower since they become low-spin less readily. (This fact has made pyrite a peculiarly important compound on Earth and perhaps at the origin of life.) The structural implications are clear from symmetry considerations as in Fig. 2.24 and Table 2.21. Other examples are found in the binding of Fe-porphyrins in electron-transfer proteins.

Table 2.21 Ligand-field stabilisation energies (LFSE)*

d^n	Example	Octahedral				Tetrahedral	
		Strong field		Weak field			
		N	LFSE	N	LFSE (Δ_{oct})	N	LFSE (Δ_{tet})
d^0	Ca^{2+}, Sc^{3+}	0	0	0	0	0	0
d^1	Ti^{3+}	1	0.4	1	0.4	1	0.6
d^2	V^{3+}	2	0.8	2	0.8	2	1.2
d^3	Cr^{3+}, V^{2+}	3	1.2	3	1.2	3	0.8
d^4	Cr^{2+}, Mn^{3+}	2	1.6-P	4	0.6	4	0.4
d^5	Mn^{2+}, Fe^{3+}	1	2.0-2P	5	0	5	0
d^6	Fe^{2+}, Co^{3+}	0	2.4-2P	4	0.4	4	0.6
d^7	Co^{2+}	1	1.8-P	3	0.8	3	1.2
d^8	Ni^{2+}	2	1.2	2	1.2	2	0.8
d^9	Cu^{2+}	1	0.6	1	0.6	1	0.4
d^{10}	Cu^+, Zn^{2+}	0	0	0	0	0	0

* N is the number of unpaired electrons. LFSE is in units of Δ_{oct} or Δ_{tet}; the calculated relation is $\Delta_{tet} = 0.45\ \Delta_{oct}$. P is the energy required to pair electrons. (See Fig. 2.26.)

B Non-stoichiometric combination of elements

2.10 Introduction

We have stressed that our interest in this book is the natural selection of elements. To some degree, and in an effort to describe ordered arrangements rather than value or other reasons for selection, we have broken with this objective since most of the systems we have described are idealised with perfect order, are stoichiometric, are in one isolated phase and have no contamination. Although this is not the condition of very many useful (observed) materials, such idealised and isolated substances are excellent for an understanding of what might happen in real systems where order is not perfect, non-stoichiometry is found, phases coexist and purity of material is very improbable. Now we must start to move towards those materials we see around us—geological, biological and man-made. We start in the opposite way from that followed above, that is, we describe first ionic rather than covalent systems of several components that have no fixed stoichiometry.

2.11 Ternary salts and silicates

When we examine ternary (or more complicated) ionic compounds, the idea that there should be stoichiometry becomes severely in error since two of the three elements as ions are likely to be quite alike in size and electronegativity and can replace one another randomly and to any degree. This is most often true for metal elements in salts, of course (see Table 2.7). An obvious case is that of the silicates of Earth, which are the basis of most of the rocks, clays, soils and sands around us and are incorporated in a range of building materials. The silicate soils are also the sources of the trace minerals, especially important for life, so that knowledge and understanding of them should never be far from our mind. Unfortunately, they are extremely complicated and of great variety.

The extensive chemistry based on H, C, O and N in organic chemistry (see Chapter 9), is matched by that of silicates where the principal unit, (Si—O) as opposed to C, builds itself into a vast variety of strings, planes and three-dimensional structures. These structures are very difficult to describe since the network is anionic and held together by more salt-like interactions, of which those with Na^+, Mg^{2+}, Al^{3+} and Ca^{2+} dominate. They form non-stoichiometric substitutional solids $A_x B_y$ (etc.) $Si_w O_q$, (see Fig. 2.27 and Table 2.22), in which some elements can substitute for Si as well as A can substitute for B. The

Geometry of linkage of SiO$_4$ tetrahedra	Si/O ratio		Example mineral	Formula
Isolated tetrahedra: linked by bonds sharing oxygens only through cations	1:4		Olivine	$(Mg,Fe)_2SiO_4$
Rings of tetrahedra: joined by shared oxygens in three-, four-, or six-membered rings	1:3	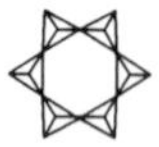	Beryl	$BeAl_2(Si_6O_{18})$
Single chains: each tetrahedron linked to two others by shared oxygens. Chains bonded by cations	1:3		Pyroxene	$(Mg,Fe)SiO_3$
Double chains: two chains joined by shared oxygens as well as cations	4:11	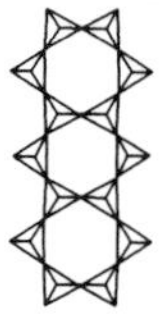	Amphibole	$(Ca_2Mg_5)Si_8O_{22}(OH)_2$
Sheets: each tetrahedron linked to three others by shared oxygens. Sheets bonded by cations or alumina sheets	2:5	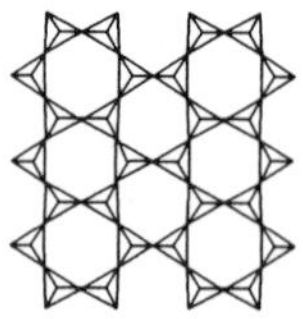	Kaolinite	$Al_2Si_2O_5(OH)_4$
Frameworks: each tetrahedron shares all its oxygens with other SiO$_4$ tetrahedra (in quartz) or AlO$_4$ tetrahedra	3:8		Feldspar (albite)	$NaAlSi_3O_8$
	1:2		Quartz	SiO_2

Fig. 2.27 Some major silicate structures. After Press, I. and Siever, R. (1986). *Earth*, 4/E, Copyright © W. H. Freeman and Company. Used with permission.

Table 2.22 Silicate minerals: types and examples

Class	Si—O formula	Examples	Typical formulae
Ortho-silicates	$[SiO_4]^{4-}$	Olivine	$(Mg, Fe)_2SiO_4$
		Zircon	$ZrSiO_4$
		Garnet	$Ca_3Al_2Si_3O_{12}$
Di-silicates	$[Si_2O_7]^{6-}$	Thortvietite	$Sc_2Si_2O_7$
Meta-silicates	$[SiO_3]^{2-}$	Beryl	$Be_3Al_2[Si_6O_{12}]$
		Pyroxene	$(Mg, Fe)[SiO_3]$
Amphiboles	$[Si_4O_{11}]^{6-}$	Glaucophane	$Na_2Mg_3Al_2[Si_8O_{22}](OH)_2$
	$[Si_3AlO_{11}]^{7-}$	Hornblende	$(Na, Mg, Al, Ca, Fe)_4[Si_3AlO_{11}](OH)$
Sheet silicates	$[Si_2O_5]^{2-}$	Serpentine	$Mg_3[Si_2O_5](OH)_4$
		Kaolinite	$Al_2[Si_2O_5](OH)_5$
	$[Si_3AlO_{10}]^{5-}$	Phlogopite	$K(Mg, Fe)_3[Si_3AlO_{10}](OH, F)_2$
Framework silicates	$[SiO_2]$	Quartz	SiO_2
	$[Si_3AlO_8]^-$	Alkali feldspar	$K[Si_3AlO_8]$
	$[SiAlO_4]^-$	Plagioclase feldspar	$Ca[Si_2Al_2O_8]$

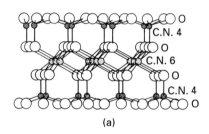

(a)

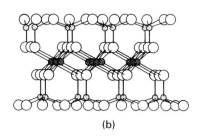

(b)

Fig. 2.28 (a) The structure of 2:1 clay minerals such as muscovite mica $KAl_2(OH)_2Si_3AlO_{10}$ where K^+ resides between the charged layers (exchangeable cation sites). Si^{4+} resides in sites of co-ordination number 4 and Al^{3+} in sites of co-ordination number 6. (b) In talc, Mg^{2+} resides in the octahedral sites and O atoms on the top and bottom are replaced by OH groups.

only valence considerations here are that the separate sums of all negative and positive charges must be equal. The really constant feature is the O^{2-} ion, which can rarely be substituted except by F^- giving oxyfluorides in a constant close-packed lattice. It is obviously impossible to define these complicated oxide phases precisely in terms of compounds of fixed combining proportions of elements. Any complex crystalline silicate structure, never mind an amorphous solid or a melt, has therefore a wide range of potential composition depending on the choice of elements from the periodic table. An example is given in Fig. 2.28 and see Chapter 4. (We shall see later that bone, a complex calcium phosphate, and other ternary phosphates are equally difficult to describe.) While earlier in this chapter it has appeared that materials could be described chemically in rather simple ways—through stoichiometry, for example—we have begun to run into severe problems with these ionic systems which will be examined further in Chapter 4 where we deal with the stability of *non-stoichiometric phases*. Furthermore, the manner of substituting A and B in tetrahedral or octahedral holes can be very ill defined so that *order* is limited although atoms are immobilised. The severity of the problems is similar to that found when we look at other classes of ternary solids more closely (Table 2.23). Two similar non-stoichiometric classes are covalent solids and alloys, which we discuss briefly in the next sections.

Table 2.23 Important non-stoichiometric solid phases

Metallic	Salts	Semimetals
Alloys	Clays	Some oxides
	Rocks	Some sulphides, e.g. Se_xS_y, or
	Cements	covalent solids, e.g. SeS, As_2S_3, GaAs, etc.

2.12 Intermediate cases between salts and covalent molecules: covalent solids A_mB_n

Just as we saw in the cases of elements in groups close to group 14 that the structures and properties of elementary substances, for example, from carbon to tin, can change, so is it the case for compounds when the degree of electronegativity difference is reduced. We can, for example, see the switch from ionic to covalent bonding in the series NaCl, MgS, AlP, SiC. Another example is given by the series of sulphides K_2S, CaS, Ga_2S_3, GeS_2, As_2S_3, SeS, where K_2S and CaS are ionic salts but As_2S_3 and SeS are obvious examples of covalent solids. Interest now settles on the 'covalent lattice' solids such as AlP in the half-way condition between a molecular unit and a salt. As the elements in these compounds become very alike, then their structural formulation becomes difficult due to the possibility of lattice substitution, so that the combination of, say, Se with S exists over a wide range of composition, Se_xS_y. This multiple possibility of substitution is extremely important since it affects the semiconductor properties of solids such as GaAs when we add a third element such as Sb or In. The properties of such non-stoichiometric solid solutions will be discussed in Chapter 4. [At the level of molecules such as C_nH_{2n+2} and polymers, for example proteins, the ways in which they can pack together is almost limitless. In Chapter 6 we look at these problems only noting here that some polymers bind together very selectively.]

2.13 Combination of like metal atoms: alloys

We have seen that the atoms of metallic elements condense in collective close-packed arrays in which cationic atoms are in a sea of electrons. Replacement of some metal atoms A by other metal atoms B to give alloys A_xB_y cannot be easily related to any principle concerning bonds in such metallic solids and stoichiometry is generally not observed in A_xB_y. Thus, combination of the vast range of metal elements generally gives rise to *alloys*, which are to be likened to solid solutions as much as compounds. They are of continuously variable composition in some cases, though more usually over limited ranges. For example, many different phases are formed in different ratio domains of Cu_xZn_y (see Fig. 4.15). It is very important to note that Fe/Ni, bronzes, steels and Mg/Al alloys also fall into this group (see Chapter 4). These materials of specifically selected elements have had essential roles in the development of man's civilised societies. The basis of their use rests on their properties; they are solid, have high melting points, are usually malleable and ductile, with thermal and electrical conductance, and some of them have exceptional mechanical strength. As expected from their structure they have no extremely stable intrinsic shape and can be fashioned into wires, rods, plates, etc.

On the whole, small metal atoms do not form alloys with large metal atoms, and metals of very different electronegativity do not form alloys. Just as in the case of multi-component salts, which formed non-stoichiometric solids and where size controlled packing, so the formation of ternary alloys also does not conform to stoichiometry rules.

We must describe these systems once again in terms of *phases* (see Chapter 4), asking only about the existence of limits to composition zones. It so happens that these limits in alloys are related to the filling of *lattice* orbitals derived from atomic orbitals so that there is indeed a connection with compound formation. Stoichiometric compound formation corresponds to the limiting case of creation of a phase with an infinitesimal composition range and, although this is unusual for metal combinations, it does occur in so-called intermetallic compounds.

Clearly, we have to abandon now the attempt to explain the nature of compounds by direct reference to atomic properties (which is called valence theory) and must try to understand better the nature of *phases*, that is, we require a *collective* or *co-operative* theory. It is often the formation of complicated phases that has decided the natural selection of chemical elements in geological (Chapter 8) and biological systems (Chapters 10–13) and in man's organic chemistry (Chapter 9) and industry (Chapter 14).

2.14 Liquids and solutions

The above cases of non-stoichiometry referred to solids but we must also consider liquids including melts. It is immediately clear that combinations of atoms in complicated liquid solutions are likely to be non-stoichiometric since there is, by definition, no requirement for really long-range order but only for containment. Thus, all natural materials in liquids or melts are open to contamination or even gross mixing, often without limits.

Obviously, we need to leave the subject of order and how things come to be ordered and consider the more general point as to which substances of interest are naturally ordered as stoichiometric compounds, as opposed to which substances are disordered, and in what phases they are found. The ordered situation is of a single state with one defined energy, while a disordered condition represents usually a summed system over many states. Furthermore, we clearly see that ordered and disordered conditions of even one substance exist together, for example, ice, liquid water and water vapour. Whereas in this chapter we have discussed especially why order is observed, we must consider disorder and the reason for its coexistence with order in the next.

2.15 Summary of order

We have now described the main types of chemical bonding and the properties of the resulting *ordered* substances or compounds; these are summarized for convenience in Table 2.24. Depending on basic physical characteristics of atoms such as size, ionisation energy and electron affinity, and orbital kind and occupancy, different types of substances are formed, ranging from close-packed metals to ionic crystals and covalent infinite networks, and from discrete molecular substances to infinite linear or branched chains. There are still other intermediate types of association such as in liquid crystals (see Section 4.4.5). A dominant feature is *co-operativity in condensed states*.

Amongst the compounds formed we observed that many had stoichiometric ratios and hence a series of discrete compounds A_mB_n could be obtained, for example, $CH_3(CH_2)_nCH_3$. Rules for the ratios, *valence* rules, were easily found and explained in many cases. To a large degree this formation of discrete compounds in series was restricted to molecules dominated by non-metals without regard to the phase they were in. Even very large stoichiometric molecules are readily observed for combinations of especially H, C, N, O, S and P in covalent compounds such as proteins and DNA. Discrete compounds were also formed when non-metals were bound to metals, for example in binary salts such as NaCl, but the identity of their molecules was lost in solids, liquids or solutions. In these combinations principles based on electrostatics showed that the variability within chemical combination of two elements is again restricted, that is, we observed particular ratios and ordered structures. However, this was not found to be a general rule where metal elements were involved. When metallic elements were combined in alloys, A_xB_y, it was usual

Table 2.24 Main types of chemical bonding and properties of substances

Nature of substance	Properties (at room temperature)	Examples: elements and binary compounds
Covalent bonding. Sharing of pairs of electrons between atoms (single and multiple bonds); directional bonding; usually low co-ordination number		
Discrete molecules	Gases, liquids or solids, depending on the intensity of intermolecular forces; relatively low boiling and melting points; non-conductors	H_2, CO_2, H_2O, I_2, C_6H_6, glycine, etc.
Atomic networks	Solid; usually high melting point; hard and insoluble; non-conductors	Graphite, diamond, quartz, SiC (carborundum), BN (borazon)
Ionic bonding. Non-directional electrostatic attraction between positive and negative ions; intermediate co-ordination numbers (usually 4 to 8)		
Ionic crystals	Solid; relatively hard and brittle; more soluble in polar than in non-polar liquids; high melting points. Non-conductors as solids but conductors as melts and, especially, in solution	NaCl, KBr, Li_2O, MgO, BaO, etc.
Metallic bonding. Non-directional attraction between positive ions and delocalised valence electrons; higher co-ordination numbers (8, 12)		
Metals and alloys	Solid (except mercury); high melting, malleable and ductile; high thermal and electrical conductance; insoluble unless reaction occurs with solvent	All metals, e.g. Fe, Cu, Ni, Ag, etc., and alloys

to find a region of *continuous* variation of $x:y$ between certain ill-defined ratios that are even temperature-dependent. Moreover, structure was now often disordered. As we increased the number of elements that we combined in this and other condensed system solids, especially ionic salts, or liquids, then continuous variation became more and more common. In particular, multicomponent liquids are miscible over wide ranges of composition, many solids also dissolve in some liquids over wide ranges and 'solid solutions' are common in complex salts. Here (and frequently throughout the book) we have studied extracted materials as well-defined separated substances in particular isolated physical states only in order to gain a basis for understanding of complex systems.

We have then described the properties of the substances as if they existed independently from any force external to them or from necessary contact with other chemicals or states of matter, and without any restriction as to the amounts present. For example, a given pure gas or liquid consists of identical atoms or molecules and we did not consider the volume of the container as either limiting or distorting its properties. The shapes of *molecules* were described as if they arise through internal interactions between atoms that are in one molecule alone, which is reproduced precisely by every other molecule in the system, and without regard to their environment. This is not our experience of the real world of objects, which have surfaces.

In such a discussion we could not handle properly important features of the real environment. We must now recognise additional features of everything that is around us, for example, the balanced coexistence of different states of matter (gas: liquid: solid) of any one substance. Thus H_2O cannot be considered in a real system as just vapour, liquid water or ice separately since below 100°C there is balance between them at any temperature we experience. Similarly, it is frequently the case that a given chemical is found in several forms, sometimes in chemical balance and sometimes not. Thus CO_2 and H_2O are in balance with HCO_3^- and H^+ while O_2 is not in balance with H_2O. We turn in Chapter 3 to a discussion of (the thermodynamics of) physical and chemical balance. This re-introduces a central problem in all our surroundings. The existance of *order*, described in this chapter, and *disorder* are obviously in some balanced condition around us, for example, the water vapour in the air coexisting with liquid water and ice. Once we have described the principles of balance in Chapter 3, we can apply them under simplifying assumptions to bulk phase physical balances, for example ice/water/water-vapour, and to chemical balances, for example $H_2O/O_2/H_2$, in Chapters 3 and 4. Balances in dilute solution systems, especially precipitation and complex association are discussed in Chapter 5. This description of *order/disorder* remains incomplete and in Chapter 6 we consider balances between components (chemical compounds) in phase systems of limited size and under the influence of fields from neighbouring phases, that is, across boundaries. This will complete the account of observed stationary conditions. Only in Chapter 7 do we introduce rates of change, kinetics, which are so apparent around us and will lead to organisation. It is obvious to us all that, as explained in Chapter 1, the major challenge we face is to understand order and disorder balances within systems that flow and change, as in geological or biological systems, and that we have called organised.

Further reading

Most good texts of general or inorganic chemistry cover the subjects discussed in this chapter in greater or lesser depth. The choice is very much a matter of individual preference. We limit ourselves to indicate just three recent books and to suggest a number of other classic texts that still have much to recommend them.

1. Atkins, P. W., CLugston, M. J., Frazer, M. J., and Jones, R. A. Y. (1988). *Chemistry—principles and applications*. Longman Group, London. Mostly Chapters 1, 2, 5–8 and 15.
2. Huheey, J. E., Keiter, E. A., and Keiter, R. L. (1993). *Inorganic chemistry—Principles of structure and reactivity* (4th edn). Harper Collins College Publishers, New York. Mostly Chapters 2, 4, 5 and 8.
3. Shriver, D. F., Atkins, P. W., and Langford, G. H. (1994). *Inorganic chemistry* (2nd edn). Oxford University Press, Oxford. Chapters 1–4, 9 and 18.
4. Pauling, L. (1960). *The nature of the chemical bond* (3rd edn). Cornell University Press, Ithaca, New York. Chapters 3, 7 and 11–13.
5. Phillips, C. S. G. and Williams, R. J. P. (1966). *Inorganic chemistry*. Oxford University Press, Oxford. Chapters 2–5.
6. Day, M. C. Jr and Selbin, J. (1969). *Theoretical inorganic chemistry* (2nd edn). Reinhold Book Corporation, New York. Chapters 3–6.
7. Pimentel, G. and Spratley, R. D. (1970). *Chemical bonding clarified through quantum mechanics*. Holden-Day Inc, San Francisco. Chapters 1–3.

8. Coulson, C. A. (1961). *Valence* (2nd edn). Oxford University Press, Oxford. Chapters 1, 2, 4, 11 and 13.

9. Cartmel, E. and Fowles, G. W. A. (1966). *Valence and molecular structure* (3rd edn). Butterworths, London. Mostly Chapters 5, 7 and 9–11.

10. Edwards, P. P. and Sienko, M. J. (1983). Metallic character in the periodic table. *Journal of Chemical Education* **60**, 691–6.

The following paper and books are recent additions to the literature, providing useful insight to the views of today.

11. Orchard, A. F. (1994). Electronegativity and ionisation energies. *Philosophical Magazine, Series B* **69**, 821–31.

12. Cox, P. A. (1995). *The elements on Earth*. Oxford University Press, Oxford.

13. Gray, H. B. (1995). *Chemical bonds—an introduction to atomic and molecular structure*. University Science Books, New York.

3

The balance between order and disorder

He who wants to have right without wrong,
Order without disorder,
Does not understand the principles of
Heaven and Earth

Chuang-Tzu (Fourth century BC),
Texts—Autumn floods, Section 17
(adapt. Thomas Merton)

3.1 Introduction

In Chapter 1 we took as a historical starting point for the discussion of all observations concerning the material objects we sense around us the four 'elements' of certain Greek philosophers and scientists such as Aristotle (Fig. 1.1). We also noted the relationship to similar thinking in Chinese philosophy (Fig. 1.5).

We have shown so far that, according to today's views, three of the classical 'elements'—air, water and earth—are more or less directly related to the physical states of materials (gases, liquids and solids). Moreover, as examined in Chapter 2, materials themselves of every kind are now known to be made from the fixed number of chemical elements of the periodic table (Fig. 1.7). These elements came together in *ordered*, often shaped constructs to make molecular gases and condensed systems, and thence all chemical material things on Earth (and elsewhere), due to what we call 'forces', which can still be said to be related to the Greek idea of underlying 'form'. The main forces of concern are those that we have labelled gravitational and electromagnetic (see Chapter 1). The fourth Greek 'element'—fire (or heat and light)—can transform solids to liquids to gases; the Greeks connected it also to underlying 'form' and therefore made it similar in kind to the other three 'elements'. Today we say instead that the uptake of *energy* (heat or light) by materials acts to increase their internal energy content and random motion, disorder, which opposes the forces that bring chemical elements together generating order. Thus 'fire' is related in part to an intrinsic energy content of materials, to weakening of bonds (decreasing potential energy), to increasing separation and increasing random motion of atoms and molecules.

Since locally and in the universe there is both energy and matter it is not clear how order and disorder come into balance and as yet in this book we do not know how to treat energy in this context of disorder and order of matter. We must try to analyse its nature as we did the binding energy of atoms in materials. We know now that one form of energy within the observed concept of fire is radiative (photon energy, $E = h\upsilon$, for example light) and another is the energy associated with rest-mass random motion, (kinetic energy, $E = \frac{1}{2}mv^2$ Section 1.6). As we stressed in the first chapter, there is a need to analyse the radiative energy of photons together with that of the motions of rest mass; indeed, we frequently describe a physical or chemical change by writing equations such as

$$X + Y \longrightarrow XY + energy$$

where the energy is not restricted to either kinetic (rest mass) or radiative (photon) energy. Now this equation has been written deliberately here with a direction of change, from $X + Y$ to XY, where XY is obviously more 'ordered'. One of the purposes of this book is to examine the way in which such directional changes have come about, particularly in chemical systems,† from the beginning of the universe. This is not just a question of forming the ordered isolated units (XY) described in Chapter 2, but of the obvious overall tendency for rest-mass particles and photons to be disordered, as seen by their dispersion in the expanding universe, which turns out to be related in part to the energy released in the above equation (see marginal illustration).

In fact, we see everywhere around us materials that coexist apparently in balance between ordered and disordered states, and where the balance responds to both temperature and pressure. An obvious example is water which is present at a given temperature and pressure as water vapour, as liquid water and/or as ice. Here the water vapour is a disordered gas except for

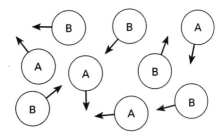

Disorder spatial and kinetic

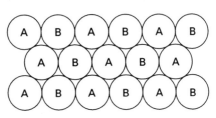

Order (see Chapter 2)

† In this book we use the word *system* to represent any particular substance and/or process, or set of substances and/or processes, that we choose to separate from all the others. More generally, a system is a part of the 'universe' but separated from the rest of the universe by definite boundaries, real or arbitrary (see Fig. 3.9). Systems can be *isolated*, when the boundaries do not allow transfer of mass or energy; *closed*, when only energy transfer is allowed; or *open*, when both matter and energy transfer are allowed.

the structure in the H_2O molecules, the liquid water is in a more limited disordered state and the ice is an ordered solid. Other materials, such as dioxygen, are found as disordered gases almost exclusively and yet others, minerals, are in essence ordered solids. In contrast, living systems are clearly only partly ordered, having within them much liquid, disordered water solutions. What is it that decides this balance between the physical states and that obviously depends on temperature and pressure?

We shall approach the problem of disorder in a simplified orthodox way, much as we have done in the description of order in Chapter 2. Thus Sections 3.2 to 3.2.4 can be passed over quickly by the reader who is familiar with the thermodynamic treatment of disorder or so-called entropy. In Section 3.2.5 we turn to one main purpose of the book, which is to understand how the natural selection of the chemical elements has generated the physical order $\rightleftarrows$ disorder balance that we observe around us, treating the question whether particular chemical elements (in compounds) favour a particular balance so that some are gases, some liquids and some solids under defined conditions. This division has dominated both inanimate and animate evolution.

When the question as to the association of particular physical states with particular groups of elements (and their compounds) has been analysed we go on to ask, in Sections 3.4 and 3.5, about the *chemical* balance in materials. An example is the balance between an ordered mineral, NaCl, and its obviously disordered ionic solution in water, and another is that between the reactant and the products in the decomposition of calcium carbonate

$$CaCO_3(s) \rightleftarrows CaO(s) + CO_2(g).$$

Here the balance is between the two ordered solids, $CaCO_3$ and CaO, and a disordered gas, CO_2, and is obviously temperature-dependent.

The natural selection of chemical elements in the formation of gases, liquids and solids is now clearly a question of balance between chemical affinities that can give either ordered structured systems with highly *co-operative internal binding*, $CaCO_3$ and CaO, or gaseous disordered molecular systems lacking co-operative interaction (CO_2). It is this selectivity that allows structured and unstructured materials to coexist in geophysical constructs and to co-operate in the formation of living organisms. Without some disorder nothing can be alive.

Once the principles of physical and chemical balance are in place, we shall consider situations where there is lack of balance between two systems, each with internal balance (Sections 3.5.2–3.6.2). In this situation one system can transfer energy to the other, doing work on it and creating new situations. Such transfer of energy and/or material generates change of physical/ chemical state and allows abiotic and biological evolution. We shall consider the energy transfer as a transfer of heat, not distinguishing between convective and radiative energy. Since, however, the source of all our energy is at present the sun we devote a separate section (3.7) to the description of radiative energy before summarising the nature of order $\rightleftarrows$ disorder balances. (Later, in Chapter 7, we will consider one other fundamental characteristic of the physical world and, indeed, of the universe, noted particularly by the Chinese—flow, which is related to rate of change (in a direction) and then to organisation, as distinct from order.)

3.2 Order–disorder balance

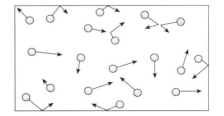

Fig. 3.1 Individual molecules in a gas have different space co-ordinates and different kinetic energies at any one time. The assembly has a vast number of arrangements of spatial and energy distributions: each is a microstate within a macrostate condition, see also Fig. 1.5.

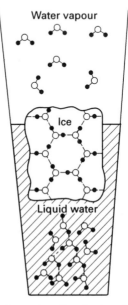

Fig. 3.2 Increasing disorder: ice < liquid water < water vapour.

In Chapter 1 we have written the balanced equation for a given system as

$$\text{Ordered} \rightleftharpoons \text{Disordered}$$

and we have described the binding energy of the ordered condition in Chapter 2, but we did not consider the energies corresponding to the disordered condition. We wish at first to consider the disordering of rest mass without reference to radiation, photons. It turns out that we must take into account: (1) the different possible spatial arrangements of the particles of the system; and (2) the various possible distributions of the energy of the system amongst its constitutive particles. To proceed with this analysis we may now consider that the pure ordered state has a certain amount of energy less than that of the disordered state, the ordered state being associated in principle with a perfect single arrangement due to forces of interaction between its units (atoms or molecules in the cases dealt with in Chapter 2), for example, a crystal, while a typical disordered state is that of an ideal gas.

The disordered condition, in which for practical purposes we consider that there are no interactions between the units of the system, may have many different spatial and thermal arrangements of these units (different microstates) corresponding to the same macroscopic state, that is microscopic arrangements that look identical from the point of view of the macroscopic properties of the systems, for example temperature, pressure and relative composition. The fact that there are many ways for the arrangement of the disordered state and only one for the ordered state means that there is always a *probability bias* in favour of the disordered condition (see Fig. 3.1).

If there are Ω possible ways of arranging the disordered system and *if the ordered and any of the disordered microstates have the same probability of being observed*, the *bias* in favour of disorder will be $\Omega:1$ for large values of Ω. What we want to obtain is a quantitative expression for the balanced relationship between what effectively is the 'ordering energy', ΔQ, and this *bias* toward disorder. We need to express disorder, therefore, in the same units as those of the ordering energy ΔQ so that we can understand balanced conditions quantitatively. An example may help. We observe frequently that some substance is in an ordered state, for example water as ice, and is in balance *at a fixed temperature* with some of the same substance in the gas state, that is water as vapour, when its vapour pressure is fixed. The molecules of H_2O in the gas phase have many arrangements open to them in respect to one another, while in perfectly crystalline ice there is but one possible arrangement in ordinary conditions (Fig. 3.2). (We also show in Fig. 3.2 the interactions that determine some degree of ordering in the liquid.) What is the quantitative characteristic of the term in energy units that favours the vapour so that there is a fixed vapour pressure?

Now, the disordered side of the balanced equation above obviously visits more microstates Ω as the volume of the system increases, that is, there are more ways of arranging the relative space co-ordinates of the molecules when the volume available, V, increases (see Fig. 3.1). A common observation is that a vapour will fill the volume open to it, but a solid will not. Clearly, in this sense, Ω and V are related. Similar arguments can be used to show the effect of temperature. Although the molecules of water in the vapour have the same *average energy* over time, the individual molecules may differ in their (kinetic) energy at any one moment and we may distribute them over a series of 'energy levels'. Several such distributions are possible. The 'occupancy' of these energy levels can be interpreted in terms analogous to the occupancy of space and in this sense the increase of temperature corresponds to an increase of volume, that is, it makes more energy levels accessible and there will be more ways to distribute the energy amongst the particles of the system. In brief, both spatial rest-mass distribution and energy distribution amongst rest mass must be considered in order to evaluate the number of possible microstates of the system.

One conclusion can immediately be derived: order is favoured by smaller volumes (hence higher external pressure) and by lower temperatures; both effects correspond to our common experience, as the example of ice/liquid water/water vapour balances shows.

To help the understanding of the order/disorder concept (only apparently simple) and to introduce its quantitative appreciation, we will consider first some small model systems for which the different possible microstates can be 'counted' ('countable systems') and we will separate, for convenience, the spatial (pressure/volume) arrangements and the energy (temperature) distribution.

3.2.1 Spatial arrangements of the particles of a system

Let us suppose that we have a system of four cells, each of which can contain a particle, for example a molecule or an atom, and that the four cells are non-identical in their location, for example relative to a plane dividing them in two parts—left and right.

The following distribution possibilities arise:

There are, then, six different arrangements or microstates, sometimes called complexions, two of which correspond to having two particles both at the left or at the right side of the plane, and four that correspond to having one particle at each side. Obviously, the probability of finding the two particles at the same side is $2/6 = 1/3$ and the probability of finding one molecule on one side and one on the other is $4/6 = 2/3$, assuming that any microstate is equally probable.

If we now double the number of cells to 8, a similar exercise will lead to the conclusion that there are 28, that is, $(8 \times 7)/2$, different possibilities (microstates), of which 12 correspond to having two molecules in the same half of the system. The corresponding probability of occurrence of any of such states is $12/28$, that is, $3/7$, but if we are interested in one particular packed state, the probability of its occurrence is only $1/28$, whereas that for the summed probability of the other states is $27/28$. This reasoning can be generalised to any number of microstates, in which case the probability of occurrence of one particular state is $1/\Omega$, whereas the total probability for all the others is $(\Omega-1)/\Omega$. If we now say that the one particular state corresponds to a perfectly ordered solid, say, ice, and all others represent vapour in equilibrium, that is, balance, then the *probability* of observing the ice structure, rather than the vapour with its multiplicity of possible arrangements, is $1/(\Omega-1)$ which is approximately equal to $1/\Omega$ for large values of Ω. Of course, we are assuming that in the calculation all of the microstates are equally probable, that is, that there is no *bias* of internal forces, although we know that they exist, favouring the formation of ice, and represented as ΔQ. (Temperature here is, of course, a constraint that can favour one physical state or another, which we consider later.)

Now, following Boltzmann (1896), we define the quantity S, called *entropy*, by the expression

$$S = k \ln \Omega$$

where k is the so-called Boltzmann's constant ($k = 1.380 \times 10^{-23} \, \text{J K}^{-1}$) and Ω is the number of microstates that are in consonance with the specified state of the system. The reason for the units of k become clear later. (Note that we are as yet considering configurational states only.) The entropy, S, is a particular measure of the number of microstates available to a system, hence a measure of its probability of occurrence, and a measure of its disorder in the sense that it has many possible equivalent distributions, whereas for the ordered form there is but one or perhaps a small number of possibilities. We now wish to show the relevance of this particular expression in the quantitative estimation of *disorder in energy units*.

3.2.2 Changes in entropy with changes in volume of a gas—configurational entropy

Let N be the number of independent cells that molecules can occupy in a volume V and N' the number of cells in a volume, V'. The values of Ω and, from it, the entropy change for the process in which N_A molecules distributed in N cells (state 1) become distributed in N' cells (state 2) (Fig. 3.3) can be calculated to be $\Delta S = k N_A \ln N'/N$ as follows. (Those readers who do not care to follow the mathematics should omit the next paragraphs in small print that show this to be so.)

It can be demonstrated by inductive reasoning (see reference 3, 'Further reading') that if the number of non-equivalent cells is N and the number of indistinguishable

Volume V with N cells

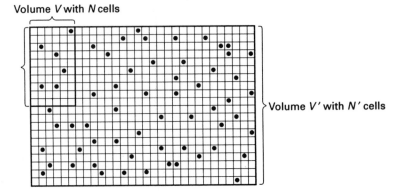

Volume V' with N' cells

Fig. 3.3 The figure shows the relationship between the number of cells, N and N', and the corresponding volumes, V and V'. If all molecules (black circles) in V' were put into V there would be a single possible ordered state.

particles A, is N_A, the number of possible states will be given by ($N!$ = factorial N)

$$\Omega = \frac{N!}{N_A!\,(N-N_A)!}.$$

We can now calculate the entropy change, ΔS, defined above, for the process in which the N_A molecules distributed in N cells (state 1) become distributed in N' cells (state 2) as

$$\Delta S = S_2 - S_1 = k \ln \left\{ \frac{N'!}{N_A!(N'-N_A)!} \right\} - k \ln \left\{ \frac{N!}{N_A!(N-N_A)!} \right\},$$

which can be restated as

$$\Delta S = k \ln \left\{ \frac{N'!(N-N_A)!}{N!(N'-N_A)!} \right\}.$$

For large values of N, N' or N_A this expression is difficult to calculate and approximate methods must be used. A useful method is the so-called Stirling approximation, according to which (for large values of N)

$$\ln N! = N \ln N - N.$$

Considering $N, N' \gg N_A$ the expression above then becomes

$$\Delta S = kN_A(\ln N' - \ln N) = kN_A \ln N'/N.$$

Here we have deduced, through a consideration of the change of the number of microstates, that there is a logarithimic relationship between the change in entropy and the change in the number of cells that can be accessed by N_A molecules.

Since the number of cells in Fig. 3.3 is, at a fixed temperature, clearly proportional to the volumes V and V' that contain them, and making N_A (number of particles) equal to Avogadro's number, that is, considering one *mole* of particles, one obtains ($R = kN_A$),

$$\Delta S_{\text{conf}} = R \ln(V'/V),$$

which shows the effect of expanding the volume on the number of (spatial) microstates of the system and hence on its entropy, which we call, in this case, the *configurational* entropy change, ΔS_{conf}. R is the gas constant of the expression $pV = RT$ as we now confirm.

For an *ideal* gas this configurational entropy change relates to the energy that is expended in the process of expansion from V to V', traditionally called work† of expansion and defined as

$$W = -\int_V^{V'} p \, \mathrm{d}V$$

(work performed by the system is given a negative sign). Since for 1 mole of an *ideal* gas $pV = RT$ (see Chapter 1), one has

$$W = -RT \int_V^{V'} \mathrm{d}V/V = RT \ln (V'/V) = -T \, \Delta S_{\text{conf}}.$$

† As will be discussed later, *work* is a *process* of energy transfer. In the present case the process is the expansion of the ideal gas, but there are other *processes* of energy transfer, hence other types of work. An *ideal* gas is used to avoid interactions between particles.

This nicely relates the kinetic theory of gases to probability arguments and to the meaning of R (or k) which is in the units of calories (joules) per mole per degree. It is often the case in textbooks that, instead of showing the source of entropy in statistical terms as here, the reader is presented with the equally correct approach from functional thermodynamics using the expression for entropy, by definition.

$$\frac{\Delta Q}{T} = \Delta S$$

where ΔQ is the heat energy equivalent of W above involved in a process such as the expansion of a gas. (ΔS is clearly here in units of energy, calories (joules), per mole per degree.) The two ways of defining entropy relate to the same fundamental property of matter—that it can be disordered by increasing its energy content, here increasing volume. This means that the larger the expansion the larger will be the configurational entropy change and the larger (more negative) the mechanical work performed by the system at the cost of an equivalent amount of energy (heat), ΔQ, transferred from the surroundings. For solids and liquids the internal configurational entropy changes are usually small since they are not affected by pressure to a significant extent. When pressure is reduced more material enters the vapour phase particularly from volatile liquids and solids, whence chemical bond selectivity is critical.

It is very important to recognise the nature of the expression $R \ln V'/V$ in relation to disorder; S does not represent a single 'snapshot' (microstate) of the gas but a particular sum of all possible 'snapshots' at V' or at V, respectively. It is a property of a macrostate, which is the observable condition. Each snapshot, microstate, is here given equal weight (or probability of occurrence). Finally $T \Delta S$ is in the same units of energy as Q so that we have achieved the objective of balancing order and disorder in the same units.

It is instructive to give some numerical values to the energies involved in this process and for this purpose we take the example of an ideal gas, 1 mole of which expands isothermally at 298 K, say, from 22.4 l to 224 l. The configurational entropy, change will be given by $R \ln V'/V = 19.2$ J K^{-1} mol^{-1} = 4.6 cal K^{-1} mol^{-1}. At a temperature of 298 K (25°C), $RT \ln V'/V = 1380$ cal mol^{-1} which is to be compared with the energy of binding together of molecules in lattices (Section 3.2.5). Before proceeding to the discussion of the effect of temperature we should notice that we have considered small molecules only in random translational motion. For larger molecules we have to add the different conformations (which contribute to the overall configurational entropy) and include the different internal rotational states that a molecule can adopt.

3.2.3 Temperature change and disorder—thermal entropy

We can now discuss energy distribution amongst particles, molecules, when we change temperature. The derivation follows the same steps as for the consideration of the effect of increase of volume but now we consider discrete (that is, 'quantised') energy levels (for example kinetic energies) *unequally* '*occupied*' by particles instead of the *equal* distribution of particles in different volume cells. Let us then suppose that there is a series of energy levels ε_0, ε_1, $\varepsilon_2 \ldots$, that are occupied by $N = N_0 + N_1 + N_2 + \ldots$ particles. The energy

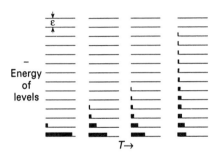

Fig. 3.4 The distribution of particles over equally spaced energy levels becomes broader as the temperature, T, increases (see Figs 1.16 and 1.19).

distribution is then described by specifying the number of particles N_0 with energy ε_0, the number of particles N_1 with energy ε_1, etc. for a total energy $E = N_0\varepsilon_0 + N_1\varepsilon_1 + N_2\varepsilon_2 + \ldots$, corresponding to a given temperature T.

There will be many different distributions (microstates) corresponding to the same total E within which we must evaluate N_0, N_1, N_2, etc. For simplicity we will consider equally spaced translational (kinetic) energy levels, as in Fig. 3.4, and will assume that the particles are non-interacting so that the total energy of the system is indeed the sum of the individual kinetic energies of the particles and there is no contribution arising from interactions between them (as in an ideal gas).

As in the case of particles occupying cells, we derive an expression for the number of microstates, Ω, corresponding to the distribution of particles over energy levels, now for a total energy E. The expression is (see references 1, 2 and 6 in 'Further reading').

$$\Omega = \frac{N!}{N_0!N_1!N_2!\ldots}. \tag{3.1}$$

The *thermal* entropy, S_{th}, is given as before, by Boltzmann's expression $S_{th} = k \ln \Omega$. Hence, if Ω is large, the entropy will be large and, if Ω is small, the entropy will be small, down to $S_{th} = 0$, when $\Omega = 1$, with all the particles occupying the lowest energy level, that is, $N_0 = N$ and N_1, N_2, etc. are equal to zero. This is the case of a perfectly crystalline substance with but a single arrangement of its particles obviously at the lowest possible temperature. We can again think of the case of the ice/water vapour, order $\rightleftarrows$ disorder, balance of probabilities making the simplifying assumptions that ice is perfectly crystalline and occupies just one microstate while the vapour has open to it a distribution of microstates given by eqn (3.1) due to the different distributions of *translational* velocities of particles.

To achieve a state of highest entropy at any one temperature, the particles of the system must spread out into as broad an energy distribution as possible, compatible with the given total energy E. (There is a severe constraint on the occupancy of higher energy levels, of course.) The analogy with a gas filling all the volume, the cells in Fig. 3.3, available in a container is obvious; the 'container' in this case is the fixing of the total energy E, but all energy levels in 'cells' ε_i will be differently occupied. (When we considered volume cells in Section 3.2.1 all cells were equally occupied.) Here we face the difficult task of determining values of occupancy, $N_0, N_1, N_2\ldots$, generally represented by N_i, at equilibrium which will maximise Ω of eqn 3.1.

This requires a rather complicated process of calculus which need not be described here (the interested reader can look into references 1, 3 and 6 in 'Further reading'), but it was shown by Boltzmann that such N_i values (for the occupancy of the ε_i state) satisfy the equation

$$N_i = \frac{Ne^{-\varepsilon_i/kT}}{\displaystyle\sum_{i=0}^{i=\infty} e^{-\varepsilon_i/kT}} \tag{3.2}$$

where N is the total number of molecules present. The sum

$$\sum_{i=0}^{i=\infty} e^{-\varepsilon_i/kT}$$

is the total occupation of the system and is called the *partition function* since it shows the way in which molecules are divided, partitioned, amongst all energy levels. Use of the occupation of the ε levels, like use of the occupation of the N volume cells in Section 3.2.2, allows us to calculate the probability, the number of thermal microstates, Ω_{th}, of any system for one temperature.

To obtain the change in thermal entropy from one temperature T_1 to another T_2, corresponding, for example, to the change of water vapour pressure over ice with temperature, we find Ω at the two temperatures. Then by an analysis, utilising expression (3.2) in eqn (3.1), we find the relationship between Ω and temperature at constant volume for a given substance (compare the calculation of the relationship between Ω and volume, V, in Section 3.2.2). For a perfect (ideal) monatomic gas, that is just for translational energy states, it has been shown that Ω varies with $T^{3/2}$, whence the thermal entropy change for one mole of gas on going from T_1 to T_2 is $\Delta S_{\text{th}(V)} = \frac{3}{2}R \ln T_2/T_1$.

Now $\frac{3}{2}R$ is the *molar heat capacity*† at constant volume, C_V, for a perfect monatomic gas and a more general treatment of expression (3.2), including all the possible energy states of polyatomic gas molecules, gives

$$\Delta S_{\text{th}(V)} = C_V \ln(T_2/T_1)$$

where units of R and C_V (and ΔS), as defined above, are in calories (joules) per mole per degree. To get the energy corresponding to the above entropy change we have to multiply ΔS by the mean temperature T, that is,

$$T\,\Delta S_{\text{th}(V)} = T\,C_V \ln(T_2/T_1) = \Delta Q.$$

(The relationship is evident if we consider that heat put in at constant volume, δQ, to change the temperature by dT is $C_V\,dT$ and that if we take this heat to change the distribution over microstates alone, that is, to change the entropy by a differential amount, dS_{th}, then, on going from T_1 to T_2,

$$\int_{T_1}^{T_2} dS_{\text{th}} = \int_{T_1}^{T_2} \frac{\delta Q}{T} = \int_{T_1}^{T_2} \frac{C_V\,dT}{T}$$

and $\Delta S_{\text{th}} = C_V \ln(T_2/T_1)$.)

The above relationships are often deduced using functional thermo-dynamics. In this alternative analysis the value $\frac{3}{2}R$ for C_V of a perfect monatomic gas is found using the kinetic theory of gases (similarly C_p is found to be $\frac{5}{2}R = C_V + R$ per mole) and hence we see clearly that $\Delta S_{\text{th}(V)} = \frac{3}{2}R \ln(T_1/T_2)$ is in the units of $\Delta Q/T$, as was stated above.

This deduction applies at constant volume or pressure and we can see that thermal entropy change per mole quite generally is given by

$$\Delta S_{\text{th}} = C \ln(T_2/T_1)$$

where C is the specified molar heat capacity.

As stated, in the general case the molar heat capacity depends upon the other energy levels available to the system apart from translational motion. The quantised energy levels of a molecule include translational, rotational, vibrational and electronic levels, and the total energy E is given by

$$E = E_t + E_r + E_v + E_e.$$

† Molar heat capacity is the amount of energy (heat) needed to increase the temperature of a mole of a substance by a degree. It can be defined at constant volume, C_V, or at constant pressure, C_p.

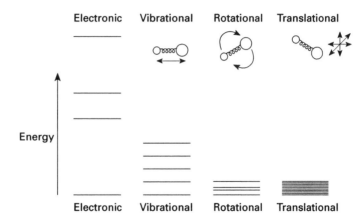

Fig. 3.5 The quantised energy levels for different forms of molecular motion.

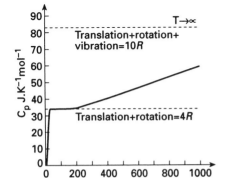

Fig. 3.6 Plot of C_p versus T for NH_3 (g) at 1 atm over the temperature range 0 K to 900 K. Note that $C_p = C_V + R$ for an ideal gas and that for polyatomic molecules C_V is higher than $\frac{3}{2}R$, due to the rotational and the vibrational modes into which energy can be distributed.

The spacings of these levels are very different and decrease in the order

$$\text{Electronic} \gg \text{vibrational} > \text{rotational} \gg \text{translational}$$

(see Fig. 3.5 and Table 3.1), so that C increases very rapidly from 0 to 10 K but remains roughly steady to 300 K after which it gradually rises (Fig. 3.6). Thus, the entropy ΔS of a gas is strongly dependent on the number of molecules present in a given volume but only weakly dependent on the type of molecule until quite high temperatures > 300 K. This is very fortunate for the ease of understanding semiquantitatively entropy changes at around 300 K in any reaction when the number of molecules changes such as in

$$X + Y \longrightarrow XY.$$

It is also fortunate that both liquids and solids in this low temperature range have very little entropy so that the entropy change on vaporisation at the boiling point ($p = 1$) is relatively fixed for all substances (Section 3.2.5). To a first approximation the entropy is related just to the *number* of molecules not in a condensed state so that we can easily handle chemical as well as physical changes. Returning to the equilibrium water vapour over ice the balance of

$$\text{Ice (ordered)} \rightleftharpoons \text{Vapour (disordered)}$$

is seen to be biased by decrease in pressure and increase in temperature since both changes increase the entropy, that is, the probability of being in the vapour rather than in the crystalline state and is chemically selective.

Table 3.1 Characteristic energies of thermal excitation of quantised molecular motion (example)

Type of motion	Particle mass (atomic units)	Typical distance (L) of motion (cm)	Energy spacing between adjacent levels[*]			Regions of radiation spectrum allowing observation of the transition
			(J molecule^{-1})	(kJ mol^{-1})	(cm^{-1})	
Translation	50	10	5×10^{-40}	3×10^{-19}	2.5×10^{-17}	
Rotation	10	1×10^{-7}	2.5×10^{-23}	1.5×10^{-2}	1.25	Microwave
Vibration	10	5×10^{-9}	1×10^{-20}	6	500	Infrared
Electronic	1/1837 (electron)	5×10^{-8}	2×10^{-18}	1×10^3	1×10^5	Ultraviolet/visible

[*] Note that the larger the mass, m, of the particle considered and the larger the distance, L, involved in the motion, the smaller is the spacing between adjacent levels, given by $\varepsilon = h^2/mL^2$, where $\varepsilon = kT$.

To give a numerical example of ΔS_{th} let us take 1 mole of gas such as carbon dioxide, which is heated at constant volume from 298 K to 373 K, $C_V = 28.5$ J K^{-1} mol^{-1} (assumed to be constant in this range of temperature). The increase of entropy for 1 mole $\Delta S_{th} = 28.5 \ln(373/298) = 6.4$ J K^{-1}, a value of the order of that corresponding to the configurational entropy change when the volume is increased tenfold. Thus we see that $CaCO_3$ must dissociate to $CaO + CO_2$ (volatile) as temperature is increased.

3.2.4 Total entropy change of a system

Naturally, the number of possible microstates of a system when both configurational and thermal effects are considered will be equal to the *product* of the number of microstates for each separate effect, say, Ω_i and Ω_j (this is obvious since any microstate of one kind may combine with all microstates of the other kind).

We may then write, for 1 mole

$$S_{sys} = S_{conf} + S_{th} = R \ln(\Omega_i \Omega_j) = R \ln \Omega$$

where S_{sys} is the overall entropy of the system and Ω the total number of microstates compatible with any one macrostate. We may then consider that for the balanced equation above (Section 3.2.1) $RT \ln \Omega$ is a measure of the switch from order (state 1) to disorder (state 2) which, in an equilibrium, we may now equal to the total ordering energy ΔQ, that is,

$$\Delta Q = RT(\ln \Omega_2 - \ln \Omega_1) = RT \ln(\Omega_2/\Omega_1) = RT \ln \Omega_2 = T \Delta S_{sys}$$

(since for the ordered state, for example corresponding to $T = 0$ K, $\Omega_1 = 1$).

We have, therefore, $\Delta S_{sys} = \Delta Q/T = R \ln \Omega$, which allows the total entropy change of the system to be determined from the knowledge of the 'ordering energy', ΔQ, determined experimentally, or from the knowledge of the total number of possible microstates of the system, Ω, calculated from statistical considerations. More commonly, we consider systems at constant pressure or constant volume and the ordering energy, ΔQ, exchanged reversibly with the surroundings of the system (see Fig. 3.9), may be at constant p, ΔQ_p, or at constant V, ΔQ_v.

Note again that $T \Delta S_{sys}$ is the energy 'used' on going to a more disordered state; hence in any change from a disordered to a more ordered state, for example, water→ice at 0°C, there is a decrease of disorder ΔS_{sys} of the water going to the solid state, ice, implying that a corresponding amount of energy is lost to the surroundings as ΔQ.

From what has been said we can also understand a very important consideration from an evolutionary (and a technical) point of view: the fact that, on expanding, a gas cools. Let us consider again a gas in a closed vessel that expands very slowly so that no heat is exchanged with the surroundings (so-called *adiabatic reversible expansion*). In this case $\Delta Q = 0$ so that $\Delta S_{tot} = 0$, that is, the entropy remains constant. However, since the volume increases, the spatial distribution broadens and the configurational entropy increases; hence the thermal entropy must decrease by the same quantity to keep the total entropy constant. This corresponds to a decrease of random kinetic energy and hence to a decrease of the average temperature. That is, the work of adiabatic expansion of a gas is done at the expense of the decrease of random

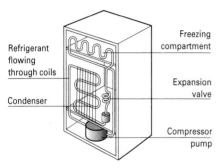

Refrigerant
flowing
through coils

Condenser

Freezing
compartment

Expansion
valve

Compressor
pump

In a refrigerator, a liquid called a refrigerant
is first evaporated (turned to gas), then
condensed by being pumped and turned
back to liquid. As it evaporates inside coils
in the refrigerator, it takes in heat. As it
does this, it cools the food.

kinetic energy of its molecules as in a refrigerator (see marginal illustration). This is in part analogous to what happens in an isolated system such as the universe; *the cooling of the universe due to expansion was and is today one major cause of the natural selection of the chemical elements into their observed physical conditions.* (They have condensed and become more ordered locally because, although as a whole the universe on expanding has become more disordered so that $\Delta S_{tot} > 0$, (see Section 3.3), the average temperature has fallen.)

(We must, however, be careful—the development of the universe cannot be analysed just in these simple terms. Although it is an isolated system it is not homogeneous in the same sense as an ideal gas in a container is disordered; there are fields acting—particularly the gravity field and the electrostatic fields between elements and molecules—and entropy is dominated by radiation, that is, of photons—at least by a factor $10^9:1$ (see Section 3.7 and Chapter 6).)

The major result of the above analysis is readily stated. We have a way of appreciating the balance

$$\text{Order} \rightleftharpoons \text{Disorder}$$

for any system of chemicals since both are now related to energy units. Chapter 2 showed how to think about *selective ordering* energy and we now appreciate *general disorder* in the same units. We can, therefore, treat balances between physical states and between chemical substances that react in these terms. Thus we shall discover the drives (but not the rates of change) in the universe that have helped to bring about observable materials and conditions today. A simple example follows.

3.2.5 Physical states in balance—latent heats

From what has been stated above it is expected that, if we measure the energy at equilibrium of a gram-mole of a liquid (or a solid) in isolation and separately from that of a gram-mole of its vapour, there will always be found, at a given temperature and *external* pressure, a difference in energy, ΔQ, between the two states that is related to the strength of the interactions between particles, atoms or molecules, in the condensate. There will also be differences in entropy, ΔS, which for a gas, relative to a liquid (or a solid), include terms dependent on volume and temperature related to $R \ln V$ and $C \ln T$. When the system is in balance, $\Delta Q = T\Delta S$ when moving from condensed to gaseous states.

Obviously, if we increase the temperature then the vapour pressure increases too and when it equals the external pressure the liquid reaches its boiling point—any further heating will convert liquid into vapour—and the amount of energy, ΔQ, necessary to change 1 gram-mole of liquid into vapour at constant p and T is called the *latent heat* (or enthalpy, see later) of vaporisation of the liquid. (It is usually defined at external pressure = 1 atm but can be determined at any external pressure.) There is a parallel latent heat of fusion for the transition of a solid to a liquid.

Insofar as liquids rather than solids are formed at a given temperature, the difference between the latent heat (actually enthalpy, see later) of fusion and the energy corresponding to the entropic change, $T\Delta S$, of the liquid relative to the solid (which is assumed to be of lower entropy) decides the physical state of a given substance. The liquid range is not easily calculated, but is of obvious

Table 3.2 Range of existence of liquid form (at atmospheric pressure)

Substance	Melting point (°C)	Boiling point (°C)	Range of existence of liquid form (°C)
Oxygen, O_2	−218	−183	35
Nitrogen, N_2	−210	−196	14
Water, H_2O	0	100	100
Ammonia, NH_3	−78	−44	33
n-Hexane, C_6H_{14}	−95	69	164
n-Octane, C_8H_{18}	−57	126	183
Benzene, C_6H_6	5.5	80	74.5
Ethanol, C_2H_6O	−114	78	192

importance; in particular note that co-operative angle-dependent interactions stabilise crystalline states, not particularly liquid states. Table 3.2 and Fig. 3.7 give some data for liquid ranges.

A question which we then can ask is: are particular physical states of matter favoured by particular chemicals?

In Chapter 2 we showed that atoms have preferences for one another, represented by two extreme possibilities. In one extreme case many covalently bonded *molecules* are formed with no polarity, for example H_2, N_2, O_2, CCl_4, C_6H_6, etc., and very little interaction of one molecule with another. (The extreme cases are the noble gases.) In the second extreme case, always involving heteronuclear species (which we call *systems* quite generally), the species are strongly polar, ionic, and form solid *co-operative* units, for example, $[Na^+ \cdot Cl^-]_m$, rock salt. In between these two there are intermediate cases, for example, of covalent molecules that are rather strongly polar, such as HF, H_2O and NH_3, where a main consideration is the *moderate co-operative* association through the formation of hydrogen bonds. A different case again is that of metals, which also form solid, *strongly co-operative* lattices, involving the metal ions and their 'free' electrons $[M^+, e^-]_m$.

As stated above, these differences are reflected in the values of the latent heats, which correspond to the energy necessary to take one gram-mole of the

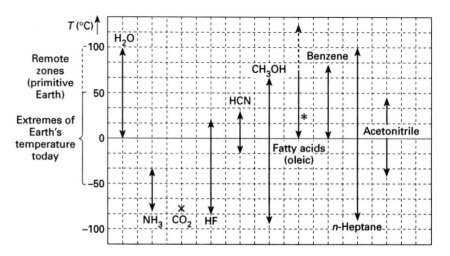

Fig. 3.7 The liquid ranges of a variety of covalent compounds. (*) Includes liquid crystal range.

Table 3.3 Data on melting and vaporisation*

Substance	Enthalpy (kJ mol^{-1}) of		Melting point (K)	Entropy (J K^{-1} mol^{-1}) of	
	Fusion	Vaporisation		Fusion	Vaporisation
Metals					
Na	2.64	103	371	7.11	88.3
Al	10.7	283	932	11.4	121
K	2.43	91.6	336	7.20	87.9
Fe	14.9	404	1802	8.24	123
Ag	11.3	290	1234	9.16	116
Pt	22.3	523	2028	11.0	112
Hg	2.43	64.9	234	10.4	103
Ionic crystals					
NaCl	30.2	766	1073	28.1	456
KCl	26.8	690	1043	25.7	389
AgCl	13.2		728	18.1	
KNO$_3$	10.8		581	18.5	
BaCl$_2$	24.1		1232	19.5	
K$_2$Cr$_2$O$_7$	36.7		671	54.7	
Molecular crystals					
H$_2$	0.12	0.92	14	8.4	66.1
H$_2$O	5.98	47.3	273	22.0	126
Ar	1.17	7.87	83	14.1	90.4
NH$_3$	7.70	29.9	198	38.9	124
C$_2$H$_5$OH	4.60	43.5	156	29.7	124
C$_6$H$_6$	9.83	34.7	278	35.4	98.3

* 1 J = 4.18 cal.

| **Gases** |
| H$_2$, O$_2$, N$_2$, F$_2$, Cl$_2$ |
| NH$_3$, CO$_2$, CH$_4$ |
| **Liquids** |
| H$_2$O, HF, HCN, C$_6$H$_{12}$ |
| CH$_3$OH, many oils, Hg |
| **Solids** |
| Metals, metal oxides, silicates (most minerals and higher molecular weight organic compounds) |

Fig. 3.8 Some substances found as gases, liquids or solids in the temperature range 273 to 323 K.

solid, or a liquid, to the state of gaseous molecular or atomic vapour, that is, from the ordered to the disordered state. The molar entropies of fusion are considerably lower than the molar entropies of vaporisation (see Table 3.3), which, as stated, are rather similar for many liquids, $\sim 100 \pm 20$ J K^{-1} mol^{-1}. This is not unexpected since a similarly large amount of disorder is generated when the same number of molecules of any liquid evaporate. Thus, we have, approximately, that the latent heat of vaporisation of a liquid (but often of a solid too) is given by $\Delta Q/T_c = \Delta S_{vap} \simeq 21(\pm 5)$ cal $°$C^{-1} mol^{-1} (Trouton's rule), where T_c is the condensation or vaporisation temperature. The largest ΔQ values are for the strongly associated liquids, for example, particularly ionic or metallic melts which do not become gaseous until very high temperatures. The smallest values are for inert gases that remain in the gaseous state even at low temperatures. A few molecular substances, mainly those giving H bonds that have intermediate co-operative energies and therefore moderate values of ΔQ, condense at close to room temperature. A particular and almost unique case is water which has a very high boiling point considering the number of atoms per molecule. Many larger organic molecules, petrol or octane C$_8$H$_{18}$, also fall in this class due to the larger number of interactions between atoms in larger molecules increasing ΔQ per mole.

This leads us to Fig. 3.8, which shows that at the relatively low temperatures of the Earth, say -50 to $50°$C, the following events happen.

1. Gases are largely formed by elements in the top right-hand corner of the periodic table alone and when in mutual combination in small molecules, that is, H, C, N, O, F, Cl and the noble gases.

2. Liquids are formed in this temperature range by a very small group of elements in compounds, largely from combination of elements in the same part of the periodic table as those elements in (1). The examples are a very few *small* molecules, for example H_2O, but a larger number of bulkier molecules, for example cyclohexane C_6H_{12}, and other organic solvents, for example carbon tetrachloride CCl_4.

3. Solids are formed by the vast of majority of metal elements and by combinations of these elements, alloys, by salts and by quite a large number of continuous covalent lattices (see Figs 2.7 and 2.29).

We now can state that the electronic structures of different elements have led them to appear in differently structured compounds with different charge distributions (Chapter 2), that is, some as molecules of low polarity and low co-operativity and some as polar co-operative solids, and that, by virtue of their electronic structures (especially polarity), atoms and molecules retain kinetic energy (entropy) quite differently at 25°C, the average temperature of Earth's surface.

The consequences of this division are profound and are the basis of the constitution of Earth and of the chemical nature of life, which is totally dependent on the gases of the atmosphere and on liquid water and oils. Gases and liquids form only a very small percentage of what is around us but they flow, a necessity for life. Thus, it is not the boiling point of water that matters so much but rather its liquid range. Later we shall see that it is, in fact, the matching of the liquid ranges of water and fatty materials (Fig. 3.7) that allowed life to appear.

The natural selection of chemical elements decided that the only liquid on Earth's surface some 4–5×10^9 years ago was water. If the temperature had fallen lower, to around $-90°C$, and CO_2 had been the only gas available, solid carbon dioxide would have formed, a very real possibility observed on other planets, for example on Venus and Mars at the polar caps. Again the vapour pressure of water largely controls the ability of the Earth's surface to maintain the temperature, still $\sim 25°C$ today, through its greenhouse effect. The weaknesses and strengths of *co-operative* energies in molecules in liquids and solids, giving a very different accessibility to a variety of states (Fig. 3.8) allowed the development of immobile minerals in rocks, liquid water and then later liquid fats and the curious semimobile protein and other structures in life (see also Fig. 3.15).

3.3 Systems out of balance

Until this section we have considered systems in equilibrium, balance, and reached the conclusion that for such systems $\Delta Q = T\Delta S$. Thus, for changes of state—of solid to liquid or of liquid to vapour—the values of ΔQ, latent heats, were referred to the transition (equilibrium) temperatures between the two states involved, that is, the melting and vaporisation temperatures.

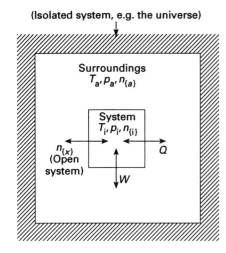

(Isolated system, e.g. the universe)

Surroundings
$T_a, p_a, n_{\{a\}}$

System
$T_i, p_i, n_{\{i\}}$

$n_{\{x\}}$
(Open
system)

Q

W

Fig. 3.9 'System' and 'surroundings'. T, p, n are temperature, pressure, and numbers of molecules, respectively; W exchange of work; Q of heat; $n_{\{x\}}$ of matter. A closed system exchanges energy but not material with its surroundings; an open system exchanges both.

More generally, however, systems are not in balance, that is, order and disorder are not in balance and undergo change as we observe all the time while looking around us. Then $\Delta Q \neq T\Delta S$, but any system for which this is so will tend to be going to the balanced state, ordered $\rightleftarrows$ disordered, by giving out energy in one form or other. (In non-equilibrium thermodynamics it is said that equilibrium is an 'attractor' state.)

We wrote initially (Section 3.1) that

$$X + Y \longrightarrow XY + \text{energy}.$$

Now, clearly, this process will not occur simply as written (without change of temperature) if the energy released remains trapped in $X + Y$ and XY since the energy changes the microstates open to *all* the components and hence the temperature and the resulting composition. To drive the reaction to completion (or to a fixed balanced state $X + Y \rightleftarrows XY$) without changing temperature we must take the energy generated away to an external system, the surroundings (Fig. 3.9), which we take to be an infinite heat sink. (This is the way observed changes often occur since the energy transferred goes to increase the entropy (degree of disorder) of external systems (surroundings) in such a way that $\Delta S_{\text{total}} = \Delta S_{\text{system}} + \Delta S_{\text{surrounding}} > 0$ which can be shown to be true for condensation on cooling even though for the condensation itself $\Delta S_{\text{sys}} < 0$. That is, the larger probability of the *overall* process dominates the smaller probability of a more ordered state forming in the system concerned. The fact that always when systems change $\Delta S_{\text{tot}} > 0$ is a statement of the so-called *Second law of thermodynamics*.)

Now, there are several ways of causing change in physical and chemical systems which are characterized by a set of independent variables, that is, pressure (or volume), temperature and composition. (Here again effects of size of a system, of fields—gravitational, electrical, etc.—and radiation are ignored until Chapter 6.) For example, as seen above, we can transfer energy ΔQ from one system to another, we can have one system doing work W on another or we can change the composition allowing all variables to change without restriction. In chemistry and biology, changes often occur in a fixed set of conditions, for example at constant volume and temperature, at constant pressure and temperature, etc. It is convenient, then, for prescribed quantitative consideration of the implications of the general inequality $\Delta Q \neq T\Delta S$, to replace ΔQ which, like W, describes changes in unprescribed (not fixed) circumstances, by other symbols for defined changes, that is, from an initial to a final state under particular physical conditions. These so-called *state functions* are the following:

ΔE—equal to ΔQ_V for processes at V, T constant, where E is called

the *internal energy* of the system.†

ΔH—equal to ΔQ_p for processes at p, T constant, where $H = E + pV$

and is called the *total energy* (or *enthalpy*) of the system.

† The internal energy of a system encompasses everything in it we can associate with energy: potential and kinetic energies, lattice energy, bond energies, electronic, vibrational, rotational and translational energies, etc. and even, if we so want, energy equivalent to its mass, $E = mc^2$. Since we are generally interested in changes in internal energy in physical or chemical processes, only a few of these terms need to be considered, usually those derived from transfer of energy (heat, work) from or to the surroundings and eventually associated with chemical change or with order/disorder equilibria. The enthalpy ΔH corresponds to this internal energy plus the energy absorbed as a result of (pV) changes, usually $p\Delta V$ work at constant pressure.

Note that changes in internal energy, that is ΔE, can be achieved by transfer of ΔQ and or work performed, W, that is, $\Delta E = \Delta Q + W$ which is a statement of the *First law of thermodynamics*.

We define next two other useful state functions, A and G, called, respectively, the *Helmholtz free energy* and the *Gibbs free energy*, such that their changes are

$$\Delta A = \Delta E - T\Delta S: \text{ suitable for discussion of changes of systems}$$
$$\text{at constant } T \text{ and } V,$$

$$\Delta G = \Delta H - T\Delta S: \text{ suitable for discussion of changes of systems}$$
$$\text{at constant } T \text{ and } p.$$

We can now deal with the non-equivalence $\Delta Q \neq T\Delta S$ in quantitative ways in prescribed conditions; ΔA and ΔG represent the distance in free (available) energy terms of the system relative to the state of equilibrium when $\Delta Q = T\Delta S$, that is, $\Delta E = T\Delta S$ or $\Delta H = T\Delta S$. We have to remember that many systems on Earth are not at equilibrium so that the energy content associated with elements in different compounds and under different conditions has to be explored using values of ΔA and ΔG of isolated substances in physical equilibrium. These functions are extremely useful to characterise the tendency (drive) for change, especially in chemistry and biology when a material is brought into contact with another material, both being only internally and initially in physical/chemical equilibrium conditions. By calculating the values of ΔA and ΔG we can know the energy that can be released or absorbed in the possible changes as well as the direction of the process which is that which makes the overall changes ΔA or $\Delta G < 0$, meaning that energy is released. (This can be shown to be equivalent to $\Delta S_{total} > 0$. A negative value of ΔA or ΔG is in the direction of stability.) A simple example is given in the next section.

3.3.1 Thermodynamics of mixing

As we have seen, entropy clearly goes to a larger value in a system of a single substance the larger the number of possible states open to it, that is the larger the number of microstates compatible with a given macrostate characterised by state variables, p, T and composition.

If more than one substance is considered, for example, two ideal gases, the number of possibilities increases (since the particles are non-identical) and the maximum is reached when the two are present in equal amounts. For example, the most probable situation for an optically active substance is its occurrence with equal quantities of the *l*-(laevo) and *d*-(dextro) forms (enantiomers).

Put in simple terms, two non-interacting gases mix and do not tend to separate since the separated state is but one possibility among many others in the mixed state. The enthalpy of mixing for non-interacting gases is obviously zero and the entropy of mixing (see Section 3.2.2) is $\Delta S = -nR(x_A \ln x_A + x_B \ln x_B)$ where n is the total number of moles and x_A, x_B are the molar fractions of A and B (Fig. 3.10). Since $\Delta G = \Delta H - T\Delta S$, the Gibb's free energy of mixing is $\Delta G = nRT(x_A \ln x_A + x_B \ln x_B)$. Obviously $\Delta G < 0$ because x_A and x_B are always smaller than 1.

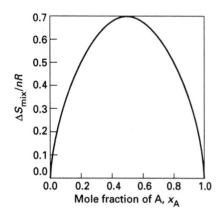

Fig. 3.10 The entropy of mixing of two perfect gases and of two dilute solutions that form an ideal solution. The entropy increases for all compositions and temperatures, so perfect gases and solutions mix spontaneously in all proportions at all temperatures.

The conclusion here is simple. In the natural selection of elements complete random mixing is dominant unless interactions select. The higher the temperature the more mixing will occur as opposed to separation into different zones or compartments. Mixing acts against the development of any selection, and this applies equally to mixing in solvents or in the gas phase. We shall discover systems that store energy because they have failed to become mixed, for example lake water and sea water. On mixing $A + B$, formation of A_n, B_n or $[AB]_n$ can *co-operatively* oppose mixing through ΔE or ΔH.

We have now described, in particular, balanced *physical* systems, and the tendency for out-of-balance physical systems to change. In order to treat separately *chemical* balance and out-of-balance we need to set reference *physical* conditions for systems around us including those not at equilibrium as is the case for H_2, O_2 and H_2O, and we need to describe chemical energy contents, that is, the distances of such systems from chemical equilibrium.

3.4 Standard states of chemical substances

In chemistry we define the physical standard state as that at $T = 298$ K ($25°C$) and $p = 1$ atmosphere (to approach the working conditions on Earth), all substances being considered in the form and the physical state they have in these conditions and at unit partial pressure† (gases) or unit concentration, for example 1 molar in solutions or, more exactly, unit activity (effectively active concentration, see references 1–3). (In each case the system considered is still supposed to be without boundaries for all substances so as to neglect any surface effects, and it is assumed that all fields are zero.)

For the state functions G, H and S we represent the standard state changes as $\Delta G°$, $\Delta H°$ and $\Delta S°$, which may refer to different reactions or processes, for example combustion (reactions with O_2), formation of compounds from corresponding elementary substances with or without changes of physical state, etc. Thus the value for the standard enthalpy of formation of liquid water, $\Delta H°$ water (liquid), is defined in relation to the corresponding values for O_2 (gas) and H_2 (gas) at 298 K and $p = 1$ atm. $\Delta H°$ water (vapour) can also be related to O_2 (gas) and H_2 (gas) directly. The same can be said for $\Delta G°$ values. Tables 3.4 and 3.5 summarize some enthalpy and Gibbs free energy standard values. By definition, the $\Delta H°$ values are zero for the elementary substances as is $\Delta S°$. For the entropies, however, the situation is somewhat different since we know, from the *Third law of thermodynamics*, that for perfect crystalline substances at absolute zero the entropy is also zero. Thus ΔS can be given an absolute reference scale if required but we shall not follow this point. It is an instructive exercise to enquire as to why particular substances have particular values in Table 3.6. Tables 3.4–3.6 can be used to discover at different temperatures which exchanges of elements can occur with output of energy, that is, spontaneously, and they will be used frequently throughout the book.

† The partial pressure of a gas A, p_A, is the pressure that the gas would exert if it occupied the total volume of the system all by itself, that is, $p_A = x_A p$, where x_A is the molar fraction of A and p the total pressure of the mixture.

Table 3.4 Enthalpies of formation, ΔH_f°, at 298 K

Compound	ΔH_f° (kcal mol^{-1})	Compound	ΔH_f° (kcal mol^{-1})
Inorganic compounds*			
H_2O (g)	-57.79	CaO (s)	-151.8
H_2O (l)	-68.32	$Ca(OH)_2$ (s)	-235.6
H_2O_2 (g)	-32.53	$CaCO_3$ (s)	-288.4
O_3 (g)	34.0	BaO (s)	-133.5
HCl (g)	-22.06	$BaCO_3$ (s)	-290.8
HBr (g)	-8.66	MgO (s)	-143.8
HI (g)	6.20	MnO (s)	-92.2
SO_2 (g)	-70.96	CoO (s)	-57.2
SO_3 (g)	-94.45	NiO (s)	-56.5
H_2S (g)	-4.81	Fe_2O_3 (s)	-196.5
N_2O (g)	19.49	Al_2O_3 (s)	-399.1
NO (g)	21.60	SiO_2 (s)	-209.9
NO_2 (g)	8.09	CuO (s)	-37.6
NH_3 (g)	-11.04	Cu_2O (s)	-40.4
CO (g)	-26.41	ZnO (s)	-83.2
CO_2 (g)	-94.05	ZnS (s)	-48.5
Organic compounds†			
Gases			
Methane, CH_4	-17.89	Ethylene, C_2H_4	12.50
Ethane, C_2H_6	-20.24	Acetylene, C_2H_2	54.19
Propane, C_3H_8	-24.82	1-Butene, C_4H_8	0.28
n-Pentane, C_5H_{12}	-35.00	cis-2-Butene, C_4H_8	-1.36
Isopentane, C_5H_{12}	-36.92	trans-2-Butene, C_4H_8	-2.40
Neopentane, C_5H_{12}	-39.67		
Liquids			
Methanol, CH_3OH	-57.02	Benzene, C_6H_6	11.72
Ethanol, C_2H_5OH	-66.35	Chloroform, $CHCl_3$	-31.5
Acetic acid, CH_3COOH	-116.4	Carbon tetrachloride, CCl_4	-33.3

* g, Gas; l, liquid; s, solid. These values are used in Chapter 8.
† These values are used in Chapter 9, Fig. 9.3.

Using ΔH°, ΔS° and ΔG° values we are now in a position to determine any ΔH, ΔS and ΔG in conditions different from those conventional for the standard state. For instance, to calculate the enthalpy change at any other temperature, T, one uses the expression

$$\Delta H = \Delta H^\circ + \int_{298}^{T} C_p \, dT$$

to obtain the Gibbs' free energy change one may calculate $\Delta H(T)$ and $\Delta S(T)$ and substitute the values obtained in $\Delta G = \Delta H - T\Delta S$. Actually these calculations are slightly more complicated since C_p varies with T and usually we proceed empirically.

It is extremely important to appreciate what has been achieved. We now have procedures that allow us to have numerical values for the driving energy for change given any mixture of chemicals in any physical conditions. In

Table 3.5 Free energy of formation, ΔG_f°, at 298 K

Compound	ΔG_f° (kcal mol^{-1})	Compound	ΔG_f° (kcal mol^{-1})
Inorganic compounds*			
Gases		*Solids*	
H_2O	−54.64	BaO	−126.3
H_2O_2	−24.7	$BaCO_3$	−272.2
O_3	39.16	CaO	−144.4
HCl	−22.77	$CaCO_3$	−269.8
HBr	−12.72	$Ca(OH)_2$	−214.3
HI	0.31	SiO_2	−192.4
SO_2	−71.79	MgO	−136.2
SO_3	−88.52	MnO	−86.8
H_2S	−7.89	CoO	−51.2
N_2O	24.9	NiO	−50.6
NO	20.72	Fe_2O_3	−177.1
NO_2	12.39	Al_2O_3	−376.8
NH_3	−3.97	CuO	−30.4
CO	−32.81	Cu_2O	−34.98
CO_2	−94.26	ZnO	−76.05
Organic compounds			
Gases			
Methane, CH_4	−12.14	Ethylene, C_2H_4	16.28
Ethane, C_2H_6	−7.86	Acetylene, C_2H_2	50.00
Propane, C_3H_8	−5.61	1-Butene, C_4H_8	17.09
n-Pentane, C_5H_{12}	−2.0	*cis*-2-Butene, C_4H_8	15.74
Isopentane, C_5H_{12}	−3.5	*trans*-2-Butene, C_4H_8	15.05
Neopentane, C_5H_{12}	−3.6		
Liquids			
Methanol, CH_3OH	−39.73	Benzene, C_6H_6	29.76
Ethanol, C_2H_5OH	−41.77	Chloroform, $CHCl_3$	−17.1
Acetic acid, CH_3COOH	−93.8	Carbon tetrachloride, CCl_4	−16.4

Chapters 4–6 we shall not be concerned with change, but will elaborate the description of systems in which change is not observed. This does not necessarily mean that they are at global equilibrium since barriers to change may maintain the physics and chemistry of the system in compartments but they will then store energy that can be used to drive change at some time. Such storage conditions can be analysed relative to standard states of chemicals. Moreover, we can appreciate which systems will tend to become more ordered or disordered. We have a basic set of free energies, heats and entropies that help us to understand the natural selection of chemical elements based on thermodynamics and where the observed materials around us, animate or inanimate, lie with respect to distance from equilibrium. Some of these points will be illustrated below. (The situation is quite different from that in Chapter 2 where we analysed effectively internal binding energies (approximately ΔH or ΔE) of isolated ordered materials and hence could not analyse balances or equilibria.)

Table 3.6 Absolute entropies, $S°$, at 298 K

Substance	$S°$ (cal mol^{-1} K^{-1})	Substance	$S°$ (cal mol^{-1} K^{-1})	Substance	$S°$ (cal mol^{-1} K^{-1})
Solid elements		**Solid compounds**		**Liquids**	
Al	6.77	CaO	9.5	Br_2	36.4
C (diamond)	0.6	$Ca(OH)_2$	17.4	H_2O	16.73
Ca	9.95	$CaCO_3$	22.2	Hg	18.17
Cu	7.97	CuO	10.4		
Fe	6.49	Cu_2O	24.1		
Na	12.2	Fe_2O_3	21.5		
S (rhombic)	7.62	SiO_2	10.0		
Si	4.51	ZnO	10.5		
Zn	9.95	ZnS	13.8		
Monatomic gases		**Gaseous diatomic molecules**		**Gaseous polyatomic molecules**	
He	30.13	H_2	31.21	H_2O	45.1
Ne	34.95	D_2	34.6	CO_2	51.1
Ar	36.98	F_2	48.6	SO_2	59.4
Kr	39.19	CO	47.3	H_2S	49.1
Xe	40.53	NO	50.3	NO_2	57.5
H	27.39	N_2	45.7	N_2O	52.6
F	37.92	O_2	49.0	NH_3	46.0
		HF	41.5	O_3	56.8
Organic compounds (gases)					
Methane, CH_4	44.5	Isopentane, C_5H_{12}	82.1	1-Butene, C_4H_8	73.48
Ethane, C_2H_6	54.8	Neopentane, C_5H_{12}	73.2	cis-2-Butene, C_4H_8	71.9
Propane, C_3H_8	64.5	Ethylene, C_2H_4	52.45	trans-2-Butene, C_4H_8	70.9
n-Pentane, C_5H_{12}	83.4	Acetylene, C_2H_2	49.99		

3.4.1 Applications to chemical systems: thermodynamic stability of compounds

Let us consider a chemical reaction (change), say, the combustion of hydrogen which gives water, that is,

$$H_2(g) + \tfrac{1}{2}O_2(g) \longrightarrow H_2O(l), \qquad \Delta G°(\text{table}) = -56.7 \text{ kcal}$$

(note that this is not the 'heat of reaction').

$\Delta G° = \Delta H° - T\Delta S°$ and, since $\Delta H° = -68.3$ kcal, $T\Delta S°$ must be positive, that is, $\Delta S° < 0$ because the total number of particles of the system (especially those in the gaseous state (g)) is reduced so the entropy change is against the reaction. This is counterbalanced by the large favourable (negative) enthalpy change, heat evolved; hence the change in the Gibbs free energy, $\Delta G°$, is negative and the reaction proceeds spontaneously as written (remember that the heat evolved increases the entropy of the environment). We may say that liquid water, $H_2O(l)$, has a thermodynamic stability $-\Delta G° = 56.7$ kcal mol^{-1} relative to its components H and O in the states they are all in at 298 K, $p = 1$ atmosphere.

We can now extend this observation and find out which combinations of elements are the most stable in the standard conditions, that is, at 298 K and atmospheric pressure, since this will be shown to be a major factor in the natural selection of the chemical elements. We choose to concentrate first on the dominance of the oxides of the elements.

Table 3.5 allows us to establish $\Delta G°$ orders for metal oxides, per mole of oxygen incorporated. $\Delta H°$ values (Table 3.4) can also be used since the $\Delta S°$ values are small and not very dissimilar and only become important at high temperatures through the product $T\Delta S$. We observe then the orders

$$Mg > Al > Si > P > S,$$

$$Mg > Mn > Fe > Co, Ni > Cu < Zn,$$

and for non-metals

$$H > C \gg N.$$

These orders define the sequence of decreasing affinity of these elements for oxygen at equilibrium in the standard conditions, but if we raise the temperature those oxides which are gases, non-metal oxides, are most favoured due to their larger $T\Delta S$, disorder.

We must not forget, when thinking of natural processes, that there can be, simultaneously, competition of some of these elements for hydrogen, when the order of affinity is

$$O > Cl \gg C > N > S \gg \text{all metals.}$$

These competing sequences of *equilibrium* chemical energies take on particular importance when discussing, for example, the formation of the Earth from a mixture of elements (see Chapter 8). *They govern a major step in the natural selection of the chemical elements based on selective tendencies of a multiplicity of reactions taken together. The drive of heat loss, corresponding to entropy gain of the environment, is the essential directing principle against the background of preferential combination due to $\Delta H°$, binding energy, (see Chapter 2).* In a qualitative way Table 3.7 summarizes our knowledge of the thermodynamic stability of some relevant types of compounds.

Above we stressed that the orders of stability corresponded to equilibrium energies in standard conditions; if the conditions are different, for example if the temperature becomes much higher, the orders will vary and this may be of deep significance in natural processes, as it is in the laboratory or man's industry. To understand these changes of chemical processes in general we have to explore quantitatively the effect of conditions on equilibrium between ordered and disordered units now within chemical systems, for example the balance position in the reaction

$$H_2(g) + O_2(g) \rightleftharpoons H_2O(l).$$

To achieve this we must define first the relationship between free energy and equilibrium.

Table 3.7 Compounds of special thermodynamical stability

	Stable	Very unstable
Hydrides	O, N, C, (S), Halogens	Heavy elements, metals
Oxides	H, C, Si, P, (S), Metals	Halogens, N, O
Sulphides	Metals, (H)	Non-metals

Brackets refer to compounds of somewhat lower stability

3.5 Free energy and chemical equilibrium

In the previous section we have seen how to calculate the free energy changes of any substance at any temperature given the value under standard conditions. In a chemical reaction, the pressure and the concentration of reactants may also be altered and will affect and even change the sign of the overall ΔG for the reaction.

Consider the chemical reaction $aA + bB \rightarrow cC + dD$ which could also occur in the reverse direction, that is, $cC + dD \rightarrow aA + bB$ in a gas or solution phase. When balance is achieved one finds that the ratio $[C]^c [D]^d / [A]^a [B]^b$, where $[X]$ represents the concentration of the element X, is constant and we denote it by K_c. K_c is called the 'thermodynamic equilibrium constant' of the reaction defined in terms of concentrations in solution. If A, B, C, D were ideal gases at constant pressure and temperature, the equilibrium constant would be defined in terms of the 'partial pressures' of the components, p_i, which according to Dalton's law are a measure of their concentration in the mixture (see Fig. 3.11). This constant is represented by K_p.

For any such reaction we write

$$\Delta G = cG_C + dG_D - aG_A - bG_B = 0$$

where the Gs are the free energies *per mole* of the products and reactants in the given equilibrium which cannot, of course, be their standard states. The free energies per mole are usually called 'chemical potentials' and given the symbol μ_i. (Note that since μ_i is a molar property it is an intensity factor, whereas ΔG for a system is, like ΔH or ΔS, a capacity factor.)†

We may then write for the equilibrated reaction above

$$\Delta G = \sum_i n_i \mu_i = 0$$

where n_i is the number of moles of the ith substance, positive for products formed, negative for reactants used up, and μ_i the corresponding chemical potential.

Naturally, the chemical potentials or free energies per mole of the various substances must be defined relative to some reference state, that is, μ° or ΔG° quantities per mole. (For pure solids or liquids at constant pressure and temperature, ΔG is effectively not variable no matter what equilibria are involved; hence $\Delta G = \Delta G^\circ = \sum n_i \mu_i^\circ$.) For gases at constant temperature although the total pressure is constant, the partial pressure of any component i is variable, dependent on the chemical equilibrium involved. The same is true for ideal solutions, when the solvent is considered to be just an inert supporting medium. In these cases we may derive, for ideal gases and ideal solutions at

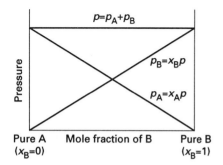

Fig. 3.11 The partial pressures p_A and p_B of a binary mixture of (real or perfect) gases of total pressure p as the composition changes from pure A to pure B. The sum of the partial pressures is equal to the total pressure. x, mole fraction.

---

† Intensive properties (or intensity factors) are those that have the same value for the system and its individual parts, for example, temperature, pressure, density, molar properties, etc. Extensive properties (or capacity factors) are those dependent on the amounts and are additive, that is, the value for the system is the sum of the values for its individual parts, for example, volume, mass, internal energy, enthalpy, entropy, etc. Note also that the quotient of two extensive properties may give a meaningful intensive property, for example, density (mass/volume), molar volume (volume/number of moles), chemical potential (free energy/number of moles), etc. All forms of work are a product of an intensive property and an extensive property (see Table 3.10).

constant atmospheric pressure and $T = 25°C$ (see Section 3.2.2) the expressions†

$$\mu_i = \mu_i^\circ + RT \ln p_i \qquad \text{cf. } \Delta G_i = \Delta G_i^\circ + RT \ln p_i$$

$$\mu_i = \mu_i^\circ + RT \ln c_i \qquad \Delta G_i = \Delta G_i^\circ + RT \ln c_i.$$

Since, at equilibrium, $\sum n_i \mu_i = 0$, one obtains in the case of gases

$$-\frac{\sum n_i \mu_i^\circ}{RT} = \sum n_i \ln p_i.$$

At standard temperature ($25°C$) the left-hand side of this equation is constant and is related to $\ln K_p$. Therefore,

$$\ln K_p = \sum n_i \ln p_i, \qquad \text{and} \qquad \sum n_i \mu_i^\circ = -RT \ln K_p.$$

This means that we may calculate the equilibrium constant K_p knowing the values of $\mu_i^\circ (\Delta G^\circ$ per mole) which have been tabulated for a large variety of substances (see Table 3.5).

Similarly, for ideal dilute solutions, we can calculate the value of the equilibrium constant K_c (or, more correctly, K_a, see references 1–3) as $\sum n_i \mu_i^\circ = -RT \ln K_c$. Most frequently, these expressions for the equilibrium constants appear in textbooks as

$$\Delta G^\circ = -2.303 \, RT \log K.$$

(Note that ΔG° is *not* the free energy change at equilibrium (which is zero); it is that difference ΔG° that corresponds to reactants and products being taken in their standard conditions, namely unit partial pressures, concentrations or activities.)

In order to appreciate the value of knowing of ΔG°, we will take as an example the oxidation of SO_2 to SO_3, namely

$$2SO_2(gas) + O_2(gas) \rightleftharpoons 2SO_3(gas),$$

and consider an infinite system in balance and in the absence of fields. From Table 3.5 we can obtain ΔG° at 298 K and a pressure of 1 atmosphere we have

$$\Delta G^\circ = 2\mu_{SO_3} - \mu_{O_2} - 2\mu_{SO_2} = -18.9 \text{ kcal.}$$

This is a fairly large negative value and hence favourable to the formation of SO_3. The equilibrium constant $K_p = (p_{SO_3})^2/(p_{O_2})(p_{SO_2})^2$ is easily calculated from $\Delta G^\circ = -2.303 \, RT \log K_p$ and we obtain $K_p = 7.9 \times 10^{13}$.

We may also calculate the free energy change for any conditions using the expression $\Delta G = \Delta G^\circ + RT \sum n_i \ln p_i$ easily derived from the considerations above. For example, starting with a mixture in which $p_{SO_2} = 0.6$ atm, $p_{O_2} = 0.3$ atm and $p_{SO_3} = 0.1$ atm, the free energy distance to equilibrium is

$$\Delta G = -18.9 + 1.356 \log(0.1^2/0.6^2 \times 0.3) = -17.9 \text{ kcal.}$$

Of course, when a reaction reaches true equilibrium $\Delta G = 0$ and no more change can occur and no more work can be done. It should be noticed again

---

† We use concentrations, c_i, rather than the more accurate 'activities', a_i, (see references 1–3) for simplicity.

that this is not the same situation as that of a system in a state that does not undergo observable change over a limited time period. There are many states far from equilibrium that look to us to be permanent and unchanging, but they do not change quickly because they are restricted from releasing energy, for example from doing work, by barriers of distance, physical barriers in general, or chemical barriers at low temperature. Such systems in apparently permanent *stationary (or steady) states*† include the universe, all living organisms and much of the Earth locally.

The interest of thermodynamic equilibria considerations is that they show the direction in which all systems are trying to go *without energy input*. Given knowledge of an actual condition of any kinetically stabilised system we can calculate through the 'distance' from equilibrium (the 'attractor' state) the maximum capacity of the system to give out 'free energy', that is, to do work. In Chapters 8 to 16 we shall be constantly considering which chemicals are in situations where they can do work, that is, we look at the possibility for change. Relative ΔG values are then work capacities and the route to equilibrium does not alter the maximum work the system can do since ΔG is a state function. The route, though, is under barrier, kinetic, control.

3.5.1 The equilibrium constant at different temperatures or pressures

The equilibrium constants calculated from $\Delta G°$ values are for 298 K and unit atmospheric pressure. To calculate constants at any other temperature or pressure we must find the equivalent ΔG for the reaction at that temperature or pressure and then use the same relation, $-\Delta G = RT \ln K_p$. Since partial pressures or concentrations are not changed by a change in pressure, both K_p and K_c are independent of pressure; only K_x, a function of molar fractions, x_i, is affected since $x_i = p_i/p$ (see Fig. 3.11). As to the effect of temperature we can deduce from $\Delta G° = \Delta H° - T\Delta S°$ that $d(\ln K)/d(1/T) = -\Delta H°/R$ and if the range is not too large and ΔH can be considered independent of T, we can derive a simpler equation correlating equilibrium constants at different temperatures, namely,

$$\log K_2/K_1 = \frac{\Delta H°}{2.303R} \cdot \left(\frac{T_2 - T_1}{T_1 T_2} \right).$$

If $\Delta H° = \sum H°$ (products) $- \sum H°$ (reactants) > 0 (endothermic reaction), the increase of temperature $T_2 > T_1$ increases the equilibrium constant; if $\Delta H° < 0$ (exothermic reaction) then the increase of temperature decreases the equilibrium constant. The changes of equilibrium with temperature are extremely important for competing processes (see Section 3.4.1) as in the chemical selections that led to the Earth's formation and composition.

---

† We define three conditions of a system. First, a *balanced state* independent from time and requiring no net input of energy or material from outside and which is unable to develop by itself in any circumstances. Second, *a stationary state*, which is apparently independent of time, but, although it could collapse to a balanced state by giving energy and/or material to the outside, it cannot do so since it is trapped by strong physical or chemical barriers. Third, *a steady state*, which is also independent of time but now remains so only as long as it receives energy and material and loses transformed energy and material equivalently. A steady state lies in a shallow physical and chemical trap and is part of a flow system.

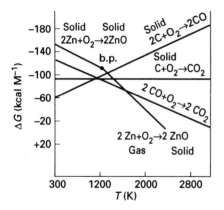

Fig. 3.12 The change in free energy, ΔG, with temperature, T, for the oxidation of Zn, C and CO; for C there are two different products. Above the boiling point (b.p.) of Zn the oxide loses stability more rapidly. Note that the more gas molecules are formed in the changes the more $T\Delta S$, disorder, favours that change. (Stability increases upwards.) The figure is part of the Ellingham diagram (Fig. 8.4).

An example where we monitor ΔG at unit pressure as we vary T is given in Fig. 3.12. Carbon or carbon monoxide cannot reduce zinc oxide until close to 1000°C, but above that temperature carbon, C, is a good reducing agent but CO is not. The critical factor, the slope of $\Delta G/T$, is the entropy ΔS which depends on the number of molecules that are gases in each reaction. This is an order $\rightleftarrows$ disorder balance in a *chemical* equilibrium. Such knowledge shows how to obtain metals from minerals, for example using appropriate extraction conditions. Its use became the material basis for man's cultural development, see Chapter 14 and lies behind selective chemical evolution.

3.5.2 Chemical change and physical change

We conclude this section on chemical change with remarks which parallel those on physical change. As we have seen, there is a condition for a given system that is described as balanced or at equilibrium when the *free energy status* of the two sides of a reaction are equal. They are not in their standard states except by coincidence. At equilibrium no change can occur spontaneously, that is, without outside interference and energy input. Clearly, the condition of balance will prevent any further natural selection of the elements. For systems out of balance we refer their condition (X) relative to the standard state condition (X°) of each reactant so as to calculate the difference in free energy (ΔG) between that condition and the balanced equilibrium state. This difference is the energy that is related numerically to the change that can take place, no matter however slowly, since G is a state function. The amount of energy released is then that which can be transferred to an outside system (the surroundings) and represents the maximum ability to do work.

In the following section we will examine two particular forms of energy and show how they are transferred from one system into another and how one system can do work on another.

3.6 Energy transfer: work

Given the multitude of ways in which energy is important in our environment and lives (see Table 3.8) we must now explain how it is exchanged (transduced) from one mode to another. We describe first the transfer of chemical energy of reactions to pressure/volume energy, leaving aside thinking about energies associated with surfaces, gravity or any effect of any other field. We shall not consider flow at all until Chapter 7.

3.6.1 Chemical energy transfer to mechanical energy (pV)

We return again to the fundamental problem inherent in the manner of categorizing observables that troubled the ancient scientists. They noticed that many materials could go up in flames (fire) and asked, naturally enough, what was the relationship between the light produced and the change in temperature caused by heat from the fire, and the material that was consumed. The obvious answer was that they all belonged to a system of

Table 3.8 Types of energy*

Radiant energy (very short, e.g. X-rays, to very long wavelength, e.g. radio waves, and associated with temperature (Fig. 1.18))

Kinetic (thermal) energy (associated with temperature and motion of masses (Fig. 1.16).)

Nuclear energy (associated with nuclear mass)

Electronic energy (associated with electrons in atoms and chemical bonds)

Electrostatic energy (due to charge separation)

Gravitational energy (due to attraction between separated masses)

Surface energy (associated with curvature of liquids)

Mechanical energy (associated with stress in condensed phases)

Chemical energy (associated with transformations of chemicals)

---

* NB. There is really a fundamental underlying unity in energy, but these divisions are useful to allow us to calculate effects observed in unlike conditions.

irreducible entities (Fig. 1.1), meaning that the fire was in the material (like some spirit) and was given to the environment on change of materials. The idea was very plausible and led, as we have seen, to the thought that matter and energy were similar in kind. The phlogiston theory was based on this idea too, which was found later to be wrong. (Actually not quite so, as Einstein has shown in the theory of relativity, but this is a deduction at quite a different level of thinking about matter.) The early scientists could not know that, when coal (carbon, C) is burnt with air (which contains molecular oxygen, O_2, but about which they knew little) in, for example, a modern coal power station, it gives a chemical change, a gas, which is also a minor component of the air, namely carbon dioxide (CO_2) that is

$$C + O_2 \longrightarrow CO_2$$

as well as light and heat, but that light and radiant and convective heat, energy, are not material though O_2 and CO_2 are. Analytically we can catch the CO_2 and even force it back to $C + O_2$, *but* this *costs* a lot of energy, taken from the environment. For example plant life exploits the sun's energy to do this conversion. It took centuries of work to overcome the difficulties inherent in dissecting such processes and the beginning of a real deeper understanding is still not 200 years old. A further example that leads us to say that food gives us energy (calories) and so enables us to do work derives from the combustion of carbon compounds, for example sugar with dioxygen. (We use 'calories' as a measure of the energy we can derive from this reaction of elements.)

$$\text{Sugar} + O_2(\text{air}) \longrightarrow CO_2 + H_2O + \text{energy(calories)}$$

$$\text{(for example, } C_6H_{12}O_6 + 6O_2 \longrightarrow 6CO_2 + 6H_2O + 688 \text{ kcal).}$$

One of the things we have to know, therefore, is the free energy, ΔG, released in the combustion of chemical compounds (Table 3.9), which can be determined experimentally, and how it is related to the work that can be done on other systems where the activity takes place in a specific environment.

(There is a different problem, mentioned before but needing to be stressed, that concerns not the formulation of chemicals or their energy states but rather the rate of their change, that is, their *kinetic stability*. Neither coal or sugar burn in the air spontaneously, but they do so in furnaces and in the human body (contrast metallic sodium, which reacts spontaneously on the

Table 3.9 Heats of combustion ΔH at 25°C

Compound	$-\Delta H$ (kcal mol^{-1})	Compound	$-\Delta H$ (kcal mol^{-1})
Carbon (graphite)	94.1	Acetone (l)	428.2
Hydrogen	68.3	Formic acid (l)	61.0
Methane (g)	212.9	Acetic acid (l)	209.3
Ethane (g)	373.2	Oxalic acid (s)	60.8
n-Hexane (l)	995.9	Lactic acid (s)	321.5
Cyclohexane (l)	933.4	Ethyl acetate (l)	535.7
Benzene (l)	781.8	Tripalmitin (fat)	7510.0
Toluene (l)	945.7	Sucrose (s)	1350.4
Methanol (l)	173.6	D-glucose (s)	671.8
Ethanol (l)	327.7	Glycine (s)	231.8

surface of water). The much faster (higher) rate of burning depends now on assistance from other substances that help the change; these substances are called *catalysts* and we will consider them and their action in Chapter 7.)

We turn therefore to the relation between the released energy (heat) to the mechanical work that can be done. The easiest example is the work done in compressing a gas from a volume V to a volume V' as described in Section 3.2.1, that is,

$$\int_V^{V'} p \, dV.$$

We wish to connect this work to an operational engine using chemical fuels. In a petrol engine of a car, for example, we burn a gas (CH_4 for the sake of simplicity) with dioxygen,

$$CH_4 + 2O_2 \longrightarrow 2H_2O + CO_2 + 212 \text{ kcal},$$

and, due to the energy released forcing a change of temperature, we gain an increase in pressure in the car's cylinders. This pressure is later partly converted into a volume increase and into the mechanical motions of the machine, but note that it is partly dissipated as heat, and so efficiency is less than 100%.

We have demonstrated here the following transformation: energy in chemicals (initially out of equilibrium) goes to thermal energy (largely change of kinetic energy), goes to configurational energy (change of volume or pressure). Such changes occur overwhelmingly in a reaction vessel, car engine or a human body, that is in a small system that is in energy contact with an external environment (a so-called closed system), and are used to perform work. Inevitably, the environment is raised in energy too by heat loss from the system. (This is one part of the classical idea of observed 'fire'—heat and light released to the environment.) We need to know how much of the energy generated, while the chemical system goes to equilibrium and does pV work, is lost *inevitably* to the environment.

Before going on we stress that we could have connected the chemical engine to other mechanical devices, such as a pump, and so developed a pressure or concentration gradient, say, across a balloon membrane. Energy from chemicals can also be converted into gradients of solution concentration using mechanical devices. In biology there are mechanical devices, similarly called pumps, to lower the concentration of sodium ions and to take potassium ions into cells, or indeed to move many chemicals around man's body. The heart, for example, is a pump of blood. All of these activities are in closed systems in thermal contact with the environment and, in effect, the machinery transfers energy as radiative or convective heat. There is then the question as to how good the machinery can possibly be.

3.6.2 Thermodynamic efficiency of transfer of energy: doing work

It is the transforming capability, efficiency of energy transfer as heat to work, that we now wish to understand. In this book generally we are concerned especially with the passage of energy from a high-temperature T_1 source, say,

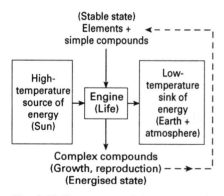

Fig. 3.13 The non-reversible sequence of events as energy flows through an engine that does work at equilibrium (reversible system or not) with its environment. The products may or may not be connected back to the input materials. In biology they are, but in man's factories they are not.

the sun, through a machine of some kind, to a sink, say, the Earth at temperature T_2, where in particular the machine can be a living, reproducing, organism or a man-made machine producing ordered chemicals (see Fig. 3.13), that is,

$$
\begin{array}{ccccc}
 & & \text{INPUT material} & & \\
 & & \downarrow & & \\
\text{(Sun) Energy source} & \xrightarrow{Q_1} & \text{Machine} & \xrightarrow{Q_2} & \text{Sink of energy (Atmosphere)} \\
T_1 & & \downarrow & & T_2 \\
 & & \text{OUTPUT material} & & \\
 & & \text{(through work } W\text{)} & &
\end{array}
$$

The interest clearly rests in what the machine can do with the energy Q_1 it gets, light (radiative heat) in the case of the sun, without generating too much energy loss, Q_2, as heat. We treat any system that converts energy to work (Table 3.10) as a cycling machine so that we also include biological systems that synthesise chemicals from other chemicals before eventually returning them to their original state.

Table 3.10 Expressions for thermodynamic work, δW, done on a system

Type	Intensive property	Extensive variation	Expression
Expansion	Pressure, p	Volume, dV	$-p\,dV$
Electrical	Potential, E	Charge, de	$-E\,de$
Gravitational	Force, mg	Height, dh	$mg\,dh$
Chemical	Chemical potential, μ	Moles, dn	$\mu\,dn$
Surface	Interfacial tension, γ	Area, dA	$\gamma\,dA$

Generally, the mechanical machine itself just cycles while it manipulates the input to the output materials. As stated, as the machine cycles, acting on particles external to itself, that is, doing work, W, it has to change and pass on energy to an external energy sink before returning to its initial state. It uses heat and is in thermal contact with the environment. The problem of *thermodynamic efficiency* does not therefore concern avoidable losses of heat but unavoidable losses, Q_2.

The thermodynamic efficiency of an engine, η, is defined as the ratio of the maximum work done to the quantity of heat extracted from a hot reservoir, Q_1, that is,

$$\eta = \frac{W}{Q_1}.$$

If an amount of heat, Q_2, is also inevitably transferred to the environment, then the maximum work the machine can do corresponds to the difference $Q_1 - Q_2$. We can then write $W_{max} = Q_1 - Q_2$ and we can easily derive

$$\eta_{max} = 1 - \frac{Q_2}{Q_1}.$$

This means that, the smaller Q_2 is relative to Q_1, the greater is the efficiency. Therefore we must minimise the ratio Q_2/Q_1. In fact, the maximum energy transfer to work is achieved only when the transfer is carried out extremely slowly (that is, effectively reversibly) so that losses are minimised. The reason for this is that when a process is carried out in minute steps the change in temperature of the system is made extremely small in each step so that we can use the equations $\Delta S = \Delta Q/T$ or rather $dS = \delta Q/T$. For any other situation, losses are greater. Of course, any reversible process is extremely slow, that is, it must go by infinitesimal steps to maintain very close to equilibrium; thermodynamic efficiency must be forfeited for speed.

Now you may ask why are we going to so much effort to explain the way in which work can be extracted by energy transfer in a machine when we have in mind problems concerning both the universe and Earth. The answer lies in the switch from systems of which we can have little understanding, such as which particular system evolved in the universe due to fluctuations, to Earthly ones where our interest must be paramount since efficiency must have affected evolution and could decide our future for we are the operators of the local (closed) machinery. The question as to whether the state of the universe with its scattering of matter formed by rapid cooling was inefficient in thermodynamic terms does not concern us since it is not known to be in competition with alternative processes. However, forms of life may well be in competition, so that survival could be a matter of thermodynamic efficiency in the sense that one form could work efficiently and grow slowly, close to equilibrium, as opposed to another that could have a fast rate of growth which is inefficient in thermodynamic terms. These two are clearly alternative and not necessarily compatible ways of living. Animals are much less efficient thermodynamically than plants. This is one factor that has caused a balance to be struck between huge numbers of forms of life, their growth patterns and their environments. Another problem is that all intensive, rapid activity heats the Earth, see Fig. 3.14, yet we may need temperature stability.

It is here that we meet the real problems. While we can *discuss* changes from one state of matter to another, for example C plus O_2 to CO_2, in terms of maximum free energy, $\Delta G°$, differences and then their ability to do work, in practice we cannot carry them out without considerable energy losses to the environment; the greater the speed the greater the loss. All cycling systems, that is, systems that use energy to maintain themselves away from equilibrium, for example very much of the activity on the Earth's surface and all living systems, must consume energy and develop entropy, that is, they dissipate energy in flowing steady states. Steady states, therefore, can only exist in systems in which energy (and materials) flow *continuously* from source (hot or high energy) to sink (cold or low energy), for example in a field and a radiation system (the sun/earth pair). We see too that in this discussion the rate of change as well as the magnitude of the overall change is important. The *rate* is not related only to order and disorder themselves; it also depends upon the way in which the material and energy uses are *organised* as we see in the production of living things and of goods. We turn to the kinetic rate problem in Chapter 7.

3.7 Radiative energy and entropy

So far, in these considerations we have only mentioned in passing radiative energy, that is, energy *not* associated with *rest* mass. We have seen that temperature can be described by random translational kinetic energy (see Section 1.6) and that heat is energy transfer due to differences of temperature and hence describable as differences in random kinetic translational energy. We also know that energy can be described in equivalent terms of quanta of radiation, $E = h\upsilon$. There is, furthermore, a parallel with light in that light quanta (ultraviolet (UV), visible) can be taken into or emitted from rest-mass electronic states, and intermediate quanta (infrared, microwave) can be taken into or emitted from vibrational or rotational modes of molecules, both resulting in rise or fall of temperature. If we then consider a set of particles at any temperature and at equilibrium, *it has to be in balance with a radiation spectrum as well as with rest-mass motions*. Any system from which mass cannot escape (isolated or closed system) must then reveal its temperature by its radiation spectrum. Moreover, any system in equilibrium with its environment is receiving exactly the same spectrum of radiation (Fig. 1.18) as it is giving out since they are at the same temperature.

Consider now an isolated system of two masses at different temperatures that are close to one another in a given environment. The system is not in balance and the mass at higher temperature emits (isotropically) quanta of high energy, which the other mass absorbs in part but, since its *internal* balance is always attained, it will also radiate (again isotropically) quanta of lower energy. Since the total energy of the system remains the same, the number of quanta in the system increases and so does the entropy, which is proportional to the number of photons (which can be treated as particles). Hence, the tendency within fixed energy in an *isolated* system is to increase entropy until balance is reached in the whole system.

The situation described is, in fact, similar to that of the sun and the Earth (Fig. 3.14 and see Fig. 7.43), which, however, remains *far from equilibrium* since energy is continuously created, at the expense of matter, in the nuclear reactions occurring in the interior of that star. Therefore, if we wish to apply the concepts of thermodynamics described in this chapter to the Earth, we need to postulate a principle of *local thermodynamic equilibrium*, according to which the macroscopic state variables such as temperature, pressure, etc. can be defined *locally* using equilibrium concepts since the molecular collision relaxation times are small enough to enable the absorbed solar radiation to be 'thermalised', that is, to be in balance with rest-mass kinetic energy. (The principle is not valid, for example, in the upper atmosphere because there are so few molecules there, that is the air density is so low, that collisions are rare and balance cannot be achieved. No unique local temperature can be defined in such conditions.)

It is this re-radiation of low-energy photons (corresponding to $T \approx 260$ K), escaping to the 3 K background radiation of the universe, that allows ordered and far-from-equilibrium organised structures to exist on the Earth, and this includes life (see Fig. 3.14). The overall entropy of the Earth plus surroundings does indeed increase, despite its decrease at the Earth level due to the formation of ordered and organised living structures. The process is permanent (for at least 5×10^9 more years) since radiant energy from the sun (at some 6000 K) is continuously absorbed and most of it is re-radiated (only about 1 per cent is retained and transformed, for example in photosynthesis, or dissipated, for example by atmospheric circulation). This photon entropy change is actually the main ordering drive, that is, the main contribution to overall entropy

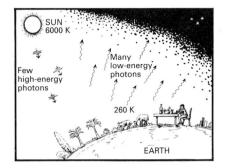

Fig. 3.14 How we make use of the fact that the sun is a hot-spot within the coldness of space. Note the order and organised character of the objects on Earth. The average temperature of Earth is greater than 260 K due to heat from Earth's core. (From Penrose, R. *The emperor's new mind*, reference 7 in 'Further reading'.)

increase; export of rest-mass entropy (for example, hydrogen) and with it kinetic energy to the surroundings (the Earth may now be considered as an *open* system) makes only a small contribution. We will come again to this subject in Chapter 8.

This principle of local thermalised dynamic equilibrium can be applied to the formation of nuclei, atoms, gaseous molecules, condensates, planets, etc., even when full equilibrium is not quite reached at any time due to the progressive lowering of the temperature caused by the expansion of the universe. In all cases the local internal energy decreases (since the products are more stable than the reactants), the local system entropy decreases (since more or less ordered objects are formed), but the overall entropy of the system plus surroundings increases due to the export of the 'heat' of the reactions as radiant energy in (local) equilibrium with the rest mass of each system. As we have seen, the free energy change is negative for such processes. In Chapters 4–6, we shall consider many aspects of local equilibria before we describe controls over rates of change in Chapter 7. We shall find in Chapter 8 that this approach is often valuable when we discuss the formation of Earth (minerals) but that it cannot be applied to living organisms.

3.8 Energy state distribution in materials: the construction of machines

In this chapter, apart from the discussion of equilibria between physical states and chemical species, we have shown that the different states of matter have molecules with very different distributions of motion, that is, very different entropies. This is obvious in the very different cases of crystals such as NaCl, liquids such as water and gas molecules such as N_2. Here it is mainly the kinetic energy distribution that is different (Fig. 3.15). There is a more subtle but similar distinction in the distribution of energy in configurational, rotational and vibrational states in condensed systems. The mechanical

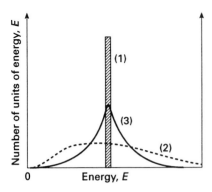

Fig. 3.15 The distribution of energy over molecules or monomer units of a polymer in: (1) a perfect crystal; (2) a liquid; and (3) a weakly co-operative material such as some proteins (but proteins can vary between the extremes). In each case the absolute value of E differs.

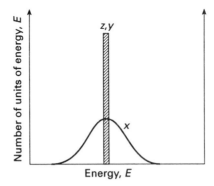

Fig. 3.16 The distribution of energy within an energised polymer such as a protein that can oscillate its component parts along x but not along y or z directions.

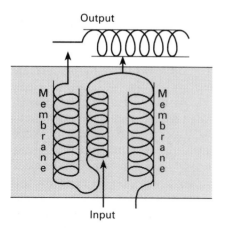

Output

Input

Rod-like helical proteins for energy transduction (see Chapters 10–13).

properties of materials, for example compressibility, depend on these energy state distributions. When many configurational states of a substance, for example a polymer, are close in energy (and barriers between them are not large (see Chapter 7)) the material is readily deformed by mechanical pressure. Such materials can be used in rubbers, springs, hinges, lubricants and so on, while non-deformable materials form virtually static structures. (Clearly, internal entropic considerations are as important as bond energies to the properties of materials.) When we consider mechanical machinery that transfers energy we need both stationary frameworks and dynamic parts, and therefore must construct them with materials of dynamic properties of different kinds for different purposes. Moreover, a machine operates differently in different directions. Thus, it can well be that a substance, in biology a molecule, must be designed so that it is more deformable in one vectorial direction than in the other two (Fig. 3.16). Such is the case for a spiral spring but it is also true of the design of complicated networks of rods, for example, in an automobile engine. We shall see in Chapter 6 that polymers have been 'designed' in biology so that they act like sets of rods that can be displaced more easily in a particular direction than in all others. Thus biological polymers can be parts of molecular machines, see marginal illustration. These vectorial considerations will be developed when we discuss flow in Chapter 7.

3.9 Summary of order/disorder equilibrium

In the above discussions we have been concerned mainly with (1) the fundamental problems of the equilibrium in materials between order and disorder; (2) the drive towards change due to absence of equilibrium (not the rate of change); and (3) the involvement within materials of energy in different forms. We have seen that we can know with precision the free energy status of any part of a system of phases and chemicals that lies behind these problems and that, once a system is in equilibrium, it cannot change unless energy is put into it. Now we are concerned in this book with the origin of the chemical objects around us, that is, the process and progress of natural selection of chemicals, and we must therefore concentrate on the physical and chemical conditions of the materials produced by the cooling of the Earth which has left the planet at approximately 25°C.

We considered the physical condition of evolving and evolved materials first and showed *that it was the co-operativity in condensates that dominated the state of particular compounds* at ambient temperature and pressure once the chemical combinations had been sorted out. It was shown that the balance order⇌disorder favours ordering of matter on cooling. The balance of ordering in materials is unevenly in favour of disorder in gases and of order in solids and only in liquids is it somewhat evenly balanced. We noted that very few substances on Earth are liquids at 25°C and the most important of these is water. In earlier historical times at higher temperatures many more materials were liquids (or gases). The presence or absence of particular liquids was and is crucial to a discussion of the continuation of the natural selection of the chemical elements, since, overwhelmingly, liquids (and gases) allow transfer of material to ordered states. The presence of the more co-operatively ordered

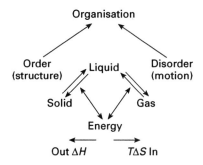

Fig. 3.17 A version of Figs 1.1 and 1.2 that now introduces the concepts of order and disorder ($T\Delta S$) with relationship to the states of matter, their structure and dynamics and energy.

materials thus gives structure and the structures give boundaries to the liquid phases that they contain. One immediate task in Chapters 4–6 is to see which elements in compounds form gases/liquids/solids or complex associations in balance so that we can see which particular properties could appear and which element transfers are possible in which temperature ranges. Obviously, geophysical, biological and man's activities take place in different temperature ranges causing different element selection (for different ends in the case of biology and man) in ordered and disordered states. The characterisation of the observable physical materials around us is still describable in terms of the central part of a diagram much like that of Fig. 1.2 (Fig. 3.17).

As well as discussing equilibrium we have considered the possibility of change, given that, locally, systems having internal balance individually are not in balance with one another. We must then pose some questions. While thermodynamic analysis establishes equilibrium conditions how much of our world is in this condition? What is the consequential state of the elements? How does a living system as well as a geophysical one relate to equilibrium conditions? If parts of our world are or are not in equilibirum, can we find out if it is possible to go against the direction of spontaneous change toward equilibrium which is known from calculation using standard states?

In the case of man and effectively in the case of biology machines are used, especially mechanical devices, to direct change and so generate organisation (Fig. 3.17). The design of machines with choice of materials (Fig. 3.16) and the way in which they utilise energy (Fig. 3.13) are critical problems that we will discuss in Chapters 10–14. As we shall see lying behind all such discussion is the natural selection of the chemical elements in the materials.

Further reading

This chapter deals mainly with some basic principles of equilibrium thermodynamics. There are, literally, hundreds of books on this fundamental branch of physics, from elementary introductions to advanced treatises, many of which could be recommended. The following are just a few references that we found useful to clarify concepts, but the selection is a matter of individual preference.

1. Atkins, P. W. (1993). *Physical chemistry* (5th edn). Oxford University Press, Oxford. Chapters 1–5, 7 and especially 19 and 20. A modern university textbook with a clear, straightforward, yet rigorous treatment of classical and statistical thermodynamics.
2. Moore, W. J. (1972). *Physical chemistry* (5th edn). Longman Group Ltd, London. Chapters 1–3 and 5. A classic university textbook, again with a clear yet rigorous treatment of thermodynamics, written in an elegant and appealing style.
3. Castellan, G. W. (1964). *Physical chemistry*. Addison-Wesley Publishing Co, Inc, Reading, Massachusetts. Chapters 8–11. Another classic university textbook, with clear definitions, full derivations and clarifying examples.
4. Mahan, B. H. (1964). *Elementary chemical thermodynamics*. W. A. Benjamin Inc, New York. A simple, short but excellent text that describes and discusses the basic principles and concepts of thermodynamics in an easily understandable style.
5. Bent, H. A. (1965). *The second law*. Oxford University Press, New York. A monograph that goes beyond its title and helps to clarify concepts frequently dealt with in a cursory and sometimes incorrect way.

6. Craig, N. C. (1992). *Entropy analysis*. VCH Publishers, New York. Another short monograph, inspired by the previous text.

The first two references also contain information on non-equilibirum and far-from-equilibrium thermodynamics. Specialised texts, see for example 11, are generally at advanced level. References 7 and 8 supplement the discussion of some of the topics of this book—entropy of radiation and organisation on the Earth.

7. Penrose, R. (1989). *The emperor's new mind*. Oxford University Press, Oxford. Chapter 7. Not an easy book for the average reader, but the sections on entropy are amongst the easiest to follow.
8. Lesins, G. B. (1993). Radiative entropy as a measure of complexity. In *Scientists write on Gaia*, The MIT Press Cambridge, Massachusetts.
9. Adams, S. (1994). No way back. *New Scientist* 22 October. This short paper gives a quick introduction to those unfamiliar with the laws of thermodynamics.
10. Williams, R. J. P. (1993). Protein dynamics. *European Biophysics Journal* **21**, 393–401.
11. Schuster, P. (1983). Irreversible thermodynamics. In *Biophysics* (ed. W. Hoppe). Springer-Verlag, New York, Chapter 8.

4

Phase equilibria

Things do not interact amongst themselves at random; they follow of necessity from their principle of order. They are integrated by an underlying cause. They are gathered together by a determining influence. Thus things are complex but not chaotic.

Wang Pi (BC Third century)
Chou Yi (Explaining the appendix of the *I Ching*)

4.1 Introduction

In Chapter 1 we maintained that all material objects on Earth are to be described in terms of atomic elements and their combinations. Since there are 90 different types of element on Earth, it might appear as if their combinations were of virtually infinite variety, especially since we can choose to put together any number of kinds of element in any quantity. In Chapter 2 we showed that, in fact, some ordered combinations or association of atoms were more stable than others due to the different strengths of bonding between atoms and molecules. Thus, there are limitations on combinations and this is one indication of the natural selection of the elements. By treating specific combinations of elements in isolation, both in fixed atomic chemical proportions, for example H and O in H_2O, and in given physical states, for example H_2O as ice, we could then discuss many features of the nature of element combinations, here of the atoms H and O. We established that some followed valence rules, which placed further restrictions on variability. We noted too that it was possible to produce more than one combination of some elements, for example H_2O_2 from H and O, although this compound is not as stable as H_2O. Each such chemical substance could also be discussed in isolation in at least three physical states—solid, liquid and gas. However, as stressed in Chapter 2, the majority of combinations did not give rise to simple stoichiometric compounds, which implies a greater opportunity for variability. In the present chapter we use the principles of Chapter 3, that is, of order$\rightleftharpoons$disorder equilibrium, to extend our discussion of the variability of chemical systems to put together all kinds of physical and chemical equilibria. This will help us to see to what extent physical and chemical equilibria, all related to chemical selective combination, have constrained the formation of the objects seen around us within the limits of the particular composition of the Earth.

4.2 Phases and chemicals in equilibrium

In Chapter 3 we looked at real systems rather than at each isolated compound in a single physical state. We observed that, quite generally, co-operative liquid and solid phases are found in balance with a vapour phase and that, frequently, one chemical is in balanced equilibrium with another. The position of balance is dependent on the energy, associated with the temperature particularly, which is put into the system. For example, water or ice have a specific balanced vapour pressure at unit atmospheric pressure which increases with temperature, but at very high temperatures the gas H_2O also exists increasingly in balance with $H_2 + O_2$. The balance point in the case of chemical equilibria depended on the relative amounts of the elements introduced (Section 3.5). We could appreciate therefore that there are constraints on physical/chemical systems based on both physical and chemical equilibria. In this chapter we shall ask of any chemical system how its condition is constrained. This is a question of the variability of chemical systems and constitutes fundamental knowledge concerning all that we observe around us and how it could have arisen especially as temperature changed.

What we should like to know, in order to explain the materials around us, is therefore simply stated: *when a system of chemical elements is being cooled* and/or its pressure adjusted, *are there fixed ways in which it has to change if it is to remain at equilibrium?* In such a case it passes inevitably through a series of precisely defined physical *phases*, by definition homogeneous, of given chemical compositions. Thus only certain phases and compositions can coexist at the particular temperature and pressure now existing on Earth. This knowledge would then allow understanding of all the things around us from initial inputs to the universe.

The simplest example is the nature of any gas. We can measure the volume, V, pressure, p, and temperature, T, while they are changed in balance. Consider a fixed amount, one gram-mole of a gas, and ask, for non-interacting molecules, are there three (V, p, T) independent variables? We know that the answer is that there are but two variables since the observed gas law $pV = RT$, where R is a constant, generates an equilibrium restriction. We can derive this equation from the fact that the kinetic theory of gases allows us to discuss p in terms of *momentum exchange* and T in terms of *kinetic energy*, which are related through mass and velocity to each other. Thus V varies strictly with T at fixed p. Note that we here assume that the gas is homogeneous and ideal. A further step forward is to consider condensation equilibria starting from a gaseous state of any substance and cooling at equilibrium without changing composition. Usually we observe that, successively, a liquid and then a solid are formed but, as explained in Chapter 3, these phases always coexist with an equilibrium vapour (gas) pressure. First then by experiment we need to discover how many phases, gas (vapour), liquid and solid (or more than one solid if this is the case) can coexist over which pressure and temperature ranges when we cool at equilibrium. To what extent is this number dependent on the number of chemical elements in the system? This is an examination of phase coexistence.

For example, water as a liquid coexists with water as a gas at all pressures and temperatures from above 0 to 100°C, but the water vapour pressure is reduced as the temperature is reduced. If we reduce it below 0°C at atmospheric pressure, water freezes to give ice, but even below the freezing point ice still has a small vapour pressure, which again falls with temperature. At a higher temperature, above the boiling point, 100°C at atmospheric pressure, no liquid water remains. All these observations can be summarized in a temperature against vapour pressure, p_{H_2O} diagram for water (Fig. 4.1) that shows where the p_{H_2O}, T regions in which ice/vapour, liquid water/vapour, or just vapour are located. This is called an *equilibrium* phase diagram since all phases are always homogeneous and in balance and the chemical condition of the water is assumed to be independent of time. At only one p_{H_2O}, T point do all the three phases—ice, liquid water and vapour—coexist; this is called a triple point. At all other p_{H_2O}, T combinations, if three phases are apparent, they will not be at equilibrium and one will disappear with time.

What we have discovered here is that at equilibrium, and assuming that the amount of water present is large and that no external fields are applied except fixed pressure† (see Chapter 6), the presence of three phases has removed completely the variability of the system. Again water in any condensed form has a definite water vapour pressure at a given temperature and fixed external pressure. We can put these findings in another way. When we cool a system at fixed external pressure then we can only obtain particular condensed phases over limited temperature ranges. The more condensed phases there are present the more *restricted* is the *variability* of the system. Now these rules are open to simple demonstration and generalisation since water is in no way peculiar in these respects. Moreover, we can treat systems in which there are two or more chemicals made of independent chemical components (elements or compounds), each one of which will be separate from all other components in behaviour, so long as each one does not exchange atoms of elements with other compounds (components) (see Section 4.3.1). The components will mix, that is, dissolve in one another, to some balanced degree within each phase they form, see marginal illustration. We now proceed to derive a general relationship for equilibrium in such a system.

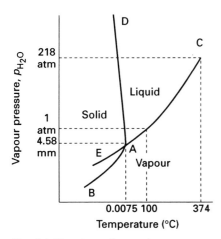

Fig. 4.1 The phase diagram for water, see Figs 1.1 and 1.2.

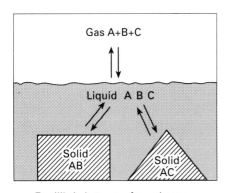

Equilibria between four phases and three components with five separate compounds present.

---

† The fixed atmospheric pressure is due to Earth's gravitational field.

4.3 The phase rule

The following is a simple deduction of the phase rule devised in 1875–6 by J. Willard Gibbs, which governs the number of ways in which a system of chemical components can be varied within a variable number of phases at equilibrium. There are only four considerations that define such systems: temperature and pressure, that is, two variables, the composition defined in terms of numbers of components, C, and the number of phases, P. (The first three are *intensity factors*, that is, they are the same for the system and its individual parts.) Here we are dealing with percentage composition (or mole fraction) not total amount, which is a capacity factor, so that the composition of each phase is defined by the number of independent components that can be varied at equilibrium. Clearly, in any one phase this is $C-1$. The *maximum* number of variables are the independent compositions of the independent phases, $P(C-1)$, plus temperature and pressure, that is $P(C-1)+2$. But the number of variables is restricted by the equilibria *between* the phases and all components are in equilibrium in all phases. Every component will distribute itself in every phase so we have for each component $(P-1)$ equilibria that prevent arbitrary changes in composition of any phase. All components have a similar set of phase equilibria so that there are a total number of $C(P-1)$ restrictions or equilibria.

Adding together the maximum number of variables and the number of restrictions gives us the real number of variables, F, also called the number of *degrees of freedom*, or the *variance*, of the system

$$F = P(C-1)+2 - C(P-1) = C+2-P.$$

This remarkably simple result gives the number of degrees of freedom due to the variables, pressure, temperature and the number of components, C, which can be varied *independently* without changing the number of phases, P. It is an analysis of the physical states in equilibrium for any number of chemical components, C. Before going further let us consider simple real cases since this knowledge is a major step in our understanding. It could help to reveal how the stable state of material that we observe around us has been obtained on cooling a given set of atoms, which must be the ultimate basis of all chemicals extant on the Earth. Obviously this is central to the enquiry of this book for it is quite possibly the process by which Earth formed. (We note in fact that this does not turn out to be the way in which our surroundings formed, but the examination of these equilibrium thermodynamic conditions proves to be extremely useful since it shows the direction of change in which all systems must go eventually, Section 3.3.)

Returning to the case of water, we see that if we have just one chemical component, which water is, as defined by the formula H_2O, then $F = 3-P$, and we may have three situations.

1. One phase only: $F=2$, that is, there are two independent variables (bivariant system). This occurs when all the H_2O is a gas.

2. Two phases: $F=1$, that is, there is only one independent variable (univariant system).

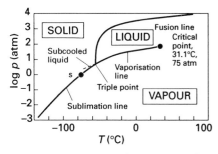

Fig. 4.2 Phase diagram for carbon dioxide (after *Gmelin Handbuch der anorganischen Chemie: Kohlenstoff*, T3/L1 (1970). *p* is here the external pressure. Notice that CO_2 sublimes at s where $p=1$ and $T = -78°C$.

3. Three phases: $F = 0$, that is, there are no degrees of freedom (invariant system).

All three situations can be seen in Fig. 4.1. The singular point of case 3 is, of course, the triple-point A; in this situation there is no possibility of changing either the temperature or the pressure without a change of the number of phases. Similar cases occur in many other systems, for example for CO_2, (Fig. 4.2). Two phases coexist in a wide range of possible external p, T pairs, but, once either p or T is fixed, the other can only have a well defined value, that is only one of the two variables p and T can be independently varied (univariant system). In every case too when the vapour pressure exceeds the external pressure only a gas phase exists.

There is a very rare peculiarity in the water phase diagram in Fig. 4.1: the slope of curve AD shows that the melting point of ice decreases as the pressure increases. As a consequence, water, like bismuth and antimony but unlike CO_2, expands on freezing and its density decreases—ice floats on liquid water. This is a direct consequence of the formation of hydrogen-bond structures in ice (see Section 2.6.1). For most substances the density of the solid is higher than that of the liquid. As stated in Chapter 2, it is not at all clear how Earth and especially life on it would have evolved if ice had been denser than liquid water.

(The continuous lines in the diagram of Fig. 4.1 refer to equilibrium situations, but in some *situations* we may have a *metastable* condition. Line AE, for example, corresponds to supercooled liquid water which tends to solidify spontaneously. Its condition is time-dependent.)

4.3.1 Chemical equilibria and components of chemical systems

The chemically distinct and independent constituents of a given system are the *components* of the system. In principle, therefore, since there are 90 chemical elements on Earth from which all possible materials are made (two of the 92 of the periodic table have disintegrated completely) and ignoring phases for the moment, there will be for the chemical system Earth (90) +2 degrees of freedom, that is, the number of elements plus p and T (NB. Fields are ignored). However, most of the 90 chemical elements are not inert so that they did not remain unreacted at the high temperatures at which the Earth was formed but came together to form compounds (see Chapters 2 and 3). We must then ask: does the formation of a compound, for example H_2O, composed of hydrogen and oxygen, increase the number of chemically distinct entities that we consider to be components of a particular system? The answer to this question is given by the specification of components as 'constituents whose concentration can be independently varied'. Now for pure gaseous water *at equilibrium*, that is, at very high temperature with no deficit or excess of two H relative to one O, we have the chemical equilibrium

$$2H_2 + O_2 \rightleftharpoons 2H_2O.$$

For this equilibrium we may write an *equilibrium constant* (Section 3.5) in terms of the partial pressures of the species involved. The constant is

$$K_p = \frac{[H_2O]^2}{[H_2]^2 \, [O_2]}$$

where [. . .] refers to concentration or partial pressure in the gas phase. Hence the partial pressure of H_2O at equilibrium is not independent from those of hydrogen and oxygen and the formation of this compound did not alter the number of chemically distinct species of the system whose concentrations can be varied *independently* and consequently the number of degrees of freedom or *variance*. Since the ratio of H to O was decided as two H to one O in this example there is also no percentage compositional variable.

We may summarise by saying that, at full chemical equilibrium where all atoms in all compounds are exchanging freely, a chemical system is constrained by the composition in terms of the ratio of the number of elements themselves, not of any compounds formed, and the only other variables are temperature and pressure if fields are ignored. (The amounts of materials have been made unrestricted, that is, phases are considered as effectively infinite.)

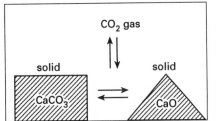

There are three phases and three compounds but only two components

More generally, we may state $C = C_i - R_e$, where C_i is the number of chemically identifiable substances and R_e is the number of reactions that can convert one into another at equilibrium. For instance, let us take a system consisting of any ratios of $CaCO_3$, CaO and CO_2, three chemically different species, see marginal illustration. However, since at high temperature there is the equilibrium, $CaO + CO_2 \rightleftarrows CaCO_3$, the number of independent components, C, is two, not three. If we take pure stoichiometric $CaCO_3$ then there is only one component whether or not we raise the temperature above 500°C, when $CaCO_3$ dissociates perceptibly into $CaO + CO_2$, or at even higher temperatures when the system dissociates to Ca, C and O, its elements. The composition ratio has been chosen as 1:1:3 and is fixed, not variable, so that the fact that it can form all sorts of combinations including CO, CO_2, CaO and $CaCO_3$ in equilibrium does not affect the degrees of freedom before we refer to phase formation.

Any equilibrium or similar relationship restricting the possibility of independent variation of concentration of any substance is, therefore, a limitation upon the number of components of a system.

Let us now return to Earth as a system of 90 elements. If all the elements were in equilibrium, both physically and chemically, that is in all compounds and in all phases, then for all of them we will have conditions of the type

$$xA + yB \rightleftharpoons AxBy,$$

or more complex equilibria involving many elements. From what we have said, no additional variable is introduced by the presence of any A/B combination; *at equilibrium*, the maximum number of degrees of freedom to be considered is still $90 + 2$. Since there must be at least one phase this number is reduced to only 2 if the relative amounts of all elements are also fixed, when $C = 1$. What this means is that at fixed temperature and pressure there are no degrees of freedom in an equilibrium system of defined composition, that is, $F = 0$ (see above). In such an equilibrium situation there should be no development with time *except as T and p change the compounds and phases present*. Evolution in chemical selection and phase change are possible, even at equilibrium, with fixed proportions of elements, but only by changes in temperature and pressure. (If the total *amount* of elements can be changed or external fields are introduced then the number of variables also changes and the system can evolve in new ways (see Chapter 6) still at equilibrium. However only by breaking the equilibrium conditions does a system become extremely variable.)

We must observe that we have had to prescribe conditions, especially temperature, such that all elements exchange between all compounds. Now this was not and is not the condition on Earth at 25°C where only some elements exchange between some compounds at equilibrium. For example, H_2 and O_2 no longer exchange with H_2O. Earth did not form in this equilibrium manner and no biological system conforms to this description at all. In the absence of equilibrium we must re-examine the definition of component to be used in the phase rule.

4.3.2 The operational definition of component

We stated in Chapter 1 that all chemicals can be defined in terms of the 90 elements made from neutrons and protons. If the fundamental equilibrium in the universe had concerned itself with the proton/neutron distribution amongst the different nuclei (see Chapter 8), this equilibrium would have led to a definition of components in terms of the ratio of neutrons and protons alone. There is today on Earth no exchange of nuclear particles (protons and neutrons) between elements at the temperatures where chemistry is concerned. We shall consider that the state of the nuclei of the elements is so frozen on Earth that an approach through neutrons and protons has no value for our objectives. The definition of the number of components is therefore related to the effective number of *independent chemical species* in the system under the conditions under consideration. We have to restart discussion from the number of elements on Earth (90) but take cognisance of the fact that the temperature on Earth was and is so low that many chemical combinations were frozen relative to one another. Since all combinations were formed from elements originally at high temperatures we must be careful in our treatment to evaluate the idea of a component according to operational (temperature/pressure) conditions. In an operational situation where a chemical undergoes no exchange of elements (that is, no reaction), we can legitimately consider it as an independent component separately from its elements. We considered H_2O in this way in Fig. 4.1 where we did not include H_2 or O_2 in equilibrium. Hence, in using the phase rule we must not reduce all combinations of elements, compounds, to the number of elements present any more than we reduce all elements to the number of neutrons and protons present. A further example illustrates the point.

Consider the solubility of a liquid organic molecule $C_xH_yN_wO_z$ in water, H_2O, and assume that there is no exchange of atoms between the two (Fig. 4.3). Beyond the solubility limit we consider that there are two phases present, water and the organic material, and two components. Thus there are two degrees of freedom, temperature and pressure, at equilibrium despite the facts that there are four elements present and that those in water, H and O, are also present in the organic molecules. The reader will see that it does not matter if the material considered is a salt that dissociates in water or an organic molecule that does not so long as there is no reaction (exchange of elements

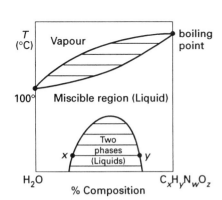

Fig. 4.3 The typical composition/temperature phase diagram for two somewhat similar liquids. The x y points, joined by a tie line, represent compositions of two liquid phases in equilibrium.

between chemical species, here between solvent and solute); the number of components, C, is described operationally as the number of independent chemical species. C will then depend on such features as temperature or on any other factor that causes equilibria to be established or lost.

The cooling of the Earth generated a large number of independent components, chemical combinations of most elements. This number greatly exceeds the number of elements and therefore greatly increases the variability of the whole planet. We need to know the numbers of 'frozen' independent chemical components and we return to the further consideration of these frozen states in Chapter 7 where we ask which kinetic barriers maintain this independence. Before we analyse this escape from element equilibria we shall consider those kinds of physical and chemical equilibria that are present in systems of the greatest interest to us in our description of what we observe around us. In this chapter therefore we shall describe the limitations on some chosen chemical systems at equilibrium applied both to chemicals that do and that do not exchange elements. Our initial concern will be with the number of phases we observe.

4.3.3 The number of phases

The phase rule equation includes P, the number of phases, as well as the number of components, C, and so, while temperatures and pressure are varied, we have to consider the possibility of the formation of separate gas, liquid and solid phases at equilibrium. This means that, in the development of the universe, of the Earth, of biological or non-biological chemical systems, these systems were not only increased in variance by formation of components but also constrained by formation of phases as the temperature and the pressure changed. Let us consider a vapour of *fixed* composition $xA + yB$, (xy), in equilibrium with any set of combinations of A and B in molecules in a gas phase. From what has been said above, $C = 1$. On cooling we could get from the system, in principle, a variety of liquids depending on the strength of the relative interactions of A with A and B with B, as well as on the affinity of A for B which determines the formation of combinations of A and B. Let us denote the liquid composition by $uA + vB$, (uv) (see Fig. 4.4). In this case $P = 2$, whence $F = 1$, which means that of the two intensive variables, temperature and pressure, only one can be varied independently—the system is univariant. Now, this is extremely instructive since it says that, *on cooling a fixed ratio of elements at constant pressure at equilibrium, not only will there appear at a given temperature, the boiling point, a fixed combination of A and B (even a compound) but also that any fixed liquid or solid composition that forms does not necessarily have the same composition as the vapour.* This must apply to the formation of Earth. We discuss the problem further in Chapter 8, but let us continue the examination of equilibria.

We have introduced no specifications concerning the ratios x/y or u/v and it follows that no rules apply except that a fixed condition links an x/y ratio in the gas to a u/v ratio in the liquid. If we change the x/y ratio, that is, the relative proportion of the elements A and B in the gas phase, then the ratio u/v will also change. This is a result of the relative A —A, B—B and A—B affinities in the gas and liquid phases. For cases of very strong preference, for example

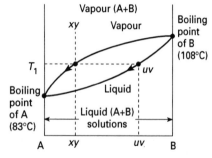

Fig. 4.4 A simple vapour–liquid phase diagram here for propanol-2 (A) and 2-methylpropanol (B); see text for further description.

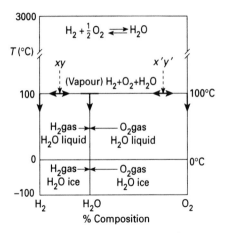

Fig. 4.5 Simplified phase diagram for H_2/O_2 mixtures at equilibrium. In fact, chemical equilibrium is only established at high temperature, for example 3000°C for $p=1$ atmosphere.

formation of a stoichiometric compound AB, only one ratio u/v in the liquid may appear for all mixtures x/y as illustrated by the case of $H_2 + O_2$ mixtures cooled down to below the boiling point of water (Fig. 4.5) with all phases and chemicals at equilibrium and at atmospheric pressure. The diagram shows that at around 3000°C there is formed H_2O which remains in a homogeneous gas phase until 100°C when liquid water forms, followed at 0°C by the formation of ice. Only at very low temperatures (not represented) do O_2 and H_2 form liquids and then solids. The variety of *mixtures* of liquid and solid phases that can be obtained is shown in the diagram, but here the trivial solubility, for example of O_2 in water (which in part keeps us alive! . . .) is ignored. (We shall return to it in Chapter 5.) In this case the binding in H_2O is so strong and specific at a given ratio that no other liquid or solid phase of any other composition is observed as a stable substance.† H_2O_2 is not stable, for example. There is no *a priori* reason why this should be so in other cases; in effect, it will be easily seen to be quite other than a general rule. On further consideration we find that combinations of elements are multiple (Chapter 2, part A) and most phases are non-stoichiometric (Chapter 2, part B) and variable in composition with temperature.

4.4 Variability of chemical systems: examples of systems of more than one component

During the considerations of the previous section we referred rather extensively, albeit simply, to systems of one component, deliberately stressing H_2O due to its outstanding importance on Earth, but we only mentioned superficially systems of more than once component—binary, ternary, etc. Of course, the more independent components that are introduced the more complicated the system becomes and the greater its variance, F. It may then be quite difficult or even impossible to visualise from extensive tabulations of values of p and T and composition or from complicated multidimensional diagrams the possible phase changes when the conditioning factors vary. The diagrams may be simplified, however, to show the separate effects of some of the variables. For many systems we can use two-dimensional vapour pressure p_x, T diagrams since the composition is fixed, for example in H_2O, Fig. 4.1. For binary systems there will be three variables, external p, T and relative composition; hence three-dimensional phase diagrams are required to give an overall view of the systems. Since these are complicated we normally use two-dimensional cross-sections of the general diagram, usually at constant external pressure or constant temperature and, less frequently, at constant composition. We have, then, simplified pressure–composition or temperature–composition diagrams, useful in particular conditions, for example at atmospheric pressure, when we plot boiling point/composition curves.

In the following section we give a few illustrative examples before taking, in later chapters, a more general approach, closer to reality, where amounts, fields and, finally, certain non-equilibrium conditions are also included. Only

† Phase diagrams can be drawn showing unstable substances but we shall not introduce this complication in this book.

later, then, will we be able to deal with the questions of shape and still later organisation that are central to the purposes of this book which is an analysis of the selection of chemicals in the real objects around us.

The classification of chemical systems considers first the number of components and the kinds and number of phases present (see Table 4.1) in terms of their variability. Formation of compounds (or intermolecular species) and polymorphism in solid phases are constraints that must also be taken into account. However, there is no need to be exhaustive since not all combinations of variables are of practical interest; the most common are systems with only vapour and liquid phases, several liquid phases, liquid and solid phases, or just

Table 4.1 Classification of chemical systems of components

Number of phases	Variance	Free variables*		
		General	Fixed p	Fixed p, T
Unitary system (one component)				
1	2	T, p	T	None
2	1	T or p	None	—
3	0	None	—	—
Binary system (two components)				
1	3	T, p, x	T or x	x
2	2	Any two	x	None
3	1	Any one	None	—
4	0	None	—	—
Ternary system (three components)				
1	4	T, p, x_1, x_2	T, x_1, x_2	x_1, x_2
2	3	Any three	x_1, x_2	x_1
3	2	Any two	x_1	None
4	1	Any one	None	—
5	0	None	—	—

* T, Temperature; p, pressure; x, mole fraction composition.

solid phases, as in minerals, ceramics and alloys. Organic systems have been less studied than inorganic systems and biological systems have not received much attention along these lines despite their obvious interest (see reference 11 in 'Further reading'). The effects of temperature and pressure are the most relevant for liquid–vapour systems, whereas for condensed states it will be mostly the effect of temperature that needs to be taken into account. Some paradigmatic cases will be considered for better understanding of the potential of the phase diagram approach to chemical systems. Eventually, we shall find reference to this treatment valuable in the examination of both geological (Chapter 8) and, in a different way, of biological species (Chapter 16).

4.4.1 Unitary systems

In Fig. 4.1 we presented the phase diagram for water within the range of its stability and we have commented on some of its characteristics. Similar diagrams can be established for many other systems with but a single component, for example CO_2 (Fig. 4.2). In the latter case the melting point of the solid increases with the increase of pressure, and a critical point (for which the liquid and the vapour have the same properties) appears at a relatively low temperature, $31°C$, and moderate pressure, 75 atmospheres. This phase diagram may be of great importance since man may need to remove CO_2 by liquefaction and store it at high pressure beneath the deep ocean. It is of immediate concern too when we examine the condition of Earth's atmosphere and the surfaces of cold planets such as Mars or Jupiter. CO_2 is plentiful but it does not give a stable liquid at atmospheric pressure and no life can exist without a liquid. Earth is a very idiosyncratic place.

At very high pressure the diagrams become more complicated; water, for example, exhibits no fewer than nine solid phases for pressures up to 10 000 atmospheres and it has been reported that at 40 000 atmospheres ice melts at $200°C$ and should be more dense than liquid water.

Table 4.2 Examples of some phase systems of concern in this book*

Liquid/liquid phase system

Water/lipid (biological cells)
Lipid/lipid (biological cells)
Minerals/metals (planet Earth)

Liquid/solid phase system

Water/salt (minerals and bone)
Lipid/fatty solids (cholesterol in cells)

Solid/solid phase system

Alloys (Cu/Zn)
Minerals (Mg, Ca, Al, oxides)

* NB. Dilute solutions are not considered in this chapter.

4.4.2 Binary systems

These cases are, of course, more complicated, but much information can be obtained from two-dimensional diagrams, usually pressure/composition or temperature/composition, although some limited use is also made of p, T diagrams at constant composition. Usually, these diagrams are restricted to a limited number of phases for practical purposes. We will consider a few examples of some important cases to give an overview of those possibilities that interest us in this book (Table 4.2). The figures show transition temperatures between phases at different compositions.

Vapour–liquid equilibria

The simplest example of vapour/liquid equilibrium is given in Fig. 4.4. It shows one possible variation of composition with boiling point. The full variety of vapour–liquid equilibria is given in Fig. 4.6 using only temperature–composition diagrams for the sake of simplicity (the corresponding pressure–composition diagrams are approximately similar but rotated $180°$ in the plane of the paper).

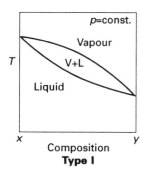
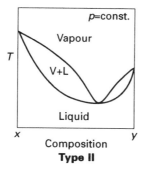
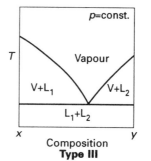
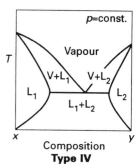

Fig. 4.6 The main types of vapour–liquid phase equilibria. Type I, water/acetone; type II, ethanol/water; type III, carbon disulphide/water; type IV, water/butanol. V, Vapour; L, liquid.

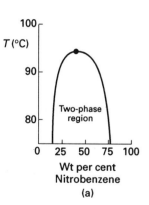

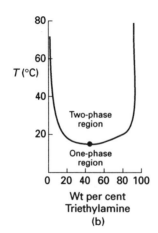

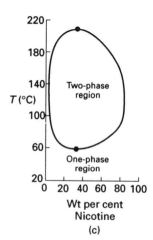

Fig. 4.7 Partial miscibility of two liquids. (a) *n*-hexane + nitrobenzene; (b) triethylamine + water; (c) nicotine + water. Critical solution temperature (CST) values are indicated (●).

A case of particular interest is that in which two liquids separate to form very different phases, for example oil and water as in Fig. 4.6, Type III. The value of two liquid phases (which will separate vertically under gravity, see Section 4.5) is that they allow dilute components to distribute between them on the basis of physical chemical selectivity. Such distributions of chemicals dominate biology and many industrial separations (see Section 5.3). The equilibrium between two liquid phases of a low degree of mutual miscibility, as in Fig. 4.6, type IV, could well correspond to the Earth's situation of limited mutual solubility of the elements of the central iron (alloy) core with the liquid silicate system immediately above it, when Earth remelted after formation by the accretion of meteorites, see Chapter 8. Thus there is quite a lot of iron in the silicates and quite a lot of other metals are extracted into the iron core from the silicate. This fractionation (see Section 8.6.5) is a part of the natural selection of the elements at a very early stage of the condensation of compounds.

Liquid–liquid equilibria

Usually two liquids form either one or two phases at all temperatures (see Fig. 4.7), but some liquid mixtures show liquid phase transitions. A particular case is that of the biological liquids formed from lipids. Two lipids may form one phase at a particular temperature but two phases in a different temperature region, and this situation may occur in mosaic zones within biological membranes (see Section 9.17). Usually, the extent of two-phase regions is reduced by the increase of pressure, and the increase of temperature also favours mutual solubility. The temperature at which complete miscibility is attained is called the critical solution temperature, CST, and it is clear from the figure that there may be both upper and lower CSTs.

As explained in Section 9.4.3, the lipids that are components of biological membranes consist of a long non-polar hydrocarbon chain and a polar head group. They may associate in different ways forming layers, vesicles and other aggregates, depending on their relative proportion to water. All these phases are treated in this chapter as different 'liquid' phases in equilibrium—we will discuss them further in Section 4.4.5.

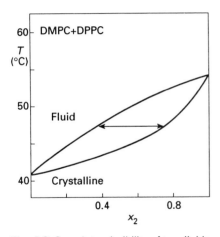

Fig. 4.8 Complete miscibility of two lipids in both fluid (liquid) and solid phases. x_2 is the mole fraction of DPPC. DMPC, dimyristoyl-phosphatidyl-choline; DPPC, dipalmitoyl-phosphatidyl-choline.

Liquid–solid equilibria

Phase diagrams involving liquid and solid phases in equilibrium are obviously more complicated than the ones previously considered, since we may now have cases of liquid–liquid partial miscibility, solid–solid partial miscibility, extensive polymorphism in the solid state, formation of compounds or intermolecular species, etc. Thus there are more transition temperatures as well as components to consider. The simplest case, naturally, is that of two substances, A and B, totally miscible both in the solid and liquid state. The corresponding phase diagrams are then analogous to those of liquid–vapour equilibria in Fig. 4.6. In many cases miscibility is not observed and some real examples are shown in Figs 4.8 to 4.14, where melting points or transition temperatures are plotted against composition. A completely miscible system is shown in Fig. 4.8 for biological lipids.

Completely immiscible solid phases generate relatively simple diagrams (Fig. 4.9, and see Fig. 4.13). The interesting peculiarity of these binary

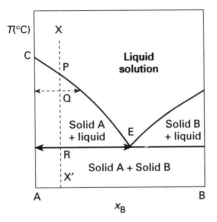

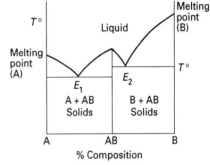

Fig. 4.9 Simple eutectic diagram for two components, A and B, completely intersoluble as liquids but with negligible solid–solid solubility. The effect of cooling follows the line X to X'. At R the eutectic E forms. x_B Mole fraction of B. Note that below P and at Q two phases separate.

Fig. 4.10 The phase diagram for formation of a compound AB that is completely immiscible with either A or B in the solid state. Note the two eutectics, E_1 and E_2.

systems is that each component lowers the melting point of the other so that, on a temperature–composition diagram such as that of Fig. 4.9, the two freezing curves meet at a minimum point, E, called the *eutectic* for which the compositions of the liquid and the solid are identical. Binary eutectics are two-phase systems in which different crystals are often intimately intermingled as expected from their crystallisation conditions—they are *not* A/B compounds. Such crystals frequently appear in natural minerals and in rocks. Stoichiometric compounds, for example AB, however, are also of common occurrence, and may themselves then form eutectics with the other components including A and B separately and other compounds. In such cases the phase diagrams become more and more repetitive as more and more compounds form (see Fig. 4.10). (Note that Fig. 4.10 is just Fig. 4.9 twice, back to back.)

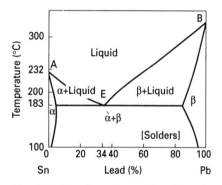

Fig. 4.11 The phase diagram for alloys of tin and lead, showing solid solutions in the regions α and β. Note the temperature scale, due to high co-operativity in condensed phases.

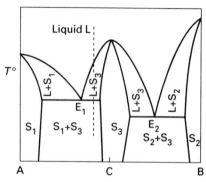

Fig. 4.12 This phase diagram may be regarded as made of two simpler diagrams: of substance A and intermolecular compound C, and of intermolecular compound C with substance B. All substances have limited miscibilities in the solid phase (see regions S_1, S_2 and S_3) but are completely miscible as liquids.

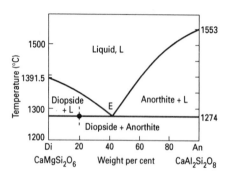

Fig. 4.13 The phase diagram for two minerals, diopside (Di) and anorthite (An), that are completely miscible in the liquid state but completely immiscible as solids. Earth's minerals formed from such melts. Note the temperature scale and compare organic systems (Figs 4.7 and 4.8).

Generally, components are not totally immiscible so that any A is partially miscible with any B. The phase diagram is now that shown in Fig. 4.11. It can be elaborated for non-stoichiometric compound formation where A and B are partially miscible with AB (Fig. 4.12). This figure requires some comment since we meet here, for the first time in this chapter, a solid phase, S_3, that does not correspond to a defined intermolecular compound, AB, as in Fig. 4.10, but to a non-stoichiometric species with a broad composition range due to the solubility of both A and B in C. In essence, however, the diagram is not so different from the simpler one. It may indeed be regarded as composed of the addition of two diagrams, one for substance A and the intermolecular compound C and the other for the intermolecular compound C and substance B, where all substances have limited mutual solubilities in the solid phase. It is an interesting exercise to follow the dashed line in Fig. 4.12 and to see that, on cooling the homogeneous liquid mixture of A and B, the solid phase S_3 separates first, followed by phase S_1 which is largely A (with some dissolved C).

The essence of equilibrium chemistry lies in these diagrams; they are of the utmost importance for understanding the conditions of the Earth today and its production from a gaseous and then from a molten state (Chapter 8), of many industrial processes and of biological systems. Many of Earth's and our chemicals were and are formed by cooling and in conditions close to equilibrium. Co-operativity between units as well as bonding dominates the processes.

The following two real examples (Figs 4.13 and 4.14) illustrate this statement. The first corresponds to the simple phase diagram of the diopside $(CaMgSi_2O_6)$–anorthite $(CaAl_2Si_2O_8)$ system, which forms a eutectic, and the second to the more complex pressure–composition phase diagram of the olivine system $(Mg, Fe)_2SiO_4$, showing the stability range of the various solid phases in equilibrium. All these silicates are present in quantity in the Earth's crust. A fine parallel example amongst alloys is provided by the Cu/Zn system (Fig. 4.15), which has more than five phases from two components, but only three of these phases can coexist at equilibrium for one composition. (The reader should check this statement using Table 4.1.)

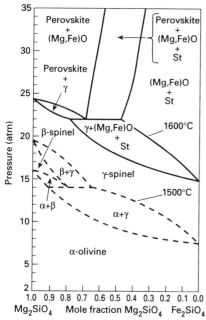

Fig. 4.14 The pressure/composition phase diagram at different temperatures of the olivine system (Mg · Fe)$_2$SiO$_4$, showing the stability fields of the various stable phases, St-stishovite (SiO$_2$). The α phase is olivine, and both the β and γ phases are forms of spinel. Notice how both the olivine to spinel and spinel to perovskite, etc. phase changes occur at greater depths of the Earth, higher pressure, in Mg-rich than in Fe-rich systems. These two sets of phase changes are believed to be important, respectively, at the 400- and 650-km depth discontinuities of the Earth's crust. The formation of Earth and even present-day changes of minerals are represented on the diagram. Note the high temperatures.

Fig. 4.15 The phase diagram for copper/ zinc alloys. Note the temperature scale.

Such diagrams have great value in industrial processes. The reader is referred to Figs 4.15 and 4.16, the Cu–Zn and the Fe–C phase diagrams, respectively, which are both essential knowledge in metallurgy. Here we limit ourselves to the comment that in Fig. 4.15, the formation of several types of brass at various percentages of Cu and Zn is shown, while Fig. 4.16 shows the formation of several allotropes of iron, of one type of steel, austenite, which can be hot-rolled and worked, of an intermolecular compound, FeC$_3$, cementite, which is hard and brittle, and of several solid solutions of iron–carbon, generally called ferrites, which with cementite, at less than 2 per cent C, give a type of high-strength steel called pearlite. (By metallurgical operations, annealing and tempering, different mechanical properties can be achieved.) Above 2 per cent C one obtains cast irons, which are hard and resistant to corrosion but cannot be conveniently mechanically worked. Unwittingly, these phase diagrams lay behind the development of the bronze and the iron ages by man (Chapter 14). They give great insight also into the formation of minerals on Earth. The components need not be minerals or alloys of course, and very similar phase diagrams can be provided for organic compounds and inorganic solutions (Figs 4.17 and 4.18). The formation of salts from the sea, such as dolomites, also falls in this class.

It is a peculiarity of the diagrams in Figs 4.13 to 4.16 that they cover temperature ranges often greater than 500 °C and pressures of up to 30 atmospheres. Such conditions were open during the formation of the Earth 5×10^9 years ago and are still present below the Earth's surface, and have again become available in the last 10 000 years through man industry. The range of physical conditions open for life's evolution is from 0–100 °C and 1 atmosphere. We now look at phase systems relevant in this circumstance.

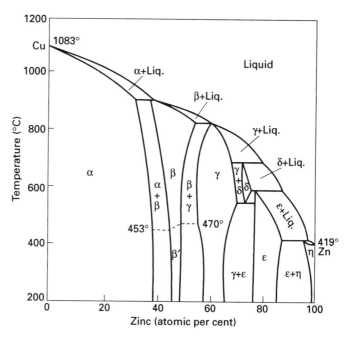

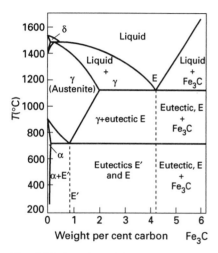

Fig. 4.16 A portion of the iron–carbon phase diagram. (After Austin, J. B. (1948). *Metals handbook*, p. 1181. American Society for Metals, Cleveland.) Note the temperature scale and the composition axis.

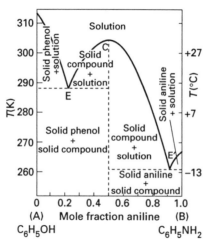

Fig. 4.17 The system phenol + aniline, illustrating the formation of an intermolecular compound. Note the temperature scale due to low co-operativity in condensed organic phases.

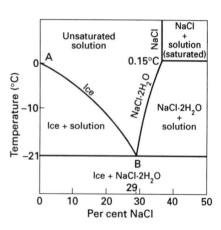

Fig. 4.18 Water and sodium chloride system. B is effectively a eutectic point. Note the temperature scale.

In biology mixtures of lipids again give us some interesting examples. In Fig. 4.8 we saw the case of two similar lipids, differing only in one —CH₂— in their hydrocarbon chains, which are miscible both in the liquid and solid states in any proportion. If the difference in chemical character becomes larger (or one lipid has a non-saturated hydrocarbon chain) then the substances are, generally, not miscible, and the corresponding phase diagram is like that represented in Fig. 4.19. The value of liquid as opposed to solid phases is that they give rise to the possibility of transport and hence organisation, so that solid lipid phases are avoided in much of biology. During the course of evolution the types of lipid molecules in biology have changed and these changes have greatly affected the temperature ranges of biological activity (see Fig. 4.19).

4.4.3 Ternary systems

Most natural and many industrial systems contain more than two components and the type of phase diagrams that we have been describing is hardly sufficient to give an overview of the phase changes occurring when the variables—pressure, temperature and composition—are changed. Multicomponent systems cannot, therefore, be handled in this way, but for ternary systems, for which $F = 5 - P$ (see above), it is still possible to use a similar type of diagram provided that some variables are fixed. For example, by restricting the analysis to the condensed states and fixing the pressure we can construct a triangular prism phase representation such as that in Fig. 4.20(a), corres-

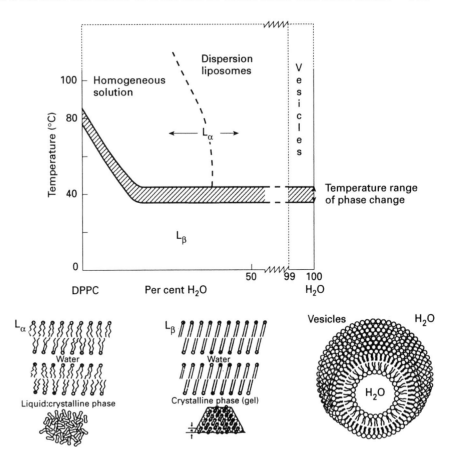

Fig. 4.19 The complicated phase diagram of a lipid, dipalmitoylphosphatidylcholine (DPPC), in water. At low temperatures the system crystallises to L_β but it changes phase *over a temperature range* to a liquid crystal, L_α, which may be dispersed in various forms according to composition. (The phase changes themselves are shown as taking place over temperature ranges; see Section 4.7.) Adapted from Sackman, E. (1982). In *Biophysics* (eds. W. Hoppe *et al.*) Springer-Verlag, Berlin.

ponding to a very simplified theoretical case in which three binary eutectic mixtures and one ternary eutectic mixture are shown. Now, if T is also held constant, $F = 3 - P$, we can use a horizontal cross-section of the prismatic diagram ABC to study the effect of changing three concentrations (Fig. 4.20(b)). Examples of inorganic and organic systems are given in

Fig. 4.20 (a) Ternary diagram of three binary eutectic systems, showing intermediate compounds. On cooling the liquid pure A, B or C only (or mixtures of two of them) appears as a solid. (Adapted from Birchenall, C. E. (1959). *Physical metallurgy*. McGraw-Hill, New York.) (b) The slice of (a) at the lowest temperature (E, eutectic). At O three eutectic solids form. Notice how phases can increase in number as the numbers of components increase.

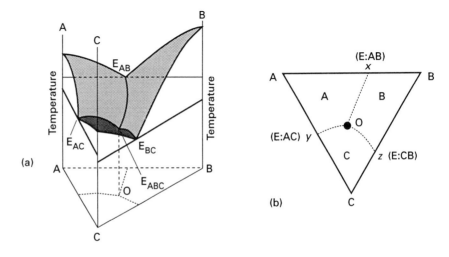

Fig. 4.21(b) and (c). The construction of these diagrams is clearly different from that of the previous cases; the apices of the triangle correspond to 100 per cent of the three components, A, B and C, and any point P within the triangle corresponds to a given fractional composition x_A, x_B, x_C, when $x_A + x_B + x_C = 1$ (see Fig. 4.21(a)). The lines parallel to any given side of the triangle (AB in the figure) represent the various concentrations of the component in the opposite vertex, (C), from 0 per cent (on the triangle side) to 100 per cent (at the vertex). Therefore, the length of the perpendicular to a given side of the triangle divided by the sum of the three lengths represents the percentage composition of the component at the opposite vertex. Triangular diagrams such as those for the important geological systems $MgO-Al_2O_3-SiO_2$ and $SiO_2-NaAlSiO_4-KAlSiO_4$, are given in Figs 4.22 and 4.23 where the different possible phases which crystallise from particular compositions are shown. Modifications with change of temperature can also be plotted. These types of diagram are extremely valuable in the description of conditions in soils.

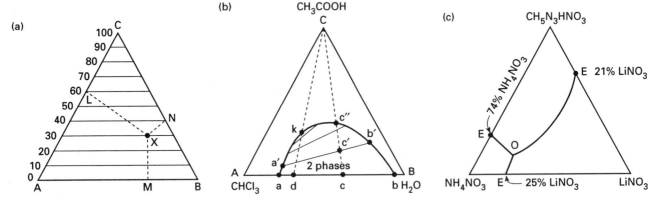

Fig. 4.21 Melt equilibria of some ternary mixtures. (a) The triangular diagram. (b) Two (three-component) partially miscible liquids. Lines a', b' (tie lines) connect concentrations in equilibrium in the two-phase region. (c) The system ammonium nitrate + lithium nitrate + guanidine nitrate forms simple binary and ternary eutectics, (E. Clarke *et al.* (1949). *J. Phys. Colloid Chem.* **53**, 225.) Compositions in weight per cent.

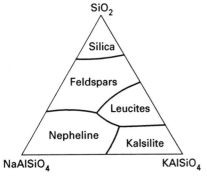

Fig. 4.22 The solid phases that appear from mixtures of SiO_2, $NaAlSiO_4$ and $KAlSiO_4$ in equilibrium at 1000°C. Many such phases are present in clays and soils.

Systems can be analysed that contain liquids as well as solids. An important biological case is the water (liquid), cholesterol (solid) and lecithin (liquid crystal) system of three components. Their limited miscibility is shown in Fig. 4.24.

4.4.4 Systems with more than three components

The only practical way of dealing with systems of more than three components is to fix some of the variables and analyse the separate effect of the others. For example, a system of four components requires three dimensions for $(n-1)$ components (the fourth is fixed once three are defined) at constant temperature and pressure. A regular tetrahedron could be used for the purpose in terms similar to those for triangular diagrams. If a further variable is constrained, for example a constant value for the mole fraction of one component, simpler diagrams can be used, but several of them for different conditions are required to have an overall understanding of the system. It is perhaps necessary to point out that for such complex systems of more than

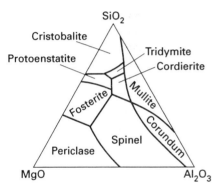

Fig. 4.23 The solid phases that appear from mixtures of SiO_2, MgO and Al_2O_3 and are at equilibrium at 1000°C. Many form in typical rocks (see Figs 4.14 and 4.15).

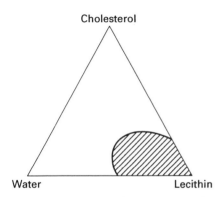

Fig. 4.24 The triangular phase diagram for water, cholesterol and lecithin showing solubility limits.

three components there is at present insufficient data to enable the construction of many useful diagrams. As we increase the number of components we must also be concerned about the number of phases that can coexist (see Section 4.6.1). The problems are not just theoretical; the formation of many objects around us has arisen from multi components.

4.4.5 Liquid crystals

An interesting case of phase formation, intermediate between the solid and liquid states, which we have not described since the phases have a complicated structure, is that of *liquid crystals* (see Fig. 4.25). They tend to occur in molecules that are asymmetrical in shape, for example, with a long flexible chain exhibiting weak interactions in combination with a rigid short unit exhibiting strong interactions. Phospholipids are typical examples, but there are others in biology, for example, some polypeptides or even proteins. In ordinary chemistry and for technical applications, other substances are used, for example *p*-azoxyanisole or ethyl-*p*-anisalamino-cinnamate (Fig. 4.26(e)).

In the crystalline state these molecules are usually lined up. On raising the temperature the energy disrupts the binding between the weakly bonded parts of the molecules but is not sufficient to overcome the order in the strongly bonded parts. We may, then, obtain two types of anisotropic melts, as shown in Fig. 4.26: the so-called *smectic* state, in which the molecules are oriented in well defined planes, and the *nematic* state, in which the planar arrangement is lost but some orientation is maintained. These two states have different viscosities and optical properties, which are put to use in many electronic devices (liquid–crystal indicators). On further heating all the order is lost and an isotropic fluid is obtained as in ordinary solid–liquid phase transitions. These changes of state over ranges of temperature are becoming extremely valuable in mechanics and in signalling. The liquid crystal state is also very important in biological systems, but mainly in membranes.

4.4.6 Summary of phase diagrams

These diagrams show that, when we examine systems of more than one component and increasingly as the number of components increases, the number of independent phases that can be formed and can coexist under equilibrium conditions through small variations in composition becomes very large. Thus if, for example, regions of the Earth are of even slightly different composition we may expect from the phase rule, and we find in practice, that the local minerals are different, even very different, in composition, and are different in phase structure. There are, in fact, thousands if not millions of species of mineral structures. (Note in passing that there are millions of living species too. Is there any connection between the constraints on variability?) The rules governing this variation are those of chemical bonding and packing compatibility. Silicates, for example, illustrate a great variety of packing (see Section 2.11). This finding is not limited to minerals but extends to mixtures of any condensed systems of components, such as alloys, organic molecules such as lipids in biology, salts in water and so on. Thus variability of phase structure opposite variation in composition is immense. (This is *not* a 'full' equilibrium

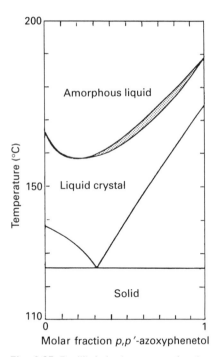

Fig. 4.25 Equilibria in the system of *p,p'*-azoxyphenetole + *p*-methoxycinnamic acid, showing equilibria between liquid crystals and amorphous liquid and the freezing behaviour of the liquid crystals (Prins, (1909). *Z. phys. Chem.* **67**, 718.) Note there is a melting *range* for a liquid crystal (shaded); see Section 4.7.

consideration, however, in that we have allowed the *real* situation of vast numbers of components, (see Section 4.3.2), and many *different local starting conditions of percentage composition.*) Naturally, the use of materials will depend on the phases that they form with particular properties.

We can ask a separate question. If we start from one local or general composition how many phases can coexist? Here the rule governing coexistence is the phase rule in all its simplicity and it clearly limits the number of coexisting phases according to the number of independent components once temperature and pressure are fixed, that is,

$$P + F = C + 2.$$

As we have seen for one component, water, the maximum number of coexisting phases was three. Generally, the number of phases, types of mineral, alloy, organic mixture that can exist independently from mutual solubility, that is, not as pure substances, is, of necessity, $C + 2$. We have seen that, operationally at $T = 25°C$, C is very large. It is, therefore, essential in the treatment of the natural selection of elements to observe that, if global physical equilibrium *with many components* had been maintained (assuming no local spatial division into compartments of different composition), a huge variety of materials could in fact have arisen around us. Variety in what we observe would then be due only to the independence of components. However, as there is also spatial separation locally (not at general equilibrium) a much greater variety still is to be expected from phase variations. If we also remove local phase equilibrium then the variety of substances observable must increase further (see Chapter 6). It is always to the conditions of equilibrium to which we must turn, however, to help us understand the complexity of materials before we examine the situation in which all equilibrium considerations are removed. As the local equilibrium zones become smaller in size we must take into account further new variables due to surface energies and electrostatic fields.

Fig. 4.26 Possible degrees of order in condensed states of long-chain molecules: (a) crystalline—orientation and periodicity; (b) smectic—orientation and arrangement in equispaced planes, but no periodicity within planes; (c) nematic—orientation without periodicity; (d) isotropic fluid—neither orientation nor periodicity. (e) Two compounds that give liquid crystals.

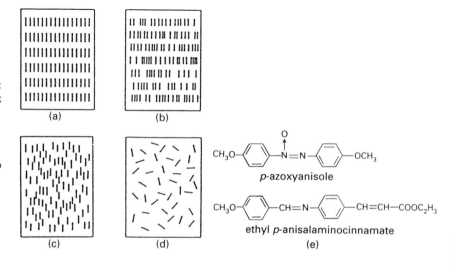

4.5 The effects of gravity

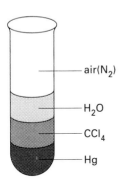

Clearly, the presence of fields requires us to introduce new degrees of freedom referring to distribution in space. We shall consider the effects of such external fields generally in Chapter 6. Here it is appropriate, however, to make some comments on one particular field—that due to gravity. Its general presence affects all systems, including those on Earth. We restrict ourselves to a few observations. It is gravity that promotes separation of phases on the basis of their density, a selective property of elements (Section 2.2), and so gives phases relative position in space, see marginal illustration. Gravitational fields have caused the formation of the planet Earth, and all chemistry of interest in this book is carried on in the general field of the Earth. The gravitational field of Earth also fixes atmospheric pressure. Almost everything that occurs in nature, from the way the phases of the Earth are layered to biological activity and then to factory or laboratory chemistry, is under the influence of Earth's gravitational field. Its neglect in the standard chemistry textbooks shows a lack of interest in our environmental surroundings as opposed to laboratory activity. To see the importance of gravity in chemistry the reader needs only to examine most separation techniques that chemists use, for example fractionation by partitioning or distillation (see also Chapter 6).

4.6 The free-energy conditions and phases at equilibrium

So far in this chapter we have described phase diagrams in which we plot one chemical property, usually composition, against one physical property, most usually temperature. These plots are then a description of regions of stability of phases but do not provide a quantitative account of their energetics. We showed in Chapter 3 that two or more phase boundaries were stable in a condition of equal free energy of two or more physical states across the boundaries, for example, ice $\rightleftarrows$ water vapour, or of two or more phases in the same physical state but of different composition, for example in the Cu/Zn phase diagrams of Fig. 4.15. Now while the ice/water vapour equilibrium was continuously adjusted with temperature below $0°C$, the Cu/Zn diagram broke up in *zones of composition*. The implication is that there are regions of physical and/or chemical composition between the observable alloys with free energies such that some regions are unstable. The appearance of a phase diagram is therefore dependent on the way the free energy on mixing (of different components) changes with temperature and composition. It is then of interest to plot observed free energy changes on mixing against composition at a given temperature to see why systems split up into very different phase patterns. N.B. The free energy of mixing in Section 3.3.1 describes only the entropy and not the heat of mixing.

Before we do so we need to relate the phase diagrams to the general treatment of thermodynamic equilibria given in Chapter 3. As we have seen in Section 3.2, under conditions of constant temperature and pressure, spontaneous change in a system proceeds from a state of higher free energy, G_1 to a state of lower free energy, G_2. Since a decrease in G, denoted by ΔG, corresponds to a change to greater stability, we have related this state function of a system to a *thermodynamic potential*.

Now, a system is defined in this chapter by its components and by the conditions to which it is subjected so that, in addition to the intensive factors pressure and temperature, we have to specify the composition in terms of the molar fractions (or percentage composition) of the components, denoted by x_i. If follows that the function G depends on these mole fractions (or percentages) as well as on pressure and temperature (and on other intensive factors, neglected for the moment, such as amounts and fields—electrical, gravitational, etc.).

An overall differential change in G, represented by dG, will then be due to the separate contributions of the effects of all the variables, that is,

$$dG = \left(\frac{\partial G}{\partial T} \right)_{p,x_i} dT + \left(\frac{\partial G}{\partial p} \right)_{T,x_i} dp + \left(\frac{\partial G}{\partial x_i} \right)_{p,T} dx_i,$$

in which the partial derivatives represent the change in G with the change in each variable when the other two are kept constant. At constant pressure and temperature we will have only

$$dG = \left(\frac{\partial G}{\partial x_i} \right)_{p,T} dx_i.$$

As stated in Chapter 3 this partial derivative, referred to a mole of material, was called the *chemical potential* by Gibbs and given the symbol μ_i. Hence for a particular component, i,

$$\mu_i = \left(\frac{\partial G}{\partial x_i} \right)_{p,T,n},$$

which in the context of this chapter represents the free-energy change of a given phase to which 1 mole of component i is added, when the temperature, the pressure and the number of moles of all other components are kept constant. Note again that the phase itself must *not* change; hence it has to be considered as effectively infinite to remove boundary effects.

Now in a system containing several phases, for example two, α and β, the conditions for thermal and pressure (mechanical) equilibrium obviously are $T_\alpha = T_\beta$ and $p_\alpha = p_\beta$. The condition for chemical equilibrium is easily derived from the expression for ΔG above and since at p and T constant one has $\Delta G = \sum_i \mu_i \, dn_i$, and because the concept of chemical equilibrium implies that there is no net change in the system, that is $\Delta G = 0$, it follows that $\sum_i \mu_i \, dn_i = 0$. Considering the two phases α and β and making a transference of dn_i moles of the component i from phase α to phase β, we have

$$\mu_i^\alpha(-dn_i) + \mu_i^\beta(+dn_i) = 0 \qquad \text{or} \qquad \mu_i^\alpha = \mu_i^\beta.$$

That is, the condition for chemical equilibrium in a system containing several phases is the equality of the chemical potential μ_i in every phase and for all components. This is the basic assumption behind equilibria between phases, the phase rule. We can therefore discuss phase stability in terms of free-energy changes with composition.

4.6.1 Stability of phases: free-energy changes with composition

We shall now look at the phase rule in a somewhat different way asking why combinations of elements can appear with very limited composition ranges, sometimes more or less exactly approaching stoichiometry, as in Fig. 4.10. Consider A and B as interacting chemical entities that could give either a continuous variable condensed phase, liquid or solid, or a stoichiometric compound. Now, we may plot, at fixed temperature and pressure, the change in free-energy content of A + B mixtures (ΔG) as we vary composition (see Fig. 4.27) assumed to apply to gas, liquid or solid phases. (The smaller or more negative ΔG, the higher is the stability.) The simplest case is that represented by curve (a) in Fig. 4.27 in which the stability is independent of composition and there is no phase separation.

In curve (b) of Fig. 4.27 it is seen that for all mixtures xA + yB ($x = 1 - y$) there is a single phase; no particular phase separates from another since drawing a straight line uv between any two possible phase compositions always corresponds to a lower stability for $u + v$ than the value corresponding to point O'—the curve is concave. The entropy of mixing is sufficient to generate this case. The convex curve (c) in Fig. 4.27 shows that all possible intermediate compositions are less stable that the separate pure phases A and B.

Now let us consider curve (d) in Fig. 4.27 which corresponds to the formation of an AB intermolecular species. From similar reasoning it is clear that, whatever the composition may be, only A + AB or B + AB can be stable and AB separates as a pure stoichiometric compound. This is a common case when the affinity between A and B is highly co-operative and/or when in a solid or even liquid phase A and B are very different (Section 2.9). As A and B become more alike then the ΔG/composition curve may become like curve (e) in Fig. 4.27, and it is easy to see that phase A' can coexist with an intermediate phase u and phase B' with another intermediate phase v and that, between u and v, the compositions are continuously variable in a non-stoichiometric phase A_xB_y. This, again, is a very common situation, particularly in solids. The law of multiple proportions is, in fact, only the extreme case of curve (e) in Fig. 4.27 taking the form of the curve (d) in Fig. 4.27 to give AB. When there are more than two elements, non-stoichiometry is the rule rather than the exception.

Now, it is obvious that ideal gases mix in all proportions and do not give rise to phase boundary limitations (see Table 4.3). In other words, the curve for two gases looks like case (b) in Fig. 4.27 due to what is called the free energy of mixing (Section 3.3.1). Two liquid phases, which are obviously at higher temperature than the corresponding solids, very frequently fall into cases (b) and (e) of Fig. 4.27 and, less frequently than for solids, they fall into case (d) of Fig. 4.27. Most liquids show limited miscibility represented by the zone u–v of curve (e).

We can decompose the diagrams shown in Fig. 4.27 into their contributions arising from ΔH and ΔS. This, in effect, is a way of illustrating effects due to bonding interactions, ΔH, (ordering) and probability summations (disorder) in the formation of phases.

In Fig. 4.28 the dashed lines represent the *total* enthalpy *before* mixing and the full lines represent the *total* enthalpy after mixing. It is clear that Fig. 4.28

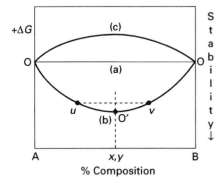

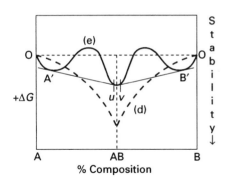

Fig. 4.27 (Above) Free energy/composition diagrams for cases of: (a) to (c) complete immiscibility and (a) to (b) complete miscibility. Below (e) the ΔG/composition curve corresponding to the phase diagram in Fig. 4.12, giving a non-stoichiometric compound in the range uv while (d) represents stoichiometric compound, AB, formation (Fig. 4.10 and see Fig. 4.5).

Table 4.3 Distribution between phases

Gases only present

Complete mixing: one phase

Liquid(s) (gas always present)

Complete mixing: one phase
Some mixing: two phases
Possibility of formation of A_nB_n complex species

Liquid + solid (gas always present)

Complete dissolution: one phase
Partial dissolution: one liquid and remaining solid
No dissolution: two phases
Possibility of A_nB_n complex species: one or two phases

Solid(s) (gas always present)

Complete mixing: one phase
Partial mutual mixing: four phases (A, B, A in B and B in A)
Possibility of formation of A_mB_n compounds: several phases

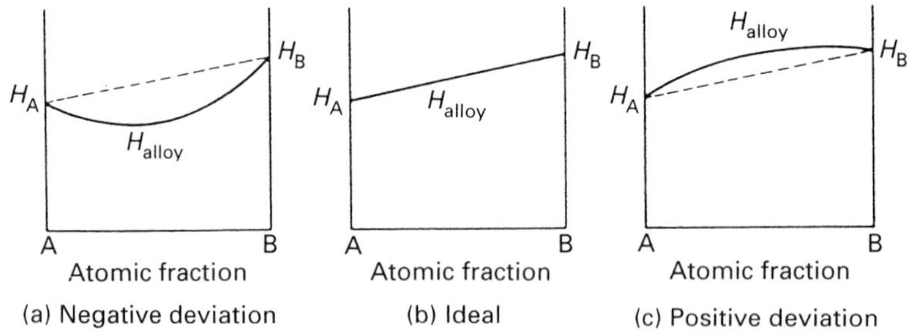

Fig. 4.28 Enthalpy-composition diagrams.

(a) corresponds to an interaction favourable to mixture (ΔH negative) and Fig. 4.28(c) to an interaction unfavourable to the mixture, whereas Fig. 4.28(b) is the ideal case when no enthalpy change occurs on mixing, that is, the bonding strength is equal for A—A, B—B and A—B interactions. In Fig. 4.29 we include the entropy as $T\Delta S$; the entropy change is favourable to the process of mixing shown in cases (a) and (b), which exhibits a maximum in the curve at 50–50 per cent composition, as expected.

Now, finally, Fig. 4.29(c) shows the result of the sum of more complex enthalpy but simple entropy changes which is similar to curve (d) of Fig. 4.27. Cases (a) and (b) correspond to favourable processes, but no phase separation occurs. In case (c) the mixing is determined by the favourable $T\Delta S$ term and the *co-operative* ΔH term. The shape of the curve is such that any unstable solution of composition Z splits up into a *mixture* of two solutions of compositions X and Y. For all intermediate compositions from X to Y such a mixture of the phases is the most stable situation. This is the case for simple (almost) stoichiometric compound formation. Less co-operative changes of ΔH with composition lead to non-stoichiometric phases.

Note, in conclusion, that the influence of the $T\Delta S$ term increases with the increase of temperature which, generally, will favour mixing (and dissolution). The lower the temperature the less will be the contribution of $T\Delta S$ and the process is dominated by the enthalpy change which can be favourable or

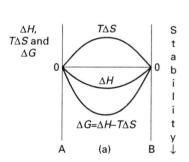

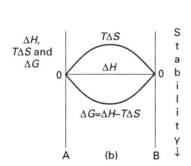

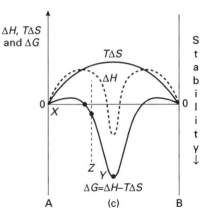

Fig. 4.29 The sum of entropy and enthalpy changes to give the free energy of combination of mixtures of A + B. (a) ΔH and $T\Delta S$ favourable—complete miscibility; (b) ΔH is zero (ideal behaviour)—complete miscibility; (c) ΔH is only favourable when A and B are nearly equal due to co-operativity. Here a compound AB of narrow composition range forms at Y. (NB. $T\Delta S$ is always favourable to mixing but is shown here with the + sign for clarity.)

unfavourable. This discussion allows us to see the formation of phases in terms of the way in which there is change of balance in order⇌disorder equilibria on cooling. Of course just as we drew a tangent to ΔG at two points in the free energy figures for two components we should draw a plane to touch ΔG for three components and so on. As the number of components increases so at equilibrium the number of phases present can increase.

We now see clearly the nature of some features that govern the natural selection of the chemical elements. The variables, up to now, remain temperature and pressure and the fractional composition (the choice of the ratio of elements). The restrictions on the degrees of freedom depend on the number of phases present and upon the affinity of A for A, B for B and A for B at various compositions which have generated independent components. (For example, in curve (d) of Fig. 4.27 there is no way in which the composition of phase AB can be varied.) *Thus, over fixed temperature and pressure ranges, chemical interactions set strict conditions to the composition of the phases present at equilibrium.* It is, therefore, the atomic characteristics (internal fields) of the individual elements that determine their natural selection in compounds (Chapter 2) and then in different physical states and particular phases— different 'compartments'—depending on the intensive variables—pressure, temperature and fractional composition. (Amounts of substances and external fields are still ignored and all restrictions in variance, number of degrees of freedom, were derived for systems of components at equilibrium.)

4.7 The co-operative character of phase transitions

In chemical systems it is not bonds between atoms in molecules alone which dominate but the *degree of co-operativity* between the units in the condensed phases.

The usual changes of state (solid to liquid, liquid to vapour, etc.) are characterised by abrupt changes of volume and also of entropy and enthalpy at the fixed point of transition. They are called *first-order co-operative transitions*. At the transition temperature (at constant pressure) the chemical potential, μ, of the two states is equal, but there is a discontinuity in the shape of the μ versus T curve for the substance (see Fig. 4.30). There is also a discontinuous change in volume since the densities of the two phases considered are not the same. There are, however, transitions for which no such discontinuities are observed, that is, in which there seems to be no discontinuous difference of volume or in entropy or enthalpy. What is observed instead is a sharp (lambda-shaped) change in slope (but no discontinuity) in the various functions such as C_p (heat capacity) versus T at the transition temperature. A change of state of this kind is called a *second-order co-operative transition* (or a λ transition); see Fig. 4.30. Examples of first- and second-order transitions are given in Table 4.4. Where there are second-order phase transitions phase diagrams must be changed so as to show regions of transitions over temperature rather than point transition temperatures (see Fig. 4.19).

Now in both cases a phase transition is a *co-operative phenomenon*. The difference between first- and second-order changes is then in the degree of co-operativity. For a first-order change it is sharp and global, applying to all atoms

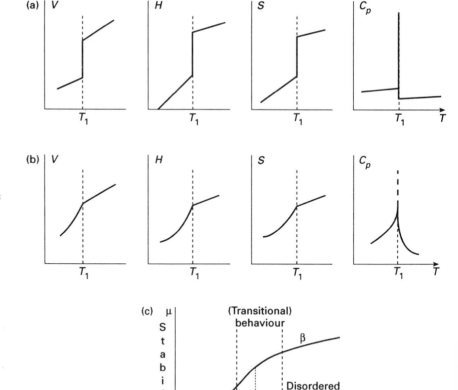

Fig. 4.30 The changes in thermodynamic properties accompanying (a) first-order and (b) second-order phase transitions. (c) The variation of the thermodynamic chemical potential (μ) close to the temperature of a second-order transition.

Table 4.4 First- and second-order phase transitions

	Transition	Comment
First-order phase transitions		
Sn	Grey–white	Change of co-ordination number
S	Rhombic–monoclinic	Change in packing of S_8 molecules
RbCl	NaCl–CsCl	6:6 to 8:8 lattice
$CaCO_3$	Aragonite–calcite	6- to 9-co-ordinate of Ca ions
Fe/Ni alloy	Fcc→bcc	Martensite transformation
Second-order phase transitions		
NH_4Cl	III→II	Rotation of NH_4^+
MnO	Magnetic transformation	Spins aligned below Curie temperature
Ti_2O_3	Semiconductor to metal	Change of axial ratio in crystal
Protein	Conformation change	Main chain re-organisation
Cell membrane + cholesterol	Fluidity change	Lipid packing altered

in the sample. For a second-order change it is local but extensive, statistically distributed, and fluctuating (Fig. 4.30(c)). It is more likely that second-order changes will occur in weakly interacting systems with many energy states (see Fig. 3.15).

Discontinuities over a temperature range have been observed in many solids, notably in alloys, in solid methane, in crystalline ammonium salts and other salts containing symmetrical ions such as NO_3^-, SO_4^{2-} and PO_4^{3-}, as well as in polymeric materials such as rubber. They are very common in liquid crystals and in biological lipids, for example in membranes. In polymers such a transition may be related to the increased disorder of vibration or rotation of side groups or motions of some segments of the molecular chains as the temperature rises to a limiting value. This often corresponds to an appreciable modification in the plastic properties of the polymer. We need to note that for many polymers, for example proteins and polynucleotides, second-order changes are usual and are of great consequence in biology (see Section 6.8.1). In this ability of continuous adjustment could lie one of the major reasons for the flexibility and sensitivity of biological objects.

4.8 Supercooled phases—glasses

The three states of matter that we have described in Section 4.2 (see Table 4.5 for a summary) are limiting situations; in effect, for some substances there are intermediate stages that have properties of both adjacent states, for example liquid crystals. Occasionally intermediate states appear, out of equilibrium but long-lived (see Fig. 4.1). *Glasses* are one of these substances—they are a compromise between crystalline and liquid states and many authors consider them to be supercooled liquids. An example is vitreous silica, which maintains in part the structure of quartz but whose bonds are of variable length; consequently, the order is altered and the material is mechanically weaker (see Fig. 4.31). On heating, glass softens gradually rather than melting sharply since co-operativity is locally diverse. Glass is not the most stable state of silica,

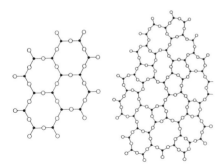

Fig. 4.31 Schematic representation in two dimensions of the difference between the structure of a crystal (left) and a glass (right). Solid circles represent silicon atoms; open circles are oxygen atoms. (After Zachariasen, W. H. (1932). *J. Am. Chem. Soc.* **54**, 3841.)

Table 4.5 Properties of bulk phases at 300 K

Gases

Disordered
Readily deformed
Very low density
Rearrangement of components easy
Allow transport and flow

Liquids

Limited local order
Relatively high density
Self-restricted volume
Rearrangement of components easy
Allow transport and flow
Communication can be generated

Solids

Definite ordered structure
Functional in construction
High density
Retain liquids and gases in confined zones
Rearrangement of components very difficult
Allow transport of electrons only (in metals)
Organisation can be generated (with liquids)

of course, but we shall meet a similar problem in the case of proteins, which have many of the irregular characteristics of glasses (Chapter 6).

4.9 Summary of equilibrium conditions of phases

There is no doubt that the understanding of the chemical objects around us is greatly aided by an understanding of bulk phase systems in equilibrium. However, such systems are restricted, that is, limited in variance or degrees of freedom, by the fixing of several factors.

1. The two more general variables—temperature and pressure.

2. The fractional composition, that is, the number of chemical components in any one system, which is connected to the chemical forces that hold elements together. As we have seen, full chemical equilibrium reduces the variance of the composition to the number of elements present, and fixed element stoichiometry or fixed relative proportions then reduces chemical composition, number of components, to one. The absence of full chemical equilibrium in practice then makes it essential to define components under *operational conditions* of temperature and pressure. (The variance at high temperature is much less than that at low temperature due to chemical equilibration.) We shall have to examine what feature of chemicals prevents exchange of atoms between them (Chapter 7).

3. The number of phases that exist through physical and chemical forces. *Each* additional phase *at full transfer* equilibrium introduces a new restriction reducing the number of degrees of freedom by one. The variance at high

temperature is greater than at low temperature, since the number of phases diminishes toward one—a homogeneous gas.

When we put factors 1, 2 and 3 together under operational conditions, and apply the phase rule, we can understand many systems we observe, at least in outline. However, the real situation is more complex for the three additional reasons.

4. The total amount of materials is a variable, a fact that we have not examined and that we shall consider in Chapter 6. Amount introduces consideration of size, shape and surfaces.

5. We have to take into account the relative position in space of bodies of substances of defined size since they exert bulk fields on one another, and the space co-ordinates of mass and charge introduce new variables. This will also be considered in Chapter 6.

6. We must consider that in some cases transfer between phases is frozen. Here equilibrium conditions are lost entirely.

There is a peculiarity about the phase diagrams in this chapter which we must not miss. There are parallel diagrams for two temperature ranges, 0–100°C (273–373 K) and $\geqslant 500$°C (770 K). In the high-temperature range we gave examples of two kinds: (1) of minerals, the liquidus–solidus curves that are of great interest for the study of the origin of the Earth; (2) of man-made products, metals (alloys) and ceramics of very recent origin. It is the subsequent cooling of materials made and equilibrated at these high temperatures ($\geqslant 500$°C) as shown in phase diagrams that has given us most of the (frozen) *ordered* (structured) objects around us. In the low-temperature range the main concern until man's intervention lay in the mutual solubility or miscibility of molecular compounds and salts with water. Moreover, this low-temperature range is the extant ambient condition of Earth so that here in phase diagrams we are often discussing disordered liquids as well as gases *equilibrated* in pairs or with ordered solids. The objects we see around us based on these materials are not strictly ordered but are often organised, especially living systems. Subsequently, man has also come to use these ambient conditions by developing organic solvents from kinetically stable chemicals. Curiously, for stability reasons much of this use of molecular substances is restricted to the same temperature range as that of aqueous solutions. At room temperature the only material that flows easily within solids is the electron, while almost any material flows in any liquid (for example, H_2O).

The universe evolved meeting these constraints at different stages, but later in time the condition of equilibrium became the exception rather than the rule due to the rapid cooling of the original atoms and molecules which has 'frozen' some chemicals locally in kinetic traps. Now, if the products of reactions are not in equilibrium with the elements in them, the variability of composition is greatly increased, almost *ad infinitum*, and the same happens to the variance. We need, then, to understand non-equilibrium systems, which are a source of chemical variety and consequently of increased possibilities. In addition, we shall have to discuss lack of equilibrium between phases, that is, restrictions on transfer, and its further consequences. In fact, the whole Earth failed to reach equilibrium in both respects (fortunately for us), and all biological systems are far from equilibrium. We need to understand how such systems far from equilibrium can spontaneously adopt energised organised structures of very closely fixed composition, for example living structures. They look like other material systems open to our observation and are certainly based on the same chemical elements, but seem to be in apparent contradiction with the principles of thermodynamics, which demand an irreversible progression toward optimally disorganised systems not then able to do 'useful' work (see Section 7.22). Yet they have a further peculiarity. They develop in species with reproducible characteristics much like any condensed phase. The secret of the stability of a condensed phase lies in the way internal co-operativity varies with composition as we see in Fig. 4.27. Is there some kind of internal co-operativity within an individual of a species? We tackle this problem in Chapter 16.

Before dealing with the problems of limited volumes of phases, fields and of non-equilibria (Chapter 6) we need to consider chemical equilibrium in a single aqueous phase. This is a particularly important subject—dilute aqueous solutions affect much of geological and all of biological systems which will be discussed in Chapters 8–13.

Further reading

The literature on phases and phase equilibria is not so accessible to chemists and biologists. Most of the more recent works are for chemical and mechanical engineers, physicists and metallurgists. Physical chemistry books do not always treat the subject in depth. The reader would however, be advised to look into three excellent classics (references 1, 2, 3) and a more recent text (reference 4).

1. Moore, W. J. (1972). *Physical chemistry* (5th edn). Longmans, London. Chapters 1–3 and 5–7. Apart from its soundness, this text is beautifully written with a literary style seldom found in scientific literature.
2. Castellan, G. W. (1964). *Physical chemistry*. Addison-Wesley, Palo Alto. Chapters 9, 11, 12, 14 and 15. Sound and more detailed than the previous reference in this particular subject.
3. Hill, T. L. (1968). *Thermodynamics for chemists and biologists*. Addison-Wesley, Reading, Massachusetts. A classic textbook on chemical thermodynamics.
4. Atkins, P. W. (1994). *Physical chemistry* (5th edn). Oxford University Press, Oxford. Chapters 1 and 6–9. An excellent modern text.

The following is a recent and comprehensive work on phase equilibria.

5. Walis, S. M. (1989). *Phase equilibria in chemical engineering*. Butterworths, London. Particularly Chapter 5, with many examples of phase diagrams.

Two good classics dealing only or mostly with phase equilibrium diagrams are

6. Ferguson, F. B. and Jones, T. K. (1966). *The phase rule*. Butterworths, London.
7. Reisman, A. (1970). *Phase equilibria*. Academic Press, New York.

Extensive compilations of phase diagrams are given in the following two works.

8. Landolt.-Bornstein: *Zahlenwerte und Funktionen*, II/2a (1960), II/2b (1962), II/2c (1964), Neue Serie IV/3 (1975). Springer-Verlag, New York.
9. Tamas, F. and Pal, I. (1970). *Phase equilibrium spatial diagrams*. Butterworths, London.

Phase diagrams for the formation of Earth's minerals can be found in

10. Brown, G. C., Hawkesworth, C. J. and Wilson, R. C. L. (eds.) (1992). *Understanding the earth—a new synthesis*, Cambridge University Press, Cambridge.

Applications to biological systems are given in

11. Williams, R. J. P. (1975). Phase and phase structure in biological systems. *Biochemica et Biophysica Acta* **416**, 237–86.

5

Equilibria in dilute solutions in water

*Water is the element of selfless contrast—it passively exists for others
. . . water's existence is, therefore, an existing-for-others . . . Its fate is to
be something not yet specialized . . . and thus it soon came to be called
'the mother of all that is special'*

Hegel, *Philosophy of nature* (1817)

5.1 Introduction

In Chapter 3 we have shown that, on bringing together volumes of two different ideal gases, A and B, they form a homogeneous mixture, which, at equilibrium, is totally randomised, that is, disordered. We also referred to the possibilities of weak interactions between A and B, which give deviations from perfect or ideal gas behaviour, and stronger ones, which may lead to the formation of gaseous compounds, A_nB_m, at equilibrium with $A+B$. An example is H_2O in an equilibrium mixture with H_2 and O_2 when the temperature is kept $>3000°C$. The mixture of the three gases is still homogeneous and disordered, but some order, structure, in H_2O has appeared. The reasons for such ordering were described in Chapter 2.

In the case of two bulk liquids similar considerations apply, but in these cases miscibility is more limited and phase separation frequently occurs (see

Chapter 4). In liquids and particularly in liquid/solid and solid/solid mixtures, we often observe a higher degree of order in the phase distribution of the two components A and B, up to the limit of pure *co-operative* compound formation, which is absent in the mixtures of gases. In all these cases the mixing of *similar mole fractions* of components only was considered, starting from two, A and B, and then going forward to larger numbers.

In this chapter we shall describe the particular case of dilute *solutions*, defined as liquid homogeneous systems containing no fewer than two components—A, B . . . S—one of which is overwhelming dominant, say S, which from now on we will call the bulk solvent. For most of our considerations this solvent will be *water*, given its dominant role on the Earth and in life, and hence its involvement in the selective development of element combinations, to be described in Chapters 8–13. Life chemistry, particularly, is very much a chemistry of dilute water solutions and most authors agree that life originated in 'some warm little pond' and developed in the sea for the first 3×10^9 years. (It is curious to notice that in Sumerian *mar* means 'womb' as well as 'sea' and in Japanese mythology the word *umi* (ocean) is homophonous with the word meaning 'give birth'. Almost every old civilisation assumed that life began in the sea and the word for sea is feminine in many languages.) We wish to understand the ways in which order and disorder (solution) come into balance in this solvent in particular since such knowledge must assist us in understanding the factors that helped to produce the objects we see around us, including those which are living. As in Chapter 4 we look especially at the restrictions on variability for specific elements due to chemical combination now involving equilibrium complex formation and new phase formation arising from insolubility.

5.2 Factors that affect solubility generally

In essence, a liquid is a rapidly fluctuating condensate of atoms or molecules that, at any instance in time, can be likened to a condensed random gas. When we add an additional component, a solute, A, to a solvent, S, there are several approaches open to us. The simplest—the ideal solution approach—is to consider each solute molecule as an independent randomly disorganised unit in a bulk medium of given viscosity (and dielectric constant). In fact, this treatment of the solute parallels exactly that of a molecule in the gaseous state so that, for ideal dilute solutions of one component A in a bulk solvent S, the molar concentration c_A† replaces the partial pressure p_A. The next step, as in the treatment of gases, is to consider deviations from ideality, due, as before, to the effect of intermolecular forces which, of course, increase with the concentration of the solute. This leads, progressively, toward a system of

† The composition of a solution is sometimes expressed by the ratio of the number of moles of each solute A, B, C, etc. to the total moles of A plus B plus C, etc., the bulk solvent being ignored. The sum of the mole fractions, say $x_A + x_B + x_C$, is, of course, equal to unity. In dilute water solutions, however, it is more convenient to express the composition in terms of the number of moles of each solute per litre of solution (molarity) or per kilogram of solvent (molality). Chemists generally use the molarity scale (M) rather than the molality scale (m) although this latter has certain advantages; for example, the molality of a given solution is independent of the temperature whereas the molarity is not (the volume varies with the temperature).

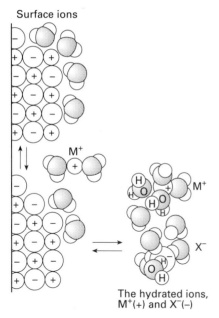

Surface ions

The hydrated ions,
$M^+(+)$ and $X^-(-)$

Fig. 5.1 Equilibrium between a precipitate MX and ions $M^+(+)$ and $X^-(-)$ in water illustrating solvation.

Molecule	Character
N_2, O_2	Non-polar
HCHO	Polar
CO_2	Quadripolar
H_2O, NH_3	H-bonding, polar
HCl	Dissociates in water (polar)
$(HCOH)_6$	H-bonding, polar
benzene ring —	Non-polar
—	Polar
CO_2H —	H-bonding

Polyvinyl alcohol

phases due to the relative importance of A–A, A–S and S–S interactions, but we will then be dealing more with questions of *miscibility*, discussed in the previous chapter, rather than of *solubility*, even if the distinction between the two, as well as the definition and classification of solute and solvent, are arbitrary. (A discussion of solubility follows very closely that of vapour pressure of a component over its solid phase.)

Usually, the interaction of A with the solvents is non-trivial and corrections have to be introduced as is the case with real gases. The interaction may even be considerable, when there will be some ordering of the solvent molecules around the solute molecules which are then said to be solvated (hydrated, in the case of solvent water; see Fig. 5.1). As stated before, in Chapter 2, the interactions between molecules are determined by intermolecular forces (van der Waals forces and hydrogen bonding, see Section 2.6) assuming that there is no chemical reaction between A and S. When only these forces are operative, the attraction between A and S molecules depends mainly on London forces or hydrogen-bond formation (Section 2.6.1). In strongly polar solvents, for example water, Keesom forces may be significant for polar solutes and ionic compounds, but Debye forces are usually not too relevant. Solubility, therefore, tends to increase with the polarity and polarizability of the solutes and solvents (see marginal illustration), although, of course, these factors also contribute to the cohesion energy of the solute and solvent units themselves. One illustrative example is that of the solubility of noble gases in various solvents (Table 5.1), and another is the solubility of various common gases in

Table 5.1 Solubility of the noble gases in various solvents

Noble gas	Polarisability ($\times 10^{24}$ cm^3)	Solubility (mole fraction $\times 10^4$) in solvent			
		Water (at 0°C)	Ethanol (at 0°C)	Acetone (at 0°C)	Benzene at (7°C)
Helium	0.202	0.177	0.599	0.684	0.507
Neon	0.392	0.174	0.857	1.15	0.809
Argon	1.629	0.414	6.54	8.09	7.84
Krypton	2.460	0.888	—	—	—
Xenon	4.000	1.94	—	—	—
Radon	5.415	4.14	211.2	254.9	638.9
		Water (at 0°C)	Ethanol (at 0°C)	Acetone (at 0°C)	Benzene (at 7°C)
Dipole moment (debyes)		1.84	1.70	2.85	0
Polarisability ($\times 10^{24}$ cm^3)		1.48	5.29	6.59	10.87

water (Table 5.2). The importance of hydrogen bonding can be seen in the general solubility of, say, short-chain amines, alcohols or sugars in water (Table 5.3) where it is seen that the more hydrogen bonds possible between the molecules of the solute and the solvent the more soluble the solute is likely to be. In these conditions even a large polymer such as polyvinyl alcohol (see marginal illustration) is fairly soluble in water and so are many proteins, polypeptides, and RNA and DNA, polynucleotides (see Chapter 9).

Qualitatively, one usually says that 'like dissolves like', that is, a given

Table 5.2 Solubility of some gases in water at $T = 0°C$

Gas	Solubility in water (mol l^{-1})
Oxygen	2×10^{-3}
Ozone	2×10^{-2}
Nitrogen	1×10^{-3}
Nitrogen monoxide	3×10^{-3}
Carbon monoxide	1.6×10^{-3}
Carbon dioxide*	7.5×10^{-2}
Methane	1.5×10^{-3}
Sulphur dioxide*	1.0×10^{-2}
Hydrogen sulphide*	2.0×10^{-1}
Ammonia*	>50
Hydrogen chloride*	>20

* These gases react with water.

Table 5.3 Effect of hydrogen bonding groups on solubility of organic compounds in water (25°C)

Compound	Formula	Solubility (mol l^{-1})
Succinic acid	$HOOC—(CH_2)_2—COOH$	0.6
Malic acid	$HOOC—CHOH—CH_2—COOH$	13.8
Adipic acid	$HOOC—(CH_2)_4—COOH$	0.1
Citric acid	$HOOC—CHOH—(CH_2)_3—COOH$	10.8
Toluene	$C_6H_5—CH_3$	0.005
Benzoic acid	$C_6H_5—COOH$	0.015
Phenol	$C_6H_5—OH$	0.9
Aniline	$C_6H_5—NH_2$	0.4
D-Glucose	$C_6H_{12}O_6$	4.6
Sucrose	$C_{12}H_{22}O_{11}$	5.2

solvent tends to dissolve substances that are similar to it: non-polar solvents tend to dissolve non-polar substances and polar solvents tend to dissolve polar substances. Extremely interesting cases arise when a solute A contains segments that are polar and segments that are not. Consider, for example, the molecule of benzoic acid $C_6H_5CO_2H$: one part tends to dissolve in organic solvents; the other in water. This dualistic character in larger molecules is the basis of the folding of many biological polymer molecules in water and of membrane formation from fatty acids. Obviously, ionic compounds, containing charged species, are not soluble in non-polar solvents as a rule but tend to dissolve in polar solvents, particularly in water, due, in part, to the favourable contribution of the energy of 'hydration' of the ions, and to the drive towards increased randomization, disorder, of the particles in the solution. Table 5.4 shows however, that the solubility of a salt, M_nX_m, decreases rapidly as the ions

Table 5.4 Solubilities in water at 25°C

Anion*	Solubility in water at 25°C (mol l^{-1})													
	Li$^+$	Na$^+$	K$^+$	Rb$^+$	Cs$^+$	Be^{2+}	Mg^{2+}	Ca^{2+}	Sr^{2+}	Ba^{2+}	Al^{3+}	Sc^{3+}	Y^{3+}	La^{3+}
OH$^-$	5.6	27.0	20.0	19	25	1.3×10^{-5}	3.2×10^{-4}	1.8×10^{-2}	8.6×10^{-2}	0.27	$<10^{-4}$	$<10^{-4}$	$<10^{-4}$	$<10^{-4}$
F$^-$	0.05	1.0	17	12	24	5.5	1.9×10^{-3}	3.1×10^{-4}	9.3×10^{-4}	1.2×10^{-2}	$<10^{-3}$	$<10^{-3}$	$<10^{-3}$	$<10^{-3}$
CO$_3^{2-}$	0.2	2.0	8.1	9.6	8.0		1.3×10^{-4}	6.2×10^{-5}	3.9×10^{-5}	4.4×10^{-5}	†	†	†	†
Oxalate	0.7	0.3	2.0	1.5	1.6	3.4	2.7×10^{-3}	6.7×10^{-6}	2.4×10^{-4}	5.0×10^{-4}	$<10^{-2}$	$<10^{-2}$	$<10^{-2}$	$<10^{-2}$
Formate	8	15	40	‡	‡		0.71	1.3	0.53	1.4	§	§	§	§
Acetate	5	6	29	‡	‡		4.5	2.2	2.0	2.9	§	§	§	§
Cl$^-$	19.0	6.0	4.6	7.5	11.0	Large	5.7	7.3	3.5	1.8	2.0		3.0	
Br$^-$	18.8	9.0	5.5	6.5	6.0	Large	5.5	7.0	4.0	3.5	>5.0		2.5	
I$^-$	12.4	12.3	9.0	7.0	3.0	Large	5.0	7.0	5.3	5.3	>5.0			
SO$_4^{2-}$	3.2	4.3	0.6	1.8	5.0	4.0	3.0	7.8×10^{-3}	6.2×10^{-4}	9.5×10^{-6}	1.0	0.6	0.2	0.04
NO$_3^-$	10.5	11.0	3.0	3.6	1.0	8.2	5.1	7.4	3.3	0.35	2.0		5.0	0.2
ClO$_4^-$	5.7	16.0	0.1	0.07	0.09	7.1	4.5	7.9	10.8	5.9	2.0			
ClO$_3^-$	48.0	10.0	0.6	0.3	0.3		7.7	9.7	6.9	1.1				

* NB. With these anions transition metal ions in salts follow a similar pattern of solubility to that of M^{2+} and M^{3+} salts in the table.
† Not stable.
‡ Very high solubility.
§ Fairly soluble.

become more highly charged, for example, amongst M^{n+} hydroxides and carbonates (see Section 5.5).

Non-polar solutes are not soluble in polar solvents, for example oils or large fatty acid soaps (lipids) are not soluble in water, so that immiscible layers or vesicles are formed even from dilute components creating new phases in which any added ions or polar molecules are differentially concentrated in the aqueous phases and non-polar molecules are concentrated in the oily phases. This is again a form of chemical selection based on atomic properties, that is, a natural selection of the chemical elements. All of these considerations are more or less straightforward, but should not be overgeneralised; for example, O_2 and CO_2 are non-polar molecules, but they are soluble to some extent in water (a fundamental fact for the existence of life); non-polar dioxane is completely miscible with water and urea is also soluble in the same solvent (due to extensive hydrogen bonding). Similarly, LiI, an ionic compound, dissolves in ether, which is only slightly polar, and anhydrous magnesium and sodium perchlorates are very soluble in acetone. In all these examples and, generally, in all matters concerning solubility, we may use the expression introduced in Chapter 3, that is,

$$\Delta G = \Delta H - T\Delta S \qquad \text{(for an isothermal process at constant pressure)}$$

where $\Delta G = RT \ln K = RT \ln(\text{solubility})$, where $K = [A]_{\text{solution}}/[A]_{\text{solid}}$ and $[A]_{\text{solid}} = \text{constant}$. K is an equilibrium constant related now to the solubility of a substance.

As we have seen, a negative value for ΔG corresponds to a thermodynamically feasible spontaneous process, here the tendency to form a solution rather than keeping the solute and solvent in separate phases. ΔS is of course linked to disorder (the solution) while ΔH is more strongly linked to order (the precipitate). We can set up standard state thermodynamic functions, for example $\Delta G^{\circ}_{\text{aq}}$, as we did in Table 3.5 for all ions and molecules (see note at the end of the next page).

5.3 Relative solubility in two liquid phases: partition coefficients

We may now consider the case of solutes distributing (partitioning) between two different and immiscible liquids, say, S_1 and S_2. The way in which an ideal dilute solute, A, partitions at equilibrium between two liquid phases S_1 and S_2 is given by

$$\text{Partition coefficient} = c_1/c_2,$$

and, to a first approximation, this ratio of concentrations is related to the ratio of solubilities of A in the two solvents.

The partition coefficient is an extremely important concept in the understanding of many systems of solute equilibria between a multiplicity of phases. It can be applied to trace element equilibrium uptake into immiscible pairs of mineral melts (see Section 8.6.5) and to the equilibration of any molecule between biological liquids, for example, membranes and water. Clearly, chemical selection based on specific characteristics of solutes and solvents is involved according to the ratios of solubilities as in Tables 5.1 and 5.5. Water in the presence of an organic solvent or lipid phase selects a special

Table 5.5 Solubility of classes of compounds in water and in organic solvents

Considerably soluble in

Water	Organic solvents
Salts, organic and inorganic	Alkanes and alkyl halides
Complex ions (salts)	Ethers
Short-chain acids, amines, alcohols, thiols, aldehydes and ketones	Long-chain acids, amines, alcohols, thiols, aldehydes and ketones
Amino acids (and many proteins)	Alicyclic compounds
Nitrogen bases, e.g. pyridine, nicotine, adenine, thymine, etc.	Aromatic compounds
Sugars (polysaccharides)	Lipids
DNA^{n-} and RNA^{n-}	Organometallic compounds

set of organic molecules that are polar or charged or that form strong hydrogen bonds, largely due to the need to overcome the relatively high affinity of H_2O for itself through H bonds. This sets biological organic chemistry (in water) apart from general organic chemistry (in non-aqueous solvents) though in biological systems we must always remember the lipid (organic solvent) membranes.

5.4 Dilute solutions containing more than one solute

The most interesting situation, for our purposes, arises when we introduce two or more solutes A, B, etc. into one solvent. Ignoring solvation effects, we are interested in any selective AB, etc. associations, orderings, which we call 'complexes' generally and in which A and B are stoichiometrically combined. In such cases we describe AB as partly dissociated at equilibrium, for example,

$$AB \rightleftharpoons A + B.$$

Obviously, we can extend the idea of association to bigger and bigger units, clusters, A_nB_m, up to the formation of a solid phase (precipitate), which can still be in equilibrium with its dissociated components, that is,

$$A_nB_m \text{ (solid)} \rightleftharpoons nA + mB \text{ (solution)}.$$

A point of great concern in this book is the selective degree to which different elements (as ions) or elements in compounds give these associated units. (The size and shape of such units will also be relevant in most cases.) Such equilibria dominate species in the sea, of course. In particular, we may well ask: what happens if we mix in water A, B, C, etc. up to, say, the 15–25 elements that exist in biological systems, allowing some of them—C, H, N, O, S and P—to combine in organic molecules such as amino acids, fatty acids, proteins, nucleotides, etc.? We would then be examining the complexes and precipitates that form at equilibrium amongst organic molecules themselves or with the elements introduced as inorganic ions such as Ca^{2+}, Mg^{2+}, Fe^{2+}, Cu^{2+}, Zn^{2+}, etc. together with negatively charged inorganic anions, for example CO_3^{2-},

NO_3^-, SO_4^{2-}, etc. These associations, precipitates and complexes, must be part of the basis of self-assembly in solution, which is so critical to the construction of living cellular systems. The analysis will be seen later to be much more complicated than the comparative treatment of the precipitation or complex formation of a single species A_nB_m (see Sections 5.5 and 5.6).

The formation of a precipitate is the creation of a new phase and, as such, it is a restriction on variance (degrees of freedom) which is selective for both inorganic and organic components. As in Chapter 4 we shall not concern ourselves with the volume (amount) of any such solutions. This means that we can ignore the nature of the containing vessel and the surface of the liquid. We are also not concerned with the effect of inhomogeneity in the liquid due to fields, for example (see Chapter 6). Our concern in this chapter is with the equilibria that relate to *chemical* interactions between components, but note that equilibria giving new associations do not increase variance; rather they decrease it when new phases are formed.

5.5 Equilibria between acids and bases in aqueous solutions

After this general digression on the factors that affect solubility in general and particularly in water, we can proceed to examine in turn the main types of equilibria that occur in aqueous solutions: formation of precipitates; formation of complexes, local order (both often included under acid–base behaviour); and oxidation–reduction reactions. The description will be kept very simple and very general; the interested reader is advised to look into the references given in the 'Further reading'. We turn first to acid–base equilibria. (NB. The general considerations can be applied to equilibria in any solvent.)

As stressed above a major concern in this book must be with the association of substances in aqueous solution. At its simplest this is a consideration of two steps

$$A+B \quad \rightleftharpoons \quad AB \quad \rightleftharpoons \quad [AB]_n$$
$$\text{(acid)} + \text{(base)} \qquad \text{complex} \qquad \text{precipitate (separate phase)}$$

where A and B may be inorganic or organic molecular species, neutral or charged (ions), small or large, from simple elements up to the size of proteins and nucleic acids.

In order to appreciate the nature of specificity in this case we shall analyse the coming together of ions and the solubility of inorganic salts (or the ease of formation of precipitates) as particular examples. We may do this by comparing heats of formation of AB (Section 3.4) with heats of hydration of A and B in the following way.

In Fig. 5.2 we plot the heat of formation of some aqueous ions, largely singly charged A_{aq}^+ and B_{aq}^-, of some ion pairs, A^+B^-, in the gas phase, and the lattice energy for the formation of the crystalline solid, $[AB]_n$, against the size of simple inorganic ions. All the heats are reduced in the figure by z^2, where z is the ionic charge. The heats of formation of the lattices are seen to be some 1.5 times greater than the heats of formation of the ion pairs due to co-operativity effects and *but for this co-operativity very few ions would form a precipitate.*

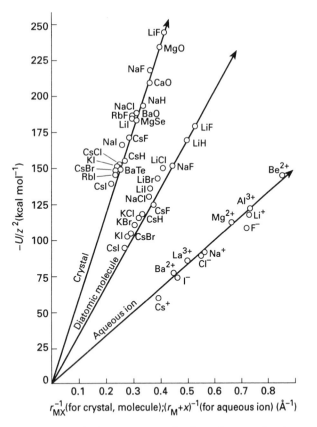

Fig. 5.2 Ionic binding energy (crystal and diatomic molecule), U, as a function of interatomic distance, r_{MX}, and ionic hydration energy as a function of effective ionic radius $(r_M + x)$. Abscissa units are Å^{-1}. The lines are theoretical and the points experimental. (See reference 1 in 'Further reading'.)

The heat of solution in standard conditions of a given salt, ΔH°_{sol}, is related to the algebraic sum of the heats of hydration of the corresponding ions, ΔH°_{hyd}, and to the lattice energy of the salt, U°, as is seen by considering the so-called Born–Haber cycle,

$$
\begin{array}{c}
\xrightarrow{\;+\,\Delta H^o_{sol}\;} \\
MX_{sol} \qquad\qquad M^+_{aq} + X^-_{aq} \\
\end{array}
$$

$$
\nearrow\; -U^o \qquad\qquad \swarrow\; +\Delta H^o_{hyd}(M^+) + \Delta H^o_{hyd}(X^-)
$$

$$
M^+_g + X^-_g
$$

Since this is a cycle

$$\Delta H^\circ_{sol} + (\Delta H^\circ_{hyd}(M^+) + \Delta H^\circ_{hyd}(X^-)) - U^\circ = 0.$$

(A parallel cycle can be used to describe the association of ions *in solution*.) Hence, the heat of the solution (or, more correctly, the enthalpy of solution) is the balance between the lattice energy and the sum of the hydration heats (enthalpies) of the ions. If it is negative, the dissolution process is favourable; if it is positive, the dissolution process is unfavourable (in heat terms). For example, for LiF the lattice energy is 243 kcal mol^{-1} and the sum of the hydration enthalpies of Li$^+$ and F$^-$ is also 243 kcal mol^{-1}, hence $\Delta H_{sol} \approx 0$ kcal mol^{-1}. The degree of solution of this salt would then be slight but for the entropy change of the process which is positive since the disordered dissolved state is more probable than the ordered crystalline state. In effect, the solubility is related to the very close balance between the free energy of formation of the solid and the free energy of hydration of the ions, both from the component ions in the gas state, that is, both enthalpy and entropy contributions are involved, but, whereas the enthalpy of solution may be favourable or unfavourable, the entropy of solution is generally favourable. (In a more complete treatment we would have to take into account the ordering of the hydration sheaths of the ions which decreases solution entropy to a greater or lesser extent depending on the charge and size of the ions.) Thus we see that, unless the lattice energy is considerably in excess of the sum of the hydration energies of the ions, ionic salts are rather soluble. This is the case for many simple salts and it is this fact that allows extensive chemistry of salts in water.

On the other hand, there is a very considerable number of insoluble ionic compounds, for example MgF_2, MgO, $BaSO_4$ and $Al(OH)_3$. Table 5.4 shows that there are commonly observed patterns of increasing insolubility.

1. Smaller anions form more insoluble salts with smaller cations. For example, the insolubility of $LiF >$ that of $LiCl >$ that of LiI.

2. Larger cations form more insoluble salts with large anions. For example, the insolubility of $BaI_2 >$ that of MgI_2.

3. More highly charged cations form more insoluble salts with small more highly charged anions. For example, the insolubility of $Al_2O_3 >$ that of $MgO >$ that of Na_2O, and that of $AlCl_3 >$ that of Al_2O_3.

A major factor dominating these orders is the nature of the packing of the ions together in lattices. Although we use a generally valid ionic radius for a given ion, in fact, its radius varies slightly according to the lattice packing (see also Section 2.9) and these small 'size' changes greatly influence lattice energy and hence solubility. Further discussion on this subject is given in References 1, 3 and 5 in 'Further reading', including the consideration of additional factors in lattice stabilisation such as covalency (see Section 2.9.4–2.9.5).

Hydrated ions of opposed charge also tend to associate in solution, due to the electrostatic attraction, to form so-called ion pairs rather than complexes (a term more commonly used for species in which there is some direct covalent bonding between A and B). A packing problem is again involved due to the hydration of the anions and cations. The affinities are once more related to the charge and size of the ions and experiment shows that, in the case of, for example, Na^+ and Cl^-, ion-pair formation occurs only to a neglectable extent while Al^{3+} and F^- ion-pairing is strong, so strong in fact that complex species up to AlF_6^{3-} can be formed in strong F^- solutions (see below). Ion pairing can be thought of as being in competition with precipitation in solutions of many anions and cations. The orders of complex ion formation follow closely those of precipitation.

Electrostatic interactions, therefore, allow considerable selection based on charge and size of ions and both Nature and man have taken advantage of this fact by preferential concentration and then precipitation, or by developing special precipitating and complexing agents, for example oxalates and ion-carriers, and other devices, for example ion channels, see marginal illustration, that allow separation of particular ions in the presence of others, such as K^+ and Mg^{2+}. The same description applies in a general way to the association of organic molecules or inorganic ions with organic molecules through electrostatics. In all systems further selectivity can arise through

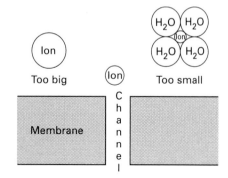

other forms of attraction such as covalence, van der Waals forces and H-bond formation, especially in water, when shapes, number of bonds and size matching are important factors (see Chapter 2). Repulsion, again due to shape and size matching, is also important both in selectivity of ion-pair formation and in precipitation; it dominates much of the selectivity of association of inorganic and organic molecules, for example between and on protein surfaces, and is always taken into account by treating *equilibrium* distances.

Following the discussion in Chapter 3, we see that in the description of the balance, order⇌disorder, in aqueous dilute solutions the disorder is at its greatest in the separated ions while the order is at its greatest in complexes and dominant in crystalline precipitates, where the interaction energies arise from all the above terms.

After this general digression on the factors that affect order/disorder equilibria in aqueous solutions we can proceed to examine more specifically in turn the main types of reactions that occur in such solutions: formation of precipitates; formation of complexes; proton–hydroxide equilibria; and oxidation–reduction reactions. The description will be kept very simple and very general; for details the interested reader is again advised to look into the references in the 'Further reading'. We shall stress selectivity of element combinations in all equilibria, since it governs much of what we observe.

5.5.1 Formation of precipitates: solubility products

At equilibrium there is a quantitative connection between the contents of a solution and the solids with which it is in contact, for example the sea and the nature of the minerals on the sea bed. This is easily seen on the tidal coasts of the ocean, especially in hot climates where salt is deposited as water evaporates. We have then the equilibrium

$$\text{(sea) } Na^+ + Cl^- \rightleftharpoons NaCl \text{ (solid deposit)}.$$

The two sides are in balance, a fact that we express by stating that the product of the concentrations of the ions in solution when solid appears is constant, that is,

$$[Na^+][Cl^-] = K_{sp}$$

where K_{sp} is the solubility product (constant).† (Here we have loss of degrees of freedom due to formation of a new phase.)

Sodium chloride is quite a soluble salt, so the sea is very salty. Calcium carbonate is much less soluble, so the sea contains much less Ca^{2+}, CO_3^{2-} and HCO_3^-. For M_iX_j precipitates generally, the solubility product K_{sp} is defined as $K_{sp} = [M]^i[X]^j$ where we express concentrations in moles per litre. In some cases it is very small; for example, that of iron (III) hydroxide, is

$$K_{sp} = [Fe^{3+}][OH^-]^3 = 10^{-39},$$

and consequently there is very little free ferric iron in the sea since $[OH^-] \sim 10^{-6}$ M. Table 5.6 gives many solubility products and Fig. 5.3 shows how precipitation of hydroxides and sulphides vary, a point that has been of extreme importance for the evolution of living organisms (Section 12.4).

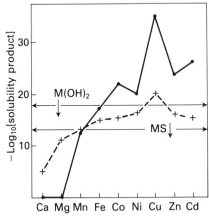

Fig. 5.3 The solubility products of hydroxides (broken line with crosses) and sulphides (solid line with solid circles), which have been of extreme importance in the evolution of life. The horizontal line indicates that solubility product which gives a precipitate at pH = 7.0 when [M] = 10^{-4} M and, for the sulphides shown, when [H₂S] = 10^{-3} M.

† Actually the concept of solubility product is strictly applicable to sparingly soluble salts only.

Table 5.6 Solubility-product constants, K_{sp}

Substance	Formula	Solubility product*
Aluminium hydroxide	$Al(OH)_3$	2×10^{-32}
Barium carbonate	$BaCO_3$	5.1×10^{-9}
Barium oxalate	BaC_2O_4	2.3×10^{-8}
Barium sulphate	$BaSO_4$	1.3×10^{-10}
Cadmium hydroxide	$Cd(OH)_2$	5.9×10^{-15}
Cadmium oxalate	CdC_2O_4	9×10^{-8}
Cadmium sulphide	CdS	2×10^{-28}
Calcium carbonate	$CaCO_3$	4.8×10^{-9}
Calcium fluoride	CaF_2	4.9×10^{-11}
Calcium oxalate	CaC_2O_4	2.3×10^{-9}
Calcium sulphate	$CaSO_4$	2.6×10^{-5}
Cobalt sulphide	CoS	8.0×10^{-23}
Copper(II) hydroxide	$Cu(OH)_2$	1.6×10^{-19}
Copper(II) sulphide	CuS	6×10^{-36}
Iron(II) hydroxide	$Fe(OH)_2$	3×10^{-15}
Iron(II) sulphide	FeS	6×10^{-16}
Iron(III) hydroxide	$Fe(OH)_3$	4×10^{-38}
Magnesium ammonium phosphate	$MgNH_4PO_4$	3×10^{-15}
Magnesium carbonate	$MgCO_3$	1×10^{-5}
Magnesium hydroxide	$Mg(OH)_2$	1.8×10^{-11}
Magnesium oxalate	MgC_2O_4	8.6×10^{-5}
Manganese(II) hydroxide	$Mn(OH)_2$	2.5×10^{-14}
Manganese(II) sulphide	MnS	3×10^{-15}
Nickel sulphide	NiS	2×10^{-21}
Strontium oxalate	SrC_2O_4	5.6×10^{-8}
Strontium sulphate	$SrSO_4$	3.2×10^{-7}
Zinc hydroxide	$Zn(OH)_2$	1.2×10^{-17}
Zinc oxalate	ZnC_2O_4	7.5×10^{-9}
Zinc sulphide	ZnS	4.5×10^{-24}

Data from *Stability constants* (1964). Spec. Pub. No. 7. The Chemical Society, London. Note the preferred use of Roman labels of oxidation states.
* The units of concentration are $(moles)^i (litre)^{-i}$

From what has been said, it is clear that solubility gives upper limits to the content of the aqueous solutions in living systems, just as much as in the sea. As a consequence, some living creatures can utilise precipitates such as $CaCO_3$ (shells), others $SrSO_4$ (acantharia), others $Ca_2(OH)PO_4$ (bones) and yet others precipitate iron oxides and even sulphides, all solutions of which are not too far from saturated locally in the sea. As stated in Section 2.9, major factors favouring stable solids are: (1) high charges on both anion and cation; (2) matching sizes of anion and cation; and (3) matching strong electron acceptor cations to strong electron donor anions, that is, increased covalent character. For detailed discussion see the references in 'Further reading'.

In this book we are particularly concerned not with isolated equilibria as in the preceding tables but with the result of mixing a large number of elements as happened during the formation of the Earth's solutions. This means that a realistic description of solubility in the environment must include very complex precipitates, for example of silicates, which we can only treat semiquantitatively. If we consider a solution made up from the metal ions of the first 30 elements in the periodic table in water in the presence of a slight excess of all anions $HX + X^-$ from the same part of the periodic table, then we find that at $pH = 7.8$ in a moderately reducing (H_2S) atmosphere solubility products will give us the results shown in Table 5.7. This would be the condition of the sea if it were in equilibrium with all the minerals available to it from the sea floor. These equilibria restrict as well the ways in which life could develop on Earth. *Precipitation determines the maximum allowed amounts of certain components that can occur together in a solution. Where there are many dilute components, many precipitated phases can coexist at equilibrium.*

Table 5.7 Elementary species remaining after mixing cations of the first 30 elements of the periodic table at 10^{-1} M, $pH = 7.3$ (25°C) with a small excess of anions as X^- or weak acids (HX)*

Soluble (10^{-1} M)	Li^+, Na^+, K^+ Cl^-, NO_3^-, SO_4^{2-}
Relatively soluble (10^{-4} to 10^{-1} M)	Mg^{2+}, Ca^{2+}, Mn^{2+} F^-, HCO_3^-, HPO_4^{2-}, $B(OH)_4^-$, VO_4^{3-}
Sparingly soluble (10^{-8} to 10^{-4} M)	H^+, Ba^{2+}, Zn^{2+}, (Fe^{2+}), Co^{2+}, Ni^{2+} OH^-, $(SH^-)(H_2S)$, $Si(OH)_4$
Almost insoluble ($<10^{-8}$ M)	Al^{3+}, Sc^{3+}, Ti^{4+}, Cr^{3+}, Fe^{3+}, (Co^{2+}), (Ni^{2+}), Cu^{2+}, (Zn^{2+}) OH^-, SH^-, H_2S, $Si(OH)_4$

* It is assumed that the anions are in slight excess in each category of solubility as X^- or HX.
(X) indicates a borderline case.

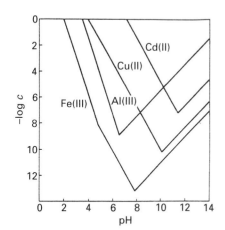

Fig. 5.4 Equilibrium of ion concentrations, c, in contact with precipitated hydroxides $Al(OH)_3$, $Fe(OH)_3$, $Cu(OH)_2$ or $Cd(OH)_2$, calculated using values for equilibrium constants for solubility and hydrolysis equilibrium. Note the effect of formation of $[FeOFe]^{4+}$ at around $pH = 4$ and the way in which complexation and precipitation can compete. Solubility increases at high pH due to hydroxy-anion formation (amphoteric behaviour).

5.5.2 Complex ion formation

While some elements combine to form the insoluble minerals leaving ions in solution in equilibrium with them, it sometimes happens that the precipitate also reacts with the excess of one of its components added to the solution to give *complex* ions. A typical example is the formation of chloride complexes of lead,

$$Pb^{2+} + 2Cl^- \rightleftharpoons PbCl_2; \qquad PbCl_2 + 2Cl^- \rightleftharpoons (PbCl_4)^{2-}.$$
$$\text{(solid precipitate)} \qquad\qquad \text{(soluble complex ion)}$$

Other common cases are those of aluminium with hydroxide and of arsenic, antimony or molydenum with sulphide ions, that is,

$$Al^{3+} + 3OH^- \rightleftharpoons Al(OH)_3; \qquad Al(OH)_3 + OH^- \rightleftharpoons (Al(OH)_4)^-$$

$$Mo^{6+} + 3S^{2-} \rightleftharpoons MoS_3; \qquad MoS_3 + S^{2-} \rightleftharpoons (MoS_4)^{2-}.$$

All these are examples of so-called 'amphoteric' behaviour (see Fig. 5.4).

It is also possible for metal ions to bind to anions or neutral molecules, such as NH_3, and remain in solution,

$$Cu^{2+} + 4NH_3 \rightleftharpoons (Cu(NH_3)_4)^{2+},$$

and they may react in solution with more complicated inorganic ions or large organic molecules, including proteins, sugars, nucleic acids, etc., which are generically called *ligands*. We can express the extent of these reactions within complex ion equilibria by defining a (thermodynamic) *stability constant K* that is related to the driving energy, $\Delta G° = -RT \ln K$,† for the reaction (see Section 3.4.1). In the case of the copper complex with NH_3 this constant is

$$K = \frac{[Cu(NH_3)_4^{2+}]}{[Cu^{2+}][NH_3]^4},$$

which gives an indication of the preference of a given metal ion, Cu^{2+}, to become 'complexed' by a given ligand, NH_3. (For the sake of simplicity, we have ignored the hydration sheaths of the metal ions and ligands.)

Much as there is a selection of partners through lattice energy (solubility products), so there is selection through complex ion formation, depending on the value of the stability constant of the corresponding complexes, usually quoted as log K (see examples in Table 5.8 and Fig. 5.5). It follows that, if we

Table 5.8 Stability constants (log K) of some complexes of biologically relevant ions in aqueous solution ($T = 25°C$)

Ion	Log K						
	Acetate (ML)	Ammonia (ML$_4$)	Glycine (ML)	Cysteine (ML)	Histidine (ML)	ATP (ML)	EDTA (ML)
Ca^{2+}	0.57		1.31			4.25	10.65
Mg^{2+}	0.55		2.22	<4		4.55	8.85
Co^{2+}	0.7	5.53	4.64	8.00	6.90	5.10	16.45
Ni^{2+}	0.84	8.12	5.78	9.8	8.67	5.22	18.4
Cu^{2+}	1.82	13.00	8.15		10.20	6.42	18.8
Zn^{2+}	1.20	9.65	4.96	9.17	6.55	5.16	16.5
Cd^{2+}	1.56	7.38	4.22		5.39	5.31	16.5

Data from Martell, A. E. and Smith, R. M. (1989). *Critical stability constants.* Plenum Press, New York.

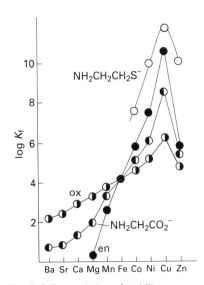

Fig. 5.5 The variation of stability constants, K, for the complexes of M^{2+} ions of the Irving–Williams series. ox, oxalate; en, ethylenediamine.

take an aqueous solution that contains a mixture of some 30 metal ions and other ions or compounds now with non-metal donor atoms, ligands (Table 5.9), we shall find that there results an equilibrium corresponding to a preferential selection of partners as given in Table 5.10. Once again these selections dominate simple ion complex equilibria in the sea but are of very much greater importance in all living organisms, where considerable amounts of metals are maintained in solution in complexes with organic ligands, that is, they are not precipitated. We shall consider this selection again in Chapters 9–13. A major point is that at equilibrium many different molecular species can coexist in the same solution limited by chemical selectivity.

The group analysis tables used formerly by all students of chemistry to analyse complex mixtures of compounds make use of both the relative strengths of precipitation reactions and complex ion formation. In fact, by the use of different complexation reagents, such as ammonia, in addition to hydroxide, sulphide, chloride, carbonate and sulphate for precipitation, we can separate all the elements by precipitation and solubilisation (see

† $\Delta G°$ is the change in free energy in given *standard* conditions ($p = 1$ atm, $T = 298$ K with all reactants and products at 1 M concentration). As before, $\Delta G° = \Delta H° - T\Delta S°$, but note that the standard conditions do not correspond to the conditions at equilibrium (the concentrations of the reactants and products are not, generally, 1 M; see Section 3.4).

Table 5.9 Selective ligands based on different donor atoms

O-donor	$^-O_2C-CH_2-CO-CH_2-CO_2^-$
N,O-donor	$^-O_2C-CH_2-NH-CH_2-CO_2^-$
N-donor	$H_2N-CH_2-CH_2-NH-CH_2-CH_2-NH_2$

O-donor
(phenolate)

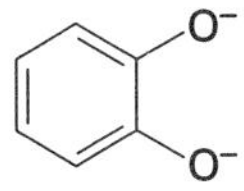

N-donor
(aromatic)

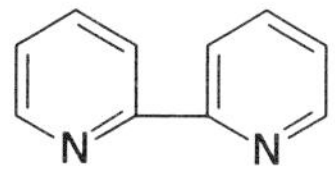

O,S-donor	$^-O_2C-CH_2-S^-$
N,S-donor	$H_2N-CH_2-CH_2-S^-$
S,S-donor	$^-S-CH_2-CH_2-S^-$

O-donor
(hydroxamate)

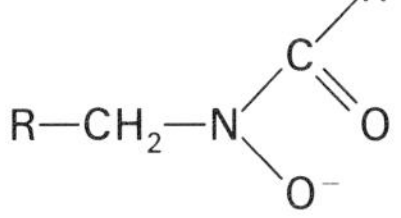

EDTA
N,O-donor

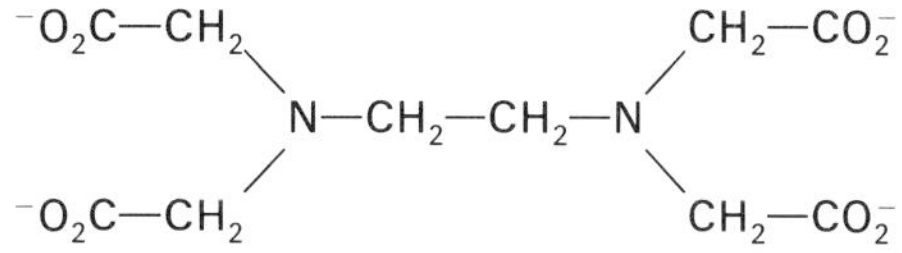

Table 5.10 Ligands preferred by different metal ions in simple co-ordination compounds

Metal ion	Ligands
Na^+, K^+	Oxygen-donor ligands, neutral or of low charge (-1)
Mg^{2+}	Carboxylate, phosphate and polyphosphate (total charge > -2); N-donation (special)
Ca^{2+}	Carboxylate, phosphate (less than Mg^{2+}), some neutral O-donors
Mn^{2+}, Fe^{2+}	Carboxylate, phosphate and nitrogen donors combined, (thiolate)
Fe^{2+} (special)	Unsaturated amines (particularly porphyrins)
Cr^{3+}, Mn^{3+}⎫ Fe^{3+}, Co^{3+}⎭	⎰Phenolate (e.g. tyrosine), hydroxamate, hydroxide ⎱Carboxylate, N-donors, polypyrroles, (thiolate)
Ni^{2+}	Thiolate (e.g. cysteine), unsaturated amines, polypyrroles
Cu^{2+}, Cu^+	Amines, ionised peptide $> N^-$, thiolate
Zn^{2+}	Amines, thiolate, carboxylate

The case of Fe^{2+} is especially complicated by spin-state changes and polymerisation of mixed oxidation states.

Table 5.11). As discussed in Section 8.5 the set of reactions is not very different from those which govern the geochemistry of land and sea. At present, the most sophisticated development by man of this procedure in analysis uses organic molecules containing chosen donor (complexing) atoms, and this also has many parallels in biological chemistry (Table 5.9 and see Chapters 9 and 12).

Table 5.11 Separation of cations into groups (organic acids, boric, hydrofluoric, silicic and phosphoric acids being absent)

Add a few drops of dilute HCl to the cold solution. If a precipitate (ppt.) forms, continue adding dilute HCl until no further precipitation takes place. Filter (1) and wash the ppt. with a little water: add washings to filtrate (2).

Residue. The ppt. may contain:
$PbCl_2$—white.
Hg_2Cl_2—white.
AgCl—white.
(silver group)

Filtrate. This must give no further ppt. with a few drops of dilute HCl. Add 1 ml of 3% H_2O_2 solution (3). Adjust the HCl concentration to $0.3N$ (4). Heat to boiling and pass H_2S through the solution until precipitation is complete (5). Filter and wash (6).

Residue. The ppt. may contain:
HgS—black.
Pbs—black.
Bi_2S_3—black or dark brown.
CdS—yellow.
CuS—black.
SnS_2—yellow.
Sb_2S_3—orange.
As_2S_3—yellow.
(copper and arsenic groups)

Filtrate. Test a small portion with H_2S to be certain that precipitation of Group II is complete. Boil down to about 10 ml in a porcelain dish and thus ensure that all H_2S has been removed (test with lead acetate paper). Add 1–2 ml of concentrated NHO_3 and boil to oxidise any ferrous salt to the ferric state (7). Add 1–2 g of solid NH_4Cl, heat to boiling, add dilute NH_3 solution until mixture is alkaline and then 1 ml in excess, boil for 1 min and filter immediately. Wash (8).

Residue. The ppt. may contain:
$Fe(OH)_3$—reddish-brown.
$Cr(OH)_3$—green.
$Al(OH)_3$—white.
MnO_2, xH_2O—brown.
(iron group)

Filtrate. Add 2–3 ml of dilute NH_3 solution, heat, pass H_2S (under 'pressure') for 1 min. Filter (9) and wash (10).

Residue. The ppt. may contain:
CoS—black.
NiS—black.
MnS—pink.
ZnS—white.
(zinc group)

Filtrate. This must give no further ppt. with H_2S (9). Transfer to a porcelain dish and acidify with dilute acetic acid (11). Evaporate to a pasty mass [FUME CUPBOARD], allow to cool, add 2–3 ml of concentrated HNO_3 so as to wash the solid around the walls to the centre of the dish and heat cautiously until the mixture is dry. Then heat more strongly until no more white fumes are evolved (12). Cool. Add 3 ml of dilute HCl and 10 ml of water: warm and stir to dissolve the salts. Filter, if necessary. Add 0.25 g of solid NH_4Cl (or 2.5 ml. of 10% NH_4Cl solution), render alkaline with concentrated NH_3 solution and then add. with stirring, $(NH_4)_2CO_3$ solution in slight excess. Keep, and stir, the mixture in a water bath at 50–60°C for 3–5 min (13). Filter and wash with a little hot water.

Residue. The ppt. may contain:
$BaCO_3$—white.
$SrCO_3$—white.
$CaCO_3$—white.
(calcium group)

Filtrate. May contain Mg^{++}, Na^+ and K^+ (14). Evaporate to a pasty mass in a porcelain dish [FUME CUPBOARD], add 2 ml of concentrated HNO_3, evaporate cautiously to dryness and then heat until white fumes cease to be evolved. White residue.
(alkali group)

After Vogel, A. I. (1979). *Textbook of macro- and semi-micro qualitative inorganic chemical analysis* (5th edn; revised by G. Svehla). Longmans, London.

The values of K_{sp} and K can be looked upon as driving energies for order related to $\Delta G°$, when we start artificially with one mole per litre of each of the components, anion and cation, in solution and go to the balanced position where no further reaction occurs but the products are also considered to be present in 1 M concentration. As shown in Section 3.5, for a balanced reaction $aA + bB \rightleftarrows cC + dD$ in which [A], [B], [C], [D] are the concentrations at equilibrium, one has

$$\Delta G = \Delta G° - RT \ln[C]^c[D]^d/[A]^a[B]^b.$$

(Note that *due to the definitions* while very small values of K_{sp} correspond to very insoluble precipitates, very large values of K correspond to very stable complexes, that is, to a great tendency to form such species.) We stress that such equilibria lie behind the appearance of locally ordered structures from random solutions. Solution chemistry is often treated in this thermodynamic fashion and this is justified since we show in Chapter 7 that, in water, ions usually combine and exchange rapidly thus coming to equilibrium. The situation in organic chemistry and organometallic chemistry of molecules in organic solvents is different in that exchange of atoms does not occur except under controlled (catalysed) conditions. We will delay treatment of organometallic chemistry to Chapter 9. In fact, all intermediate conditions of exchange exist, and in biological chemistry we shall see how this is exploited in control and catalysis.

5.5.3 Selectivity of precipitation and complex ion formation

In previous books (see references in 'Further reading'), we have analysed in detail the factors affecting the stability of binding of metal ions to inorganic anions and organic molecules. The selection rules are as follows.

1. Organic ligands, for example macrocycles, or inorganic ligands, for example silicates, can be devised so that they select for cations on the basis of their size, and either a small cation or a large one may be preferred depending on whether the cavity is small or large. In biological or mineral systems selection can be of either $Na^+ > K^+$ or $K^+ > Na^+$ and again of $Mg^{2+} > Ca^{2+}$ or $Ca^{2+} \gg Mg^{2+}$. Clearly, charge is also an effective differentiating factor too. These two factors, size and charge, greatly influence packing in lattices and therefore co-operativity and selectivity.

2. Ligands can be devised such that they favour the cation with the greater demand for electrons, which corresponds to the energy necessary for the ionisation of the metal atom to the metal ion. A well known series (the Irving–Williams order) giving the relative stability of divalent ion complexes is

$$Mn^{2+} < Fe^{2+} < Co^{2+} < Ni^{2+} < Cu^{2+} > Zn^{2+}.$$

The differences along this series are increased as the ligand donor atoms become better electron donors, for example $S > N > O$ (see Fig. 5.5). This series is common to much of inorganic chemistry including solubility product series (Fig. 5.3). (Note that only cations of similar size give series that are independent of donor atoms.)

3. We must add to these two rules, concerning the contribution of central field or isotropic terms, the contribution of other interaction energies due to angular polarisability. These terms, often called *ligand field energies*, depend on the geometry around the metal (symmetry) of arrangement of the donor atoms. For an octahedral field, that is, an octahedral arrangement of donor atoms (Fig. 2.24) they act in the order

$$Zero = Mn^{2+} < Fe^{2+} < Co^{2+} < Ni^{2+} > Cu^{2+} > Zn^{2+} = Zero$$

(see references 1 and 3 in 'Further reading'). The gain in lattice energy due to the effect can be appreciated from Fig. 2.25 but it also applies to complex ions, for example hydrated ions, as shown in Fig. 5.6.

Assuming the gas-ion ground electronic state is not maintained, an ion, Fe^{2+} in particular, can be forced into a different electronic state by strong ligand donors whence it may gain considerable further ligand-field energy (Fig. 2.26). Examples are the formation of FeS_2, pyrite, which has a stoichiometry not seen in very stable forms of Mn^{2+}, Co^{2+}, Ni^{2+} or Zn^{2+} with

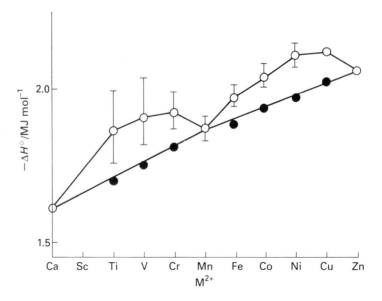

Fig. 5.6 The hydration enthalpy of M^{2+} ions of the first row of the d block $\Delta H°$. The filled circles show the trend when the ligand-field stabilisation energy has been subtracted from the observed values. Note the general trend to greater hydration enthalpy (more exothermic hydration) on crossing the period from left to right. The ligand-field energy here includes covalent radial terms due to $3d$ orbitals. Values are relative to $\Delta H°(H^+) = -1.09$ MJ mole^{-1}.

S^{2-}, and of the stable complex of Fe^{2+} in special chelates such as that of porphyrin. The peculiarity of pyrite could have been an initiating factor in the origin of life.

4. Finally we must observe that long-range effects, that is, effects not derived from the direct influence of bonding, can influence the donor/acceptor atom interaction. For example, steric hindrance, electrostatic forces, H bonds, etc. all affect stability. These effects may dominate in large molecules, such as proteins, which fold.

5. As a consequence of all these interactions, metal ions have preferences for specific donor atoms and geometries in their binding, for example Zn^{2+} and Fe^{3+} prefer tetrahedral ion sites (see Table 5.13). Ligands can be devised to satisfy these selective requirements, but, more generally, it is the metal that imposes the geometry on a set of *small* ligands. Some multidentate ligands of rigid geometry, for example some proteins, do impose stereochemistry on the metal ion, contrary to its geometric preference, and this imposed geometry, called an entatic state (from the Greek for 'under strain'), generates catalysis in many enzyme reactions (see Chapter 7).

It is the combination of rules and factors 1 to 5 that decides the selectivity of association; this has led to several classifications of acids and bases (acceptors and donors). One that is commonly used is that of 'hard' and 'soft' acids and bases, given in Table 5.12. Softness is clearly related to electronegativity which is dominated by factors under 2. For ions, as explained in Section 2.9.4, we shall often prefer to use in this book a two-parameter classification based on the acceptor power of a cation that is a measure of electronegativity (hard and soft), namely the ionization potential of M^{2+} together with the charge/ion size ratio, z/r (Fig. 2.23). The Irving–Williams series for complex ion formation derives from these factors (see borderline hard/soft ions in Fig. 2.23) as does precipitation by hydroxide (hard) or by sulphide (soft), or exchange of oxygen for sulphur, etc. Similar plots are readily obtained for other oxidation states, but, while they help us to get a quick overall view of stability constant trends, a

Table 5.12 Hard and soft acids and bases

Hard (class a)

Acids H^+, Li^+, Na^+, K^+, Be^{2+}, Mg^{2+}, Ca^{2+}, BF_3, BCl_3, $B(OR)_3$, Al^{3+}, $AlCl_3$, $Al(CH_3)_3$, Sc^{3+}, Ti^{4+}, VO^{2+}, Cr^{3+}, Fe^{3+}, Co^{3+}

Bases NH_3, RNH_2, N_2H_4, H_2O, OH^-, O^{2-}, ROH, RO^-, CO_3^{2-}, SO_4^{2-}, ClO_4^-, F^-

Borderline

Acids Fe^{2+}, Co^{2+}, Ni^{2+}, Cu^{2+}, Zn^{2+}, Rh^{3+}, $B(CH_3)_3$, R_3C^+, Pb^{2+}, Sn^{2+}

Bases $C_6H_5NH_2$, N_3^-, N_2, Br^-, Cl^-

Soft (class b)

Acids Cu^+, Ag^+, Au^+, Cd^{2+}, Hg^{2+}, Pt^{2+}, Pt^{4+}, MoO_2^{2+}, Pd^{2+}

Bases H^-, R^-, C_2H_4, C_6H_6, CN^-, CO, SCN^-, R_3P, R_2S, RSH, RS^-, I^-

* Hardness is defined (see textbooks in 'Further reading') as $\eta_M = \frac{1}{2}(I - A_e)$ where I is the ionisation potential of the state of M concerned and A_e is its electron affinity. I dominates for positive ions, but in aqueous solution must be considered relative to z/r. (Data from Ahrland, S., Chatt, J., and Davies, N. R. (1958). *Quart. Rev.* **12**, 265–76; Pearson, R. G. (1963). *J. Am. chem. Soc.* **85**, 3533–9.)

Table 5.13 Table of selectivity for Cu^{2+}, Zn^{2+}, Mn^{3+} and Fe^{3+} (selection by adequate combination of donor atom and geometry of site)

Geometry of site	Donor atom			
	Phenolate RO^-	Ionised peptide $\rangle N^-$	Thiolate RS^-	Nitrogen $\rangle N$ or $\rangle N$, O^-
Tetrahedral	Fe^{3+}	—	Zn^{2+}, Fe^{3+}, Cu^+	Zn^{2+}
Tetragonal	Mn^{3+}	Cu^{2+}	Cu^{2+}	Cu^{2+}
Octahedral	Fe^{3+}	—	$(Mo(V))$	Mn^{2+}, Fe^{2+}

more detailed description is required if a deeper understanding of the selectivity factors that operate in chemistry, geology and biology is to be attained. We must introduce, as stated above, stereochemical properties of ligands, preferred symmetry of sites, ligand-field effects, etc. for different metal ions (Table 5.13) in order to have an idea of the thermodynamic principles that govern selection by complex formation. (To these principles we must then add kinetic factors which are of overwhelming importance in real natural selection processes; see references in 'Further reading'.)

5.6 Competitive equilibria

It is now necessary to introduce a more realistic quantitative discussion of association in solution where we do not deal with an isolated acid/base, $A + B \rightleftharpoons AB$, equilibrium but where competition exists between large numbers of donors and acceptors as in the sea or the human body. We wish to concentrate on problems in water where there is always one main acceptor, H^+, in competition with A, and one main donor, OH^-, in competition with B. The competition applies both to considerations of precipitation and complex formation. It is conventional to use two different definitions of acids and bases in this context. According to the definition of Brønsted and Lowry an acid is any substance that gives hydrogen ions to the solvent, for example, forms H_3O^+ in water, while a base is a substance that removes hydrogen ions from the solvent, for example, gives OH^- ions in water. The Lewis definition of acids and bases as electron-pair acceptors and donors, respectively, is more general and includes that of Brønsted and Lowry, but the latter is more convenient in aqueous solution when equilibria involving the proton and the hydroxide ion are very frequently considered. We therefore use the Brønsted–Lowry approach first.

5.6.1 Proton and hydroxide ion binding

In pure water, at 25°C, the molar concentration of hydrogen ions is approximately 10^{-7} M and the product $K_w = [H^+][OH]^-$, the so-called ionic product of water, is 10^{-14}. This is related to the acid dissociation constant of water

$$H_2O \rightleftharpoons OH^- + H^+ \text{ (or, more adequately, } 2H_2O \rightleftharpoons H_3O^+ + OH^-),$$

for which $K_a = [OH^-][H^+]/[H_2O] = 10^{-16}$ since $[H_2O] = 55.5$ M.

Any other substance, HA, added to water, will appear acid if in molar concentration it gives a measurable concentration of H_3O^+ ions greater than 10^{-7} M. This will happen only when its concentration is high and its dissociation constant $K_a = [A^-][H^+]/[HA]$ considerably exceeds 10^{-10}. Similarly, the classification as a base in water is restricted to those compounds that considerably exceed in *basic* dissociation constant, K_b, the value of 10^{-10} (for reactions such as $A^- + H_2O \rightleftharpoons AH + OH^-$, when $K_b = [AH][OH^-]/[A^-]$). Clearly, for any acid AH and base A^- one has $K_a \cdot K_b = K_w$. AH and A^- are said to be a conjugate acid/base pair. An immediate consequence is that, if an acid is strong, that is, tends to donate hydrogen ions to water, its conjugate base is weak, that is, does not tend to remove hydrogen ions from water, and the weak base is available for weak binding to metal ions. For example, $HClO_4$ and HCl are very strong acids, and ClO_4^- and Cl^- are very weak bases. The reverse is also obviously true; for example $C_2H_5O^-$, ethanolate, is a very strong base and C_2H_5OH (ethyl alcohol) is such a weak acid in water that it is not considered as such and is hardly able to bind to even the strongest Lewis acid, for example Hg^{2+}. The best donors formed from HL are moderately strong bases.

There are also substances that donate (or remove) more than one proton to (or from) water in a rather wide range of acid (or base) strength. An example is phosphoric acid, H_3PO_4, which can donate three protons to water, but the successive ionisation constants decrease rapidly ($pK_{a1} = 2.1$; $pK_{a2} = 7.4$; $pK_{a3} = 12.3$ where $pK_a = -\log_{10} K_a$). The conjugate base PO_4^{3-} will exhibit the opposite behaviour. Consequently, in aqueous solution at pH = 7, the species $H_2PO_4^-$ and HPO_4^{2-} will be present simultaneously in approximately equal concentrations. These equilibria are of fundamental consequence to the properties of phosphate and its esters in biology and to their ability to bind and/or precipitate Mg^{2+} and Ca^{2+}, for example in polyphosphate complexes and in bone precipitates. In aqueous solution we need to consider only the two main types of Lowry–Bronsted acids listed in Tables 5.14 and 5.15 where their acid dissociation constants are given.

Table 5.14 Ionisation constants (pK_a) of non-metal hydrides HA and H_2A

Hydride	Ionisation constant, pK_{a1}
HF	3.18
HCl	-7
HBr	-9
HI	-11
H_2O	15.7
H_2S	7.0
H_2Se	4
H_2Te	3

Table 5.15 Acid dissociation constants of oxyacids

	pK_1	pK_2	pK_3
H_4GeO_4	+8.6	+12.5	
H_3AsO_3	+9.2	+12.0	
HClO	+7.2		
(H_3BO_3*	+9.2)		
H_6TeO_6	+7.8	+11.2	
H_3PO_4	+2.1	+7.4	+12.3
H_3AsO_4	+2.3	+6.9	+11.8
H_6IO_6	+1.6	+7.0	
H_2SO_3	+1.9	+7.2	
H_2SeO_3	+2.6	+8.0	
H_2TeO_3	+2.7		
H_2CO_3 (true)	+3.6	+10.2	
H_2SO_4	−3.0	+1.5	
HNO_3	−1.4		
$HClO_3$	−1.0		
$HClO_4$	very strong		
	~ −10.0		

* H_3BO_3 is included in parentheses as it reacts with alkali to give B(OH)$_4^-$ rather than by losing a single proton.

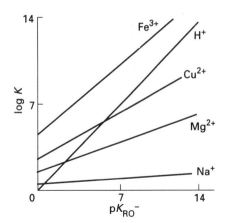

Fig. 5.7 An idealised plot of log K for complexes of Fe^{3+}, Cu^{2+}, Mg^{2+}, Na^+ and H^+ (log $K = pK$) with ligands binding through RO^-, for example, SO_4^{2-}, RCO_2^-, RO^- and OH^-. Competition between a fixed $[M^{n+}]$ and H^+ varies with pH. When pH = 7, H^+ does not compete for ligands with $pK_a < 7$.

Very commonly, in chemistry and particularly in biology, we are also interested in the acidic properties of protonated nitrogen bases, for example NH_4^+, $R—NH_3^+$, $C_3H_4N_2H^+$ (imidazolium), etc., or in the properties of the conjugate bases NH_3, $R—NH_2$, $C_3H_4N_2$, etc. Again, pK_a measures the binding strength of the proton.

Now the ability of series of related bases to bind protons (by electron-pair donation) often parallels the ability to bind metal ions, Lewis acids, and we frequently observe an approximately linear relationship between stability constants of metal complexes (log K) and ionisation constants of acids (pK_a) for families of closely related ligands, that is,

$$\log K = apK_a + b.$$

This implies that unless $a > 1$ or b is large due to special features of the metal ion not present in the proton, single weak acid centres will not bind metal ions (see Fig. 5.7). For most metal ions of low charge ($< 3 +$), a, in the above equation is < 1.0 so that a ligand with $pK_a \geqslant 7.0$ will not bind effectively at pH = 7.0. However cations of charge $> 3 +$ may well bind to such ligands, for example phenolates. The greatest contribution to log K is then the so-called chelate effect—the resulting increase in stability from the ability of a metal ion to bind from two to six groups of a polydentate ligand co-operatively while the proton binds but one at a time (Fig. 5.8).

Small highly charged cations, for example Fe^{3+}, Al^{3+}, Cr^{3+}, etc., also give acidic solutions due to the ionisation of molecules of water from their hydration sheaths, for example,

$$M(H_2O)_6^{3+} + H_2O \rightleftharpoons M(H_2O)_5 (OH)^{2+} + H_3O^+.$$

This process is commonly referred to as hydrolysis, but it is really a normal acid–base reaction. (It can also be looked upon as hydroxide binding, see below.) The hydroxo species thereby formed can frequently polymerise

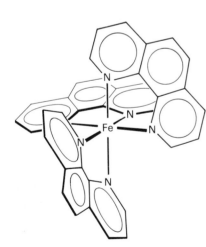

Tetragonal Cu^{2+} bis ethylenediamine

Tetrahedral Zn^{2+} bis ethylenediamine

Octahedral Ni^{2+} bis ethylenediamine dihydrate (high-spin)

Octahedral Fe^{2+} tris 1,10-phenanthroline (low-spin)

Fig. 5.8 Examples of chelate formation where a metal binds to several ligands displacing (many) protons from the protonated ligands.

through the formation of bridging groups or by condensation involving loss of water between two hydroxyl groups, that is,

$$(H_2O)_4M \begin{matrix} OH^- \\ \diagup \quad \diagdown \\ \diagdown \quad \diagup \\ OH^- \end{matrix} M(H_2O)_4 \text{ or } (H_2O)_5M - O^{2-} - M(H_2O)_5.$$

Eventually, these mechanisms may lead to the formation of clusters (see Table 5.16). Alternatively they give rise to oxo-ions.

$$(H_2O)_2M(OH)_4 \longrightarrow (H_2O)_2MO_2^{2+} + 2H_2O$$

(for example where M = V or Mo). (Note that if H_2S is present S^{2-} or $-SH^-$ bridged clusters may also be formed with some metals, for example Fe.)

Table 5.16 Multinuclear complexes using O or S donors*

Type†	Metals believed to form such complexes‡
Cationic complexes	
Me—OH—Me	Be(II), Mn(II), Zn(II), Cd(II)
Me $\diagup$OH$\diagdown$ Me (OH bridges)	Cu(II), Fe(III), Hg(II), Sc(II), UO$_2^+$
Me—Me—Me (double OH bridges)	Hg(II), Sn(II), Pb(II), Sc(III), Si(IV), Al(II)
Me $\diagup$S$\diagdown$ Me $(MeS_2)_n$	Fe(II, III)
Me$_4$S$_4$	Fe(II, III), Ni(II)
Varied	Be$_3$(OH)$_3^{3+}$, Bi$_6$(OH)$_{12}^{6+}$, Pb$_6$(OH)$_8^{4+}$, Al$_7$(OH)$_{17}^{4+}$, Al$_{13}$(OH)$_{34}^{5+}$, Mo$_7$O$_{24}^{6-}$, V$_{10}$O$_{28}^{6-}$

Polymeric oxo complexes of metals, metalloids and non-metals

Type	Examples
O$_3$XOXO$_3$	Cr$_2$O$_7^{2-}$, S$_2$O$_7^{2-}$, P$_2$O$_7^{4-}$
O$_3$X(O$_4$X)XO$_3$	P$_3$O$_{10}^{5-}$
(XO$_3$)$_n$	(PO$_3$)$_n^{n-}$, (SiO$_3$)$_n^{2n-}$, CrO$_3$(s)
(X$_2$O$_5$)$_n$	P$_2$O$_5$(s), (Si$_2$O$_5$)$_n^{n-}$
(XO$_2$)$_n$	SiO$_2$(s)

* Modified from P. Schindler (1968).
† Charges are omitted; the structural arrangement given, although plausible, is hypothetical.
‡ The list is not complete.

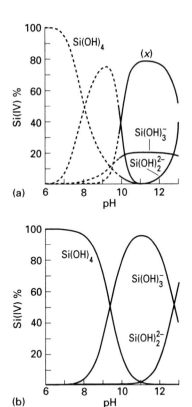

Fig. 5.9 The distribution of mono- and polynuclear species of Si(IV) in water at 25°C illustrating the effect of change of concentration; (a) [Si] = 0.1 M; (b) [Si] = 10^{-5} M. The dotted lines in (a) refer to viscous gels, neutral and anionic species, and curve (x) to a possible tetrameric anion.

Not only these metals, but at least 40 elements give polymerised oxide species in aqueous solution (Table 5.16) and Fig. 5.9 gives an example from the chemistry of silicon showing the influence of concentration. In general, polymeric species are formed only by *weak acids* or *weak bases* in which the central element is in an oxidation state (see Section 5.7) equal to or greater than III. The explanation of this behaviour rests on the overall reduction of charge in the polymerised species, which is energetically favourable. We must look more generally at the way in which the hydroxide ion (from water) removes Lewis acid acceptors (metal ions) from the possibility of binding to donors.

It is a feature of most waters on Earth that they have a pH = 7.0 ± 1.5 and this range is critical to much of life. Thus [OH$^-$] is approximately equal to [H$^+$]. This equivalence arises from the relative abundance of acidic and basic components in waters. Major contributors are the amphoteric HCO_3^-, the non-metal acids HCl, H_2CO_3 and the basic oxides of silica, alumina and magnesium. Thus, while H$^+$ binding removes many basic centres from combination with metal ions leaving protonated centres that are extremely important in organic structures (H bonds) and reactions (H$^+$ catalysts), hydroxide removes many metal ions from combination almost entirely as precipitates. Associated structure (order) in minerals depends greatly on the second, as in hydroxide/oxide lattices, while structure in biology is dependent on the first, as in organic molecules (contrast aluminosilicates and protein associations).

We must now increase the number of competing equilibria so as to increase understanding of the selectivity of all these associations under the real conditions in the sea or in biological solutions.

5.6.2 Conditional, apparent or effective stability constants

The stability constant of a metal complex is defined as $K_{ML} = [ML]/[M][L]$. If this is the only complex formed, then the total concentration of the metal M is $C_M = [M] + [ML]$ and the total concentration of the ligand L is $C_L = [L] + [ML]$.

This means that

$$K_{ML} = \frac{[ML]}{[M][L]} = \frac{C_M - [M]}{[M][L]}.$$

In actual practice there may be other complexing agents present that may complex M; hence $C_M \neq [M] + [ML]$. Furthermore, L is frequently derived from a weak acid and has a tendency to bind the proton.

In such circumstances the stability constant K_{ML} is not the constant that should be used to characterise the conditions since it does not represent the real equilibria and may lead to the wrong conclusions if not duly corrected. The correction is made by defining a conditional, apparent or effective stability constant that is valid for the exact prevailing conditions. These constants, $K_{M'L'}$, are defined as

$$K_{M'L'} = \frac{[ML]}{[M]'[L]'}$$

where [M]' is the total concentration of metal not bonded to L and [L]' is the total concentration of the ligand L not bonded to M.

Hence $C_M = [M]' + [ML]$ and $C_L = [L]' + [ML]$.

On the other hand, if A is any other ligand present

$$[M]' = [M] + [MA] + [MA_2] + \ldots + [MA_n] = [M]\{1 + \sum \beta_i [A]^i\}$$

where the β_i are overall stability constants of any MA_i species.

Putting

$$1 + \sum \beta_i [A]^i = \alpha_M,$$

then

$$[M]' = \alpha_M \cdot [M].$$

Similarly, if there is any other metal in solution, say, N, that forms complexes with L, or if ligand forms H_iL complexes with the proton p, then

$$[L]' = [L] + [HL] + [H_2L] + \ldots + [H_pL] + [NL] = [L]\{1 + \sum \beta_i [H^+]^i + K_{NL}[N]\}$$

where the β_i are the overall formation constants of the H_iL species and K_{NL} is the stability constant of the NL complex.

Putting, as above,

$$1 + \sum \beta_i [H^+]^i = \alpha_L,$$

then

$$[L]' = \alpha_L \cdot [L] + K_{NL} \cdot [N][L],$$

which, if there is no other metal N, reduces to

$$[L]' = \alpha_L \cdot [L].$$

In such conditions

$$K_{M'L'} = \frac{[ML]}{[M]'[L]'} = \frac{1}{\alpha_M \cdot \alpha_L} \cdot \frac{[ML]}{[M][L]} = \frac{K_{ML}}{\alpha_M \cdot \alpha_L}.$$

The values of α_M and α_L are generally large and positive; hence $K_{M'L'}$ is smaller than K_{ML}. In effect, $K_{M'L'}$ may be so small that in practice the complex ML is not formed due to the competition established with other ligands or metals, or due to the effect of increased acidity, that is, higher $[H^+]$. It is not absolute stability constants that define the associations that appear but the conditional constants in the given media.

As an example we take the real complexing power of a given ligand towards a given metal M in a given medium, usually water, at varying pH values. We may say that the effective stability constant in such conditions is very approximately given by

$$\log K_{M'L'} = \log K_{ML} - \log \alpha_L \simeq \log K_{ML} - \log K_{HL} + pH \simeq \log K_{ML} - pK_{HL} + pH.$$

For instance, the stability constant of the Ca · EDTA complex is $\log K_{ML} = 10.9$. However, at pH = 2 the conditional stability constant ($\log K_{M'L'}$) is only about 2.6 since $pK_{HL} \approx 10.3$. This means that, unless the pH is of the order of pK_{HL}, the stability constant of the complexes formed with L are substantially reduced (see Fig. 5.10). Thus, selective binding is here strongly dependent on pH. The discussion can be extended to hydroxide competition (precipitation) for M and then generally to all acids and bases. We must consider that the observed combining arrangement of the some 20 essential elements, which develops from the simplest forms in which they could occur together in water, is due to

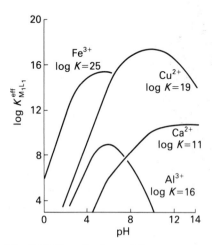

Fig. 5.10 The effective stability constants, K^{eff} or $K_{M'L'}$, at different pH values for EDTA complexes of four metal ions. K is the absolute binding constant when there is no competition from H^+ or OH^-.

intense competition based on the above equilibrium considerations. Solubility, itself often reflected in availability, decides the concentration terms in the above competition amongst soluble species, while the stability constants reflect other affinities. There is no doubt that the building of structures (order) is competitive between the existing mineral and the organic (living) components. This competition lies at the very heart of what we observe. We must, therefore, include a few notes on organic molecule association in water, as well as on inorganic combinations seen in seawater or inorganic–organic combinations which are used in biological systems and chemical analyses.

5.6.3 Organic molecule associations

The principles that apply to association between two organic molecules are exactly parallel to those described for inorganic species. It is clear that simple electrostatic interaction such as that between R^+, an organic cation, and R^-, an organic anion, does not generate association. For example, $[N(CH_3)_4]^+$ does not associate to any relevant extent with $R—CO_2^-$, $ROSO_3^-$ or $R—OPO_3^{2-}$ (compare Na^+ with Cl^-). It is only when their structures carry larger numbers of charged units that electrostatic energies between polyanions and polycations give rise to association generating a reduction in disorder. Thus, a wide range of simple anions, for example carboxylates and phosphates, form soluble organic salts with simple and ammonium-type cations. Many of these salts are the basic units of biological metabolism. By way of contrast the formation of complexes of small organic molecules with large proteins, which have many centres for binding, becomes the basis of many biological associations and reactions especially in enzymes. All such associations are treated by competitive equilibrium just as in the above examples of inorganic species. Of course, longer hydrophobic organic chains aid association as in the case of lipids (see Section 9.4.3). Since all species are in competition, surfaces have to be constructed to give selectivity if specific shapes are to evolve and this turns us to the properties of polymers.

5.6.4 Organic polymers in solution

Organic polymers, especially biological polymers, are of very great importance in the development of the natural selection of the chemical elements, as we shall see in Chapters 9–13, and their formation at equilibrium must be considered first. The condition of polymerisation for a monomer A at equilibrium is written generally

$$n\text{A} \rightleftharpoons \sum \text{A}_n.$$

Many inorganic polymers, such as polymeric hydroxides, equilibrate quickly, that is, in times less than hour; they often *cross-link* and, even at low molecular weight, they precipitate. Examples are the mixed hydroxides and oxides forming clay minerals.

Amongst organic monomers there is frequently found to be association, rather than polymerisation, of the more hydrophobic tails of molecules, for example of fats, which leads, at equilibrium, to the formation of small separate 'phases', vesicles. As we have seen, this association is due to van der Waals weak interactions. We describe these small units in Section 6.2.2 and see in

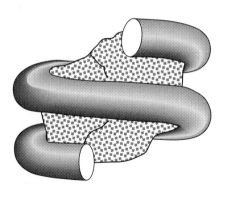

Two turns of DNA
around eight histone proteins

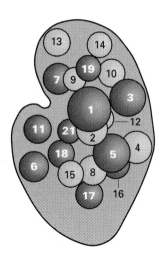

Twenty-one proteins in a
bacterial ribosomal
particle. Un-numbered
shaded space is RNA.

Fig. 4.19 that they can remain in equilibrium even when dispersed at different degrees of association.

While discussing polymerisation we refer next and more generally to covalent polymers in (aqueous) solutions in which there is a great variety of species, A_n. Equilibrium through exchange of A with any A_n or of any A_n with any other A_m is now not achieved over long periods of time. These polymers are usually not stable in the thermodynamic sense but, because they do not exchange units, they have to be treated operationally as separate components (Section 4.3.2). Organic polymers made by man are also often of this type and are often very insoluble in water, for example many plastics. We return to them in Chapter 14. Biological polymers such as proteins and DNA/RNA are very different from these man-made polymers in that they have fixed complex sequences, composition and molecular weight, and when in water they are then components. We treat them as individual molecules in water, but they can associate in equilibria of the kind, see marginal illustration

$$\text{Polymer } 1 + \text{polymer } 2 + \text{etc.} \rightleftharpoons \text{complex unit.}$$

Some of these complex units are extremely complicated, yet many remain soluble in water while some go into lipid membranes. They then form non-repeating lattices, unlike an inorganic crystal, since the polymers 1, 2, etc. have unique matching surfaces so that they can come together in highly specific forms, overcoming competition. The treatment of the system as one at equilibrium may seem artificial, but it does assist in an understanding of their assembly. We return to discussion of these and other systems later in this chapter and in Chapters 6 and 7 where we look at a single large polymer in terms of the phase characterisations usually applied to bulk components; that is, a protein molecule or DNA can be likened to a small phase dispersed in water (Section 6.8.1) when it folds and then to a collection of small phases in associative equilibrium, leading to a mosaic of 'phases'. The mosaic of phases (proteins) is stabilised by co-operative interactions based on matching of surfaces so that in principle, perhaps even in practice, an organism self-assemblies to give a particular shape.

While above we have utilised well-known and easily analysed systems of small organic molecules binding together and of metal ion binding to ligands, the principles are general and can be extended to any mixed system of substances, for example the binding of proteins to nucleic acids or of either or both to anions and cations. In fact, these are general principles of *thermodynamic* control over *self-assembly* at equilibrium using acid–base equilibria, which we now summarise.

5.6.5 Summary of solubility and complex association equilibria

This consideration of acid–base equilibria of the kind

$$n\text{A} + m\text{B} \rightleftharpoons \underset{\text{complex}}{\text{A}_n\text{B}_m} \rightleftharpoons \underset{\substack{\text{separate phase} \\ \text{(precipitate)}}}{\text{A}_n\text{B}_m}$$

is given so as to stress three points:

1. No matter what the complications of *the equilibria* in solution, the chemical variables are only the components A, B, etc., temperature and pressure. As stated in Section 4.3.2, binding of A to B does not increase the way in which a system can be varied *at equilibrium*. The decisive factor for variance is the relative contribution to the composition of those species of A and B, inorganic or organic, that are not in equilibrium, that is, the components. Increase in components increases the variance of systems. Increase of the presence of precipitates, phases, decreases variance.

2. There is great selectivity between reactions of elements as simple ions or as complex charged molecules (inorganic species) within the equilibria.

3. Given that there is a vast range of organic components and many inorganic components, diversity of combination at fixed temperature and pressure is a matter of relative amounts of each in the system. This diversity of association is then under local concentration control.

These conclusions are independent of the size or nature of A and B.

Now we face the difficult task of considering, at equilibrium in water, the vast range of possible products from a mixture of some 30 inorganic elements (the most common) (see Table 5.7) and organic components of a considerable variety of donor/acceptor kinds. A few firm conclusions are highly important at pH=7.0–8.0 (in life and in the sea):

1. Very small anions and cations of high charge, >2, tend to remove one another as precipitates. In effect, few of the possible trivalent or more highly charged elementary cations remain available; for example, Al, Sc, Ti, Cr and even Fe are precipitated with O^{2-} or bound as hydroxy-species. (However stronger bonding still allows B and Si to become available as borate and silicic acid.) Thus oxide (hydroxide)-based lattices dominate the Earth's soil chemistry. The concentrations of these insoluble cations in water are very limited and invariant. (NB. Al, Sc, Ti and Ga are not used in biology.) A few selective anion donor ligands can solubilise these elements.

2. The presence of sulphide which, unlike OH^-, can be varied at fixed pH, can lead by oxidation/reduction and precipitation to virtual removal of some selected divalent metal ions, for example Cu. However, soft organic donors solubilise some of these ions; for example N/S donors can solubilise Fe, Co, Ni, Cu, Zn and Cd as divalent ion complexes. There is then competition in sulphide media but the last three cations tend to precipitate.

3. Competitive proton binding prevents many saturated nitrogen and carbon donors from binding to most cations. There is little natural organometallic chemistry possible (except that developed by man in very different solvents from water).

4. Association of organic donor ligands with organic acceptors is based on principles different from those governing metal acceptor/organic donor interactions and is due to the effects of a multiplicity of weak interactions on large surfaces. The association often depends on the proton (or the hydrogen atom) in bound form, that is H-bonding between N—H and O—H groups. Together with hydrophobic surfaces, essentially —CH_3, interactions they dominate organic (biological) ordering.

5. In the absence of precipitation the limits of concentration of complexed and

free species is not controllable by thermodynamic factors but must be controlled by constraints imposed by availability or supply.

Finally, note that it is the pressence of very many components, now especially organic, that allows the solutions in biology to have a great number of species present simultaneously even at equilibrium.

5.6.6 Acid–base equilibria and components in solutions

When describing acid–base behaviour under the title of effective association constants, equilibration is assumed between all species. Thus, it is important in any given system to see the relationships between all the available species for which there is equilibration, where equilibration means that exchange of elements between species is fast in relation to the activity under discussion. An example gives a view of the complexity. Consider the proton and the magnesium ion interacting with all possible phosphates and polyphosphates in biological systems (for example, adenosine triphosphate, ATP; Fig. 5.11).

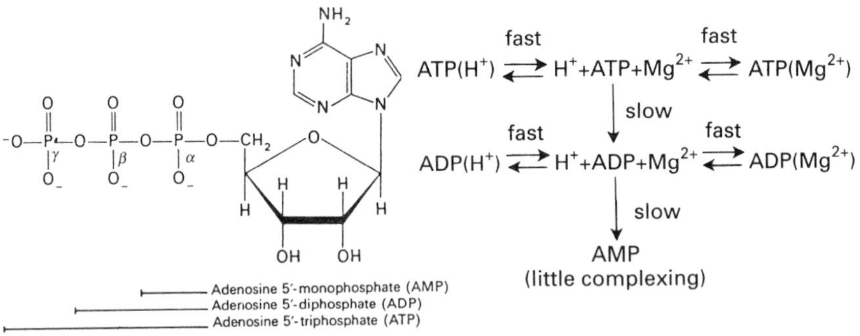

Fig. 5.11 The reactions of adenosine triphosphate (ATP) in equilibria ($\rightleftarrows$) with inorganic ions and reacting slowly ($\rightarrow$) to the diphosphate (ADP) and the monophosphate (AMP). ATP, ADP and AMP are independent components.

These phosphates are involved in metabolic pathways with enzyme-catalysed reactions with turnover rates of the order of 10^{-3} s. Protons and magnesium exchange at a rate in excess of this from phosphate complexes. Thus, in a cell there is an equilibrated system of Mg^{2+}, H^+ and phosphate species RPO_4^{2-} in the time-scale of metabolism. In thermodynamic terminology, provided the ratio of amounts of the elements does not change, there are two components Mg^{2+}, H^+ plus any number of components RPO_4^{2-}, which are slowly exchanging amongst themselves. Each RPO_4^{2-} is, therefore, a component, but the presence of the fast exchanging complexes of Mg^{2+} and H^+ with these phosphates does not change the number of components (Section 4.3.2). We shall need to consider very carefully which species are in equilibrium so that the variables in an aqueous solution are known (under the reaction conditions) from a thermodynamic point of view; immediately it is apparent that in a biological cell this must be a very large number.

5.7 Equilibria between oxidation states

The equilibria that we have considered so far involve just one substance in different physical states or acid–base reactions of the types $A + B \rightleftarrows AB$ or

$AB + CD \rightleftarrows AC + BD$ in which elements are exchanged. There is, however, another possibility which is the transfer of electrons, for example,

$$2Fe^{3+} + 2I^- \rightleftharpoons 2Fe^{2+} + I_2.$$

In this reaction the ferric ion, Fe^{3+}, captures one electron and goes to the ferrous state, Fe^{2+}, and each iodide ion, I^-, gives up one electron and goes to the molecular state, I_2. As already discussed in Chapter 2, the substances that have a tendency to give up electrons are called *reductants* or reducing agents and those that have a tendency to accept electrons are called *oxidants* or oxidising agents. The intrinsic tendency to give up or accept electrons is independent of the particular reaction and we may express this fact by considering each element, free or combined, separately; that is, for the case above

$$Fe^{3+} + 1e \rightleftharpoons Fe^{2+},$$

$$I^- - 1e \rightleftharpoons 1/2\ I_2$$

which can be written, in general terms, as

$$\text{Oxidised form}_1 + n_1 e \rightleftharpoons \text{reduced form}_1.$$

Since electrons cannot exist free in solution, such a reaction has no physical meaning (although it reflects what happens in so-called electrode reactions taking place in electrochemical cells where a solid metal—the electrode—exchanges electrons with an ionic species in a solution with which it is in contact, see Fig. 6.17). This means that in order to be observed this reaction needs another system, which we can represent as

$$\text{Reduced form}_2 - n_2 e \rightleftharpoons \text{oxidised form}_2$$

so that the overall reaction will be

$$n_2(\text{oxidised form}_1) + n_1(\text{reduced form}_2) \rightleftharpoons n_2(\text{reduced form}_1) + n_1(\text{oxidised form}_2),$$

as seen above in the reaction $2Fe^{3+} + 2I^- \rightleftarrows 2Fe^{2+} + I_2$. This is a *reduction–oxidation reaction*, more simply called a *redox* reaction.

All redox reactions can be written in terms of two half-reactions using electrons as above. A more complicated example, of geological and biological relevance, is

$$O_2 + 4Fe^{2+} + 4H^+ \rightleftharpoons 4Fe^{3+} + 2H_2O$$

which results from the combination of the two partial ('half') reactions

$$O_2 + 4H^+ + 4e \rightleftharpoons 2H_2O,$$

$$4Fe^{3+} + 4e \rightleftharpoons 4Fe^{2+}.$$

The practical result of such redox reactions is the change in the so-called *oxidation state* of some of the elements, free or combined, involved in them. The oxidation state is usefully expressed by a number that must not be confused with valence discussed in Chapter 2, Sections 2.4.2 and 2.4.3. Carbon is four valent in CO_2 and CH_4 but the oxidation states are $+IV$ and $-IV$, respectively. (An *oxidation* corresponds to the increase of the oxidation state and a *reduction* corresponds to the decrease of the oxidation state no matter what the valence.) The numerical assignment of the oxidation state to any element is empirical and straightforward and is given a Roman numeral. When elementary forms are involved, it is *zero* for the free element (for example, Fe, Ca, K, O_2, N_2, I_2, etc.) and equal to the charge when in an ionic form (for example, $Fe^{3+}(+III)$, $I^-(-I)$, $H^+(+I)$, $O^{2-}(-II)$, etc.). In other cases, the oxidation state can easily be derived using the following simple rules, treating elements as if they were ions in combination.

1. The sum of the oxidation states of the elements in any combined species is zero if the species is neutral, or equal to its charge if it is an ion.

2. The oxidation state of hydrogen as a proton is $+I$. The exceptions are the hydrides of electropositive metals (for example, NaH, CaH_2, H_4LiAl, etc.), when it is $-I$ and, of course, the free element (H_2, oxidation state zero).

3. The oxidation state of oxygen is generally $-II$. The exceptions are the free element (O_2, O_3), when it is zero, and peroxides such as H_2O_2 when it is $-I$.

4. In covalent compounds the oxidation state of each element corresponds to the charge that the element would have if each shared pair of electrons were assigned completely to the more electronegative of the elements sharing them. If these elements have similar electronegativity the pair of electrons is split between them. For example, in ICl we have $I(+I)$ and Cl $(-I)$, in SO_4^{2-} we have $S(+VI)$ and $O(-II)$ and in FeO^{3+} we have $Fe(+V)$ and $O(-II)$, whereas in B_4C_3 or SiC, in which the elements have very similar electronegativity, B, C and Si can all be considered as being in the zero oxidation state. In more complicated cases we can refer to the average oxidation state, as in $CH_2OH \cdot CH_2OH$ or CH_3CHO, where that for C is $-I$, or we can consider functional groups separately, as in $CH_3 \cdot COOH$, where C in CH_3 is in the oxidation state $-III$ and in COOH is in the oxidation state $+III$ (see Table 5.17).

These rules can be extended and used in practical terms for many reactions in which halogens, oxygen or sulphur are bound to a given central atom. Hence, for each halogen atom, X, bound to a central element, the oxidation state of this element increases by $+I$, and for each oxygen or sulphur atom bound the oxidation state of the central element is increased by $+II$. Similarly, the binding of less electronegative elements decreases the oxidation state since these give electrons to the central atom.

Table 5.17 Oxidation states

Nitrogen compounds				Sulphur compounds			Carbon compounds	
Substance	Oxidation states			Substance	Oxidation states		Substance	Oxidation states
NH_4^+	$N=-III$,	$H=+I$		H_2S	$S=-II$,	$H=+I$	HCO_3^-	$C=+IV$
N_2	$N=0$			$S_8(s)$	$S=0$		HCOOH	$C=+II$
NO_2^-	$N=+III$,	$O=-II$		SO_3^{2-}	$S=+IV$,	$O=-II$	$C_6H_{12}O_6$	$C=0$
NO_3^-	$N=+V$,	$O=-II$		SO_4^{2-}	$S=+VI$,	$O=-II$	CH_3OH	$C=-II$
HCN	$N=-III$,	$C=+II$,	$H=+I$	$S_2O_3^{2-}$	$S=+II$,	$O=-II$	CH_4	$C=-IV$
SCN^-	$S=-I$,	$C=+III$,	$N=-III$	$S_4O_6^{2-}$	$S=+2.5$,	$O=-II$	C_6H_5COOH	$C=-II/7$

In acid–base reactions, clearly, the oxidation states of the elements must not change. Thus, in the reactions (C oxidation states given)

$$CH_3-COOH+HOCH_3 \longrightarrow CH_3-COO-CH_3+H_2O$$
$$-III \quad +III \qquad -II \qquad -III \quad +III \quad -II$$

$$\text{and } Fe^{3+}\cdot H_2O_2 \longrightarrow Fe^{3+}\cdot O_2H^- + H^+,$$

the oxidation states of C, H, O and Fe do not change.

We wish to describe quantitatively the equilibria between oxidation states of elements including the elemental state itself since (1) there will arise opportunity for the natural selection of the elements by use of different binding in different oxidation states; and (2) the variance of a system depends on the rate of exchange of electrons between components, for example Fe^{2+} and Fe^{3+} may be considered as two separate components or as in equilibrium with the potential energy of electrons, e, which means that iron contributes one component only.

5.7.1 Oxidation–reduction potentials in water

In order to understand how elements bind in given oxidising or reducing conditions we need to discover the relative stability of different oxidation states. We start from the stability of the hydrated ion M^{2+} relative to the solid metal, which can be understood in terms of the thermodynamics of the theoretical process,

$$M+H_2O \rightleftharpoons M^{n+}+n\cdot e.$$
$$\text{(Solid)} \qquad\quad \text{(aqueous solution)}$$

Just as for any other equilibrium system we can assign to it an equilibrium constant and a free energy change ΔG° in standard conditions using e, the electron, as if it were at an independent concentration in the solution, representing reducing potential.

By convention, we write the reaction above as a reduction so that

(Ions) oxidised form $+n\cdot e \rightleftharpoons$ reduced form (metal atoms).

For this *theoretical* process the equilibrium constant, K, is given by

$$K=[\text{reduced form}]/[\text{oxidised form}]\cdot[e]^n$$

and $\Delta G^\circ = -RT \ln K$.

There is an important distinction, however, since we are here relating a particular chemical change, oxidation state change, to the free energy of electrons. Hence we can link chemical changes directly to electrical potential energy (volts) instead of to heat (joules). It is then usual to express the value of ΔG° (all concentrations at 1 M) in terms of the so-called standard electrode reduction potential E° as $\Delta G^\circ = -nE^\circ\cdot F$, where n is the number of electrons involved in the reaction and F is the Faraday constant (96 500 coulombs per mole of electrons). When E° is expressed in volts, ΔG° is obtained in joules.

Obviously $E^\circ = RT/nF\cdot \ln K$ and negative values of E° correspond to negative values of $\ln K$, that is, to $K<1$, which means that there will be a pronounced tendency for the metal (reduced form) to give ions M^{2+} (oxidised form) in solution plus electrons that will be available to reduce the oxidised form of any other similar system present. Large negative values of $E°$ correspond, therefore, to strong reducing agents (in the reduced form) and, similarly, large positive values of $E°$ correspond to strong oxidising agents (in the oxidised form).

A more general equation, for conditions different from those of the coventional standard states is called the Nernst equation and has the form

$$E = E° + \frac{RT}{nF} \ln \frac{[\text{oxidised form}]}{[\text{reduced form}]}$$

$$= E° + \frac{0.059}{n} \log \frac{[\text{oxidised form}]}{[\text{reduced form}]} \quad \text{(at } T = 298 \text{ K)}$$

where E is called the 'redox potential'. This is obviously the same equation but in different units as that given for ΔG in terms of $\Delta G°$ in Section 3.5.

We now need a reference for the redox potential since we cannot measure directly the free energy of electrons. The standard hydrogen potential in water is normally used corresponding to the half-reaction

$$\underset{(1 \text{ M})}{2H^+} + 2e = \underset{(1 \text{ atm})}{H_2}$$

which is conventionally given the value $E° = 0$. (This is an equivalent procedure to the determination of thermodynamic $\Delta G°$ values with reference to standard state conditions in Section 3.4.)

For any system, say $M^{2+} + 2e = M$, we have the free energy difference, now the difference of electrode potential of the coupled redox system, namely

$$\underset{}{M} \rightleftarrows \underset{1 \text{ M}}{M^{2+}} \qquad \underset{1 \text{ M}}{H^+} \rightleftarrows H_2(1 \text{ atm}).$$

This will give us $E°(M)$ for this system and we can calculate the redox potential E in any condition using the Nernst equation for equilibria. $E°$ values like $\Delta G°$ values give a scale from which equilibrium concentrations are calculated.

Elements such as Na, K, Mg, Ca and Al have very negative standard redox potentials (Table 5.18(a)) so that in the metallic state (reduced form) they are such strong reducing agents, so unstable in water, that we shall not need to consider this metallic state in aqueous solution (and neither, of course, in biology). However, these elements are extremely important, especially Mg and Al, in man's industry. Large amounts of energy are necessary to obtain them from oxides (Chapter 14).

Amongst the transition metals, the following also reduce H^+ in water, (themselves going to M^{2+}) in the order

$$\text{Mn} > \text{Fe} > \text{Co} > \text{Ni} > \text{Cu} < \text{Zn}.$$

Their electrode potentials decrease from Mn to Cu. Note that the tendency for reaction of these metal ions to go to metal still observes the order of the Irving–Williams series and the order of ionisation potentials (Fig. 2.23), that is, the tendency of M^{2+} to be covalently bound. Copper prefers to stay bound to electrons (of copper), that is, to remain in the (covalent) metallic state, rather than to become the hydrated ion.

Although the discussion has been given for metals, obviously it applies to

Table 5.18 Redox potentials of some ion systems at pH = 0, $T = 25°C$

(a) Metal ion systems

Element	$E°*$	n	Main mineral source	Order in which metal was used by man†
K	−2.92	+1	KCl, KCl·MgCl$_2$	
Na	−2.71	+1	NaCl	
Ca	−2.87	+2	CaSO$_4$, CaCO$_3$, Ca$_3$(PO$_4$)$_2$ CaCl$_2$ (Solvay process)	
Mg	−2.36	+2	Mg salts, MgCO$_3$	5/6
Al	−1.66	+3	Al$_2$O$_3$	5/6
Mn	−1.18	+2	MnO$_2$	
Cr	−0.74	+3	FeO · Cr$_2$O$_3$	5
Zn	−0.76	+2	ZnS	3
Fe	−0.44	+2	Fe$_2$O$_3$, Fe$_3$O$_4$	4
Co	−0.28	+2	CoAsS, Co$_3$S$_4$	5
Ni	−0.25	+2	Sulphides	5
Sn	−0.14	+2	SnO$_2$	3/2
Pb	−0.13	+2	PbS	3/2
Cu	+0.34	+2	Metal, sulphide	2/3
Hg	+0.85	+2	HgS	3
Ag	+0.80	+1	Metal, Ag$_2$S, AgCl	2
Au	+1.7	+1	Metal, tellurides	1

(b) Some non-metal ion systems

Element	$E°‡$	n	Mineral source
O	+1.23	+2	Water (H$_2$O)
S	+0.14	+2	FeS
Se	−0.40	+2	—
Te	−0.60	+2	—
F	+2.87	+1	Na$_3$AlF$_6$
Cl	+1.36	+1	NaCl
Br	+1.07	+1	Seawater
I	+0.54	+1	Seawater

* $E°$ for the reaction M^{n+} (aq) + $ne \rightleftharpoons$ M where n is given in the next column.
† From Chapter 14.
‡ $E°$ for the reaction X$_2$ (aq) + nH$^+$ + $ne \rightleftharpoons 2$H$_n$X (aq).

non-metals too. One example is the hydrogen electrode but the analysis is general (see Table 5.18(b)).

We turn next to the oxidation–reduction potentials between different oxidation states of one element in aqueous solution which correspond to the free energy in volts for a given reaction. We consider the metal elements first, where

$$M^{(n+1)} + e \rightleftharpoons M^{n+}.$$

Once again we use the standard reference reaction

$$H^+ + e \rightleftharpoons \tfrac{1}{2}H_2 \qquad E° = 0.0 \text{ volts}$$

to obtain the redox potential for the metal redox potential. We can then use the

the Nernst equation to determine E at any pH or metal ion concentrations, that is,

$$E = E^\circ + 0.059 \log[M^{n+1}]/[M^{n+1}].$$

Consider redox changes from M^{2+}. The case of going from M^{2+} to M^+ in water, which is dominated by the ionisation energies given in Fig. 2.23, follows the order

$$Mn < Fe < Co < Ni \ll Cu(\gg Zn).$$

In fact, the state of copper ions in the absence of oxygen is Cu^+. This means that in geological systems and in man's industry copper can be separated (isolated) from the other elements by reduction either to Cu^+ or to copper metal. This feature of copper chemistry initiated man's development of metal tools but we shall propose that it was also involved in the beginning of multicellular evolution; soluble copper salts arose with the advent of O_2.

Now, these elements can also be oxidised to higher states M^{3+} in the order of ease of oxidation, namely,

$$Mn < Fe > Co > Ni > Cu > Zn.$$

This means that iron can be separated from all these elements by using the stability of the Fe^{3+} state. This is used both in man's analytical and biological separations. Parallel orders can be given for the oxidising powers of different oxidation states of non-metals. Now the conventional redox potentials are referred to pH $= 0$ (1 M H^+) but the pH of the sea and in biology is close to pH $= 7$. We need to look at redox reactions at this pH, but before doing so the stability of water itself to reduction (H_2) and oxidation (O_2) must be examined.

5.7.2 The H_2/O_2 potentials in water at pH $= 7.0$

Water is the solvent for all the above reactions, but we have to remember that it too is open to reduction by electrons or it can give electrons in the reactions

$$\text{(a) } \tfrac{1}{2}H_2O + e \rightleftharpoons \tfrac{1}{2}H_2 + OH^-$$

$$\text{(b) } \tfrac{1}{2}H_2O \rightleftharpoons \tfrac{1}{4}O_2 + H^+ + e.$$

Reaction (a) can be rewritten as

$$\text{(c) } H_3O^+ + e \rightleftharpoons \tfrac{1}{2}H_2 + H_2O$$

using the acid–base reaction $H^+ + OH^- \rightleftharpoons H_2O$. Now (c) for $[H_3O^+] = 1$ M that is, for $[H^+] = 1$ M and for H_2 pressure $= 1$ atmosphere is the standard reference potential chosen above, $E^\circ = 0$, whence we can calculate E at any pH using the Nernst equation, that is, $E = 0 + 0.059 \log[H^+]$. Thus, the decomposition potential for H_2O to H_2 (1 atm) at pH $= 7$ is -0.42 volts. It is easy to see that thermodynamically a reducing system of potential < -0.42 volts is hard to maintain unless there is a barrier to H_2 release (a so-called overpotential). It is conventional in many situations to use a reference potential, E_H of -0.42 volts to indicate possible reductions that will not release H_2 easily from water. This gives an indication of stability (thermodynamic not kinetic) in water at pH $= 7$. Of course, biological systems must not approach $E = -0.42$ volts too closely.

The stability of water to oxidising potentials written above is given by (b),

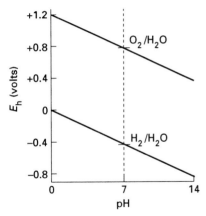

Fig. 5.12 The potential of the water redox system at 25°C, H_2 pressure = 1 atm and O_2 pressure = 1 atm, respectively. E_h stands for the redox potential relative to the standard hydrogen couple $H^+ + e \rightleftarrows \frac{1}{2}H_2$ at pH = 0 and H_2 pressure = 1 atm.

and again at pH = 7 and 1 atmosphere of O_2 we can calculate the redox potential as $E = +0.81$ volts. This voltage represents an upper thermodynamic limit to the stability of oxidising systems (thermodynamic not kinetic) in the sea and in biology. (At pH = 0 the value of $E°$ for O_2 is $+1.23$ volts.)

A graph of redox potential limits of water stability against pH is shown in Fig. 5.12. Such a graph can be drawn for any element and we now do this for both metals and non-metals. They show selective behaviour of elements in response to the presence of the *electron* at various potentials.

5.7.3 Metal ion oxidation states at pH = 7.0

In Fig. 5.13 we plot the free energies in volts of the oxidation states of several metals at pH = 7.0 against the hydrogen electrode potential at pH = 7.0. The slope of a line between any two points is the redox potential. The diagram shows that, in the presence of dihydrogen at 1 atmosphere, many oxidation states are unavailable including Cu^+, Cu^{2+}, Ni^{2+}, Co^{3+}, Mn^{3+}, Fe^{3+} and VO^{2+}.

Fig. 5.13 Oxidation state diagram for some metals at pH = 7 (*nE* versus *n*) taking into account the formation of hydroxides where *n* is the oxidation state and *E* the electrode potential from the element to that state. Note that the reference potential is $Pt \cdot H_2/10^{-7}$ M H^+, that is pH = 7.0 shown top left and not that of the standard H_2/H^+ one molar electrode.

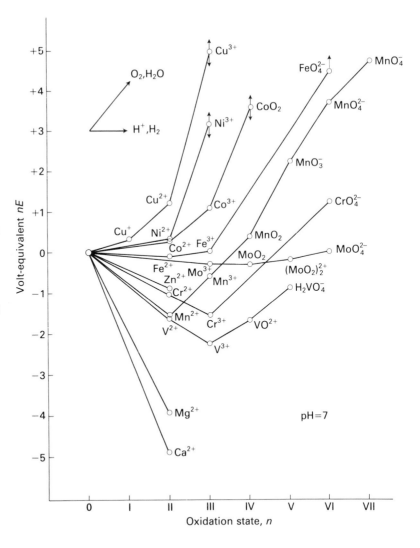

The available states include, for example, all divalent ions (Mn^{2+} to Zn^{2+} except for Cu^{2+} and Ni^{2+}), and states of Mo up to MoO_2. The restriction by redox reactions is obvious. In the presence of O_2 at 1 atmosphere the states that are stabilised are Cu^{2+}, Ni^{2+}, Fe^{3+}, Co^{3+}, MnO_2, CrO_4^{2-}, MoO_4^{2-} and $H_2VO_4^-$. The highest oxidation states are not always reached, for example, those of Cu, Ni, Co, Fe and Mn, although they may well be important as intermediates later in catalysis (Chapter 7).

In the presence of dioxygen, as opposed to dihydrogen, the peculiar ease of oxidation of iron (rusting) is well known to man and is of inestimable importance in geochemistry and biology. In biological and geological systems the special chemistry of iron as Fe^{3+} allows it to be stored as $Fe(OH)_3$ in ferritin or to be made into magnets (Fe_3O_4) and it does allow special inorganic clusters to be generated such as Fe_2O^{4+} and Fe_nS_m, where $n = 2$ to 6 and $m = 2$ to 6, which are valuable in catalysts, either in biology or industry. Biology uses these units as catalysts, for example, in electron transfer reactions, while man more often uses iron metal. A further development of the particular chemistry of this element is its easy separation as Fe^{3+} from the other available transition metal elements while they remain as M^{2+}. In biology this has allowed iron to be associated with specific functions and connected to genetic expression in special ways. Note, however, that the solubility product of $Fe(OH)_3$ is very low, so that this ion is practically removed from aqueous solutions unless strongly complexed. This removal, which was a consequence of an oxygenated atmosphere, was (and is) one of the major problems that living organisms faced (and face).

In the context of this chapter, the importance of the different oxidation state stabilities is that elements can be selectively bound by reaction under specified reducing or oxidising atmospheres. Just as pH affects competition so does the ambient redox potential.

5.7.4 C, N, O and S oxidation states in aqueous solutions

The oxidation states of C, N and S can be written X(n) where n goes from $-IV$ to $+IV$ for carbon, $-III$ to $+V$ for nitrogen, $-II$ to $+II$ for oxygen, $-II$ to $+VI$ for sulphur (see Fig. 5.14). (Note that all oxidation states are written in Roman numbers.)

As might be expected, the redox states of oxygen complexes (XO) fall in the order of oxidising power

$$F(O) > O(O) > N(O) \gg C(O)$$

$$Cl(O) > S(O) \gg P(O) > Si(O)$$

while the reducing power of HX falls in the order

$$H_2 > C(H) > N(H) > O(H) > F(H),$$

$$Si(H) > P(H) > S(H) > (H_2) > Cl(H).$$

As mentioned before, the oxidation states of Si and P in aqueous solution are mainly confined to states $+IV$ and $+V$, respectively. This non-metal chemistry, together with the above metal chemistry, dominates much of geology and biology.

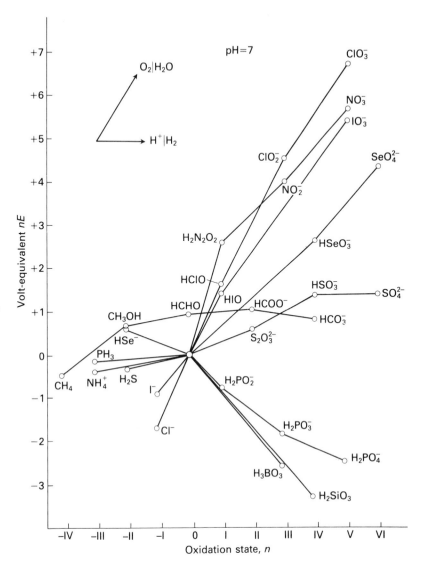

Fig. 5.14 Oxidation state diagram for non-metals at pH = 7 (nE versus n) taking into account protonation reactions. Note that the reference electrode potential is $Pt \cdot H_2/10^{-7}$ M H^+, that is pH = 7.0 shown top left and not that of the standard H_2/H^+ one molar electrode.

Thus, the thermodynamic energetics of redox reactions in water can easily be established, but we must be aware that all this chemistry is very frequently under kinetic control. Man, to a greater degree than geological or biological systems, can avoid the circumstance of working in reactive water, which forces chemicals toward equilibrium and hence he has opened up possibilities unknown before in the natural selection of elements, but his organic and metalloorganic chemistry, like biological chemistry, is under kinetic control too.

5.7.5 Oxidation states, components and variance

Since we are trying to understand the variables in a chemical system the question now arises as to which oxidation states are components in the sense that they do not exchange atoms, for example, of O or H, or exchange electrons. We readily accept CO, CO_2 and O_2 in the air as separate variables despite the fact that at high temperature there would have been an equilibrium

$$2CO + O_2 \rightleftharpoons 2CO_2$$

when there would have been two, not three components. This equilibrium is not *operational* at ambient temperature. For many purposes the different oxidation states of non-metals give rise to separate components, for example, NH_3, N_2 and NO_3^-, or H_2S and SO_4^{2-}, and, in fact, this is also the case for some oxidation states of metals at 25°C. However, at somewhat higher temperatures or if we consider equilibrium as being attained over very long periods of time, then virtually all metal oxidation states equilibrate. When we describe the formation of the Earth, at > 1000°C, we shall assume equilibrium between all metal and non-metal oxidation states (Chapter 8). When we describe the case when Earth had cooled below 100°C we must start from almost the opposite position. Thus, on cooling, the number of possible components greatly increased since oxidation states of metals and non-metals can combine in many ways. In the description of biological systems at around 25°C we shall have to be very careful with the definition of the components, since these control the variance of the system. On what time-scale is exchange slow?

5.8 Combined acid–base and redox equilibria

In the previous discussion of types of reaction in aqueous solutions we have been concerned with equilibria in which the two sides of a simple chemical equation are balanced. This applies to $\log K_{sp}$, $\log K$ and $E°$. The next step in this analysis is to combine all possible stable states involving, say, one metal, with all possible non-metal partners in just one diagram. There are many ways to do this, in simple or more complicated pictures, but the objective is to show, for example, how the relationship between one metal and several non-metals changes as the oxidation/reduction state of the system is adjusted and the pH varied. Before giving some examples of such diagrams it should be realised that the potential of redox couples in which one of the members contains bound oxygen is, by necessity, pH-dependent since the proton appears in the equilibrium reactions.† This is extremely important since water itself may be oxidised or reduced above or below its range of stability as shown again in Fig. 5.15. Now, in the same figure we show the ranges of existence of some sulphur species, from which we can conclude, for example, that in reducing media, say $E < -0.2$ volt, and up to pH ~ 7, SO_4^{2-} cannot be formed and H_2S is predominant. This is relevant information, for example in the study of prebiotic and primitive anaerobic sulphur cycles (see Chapter 15).

In the following figures we give more complicated diagrams showing the stability regions (in the presence of water) for various iron species (Fig. 5.16), manganese species (Fig. 5.17) and zinc and copper species (Fig. 5.18). The way these diagrams were established is described in references 4 and 5 of 'Further reading'. In one sense they are 'phase' diagrams in which two

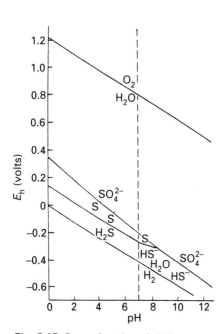

Fig. 5.15 E_h as a function of pH for selected sulphur species in the stability range of water. The symbol E_h means that the potential is pH dependent. S is elemental sulphur.

† It is common in biological texts to give $E°'$ at pH = 7 by adding to $E°$ the appropriate $RT/F \ln . 10^{-7}$. This allows direct comparison of redox systems at the cytoplasmic pH.

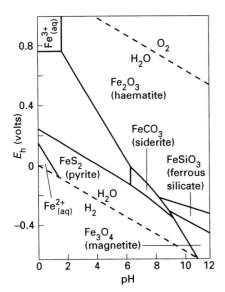

Fig. 5.16 E_h–pH diagram for FeS$_2$, FeCO$_3$, Fe silicates, and Fe^{n+} (aq). The non-metals and Fe are at equilibrium with the atmosphere and with the precipitates. After Garrels, R. M. (1960). *Mineral equilibria.* Harper and Brothers, New York; Garrels, R. M. and Christ, C. L. (1965). *Solutions, minerals and equilibria.* Harper, and Row, New York; and Brookkins, D. G. (1988). E$_h$–*pH diagrams for geochemistry.* Springer Verlag, Berlin.

variables, pH and E, are changed at equilibrium assuming the presence of fixed concentrations of H$_2$S and CO$_2$.

This type of representation, which refers to equilibrium conditions, gives us an easily understandable introduction to the essential background against which we have to appreciate real systems, for example, the sea. Above all, it shows the most stable condition and hence the direction in which unstable, out-of-balance, systems will tend to go. Of course, the exact path of change is kinetically controlled (Chapter 7).

5.8.1 Redox potentials and complex ion formation

The description of redox potentials in solution given in Figs 5.15–5.18 concerns the interaction energies of the central element in two oxidation states when bound to inorganic components of the present or early atmosphere, including H$_2$S, and in equilibrium with precipitates. Thus we have described the effects of the presence of H$_2$O as OH$^-$ and O^{2-}, of H$_2$S as mainly HS$^-$ and of CO$_2$ as CO$_3^{2-}$. (Only small effects are due to chloride binding.) Now there are major features of organic and biological chemistry involving metal ions that depend upon the complexes formed between organic donors and inorganic acceptors. We have described in outline the stability of the individual associations but it is the difference in strength of binding of two oxidation states of the same element that governs the redox potentials. Thus, the redox potential of a couple M^{3+}/M^{2+} that bind to an organic ligand with binding constants K'_3 and K'_2, respectively, at pH = 7 is given by

$$E = E^\circ - \frac{RT}{F} \ln \frac{K'_3}{K'_2}$$

where K'_2 and K'_3 are the effective stability constants at pH = 7.0 (see Section 5.6.2). The ratio K'_3/K'_2 can be adjusted by choice of ligand so that E is either more oxidising if $K'_2 > K'_3$, or more reducing if $K'_3 > K'_2$. In this way, for example the redox potential of Fe^{3+}/Fe^{2+} in water at pH = 7 can be changed from + 0.15 (aqueous ions at pH = 7) and put by complexing to ligands to between

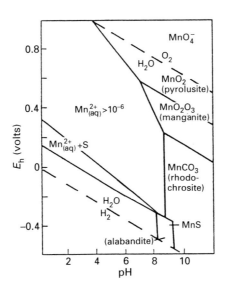

Fig. 5.17 E_h–pH diagram for some manganese species. (Source: Garrels, R. M. (1960). *Mineral equilibria.* Harper and Brothers, New York; Garrels, R. M. and Christ, C. L. (1965). *Solutions, minerals and equilibria.* Harper and Row, New York.)

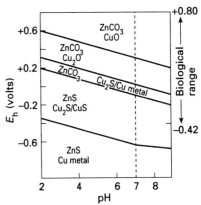

Fig. 5.18 An E_h–pH diagram showing the stability zones of zinc and copper species at 25°C and 1 atmosphere.

-0.42 (when H_2 is liberated) and $+0.4$ volts. More specifically in biological systems the sulphur complexing of rubredoxins (Fig. 5.19(a)) can give Fe^{3+}/Fe^{2+} potentials at least in the range -0.4 to 0.1 volts while the Fe^{3+}/Fe^{2+} potentials of porphyrin complexes of haem proteins (Fig. 5.19(b)) vary from, say, -0.3 to $+0.4$ volts. The factors controlling these and other redox potentials are readily understood from the description of stability constants in Section 5.5.2. It is the skilful use of binding differences that allows many oxidation states to be extremely valuable in catalysis. The design of ligands to give optimal value to redox reactions is often described as giving *fitness* to function. This topic is thoroughly covered in our previous book, reference 2 in 'Further reading', and will be mentioned again in Chapter 16.

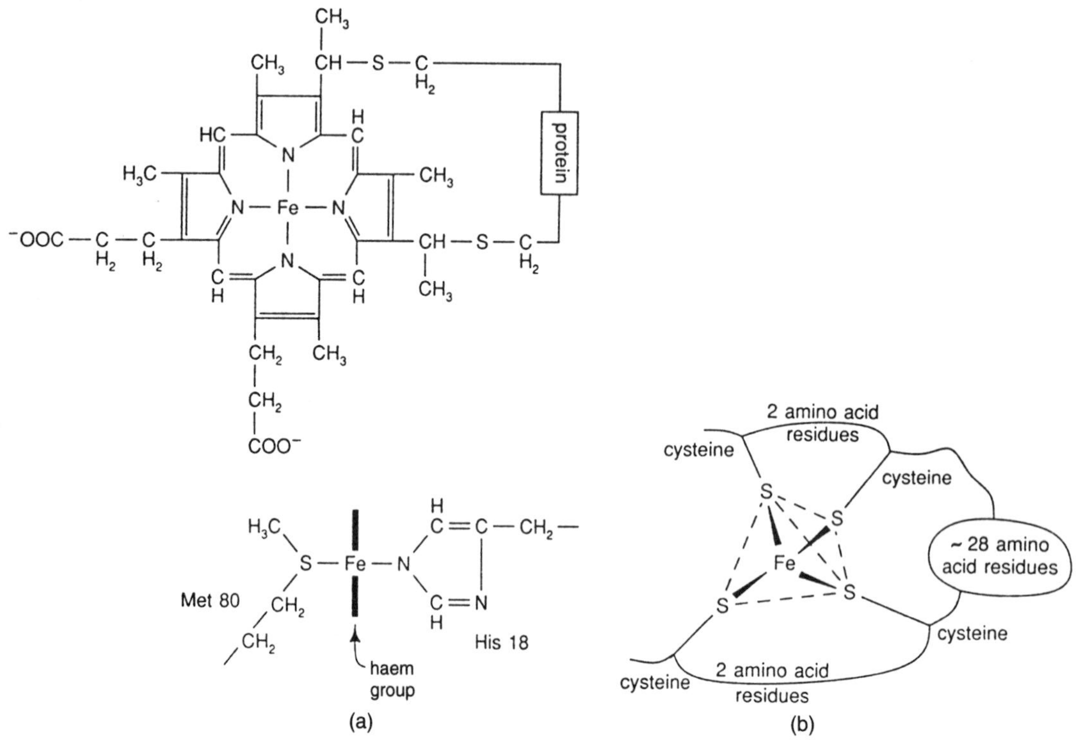

Fig. 5.19 (a) The haem group and iron co-ordination in a protein, cytochrome *c*. (b) The co-ordination of iron in a thiolate centre of a protein, rubredoxin.

5.8.2 The change from the H_2S/S_n to the H_2O/O_2 potential (or vice versa)

Later in this book we shall have occasion to consider the slow change from reducing conditions to oxidising conditions when the level of dioxygen increased in the Earth's atmosphere due to the photosynthetic activity of cyanobacteria some 3 to 1×10^9 years ago. This has had a large impact on the chemistry of the Earth's surface and completely changed the biosphere through changes in oxidation states of many elements. Today we can appreciate a similar (and more rapid) phenomenon but in the opposite direction, from oxidising conditions to reducing conditions, for example in polluted rivers and lakes when dissolved oxygen is consumed and microorganisms resort to other terminal oxidants, namely NO_3^-, SO_4^{2-} and CO_3^{2-} (the smell of NH_3 or H_2S gives us an indication of the advance of pollution).

To appreciate these trends (in one direction or the other), we show in Fig. 5.20, by using arrows, the sweep of the redox change so that it can be viewed against Fig. 5.21. It should immediately be noticed that there are changes in the oxidation state of non-metals (as above) but also of metals (we have mentioned iron) and there must be a correlation between the redox states of all elements at equilibrium at a fixed E. For example, it can be seen that in reducing conditions many metals especially in low oxidation states can bind to sulphide without the problem of oxidation of either. In oxidising conditions there is the possibility of sulphur polymerisation (frequently dimerisation) and metal oxidation must also be considered. The corresponding equilibria are

Reducing conditions

$$n\text{RS}^- \quad + \quad \text{M}^{n+} \Longleftrightarrow \text{M}^{n+}(\text{RS}^-)_n \Longleftrightarrow \text{(precipitate)}$$

$$\frac{n}{2}\,\text{R–S–S–R} \qquad \text{M}^{(n+1)+}$$

Oxidising conditions

In such conditions the metals that are able to remain bound to sulphide groups are copper (Cu^+ especially) and molybdenum. In many other cases the metal sulphide bonds are not stable in oxidising conditions. We return to this problem when we consider both geochemical and biochemical evolution.

A dominant feature of this chapter is that it has shown that the natural selection of the chemical elements for partners follows patterns in solution very like those found in compounds in Chapter 2. Thus dilute organic oxygen

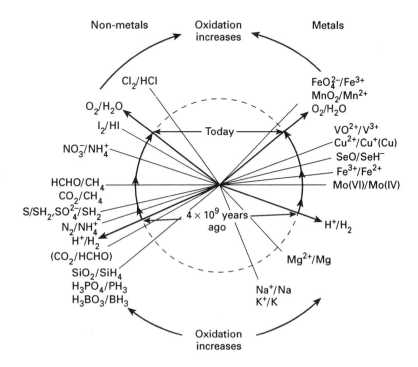

Fig. 5.20 The sweep of rising redox potential at pH = 7.0 with the evolution of Earth's surface showing changes imposed on metals from lower to higher positive oxidation states and on non-metals from higher to lower negative oxidation states.

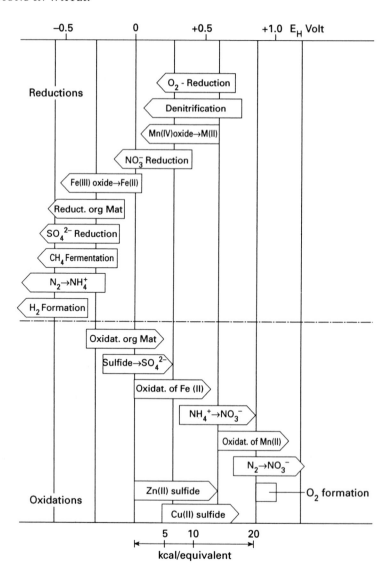

Fig. 5.21 Sequence of microbially mediated redox processes. (Adapted from Stumm and Morgan see references.)

donors parallel bulk oxides in their selectivity relative to dilute organic sulphur donors which parallel bulk sulphides. In a general way this means that *at equilibrium* atomic and/or ionic properties of elements, that is, charge, oxidation states, electronegativity (ionisation potential and electronic affinity), and size govern behaviour in both inorganic and organic media. In the broadest sense the same thermodynamic factors must lead to closely parallel selective associations of metals and non-metals in man-made chemical (analytical), geochemical and biological solids and liquids.

5.9 Summary: variance in solutions

In Chapter 4 we described the factors that restricted the variance (number of degrees of freedom) of chemical systems in terms of the phase rule. We observed particularly the number of equilibria involving partition of *bulk component chemicals* across phase boundaries with unlimited amounts of each phase present. These equilibria restricted variance. We also looked at chemical equilibria between bulk species, which are again restrictions on variance. As we have seen in the same chapter, in conditions where all elements can exchange freely between all compounds and phases, the number of elements present corresponds to the number of independent components and this is the limit of variance at fixed pressure and temperature. Such a condition existed at very high temperature before Earth formed. On Earth, at very modest temperatures, we noted that the elements do not exchange between some classes of compounds, especially organic compounds, so that variance, through the number of components, greatly increased. To describe the situation we considered that the effective number of *bulk components* was not the number of elements in a system but the number of species that could be varied independently, that is, were not in equilibrium with one another. (We could specify the equilibria concerned only in a limited manner (Section 4.3.1).)

In this chapter we have considered not phase equilibria in general but chemical equilibria between dilute species in one phase, especially in water since it is in this phase that the natural selection of elements has given rise to life. We observe again that organic compounds soluble in water do not exchange their atoms with the H and O atoms of H_2O or with other organic molecules or with $CO_2 + H_2O$ so that the organic molecules are separate components (even though they are now dilute) and can vary quite independently. (The exception of organic protonic acid equilibria was stressed.) Synthesis and degradation of organic compounds are not considered in this discussion and, effectively, we can add as many of them to water as we like and still not interfere with the behaviour of any other organic component by equilibrated element exchange. The inorganic compounds in solution are usually different in one very important respect: in general they exchange elements with water in their co-ordination sphere, for example many exchange hydroxide or oxide, and we discussed the corresponding hydrolytic equilibria based on variation of pH. Thus, in water, the concentration of, for example Al^{3+} in water when $Al(OH)_3$ is present as a precipitate, is not an independent variable from that of H^+ which varies with OH^-. The number of equilibria reduces the variance as does the number of phases involved in the equilibria, see marginal illustration.

To the physical exchange equilibrium generated by the formation of new phases, analysed in Chapter 4, there is now added, therefore, a vast range of other chemical equilibrium restrictions in any one phase which are immediately obvious in the metal complex ion equilibria described above. More generally, in any co-ordination chemistry reaction

$$A + B \rightleftharpoons AB$$

for example,

$$Cu^{2+} + 4(NH_3) \rightleftharpoons Cu(NH_3)_4^{2+},$$

there are three chemical species but only two independently variable components of the system—the concentrations of Cu^{2+} and NH_3 in the example given. Likewise, if two organic molecules act as components and

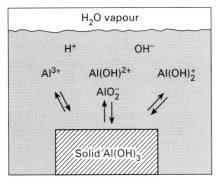

The number of compounds present is large but the number of components present is small (see text)

enter into equilibria, then, labelling them A and B as above, the molecular complex AB is not independently variable. This could apply to protein assembly or, in fact, to the whole of the assembly of an organism if it were an equilibrium system, which it may be in fair part (Chapters 10–13).

Inorganic elements, especially as ions in water, can therefore be treated frequently in terms of the elements themselves rather than in terms of their more or less complicated species when describing components. Thus Fe^{3+}, Mg^{2+}, Cl^- and OH^- are components in equilibrium, and complexes formed between them are not components, which contrasts with the absence of equilibria between, say, CO, CH_3OH, CH_3COOH, $C_6H_{12}O_6$ and so on. The distinction is explained in Chapter 7 in terms of rates of exhange due to types of bonding described in Chapter 2. Even if the oxidation state of an inorganic element is changed, this does not mean that a new component is necessarily introduced. Thus, although NO_3^-, H_2O, O_2 and N_2 do not equilibrate, there may well be equilibrium between Fe^{2+}, Fe^{3+}, H^+ and O_2 in the sea. The concentration of different oxidation states of inorganic elements in solution is often restriced by such equilibria, unlike the presence of different oxidation states of H, C, N, O, in particular, and P, S, Cl to a lesser degree. Within these considerations any direct link from an inorganic element to carbon does, however, generate a new component since the resulting compound does not exchange carbon or metal atoms with other molecules—organometallic chemistry is to be likened to non-exchanging organic chemistry rather than to exchanging inorganic chemistry. A little thought shows that we are, in fact, always introducing *operational conditions* when we allow an *absolute* distinction between non-exchanging systems on one hand and exchanging systems at equilibrium on the other. *The operational distinctions concerning variance have to be made in terms of rates of exchange within time intervals of concern.* This consideration is critical to an understanding of biological processes.

There are then two important conclusions to be derived.

1. Any chemical system has a given variance (number of degrees of freedom) opposite composition. The composition is defined by the concentrations of the independent components, but formation of a new phase at equilibrium, for example a precipitate, restricts variance (or degrees of freedom).

2. Due to differences in local availability of elements, pH and redox conditions, there is a vast possibility for chemical variation in local aqueous compartments, much as there was in silicate minerals (Section 4.4); note that biological systems are based on components in aqueous solutions contained locally in an organism.

Given the huge variety of organic components in balance with a variable supply of inorganic ions, we can only expect a continuous variation in the composition of solutions as we indeed observe in the lakes, seas and oceans. Once again, however, we note that in biology one does not observe a continuous series of variation in chemical and physical properties. Biological systems are broken into very narrow chemical composition and structure ranges within species, members of which are almost invariant. How is this managed? How is it possible for an organism to feed on a great variety of components (which is the equivalent of mixing them in water) and yet remain of fixed composition?

We shall have to search for an explanation of the anomaly of biological systems and our first enquiry must be to ask whether the examination of bulk phase considerations in Chapter 4 and 5 has led us astray since we have taken no account of *amounts* of material (only of percentage composition), of surface, or of shapes, or of the influence of fields. These factors will be the subject of Chapter 6. Especially we have to consider situations in which surfaces of phases (even as small as proteins) interact so as to give *co-operative assemblies*. We shall find that these considerations, interesting though they are, fail to explain biological (and some other) systems and we shall have to examine the way chemicals behave in time as well as in position with respect to one another.

Further reading

Most of the subjects discussed in this chapter have been dealt with at length in our two previous books, to which the reader is referred.

1. Phillips, C, and Williams, R. P. (1966). *Inorganic chemistry.* Oxford University Press, Oxford. Chapters 7, 9, 14, 26 and 32.
2. Fraústo da Silva, J. J. R. and Williams, R. J. P. (1993). *Biological chemistry of the elements—the inorganic chemistry of life* (reprint). Oxford University Press, Oxford. Chapters 1, 2 and 4.

Detailed treatment of reactions in aqueous solution can be found in many other standard texts; the following cover the subject from various complementary points of view.

3. Huheey, J. E., Keiter, E. A., and Keiter, R. L. (1993). *Inorganic chemistry.* Harper Collins College Publications, New York. Chapters 9, 10, 11 and 13. An up-to-date university textbook with the relevant basic information.
4. Stumm, W. and Morgan, J. J. (1981). *Aquatic chemistry.* J. Wiley and Sons, New York. Chapters 2, 3 and 5–7. A comprehensive treatise on reactions in aqueous solution.
5. Laitinen, H. A. (1960). *Chemical analysis.* McGraw Hill Book Co, New York. Chapters 1, 2, 3, 6, 7 and 15. The theory and the practice of chemical analysis at an advanced level.
6. Buffle, J. (1990). *Complexation reactions in aquatic systems—an analytical approach.* Ellis Horwood, New York. Chapters 1, 2, 5 and 6.
7. Siegel, H. (ed.) (1984). *Metal ions in the environment. Vol. 18. Circulation of metals in the environment.* Marcell Dekker, New York.

6

Limited phases, fields and compartments

There is no end to the weighing of things, no stop to time, no constancy to the division of lots, no fixed rule to the beginning and end. Therefore great wisdom observes both far and near, and for that reason recognises small without considering it faulty, recognises large without considering it unwieldy, for it knows that there is no end to the weighing of things.

Chuang-Tzu (*Fourth century* BC):
Texts–*Autumn floods* (Section 17)

A lump of clay is illusory; it is only its particles of earth that are real. Therefore whereas each such particle has its distinct substance, it is only through their aggregation that the lump of clay becomes formed and thereby comes to contain all the destinations of those many particles.

Chih-K'ai (AD 538–597)
Methods of cessation and contemplation

6.1 Introduction

In this chapter we wish to approach more closely the nature of the observable systems around us be removing in steps the assumptions that were employed in our discussions of equilibria in Chapters 4 and 5. These assumptions were that: (1) the phases had no surfaces, but as soon as we limit the total amount of a system, which we must do now, surfaces become important—the real world around us is constrained by divisions of space with surfaces; (2) there were no fields acting on a phase; however, as soon as we consider contact or a region between phases, that is, phases with surfaces, we must consider fields— gravitational, mechanical and electrical; (3) there was equilibrium transfer of material and energy between phases, but in the real world there are boundaries between compartments† containing phases that prevent at least some transfer; for example, the presence of semipermeable membranes is the basis of living constructs. Clearly, these three additional features, surfaces fields, and transfer of matter or energy, limit the applicability of the previous discussion of both bulk phases (Chapter 4) and of dilute solutions in isolated phases (Chapter 5). We must, therefore, look first at the effects of limited total amounts of material and the presence of fields on the variance of systems, as predicted by an extended form of the phase rule. We need then to describe the effects of barriers to transfer between compartments, that is, removal of the transfer equilibria for both bulk and dilute components. We shall find that, when we add all these factors together, substances around us cannot but be extremely diversified and far from even local equilibrium. However, as we explained in Chapter 3, it is only by reference to equilibrium that we can gain the basic understanding necessary for the appreciation of out-of-equilibrium systems and the directions of change. One of the major considerations in Chapters 4 and 5 was to establish the limits of variance of systems, that is, the number of degrees of freedom, from the use of equilibrium considerations in the phase rule. We, therefore, start the discussion in this chapter with a brief summary of the position we established in those chapters.

As stated in Chapter 4, if chemistry had remained at equilibrium in a uniform expanding and evolving universe, the naturally occurring 90 chemical elements would have formed a unique set of chemical compounds in particular phases, which would have changed predictably with further cooling while still being at equilibrium—the chemical elements would have been the only independent components in the separated phases. Given, in addition, that there was a fixed ratio of the amounts of the elements, the system would have been completely defined even until the present day. Our observations showed that this did not happen and that the expansion and cooling led to a large number of components out of equilibrium. We then found it useful to consider not the number of elements as the chemical components but, in an *operational* treatment of the behaviour of matter, the number of independent combinations that did not exchange their elements, for example H_2O and CH_3OH had to be treated as separately variable components. Consequently, we examined in Chapter 4 very many systems that we discussed as having $(C-1)$ component variables, defining the system by *percentage* composition. As far as the number of phases, P, that formed were concerned, we considered that all the components were at physical and chemical equilibrium *between* all phases

---

† The distinction between a phase and a compartment made here is that, while between phases there is transfer of components and energy at equilibrium (see the derivation of the phase rule in Section 4.3), compartments are enclosed by a barrier that does not allow equilibration of one or more components or energy. A compartment may well contain many phases. In this chapter we shall not introduce compartmental barriers until Section 6.3.

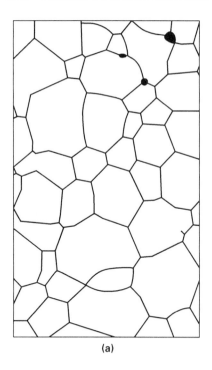

(a)

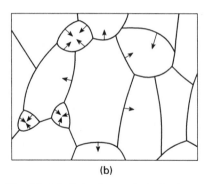

(b)

Fig. 6.1 (a) Grains in a typical mixed mineral. (b) Drop or grain growth mechanism via contact. The boundaries move toward the centre of curvature (arrows). As a result, the small units eventually disappear. Contact is not necessary since growth can occur through the vapour or liquid phases.

(hence P-1 only could be varied independently), and that any phase of a given substance was of sufficient size so that its surface was not significant. Thus, systems were examined as if neighbouring phases and their fields were absent, though some attention was paid to gravity in Section 4.5. We adopted this approach to bulk components and phases in the consideration of phase equilibria using the phase rule in Chapter 4, and included the further interactions of minor components in one phase and their partitioning between bulk phases, solids and liquids, in Chapter 5, but again we were not concerned with the volumes of the phases. We thereby came to appreciate the possible variables of chemical systems under these equilibrium constraints which could have decided the selection of the chemical elements in the observed geochemical and biological structures around us. Unfortunately for theoretical appreciation, but fortunately for our existence, the resulting descriptions do not match the variety of the systems around us, useful as they are for gaining understanding. When we look at geochemical or biological systems, ignoring any changes they are undergoing, it is very clear that, while the operational description of components developed in Chapters 3 to 5 may often be a useful approximation, the description of observed phases is not. First, as stated above, most of the space around us, except that which is gaseous, is made up of small particles close to or adjacent to one another (see Fig. 6.1(a)). In the description of these particles we cannot avoid the fact that they have limited sizes, volumes and shapes; thus their surfaces must be important.

There are, therefore, extra steps that we must now take in order to make the treatment of systems more realistic, still using local equilibrium considerations (Table 6.1). The first of these is to consider that the *total amount* (size) of any system of bulk components, not only the percentage amounts of components, is a variable in the same sense that temperature and pressure are variables. Obviously, this increases the number of degrees of freedom by one. A limited amount, volume, of material also has a *shape* and it is a matter of great interest to consider shape as another possible variable since all around us we see shaped material. The surfaces, related to shape, are of equal consequence for a system's properties as is its bulk. We, therefore, ask the question: do variations of shape and surface properties at equilibrium generate independent variables within limited phase systems? We take the simplest case of one component first.

Table 6.1 Constraints and restrictions at equilibrium

Constraints	Restrictions on
Pressure (high	Volume
Temperature (low)	Translational energy states, etc.
Atomic binding (compound formation) *	Atomic degrees of freedom in molecules or to exchange
Phase formation (condensed)	Molecular or atomic degrees of freedom in bulk
Amount	Surface/volume ratio; shape
General fields	Position (of mass or charge); shape

* This includes chemicals not in atom exchange with one another, that is, operational components at the temperature of the system.

6.2 One-component systems of limited volume

6.2.1 Sizes of phases

As stated, in real systems condensed phases have a size and this means, for example, that, since the *co-operative* energy is necessarily lower at the surface, the average energy per mole of the total phase is less than that of the bulk. Small phases are intrinsically unstable, so that small grains or crystals grow toward one large one and small droplets grow towards one liquid phase (Fig. 6.1(b)). Again the solubility in a solvent and the vapour pressure of a small phase, for example of small crystals or small droplets, exceeds that of a larger phase structure (Table 6.2). There is, therefore, a new degree of freedom in the phase rule for limited volume systems, even for one component with a condensed phase present. The rule of Section 4.3,

$$P + F = C + 2, \tag{6.1}$$

becomes

$$P + F = C + 2 + 1 \tag{6.2}$$

where the new physical degree of freedom is due to the required consideration of the total amount of material in the system. The next problem of concern is whether the shape of a limited phase is or is not an independent variable.

Table 6.2 Vapour pressure p of water droplets (water at 20°C, $p_0 = 2.3 \times 10^{-2}$ atm.)

r(mm)	p/p_0
10^{-3}	1.001
10^{-4}	1.011
10^{-5}	1.114
10^{-6}	2.95

6.2.2 Shape of a limited phase alone and in contact with a bulk phase

The shape of a limited volume of a liquid phase is clearly decided by the contact it makes with other phases. We consider first a small liquid phase interacting with a bulk phase. In the absence of fields, that is, in an isotropic environment such as in a gas phase, a small volume of liquid forms a sphere. On contact with a much larger, for these purposes infinite amount, of a second liquid or a solid, (effectively a uniform field, see below), which gives a horizontal surface, the liquid forms a spread zone and its shape (curvature) is reflected in contact angles due to 'wetting' or surface energies of interaction (Fig. 6.2). Examples

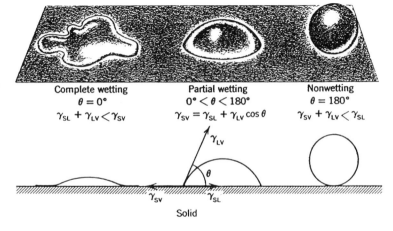

Fig. 6.2 Wetting of a substrate by a liquid, showing balance of forces, γ and the wetting angle, θ. S = solid, L = liquid, V = vapour.

Complete wetting
$\theta = 0°$
$\gamma_{SL} + \gamma_{LV} < \gamma_{SV}$

Partial wetting
$0° < \theta < 180°$
$\gamma_{SV} = \gamma_{SL} + \gamma_{LV} \cos \theta$

Nonwetting
$\theta = 180°$
$\gamma_{SV} + \gamma_{LV} < \gamma_{SL}$

γ_{LV}

θ

γ_{SV} γ_{SL}

Solid

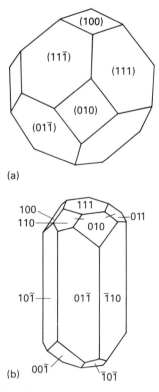

(a)

(b)

Fig. 6.3 The crystallography of: (a) a conventional cubic crystal Fe_3O_4; and (b) a crystal of Fe_3O_4 from a magnetotactic bacterium.

of the two cases are: (1) in the absence of effective fields the Earth, after remelting, took on its final form, closely related to that of a sphere; again, soap bubbles when formed in the air are spherical; (2) drops of some liquids or of oils or soaps spread on water. Extremes of drop shape extend from those shown in Fig. 6.2 to those which form a film. (Such spreading of oily liquids on water may have determined the initial form of the lipid membrane materials of biology.) These observations show that the shape of a small volume of liquid depends on its surface energy.

The shapes of ordered solids depend on surface energy too. In effect, the crystallisation of a solid from a bulk liquid or gas can generate a large variety of shapes, all consistent with the same packing internally; the implication is that the surface energy decides the crystal shape and is a property dependent on the phase in which the crystal sits—it is a common experience that different crystalline forms of the same substance are obtained from different solvents (Fig. 6.3). When there is more than one phase, such as small crystals in a solvent and droplets in a cloud, they are unstable in a thermodynamic timeless sense but can persist for very long periods. We often treat each drop or crystal as being in a permanent *stationary* state (see Fig. 6.5) as if it were at equilibrium. This is a useful treatment for many purposes in this chapter since it allows us to continue the use of the phase rule. There is then, apparently, a thermodynamic most stable state with a given shape in a given surrounding and in the next section we shall show why this is so.

As we have seen (Section 4.4.5), liquid crystal phases for example, of phospholipids in aqueous solutions, fall between liquids and solids, but the description of their shapes (see Figs 4.19 and 6.4) follows from the above, being decided in the first place by contact with a bulk phase, gaseous or condensed. Looking around us we observe the shapes of a vast variety of limited systems in contact with bulk phases, geological, biological and man-made, which are decided in large part by that contact.

6.2.3 Surfaces and shapes of limited phases as variables

The fact that droplets of different sizes and parts of a droplet on a surface have different curvature and areas implies that the local energy, related to curvature, is everywhere different. The different faces of a crystal have different energy and area too. Even so the individual drop or crystal as a whole must be as considered to be *at equilibrium*. This means that there must be a restriction making shape and size (amount) mutually dependent. (Remember

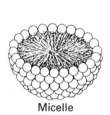

Micelle

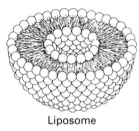

Liposome

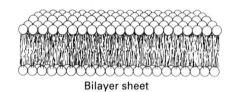

Bilayer sheet

Fig. 6.4 Cross-sectional views of the three structures that can be formed by phospholipids in aqueous solutions: spherical micelles with hydrophobic interiors; sheets of phospholipids in a bilayer; and spherical liposomes comprising one phospholipid bilayer (see Fig. 4.19).

that each equilibrium is a restriction: Section 4.3.1.) The restriction here derives from the fact that, at equilibrium, the product of the area, A_1, and the energy per mole, E_1, of one face or surface region, must be equal to the product of every different area, A_2, and its energy per mole, E_2; otherwise the phase would change shape, that is,

$$E_1 A_1 = E_2 A_2 = E_3 A_3 \ldots .$$

Furthermore, all faces must be in equilibrium with the interior and with the vapour pressure or solution concentration or else change would again occur. It follows that the shape of a phase at equilibrium is decided by the amount of material in the phase, given the contacts it makes (see Section 6.2.4, which concerns fields).

Because it is not relevant to the discussion we have not mentioned the size of the phase specifically. We can treat crystals, droplets and vesicles of monolayers or bilayers of lipid molecules (Fig. 6.4) or more complex regions of cells (see Fig. 6.8) so long as we have reason to suppose that the limited regions are individually homogeneous and at equilibrium with one another across the boundaries. The problem is not made more difficult by the inclusion of many phases containing many components. The total amount of material in the whole system introduces but one new variable since all the surfaces of all the phases and all their individual volumes (amounts) are constrained at equilibrium by this amount and by the phase rule variables, P, C, T and p.

6.2.4 Fields between limited phases in equilibrium

Once we have introduced limited volumes of systems we must consider how they interact. They clearly do so through fields of many kinds according to the material in them and the distance between them. The most obvious are gravitational and electrical fields, which develop through differential distribution of mass and charge in space. The charge distribution arises through the different electron affinities of different bulk materials and surfaces. While knowledge of gravitational fields has existed effectively for thousands of years, bulk electrostatic fields were only discovered at the end of the seventeenth century through the differential charging and discharging of two bodies made from ebony and fur that had been rubbed together so as to remove negative charge (electrons) from one while leaving positive charge on the other. This is not a case of ion charge separation in dilute solution across a membrane (see Section 6.4.2), but of charging bulk surfaces of phases. In fact, all surfaces are charged to some degree at equilibrium.

A further field, which is due to contact between phases and clearly affects shape, is mechanical stress, due to the application of pressure to the boundaries which may have connecting elastic filaments. We see this when we place objects on top of one another. The stress equilibrates with the strain introduced as in a rubber ball carrying a weight.

There is now an extra difficulty in the description of variables, namely the *relative* ways in which the phases occupy space since space is no longer homogeneous. The influence of one phase upon another will depend on the fields generated by and exerted on one another mutually, which themselves depend on the *vectorial distance* between the phases. Thus the *directions* of phases in space become important, giving rise to equilibrium bulk shape of many interacting phases as well as shapes of individual phases. At long distances there are, as we have seen, the two long-range classical forces, gravitational and electrical; at short distances we meet mechanical stresses, short-range interactions of molecular dimensions, and repulsion between two or more limited phases. It is likely that these interactions help to define biological shapes. We deal with all these effects of the phases on one another by introducing further new degrees of freedom, the parameters characterising fields, into the phase rule, which has now to include the effect of all the different fields at equilibrium. (Of course, we must only consider systems in which phases are in fixed positions, in balance of some kind, since we will not discuss change yet.)

6.2.5 Assemblies of limited phases

We shall assume now that space is occupied by many small phases that interact with one another at equilibrium in a variety of fields. A field of any kind is described by its intensity and co-ordinates in space, but there are two very different situations.

1. *Phases in contact*, with fields acting between them, that undergo no change, that is, a stationary state. This is a truly time-independent spatial situation and we can first examine cases where the two or more phases are in equilibrium but attract one another, for example, two liquids in a gravity field or two oppositely charged but electrically stable materials. The phases are in transfer equilibrium. (We will not discuss the case where barriers prevent transfer of material or energy across boundaries until Section 6.3 and the reasons for such loss of transfer equilibrium are given in Chapter 7.) This is a very conventional approach to many systems, for example in chemistry and biology. It is a system-imposed positional distribution of phases that are in static balance (see Fig. 6.5). A typical example is the separation of solid phases of the Earth, or of liquids in a separating funnel.

2. *Phases not in contact* but at set arbitrary distances, with fields acting between them. It is obvious that such systems should collapse to case (1), but we know or assume that they do not do so quickly. For example, the Earth (a not so small but nevertheless limited region in the universe) is to all intents and purposes at a fixed distance from the sun (here a limited phase). The reason there is no rapid collapse in this case derives from the angular momentum of the Earth (see Sections 1.4 and 1.8). (We will not examine this angular momentum yet.) The system is in an effective stationary state and, as far as Earth is concerned, we can ignore its motion and just treat its position relative to the sun as fixed for many purposes though we have to remember that the gravitational field is constantly present.

Notice that in case (1) attractive and repulsive fields are in balance, while in the example given for case (2) there is no repulsion but only its equivalent due to relative motion. (In fact, in this second situation energy must be applied if there is any friction (Chapter 7).) We shall assume that the chemical contents of each phase are in *true* local equilibrium in fixed fields and with the content of any other local phase.

There is a further situation of great interest, a steady state in which the

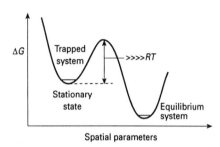

Fig. 6.5 A stationary state, which is not an equilibrium state, is defined by a system that is not at equilibrium with its surrounding material and is not gaining or losing material or energy. It is trapped behind a large energy barrier.

condition of a local system is fixed with obvious constant flow of energy and/or material through it. This *steady state* is described in Chapter 7 and is not treatable by considering equilibria, even though we can ask about energy content of the different components present in the steady state.

6.2.6 Total variables in phase systems of limited volume

Since we are discussing the variables in chemical systems of phases and components at equilibrium we need to know how many new variables have now been introduced since each variable will give rise to an axis in a phase diagram much as we have an axis for T, p and percentage composition of components in the diagrams of Chapter 4. As we have seen, limiting the total amount of the phases introduced one new variable, but shape by itself introduced no new variable since it was determined by the quantities of the components. However, the limitation on size forced us to consider boundaries and therefore fields. Each field introduces three new degrees of freedom, magnitudes connected to the three co-ordinates of position, and they define the variable *external space*, since each field is a vector acting on combinations of local systems. (We have deliberately not limited ourselves by considerations of restrictions on the fields due to the chemical composition or size of the compartments.) These variables are interactive with amount in creating total shape. This applies to the universe or the planetary systems now treated as in 'equilibrium'; all planets of the solar system interact with one another and with the sun (Fig. 1.20), giving the system a shape. The fact that fields are also variables in the description of chemical phases becomes obvious when we consider equilibrium under gravity—we know, for instance, that dense bodies settle on Earth into an equilibrium or a layer pattern. When bodies are small electrical fields can dominate assembly; for example, in electrical settlers particles of opposite charge can be separated. The importance especially of electrical fields becomes even greater for still smaller particles such as colloids, macromolecules and cells. (As we shall see, there is no way in which shapes of living biological species could be assembled except through consideration of such fields.)

6.2.7 Factors governing phase interactions in fields: chemical selectivity of surface interactions

As we have seen, the shape of two surfaces on contact is dependent on chemical surface energies and thus on their chemical compositions, that is, ultimately the chemical elements in the components of the phases and the fields they generate. The reasons are obvious enough. In the case of single crystalline solids the crystal shapes fall into certain symmetries (Fig. 6.6) due to the long-range packing of atoms or molecules internally (see Chapter 2 and reference 6 in the 'Further reading' of that chapter). There are then good reasons why surfaces of crystals of a given chemical can be of one shape, for example perpendicular to one another as in some cubic or orthorhombic crystals, while this is not possible for other chemicals that crystallise in rhombic symmetries. Within each one of the variety of symmetries there is

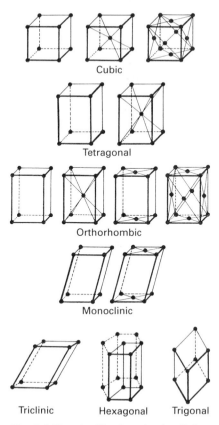

Cubic

Tetragonal

Orthorhombic

Monoclinic

Triclinic   Hexagonal   Trigonal

Fig. 6.6 The classification of unit cells in terms of the 14 Bravais lattices.

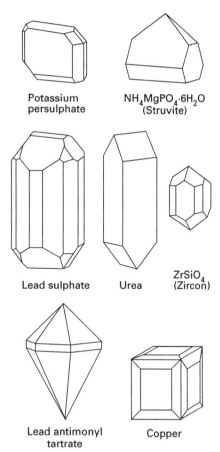

Potassium
persulphate

NH₄MgPO₄·6H₂O
(Struvite)

Lead sulphate

Urea

ZrSiO₄
(Zircon)

Lead antimonyl
tartrate

Copper

Fig. 6.7 Examples of some observed crystal shapes.

allowed a set of different shapes of crystals, these shapes being related by the main symmetry elements† (Fig. 6.7).

Thus external contacts, other chemicals and fields, decide the shape observed by changing surface energies, but within a limited range of possibilities. (Impurities, effectively new components, can change the shape too, since they can alter the surface energy.)

Shapes of liquid surfaces, their curvatures (see Fig. 6.2), are also decided by their chemical nature, but now it is the *contact* interaction with fields and with the medium that dominates the shape since there is no internal structure. Any shape is, therefore, possible. This is true of amorphous materials too because there is no imposed internal symmetry. It is clear that it will be the fine detail of the particular chemicals in a surface that decides the nature of the surface interaction, both through chemical forces and general fields, for example the charging of surfaces with electrons.

Bulk shape, that is, of combinations of phases, is seen to be a property of local surface interactions based on the energies discussed in Chapter 2 and bulk fields. A major consideration in the discussion of shape must ultimately be related to the selected chemical elements usually in compounds, in the space described. Therefore the shape of combinations of phases is not an independent variable no matter how complicated the equilibrium system. Does this apply to the shape of a biological organism (Fig. 6.8; see Section 13.7)? Or do we have to think differently about *organised* systems of phases in biology? Species, like small phases, are associated with shapes.

6.2.8 Matching of shapes of elastic substances at equilibrium

The shape of a condensed phase provides steric conditions, notably repulsive exclusion, for selective interaction with another phase at short range. The strength of the interaction, as in the packing of molecules in crystals, depends

Fig. 6.8 Transverse section of a typical plant root. Compare Fig. 6.1 but notice that this system is highly organised and reproducible. The shapes however are not exactly reproduced e.g. a plant leaf.

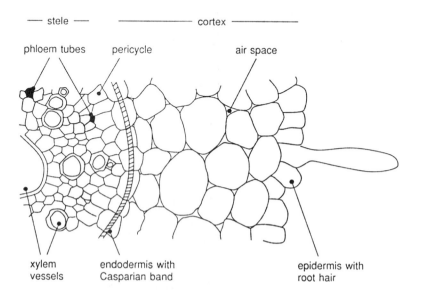

— stele — ———————— cortex ————————

phloem tubes   pericycle   air space

xylem
vessels

endodermis with
Casparian band

epidermis with
root hair

† The class of symmetry to which a crystal belongs is defined by the arrangement of the structural units about each point in a 'Bravais' lattice Fig. 6.6.

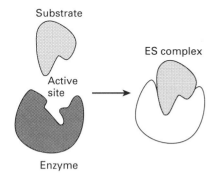

(a) DIE AND MOULD

Substrate

Active site

ES complex

Enzyme

(b) and (c) INDUCED FIT ± MOBILITY

Substrate

Active site

ES complex

Enzyme

Fig. 6.9 Three mechanisms for the interaction of two bodies of any size, for example an enzyme and a substrate. (a) In the lock-and-key or die-and-mould mechanism, the substrate fits directly into the binding site of the enzyme. (b) If binding occurs by induced fit, the substrate and enzyme surface undergo a conformational change that appropriately positions the substrate for catalysis. (c) In the case of hand-in-glove fitting the two more mobile units induce changes in one another but keep their mobility in the combined unit, as for (b).

both on attraction and this repulsion (Fig. 6.9). There is likely to be but one most stable resultant and we can imagine that to avoid repulsion and yet get maximum advantage from attraction between *two rigid bodies* they must give a precise die-in-mould fit both in physical and chemical terms. For more elastic bodies we can consider *induced fitting* of surfaces and for two even more fluid bodies fitting can be of the *hand-in-glove* kind, maintaining fluctuations in both chemicals or phases (Fig. 6.9). All such fittings result from the drive to 'equilibrium' stability. Such considerations apply equally to the building of stone walls with mortar, the setting of concretes in moulds, the assembly of cells (Fig. 6.8) and the association of large polymers in biological self-assembly. The compartments, down to the level of very large molecules, are not rigid nor are they extremely adjustable. Again, all degrees of behaviour between that of crystals and that of liquids are possible. In these circumstances mechanical strain induced by stress becomes relevant. We shall see that application of considerations such as these are very important in the discussion of surface recognition of small phases as well as of large molecules. The principles of equilibrium enzyme–substrate, antibody–antigen or cell–cell recognition are the same as the fitting of larger bodies to one another.

It may be thought that we are going to a great deal of effort to describe assembly, shape of phases and shape of whole systems. Shape, however, is peculiarly apparent around us everywhere and in biology is unique to a species, so unique that it has been used in taxonomic classifications. Thus, the refinement of chemical element use in biology must be related to the shape, now self-assembly, of the organism in some way (see Chapters 10–13). (NB. Particular shapes also have value (use) in biological systems and in man's equipment.)

In passing we observe that, under the influence of fields and in equilibrium situations, a phase is not homogeneous in charge or material, but is in free energy balance. The picture of a phase in a field is then of an equilibrium distribution of chemicals up to the boundary and including the surface. In many cases the surface composition controls properties so that we have to distinguish a limited phase in a field quite carefully from the homogeneous phase described in Chapter 4.

6.3 Further diversification through barriers: compartments

6.3.1 Storage of free energy between compartments: an ability to do work

We need now to look once again at the real observable world around us and to remove the third assumption that lay behind descriptions of it based on equilibria discussed in Chapters 4 and 5 and even in the preceding sections of this chapter (see Section 6.1). In most real systems there is a division into zones, which we shall call *compartments*, between which the transfer of stored energy and of some or all materials is inhibited or prevented even while each compartment may have equilibrium phase conditions within it. We must describe the difference in free energy across barriers that can divide phases, or split what was a single phase into compartments (Table 6.3). In these cases the compartments will store energy relative to one another and can do work. (Even when they are of equal free energy, perhaps in different forms, they could gain stability by mixing and do work.) By allowing different

Table 6.3 Some modes of energy storage in compartments

Compartments	Storage	Barrier
Reservoir versus ocean	Pressure under gravity	Dam
CH_4/O_2 versus CO_2/H_2O	Chemical energy	Chemical bond
Centre of Earth core, Fe/air, O_2	Chemical energy (and heat, temperature)	Rocks
Biological liquids/external chemicals	Pressure, chemical energy	Lipid membranes
Metal plates in electrical cell	Electrical potential	Air Plastics
Aqueous solutions	Electrolytic gradients	Plastics Lipids

compartments to store energy or material of different kinds they can, however, be in effective balance in a *stationary state* relative to one another but in a trapped situation. We shall give first some examples of energy storage between concentrated bulk components which form the compartments and then turn to the more general cases of dilute components separated by barriers between compartments, now often of the same solvent, for example, two aqueous solutions. If we look back at the phase rule, even modified to take into account limited sizes of systems and the presence of fields, we see that it cannot be applied globally to these cases since the compartmental barriers prevent exchange of either material or energy, when the $C(P-1)$ equilibria do not operate. Variance is now greatly increased. We stress again that it is only through differences in equilibrium descriptions of components in individual compartments that we can understand the out-of-balance in such systems to some degree. In this chapter we continue to discuss stationary conditions and do not analyse flow of any kind between compartments. For the most part we assume global *thermal equilibration*, that is, the system is assumed to be closed, and light and heat are in balanced exchange.

6.3.2 Pressure and concentration gradients between compartments

As stated above, rather than analyse further phase concepts in general equilibrium terms it is now appropriate to look at examples of localised chemical systems in which some part of the system concerned is held away from true equilibrium but is in a fixed state relative to another, Fig. 6.10. A simple example is an inflated balloon surrounded by air where the gases inside and outside are at different pressures, p_1 and p_2. We can discuss the relative energy of any component in terms of its local partial pressure on either side of

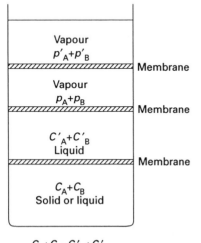

Vapour
$p'_A + p'_B$

Membrane

Vapour
$p_A + p_B$

Membrane

$C'_A + C'_B$
Liquid

Membrane

$C_A + C_B$
Solid or liquid

$C_A + C_B = C'_A + C'_B$

$p_A + p_B = p'_A + p'_B$

Fig. 6.10 The barriers illustrated do not have a total pressure or concentration difference across them but a selective chemical gradient in each case.

such a membrane. There is a free energy difference, so it can be said that the whole system has a free energy storage that can be used to do work (see Section 3.6). The free energy storage is related to $RT \ln(p_1/p_2)$ or $RT \ln(c_1/c_2)$ in the case of components in solution for each component between each pair of compartments and may be described by considerations of mixing of the two compartments (Section 3.3.1).

A second case involves the differential electrical charging of the two compartments when the energy stored is just that in an electrical condenser. Note that we are considering distributed charge in two bulk media and not charge due to specific ions diluted in the media (see Section 6.4.2).

6.3.3 Chemical potential gradients between compartments

Two elements A and B can be held in different chemical forms in different compartments, say, as pure elements and in a compound AB. In this case and assuming they are all, elements and compounds, in their standard states (note, not at equilibrium), the standard-state free-energy difference across the boundary is

$$\Delta G = \Delta G_A^\circ + \Delta G_B^\circ - \Delta G_{AB}^\circ$$

where A and B are in one compartment and AB is in another. A typical case would be of the gases, all at unit pressure, $N_2 + H_2$, held separately from NH_3 by a membrane which is impermeable to all three. To a large extent the Earth is constructed in such a fashion, in that chemicals are held in local pockets (see Chapter 8). This is also true of biological systems and of man's industrial plants. In real systems the ΔG° values must be corrected for concentrations other than 1 mole.

6.3.4 Storage of mechanical energy (tension)

When stress is imposed on a material due to a pressure differential applied, say, to a membrane that separates two compartments, it takes up energy due to tension, in part due to the adjustment of electrostatic charge interactions at the level of chemical bonds in the material. Additionally, however, the constraint on its volume causes a loss of configurational entropy, so that there is a storage corresponding to $-T\Delta S$ assuming that the temperature keeps constant. The last situation applies, for example, when pressure is applied to elastic disordered substances such as the rubber of a balloon (see Fig. 6.11). The tendency of stretched rubber is then to go back to the non-stretched situation by releasing the stored energy $T\Delta S$. In a metallic spring it is considered that mechanical energy is generated and stored by the slipping of planes of metal atoms over one another, thus weakening binding by increasing bond lengths on the convex side of a turn and weakening them by contraction on the concave side. The slipping of strands of a rope, usually helically coiled, is similar in that the twist is tightened while the rope is extended. Tension in protein systems, for example, in muscle fibres, is not dissimilar. Here chemical hydrogen bonds may slip over one another while the proteins are energised relative to a stable ground state. The picture can be reduced to the single large molecule level when, for example, application of force causes individual

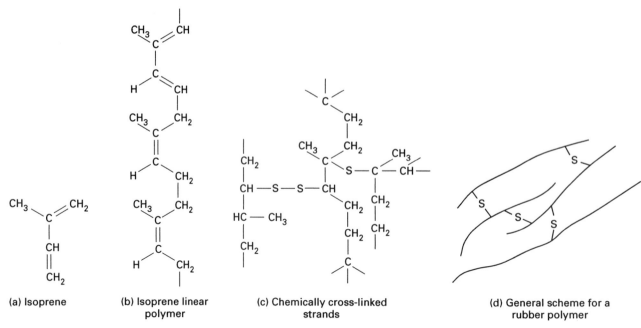

Fig. 6.11 The nature of a cross-linked rubber formed from isoprene units, (a). (b) Isoprene linear polymer. (c) Chemically cross-linked strands in rubber treated with sulphur. (d) General scheme for a vulcanised rubber polymer, see (c). The rubber is able to take up stress on compression, for example in a motor car tyre.

domain protein helical structures to move away from energy minima. Again in all these cases the tendency is for the system to go back to the initial unstressed state by releasing energy (increasing entropy) while doing mechanical work. Note that these examples imply physical contact between compartments since they require the action of external force to be mechanically transmitted.

6.3.5 Energy storage in a gravitational field

There is an obvious difference between a freshwater pond or a reservoir and the sea. The sea cannot sink further toward the centre of Earth since it is constrained by gravitational forces, it being of lower density than minerals. (Ice floating on water is another example of true equilibrium under gravity.) A water store high up on a hill represents a water compartment of higher potential gravitational energy than the sea. The lake or reservoir can be treated as a separate compartment with an internal equilibrium of its own and retained by a structural barrier. The difference in ΔG can be used in hydroelectric power production. (The way in which such compartments permit different chemistry from that of the sea is also extremely important since the lakes may have allowed the development of vertebrate life (see Chapter 12) due to the differences in minor components in these compartments.)

In passing, notice that there is a third very different water compartment, the flowing stream, which cannot be described by such pseudoequilibrium, stationary-state descriptions (see Chapter 7) since its essence is flow. It is, however, in a steady state (Fig. 7.25).

6.3.6 The natural selection of chemical elements in compartments

Now all of these examples are of chemical potential differences in systems far from equilibrium due to compartment formation, but we can use ΔG, a measure of the drive to equilibrium that relates to the free energy content difference between the compartments, to point out again the connection to the natural selection of the chemical elements. It is not accidental that energy is stored on Earth in certain compartments within concentration gradients, fields and chemical forms; it is due to the nature of the substances, chemicals, that particular elements could form those particular compartments, energised yet protected. Similarly, only certain materials can form the barriers between compartments. For example, only oils and lipids, C, H, N, O compounds (see Chapter 9) form membranes to separate compartments with different aqueous concentrations in biology and only certain soluble substances, salts such as NaCl which dissociates into Na^+ and Cl^- ions, can give rise to considerable electrolytic potentials between neighbour water compartments (Section 6.4.2). Because of the protection from one another the energised compartments have survived (see Chapter 7), and precisely because they survive over long periods of time we can handle them as if each one has an internal free energy that is independent from those of its neighbours except when under the influence of a field, for example, an electrical potential difference. We see that if man or a biological system wishes to store energy in particular forms (or to make constructs) there must be an organised selection of chemical elements in compartments and a choice of material to form a suitable barrier. Whether the observed compartmental system is biological or not, there must have been an energy input to it (see Section 6.4). This energy in any stationary state of any compartment is calculable using the equilibrium data obtained in Chapter 3.

6.4 Energy storage by dilute components in compartments

Just as in Chapter 4 we described bulk phase equilibria before describing dilute components in fixed bulk phases in Chapter 5, so in this chapter we shall now go forward from bulk phase considerations to those of *dilute* solutions in two or more small compartments in contact.

6.4.1 Osmotic pressure

Consider two ideal solutions separated by a so-called semipermeable membrane, that is, some boundary that allows the passage of the molecules of one component, the solvent, but not the molecules of the others, the solutes. Let the two solutions be at different concentrations. Since the transport of the molecules of the solutes is not possible, the system will be out-of-balance in respect to the distribution of the solutes between the two compartments (see

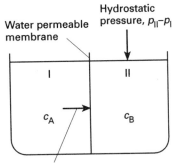

Water permeable membrane

Hydrostatic pressure, $p_{II}-p_I$

I

II

c_A

c_B

Osmotic pressure= hydrostatic pressure required to prevent net water flow

Fig. 6.12 Osmotic pressure. Here, solutions A and B are separated by a membrane that is permeable to water but impermeable to all solutes. If c_B (the total concentration of solutes in solubtion B) is greater than c_A, water will tend to flow across the membrane from solution A to solution B. The osmotic pressure between the solutions is the hydrostatic pressure that would have to be applied to solution B to prevent this water flow and is given by the van't Hoff equation as $\pi = RT(c_B - c_A)$.

Fig. 6.12). However, if the solvent can pass through the membrane until the equilibrium balance is reached, the chemical potential, μ_s, must be the same in both solutions of the membrane when

$$\mu_s(I) = \mu_s(II).$$

From the definition of chemical potential (Section 3.5) we have

$$\mu_s(I) = \mu_s^0(I) + RT \ln x_I.$$

$$\mu_s(II) = \mu_s^0(II) + RT \ln x_{II}$$

where x_I and x_{II} are the molar fractions of the solvent, a convenient way to express concentration in this case since, for example, x may vary up to 1, corresponding to the pure solvent. It is clear that the stationary state can be achieved only if the pressures to which compartments I and II are subject are different (called p_I and p_{II} in Fig. 6.12), that is

$$\mu_s^0(p_I) - \mu_s^0(p_{II}) = RT \ln \frac{x_{II}}{x_I}.$$

The difference $p_{II} - p_I$ is called the *osmotic pressure* and is usually denoted by Π which can be shown to be equal to $RT \cdot \Delta c$ where Δc is the difference in solute concentrations. Such a compartmental system stores energy related to a pressure (a concentration difference) and is utilised in plants, especially to assist flow. Now, the osmotic pressure can be replaced by adding to one or other of the solutions, inside or outside the membrane, a concentration of a second chemical, for example, NaCl outside, KCl inside (Fig. 6.13), so as to give osmotic balance. The system stores energy relative to a mixed solution of both salts. We give this example since it is essential for the stability of a living cell, which is a solution of concentrated chemicals within a relatively weak membrane, to manage osmotic pressure (Fig. 6.14). The separation of inorganic ions, Na^+, K^+ and Cl^-, largely overcomes the destructive effects of osmotic swelling due to the organic content of the cell. The mode of overcoming this problem has permitted life as we see it (Section 10.4.1) and will be seen to be a selective use of the chemical elements, here making no distinction between so-called inorganic and organic species.

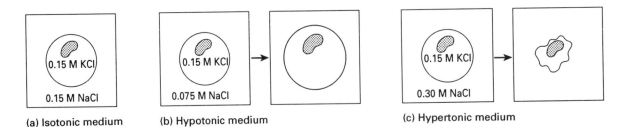

(a) Isotonic medium   (b) Hypotonic medium   (c) Hypertonic medium

Fig. 6.13 Animal cells respond to the osmotic strength of the surrounding medium. Sodium, potassium and chloride ions do not move freely across the cell membrane, but water does (arrows). (a) When the medium is isotonic, there is no net flux of water into or out of the cell. (b) When the medium is hypotonic, water flows into the cell until the ion concentration inside and outside the cell is the same. Here, the initial cytosolic ion concentration is twice the extracellular ion concentration, so the cell tends to swell to twice its original volume, at which point the internal and external ion concentrations are the same. (c) When the medium is hypertonic, water flows out of the cell until the ion concentration inside the outside the cell is the same. Here, the initial cytosolic ion concentration is one-half the extracellular ion concentration, so the cell is reduced to one-half its original volume. (After Darnell J., Lodish, H., and Baltimore, D. (1990). *Molecular cell biology*. Scientific American Books, New York.)

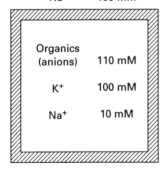

Fig. 6.14 An idealised description of cellular osmotic balance. Note the electrical balance here.

6.4.2 Electrical potential differences

As well as a difference in concentration there can be an applied or internally generated electrical field difference, ψ, acting on two compartments across their boundary, in this case due to differences in charge distribution of a minor component (Fig. 6.15). Any other charged species, any ion, even in equal concentration on both sides of the boundary, feels the difference in potential, ψ, so that the free energy difference between compartments derived from the electrostatic field is

$$\Delta G = zF\psi$$

where z is the charge of the ion and F is the Faraday constant, 96 500 coulombs per mole, so that, when ψ is expressed in volts, ΔG is obtained in joules, Section 5.7.1. The data in Fig. 6.14 are those for an idealised biological cell treating the Na^+ energies only.

In biological systems we note that biological cell membranes have a potential across them and this is the basis of, for example, stationary energy storage to be used later in nerve messages. Such an electric field is generated by *all* known biological cells (in which we consider inside and outside as separate compartments), since the distribution of oppositely charged ions is not equal. Biology evolved these electrical fields locally (across cell membranes) some 3–4×10^9 years ago by the forced separation of ions such as H^+, Na^+, K^+, Ca^{2+}, Mg^{2+} and Cl^-. In a later development a sodium/potassium pump evolved (Fig. 6.16) where the number of ions pumped is $3Na^+$ for $2K^+$, leaving excess ($+$) charge outside and a corresponding ($-$) charge inside. All charged particles in the system are energised by the field produced independently from the chemical involved. Such fields are not commonly met in everyday

Fig. 6.15 Transmembrane forces acting on Na^+ ions. As with all ions, the energy of transfer of Na^+ ions across the plasma membrane is governed by the sum of two separate forces generated by the membrane electric potential and the ion concentration gradient. In the case of Na^+ ions, these forces usually act in the same direction in biological cells.

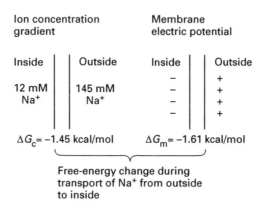

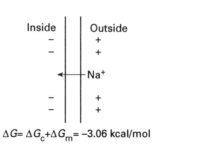

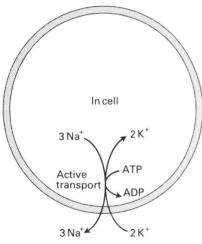

Fig. 6.16 The pumped movements of sodium ions out of and potassium ions into the cell. ATP hydrolysis (chemical energy) is required for the movements of the ions against their gradients. The actual coupling is of two K^+ for every three Na^+ so that the pump is electrogenic, that is, it develops charge and chemical gradients. We describe a resting state in terms of the free energy status of the inside and outside of the cell considered separately.

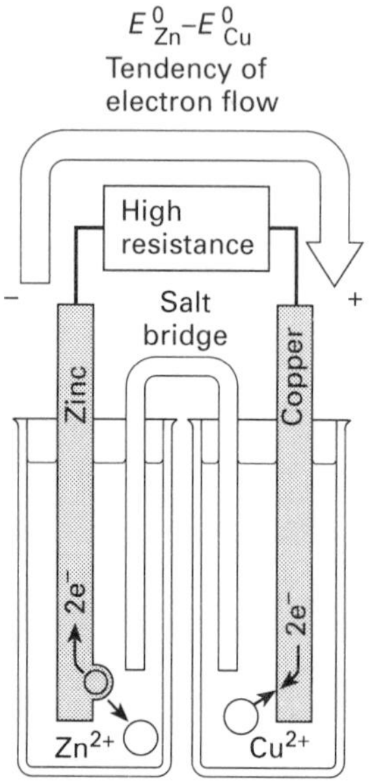

chemical systems, but are an essential feature of the electrochemistry of all biological organisms.

The potential energy in a field can be interconverted ('transduced)' with all the other forms of energy. Summing up electrical field and concentration differences we have a free energy difference for a component between two phases given by

$$\Delta G = RT \ln\left(\frac{c_1}{c_2}\right) + zF\psi.$$

We shall see how their ΔG is used in biological systems to pump required materials selectively into cells.

There are many other kinds of electrical storage, for example, batteries (man-made constructs for stationary electrical energy storage), that can be described in terms of such charge storage, free energy differences, between compartments. The relationship of electrical cells to the single electrode potentials described in Chapter 5 is one example. We showed there that the potential of, say, a zinc electrode and a copper electrode were different. Thus we can make a cell as shown in Fig. 6.17 based on principles similar to those in Section 5.7 for metal and hydrogen electrodes at equilibrium. The potential difference between the two standard potentials is 1.08 volts. This potential represents a difference between the two metals in their ability to accumulate bulk electronic charge opposite a solution of their own ions and is written generally as

$$\Delta E = \Delta E^\circ + \frac{RT}{2F} \ln\left(\frac{c_{Cu^{2+}}}{c_{Zn^{2+}}}\right).$$

There is an alternative cell in which the metal electrode is always, say, copper, but now the solutions A and B in the two compartments contain both Cu^{2+} and Cu^+ ions but in different ratios. The redox potential in any one solution (A) is given by

$$E = E^\circ + \frac{RT}{F} \ln\left(\frac{c_{Cu^{2+}}(A)}{c_{Cu^+}(A)}\right)$$

and a similar expression applies to solution B. The difference of potential between solutions A and B can be realised across an organic membrane that does not permit the diffusion of Cu^+ or Cu^{2+} or, indeed, of any ion. In fact, this type of potential can be developed by any redox pair of chemicals. For example, let us take any hydride XH_2 and its oxidation product $X + 2H^+$; the potential in either isolated compartment is

$$E = E^\circ + \frac{RT}{2F} \ln\left(\frac{[X][H^+]^2}{[XH_2]}\right)$$

and can be changed from compartment to compartment by changing [X], [H^+] and [XH_2]. Clearly, a potential difference between sensing electrodes can be written down for any such pair of redox-related substances (see Table 5.18(a)). We shall see that these electrical modes of storing energy are major contributors to biological energy transduction (Section 10.4.6).

Of course, this energy store has to be made by the input of energy which has

$E_{Zn}^0 - E_{Cu}^0$
Tendency of electron flow

Fig. 6.17 The relative electric potentials of electrodes and the tendency of electron flow. The bridge is a KCl solution in an agar gel. Here we do not discuss rate of flow (see Chapter 7).

been described as pumping. As is true for all compartments, energy is required in order to form them.

We now return to the more general consideration of free-energy differences between compartments that will affect both major and minor components and especially to consideration of temperature differences.

6.5 Equivalences of free energies

All of the above stores of energy, ΔG, arising either through the absence of equilibrium in chemical bonding, concentration gradients or the presence of various fields between two compartments, can be expressed in equivalent energy units and shown to be interconvertible (Table 6.4). In essence, we arrive at the ΔG value between two systems from a consideration of equilibrium balance in each compartment as described in Chapter 3. The out-of-balance is expressed in units of kilocalories (kilojoules) or volt-equivalents. Out-of-balance is thus expressed in units of heat energy or electrical potential energy. In fact, each out-of-balance can be demonstrated to be related directly to heat and hence thermal (kinetic) energy by connecting the two compartments that are out of balance (Table 6.4) to heat generation, for example in a resistor, or to light by connection to a light bulb. This is a simple demonstration of the ability to do work expressed within the free-energy difference ΔG. The question arises as to how these compartmental, energised systems arose.

Table 6.4 Transduction of energy

ΔG source	Transduction process	Example
Chemical compounds	Mixing (+catalyst)	C/H compounds + O_2
Electronic store (battery)	Resistor (wire)	House light and heat
Electrolytic store	Resistor (aqueous medium)	Nerve currents, body heat
Sunlight	Plant leaf, atmosphere	Photosynthesis, Earth's temperature
Gravity store water in reservoir	Dynamo to electronic store	Power station

Geochemically, they have arisen very largely through rapid cooling of the universe in local regions which generated traps (Fig. 6.5). Biologically, they arose and still arise through local trapping of the energy of the sun. We must show that radiative energies, light or heat, can be used in reverse fashion, that is, to make compartmental separation of charge or of chemical potential. In this case, as in the above, the necessary feature is an appropriate means of connection and conversion between ΔG forms, that is, a ΔG transduction system from out-of-balance of energy (photon energy related to temperature) between the Earth and the sun. (Remember we are describing systems as in stationary not flowing states.)

6.6 Radiation energy stored in compartments

The sun and the Earth are compartments that are far from equal temperature and radiation equilibrium. It is extremely difficult to construct barriers against the flow of this radiative energy so that a given volume maintains a fixed temperature without loss or gain of energy from the environment. Both kinetic energy and photons are readily transferred through many media. The bodies we see around us are, in effect, at fixed temperature on a limited time-scale, for example, the sun (all stars) and the Earth (all planets), and we can treat the states as *stationary in temperature* for practical purposes so long as we restrict ourselves to a limited period of time. In fact, all such bodies radiate and absorb simultaneously, so that planets maintain temperature through a *steady-state* flow of energy from the sun and a *steady-state* release of energy to the universe (plus a *steady-state* flow of heat from the centre of the planet to the outside). This *flow* of energy is treated in Section 7.19. It creates many of the *apparently stationary states* that are of elevated free energy, $+\Delta G$, in terms of their temperature. These steady states, apparently stationary states, also contain local components energised in fields behind barriers. It is their free energy content that we shall have to connect to the photon flux (see Section 7.19).

6.7 Limited phases and compartments in biological systems

We have shown in Sections 6.1 and 6.3 that many of the considerations that we applied under the condition of general equilibrium in Chapters 3 to 5 can be used to discuss local conditions in limited phases provided that we make certain assumptions:

(1) fixed temperature, even though over long distances there are temperature gradients;

(2) fixed pressure, even though over very long distances there are pressure gradients;

(3) fixed mass and charge contents of phases so that there is a defined set of fixed fields; they impose fields at a distance on one another, and also, upon contact, interact through mechanical forces;

(4) fixed volume and element composition of phases by fixing the amount of elements; effectively we assume there is no net transfer of material.

Using these (often hidden) assumptions we have deduced many valuable relationships. These relationships are constraints on limited systems at 'equilibrium'; for example, the disposition in space of all phases is fixed, but we have still left as variables:

(1) a large number of components, as in Chapter 4;

(2) a large number of local phases. (Even this equilibrium system of phases has a much greater variance than we described in Chapters 4 and 5.)

By relaxing from total equilibrium to local equilibrium in compartments only (Section 6.3 onwards) the system as a whole becomes very much more underdefined. Thus the Earth and its contents look increasingly accidental, dependent on a fast rate of cooling. Clearly, the situation is made increasingly more difficult the more space is split into smaller and smaller units. This is exactly the apparent construction of biological systems for which, in place of a phase where the position of individual molecules can be interchanged, many molecules are large and assembled (positioned) adjacently (Fig. 6.8), presumably by fields. These large mosaic assemblies of large molecules, each one of which can be likened to a very small phase (see Figs 6.18 and 6.19), are extremely important and we must attempt to describe their features as we have described small phases. These systems, like many composed of minerals, appear to be random mixtures of immense variety. (That they are well defined in shape within species shows this is not so, a problem we turn to in Chapter 7). We shall also have to find how such energised compartments could have arisen in the bulk water phase of the sea.

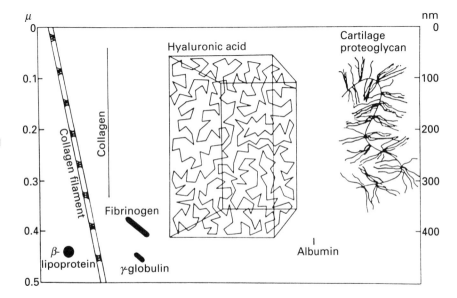

Fig. 6.18 An indication of the shape and size of 'molecular phases'.

6.8 Very small phases: large molecules treated as phases

In the above consideration of *size*, there were no definite limits. It is then legitimate to discuss assemblies where the size is no longer that associated with condensed matter but comes down to considerations of large molecules (Fig. 6.18). We can only summarise some features that are self-evident and are of great consequence in biology.

Consider a molecule made of monomer units and which can form a covalent polymer A_n. As well as having all the thermodynamic features of any molecule, it can be treated as a small phase. In effect, the polymer can be of two extreme types (or of any in-between kind):

1. It can fold into a fixed structure, when it resembles a single crystal and has a shape, a net charge, a surface charge, etc. Hence, we treat it as a small solid

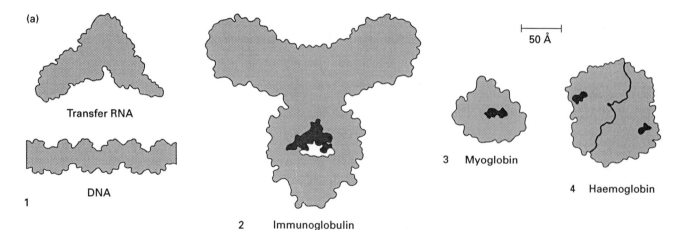

(a)

Transfer RNA

DNA

1

2   Immunoglobulin

50 Å

3   Myoglobin

4   Haemoglobin

Fig. 6.19 (a) Shapes and sizes of some biological polymers. The shapes are functional much as is the shape of an organism. Scale bar, 50 Å. Haemoglobin has a molecular weight of 64 000 daltons. (b) A more detailed view of the way in which amino acids in proteins are close-packed such that surfaces are unique to sequences and individual hot spots (shaded) can arise (see the entatic state in Section 10.8). The 1 cm scale bar equals 50 Å.

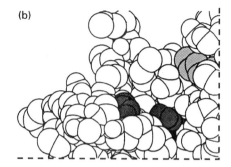

(b)

phase carrying a field and an idiosyncratic chemical surface, for example a protein (Fig. 6.19(a) and (b)). The molecule has a 'melting' point.

2. It may not fold in any preferred shape but visit many configurations and carry a variety of charges in equilibrium, so that in one sense it is like a liquid phase, for example, rubber in a ball.

The balance between conditions (1) and (2) is exactly like that of an equilibrium between a crystal solid and liquid. The balanced state is described by the co-operative interaction energy of the fold in relation to the configurational entropy of the ensemble. In fact, we can describe the polymer, for example, a protein, as having a melting point, like a co-operative solid; however, the co-operativity is not of an infinite number of identical atoms or small molecules and needs further examination. (Assembled proteins, as co-operative molecules phases, lie behind the secrets of biological systems as we shall see in Chapters 10–13.)

6.8.1 'Phase' behaviour of proteins and nucleotides

A folded protein (or DNA(RNA)) is a peculiarly small (very small) co-operative 'phase', no matter how big it is. We are used to phases of *repeating* identical units as in ice, quartz and innumerable inorganic and organic compounds, but a protein is different in that it is usually a folded structure of non-repeating units in a *linear* covalent sequence. It behaves, therefore, in many ways more like a glass than a simple crystal since it undergoes changes of structure locally at different temperatures. There are local regions in which co-operativity is continuous and high and here the major binding factor is continuity of H bonds, as in Figs 9.18 and 11.8. These regions of extensive co-operativity generate a melting curve that closely resembles that of a crystal. On the other hand, if the H bond networks are different in extent and locally separated within one protein, then the protein melts progressively over a relatively wide range of temperature describable by domain dynamics. All kinds of melting can be expressed by a co-operativity coefficient, n (Fig. 6.20(a)). We expect, therefore, that proteins individually can change virtually by a first-order phase transition or by a second-order phase transition (Section 4.7), depending upon the degree and extent of co-operativity. This discussion also applies to an ordered phase of different chemical components, for example of different proteins.

The distinction we are making between melting with a co-operativity factor of infinity and one of lower value n is related to the distinction between an all-or-nothing switch and a more gradual switch of shape with temperature (Fig. 6.20(b)). An all-or-nothing switch is not such a useful structural change in a machine or living organism as it tends to break the structure while releasing much energy for an infinitesimal change in temperature (or other condition). A much more useful switch is a co-operative change in which n is still large, but not infinite. We shall see that biological systems operate over very short ranges of temperature, pressure and composition with rapid but not discontinuous changes of structure. The great advantage of proteins is their distribution of energy states due to the peculiarities of their co-operative interactions (see Fig. 3.16). These properties will allow us to consider many proteins as components of molecular machines.

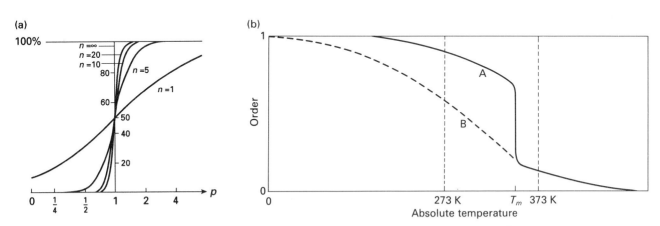

Fig. 6.20 (a) An illustration of the way in which control of any change is dependent upon the degree of co-operativity in a system $X + nX \rightleftarrows X_{n+1}$. The vertical axis records percentage change while the horizontal axis is an intensity measurement of an applied variable, for example, log p, $1/T$, etc. (b) The variation of order with temperature in two classes of protein, where order is defined as unity for a protein that occupies only one, the most stable, conformation and is zero for a random coil. In class A there is increasing but very local disorder with temperature until the onset of melting. In class B, disorder increases continuously and is more widespread at all low temperatures. A third class of protein (not shown) will have very little order at all temperatures of biological interest. Case A approaches a simple first-order phase transition (see Fig. 4.19).

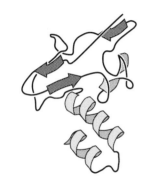

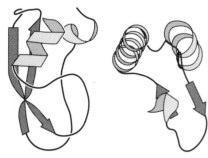

Fig. 6.21 Diagrams for the folds of a miscellaneous set of small proteins. The backbones of these polymers fold into (i) helices that are clearly seen and (ii) almost straight strands, shown as arrows which come together to give β sheets (see Chapter 9). The number of different folds known is considerable—perhaps 50–100—but the surfaces generated are of almost infinite variety since each helix or strand may have more than 20 amino acids, which can be varied in sequence among 24-amino-acid types (see Fig. 6.19(a) and (b)). (After work of J. Thornton.)

6.8.2 Ordered sequences in polymers and assemblies

Every biological polymer has an ordered sequence and as such it is a single component. From a few monomers, for example amino acids (or nucleotides), millions of components can be made by a change of sequence in proteins (or nucleic acids). The operational description is exactly that of an operational component based on small molecules, for example CH_3OH and H_2O, as opposed to that based on elements H, C, O, etc., which we used in Chapter 4 while discussing the phase rule. Here individual proteins (or DNA/RNA) not just the amino acids become the components. These components will have different surfaces or shapes (small phases) when they fold and combine (see Fig. 6.21). If we combine several of the same protein molecules together then they can give a (self-assembled) larger unit (Fig. 6.22). On combining different polymers, which are both components and phases simultaneously, that is, many components with no relationship to one another, then a huge variety (the term infinite comes to mind) of co-operative (equilibrium) shapes can be developed. Although we allow that the contents of any protein are not in equilibrium with the contents of any other (no exchange of amino acids) it can well be that their assembly forms an 'equilibrium' shape from these polymer components following the discussion in Section 6.2.5. This is one way of achieving a protein self-assembly. To all intents and purposes we are discussing the phase diagrams of Chapter 4 but with a huge number of components which are themselves minute phases.

If it should be that there is a preferential energy of coming together of some polymers with some others *but one of these polymers* is a very large one and it is needed at the centre of the assembly, its *sequence* will largely decide the assembly. Then we could again use figures like those in Chapter 4 for phase equilibria to consider the possible shapes that could evolve at equilibrium given the component polymers. Elaboration might lead to thinking of the very large DNA gene sequence or an equivalent RNA sequence as the component that had the information to generate the equilibrium shape in this way. This is observed in viruses. It so happens that the 'order' seen in a biological system, though of unique shape, is not in general maintained in this equilibrium manner. The value of the discussion, however, is that it indicates that we need to discover restrictions on shape-maintained reproducibility in biology based on a *single* (DNA) sequence if we are to explain biological shape based on vast numbers of apparently independent protein components, each of which acts as if it is a phase!

(While drawing these conclusions we have to observe that the formation of a unique sequence is highly improbable. In fact, the most probable situation is that every sequence should have equal weight when we could only expect every possible shape in a continuously variable system of polymers. This has not happened—there are, therefore, additional underlying principles in biological evolution that we must uncover.)

6.8.3 An example of equilibrium self-assembly: viruses

Figure 6.22 shows how a special protein can form a body of relatively high symmetry by association of many identical units. These bodies resemble the vesicles formed by lipids, or single crystals. There are many such multi-

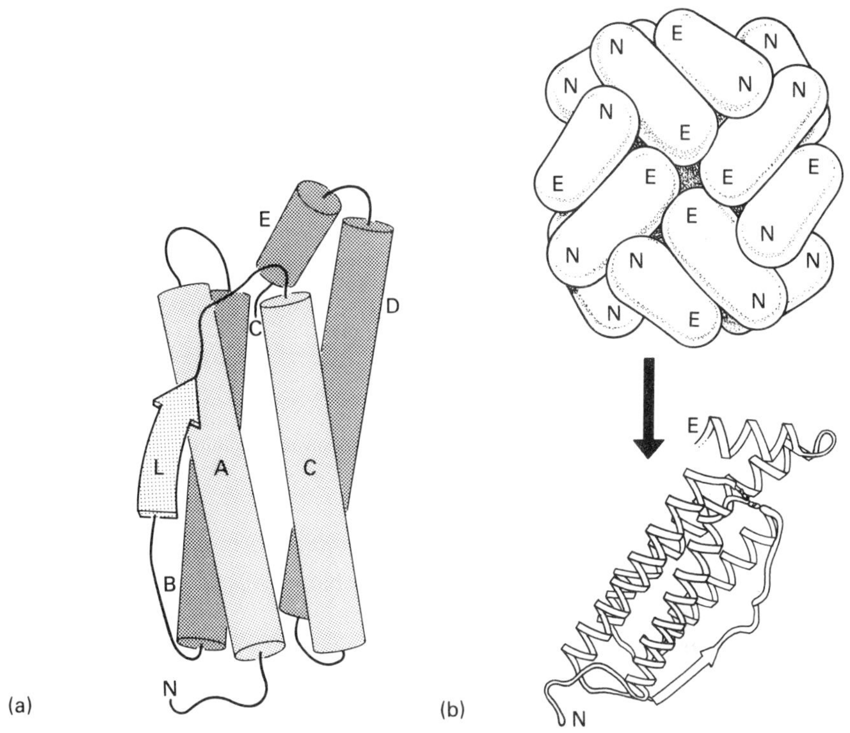

Fig. 6.22 (a) Schematic representation of the ferritin subunit. Four long helices, A, B, C and D, of one protein form a bundle with the fifth helix, E, lying at an acute angle to the bundle. There is only one region of intermolecular anti-parallel β pleat in the loop L. The way in which the subunits self-assemble is shown in (b) where the subunit, X, is illustrated separately. (After P. Harrison.)

subunits in biology and it would appear that their formation is controlled by selective surface recognition at *local* equilibrium and that the whole is driven to a composition limit by co-operative interaction forming an enclosed polyhedron or vesicle.

Figure 6.23(a) shows a more complicated example of assembly at equilibrium in which as many as 180 proteins form a floret structure around an RNA. The association of the proteins with RNA is controlled by the oppositely charged electrostatic surfaces of the polynucleotide (negative) and the protein (positive). A variety of such viruses is known and an interesting variety of geometric forms has been observed (Fig. 6.22(b) and (c)). The hollow protein bodies which form contain central RNA or DNA. The rules for such packing to give highly symmetrical shapes have been known for literally thousands of years. The number of units used to form points of a surface on a *regular* figure are 4, 6, 8, 13, 20. The resulting icosahedra are shown in Fig. 6.24. They are the same polyhedra as can be formed even from single atoms and, in fact, are observed for elements, for example, in the fullerenes (see Fig. 2.7). They are, in effect, some of the allowed ways of packing identical objects in crystals. In large part living systems build with non-identical units. (In passing note that viruses are not alive and may be equilibrium structures.)

6.9 Summary: the incommensurate increase in variables

The number of variables that could have applied to physical development within chemical selection on Earth now appears extremely formidable.

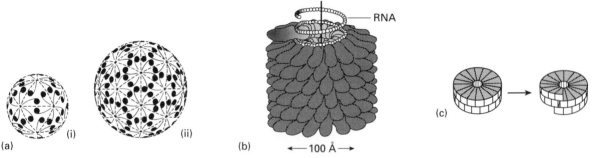

Fig. 6.23 (a) The tomato bushy stunt virus. Icosahedral surface lattices showing the packing of (i) 60 strictly equivalent subunits and (ii) 180 quasiequivalent subunits. Note that all of the tail-to-tail contacts in part (i) are in rings of five, whereas some of these contacts in part (ii) are in rings of five and other in rings of six. (After Harrison, S. C. (1978). *Trends Biochem. Sci.* **3**, 4.) (b) Model of a part of tobacco mosaic virus (TMV), showing the helical array of protein subunits around a single-stranded RNA molecule. (After Klug, A. and Caspar, D. L. D. (1960). *Adv. Virus Res.* **7**, 274.) (c) Schematic diagram showing the conversion of a TMV protein disc into the helical lock-washer form. (After Klug, A. (1972). *Fed. Proc.* **31**, 40.)

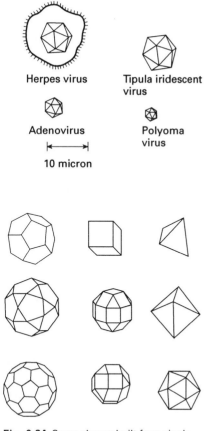

Fig. 6.24 Some shapes built from single units, for example, triangles or squares, and some built from combinations of two units.

1. Temperature and pressure can assume any value from -50 to $+100°C$ locally.

2. There are 90 stable elements on Earth (and 92 in the universe).

3. The 90 elements generated a huge variety of 'frozen' inorganic chemical components on Earth through rapid cooling. The synthesised components are not in atom exchange equilibrium. They formed many local solid compartments, the minerals.

4. There resulted a vast number of solid compartmens in contact with very few isolated liquid compartments, lakes, rivers and the sea, etc., made from one liquid, water. There remained one major gas compartment, the atmosphere.

5. By gain of energy from the radiation field of the sun or from the centre of the Earth a vast number of new components was generated, creating amongst other things new organic molecule compartments using water-insoluble oils, lipids.

6. Once formed, some of these small molecules generated frozen fixed linear-sequence polymeric components.

7. Some of these components self-assembled in large organic molecules, small phases or compartments.

8. The compartments referred to in (3) and (5) gave rise to new aqueous compartments by insulating zones of water from one another which again isolated concentrations of dilute components from one another. All these new aqueous compartments created further variables.

9. Long-range electrostatic fields as well as chemical electrostatic fields and gravity affected the way in which compartments assembled together.

10. Components from (5)–(7) under the influence of (8) and (9) came together collectively and selectively but not in physical or chemical exchange equilibria, to form living cells.

All of the above are subject to change with time but this is part of a separate discussion.

It appears that the resultant of this evolutionary sequence of events, determined in part by thermodynamically controlled chemical selectivity and in part by accidental fluctuations, could be little less than an almost infinite variety of almost fortuitous systems. Yet, in all of this chaos, a highly organised form, life, appeared *in separated forms*, reproducible species (Fig. 6.25). To appreciate this appearance it is clear that we need to discover quite new

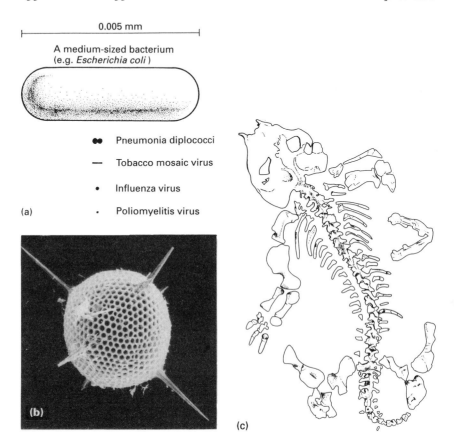

Fig. 6.25 (a) Comparative sizes of bacteria and viruses. (b) A radiolaria ($\times 10^4$). (c) The skeleton of a small lizard—life-size.

restrictions on variables. Since we have exhausted thermodynamic analysis these have to be based on *rates of change*, *kinetics*, (to which we must turn next in Chapter 7). We shall see that it is only the nature of a liquid that can allow this development since 'organised' change demands controlled diffusion (see Section 7.10). The only liquid which became available at first on Earth was water. Thus the controlled cooling of Earth, which separated out metallic iron in a protected compartment, has allowed water to appear in multitudes of small *compartments* at below 100°C. This is absolutely critical to all our discussion of life. Equally, our description of organic chemistry is extremely limited to small compartments, stable between 0 and 100°C. Finally, man's compartments, for example, electric wires, depend on heating to temperatures > 1000°C and then cooling. Only when all the kinetic principles concerning the value of compartments and exchange between components at limited rates, to be described in Chapter 7, together with the thermodynamic principles discussed in Chapters 2–6, are in place, can we attempt to analyse the evolution of the living objects we observe around us, starting again from the big bang. Likewise only then shall we be able to discuss man's modern industrial world. Part II of the book deals with this natural selection of the chemical elements in observed systems and their evolution. While we are describing these developments perhaps one thought should be uppermost in our minds. *No matter how complicated the components in any condensed phase or compartment at internal equilibrium it is a co-operative unity.* We sense that this is equally true of a living organism and our task is to see how this can be true since in no way can we believe that it can be called an equilibrium unity.

Further reading

The topics discussed in this chapter are found dispersed in many texts on thermodynamics and general physical chemistry although the approach is different from that used here. The reader can gain further insight by consulting the references indicated for Chapters 3 and 4, which are essentially the same as for this chapter and the following.

1. D'Arcy Thompson (1961). *On growth and form*. Cambridge University Press, Cambridge.
2. Van Vlack, L. (1973). *Materials science for engineers*. Addison-Wesley, Reading (Mass.).
3. The structure and properties of materials (1964). John Wiley and Sons, New York. Volumes I–IV and especially Vol. I. Moffat, W. G., Pearsall, G. W., and Wulff, J. *Structure* and Vol. II. Brophy, J. H., Rose, R. M., and Wulff, J. *Thermodynamics of structure*.
4. Hoppe, W., Lohman, W., Markl, H., and Ziegler, H. (eds). (1983) *Biophysics*. Springer-Verlag, Berlin. Chapters 5, 6, 12, and 13.
5. Williams, R. J. P. (1993). Protein dynamics. *European Biophysics Journal*, **21**, 393–401.

7

The evolution of kinetic control
and of organisation

Tao generates the One.
The One generates the Two.
The Two generates the Three.
The Three generates all things.
All things have darkness (chaos) at their bulk
But strive towards the light
And the flowing power gives them harmony.

Lao Tzu (Sixth century BC): *Tao Te Ching.*
Transl. Richard Wilhelm/H.G. Ostwald†

(English version, adapted)

A Kinetic principles

7.1 Introduction

The conclusion from the analysis of the abiological material around us in Chapter 6 is that, in both its physical and chemical conditions, the variance of inanimate Earth became incalculably large due to the prevailing conditions

† The One is unity (oneness); the Two is duality with its partition into yin and yang; the Three, the 'flowing power', is the unifying medium of the two dualist powers.

that inhibited the generation of full chemical equilibria, thus giving rise to many components, and to the inability to maintain equilibrium transfer of components between phases. In effect, most systems on Earth broke down into compartments (Section 6.3) in and between which both failures to equilibrate were due to barriers to change produced by the rapid fall in temperature. However, we were able to treat most of these systems locally through the *thermal equilibrium* considerations described in Chapter 6. Now around us we see millions of living individuals in species, which are *reproducible* almost identical objects, closely, if not exactly, invariant in chemical composition as well as in physical appearance, even during growth. They can be compared with a set of identical single crystals. Our arguments in the previous chapters led us to conclude that internally individual living systems, organisms, are therefore effectively invariant in components and transfer between their compartments, which is the condition of equilibrium. Clearly, however, they are not at equilibrium since they are developing and expanding flow systems. It must then be to considerations of flow and development, that is, to rate of change of, first, components and, second, of compartments, to which we must turn to understand the nature of this constancy of living reproducible objects. We shall ask the question: is there any true parallel between the growth of living organisms with a fixed pattern and the expansion of phases of inanimate objects (crystals) that can occur on cooling at equilibrium? Throughout this enquiry we are seeking an explanation of the objects around us in terms of the natural selection of the chemical elements, now related to their ability to change at selected rates. We shall not attempt to come to any firm conclusion until Chapter 16.

7.1.1 Change, time and flow

Until this chapter we have scrupulously avoided discussion of the *rate of change* of the substances we see around us. Thus, in Chapters 2–6 we have described stable systems in equilibrium or at least in stationary states, extending the discussion to small amounts of material in separated compartments at *local* equilibrium and including the effects of fields. In those discussions *time* was not a variable; yet it is clear that, in fact, everything around us changes with time. We enter the realm of one of the great philosophical problems of which the Chinese, rather than the Greeks, seem to have been most aware in their thinking about 'flow'.

As with many major underpinning parts of our experience such as the real nature of matter, energy and fields of force, we cannot expect a thoroughly deep understanding of time. In effect, one way to approach time is to consider it to be measured by the rate of change of position of rest mass, *motion*, which is written dx/dt, corresponding to a velocity, v, here along an x-axis. In Chapter 1 we drew attention to the fact that there were two types of motion of rest masses: (1) *random* motion, which we equated with a scalar kinetic energy and temperature; and (2) *anisotropic* motion in a direction, which we associate with a vector, momentum, mv, and *flow*. It is the flow of rest mass that gives us a sense of time. If we start from the big bang at time zero then there has been

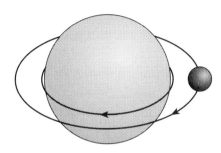

Methods of counting time

Equilibrium (time zero)
(time-independent)

↓

Rapid cooling gave
stationary states of
many components in
thermal balance
(time-independent)

↓

Further cooling gave
thermal out-of-
-balance and *physical
flow* in steady states

↓

Further cooling gave
thermal out-of-
-balance plus limited
chemical exchange in
steady states

↓

Energy trapping (heat)
Developing systems
Evolution of life

continuous radial flow and its continuation is a measure of time to the present expanded state of the universe. Local time we measure by the movement of the Earth round the sun, the yearly period, by the moon's movement round the Earth, the old month, and by the rotation of the Earth on its own axis, the day, see marginal illustration. (There are correlated measures of time due to the expansion (stretching) of the wavelength of the original photons in the universe, that is, of radiation energy, which can be equated with a fall in the temperature to about 3 K. In this way the background noise (temperature) reflects the age of the universe.) Time is then a fundamental concept associated with matter, momentum and radiation and created with them. We shall not go further into this rather metaphysical analysis except to say that we shall accept that, at time zero and zero momentum, an event took place, the big bang, which contained the possibility (a possibility which came about) of the development of energy, matter and flow, and eventually our life on Earth. A dominant feature is then *flow* due to the initial explosion and its consequences, while equilibrium considerations are but a help in understanding the direction the overall flow must take. Within the initial overall radial flow turbulence developed, and with it local flow within almost isolated systems which was and is not radial, although radial motion was also maintained within the universe as a whole. The separated systems have then undergone different rates of internal change. The question is what controls these rates of change, the central problem of *kinetics*, an understanding of which would permit us to follow the natural selection of the chemical elements from the big bang to the activities of local Earth and even those within ourselves. This chapter is concerned with the nature of such kinetics in the first place, and of its consequences subsequently. We shall find that, although the overall flow in the universe is increasingly disordered in a general sense due to turbulence, local fields and flows came about that are quite different and became *organised* into patterns. This evolution of systematic flow with selection of chemical elements reaches its most sophisticated state in living organisms and it is the general nature of such systems that we attempt to uncover in this chapter. The flow patterns can only be maintained by the constant use of energy, that is, production of entropy (for example, in the universe or in the solar, planets/sun system, light goes to heat (Section 3.7)) and of constraints (fields, self-developed from inhomogeneity or turbulence) imposed upon the flow. We stress that, unlike considerations in thermodynamics where we concentrate on the time-independent condition of a system, Sys, we are now concerned with the controls over a change of systems with time, that is, dSys/dt. Here interest can be in the change of space co-ordinates of Sys, transfer, or of its chemical condition since both of them may change. We commence this analysis of the switch from full equilibrium and stationary-state situations to *steady states* and then to *flowing development* through treatments of the natural selection of the chemical elements from the beginning of time (see marginal illustration). As stressed above, much of our emphasis must now be upon living systems since they are the objects around us that are furthest removed from the considerations developed so far and show remarkable natural selection of the chemical elements. Since it is the kinetic properties of chemical elements that now dominate in *functioning* living systems, we shall need a vocabulary that addresses the *functional* value of chemical elements and compounds.

7.2 The initial development in the natural selection of the atomic elements

We must begin our consideration of the timed natural selection of the elements, as stated above, from just after the big bang when a state of energy/matter that was homogeneous and at equilibrium, that is, in a uniform thermalised state, began to expand and cool with time. (NB. There was and is no gravitational equilibrium.) These immediate steps after the big bang involved processes at immense, uniformly changing temperature and pressure and, insofar as time was time in our sense, the initial major events and the initial kinetics took place in extremely short periods, even down to 10^{-43} sec, the so-called Planck time. We can have no feel for such events.

The beginnings of the formation of light elements, say, of H and He and their isotopes, were also very rapid, and possibly the reactions were so fast that these light nuclei formed in equilibrium with radiation and the bare neutrons and protons well within 1 minute so that the universe remained homogeneous (Fig. 7.1). Some 10^5 to 10^6 years later, the decoupling of matter and radiation

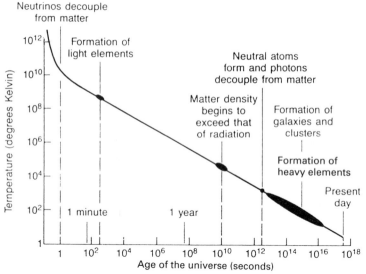

Fig. 7.1 Time evolution of the average temperature of the universe, from the moment it was 1 second old to the present day. As the universe expands, both its average temperature and density fall steadily. (Adapted from Riordan, M. and Schramm, D. N. (1991). *The shadows of creation.* W. H. Freeman and Co., New York.)

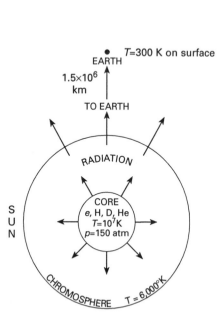

Fig. 7.2 The relationship between the atomic energy source, the sun, and Earth. The system is understood in essence but arose through a chaotic fluctuation. The reactions in giant stars larger than the sun produced the chemical elements of Fig. 7.3.

(photons), as well as subsequent fluctuations broke up this homogeneity— galaxies, nebulae and stars formed. Since that time *local* kinetic limitations have governed the formation and evolution of even quite small nuclei in the universe, for example, C, N and O. The reactions (Figs 7.2 and 7.3) took place (and still take place) in isolated giant stars as they formed (and form) due to further fluctuations in the initial gas (see Sections 8.2–8.4). Space became quite inhomogeneous. In turn, the explosion of these stars and further cooling allowed formation, locally and disproportionately, of larger nuclei and then in turn elements, gaseous compounds and finally co-operative condensed systems including the Earth (see Fig. 1.21). The formation of planets, like

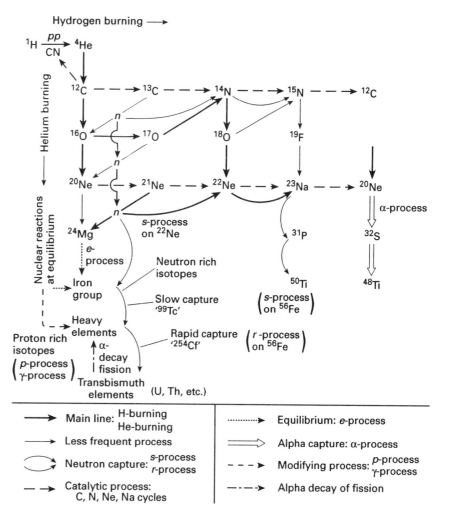

Fig. 7.3 The formation of the elements. (After Burbidge, E. M., Burbidge, G. R., Fowler, W. A. and Hoyle, F. (1957). *Rev. mod. Phys.*, **29**, 547.)

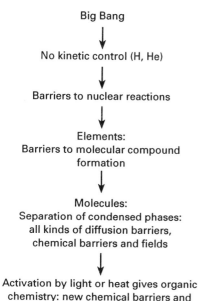

Fig. 7.4 Evolution of kinetic bariers. NB. The processes are accompanied by the fall of temperature and pressure in the universe as a whole.

earlier steps to inhomogeneity, was again caused by local fluctuations. All stages except the first were out of equilibrium (in the sense of a uniform system) and forever falling into kinetic traps (see Fig. 7.4), from many of which it is difficult to escape (see Fig. 6.5). The lifetime of atoms and molecules in these traps is the subject of kinetics which has both chemical and physical features. The lifetimes are affected by the temperature and the fields, which here include chemical interactions in which the atoms or molecules are trapped so forming the components and phases of the phase rule as well as compartments (see Chapters 2–6).

To see the consequences of these developments for the natural selection of the elements within chemistry we must come to the local conditions on Earth as it evolved some 10×10^9 years after the big bang and some 5×10^9 years ago. We must also note that on this time-scale the sun is not everlasting (in fact it could die as a star in the next 5×10^9 years) and that chemical changes of real sophistication occur only in the temperature range of 3000 to 200 K (Table 7.1) where nuclear transformations no longer occur. We shall not be concerned again with nuclear transformations until Chapter 8.

Table 7.1 Development of kinetic barriers

Reaction temperature	Barrier
Temperature falls to 10^{10} K	Nuclear reaction barriers
Temperature falls to $<10^4$ K	Chemical bonds
Temperature falls to 10^3 K	Phase and compartment boundaries, diffusion restriction
Temperature falls to 3×10^2 K	Organic phase boundaries, membranes

7.3 Factors affecting reaction rates

Two major limitations, kinetic traps, on reaction rates of chemical elements in a homogeneous phase on Earth are obvious. The first is that, generally, in order to react, particles, atoms or molecules, must come together and collide, that is, reactions are mostly of the kind

$$A + B \longrightarrow AB \longrightarrow products.$$

There are, then, limitations on collision to be considered. In such cases the reaction rate in a single gas phase depends directly on the partial pressures (relative concentrations) of A and B, and on temperature which controls the rate of diffusion. We express this by writing for the rate of reaction $v = k[A][B]$ for simple bimolecular processes,† where v is the rate of the change and k is a temperature-dependent constant, the so-called *rate constant*, which contains the collision-frequency. Spontaneous decay such as radioactivity is not of this kind, but most other reactions that are significant are temperature-dependent; in other words, they depend upon collisions. Before the Earth formed, the fact that the universe was expanding reduced progressively the chance of collision between particles of matter, while co-operative condensation, mostly as solids, further slowed down reactions by reducing diffusion within any condensed phase so making individual elements, to various degrees in compounds, inaccessible. (On Earth water became virtually the only liquid available for diffusion of solutes.) We shall call all such *diffusion* limitations to reaction *physical* barriers, which generate *compartments* (see Chapters 4 and 6).

The second limitation on reaction rates is the fact that many potential reactants, once cooled to around 0–100°C, the temperature range of the Earth's surface and of liquid water, are usually in quite stable, chemically bound, molecular (or condensed) states, even though they are not as stable as potential products in which element partners are exchanged, that is, they are *kinetically* but not *thermodynamically stable* (see Chapters 3 and 4).‡ This can

---

† Generally, the expressions of the rate of reactions are more complex and must be determined experimentally. For example, for the reaction $H_2 + I_2 \rightarrow 2HI$, the expression $v = k[I_2][H_2]$ is observed, but for the apparently similar bimolecular reactions $H_2 + Cl_2 \rightarrow 2HCl$ or $H_2 + Br_2 \rightarrow 2HBr$ the rate expressions are quite different, meaning that the pathways (mechanisms) of these three reactions differ considerably.
‡ In Chapter 4 we defined a component as a chemical that did not exchange atoms. In other words, we assumed infinite kinetic barriers between components.

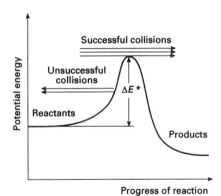

Fig. 7.5 The activation energy barrier: the activated complex corresponds to the maximum potential energy ΔE^* in cal/mol.

be observed, for example, when we mix methane and dioxygen. There is no reaction, even though the molecules collide, until we create a spark that 'excites' (energises) the molecules of these gases. There is then a chemical bond barrier to change in CH_4 and/or O_2 and we may draw a diagram (Fig. 7.5) to indicate that there is an energy hill, ΔE^*, to be climbed within the collision complex. Such energy can be put into molecules, for example by increasing their temperature (Section 3.2.3). It has been found that the reaction rate constant in a collision complex, k in the above equation, is given by $k = Ae^{-\Delta E^*/RT}$, where A is the so-called frequency factor related to the velocity of the molecules and to the effective cross-section for collision. Since k increases in proportion to $e^{-\Delta E^*/RT}$ (Fig. 7.5), it is immediately obvious that temperature is an exceedingly important factor for reaction rate through this exponential term. As stated above, the universe is cooling locally at different rates, so that, while the giant stars go on producing elements at very high local temperature and no doubt planets form at somewhat lower temperatures, on Earth the temperature is so low (in relative terms) that change of chemical partners is usually slow. ΔE^* is also widely variable, so that the range of rate constants for chemical processes is very large, (Table 7.2). We shall include all barriers ΔE^* to changes of bound states of elements under kinetic *chemical* barriers.

Table 7.2 Grouping of ions according to water exchange

Fast ($\sim 10^9$ s^{-1})	Medium ($\sim 10^5$ s^{-1})	Slow (~ 1 s^{-1})
Na^+, K^+, Ca^{2+}	Mg^{2+}, Ni^{2+}	Be^{2+}
Cl^-, Br^-, I^-	Ga^{3+}	Al^{3+}
NO_3^-, SO_4^{2-}, HPO_4^{2-}		Fe^{3+}, Sc^{3+}
Zn^{2+}, Cu^{2+}, (La^{3+})		M^{4+}
Organic molecules		

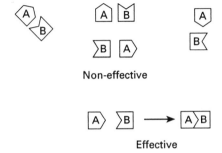

Fig. 7.6 Non-effective AB associations (above) and effective association leading to reaction (below). Collision has occurred in all cases.

There is a further factor affecting the reaction constant k, namely, the relative probability of occurrence of the required state of association, AB, in which A and B are correctly oriented for reaction, that is, there may be only one effective particular association of A and B, when they come in contact (Fig. 7.6) that allows the rearrangement of bonds leading to the products of the reaction. This can be taken into account by introducing a 'steric' factor, P, related to the 'effective' collisions, which is independent of temperature and pressure, and we may, therefore, write the expression above as:

$$k = \text{collision frequency} \cdot P \cdot e^{\Delta E^*/RT}. \tag{7.1}$$

An alternative to this so-called 'collision theory' is the 'activated complex theory' of chemical kinetics based on concepts of statistical thermodynamics (Chapter 3), in which the steric factor and indeed the activation energy ΔE^* (per mole) are introduced from general thermodynamic principles. In this theory we may label the probability of a correctly organised transition state (Fig. 7.6) in the collision complex, relative to the probability of any other form of associated AB cluster, by an entropy term ΔS^*. The rate constant is then given by

$$k = Be^{-\Delta H^*/RT} e^{\Delta S^*/R} = Be^{-\Delta G^*/RT} \tag{7.2}$$

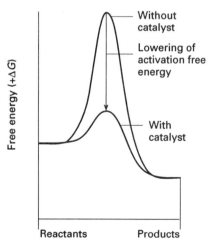

Fig. 7.7 The effect of a catalyst on the activation free energy, ΔG^*.

where B is a function of temperature and pressure related to the collision frequency. Note that the exponential terms are Boltzmann factorisations as was described for the occupation of energy states in Section 3.2.3. (For many reactions, and particularly for reactions in solution when pressure and volume are constant, $\Delta H^* = \Delta E^*$, so that this result coincides with that obtained using the collision theory when the steric factor is seen as an entropy factor dependent on the number of 'effective' configurational microstates relative to the total microstates of AB association.)

This approach (for a rigorous treatment see reference 1 in 'Further reading') allows us to consider the selection of the chemical elements for control of reaction rates on the basis of how the three terms—collision rate, ΔE^* and ΔS^*—are dependent upon the properties of elements, free or in compounds, in various states of matter. Thus, control over change is clearly exerted by three factors: (1) by the physical barriers to collisional accessibility of one atom or molecule to another; (2) by supply of energy to overcome chemical bond barriers, ΔE^*; and (3) by controlling distribution over configurational states so as to lead to the required energy distribution or orientation to overcome adverse ΔS^*. Catalysts (enzymes in biology), which we will discuss later, are substances, metals or compounds, that help to overcome one or all of these barriers (Fig. 7.7). There is a very large variety of processes that we shall have to consider in order to cover all the control factors within kinetics (Table 7.1).

7.3.1 Kinetics and natural selection of chemical elements: functional value in living systems

When we describe the conditions of objects around us based on thermo-dynamic reasoning we measured the *possibility for change* by the distance from equilibrium, ΔG, of all the elements free or in compounds. This equilibrium approach is one major consideration in assisting our understanding of the natural selection of the chemical elements amongst themselves. Since the evolution of the universe did not bring matter (here chemicals on Earth) and energy to equilibrium, and Earth remained warm, there was and is inevitably constant change. As yet we have made no quantitative analysis of the dependence of the rate of change upon selective properties of the elements. In fact, abiotic change, chemical as well as physical, goes on continuously, and in most respects it is seen as a flow toward equilibrium. While releasing energy and increasing entropy generally, we can treat this change in relatively simple ways using the above approach to kinetics and our knowledge of the energetics of the condensed states of matter and of chemical bonds as described in Chapters 2 and 3. We observe, however, that change at the low temperature of Earth has had one curious consequence: it has produced living systems that *develop* and *cycle* in persistent fixed patterns both in space and in chemical composition, not only not at equilibrium with their environment but not even related in a simple way to change toward equilibrium. The kinetics of both physical motion and chemical combination of any element in life, summed under the expression dSys/dt, have become systematised in a unique and extremely *controlled* way utilising absorbed energy and decreasing entropy internally.† We wish to understand the persistence, the *survival*, not now the

† The entropy of the environment increases disproportionately so that the second law of thermodynamics is obeyed.

thermodynamic stability, of these transforming cycles of the selected chemicals that we observe in life. We are forced to treat the development of life as an overall process having internal physical motion and going in particular observed directions, that is, flowing, not toward equilibrium but so as to secure *survival*. In this sense the chemical elements must have been selected to be *useful for this survival* and we shall employ the term '*functional use*' or '*value*' as a factor associated with an element or compound that helps to bring about kinetic persistence in a living system. The term '*fitness*' for survival will also be used. We shall see that the chemistries of distinct elements are so different that to optimalise living systems (or indeed kinetic schemes parallel to them developed by man) elements have been selected naturally for different functions related to but not obviously deducible from the time-independent physicochemical properties that we discussed in Chapters 2–6. This selection has a parallel with the natural selection of physical shape for function, as is often described in Darwinian analysis of *fitness for survival*.

While we shall often illustrate the principles of biological selection of elements for different functions, such as electronic conduction, ionic conduction, structural control and acid–base, redox and radical reaction catalyses, the same principles apply to element selection in man's activities in his externally built organisation for his own survival. However, although the principles of the kinetics of message transmission, message reception and effector† action, and of catalysis of chemical change are the same for man and biological activities, the materials used are different. This leads us to present an account of kinetics designed to highlight functional control over change by the best of all kinetic means in different circumstances using different chemicals. We also observe around us kinetic control such that change can be stopped or started usefully. In biology, the laboratory or in industry this requires a set of modules so that supplies of material, that is, chemical elements, and energy lead in a systematic way to a desired product or construction. The sequence of events is

Instruction (information control centre) → Message transmission system →

Effector system → Reaction system with catalysts → Construction.

At each step there must be physical or chemical barriers that are manipulated; thus it is the analysis of the adjustment of barriers to transfer of components (elements) and energy between compartments, as well as to reactions of chemicals, to which we must turn. In man's equipment the construction of physical barriers is again different from that in biology and again uses very different chemicals. We treat the barriers to chemical change first in part B of this chapter and those to physical transfer in part C before putting all the kinetic factors together in part D.

---

† The effector is the unit which on receiving a message adjusts the parameters of the system, for example a magnetic coil in a door bell system.

B Chemical kinetic controls

7.4 Chemical barriers to reaction

We come now directly to the limitation on rate of *chemical change* mentioned in the introduction to this chapter—movement in and out of kinetic traps based on chemical bond energy rather than on physical barriers to diffusion. We describe this condition by saying that the associated reactants, $A \cdot B \cdot C$, are not in equilibrium with the product D, and that in order for the reactants to go to products they have to pass over a barrier, ΔG^*, the summed activation energy and entropy barriers (see Fig. 7.5).

7.4.1 Molecularity and energy requirements

There are two different considerations concerning transformation barriers within chemical kinetics that do not involve restrictions on diffusion. First, there are overall schemes related to the number of units in reaction intermediates, for example, unimolecular, A by itself, or bimolecular, AB, etc reactions. We consider these associations together with various sequential mechanisms where a series of ordered steps is demanded, for example A must react with B before C can react with AB in the associated units. Second, we must consider why a rate constant of a chemical step of given molecularity has its particular value, k (eqn (7.2)), which is related to the energetics of chemical transformation of substances, ΔG^*, and particularly to ΔE^*. When we are considering either the molecularity of reaction schemes or the energy requirements to cross certain chemical bond barriers, we need to analyse why one element behaves differently from another in a kinetic context, just as we considered earlier differences in their thermodynamic properties, if we are to understand what is to be seen around us. The medium in which a reaction occurs, that is, in a compartment or barrier material, is as important as the elements involved directly in the reaction. The overall rate expression (Section 7.3) can be written

$$\text{Rate of reaction} = k' \, [A \cdot B \cdot C \ldots]$$

where $[A \cdot B \cdot C \ldots]$ is the associated complex resulting from the preformed *effective* association of A, B, C, etc. (see eqn (7.2)). We need now to examine k',[†] but we must indicate each time the molecularity of the step under discussion, that is, A alone, A · B, or A · B · C, etc. We start the discussion from the simplest chemical change, unimolecular electron transfer, in which the electron moves in a chemical matrix. Here we need to know the chemical barriers to the flow of electrons.

7.5 Chemical selection of components for rate control

7.5.1 Electron transfer

Electron transfer, not element transfer, is the basis of electronics and, in general, as stated in Chapter 2, it must be based on metal elements or at least elements (or compounds) rather close to metals in their properties, namely semiconductors. The best conductors are bulk metals such as copper (in wires), in which there are no substantial barriers to conduction (the resistance of Cu is slight), and the best semiconductors are Si and Ge, lying next in the periodic table to Al and Ga (metals), respectively (Fig. 7.8). In semiconductors, barriers, ΔE^*, are present and conduction is an excited state process. Alloys can also be used, of course, and, oustandingly, Ni, Cr and Fe are used in combination in some wires, while the semiconductor (alloys) are isoelectronic with Si (AlP) or with Ge (GaAs). Note that insulators, which have very high, virtually infinite barriers to electronic conduction, are made from organic plastics, C, N, O, H compounds, or inorganic non-metal substances such as SiO_2 (glass) and Si_3N_4 (ceramics). Clearly, the best insulators are made from elements from the right-hand side of the first row of the periodic table. The

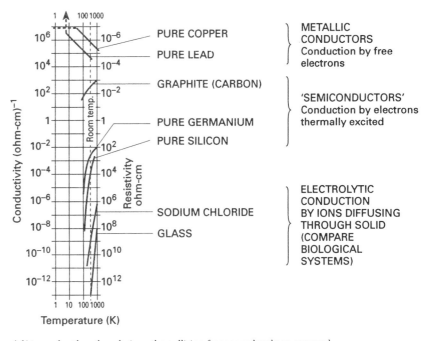

Fig. 7.8 The electrical conductivity of some representative substances. Notice that logarithmic scales are used for both conductivity and absolute temperature. (After Purcell, E. M. (1965). *Berkeley physics course*, Vol. 2. McGraw-Hill, New York.)

[†] k' is used rather than k since the collision frequency has been removed.

barriers to electron motion generally increase in elements as we go from left to right and from bottom to top of the periodic table (Section 2.2.2). These principles apply to man's bulk electron-transfer materials, of course, but increasingly interest revolves around miniature devices where some of the same elements or compounds are used, for example in computers. In biology miniaturisation is taken to its final molecular limit, but biology cannot make metals or the above semiconductors, which are high-temperature products, so that it is forced to use single metal and heavier non-metal atoms in combination within polymers—a kind of composite (Fig. 7.9). The metal

Fig. 7.9 Two examples of electronic conduction in biological matrices. (a) In photosynthesis dioxygen plus a set of reduced carbon compounds, $[CHOH]_n$, together with pH and potential (ψ) gradients are generated by the action of light. The light forces a flow of charge electrons through the membrane using Fe and Cu proteins. (b) In respiration the dioxygen and reduced carbon compounds $[CHOH]_n$, are used to create across another membrane a pH and potential gradient while remaking CO_2 and H_2O. Electrons are again moved within the membrane using Fe and Cu protein 'wires'.

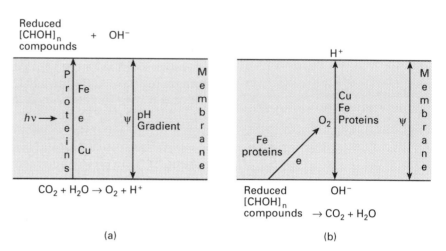

atoms used are iron, copper and manganese ions, and the main non-metal is sulphur, all of which are incorporated in polymers (proteins). It is also possible to use certain unsaturated organic molecules (compare graphite in Fig. 7.8). The ways in which biological systems generate electron conduction paths are also described in Chapter 4 of our previous book, *The biological chemistry of the elements* (see reference 1 in 'Further reading'), where the theory of their properties is outlined as well. In essence ΔE^* is reduced as much as possible. Figure 7.9 shows a typical electron transfer chain of atoms (ions) in a biological membrane. In summary, the choice of components (and compartments) is based on those elements that give ions or molecules, usually of metals associated with low ΔE^* for electron transfer within specially synthesised constructs, made from non-metals, namely proteins. (The design of compartments is described in Section 7.10.5.)

Immediately we see that the natural selection of the special chemical elements for electron transfer in biology is different from that made by man. Both are based on abundance, availability and functional property, but using different abilities to manipulate elements into optimal chemical combination.

Whereas there are many electronic devices for control in man's electronic circuits (so-called electronic *switches* or *gates*, Section 7.10), the modes of control in biology are often based on the structural changes of certain polymers, proteins, that have internal dynamics, that is, they are mechanical devices. In a device based on single atom centres, electron transfer passage through a centre depends on: (1) the energetics, that is, the relative redox potential of sites, for example of Fe^{2+}/Fe^{3+} couples; (2) the distance of separation between centres; and (3) conformational relaxation energies and rates on change of charge at a centre. All can be manipulated conformationally or mechanically (allosterically) in *proteins* so as to produce a *gated* current. It is, therefore, merely an exercise to convert, in outline, the ideas of flow of electrons in man's devices based on wires to the molecular circuits of electron flow in biology. These flows can be managed in feedback circuits as well (see Section 7.8.2). Finally, through conformation change there is also coupling between electronic and electrolytic circuits (see Chapter 10), which is essential to energy capture in these biological circuits. (NB. Biological electron flow is overwhelmingly linked to oxidation–reduction reactions for energy capture as in man's photo or fuel cells.)

Now, while man also uses electronics extensively for communication, this is a minor part of biological communication systems. In biology the transfer of charge for communication predominantly uses electrolytics, ions in water. As we shall see, all the components familiar in electronic circuits can be built in electrolytic circuits. Ion flows have the disadvantage that they are slow but the advantage of enabling great diversification of curent carriers. The medium for transport is now water and communication kinetics in water dominates biology. We are concerned next, therefore, with the chemical barriers, ΔE^*, to flow of ions in water. (Note that in man's chemistry other solvents with different kinetic properties often dominate.)

7.5.2 The nature of water: a very special solvent for transport

The majority of biological reactions occur in water, while much of man's organic and some inorganic chemistry utilises organic solvents. In water ionic processes are encouraged and those elements or compounds, especially ions, that are able to act in this solvent usually cannot enter a lipid organic (membrane) phase (Section 5.2). (We shall consider later (Section 7.10) that ion movement can be guided by constructing aqueous compartments shaped and surrounded by organic membranes.) In order to react in water ions must break the solvation barriers around themselves (see Fig. 5.1) so that the energy of water exchange is a governing impediment, ΔE^*, to ion reaction rates. The reaction rate is therefore not close to diffusion limits except for a very few simple ions, for example Na^+, K^+ and Ca^{2+}, but, of course, hydration does not restrict rates of reaction of organic molecules greatly since their hydration is weak. (For such molecules it is the covalent bond energies of H, C, N and O that control reactions since these bonds are of low polarity (see Section 7.10.1).) Finally, water and hydroxylated molecules are excellent conductors of H^+ and this unique ion has come to play a central role in bioenergetic reaction circuits. We next consider *unimolecular ion diffusion rates*.

7.5.3 Selected chemical rates of diffusion in water and ion channels: solvent exchange and message transmission

Soluble small molecules and ions diffuse in bulk water relatively rapidly and their movement is not affected by gravity. (Slower rates of diffusion are found for other ions and large molecules such as proteins and DNA. For these molecules gravity has an effect and especially large particles may settle in the gravitational field. This settling is also common for all kinds of large molecules and debris at river mouths, for example, so creating delta soils.) The rate of diffusion is now given by the *unimolecular* rate at which water molecules can move to create space for drift of molecular species or ions, for example in an electric field. The rate of diffusion of water molecules themselves is, in exchange terms, somewhat greater than 10^9 per second (Fig. 7.10). This is then the fastest rate of diffusion of ions such as Na^+, K^+ and Cl^- (Table 7.2) as well as of molecules such as NH_3. There is effectively no hydration barrier to the migration or reaction of these ions or molecules. Such rates of diffusion of these most mobile ions do not limit in practice many processes in chemistry or biology, but they do limit the fastest electrolytic conductance of solutions to rates of charge transfer at least 10^5 times slower than electronic conduction in a wire.

There is a special conductance that is 100 times faster associated with the proton since it can move from one water molecule to another through rotation and H-bond making and breaking (see Section 5.6.1 and Fig. 7.11). This makes it possible for biology to use H^+ migration in a unique way (see

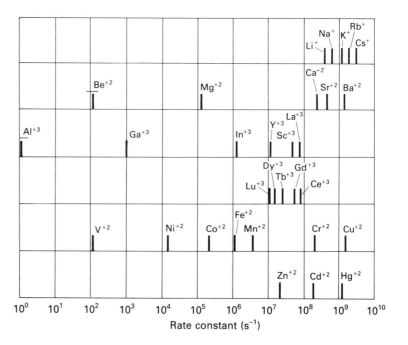

Fig. 7.10 Characteristic rate constants (s^{-1}) for the substitution of inner-sphere H_2O of various aquo ions. NB. The substitution rates of water in complexes $ML(H_2O)_n$ will depend on the symmetry of the complex. (Reproduced with permission from Frey, C. M. and Stuehr, J. (1974). Kinetics of metal ion interactions with nucleotides and base free phosphates. In *Metal ions in biological systems*, Vol. 1 (ed. H. Sigel). Marcel Dekker, New York.)

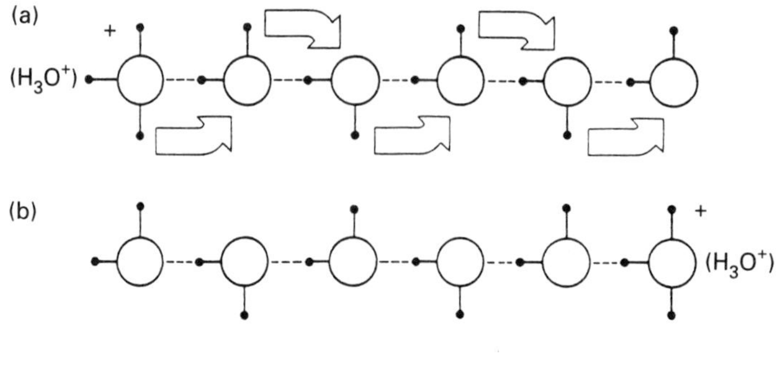

Fig. 7.11 The mechanism of conduction by hydrogen ions in water. (a) Initial state and transfer step; (b) final transfer states.

Table 7.3 Proton and electron transport

	Hop distance (Å)	Site	Activation energy
Protons	0.5–1.0	Bases	pK_a change, relaxation, rotational diffusion
Electrons	$\leqslant 15$	Metal ions, aromatics	Redox energy, ΔE, relaxation

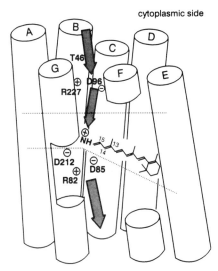

cytoplasmic side

Fig. 7.12 The proton channel in bacteriorhodopsin protein where D = aspartate, R = arginine, and T = threonine, which are amino acids in the protein chain capable of proton transfer. A similar channel in a membrane can be built for other ions leaving a larger pore. There is a suggestion that some channels are built from multiple protein β domains rather than from the protein α helices shown here as cylinders. The rhodopsin molecule is shown.

Table 7.4 Permeability ratios for two types of channels

Type of channel	Permeability ratio		
	Li$^+$	Na$^+$	K$^+$
Sodium	0.95	1.00	0.05
Potassium	0.02	0.02	1.00

Table 7.3) and so separate H$^+$ diffusion from all other ion motions (Fig. 7.12). In fact proton as well as electron diffusion in membranes contribute to energy transduction in the systems of Fig. 7.9. Selective proton diffusion in this manner is known in some inorganic materials too and is especially important in bone (see Chapter 13). Protons, like other ions, move by series of short steps, of course, and are therefore unlike electrons (Table 7.3). Protons are not useful for general electrolytics in biology, however, since the proton concentration is only 10^{-7} M (compare Na$^+$ at 10^{-1} M).

Now, aqueous spaces, compartments, need to be connected in a controlled way to allow selective flow. This is brought about by wires and salt bridges in chemical cells (Fig. 6.17) and by channels in biological membranes. The physical nature of the channels is described in part C of this chapter. Chemical selectivity of ion migration in channels, that is, ΔE^* to passage, is based on ion size and charge (Table 2.16 and Table 7.4). Ions, cations or anions, pass through the molecular channels formed with polar surfaces and of the same diameter as the ions themselves (see Fig. 7.12) but only by releasing water of solvation. The selection of the channel, based on weak interactions, can then be confined to a particular ion through its slight preference for the polar surfaces of the channel over hydration. Since the hole has a fixed geometry but water of hydration is flexible and collapses around the ion, the hole is able to accept the bare ion that fits it while it excludes larger ions, Section 5.5. It selects against still smaller ions because they preferentially retain the adaptable hydration shell, causing them to be, in effect, larger ions. These are examples of chemical selective barriers to diffusion, now dependent on ΔE^*, so that there are Na$^+$ channels, K$^+$ channels, etc. This selective permeability can also be found in minerals, for example zeolites (see Fig. 7.18), a solid-state device in this case. There can be such weak interaction between the hole and the ions that the hole or channel in a membrane can allow diffusion rates almost as fast as in free water for an ion of a particular size.

The selection of Na$^+$, K$^+$ and Cl$^-$ as the major ions for biological currents is obviously due to availability, the solubility in water of their salts and unique chemical characteristics, namely their large radii and low charge, so that they do not tend to stick to surfaces of channels or charges of opposite sign, that is, ΔE^* for any exchange is very low. Ca^{2+} channels act in a somewhat different way using the higher charge to give selective but weak binding to anion groups in the channels. In further examples, copper(I) ion movement is selected in a channel by binding to sulphur-containing side chains in the channel while anion channels have cationic channel surfaces. Thus, selective activation energy controls, that is, of ΔE^* for motion, allowed the development of the use of charge carrier selectivity. The opening and closing of the channels can be managed by voltage (physical field switch) or conformational, chemical (receptor recognition) gating. In conclusion, specific elements are naturally

Table 7.5 Classes of small information carriers

Simple ions	Complicated ions*	Molecules (transmitters)
Na^+, K^+	cAMP (P)$^-$	Acetylcholine (+)
Ca^{2+}	cGMP (P)$^-$	Glutamate (2−)
Cl^-	ATP (P)$^{4-}$	Adrenaline (+)
Zn^{2+}	Haem (Fe)	8-OH Tryptamine (+)
Fe^{2+}		NO (CO)
Mn^{2+}		Sterols
Mg^{2+}		Amino acids (+ −)
		Prostaglandins

* cAMP, cyclic adenosine monophosphate; cGMP, cyclic guanosine monophosphate; ATP, adenosine triphosphate.

selected as the major components for charge-carrying (Table 7.5), while quite other elements form the lipid barriers (C, H, O, P), the protein channels (C, H, N, O, S) and the bulk of the aqueous compartment (H_2O) (compare the discussion of selectivity in Sections 5.5.2 and 5.5.3).

7.5.4 The extension of the selectivity of chemical steps of diffusion: message reception and triggering

Ions like Li^+, Na^+, K^+ and Cl^-, from groups 1 and 17 of the periodic table, bind so weakly to any organic or inorganic surface that they do not affect structures markedly. These ions can readily carry current unimpeded by the presence of negative charge (or positive in the case of Cl^-) not only in channels but also on and around surfaces and in solutions containing other ions; similarly, the surfaces and other ions or charged species, for example proteins, are little affected by their presence in water. Natural selection then leads, as stated, to the 'choice' of uni-charged inorganic or organic ions as the major current carriers, H^+, Na^+, K^+, Cl^-, R^+, R^-, where R^+ and R^- stand for charged organic species, for example protonated amines and free carboxylates, in aqueous organic matrices. All the organic ions have additional properties, namely, specific recognition by binding, and in this respect are related to the inorganic ions M^{2+}, which allows message reception on binding by considerable conformational changes of their selective protein receptors.

The group 2 ions, Be^{2+}, Mg^{2+} and Ca^{2+}, bind more strongly to negatively charged surfaces than monovalent ions, M^+ (Chapter 5). They are, therefore, relatively poor current carriers, but, on release from a storage site, they will interact in the order $Be^{2+} > Mg^{2+} > Ca^{2+}$ with such simple sites as $R-O^-$ and $R-CO_2^-$ on nearby surfaces and cause conformational changes to the flexible molecules they bind on such surfaces (Fig. 7.13). However, they can subsequently exchange from contact relatively readily due to their not very strong binding and be pumped away. In other words, exchange of these bound ions is moderately fast. This allows Mg^{2+} and Ca^{2+} to act as rapidly reversible triggers of reaction or as conformation controls of macromolecules in biology. (Be^{2+} is not present since beryllium is a rare element and we shall not consider it again).

In a parallel manner, complicated charged organic molecules, for example,

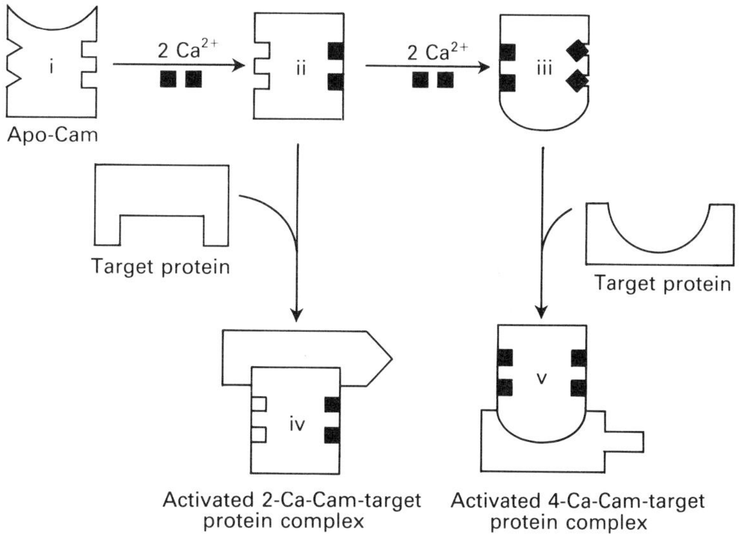

Fig. 7.13 A schematic representation of the sequence of binding steps where the calcium binding protein calmodulin, Cam, when bound by calcium adjusts its own conformation and then the conformation of a target protein. Two calcium levels of triggering for different targets are represented.

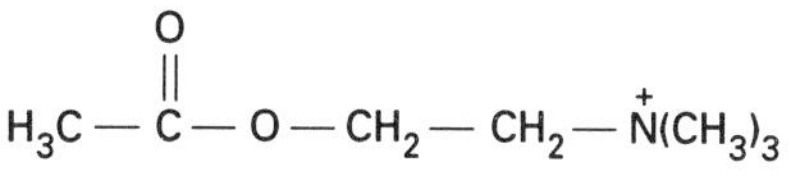

Acetylcholine

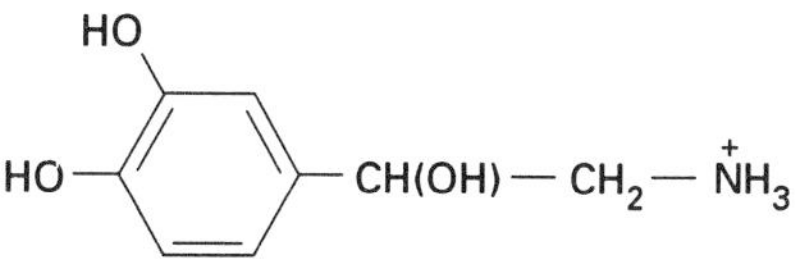

Adrenaline

acetylcholine and adrenaline, Table 7.5 and see marginal illustration, and phosphates, which bind quite strongly but exchange relatively easily, can be used as triggers. They are then called *transmitters* (Table 7.5). (It is often the case that the organic transmitters are removed by local catalytic (enzymic) destruction rather than by removal to a chemical store, i.e. by pumping as for Ca^{2+}.) A common example is that of acetylcholine in nerve transmission, which is destroyed by hydrolysis catalysed by acetylcholine esterase, an enzyme. Thus, as stated, H, C, N and O compounds have a different potential role in transmission since unlike Na^+, Ca^{2+} or Cl^- ions they are open to control by catalysed synthesis and destruction as well as by on/off rate control. NB. Man often manages triggers by very different devices in bulk electron conductors, for example by generation of electric or magnetic fields.

We consider next a group of elements and molecules that bind even more strongly and exchange even more slowly (Tables 7.5 and 7.6). These are useful as triggers for sustained transduction. The best examples are ions, such

Table 7.6 Idealised binding and rate constants

Ion	k_{on} (s^{-1})	k_{off} (s^{-1})	K	Function
Na^+, K^+, Cl^-	$>10^9$	$>10^6$	$<10^3$	Electrolytic messenger
Ca^{2+}	10^9	10^3	10^6	Mechanical trigger
Mg^{2+}	10^5	10^2	10^3	Phosphate transfer
Zn^{2+}, Fe^{2+}	10^8	10^{-2}	10^{10}	'Hormone' communication
Cu^{2+}	$>10^8$	$<10^{-7}$	$>10^{15}$	No exchange
C, H, N, O	Covalent, enzymic control			Structure building
HPO_4^{2-} (RPO_4)	10^9	10^6(?)	10^3	Trigger
Protein PO_4 (phosphorylation)	Slow	Slow	Weak	Kinetic control faster than C, N, O bond breaking
	Covalent, enzymic control			
H^+	$>10^{10}$	Slow to 10^{10}	Huge range	Catalysis, energy store
OH^-	$>10^{10}$	Slow to 10^{10}	Huge range	Catalysis

17-β-oestradiol

Oestrone

Space-filling model of 17-β-oestradiol

as Fe^{2+} and Zn^{2+}, and hormones, such as sterols, see marginal illustration, which remain bound for long periods before exchange. (Should Fe^{2+} and Zn^{2+} be considered to be hormones in such functions?) We observe again that slower on/off organic hormone reactions may be limited by catalysed steps as for faster organic transmitters like acetylcholine. Any hormone, as it is organic, may be made and destroyed by enzyme action unlike an ion such as Zn^{2+}. In the case of HPO_4^{2-}, an anion messenger, the *combination* and the *removal* of the activating moiety is also often accomplished by exchange in 'covalent' steps from organic molecules governed by the enzymic reactions phosphorylation and dephosphorylation.

It is noticeable how the natural selection of simple metal elements is now related to functional value based on energies of binding and activation, ΔE^*, of exchange, which are different for different elements. The selection is different from that for organic compounds (Table 7.6) in the mechanism of handling, for example by channels and pumps as opposed to synthesis and degradation. This is obvious enough for the light non-metal elements, C, H, N and O, from a consideration of their exchange properties, but it extends in a variety of ways to selective use of heavier non-metal elements, such as phosphorus in phosphates and iodine in thyroxine, which depends on their reactivities in complex organic molecules. It is clear that organic plus inorganic molecules can be used to increase greatly the diversity of signalling systems. (Contrast man's computers, and other devices for communication, which can only use electrons, but note how man continues to improve computers by selecting new bulk chemical materials.)

Finally, there are ions, such as Al^{3+}, Fe^{3+} and Cu^{2+}, which exchange extremely slowly in many complexes and are, therefore, located in space almost permanently by their binding. As a consequence, some of these elements, for example Fe and Cu, are, as we have seen, extremely valuable (as the ions) for making fixed electronic circuits in biology using oxidation state changes, for example Fe^{3+} to Fe^{2+} or Cu^{2+} to Cu^+, but of little general use as carriers of information. Many such very slow-exchange elements can be troublesome in that, if allowed into a system, they form almost permanent blocks for the diffusion of the more mobile ions, that is, ΔE^* for the exchange of these ions is too large. Al^{3+} and perhaps most trivalent inorganic ions fall in this category and they are often prevented from entering biological systems. In fact, biology does not use any group 3 element except to some small degree anionic boron, $B(OH)_4^-$. Be^{2+}, a group 2 element, also falls in the same category, as do ions such as Ti^{4+} and Ge^{4+}.

7.5.5 Diversification of element use through non-exchanging binding

If a given element, say iron, in all its bound forms, were in relatively fast chemical exchange, then there would be just one iron pool and we might consider it to be in a condition close enough to equilibrium, that is, to be but one chemical component. While to increase diversity the pool can be split into compartments (Chapter 7.10), a further increase in sophistication can be achieved if the element, here iron, is incorporated into different small molecular units from which it does not exchange. An example is given by

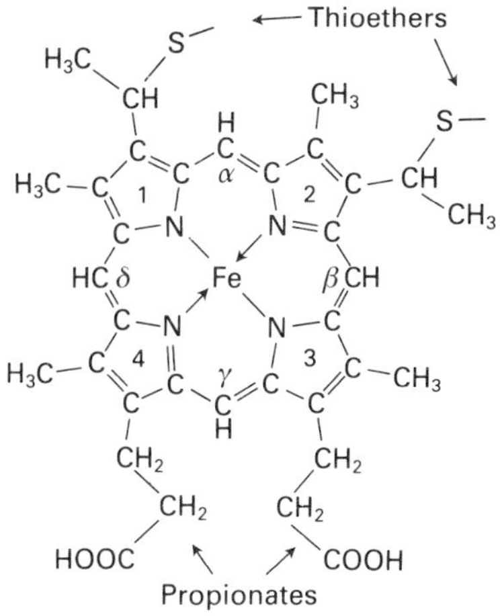

Fig. 7.14 The formula of protoporphyrin bound here through thioether links as in cytochrome *c*.

Table 7.7 Some elements acting in a dual kinetic capacity

Element	Forms
H	NADH, NADPH
P	ATP, cAMP
Fe	Fe^{2+}; Fe haem
Ni	Ni^{2+}; F-430
Mg	Mg^{2+}; Mg chlorophyll
S	Glutathione; coenzyme A

haem iron (Fig. 7.14) where the metal ion is locked into a complex and effectively does not exchange. Haem Fe behaves then as a new 'component', disconnected from free Fe; thus, in biology, haem can be used as a quite separate diffusing signalling device of the same character as organic molecules. In Table 7.7 we give other examples of the diversification of element use. Of course, the prime example is the use of C/H/N/O compounds, but it is intriguing that biology extended the diversification very early in evolution to Ni and Mo as well as Fe, while cobalt seems to belong to but one distinct pool, vitamin B_{12}.

The operational definition of component we are using is based on time-scale, of course, since eventually Fe is liberated from haem by its catalysed destruction (compare organic chemical signals) but just because the interconversion between Fe and haem is so slow the number of degrees of freedom of the system has increased (see the analysis in Section 4.3.2). Thus, chemical kinetic barriers extend the realm of possibilities, that is, the number of degrees of freedom, by increasing the number of components. Yet again and again we return to the problem that these so-called components are present in fixed ratios in living systems. What controls their levels?

Controlled and selected chemical barriers to diffusion/reaction are not only needed for signalling; they are also essential for the distribution of molecular fragments (food), for transport of material (for example, for building), for catalysts and for conversion of chemicals to gain energy and to do work. We have then to discuss exchange in more stable covalent systems, that is, in compounds of group 14 and onwards of the non-metal elements. This is the subject of molecular recognition and enzyme catalysis described in biochemical textbooks. Before proceeding we remind the reader that each and every one of the transduction activities must be seen in the physically restricted compartment appropriate to it (part C of his chapter). It is also a useful exercise to describe in part C the parallel equipment used by man. [NB. Man has but one carrier in metal circuits, the electron, but biological system have perhaps 100 with different diffusion and recognition properties, yet biology has but one bulk medium, H_2O.]

7.6 Slow exchange, structures and the requirements for catalysts: construction and reaction systems

Table 7.8 Selection of major elements in construction

Biology

Proteins and polysaccharides (H, C, N, O)
Minerals (Ca, C, O, P, Si)
Lipids (H, C, N, O, P)

Industry

Metals (steel) (Fe, Cu)
Minerals (concrete) (Ca, Si, O, Mg, Al)
Plastics (H, C, N, O)

Chapters 1–6 have descibed the selectivity associated with chemical binding at equilibrium. In this chapter we have moved to a description of the selectivity within chemical dynamics, ultimately allowing completely different functions for different elements within an organisation. As we have seen, in a biological as in any other organisation all elements must flow in a highly selective way; for this purpose many of them must also be incorporated into constructs (Table 7.8), which may have to be dismantled to allow development, so that each flow can be channelled and order created in semipermanent structures, polymers, especially proteins. While the final product must not disintegrate rapidly, it must be built at considerable speed and be at least open to slow degradation. Again, to utilise structure flexibly it must be adjusted rapidly. Thus, we must observe which elements can have long lifetimes in compounds,

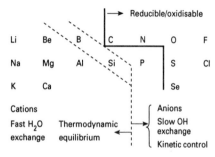

Fig. 7.15 The division of part of the periodic table by redox activity between the H_2 and O_2 potentials and by rates of H_2O or OH^- exchange.

high ΔE^*, how their synthesis and degradation reactions can be catalysed, and how they can generate flexibility for mechanical devices. In essence, Fig. 7.15 shows that exchange is slowest for H, C, N and O combinations. We need, therefore, to discover how these elements or fragments of their combinations can be moved about and how their chemical bonds can be made and broken in a controlled way using chemical catalysts so as to make structures and then degrade them. We do not need to be reminded that some of man's constructions use similar chemicals (plastics) together with extra high-temperature materials, metals and minerals. However, the demand for permanence is very different in biological self-assembling systems. (We have stated above that catalysed synthesis, transfer and degradation of small organic moieties can also be the basis of messages and transmission.)

7.6.1 Kinetics of organic compounds in traps

Around us there are millions of living species that are based on thermodynamically unstable but independent organic compounds, which we have earlier called components. In effect, the natural selection of chemical elements in life has developed inevitably as the natural selection of *kinetically stable* chemical compounds largely based on the four elements H, C, N and O, although in some of the compounds P and S are also involved. In this chapter we are not concerned with living systems as such, yet we must observe that the organic compounds of life are a special set of H, C, N, O compounds, selected for their particular survival strength, which has to be considerable but not permanent. (We shall analyse more deeply the nature of survival strength, the time of turnover and reproduction, which is again a consideration of activation energy, ΔE^*, in Chapter 16.) Without life the number and variety of organic compounds on Earth would undoubtedly be very different. Additional kinetically stable organic compounds have been synthesised by man in the last 200 years and they represent a second selection, in this case for man's functional use. We turn to their description in Chapters 9 and 14, but in a way they represent little more than additions to living biological chemistry since they are 'designed' to 'improve' the lifestyle of man. (But are there hidden risks here?)

Thus, it appears that, as far as we can judge today, it is the very nature of the kinetic barriers to reaction in hydrogen, carbon, nitrogen and oxygen compounds that generates the long life of biological energised atomic assemblies, yet they show turnover. However, because of the stability of their bonds organic elements within synthesised units cannot be transformed without catalysts which in turn cannot be made from these elements since, as we show next, bonds in catalystic action must be broken rapidly. In fact, the H, C, N, O compounds are not a self-sustaining set in a reproducing or developing system since they are not reactive enough. For this reason, amongst others such as the construction of communication networks, see above, life could not be based on them alone.†

---

† The full implications of continuous degradation and, therefore, of the need for continuous renewal and the possibility of future improvement do not seem to have hit man's thinking about organisation yet. The very nature of organisation is that it is unstable and must adapt to and use instability, not create permanence.

7.7 Chemical change in organic molecules

We shall not attempt to describe in different molecularity terms the exchange reactions of solvent water around an ion and organic covalent bond breaking. The principle is simple enough: if a chemical species with a defined geometry is to change partners, then it must either lose a partner first to allow entry of another, a unimolecular process, or it must accept the new partner first and then lose its previous partner, a bimolecular process. Both mechanisms require activation energy and redistribution of material in space (often solvent space). In order to see selectivity in these processes for different elements we first notice the changes across the periodic table in bond strength and character (see Fig 7.15). Consider the strength of the bonding of OH to X in aqueous solution in the series

Li	Be	B	C	N	O	F
Li(OH)	Be(OH)$_2$	B(OH)$_3$	CO(OH)$_2$	NO$_2$OH	(HO)$_2$	FOH.

The exchange of OH, as an anion OH$^-$, decreases very rapidly to that of the (covalent) C, N, O and F compounds, which do not show this exchange perceptibility. In contrast, Li$^+$ exchange of OH$^-$ is very fast. The order follows bond energy from Li to C, which remains high to F, and decreasing polarity. As stated already, precisely because organic elements do not exchange atoms easily they can be used quite differently from the inorganic elements, especially in water. Now a catalytic centre must exchange substrate rapidly, so we must examine the scheme

Reactant: organic system (in) $\longrightarrow$ Catalyst bed $\longrightarrow$

Product: organic system (out)

where the chemicals involved can be written

Reactants, A + B $\longrightarrow$ Cat · A · B $\longrightarrow$ Products, C.

It is clear that we need to understand Cat · A · B in a chemically specified way. Obviously, the catalyst, Cat, must increase the rate of exchange of atoms (or fragments) in a selective manner, and can do so, of course, only if it itself undergoes fast exchange of groups that are part of or all of A or B and of products. The scheme is then

Kinetically stable reactants $\longrightarrow$ Fast catalysed exchange $\longrightarrow$

Kinetically stable products.

As stated before, from the nature of the elements of the periodic table catalytic fast exchange centres generally cannot be made from H, C, N, O organic chemicals and hence they must be made from metal ions, such as Mg, Zn, Cu and Fe, and non-metals such as P, S and Se in small molecules. We shall see that a second essential use of inorganic chemicals in biology, on top of communication, is therefore in catalysis. We shall consider a variety of examples.

7.7.1 Acid–base reactions: hydrolysis and condensation

Consider the reversible reaction of ester formation,

$$RCOOH + ROH \rightleftharpoons RCOOR + H_2O.$$

The carbon atoms remain associated with the same numbers of oxygen and hydrogen atoms—there is no change of oxidation state—but the atoms are rearranged. As we have seen above, there is a need for catalysis which arises from the stability of the different carbon–hydrogen, oxygen–hydrogen, and carbon–oxygen bonds. A catalyst works in both directions of a reversible reaction equally, that is, in synthesis and degradation. To get reaction, the

R′—OH group must attack the $\diagdown$CO group of the acid, and naively we can see

that this will be assisted by higher energy states, such as

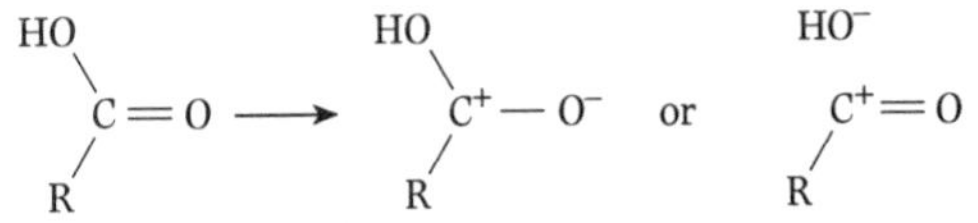

Unimolecular
intermediate

when we can write that C^+ is attacked by the negative polarity of the oxygen atom of R′OH to give

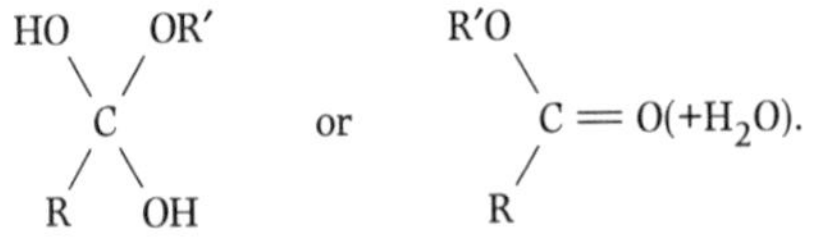

Biomolecular
intermediate

(In such schemes a positive centre is called an electrophile while a negative centre is called a nucleophile in organic texts, see Chapter 9.)

Overall loss of water (condensation) gives the product in both cases. The pattern is the same for all condensations, for example synthesis of biological polymers. Immediately we observe that catalysts must act by binding a molecule so as to favour vulnerable states (Fig. 7.7), the intermediates, and transition states. One way to achieve this is to induce strain in the substrate simply by deforming bonds using binding energy to a surface to induce that strain. This induced strain is, in principle, independent of the chemical nature of the binding in that it is just a mechanical effect. To strain a molecule, a large number of matched contacts is useful, and surfaces of proteins (large molecules, Fig. 7.16) or silicates, for example zeolites (see Fig. 7.18), are ideal for straining *large* substrates. (The entropy of the substrate is also lost on binding the catalyst surface.) If there is an electrostatic component of the binding it can be used to help induce charge separation as well as to induce mechanical strain. Such a catalyst is an acid–base catalyst since acid and base centres are distinguished by their affinity for negatively charged atoms (acids are positive centres) or positively charged atoms (bases are negative centres). Clearly, a nearby proton or a metal ion with high electron affinity, for example Mg^{2+} or Zn^{2+}, will be a good acid, and a nearby anion such as hydroxide or

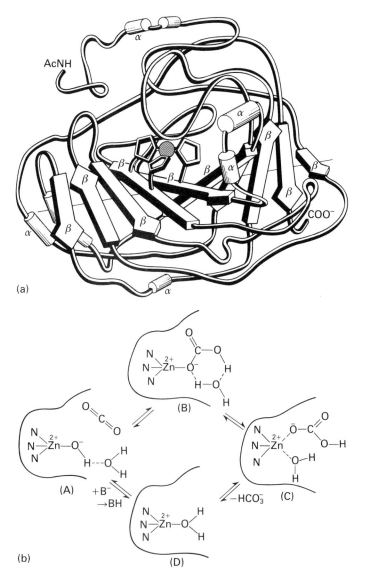

Fig. 7.16 (a) Secondary and tertiary structure of human carbonic anhydrase C. The cylinders represent the α-helices and the arrows the strands of the β-structures. The zinc atom together with its histidine ligands is also shown connected to the β-sheet. (After Linkskog, S. *et al.* (1971). In *The enzymes*, Vol. 5 (ed. P. Boyer), Chapter 21. Academic Press, New York.) (b) A proposal for the pathway of formation of bicarbonate from carbon dioxide using carbonic anhydrase as catalyst. The function of a base, B⁻, is not clear. Zinc is held in an 'ideal' (entatic) condition for the reactions.

thiolate, will be a good electron donor or base (Section 5.5). In order to control condensation and hydrolysis, biology has developed a series of such acid–base catalysts within proteins, that is, enzymes. Man uses similar centres—in silicates, for example. We shall see that the useful character of inorganic elements in these catalysts arises especially where the substrate is a *small* molecule (water is one such molecule), when strain is hard to induce or when the attack needs to be indiscriminate and fast. The metal ion is a site of particularly great attacking power which is useful in both synthesis and degradation (Table 7.9).

Now, acid attack depends on the electronic deforming (polarising) power of an ion as well as on its electrostatic binding (it is permissible to refer to this as partial covalent bond formation). In Section 5.5.3 we observed that the Irving–Williams series, that is,

$$(Mg^{2+}, Ca^{2+}) < Mn^{2+} < Fe^{2+} < Co^{2+} < Ni^{2+} < Cu^{2+} > Zn^{2+},$$

Table 7.9 Some specific metal-ion catalyses

Small molecule reactant	Metal ion	Examples
Glycols, ribose	Co in B_{12}	Rearrangements, reduction
CO_2, H_2O	Zn	Carbonic anhydrase, hydration
Phosphate esters	Zn, Mg	Alkaline phosphatase
(RNA)	Fe, Mn	Acid phosphatase
N_2	Mo(Fe)	Nitrogenase
NO_3^-	Mo	Nitrate reductase
SO_4^{2-}	Mo	Sulphate reductase
CH_3, H_2	Ni(Fe)	Methanogenesis, hydrogenase
$O_2 \rightarrow H_2O$, NO, N_2O	Fe	Cytochrome oxidase
Oxygen insertion (high redox potential)	Fe	Cytochrome P-450
SO_3^{2-}, NO_2^-	Fe	Reductase
$H_2O \rightarrow O_2$	Mn	Oxygen-generating system of plants
H_2O_2/Cl^-, Br^-, I^-	Fe(Se)	Catalase, peroxidase
H_2O/urea, CH_3CO-	Ni	Urease

was a series of decreasing radius (increasing electrostatic effect) and increasing electron affinity, polarising effect, or covalence. This order is the observed order of acid catalysis by these ions in model acid–base reactions (see Fig. 7.17). In biology, however, the order is never this one (see Fig. 7.17) even when all metal ions are available and bound using *in vitro* synthesis in the same enzyme site. It is even different in different cell compartments due to control over concentrations of these cations, and in different enzymes due to selective incorporation and control of conformation, as we shall see in the following sections. This is all part of selection for fitness for kinetic function. The power of biological polymers to utilise elements optimally is made clear here.

Amongst bases, the common attack through —OH, —SH or —NH may give rise to relatively longer-lasting 'covalent' intermediates; the problem then is not just attack, but the ability of leaving to give products. In both respects proteins are 'designed' to assist the organic catalytic acts. The bases chosen by biology are not always conventional, but are often the conjugate base of a weak proton acid that is especially reactive within the protein (enzyme) matrix; a good example is the —OH group of an alcohol, for example serine, as in many proteases.

This analysis of acid–base catalysed reactions has avoided the detailed description of the precise way in which a base like RS^- or an acid like Zn^{2+} can polarise an ester bond. The obvious mode is the formation of an intermediate so that in the reaction the base RS^- actually becomes acylated and the acid Zn^{2+} becomes attached to the ester. (Alternatively, the zinc ion can generate an attacking base such as OH^- see Fig. 7.16(b).) The required property of the catalyst is that it must also be a good leaving group, that is, it must form unstable bonds. We see that selection of an element for functional value—here for acid–base attack—is usually made first by availability; this is a result of natural selection of the elements in the universe and then on Earth. Second, the choice is dependent on radial electronic properties of atoms (see Section 2.2.2), and, third, use is made of the stereochemical angular dependence of the electronic properties of some electronic configuration (Section 2.9.5). All these properties are optimised in the best acid–base catalysts, whether they are

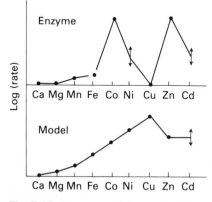

Fig. 7.17 (Lower curve) The order of hydrolysis of a simple ester by a series of Lewis acids and (upper curve) the order of hydrolysis by the same Lewis acids substituted in a zinc enzyme, carbonic anhydrase. Obviously, Lewis acid strength is not dominant in enzymes and we must look for other fitness criteria, see the entatic state.

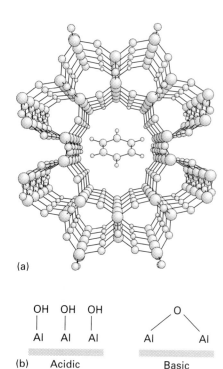

(a)

(b)   Acidic   Basic

Fig. 7.18 (a) A view through the channels in Theta-1 zeolite with an adsorbed benzene molecule in one of the channels. (Reproduced by permission from Dyer, A. (1988). *An introduction to molecular sieves*, John Wiley, Chichester.) (b) Catalytic sites on the surface of a zeolite.

Table 7.10 Major catalytic elements

Man	Biology
Acid–base	
Al, Si, Mg, H	Zn, Mg, H
R_3SiO^-, RO^-	RS^-, RO^-, R_3N
Redox	
Fe, Co, Mo, Ni	Fe, Co, Mo, Ni
Cu, Pt, Pd, Rh	Cu

simple ions or complexes in clays, synthetic zeolites (Fig. 7.18) or enzymes (Fig. 7.16). The catalytic properties are clearly most readily enhanced on surfaces that also may absorb the molecules to be attacked in stereospecific ways. We see again and again the value of large stable molecules or structures containing heavier atoms as catalysts since their structural strength, fold, can force advantageous properties on these adaptable catalyst centres. (Of course, man is not restricted in the same way as biology in his choice of catalyst (Table 7.10 and Chapter 14).

There is a final reasoning behind the choice of metal for acid–base catalysis in biology in a dioxygen atmosphere. It is much safer to use an acid–base (Lewis) catalyst (see Section 5.6) that is not open to redox reactions since redox activity in the cytoplasm, close to DNA, can lead to mutations which in turn can lead to cancer. The choice in eukaryotes is of zinc—an example of natural selection based on a multitude of functional factors.

Much of the above description of reaction systems concerns degradation but the materials and catalysts are equally essential in the synthesis of the constructs of biology—polymers. While the description of attack is clearly satisfactory for any acid–base hydrolysis or an exchange reaction, it does not permit the reaction to go uphill which requires, in addition, energy. For example, all esters and amides are unstable in water (see Section 9.13). Before dealing with this further complexity of reaction mechanism we must consider a different type of reaction—oxidation/reduction—which is deeply concerned in energy production by biology or man.

7.7.2 Oxidation/reduction reactions: further reaction systems

The rate of electron transfer to a metal, which is a redox step, was discussed in Section 7.5. Since it is general practice to include many atom transfers under oxidation and reduction as well as electron transfers we turn to the kinetics of such reactions. In order to facilitate discussion we defined in Chapter 5 an oxidation state such that increase in positive charge *or* increased combination with oxygen, sulphur or halogen atoms increased oxidation state. For example, the oxidation of iron can be written in several ways

$$Fe \longrightarrow Fe^{2+} \longrightarrow Fe^{3+} \longrightarrow Fe^{4+}$$
$$Fe \longrightarrow Fe^{2+} \longrightarrow Fe^{3+} \longrightarrow FeO^{2+} \ (FeS^{2+})$$
$$Fe \longrightarrow Fe^{2+} \longrightarrow Fe^{3+} \longrightarrow FeCl_2^{2+} \ (FeI_2^{2+}).$$

Oxidation state 0   II   III   IV

Similarly, reduction can be written as of negative charge or inclusion of hydrogen atoms, so that, even for a non-metal, we write

$$NH_3 \longrightarrow (NH_2)_2 \longrightarrow N_2 \longrightarrow NO \longrightarrow NO_2^- \longrightarrow NO_3^-.$$

$-III$ $-II$ 0 $+II$ $+III$ $+V$

There are thermodynamic relationships between oxidation states, as discussed in Chapter 5, but in this chapter we are concerned with the kinetics of such changes. Given that the barriers to electron transfer have already been described (Section 7.5.1), our concern here is with atom transfer in reactions such as

$$FeO^{2+} + RH \longrightarrow ROH + Fe^{2+}$$

or

$$SO_4^{2-} + MoO^{2+} \longrightarrow SO_3^{2-} + MoO_2^{2+}$$

and

$$QH_2 + 2Fe^{3+} \longrightarrow Q + 2Fe^{2+} + 2H^+.$$
$$\text{Quinol} \qquad\qquad \text{Quinone}$$

The critical barriers are the ability of such atoms as C, S and Fe to lose or gain O (S, Cl, etc.) or lose or gain H, either in one-electron reactions or in two-electron reactions, where loss/gain of $H\bullet$, $OH\bullet$, or $Cl\bullet$ are radical reactions (see below) while loss/gain of H^-, O, S or Cl^+ are two-electron reactions. The catalytic atoms best suited to different redox processes are given in Table 7.11. (Note that H^+ changes, which have already been treated under acid–base reactions, do not change oxidation states. However, it is obvious that two-electron switches involve similar (not identical) considerations to those of H^+ changes, that is, they are a special case of an exchange reaction.)

Table 7.11 Selection of elements for kinetic value*

Acid–base (substitution)	Electron transfer	Atom transfer
H^+, OH^-, M^{2+} M^{3+}, RO^-, RS^-	Fe, Cu, Mn low-valent metals generally	(H^-, O)/two-electron C, N, S, Se, Mo, V, Mn (Fe)

* Under acid–base reactions we are mainly concerned with the substitution of one atom for another in a two-electron covalent bond without oxidation state change, while atom transfer involves the change in oxidation state of elements by two.

7.7.3 Two-electron changes

The ability to give or take an O atom or an H^- ion is related to the capacity of the binding group to stabilise an incipient pair of electrons or to lose such a pair of electrons, respectively, which are both related to thermodynamic considerations of redox potential, but other factors are also important. Consider the rate of reaction of M_1O going to M_1; there will be changes in geometry (relaxation) around M_1 and changes in polarity. Thus, a readily deformable metal or one which can enter a bimolecular reaction path and/or a metal that exchanges groups easily (even by a unimolecular route) is to be preferred as a transfer reagent for O (or H^-). Since a reaction partner M_2 is required we need to consider the intermediate M_1—O—M_2 (or M_1—H^-—M_2) where the product is OM_2 (or H^-M_2). Thus, independently of oxidising strength in a thermodynamic sense, there are factors assisting rate of transfer associated with the ability to accept or donate O atoms (or H^- ions). The demands are just the same as those of acid–base exchange with the additional proviso that the element involved can undergo oxidation–reduction steps. The order of the rate of exchange of O is, for example,

$$SO_4^{2-} < MoO_4^{2-}$$
$$ClO_3^- < BrO_3^- < IO_3^-$$
$$CO < GeO < SnO$$
$$O{=}O < S{=}O < Se{=}O.$$

The general rule is that heavier elements exchange atoms more rapidly. It is no surprise then that heavy metals from transition metal periods—Cr to Cu, or Mo to Ag, or W to Au—are all valuable in O-transfer catalysis. To a degree heavy non-metals, for example Se, and unsaturated organic molecules, can also be useful. Coenzymes are based on all three types of transfer centre.

In H^- transfer the best elements for catalysis must, of course, combine with H^-. In an organic solvent man can use Li—H, B—H and Al—H which are much more reactive than C—H due to polarity. In water and in enzymes, transition metal elements are most useful, now found in low oxidation states, for example Ni and Co, together with unsaturated organic molecules such as pyridines. No longer curiously, Co, Ni, Mo, W and Se are found to be essential even in primitive biological systems. As in all other cases of redox catalysis where we need facile reactivity in biology, transition metals appear to be essential. Notice that there is no difference in principle between the transfer of oxygen atoms or CH_3^+ and CH_3CO^+ ions. Again there is no difference in principle between transfer of H^- and CH_3^-. We do not need to describe group transfer in general, but we shall need specific catalytic carriers for many different groups in biology in order to separate pathways.

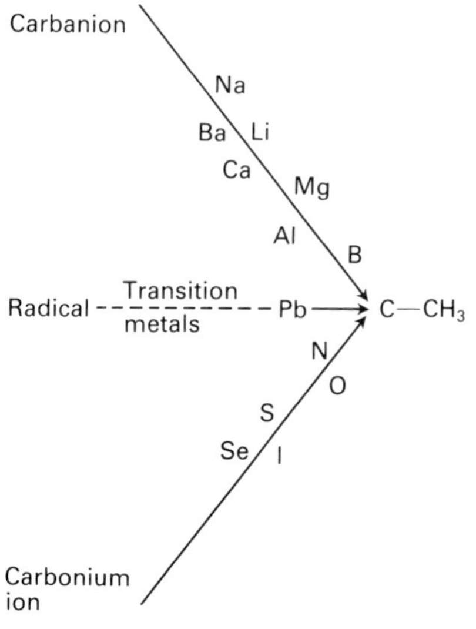

7.7.4 Free-radical reactions

The major part of the above mechanistic account of the central reactions of chemistry, except for the discussion of electron transfer, concentrates on two-electron reactions, for example the transfer of ions such as H^-, CH_3^+, CH_3^-, OH^- and H^+, of or units such as CO and O_2, inserted into molecules. There remains the possibility of radical transfer, for example of $CH_3^\bullet$, from some elements (see Fig. 7.19). We see that as polarity is removed from the substrate so $CH_3^\bullet$ transfer (and similarly $H^\bullet$, etc. transfer) becomes more probable. This is not to say that radical pathways become generally easier; it is rather that ionic pathways become very difficult. Clearly, radical pathways will be assisted by an intermediate catalyst if the central catalytic atom can undergo one-electron reactions easily. In this case, there are three possibilities:

(1) the use of transition-metal ions;

(2) the use of aromatic molecules such as quinones and flavins and, with more difficulty, tyrosine and tryptophan;

(3) the use of heavy elements, where ionization energy differences are small, for example lead.

Rather than elaborate here, we simply state that certain reactions, such as those of H_2, R—$(CH_2)_n$—CH_3, terpenes and even sugars, that is, molecules obviously not open to ionic reactions, often do take place more readily through radical pathways. The usual metal ions involved in the catalysed steps are Mn,

Fig. 7.19 The relative control over the type of leaving group derived from CH_3 or H, that is, X^+, X^- and $X°$. Carbanions obviously come from A-group metal and carbonium ions from S, Se, I methyls. Radicals, $X°$, can be generated by elements more able to undergo one-electron reactions and they may be the lowest pathway of dissociation for atoms of closely similar electronegativity, for example H_3C—H.

Fe, Co, Ni and Cu, while Mo and Se are more usually employed in two-electron reactions. The situation will be discussed in several later chapters.

We observe again the need to consider other factors than just catalysis. A biological system, especially DNA, is very vulnerable to free radicals; thus, cytoplasmic catalyses must not release products that are free radicals. Reactions are then controlled in space and in part this is achieved by localisation of the catalyst in a cell. Man achieves the same objective by initial separation of materials. Having dealt with the main kinetic mechanisms we need now to make some general points that apply to most, if not all the different reaction paths mentioned.

7.8 Inorganic elements in organic and biological chemistry: summary of kinetic aspects

From the above analysis we conclude that the use of heavier inorganic elements, that is, elements other than H, C, N and O, to transform organic chemicals made from these four light non-metal elements is unavoidable. Thus, biological transformations cannot be sustained without a battery of selected inorganic catalysts. The situation in man's organic chemistry is precisely the same (see Chapters 9 and 14). This arises from the facts that many metal inorganic centres are indestructible and that the rate of almost every metal 'activated' single-step process within bound systems is faster in both substitution (or exchange) and oxidation–reduction. In part, this is due to the electronic properties of metals—they carry local high charge M^{n+}, and consequently have high electron affinity and often are able to change their charge readily—and in part due to the flexibility of the stereochemistry of metals. Heavier non-metals, for example S and P, increasingly share these properties. Additionally, many of the simplest inorganic units bind weakly but are equally indestructible, for example Na^+, K^+, Mg^{2+}, Ca^{2+} and Cl^-, and, in biology, HPO_4^{2-}. These ions have no redox chemistry of interest and are weak Lewis acids, but they exchange co-ordination partners very rapidly. Thus, these ions are the best (fastest and simplest) carriers of information, and are without catalytic risks, as we have seen (Table 7.12). These very properties make the vast majority of heavier more electronegative inorganic elements of low value in material constructs but of extreme importance in organisations. On the other hand, it is the covalent bonding of the light non-metal elements

Table 7.12 Major elements in communication

	Man	**Biology**
Electronic	Si, Fe, Cu	Fe, Cu
Light	Liquid crystals	Chlorophyll (Mg^{2+})
Electrolytic	Li^+ cells	H^+, Na^+, K^+, Cl^-, Ca^{2+}

Mechanical	Organic polymers
	  C, H, N, O, (S)
	  Metals
		Organic polymers
		  C, H, N, O, (S)

Chemical	?	Very many

that allows kinetic stability and dynamics about single bonds, here of H, C, N and O chemistry, to dominate structure of biological machines, both in relatively rigid (sheet-like) and moving-part (rod-like) units, see Section 6.8.2. Immediately we see that a combination of heavy elements as catalysts with light elements in constructs plus a communicating system of many small units can, in principle, allow controlled development so long as energy is supplied. The essence of life is then the exploitation of the co-operative functional value of many elements, at least 15 in fact, in flow systems such as that in Fig. 7.20, which returns us to Section 7.3.1. These are general principles, but the full use of compartmentalised space greatly increases the selectivity and value of all the elements. In the last 200 years man has begun to remake organisation using a different approach from that of biology, that is, using different elements (see Chapter 14). We must, therefore, ask: are the two approaches compatible (see Chapter 16)?

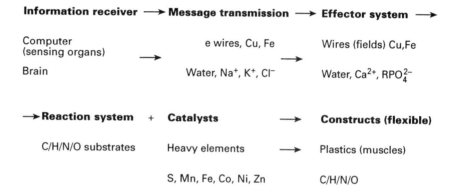

Fig. 7.20 The comparative functional value of the elements.

7.8.1 Kinetic selection of the elements for survival

Just as we saw in Chapter 2 and later in Chapters 3–6 that there were selection principles that governed the way elements came together through preferential *thermodynamic equilibrium energies*, so we see now that there are chemical kinetic reasons for selecting particular elements for special functions (Tables 7.8, 7.10 and 7.12). In Chapter 8 it will become clear that the selective thermodynamic drive to *order* via equilibrium has led to much of what we observe in the mineral world. Equally, the kinetic drive to *organisation* in living systems (and in man's organisations) has demanded the co-operative use of particular elements in highly selected ways in order to increase *survival strength*. The selection concerns not just the structures and solution kinetics and catalysts but the elements best suited for communication networks, including the electron (see Table 7.12). We shall return to an analysis of these points after we have given a description of living systems (Chapters 10–13) and man's recent activities (Chapter 14). First we must examine in greater detail the problems of the use of a diversity of chemicals, components, in an organisation.

Even though the selection of elements may be appropriate to particular functions, the use of many elements in many different non-exchanging compounds must mean that the total system has a very large number of apparently independent components that increase its variance. It is not at all obvious how this variance can be reduced in order that a particular living species may behave in development in a precise (uniqe) manner as regards chemical composition within a flowing system of chemical exchange. In a curious way the maintenance of constant chemistry is a necessary feature in both a factory and a living organism. A very simple point is that somehow an organism, for example, man, can take in a variety of chemicals, food, and yet produce a particular chemical system.

7.8.2 Allosteric control of reactions: feedback to catalysts

We return to the scheme presented in Section 7.3.1, namely,

$$\text{Control centre} \xrightarrow{k_1} \text{Transmission} \xrightarrow{k_2}$$

$$\text{Effector} \xrightarrow{k_3} \text{Catalyst} \xrightarrow{k_4} \text{Product,}$$

where we have now introduced element-selective chemical rate constants, k_i, in each step. We have seen how the transmission, the effector and the catalyst systems work to generate products utilising rates of catalysed chemical change, but to produce a unique system cell reactions at all stages must be controlled. For example, in order to limit the amounts of products we have to switch production on and off. Rates of chemical steps should then be controlled by altering the values of the rate constants as the product is made (Fig. 7.21). This is brought about if, for example, at higher product concentrations the catalyst can be switched off by inhibition due to this higher level of product. So long as the whole system supplies materials and energy (if the reaction is uphill) and the products do not react further too quickly or diffuse away, then feedback by the product to inhibit the catalyst produces a steady state. We elaborate this scheme in Section 7.8.4. Thus, amounts of elements in products generally can be regulated by feedback kinetics, but not here by thermodynamics.

One possible method for feedback to act on a catalyst, which in biology is a polymer, for example a protein, is to alter its catalytic ability by change of its conformation with increased binding of product as it accumulates (see Fig. 7.22). Here we observe that in divided pathways of reaction the balance between different products could also be managed and different pathways favoured in this strategy. We return to this possibility in Section 7.15, while noting that a flexible catalyst, for example a co-operative protein, is required in a molecular system. The great advantage of the energy distribution in protein constructions, described in Fig. 3.16, is increasingly apparent.

We now have a switch, called an *allosteric* change, of the system, that is,

State A $\rightleftharpoons$ State B

Catalytic $\rightleftharpoons$ Not catalytic

Positive message   No message

Factor bound   Factor unbound

and the possibility of control is obvious (Fig. 7.21). An all or none switch in a protein to be likened to a first order phase transition may not be as desirable as an adjustable switch, a second order change (see Chapter 6).

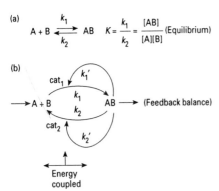

Fig. 7.21 Schematic illustration of how a constant state is maintained: (a) at thermodynamic equilibrium; and (b) by feedback from product to catalysts: the constant levels of A and B are based on supply of material (and frequently of energy) and the rate of use of AB.

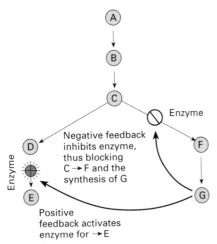

Fig. 7.22 Allosteric feedback regulation. Compound G inhibits the enzyme for the conversion to F, blocking that reaction and ultimately its own synthesis, demonstrating negative feedback by allosteric modulation. Compound G also provides positive feedback of the enzyme catalysing the step from D to E, changing that enzyme to a form that will catalyse the reaction. (After Purves *et al.*, see reference 2 in Further reading.)

These kinetic controls must be kept in mind all through the discussion of biological cells and then of man's modern industrial plants where macro devices replace micro devices in control. Of course, in biology we shall have to see how the *production* of the catalytic polymers is also controlled and is part of the total scene. There are also mechanical switches yet to be described and it is immediately obvious that physical flow controls can be knit with chemical transformation controls (see part C of this chapter)

7.8.3 The effect of temperature

Up to now we have only briefly mentioned the variables temperature and pressure while describing kinetics. However, through the expression

$$\text{Rate constant} = A \, e^{-\Delta E^*/RT},$$

each rate constant (like each equilibrium constant) is temperature-dependent. Thus, a feedback reaction system is dependent on temperature as are the phases of the phase rule, and so, if the kinetic system is to be defined, that is, if organisation (not now just order) is to be maintained, then the system must have a feedback temperature control. Man, for example, is controlled at $36.5 + 1.5°C$. Feedback thermostats are well-known, both in laboratories and factories, and obviously are operative in man's body (see Section 7.13).

7.8.4 Control of energy supply: synthesis

In the description of chemical change, for example in a biological metabolic cycle, we need to see that it can be given a pathway just as the flow of material in space can be given direction by mechanical gravitational or electric fields. The pathway of chemical change follows very similar principles, now in time, not space, governed by the second law of thermodynamics, that is, the processes require entropy production (heat loss), see Section 3.6.2. Thus, all processes need free energy sources to do their work. At the beginning of any chemical pathway there has to be an energy source, which could be, for example, the difference in free energy between $C_6H_{12}O_6$ (sugar) and $3CH_3COOH$ (acetate). The process can only run backwards to synthesis and become energy-consuming if it is coupled to an even greater energy-producing change. The use of energy in either case must be connected to a feedback mechanism if the whole system, energy input and material transformation, is to be controlled. Thus, once enough sugar is produced the input of energy to sugar synthesis is stopped. The principles can be applied to any source of energy, for example a fuel, an electric battery or radiant energy, and any number of energy-requiring processes can be connected by energy transfer between them. These energy stores, generating fields, must constantly be regenerated so that flow is maintained. Since the steps of a pathway can be selected by the catalysts, for example recognition of A only by Cat(1) and so on, the sequence of reactions follows in time. The importance of control over energy supply will be stressed again and again in Chapters 9 to 13.

7.8.5 Control over molecularity

As mentioned in Section 7.3 there is often a need to control the molecularity of a reaction in a catalysed act. Consider the desirable reaction

$$A + B + C \longrightarrow ABC \text{ not } ACB.$$

There may not be a way to control ΔE^* for any reaction, hence A, B and C could come together in all possible ways through ternary collision as could any pair. Since we need to *arrange* the final sequence ABC (not ACB) we must introduce some barriers and we can do so in two ways;

1. Use a catalytic surface that is a template and that recognises B in the correct orientation for reaction only after A is bound, and recognises C in the correct orientation to be linked only after AB is bound. This is an entropic restriction as well as a ΔE^* restriction. DNA is made to act in this way when producing RNA.

2. Alternatively, the catalyst is a mechanical device that does not recognise either B or C at all initially. On binding A the catalyst changes structure to recognise B, and then changes again to recognise C. This is called required order of binding. If this requirement is coupled to an energised pumping action, the reactants A, B, C . . . , etc. move along a reaction pathway in space as on a manufacturing line.

Clearly, space can be organised and ΔE^* can be manipulated by suitable machinery of considerable complexity. The nature of man's pumps and those in biology is such that each pump works only after it is primed by the presence of materials to be transferred. We imply that a biological cell has units which act like components of a machine, see marginal illustration and Section 3.8.

Machine components

Man-made	Proteins
Platforms	β-Sheets
Rods	Helices
Hinges	'Random' stretches
Rising hinges	Helical pairs
Valves	Multiple helices
Wheel	Flagellum base
Pump, i.e.	Molecular pump
platform + rising	β-Sheets + helical
hinges + valves	pairs + multiple
	helices

See Table 7.19.

7.9 Chemicals and self-assembly of equipment

The final step of the chemical description of a living system is the control over structure. This requires a management of the synthesis, as above, and *positioning* of resultant polymers of many kinds using group transfer to construct large units. We return to group transfer below and to energy input in biology in Chapters 10–12. It is the combination of acid–base and redox catalysis and energy input to synthesis and transfer, both under feedback control, that allows assembly of proteins and membranes and hence compartments and thereafter of biological objects. The internal activities of self-assembling systems must be linked to the structures they build, which requires consideration of the style of feedback acting on the enzymes E_1 and E_2 (Fig. 7.22). Man's apparatus also has to be constructed in a laboratory or factory and we discuss these problems of transporting and transforming chemicals in Chapter 14. Thus the apparatus itself is fundamental to organisation and the use of space, physical control, and must be given equal weight to that of chemical control. Finally, there is a clear need for a plan of all the features if an organisation is to be produced (see Chapter 11). We turn in the next sections of this chapter to a description of physical controls over variance in systems which will involve us in a discussion of flow.

C Physical barriers

7.10 Physical controls over reaction rates

7.10.1 Introduction

The limitations imposed by chemical barriers over exchange of elements between components have now been described as has the value of catalysts in overcoming these barriers. This discussion has allowed us to see that the number of independent components can be reduced by control over catalytic action, especially that due to the feedback interaction of products on the catalysts (Fig. 7.21). *Now all such chemical reactions must be held within physical barriers to control product concentrations since only physical barriers prevent dispersion.* It is, however, necessary to have some link between these compartments if an organisation of several compartments is to be maintained in a functional fixed pattern. We must then turn to the controls over limited diffusion between compartments within constructions. Much of the concern here is also chemical in that compartmental barriers are made from chemicals, but we are no longer discussing change in chemical bonding and must concentrate on transfer of material.

The two major ways of controlling reactions by physical means are:

(1) limitations of diffusion to within one compartment, effectively a phase, since no change is possible to its internal equilibria;

(2) limitations on the ability to cross boundaries between compartments.†

We will examine these physical limitations in turn.

7.10.2 Diffusion in one phase in a single compartment

In a cooled system, a given substance A can become virtually inaccessible to a gas B through condensation of A such that there is a very limited vapour

---

† In Section 6.3 we defined a phase as a zone within a boundary, equilibriated physically and chemically with its surrounds. A compartment is a region from which there is no equilibrium exchange with surrounding zones. Within the compartment there can be limited systems of equilibrated phases. In this chapter we are overwhelmingly concerned with ways in which limited communication can be made between compartments. (All structural barriers are co-operatively formed from individual units.)

pressure of A and/or solubility of B in A. The diffusion of A or B is then limited to a particular phase. At ambient temperature (300 K) the rates of diffusion are zero in any solid and rapid in gases, while rates of diffusion in single liquids are controlled, in bulk terminology, by solubility (concentration) and viscosity. Effectively we have considered viscosity in molecular terms through the rate of exchange of neighbouring solvent molecules (Section 7.5.2), which required a discussion of a chemical barrier ΔE^* (Fig. 7.5) since not all movements of solvent molecules create space for motion of the molecules of the solute without application of energy to solute–solvent bonds. A barrier around a phase, even a surface or a membrane, can also provide complete restriction to diffusion within the given phase. This allows the creation of a homogeneous single phase in or outside a containing vessel. An example is the silicate mantle of the Earth, which, acting as a kind of very thick membrane, prevents reactions of the dioxygen-rich atmosphere with the iron core. Here dioxygen and iron fail to come into contact, except in volcanoes, since neither one diffuses in the lattice of the silicate mantle—both are closed separate single phases. In fact, any containing vessel in a chemistry laboratory or factory, usually made of silica glass or steel, is a similar absolute diffusion barrier, rather like every containing outer case of an organism in biology (Tables 7.13

Table 7.13 The nature of containers

Condensed systems in planets
Steel vessels and pipes in factories and houses
Glassware and pots in laboratories and houses
Outer skin in biology

Table 7.14 The sizes of the compartments of biology*

Compartment	Scale (m)
Whole organisms	Up to 20
Organs	Up to 1
Cells	Up to 10^{-2} ($= 1$ cm)
Vesicles	Up to 10^{-3} ($= 1$ mm)

* The lowest limit for the size of a compartment is some 10^{-6} m ($1 \ \mu$m), the size of some small vesicles.

and 7.14 and see Chapter 14). The essence of these diffusion barriers is that they generate *kinetically closed systems*. Naturally, in a closed system reactions can be carried out largely under controlled conditions—for example, dioxygen-free in the gas phase, or in a chosen solvent, or in the presence of a particular catalyst. These are common practices in chemistry laboratories and factories, as they are also in biology. In Chapters 3–6 we have already considered the relative energies of chemicals in phases within a compartment where no exchange of elements could occur to the outside. This led us to describe *local* equilibria in terms of independent chemical components retained in mutually independent phases, parts of space. Clearly, there is no flow, that is, no motion in a direction from or to such a compartment.

7.10.3 Restricted diffusion across boundaries between compartments

We now proceed to examine resultant properties where there are two or more arrangements of elements (in compounds) in space and which are *restricted*, not completely prevented, from mixing. That is to say we need to engage in a discussion of physically controlled flow or *rate of transfer*. The difference in compartmental free energies is then a driving force for change (see Section 6.3). To consider change we now examine the possibility that *entry to and exist from* any compartment can be *controlled* by inserting in the containing vessel specific leaks, channels, pipes, pumps, filters, grids and so on. In the natural selection of elements for reaction (which becomes deliberate *choice* for man) such limitations and controls of intercompartmental diffusion took on a more and more dominant role with time as separations, effectively new compartments, developed from a single universal 'gas phase' of the big bang to liquids to solids and to membrane-confined systems. Of course, if confinement in a compartment is to occur then it must either be brought about by the accidental change away from equilibrium by turbulence or by deliberately placing materials (elements or compounds) selectively (and in non-equilibrium conditions) into a compartment. The latter corresponds to an uphill diffusion or movement that requires energy. Man's chemistry depends on such deliberate manipulation of materials in obvious ways, but so does the chemistry in biology (when 'deliberate' assumes a different meaning, namely 'useful for survival') where, quite comparably, energy is used to select elements and compounds to be transported into separate compartments or to form them (see Table 7.14). All such compartmental activities depend on limitations on diffusion, and it is easy to see (in a laboratory, for example, but the same is true in biology) that from these compartmental systems, with limitations on diffusion between them, great advantages accrue in selectivity of reactions. Here the sizes of biological compartments are worth noting too (see Table 7.14), but the major feature is the physical design of connections, which includes choice of materials (Table 7.15), since the materials by their very nature limit the way in which substances can be transferred.

7.10.4 Diffusion in media formed by different elements

First and foremost, we must consider which elements can diffuse in which media. Returning to the contents of Chapters 2–6 and Section 7.5 we note that there is a major distinction between metal, salt and continuous covalent non-metal phases and those of non-metal molecules. At temperatures where chemical reaction rates are relatively slow, that is, 200–400 K, the first group of elements and compounds are overwhelmingly solids, melting at temperatures around 1000 K, while in the second group they are gases or liquids in this low-temperature range. This means that only the second group, of non-metal molecules, can allow materials to diffuse and react at such low

Table 7.15 The constructs for specific transfer centres

Component	Transfer centre		Construct*		Example	
	Biological	Man-made	Biological	Man-made	Biological	Man-made
e	Metal ion	Metal ion	Protein	Oxide lattice	Cytochrome chain	Semiconductors
H^+	Acid–base groups	Silicate	Proteins, hydroxyphosphates	Silicate	Channels, bone	Zeolites
Na^+, K^+	Carbonyl groups	Silicate	Proteins	Silicates	Channels	Zeolites
Ca^{2+}	Carboxylates		Protein		Channels	

* Note the value of water in a construct in the last three cases.

temperatures. In other words, the physical kinetic traps of the first group of elements are due to their state as *co-operative* solids, not their uni-dimensional pairwise atom–atom bond strength, while the non-metal molecules are dominated by such pairwise bonding (Section 2.4). Consequently, the high melting point solids, for example most metals, metal oxides or sulphides, are, for the most part, not soluble in any solvents. Overwhelmingly then, most metal elements, in the elemental form or in naturally occurring compounds, are prevented from rearrangement since their atoms cannot diffuse. Only on melting at temperatures around or above $1000°C$ can diffusion and structural change occur readily. This is, in effect, the basis of geological change and man's development of heavy industry (see Chapter 14). Man also builds with this material his apparatus, structures and machinery, that is, substances developed by using high temperatures to melt minerals. In contrast, small (linear) molecular compounds, largely organic compounds, that is combinations of H, C, N, O, as well as of halogens, despite considerable pairwise bond strength, are capable of diffusion since they are freely mobile in gases or in liquids, into which they may dissolve mutually. It is also the molecular nature of these compounds, now especially polymers, that allowed biology to develop low-temperature (semi-mobile) structures at $0–100°C$ in water (Fig. 3.15). The essence is a low strength of co-operativity, Section 3.2.5.

7.10.5 The nature of containing vessels: boundaries

Before we consider the use of containment of chemicals in compartments it is useful to look in more detail at the differences between biological and man's equipment for control over diffusion, that is, for containment. As we have stated, the essence of biological containment is the use of light non-metals to make molecules, lipids and organic polymers, that is, organic materials, into containing walls and membranes. These containers are semifluid. Minerals from salts can be incorporated into these biological organic structures to make more rigid composites, for example shells and bones (see Section 7.14). All these materials are made in solution at low temperature, and in Chapters 10–13 we examine the molecular kinetics of their synthesis, degradation and transfer in solvents at low temperature, especially in water.

Man's materials for containment include some substances that parallel the biological polymers, for example, plastics and composites, but they are cross-linked materials, not made on a single-molecule level. Much more significantly, as stated, man uses inorganic materials that can only be made and shaped *in melts* at high temperature, for example metals, alloys and ceramics (glasses). At low temperatures they are longlasting. Of course, they have enormous advantages as absolute containing vessels for the isolated handling of reactions in gases and solvents even at high temperatures or pressures, but they cannot be handled as *molecules*; hence they lack the ease of building and rebuilding that is possible for non-metal *molecules* in solvents, as demonstrated in biology. In a sense man attempts to make permanent equipment. The lack of fluidity of man's materials for structure and the fact that they only allow mobility of electrons or more rarely of protons within themselves at temperatures around $0–100°C$ means that they are limited in many ways for the manufacture of microscopic-scale communicating equipment, though open to macroscopic construction in other ways not available to biology.

By way of contrast, biology has learned to place in its membranes *mobile molecular* channel devises—not just for the transport of electrons and protons (Fig. 7.23), but also for the transport of a vast range of larger substances up to proteins. The entrances and exits are selective and controlled. The parallel is with macroscopic rather than microscopic laboratory and industrial devices for transfer, for example selective filters. This biological equipment, like the membrane barriers themselves, is not 'permanent' and is open to constant physical (and chemical) change, development. We now consider some examples of controlled diffusion of different particles.

7.10.6 Control of diffusion in inorganic phases

We shall be concerned with diffusion in the temperature range 200–400 K rather than at temperatures > 1000 K. The dominant diffusible material in solids is the electron, as described in Section 7.5.1. The motion of electrons, electric current, is obviously physically controlled in all man's electrical circuits by retaining electrons in a structured, *shaped* 'compartment', a metal wire, surrounded by an insulating compartment, often organic in nature, which is then constructed with controls, *switches*, for making and breaking the continuity of the wire, *resistances*, to change the ease of the conduction, and *condensers*, as storage devices. There is also a variety of ways of making electronic circuits that do not use such simple free flow of electrons in bulk metal wires but use instead metal atom centres spaced short distances apart in an otherwise insulating material (Section 7.5.1). These materials, often with properties close in character to the so-called insulator/metal transition, are called 'hop' semiconductors. Biology uses such materials to make circuits (Fig. 7.23) as we discussed in Section 7.5, but only over (molecular) series of hops up to distances < 100 Å. The feature we stress here is the disposition in

Fig. 7.23 When photosystem II is activated by absorbing photons, electrons are passed along an electron acceptor chain and are eventually donated to photosystem I and finally to NADP⁺. Photosystem II is responsible for the photolytic dissociation of water and the production of atmospheric oxygen. This pathway is sometimes referred to as the Z scheme because of its zigzag route, as depicted here, but the two arms are in fact remote in space. (After Villee *et al.*, with permission, reference in Chapter 13).

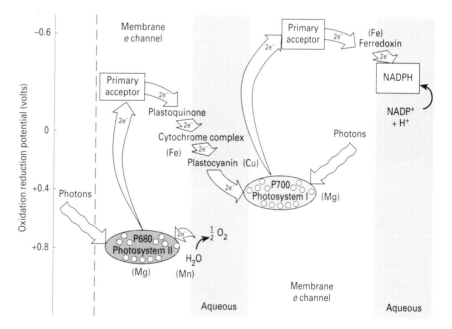

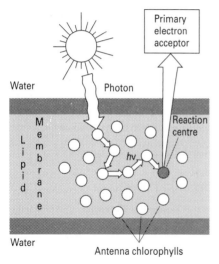

Fig. 7.24 The photosystem traps light energy. The many chlorophyll molecules (open circles) in the unit are excited by photons and transfer their excitation energy, hv, to the specially positioned chlorophyll molecule (shown shaded) at the reaction centre. The dark borders of the membrane are phospholipid headgroups. (After Villee *et al.*, with permission, reference in Chapter 13.)

space of units in compartments that direct flow, and also the physical control over flow elsewhere by a barrier. The whole system takes on a shape through co-operative self-assembly.

As far as diffusion of energy is concerned, light can be guided down man-made light pipes, optical fibres, made of glass, for example, while biology uses a special complex 'system', a series of molecular pigments in a lipid/protein membrane, in green leaves for example, to guide light physically to an absorbing target (Fig. 7.24). The material in which the inorganic or organic transfer centres are incorporated is of as great an importance as the distance between them. All these uses of zones or compartments limited by barriers give absolute control over diffusion, so that flow of electrons or energy can be mutually controlled in space. Notice that we use names such as *switches* for macroscopic devices to indicate that diffusion in these extended compartments is under *all* or *none* control, and we use the name *gates* for a more gradually imposed restriction. The implication is that mechanical as well as magnetoelectrical devices can be used to control the flow by restricting diffusion. (Both can also be coupled to chemical controls (see Section 7.5).) In the microscopic circuits of biological systems parallel devices are used based on electrical or mechanical controls at the molecular level, as mentioned in Section 7.8.2. Biological systems build components for circuits as well as for mechanical machinery from proteins.

7.10.7 The abiotic flow of water

By far the major medium for both movement of chemicals and for diffusion of ions and molecules, not electrons, and which is formed abiotically, is, of course, water. The bulk circulation of salts around the oceans, giving a redistribution of both organic and inorganic compounds, is of immense proportions and is linked to run-off from the land which mixes with the ocean flow. However, even the Earth's water as a whole is not a homogeneous phase, since, through complete or partial physical isolation of its compartments in two principle ways, the several zones of it are out of equilibrium. The first isolation, by the rocks of the crust, allows vast storage of ground water and of water in lakes and rivers to be separated from the water of the sea (but eventually flowing to it). This water carries with it dissolved materials from the crust, while the sea has direct access to the mantle and its changes. The downhill flow of rivers is spatially controlled by the structure of the land, of course. (Note that a lake above sea level *lies behind* an obvious physical barrier to gravitational flow (see Fig. 7.25 and Section 6.3.5).) The second, now partial, isolation is due to the depth of the sea and the unstirred nature of many waters which allows a gradient of conditions from top to bottom, with the bottom having access to the mantle rocks, deep vents and their products, and the top having access to the atmosphere.† Due to temperature differences the oceans also circulate in layers carrying elements with them, for example, the Gulf Stream. The limitations to free mixing in the water compartments can

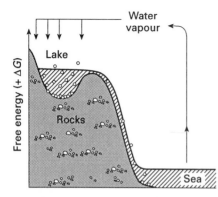

Fig. 7.25 The obvious physical barriers to water flow are the banks of lakes and rivers and man's dams.

---

† Effectively, the top layer of a water column can provide a considerable diffusion barrier to gases, for example O_2 and CO_2. Dioxygen is also removed by reaction largely by the oxidation of organic material and CO_2 equilibrates with dissolved carbonate. Biological consumption of O_2 is extremely important too, but this removal of O_2 is due to more 'deliberate' selection of a living chemical zone and biological systems are themselves selected for these zones.

therefore be due to streaming as well as static divisions. In fact, the different layers of the deep ocean only mix thoroughly on time-scales of hundreds of years, which affects, for example, the rate of CO_2 equilibration between the atmosphere and the sea and therefore greenhouse effect considerations (see Section 15.8). It is due to this circulation that the deep ocean is not dioxygen-free. A very interestingly less usual case is the difference in O_2 and that of other chemicals in layers of stagnant water. Here the bottom layers may mimic the conditions of early Earth in that they are anaerobic. In all the above, physical conditions provide barriers that allow only limited mixing of compartments in specific spatial patterns, see Chapters 8 and 15.

We must keep stressing the importance of the major flowing solvent—water (Sections 2.6 and 6.9)—since it is critical for many inorganic and organic chemical changes, and, of course, for life over a time-scale of millions or even billions of years. Not only does it flow itself but it allows distribution of a limited set of chemicals soluble in it. The fall in temperature to around 300 K locally on Earth largely removed many abundant elements, such as Fe, Si, Al, into inaccessible immobile solid phases while leaving H, C, N and O as elements in circulating gases available in large quantities, and Na, Mg, P, S, Cl, K and Ca as ions able to circulate in and with water. In addition, the flows eroded the rocks slowly and produced over long periods of time low-lying deposits that formed sands and sedimentary rocks. Later the sands allowed invasion of the land by plants and thus soils were created. The initial constructs controlling water circulation resulted largely from mineral condensation and flow as seen in the formed shapes of land masses.

7.10.8 The control of diffusion by organic phases

During the course of time a second physical compartmentalisation of the liquids on Earth occurred at temperatures below 100°C. As stated before (see Section 6.3), the production of oils, no matter how produced, generated a hydrophobic separate 'organic' semiliquid phase, and the agitation of water generated membranous structures formed from oily substances (Section 9.4). As a consequence, zones of water inside and outside oily membranes became separate, and different compartments, including eventually biological systems, evolved, all out of equilibrium. (Note that it could be that some silicates and sulphides also produced membranes.) The oil, fatty or lipid layers, presented enormous new physical barriers to the diffusion of many water-soluble, especially ionic, chemicals while allowing easy diffusion of hydrophobic compounds, particularly small neutral molecules, for example CO_2, N_2, etc. This development is every bit as important as the creation of coded polymers for biology since it provided a way of physically controlling space and directing flow in tube-like constructs that we may compare with wires in electronic circuits, or with metal or ceramic pipes for man's gas and water distribution.

Just as it is possible for man to make switches and stores in a variety of circuits made from metals, so it is possible for biological systems to make controlled molecular channels (switches) out of proteins in the confining organic outer membranes or in vesicle membranes storing dischargeable ions or molecules inside cells (condensers); see Fig. 7.26. We consider the construction of these *molecular* devices in Chapters 10–13, and we must see

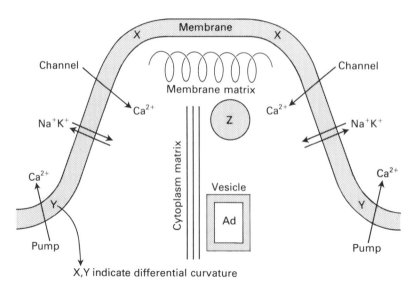

Fig. 7.26 The link between membrane-associated protein structures, which are positioned, calcium and vesicle compartments. The diagram indicates that the whole is a unit of activity in that tension, curvature, positioning of pumps and channels and movement of vesicles all depend on the activity of a cell in an interactive manner (see text). Cell shape is a steady-state property. Ad, Adrenaline. Z = calcium containing reticulum.

Table 7.16 Communication modes

Common to man and biological systems

Electrolytics (especially in biology)
Sound
Mechanical devices

Man only

Electronics in factories, laboratories, houses
Electromagnetic radiation in many systems

Biological systems only

Organic transmitters and hormones

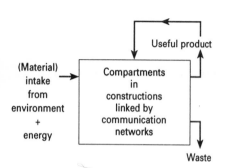

Fig. 7.27 The central box is a control centre, and we can consider that there is a parallel function between a computerised man-made system and a living system.

that not only was initial physical isolation of out-of-equilibrium compartments generated but that it was later possible to advance and make controlled physical communication between them. As we shall see, such a 'communication network' is necessary for *organisation* if an organism is to survive.

The problem that we now face is in one sense simple. The universe, Earth, organic chemicals and biological systems evolved through the continual increase in the number of smaller and smaller distinct compartments separated by barriers and out of equilibrium. They can only react with one another, bringing about organised change, when brought into communication. We have described, given the limitations of natural production of elements, which elements can give rise to which kind of phases with what kind of diffusion barriers, and we have considered which elements, compounds or forms of energy can diffuse most freely in which medium, for example which media will support electron diffusion and which will support diffusion of ions or molecules given the limitations of the physical states of the components that evolved. It is the structure of these materials in space that allowed modes of communication (see Table 7.16). Man and biology have developed a huge variety of compartments in order to allow activities of these kinds as illustrated in Fig. 7.26. The difference between biology's and man's methods for generating compartments and making communication (Tables 7.13 and 7.15) is based on the modes of preparation of materials open to them. Biology is confined to certain compartment-making materials and hence to compartments stable in the range of 0–100°C and pressure close to 1 atmosphere; industry has a much wider range of options. Thus the modern electronic computer is materially very different from the electrolytic nerve system of the brain but there are many control principles in common (see Fig. 7.27). Again the flow of water in tubes and pipes in man's industry has its parallel in veins and arteries in animals. In all cases, to become truly functional, *the different compartments have to be shaped for the function and communication has to be constructed.* We know how man does this, but how has biology produced its *self-assembling* shaped units? Once formed, how does biology allow controlled communication between such units? Finally, how can an organism develop? First we must see why compartments are so valuable.

7.11 Functional advantages of an increase in the number of (communicating) compartments

When we consider the development of universal, geological or biological systems, we can only be concerned with those systems that have generated permanently observable properties, historical traces, or those that have survived until today, for example in living systems. Thus stability (thermodynamic) and kinetic survival together dominate our rationalisations. In Section 7.3.1 we developed the idea that the systems that we observe are in fact of three different kinds: (1) unchanging, even *at local equilibrium*; (2) accidentally enclosed *in local kinetic traps* but drifting slowly to local equilibrium; (3) *organised in connected series of kinetic traps* (compartments) by flow and even moving away from (local) equilibrium. The last case required 'designed activities' so that both energy and materials were put to 'purposeful' (efficient) use in the drive to maintain the activities over long periods of time. We, therefore, referred to the *survival* of systems in complex kinetic traps, organisations, as being such that elements were used functionally. (Acting functionally means 'acting to the common value of survival of a whole non-equilibrium system'. Such an expression has anthropomorphous origins but its sense is clear.) The drive to maximum disorder (see Chapter 3) has to be prevented if any such organisation is to survive. As we now see, the two ways to avoid the drift to maximum disorder are by: (1) *physical barriers* preventing phase equilibria; and (2) *chemical traps*. Only by putting the two together can a flowing system of materials survive. Thus, at least one isolated compartment is required to prevent general diffusion, but this arrangement allows mixing and potential reaction of all its chemicals. Many physically separated compartments allow greater survival strength in that they trap different elements and activities locally, for example to remove risk of unwanted reactions (see Table 7.17). However, the different compartments usually have many common needs for energy and materials if they are to survive. Thus, it is necessarily through *communication* between the many compartments that increased survival is achieved. *One of the great steps forward in the natural selection of chemicals, including biological evolution, was and is survival through increased co-operative compartmentalisation.* Within the development of living systems this is seen in the switch from basically one aqueous solution, the sea,

Table 7.17 Elements concentrated in some biological compartments*

Compartment	Element concentrated
Extracellular fluids	Na^+, Ca^{2+}, Cl^-
Intracellular vesicles	Ca^{2+}, Sr^{2+}, Ni^{2+}
Mineral phases	Ca^{2+}, CO_3^{2-}, HPO_4^{2-}, Si, Fe^{3+}
Cytoplasm	K^+, Mg^{2+}, Zn^{2+}, HPO_4^{2-}, Fe^{2+}, S^{2-}
Vacuoles	Ni^{2+}, Mn^{2+}
Mitochondrial membranes	Fe^{2+} (often), special C/H compounds

* NB. These separations are managed by atom-by-atom or molecule-by-molecule movement.

to two or three aqueous compartments in prokaryotes, to multicompartments in eukaryotic cells and then to organisms with every-increasing numbers of linked compartments in cells. This switch was followed by further compartmentalisation, cell/cell differentiation, then by the development of truly multicellular systems and finally, in one system of organisms, of regional organs. In any organ there are three layers of compartmental element selection: (1) within the organs in the organism as a whole; (2) within differentiated cells; and (3) within vesicles or organelles within a cell. In all cases, lipid membranes have partially isolated the compartments and elements have been placed differentially in them. This is natural (biological) selection of chemical elements at its best, but it is always within *organised activity*. Communication is then maintained by controlling diffusion to and through all membranes. Of course, there is a last step in which the separate activities of *different organisms*, now treated as different compartments, linked by 'diffusion', act as co-operative units in symbiosis or in an ecosystem, communicating among themselves (Table 7.18). Man with his industrial developments, especially in the last 200 years, is following closely this pattern of biological evolution. More and more often, his increasingly compartmentalised actions require more and more communication (generating complexity) for survival strength (Fig. 7.28).

This principle of advantageous compartmentalisation is used, for example, in a set of man's factories that make components only for later assembly in a central plant (see Chapter 14). The activity requires transport, directed movement, and intercommunication if it is to survive. It is clear that the evolutionary steps of biology or man's manufacturing capability have to be seen against the background of the drift away from random mixing or complete isolation of compartments to organisation using kinetically strictly limited diffusion through *organised* barriers plus transport systems. At the same time, for integration of activity, certain compartments must be shaped so

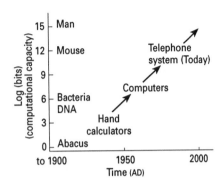

Fig. 7.28 The information storage capacity of a variety of living organisms and human technological creations.

Table 7.18 Levels of biological organisation

Level	Examples	Nature of compartment
Molecular building blocks	Amino acids, nucleotides, sugars, fatty acids	None; single molecules not treated by these considerations
Macromolecules	Proteins, DNA, RNA, starch, lipids	Just open to co-operative 'phase' treatment, e.g. melting
Macromolecular complexes	Ribosomes, enzyme complexes, membranes	Smallest 'phases'
Bacterial cells, cellular organs (organelles)	Cell nuclei, mitochondria, chloroplasts	Single compartments isolated by membrane
Higher cells Tissues	Xylem, phloem; blood, smooth muscle, bone	Single compartments of complex shape
Organs	Stem, taproot; kidney, heart	Multiple compartments
Organ systems	Root system; circulatory system, digestive system, nervous system	Multiple compartments
Higher organisms	Animals, plants	Multiple organs
Communities	Ant colonies, man's societies	Company units, of single animals

Compartments

Endoplasmic (Ribosome)   Cell membrane
reticulum

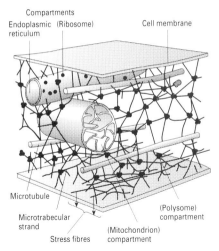

Microtubule

Microtrabecular
strand   (Polysome)
                          compartment

Stress fibres   (Mitochondrion)
                 compartment

Fig. 7.29 Elements of the cytoskeleton. The cytoskeleton consists of networks of several types of fibres including microtubules, microfilaments (shown as stress fibres), intermediate filaments (not depicted) and perhaps microtrabecular strands. The cytoskeleton forms the shape of the cell, anchors compartments, and sometimes rapidly changes shape during cellular locomotion (see Chapter 12). (After Villee *et al.*, with permission, reference in Chapter 13.)

▲ Neurofilaments
■ Actin in vesicles
○ Neurotransmitters in vesicles
N Nucleus

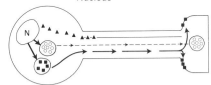

Fig. 7.30 The transport processes taking place in a normal neuron illustrating vesicle incorporation and motion along a fibre within a cell—a tramway. Neurofilaments are produced in the nucleus and incorporated into the axon to form part of the vesicle transport system. Actin and the neurotransmitters are carried on this system in vesicles to the synapse and the growth cone where the actin is incorporated and neurotransmitters are stored. See also Table 7.20.

as to connect with others and to allow communication through controlled rates of transfer of materials, that is, flow of elements as well as flow of *energy and information*. In biological systems shape is created and is an integrated part of the whole organisation.

In the above we have stressed the separation of elements in *aqueous* zones, but we must not forget other small compartments, as described in Chapter 6. For example the membranes and the networks of internal cellular protein filaments in cells (Fig. 7.29 and see Chapter 11) or the equivalent structures in man's constructs can be treated as separate compartments (see Section 6.8). Each lipid-confined compartment can have localised chemical content control over activity due to the disposition of specific catalytic activities. Each filament can also hold different activities along its length. Here advantages also accrue from relatively rigid surfaces. Thus, the membrane or the filament can be used to locate non-diffusing catalyst centres, much as in industry a solid-state catalyst is often preferred over soluble (homogeneous) catalysts since catalysts then do not diffuse with reactants or products. The exact location of catalysts can be organised provided the reaction vessel, the membrane, the filaments (or scaffold) and the cell have shape. An external scaffold is, of course, necessary for cell/cell (reaction vessel/reaction vessel) organisation. It is easy to see, in principle, that the organisations in membranes and in filament scaffolds should be linked (Fig. 7.29) so that, while diffusion barriers act between compartments, reaction-diffusion limitations are brought about by organisation of activity even within an apparently defined *single* compartment. Effectively, a surface A placed close to a surface B can transfer reactants to it avoiding general random diffusion to the bulk. In addition, there can be a localised current or transfer (Fig. 7.30), which may reduce to two- or even one-dimensional diffusion. In this way a circuit of physical transfer of material is built up just as in an electric circuit, although diffusion boundaries are less well defined. Thus, a further advance in biological evolution was *shape*, morphogenesis of zones, in all compartments down to cell, vesicle and filament levels. As we have seen in Section 6.3, shape, except for that of a sphere, implies external and/or internal control, for example by mechanically adjustable scaffolding (effectively a mechanical field) or by electric fields. Now, just because the frameworks of biology are flexible, including filaments in and between compartments (cells) and because they have to be rebuilt or to develop in time, tension in them and their semifluid shapes must be related to the metabolic activity. Thus mechanical, just like electrical, communication is always linked to chemical activity, energy or ion transfer.

A final refinement of controlled diffusion is seen in the use of a swinging arm of one centre in directed transfer, which comes close to a solid-state mechanical transfer (diffusion) device such as a crane in man's industry. A good example is the transfer of $CH_3CO—$ groups in the synthesis of fatty acids, but the principle probably applies to many polymer syntheses in biology. Some mechanical processes and devices are described in Tables 7.19 and 7.20. As can be seen, man has developed devices parallel to those of biology, but not on a molecular scale; the scale of man's operations is as yet more like that of whole large organisms.

The question now arises as to how *all the components within compartments can be active together in a flowing system so as to produce a co-operative unity or organism.*

Table 7.19 Mechanical machinery (frameworks and moving parts)

Metals, plastics or ceramics in industry and the laboratory

Proteins and composites (bone) in biology

Moving parts: hinges, ball-and-socket joints, etc.; helix/helix contacts (biology)

Elastic material: random polymers, e.g. rubber

Table 7.20 Bulk phase transfer procedures

	Procedure
Laboratory	Distillation, extraction, chromatography, etc.
Industry	As for laboratory plus spray drying, fluid beds, etc.
Biology	Exocytosis, endocytosis; vesicle transport on tubulins (Fig. 7.30); blood circulation

7.12 Fields and flow

There is a situation that we have not yet introduced specifically in our discussion of diffusion, which has been treated as equally possible in all directions (isotropic flow). However, the water vapour of the atmosphere and the water in the sea and lakes, for example, are positioned relative to one another by the gravity field of the Earth, as is the sea in relation to the Earth's core and the mountains. The compartments are in fields in communication but are not at the same free energy, so we must consider in them not only random diffusion but also vectorial flow. (Compare the discussion of equilibria in Chapters 3 and 6, before and after the introduction of surfaces.) Again, if diffusion in the atmosphere were strictly random and the gravity field absent, the atmosphere would disperse into the universe since there would be a positive outward pressure. In fact, although most of the light gases, for example H_2 and He, which are plentiful in the universe, are not retained by the Earth's field, the heavier gases O_2, CO_2 and N_2 are retained to some degree in the atmosphere in a graduated pressure column by the gravitational field. This atmosphere is not at a fixed temperature either vertically or horizontally due to the anisotropic radiation field and it therefore flows. (Note how elements are selected even in space by gravity fields.)

While condensation of a large new compartment such as Earth created a new gravitational field and this developed with the land masses, the condensation of new small local compartments, vesicles and polymers, created new fields of which the most important is an electric field due to ionic charges on their outer surfaces. This creation of an electrostatic attractive force acting on counterions of any neighbouring surface or on free counterions in a surrounding phase leads to additional preferential diffusion (see Section 6.4.2). Now, all the ion atmospheres will try to reach a local equilibrium distribution between that due to *random* diffusion and that due to the newly created field, and we observed that this static situation does arise in many cases. There is, however, a further circumstance, essential to the discussion of flow, which is one in which a homogeneous or heterogeneous phase system is subject to a maintained (constantly regenerated) non-equilibrium distribution. The use of energy to continuously produce material in a field gradient and with it a neutralising flow is, in fact, of the essence of continous water flow on Earth and of water-powered production of electricity by man. Much of the generation of biological flow is due to a parallel movement of material but there must also be continuous production of an electrostatic field generating ion flow (Section 7.17 and Chapter 10). Now, such considerations do not just apply to gravitational and electrostatic fields but can be generalised to concentration gradients, temperature gradients and mechanical stress gradients (Table 7.21 and see Chapter 6). All these fields

Table 7.21 Types of gradient causing flow

Field	System
Gravitational	Layering of Earth, sea and atmosphere Separations in chemistry Biological polarity
Chemical concentration	Storage across boundaries
Electrical	Storage of charge (many electrochemical cells and all biological membranes)
Mechanical tension or pressure	Opening or closing valves Disruption of biological membranes
Radiation causing a temperature gradient	Change of physical state Convection currents in all materials

generate vectorial flow which attempts to remove them. To keep a *steady state* they must be maintained by energy and/or supply of material. If all this is done within a controlled structure, then by its very nature the combined maintained fields and flowing material also have observable 'shape'. Maintained shape and steady-state kinetics become linked. (Again we can compare the description of shape at equilibrium in Chapter 6.)

We shall now develop a picture of a multitude of compartments, some of which generate fields acting to limit diffusion and flow within certain structures. The first need is an energy supply to make a field. Just as a field is created in an electronic circuit by a battery so it is necessary to have energy from an ion pump to drive ion flow. The sources of power are described in Section 7.17 and they are used to generate flow. The motions in these cases are called currents—electronic and electrolytic, respectively. In electrolytics, liquids or solids in which ionic salts are not soluble, for example lipids in membranes, help to control flow and are effectively the constructions and insulators. We have already compared these systems with man's electronic wire circuits. (The parallel with the flow of water in a river bed is also clear, and here again we describe rate of flow in terms of a current.)

7.12.1 Electronic and electrolytic circuits

The easiest way to examine quantitively the rate of flow in an electrical circuit, compartments in series, is through the equation known as Ohm's law, namely

$$\text{(Electron flow) current } (i) = \frac{\text{Driving field}(V)}{\text{Resistance}(R)}$$

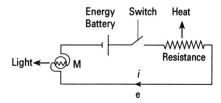

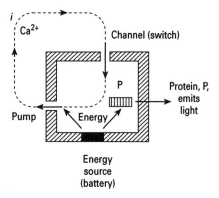

Fig. 7.31 A simple man-made electrical circuit giving out heat and light.

Fig. 7.32 A biological electrolytic circuit using a calcium current. The protein is aequorin giving out light and heat.

Fig. 7.33 An illustration of the way in which man's electronic circuits based on metals or semiconductors are related to biological electrolytic circuits. The insulator in the latter is the membrane, the switch is a gated channel, and the condenser is a vesicle store. Flow of charge is of the essence of biology. S, switch; B, power supply; C, condenser; e, electron; M, metal cation. M_1^+ entry triggers C which triggers M_2^+ entry or exit.

The current is the rate of transfer of charge, the driving field is the charge gradient (V) and the resistance (R) is the barrier met by the moving charge (Fig. 7.31). (A lamp, M, in the circuit is just a resistance.) The system is made more complex by introducing storage units (capacitors) into the circuit. The storage elements allow greater flexibility in the control over flow of charge. In biology the current is electrolytic, but the same relationships hold, and devices similar to those in electronic circuits can be built into electrolytic circuits (Figs 7.32 and 7.33). Since membranes can be given shape (by protein filaments) and ions of particular elements collected in local spaces, the circuitry of biology based on selected elements (in compounds) within compartments has evolved and diversified in space as fixed shapes. The circuitry is vastly more complex than the electronic circuitry of man and is used for transfer of all kinds of elements in selected ways, often by initial concentration in compartments (see Table 7.17). (The basic requirement for directed flow in all cases is the maintained existence of localised zones out of equilibrium with one another, that is, energised, but connected by a linking controlled compartment allowing diffusion under a differential driving force, a field (Tables 7.21 and 7.22). Now, control of the *direction* in which material and/or energy flow does not of itself control the extent, that is, the rate of flow at a desirable and perhaps adjustable level. This is achieved by the control of the power, the energy input in unit time. Again, different flows that may occur in different parts of the structure are not necessarily linked. In most

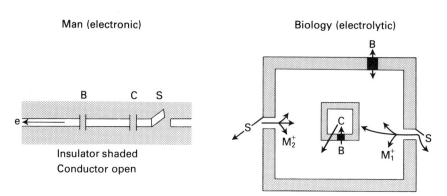

Man (electronic)

Biology (electrolytic)

Insulator shaded
Conductor open

Table 7.22 Some flow networks used in transport

Flow	Field
Water in streams	Gravity
Water in pipes	Gravity, mechanical pumps
Blood (water) in organisms	Mechanical (the heart; gravity)
Food in digestive tract	Mechanical pumps
The tides	Gravity
The atmosphere	Temperature (pressure) gradient
Electrons in metal wires	Electrical potential
Ions in water: electrophoresis, electrolysis, electrodialysis	Electrical potential (nerves)

circumstances, however, it is advantageous if the controls are self-adjusting and co-ordinated, which is the next and, as it turns out, critical advance in the *physical* management of kinetics—*feedback control*, just as it was so in chemical feedback management (compare management of flow in channels and pumps with management of catalytic rate constants (Section 7.8.2)).

7.13 From simple flow to feedback control

All the kinetic systems that we described in Sections 7.1 to 7.6 are linear in time in the sense that a later event in a series of motions does not interfere with or interact with earlier steps. Such a system is expressed by

$$\text{Rate of output} \propto F(\text{input}).$$

(Ohm's law is one such expression.)
    The generalised representation is

$$\text{Input} \longrightarrow \text{rate of motion} \longrightarrow \text{output} \qquad (7.3)$$

so that the output is linear with the supply. Now let us consider the situation where the output, material of concentration [O], can reduce the turnover rate in proportion to [O], that is, [O] feeds back on an earlier step, such that

Input, I $\longrightarrow$ rate of motion $\longrightarrow$ output, O: $\longrightarrow$
   F F'

The rate of production of O at a time (t) now becomes

$$\frac{\mathrm{d}[\mathrm{O}]_t}{\mathrm{d}t} = F[\mathrm{I}]_t - F'[\mathrm{O}]_t$$

where F and F' are some particular 'transfer' functions (*affecting* flow constants) that do not need to be considered yet. It is easily seen that as [O] increases the rate of change of [I] decreases until for a given $[\mathrm{I}]_t$ there is a relationship

$$F(\mathrm{I}) = F(\mathrm{O}) \qquad \text{and} \qquad \frac{\mathrm{d}[\mathrm{O}]}{\mathrm{d}t} = 0; \qquad \text{hence } [\mathrm{O}] = \text{constant}.$$

Thus we have a feedback system that couples the flow in of I and flow out of O in a steady state. We can represent this by a general diagram

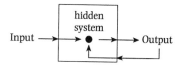

where ● is a feedback control point of flow in a compartment of the system.
    Now the control can be made in a slightly different way by having a series of

internal barriers to transfer steps that interact sequentially with one another. The simplest case is

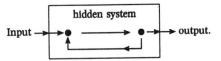

Later stages of the hidden process may slow (or accelerate it in a feed-forward system) where the feedback (or feed-forward) is internalised as shown in the next diagram. In such a way, in a message pathway of many steps, later hidden steps, may inhibit (or perhaps intensify), through their products, earlier steps. So far the feedback is simply controlling I flow relative to O. Let us extend the feedback of the type represented below to a second flow pathway, U goes to V. It is now possible to link the two processes in mutual feedback as in

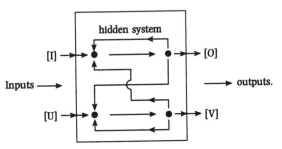

In such a cross-talk network the relationships between I and O, and U and V separately become interrelated, so that in the steady state there is *co-operative* [I], [O], [U], [V] flow relationship. It is clear that this can be extended to any number of hidden units and any number of inputs, including energy, for example light, as well as materials, electrons or ions (see Table 7.21). We can consider, for example, that there is an input of energy, in the form of light or chemical energy, such that the input of material is driven. On observing any such flow system in operation it is not always clearly known what are the inputs or the outputs, never mind what is the hidden system. As of today no living organism is fully characterised by input, output or hidden systems, but a living organism surely conforms to such a pattern. (In industry similar computer controls are understood.) Note that the feedback to the input step may control the input itself even at the level controlling whether electrons, ions or light are admitted. We have to enquire for any particular system not just what are the materials I, O, U and V, but how energy is carried and how information is fedback in the hidden system. Finally, we need to see how the necessary physical construction of compartments is made (part D of this chapter), which is far from obvious in biology.

7.13.1 Feedback electronic circuits: an illustration

In Section 7.5 the simplest electronic circuit was described. The development of this as a feedback circuit is illustrated in the doorbell arrangement shown in Fig. 7.34. The circuit, on closing, that is when keeping the bell-button in contact, passes a fixed current that works the bell by mechanical action through energising a magnetic coil. The same mechanical motion breaks the

(a)

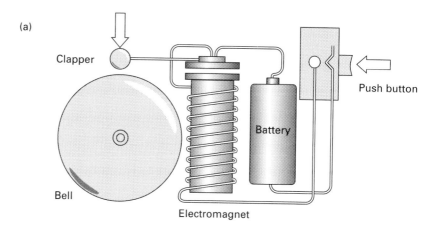

Fig. 7.34 (a) A simple example, the doorbell, with a physical feedback based on transfer control—here of electrons. When you push the button of an electric bell, current from a battery flows through an electromagnet. This becomes magnetic and attracts a clapper, which hits the bell and makes it ring but breaks the circuit. (b) A circuit diagram of (a). (c) A representation of a doorbell circuit in terms of observable and hidden parts, all of which are compartments.

(b)

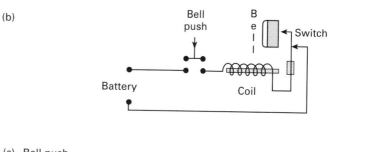

(c)

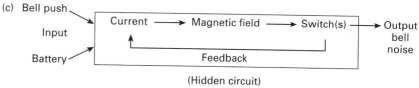

circuit thus cutting off the current. The de-energising of the coil remakes the circuit and the current flows again repeating the process until the bell push is released. Hence, the magnetic coil tells the battery through feedback induction of a magnetic field (by the current) when current is permitted to flow. (If the response is to be graded, then the all-or-none function of the bell push can be connected to a variable resistance that will allow current to flow in proportion to the resistance but it will never rise above a certain level at which the bell rings and the circuit is broken.) In these electronic devices the current is provided by a chemical battery or by the mains, the wires are copper, the magnetic bar is iron and so on. Each chemical in each compartment is selected for its particular functional physical purpose, here by man. Physical feedback circuits of enormous complexity have been developed in computers, illustrated in Fig. 7.37. In biological flow chemical selection of current carriers is required as well as physical control. We return to the way in which chemical selection is managed after we have completed the description of selection of *physical* kinetic control with one biological example and another of relevant environmental concern, the water cycle.

7.13.2 Physical fields and feedback: a biological example

Consider any animal, such as a man, standing in an upright position. He is out of equilibrium in the gravity field, that is, in relationship with the adjacent compartment, the Earth, on which he stands. If his centre of gravity is exactly placed he uses no energy in keeping this rigid metastable state (Fig. 7.35). When moving around this condition is broken and muscles need energy to keep him from falling, but the muscles do not know this unless they are told. In the animal's head are balance organs that detect the deviation from the metastable condition and constantly inform the muscles to act this way or that so as to try to return the body to the upright condition. The morphological state of the animal is then dependent upon a communication network since it does not have a rigid frame based on a single compartment. The requirements for maintenance are several compartments that do not mix but communicate:

(1) a sensing apparatus;
(2) a message system to connect the sensing to activity, that is a nervous system;
(3) an energy-utilising machine that is activated by the nerves, that is muscles, to keep balance in the correct fashion;
(4) bones and connective tissue that position (1), (2) and (3) flexibly.

Thus the morphology of the body is in (automatic) feedback relationship with the Earth's field in an attempt to maintain the metastable state that uses least energy during movement. This balancing activity is, therefore, different from the balancing of a rigid object in a static metastable condition since the very flexibility of the animal body makes it necessary to counter fluctuations of any kind by using restoring energy. The animal needs the sensing equipment to be as high above ground as possible to increase sensitivity and, therefore, its sensing equipment is far from the restoring muscles that act on every joint right down to the ground. (An upright animal can also see more of its surroundings.) This is why at least the four different kinds of compartment listed above are needed by the animal to maintain posture: the sensor; the message carrying system; the muscles; and the bones and connective tissue. (Note how spatial position is knitted into the organisation.) The reader may wonder what this (physical) physiology has to do with the natural selection of the chemical elements. The answer is that different elements fill best the roles of sensor, messenger, muscle activator, muscle and bones, just as different elements are used to advantage in physical electronic equipment. The gravity sensor in the animal's body should have high density and therefore should be a solid made from a heavy element; since the communication is through electrolytics the message carrier must be able to travel freely in water unimpeded by organic surfaces; the muscle must be made of an elastic material able to receive a messenger chemical, a trigger (induced by the electrolytic current) which must bind to muscle substance so as to change its activity while making energy available to it; finally bones must be relatively rigid. Without further thought we notice that the sensor is made of *dense crystalline* $CaCO_3$, the message carriers are NaCl and KCl in *ionic solutions*, mechanical flexible devices are made from *proteins*, using the elements H, C, N and O, the trigger is the Ca^{2+} ion in solution, and bones are made from calcium

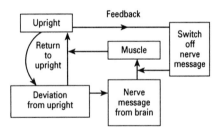

Fig. 7.35 A schematic diagram of the balance system of man with feedback control over muscle activity, so as to conserve energy.

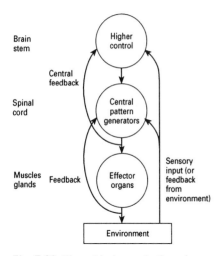

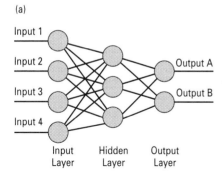

Fig. 7.36 Hierarchical organization of neuronal circuits governing the control of movement. Note the nesting of feedback loops. (After a figure in Shepherd, G. M. (1988). *Neurobiology*, Oxford University Press, Oxford.)

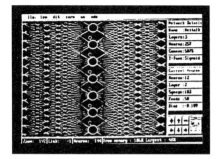

Fig. 7.37 (a) A simple crossed feedback control and (b) a more complex example of a network, for example on a chip.

hydroxyphosphate. (Note that abundance and availability are relevant in the choice of the functional elements.) We return to this topic of selection of elements for physical functional value in Chapters 10–14, but note the exact parallel with the description of chemical functional value in Sections 7.3 and 7.8.1.

The most evolved type of organisation of this kind is called a *neural network* (Fig. 7.36) and is common to computers (Fig. 7.37) and the brain. The computer uses spatial distribution of electrons both in charge carrying and in charge storage, and it uses electric power. The brain or any nervous system has to use electrolytics and its primary carriers are then ions. Immediately we see again that those elements that give soluble salts (of *indestructible* ions) are the safest and fastest current carriers in the hidden circuits of biology (see Section 7.5.3). Thus Na^+, K^+, Cl^- and, to some extent, Mg^{2+}, Ca^{2+}, SO_4^{2-} and HPO_4^{2-} became the chemically selected elements for one functional use, information transfer, without which organisms would have remained at best inefficient and at worst unable to survive. The way in which other chemical signals such as organic ions or inorganic ions complexed with organic molecules flow within biological systems using similar feedback networks is different in that they are linked to *chemical* transformations, synthesis and degradation. They have been described in Sections 7.5–7.7. The energy source for the flow is also controlled and, as it too involves chemical change, we reserve discussion until Section 7.17. In addition, we must observe that the construction of the framework of the computer and the brain needs quite different elements to make components as well as wires (Cu, Fe) or membranes (H, C, H and O) and that these elements have to be transported in a directed way to create the structure or compartment. The way in which man does this is clear from everyday activity (Chapter 14), but biological systems are self-assembled, disassembled, and reassembled by some means into which we shall enquire in Chapters 10–13. Structure itself is a dynamic feature in life and herein may lie the extraordinary character of the brain since it is linked to diverse current carriers and to the *growth into patterns* of nerve junctions.

7.13.3 Geological control and feedback: the physical water cycle on Earth

It is useful here to see how the system of water flow on the surface of the Earth arises. We can, if we wish, look upon the sea, the ice caps, the lakes and the water vapour of the air as separate local compartments, each in local stationary states describable by the appropriate equilibria (see Fig. 7.25). Because each has its particular location in the gravity field of the Earth (and of the moon) as well as in the radiation field of the sun, each compartment has a local free energy ΔG, not the $\Delta G°$, of water at 298 K and unit atmosphere pressure. We can treat these differences in free energy between the homogeneous (by definition) compartments as able to do work and hence we can understand the possibilities for change (Chapter 6). In this chapter we are interested not in the potential for doing work, that is, differences in ΔG among compartments, here due to physical conditions of the individual compartments, but in maintaining *rates of transfer* (rates of doing work, flow). This flow concerns not just the compartmentalisation but also the links and barriers between compartments and the energy needed to sustain the connected systems of different ΔG in a surviving condition. We have discovered that for such sustained survival of a flow system at constant rate there has to be controlled feedback on transfer rates. In the case of water circulation, it is easily seen that the source of energy is the sun. The evaporation from the sea leads to condensation as clouds in the colder higher air which reduces further evaporation by the sun's energy, a feedback mechanism. Moreover, the rise of the air is forced to be greatest over the high land to which it moves where the water precipitates as rain, removing the clouds, and is then returned via rivers on the land to be reheated in the sea. We could go on to see that the flow will always create snake-like patterns of rivers on the land and that these develop in recognised ways. (A peculiarity is present in the flow of the water vapour in the atmosphere: there is no apparent external structure, yet clouds have shapes (which can be classified) and their flow is directional.) The whole system of water flow is maintained by the constant energy supply from the sun controlled by feedback interaction with evaporated water. We shall need to find sources of energy for all such flowing steady states (see Tables 7.22 and 7.23). It is equally possible to consider the temperature of Earth's surface as in a feedback system taking water to be a greenhouse gas.

Table 7.23 Circuit components in message systems

Transmission	Carrier	Power
Electronic	Electrons	Charge storage (fields)
Electrolytic	Ions	Ion gradient (fields)
Chemical	Molecules	Molecule gradient (gravity)
Mechanical	Stress	Strain (compression)
Radiative	Photons	Emitter

7.14 Physical control of diffusion and variance of systems

In Chapter 4 and again in Chapter 6 we saw that, as we multiplied local division of independent compartments and independent components, the variance of systems became very, very large indeed, although this variance was limited by any equilibria existing in each 'phase' and between 'phases'. We have now removed all the equilibrium constraints on spatial co-ordinates and have allowed continuous change, flow, so that at first sight there are no constraints on concentration or position in space, components or compartments, except through haphazard development of fields. We might, therefore, have expected almost every kind of physical appearance around us. In agreement with this expectation, when we look upon the shapes of rocks and mountains and then at river systems and clouds we can perceive no reason for their particular appearances and a continuous range of systems (only limited in some respects) as well as continuous modification is observed. It is only when we look at biological objects that we see that here physical appearance has to have very strong constraints since shape and shape change is repeated time and time again in cycles within any species—we do not observe continuous gradation between species nor much variation of development within species, that is, physical form has been almost reduced to invariance in a growing, dynamic, not an equilibrium body of a given species. We conclude that physical transfer in biology must be controlled by rules parallel with equilibrium rules. The equilibrium time-independent rule was that in a system G was a minimum, $\Delta G = 0$ for exchange or transfer, and using this condition we deduced the simple expression of the phase rule, although it became somewhat more complex involving temperature, pressure, amounts of materials and fields. It was of great value where transfer was between *phases* in equilibrium, but was of little use when 'phases' were out of transfer equilibrium, in which case we called zones divided by barriers *compartments*. Thus we can expect the phase rule to be particularly valuable in certain situations, for example, at high temperatures or in dealing with solubility, but does it have a value in the description of biological systems?

The phase rule also implies that the size and shape of a phase at equilibrium are mutually invariant and any system for which this is not true must be out of equilibrium and will tend to evolve to achieve it. We have then the task of explaining the constancy of biological shape which can now only be based on kinetics. What can we describe in kinetics that parallels the rules of removal of variance by equilibria and is as general as this rule but now for unchanging steady-state systems in which there is continuous flow, for example a living organism? Here shape is maintained quite closely, not quite exactly, even with development. We must consider that this is an unavoidable consequence of *constrained self-organisation*.

Since we are dealing with flow and inputs to flow there must be limitations to the rates of transfer (of matter and/or energy) to control flows. It is obvious that controlled rates of flow require the maintenance of fields (that is, sustained energy supply) and supply of material in a system with feedback regulation so that the output can influence the initial and/or hidden inputs. This is of the essence of transfer systems involving a series of out-of-equilibrium compartments or phases through which energy and material continue in steady flow. Following the description of feedback already given, we can represent this situation schematically as

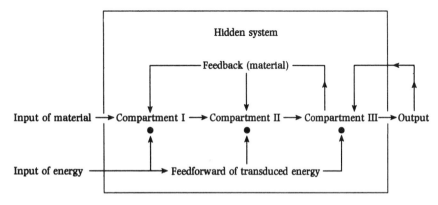

Such a system can be shown to remain constant (steady state) or to change if the input is programmed, but only to set *stable* patterns. An excellent example is the transfer of water from the sea to clouds to lakes and rivers and back to the

sea as already given. Although we described this flow as producing random patterns, their maintenance, once established is not accidental. In this case nothing changes over long periods of time, although levels can fluctuate, since mountains have a fixed shape that is imposed on the water flow. The system is not truly fixed, and will develop slowly and to some degree without control. All its steps can be characterised by rate constants, but the system lacks any initial *planned organisation* (to initiate it), has a certain haphazard or random (initial) character and, within limits, undergoes wild fluctuations.

If we alter the diagram somewhat and introduce a plan, we now have the scheme

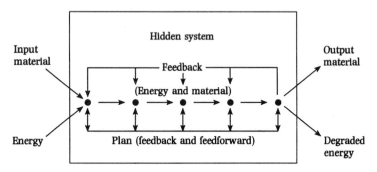

Here, in addition to a simple feedback there is a programme of instructions, a plan of action. The plan is used to co-ordinate all activities so that they become *strictly* co-operative. Such programmes are used today in computers to look after a chemical factory, for example an oil refinery, varying the output by altering the operational conditions of various compartments changing, for example the input of selected stock, the conditions of cracking and fractionation, etc. (Section 9.4.2). Thus, instructions to both input materials and energy control the factory (the hidden system). In a parallel fashion a biological system adjusts material input and energy utilisation to make structures and to maintain activity. There is no more to it than complex physical (and chemical) feedback control over flow according to instructions. Instructions come from the plan, genetic material (DNA) which itself in biological systems is in feedback communication, through synthesis, with the hidden system. Before we tackle the nature of the plan within a biological species we need to combine the physical with the chemical changes described in Sections 7.4 and 7.9 and to introduce the flow of energy since ultimately the plan must be self-producing.

7.15 Physical feedback control in relation to chemical component feedback

We return now to the description of feedback control acting on a series of chemical components, that is, within a pathway

$$A \longrightarrow B \longrightarrow C \longrightarrow \ldots \text{etc.} \longrightarrow X.$$

The feedback and feedforward within $B \rightarrow C$, etc. must determine A and X in all respects. Hence, we need to bring the set of components A, B, C, X and so on into a fixed kinetic relationship so that B, C, and so forth depend on the rate of

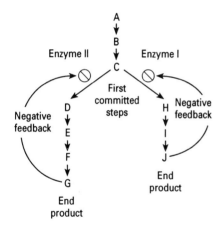

Fig. 7.38 Feedback in metabolic pathways. The reactions C to D and C to H are the first committed steps in the branch pathways leading to end products G and J, respectively. These end products can block the first committed steps by acting as allosteric inhibitors of the enzymes catalysing them. Thus, if the level of product G builds up beyond what the cell needs, its production can be slowed without slowing the production of J, and vice versa. (Adapted, with permission, from Purves *et al.* reference 2 of Further reading.)

supply of A and on the kinetic constants k that describe the chemical as well as the physical sequence of steps

$$A \xrightarrow{k_1} B \xrightarrow{k_2} C \xrightarrow{k_3} \dots X.$$

These constants, k, can be controlled by *catalysts* (see Section 7.8), which may themselves be put under feedback kinetic control of products. This is a chemical problem, specific for each system (Fig. 7.38). The catalysts are then the chemical correlate of the physical controls over transport (Fig. 7.39). The series must be such that only the supply of A can be varied. If the system is made cyclic in material however, that is,

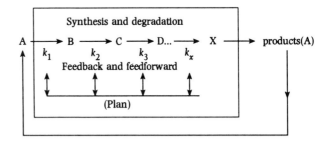

in which feedback control is maintained, then it will be invariant and will flow without change *given a fixed energy input.* As stated, any such linear series can be made of chemical components in connected compartments and can be linked to any and every other series of linear transformations in a larger system with physical restrictions (Fig. 7.40).

It does not take much effort to realise that a biological system is very closely of this kind. The chemical elements C, H, N and O in such forms as CO_2, N_2 and H_2O and some 20 others move through an energised processing only to return to their original state. It is the peculiarity of the feedback control and plan that produces $B \rightarrow C \rightarrow D$ that is so mysterious! In part D of this chapter we will attempt to put together all the contributing features of any reproducible organisation: physical and chemical barriers within structure, supplies of energy and material, and a plan, all self-made. It is this total feedback that we consider to dictate shape. We see that a sufficiently maintained (planned) feedback organisation through which there is flow has in common with a stationary *equilibrium* small system a fixed pattern but that the flow must be contained by an apparatus or structure. It is to the structure that we turn next, and then to sources of energy to drive the systems.

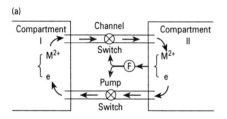

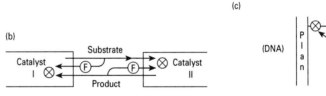

Fig. 7.39 Illustrative parallel examples of feedback in: (a) a physical circuit; (b) a chemical circuit; (c) an information circuit for expression of catalysts (enzymes). F, Feedback to control X.

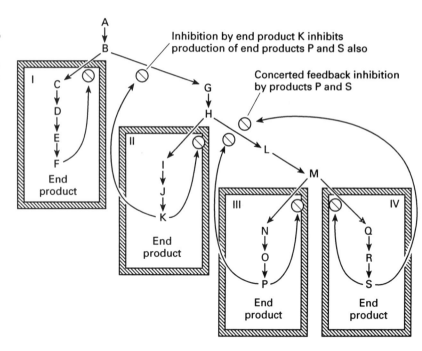

Fig. 7.40 A further example of feedback inhibition, now between compartments I to IV. A variety of feedback controls come into play in branching metabolic pathways. Inhibiting the B to C reaction, for example, shunts reactant B into the B to G step. In some cases in which a single compound (M) gives rise to more than one end product (P and S), both end products can join in concerted inhibition of the enzyme catalysing the first reaction committed to the production of M; each end product may also inhibit its own branch pathway. Occasionally, an end product such as K will inhibit an early step in its own synthesis (B to G in this example) and in so doing will slow down or stop production of products P and S as well. NB. Note the presence of diffusion barriers. ⊘ = Enzyme or Pump with chemical or physical feedback. (Adapted, with permission, from Purves *et al.* reference 2 in Further reading.)

D Organisation

7.16 Linking metabolic change to creation of structure

Time and again in the above discussion we have seen how steady-state flow and chemical transformation are maintained within structure. There is then the problem of structure and its origin, which, given an energy supply, will complete the description of organisation. There is no problem, of course, in seeing how geochemical flow or man-made organisation of flowing systems are generated. However, there is a distinction between these two in that the man-made object is designed for a functional purpose. We know how we set about organisation with a plan based on knowledge of requirements. The surprising fact is that biological systems appear to construct in exactly this way so that all their activity seems to be planned for functional value, including their own development. Biological structures are not permanent but arise in succession from single cells (see Section 13.6) so they are not to be included under discussions of equilibria or of stationary states but under steady states that follow a flowing patterned development. Clearly, we have to turn to the kinetics of how such structures could be assembled and, if need be, disassembled. In order to proceed to this end we outline first the building, self-assembly and growth of crystals of static structure. The problem is not without direct relevance to self-assembly in biology since biological systems do make crystals and from crystals make many materials with designed shapes (see Table 7.24). Again they do assemble composites such as ferritin (Fig. 6.22) and viruses (Fig. 6.23) that are clearly related to crystals in their mode of construction. In fact, many observed biological assemblies have a mathematical form not unrelated to that expected from direct considerations of the most stable way of filling space, which is exactly the way in which crystals grow (see Chapter 6). On the other hand, all the biological living systems develop from a single cell in what looks like arbitrary association in the embryonic stages (see Section 13.6).

Table 7.24 Major biominerals

Mineral	Forms	Functions
$CaCO_2$	Calcite, aragonite, vaterite, amorphous	Exoskeleton, eye lens, gravity device
$Ca_2(OH)PO_4$	Apatite, brushite, octa calcium phosphate, amorphous	Endoskeleton, calcium store
$CaC_2O_4(2H_2O)$	Whewellite, weddelite	Calcium store, deterrent
Fe_3O_4	Magnetite	Magnet, teeth
$FeO(OH)$	Goethite, lepidocrocite, ferrihydrite	Iron store, teeth
SiO_2	Amorphous	Skeleton (plants, protozoa), deterrent

7.17 Nucleation of assemblies

A crystal forms because of the thermodynamic drive to equilibrium where the disorder of the solution is overcome by the order in the crystal. The crystal is stabilised by the co-operative interaction due to electrostatic fields between atoms or molecules. However, the development of an ordered crystal has an activation energy due to the low stability of any surface (Fig. 7.41). If the crystallite is very small, say, 100 atoms, then the ratio surface/volume is

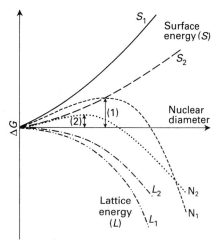

Fig. 7.41 The dependence of the free energy of the surface and the crystal interior on the size of a nucleating growing crystal. Case (1) has a high opposed surface energy despite a high crystal energy while case (2) has a less opposed surface energy and a low crystal energy. Kinetically, (2) crystallises before (1) due to the smaller activation energy barrier to growth shown by ↔.

relatively large and the crystallite tends to redissolve rather than to grow. The barrier to crystallisation is the free energy of nucleation which relates, to a high power, to the excess concentration of material in solution over the solubility, especially in the cases of highly co-operative crystals, for example salts. Nucleation of crystals, therefore, takes place most easily on preformed surfaces or, at least, from preformed sites on surfaces. The problems of initiating a crystal assembly are described in Chapter 20 of our previous book, reference 1 in 'Further reading'. Here we draw attention to the fact that *large molecules are effectively surfaces* that can assist nucleation, while small molecules in general are inhibitors. Thus, in the catalysis of crystallisation, proteins that bind only one or two units of a crystal, say of $CaCO_3$, can initiate crystal growth. Examples will be given later. It may well be that around the initial point of growth on a protein where there may be but one unit, for example a $CaCO_3$ ion pair, the surface energies of the neighbouring sections of the protein will assist further development of $[CaCO_3]_n$ provided that the protein surface disrupts the local stability of the solvent and stabilises the crystal surface. In this case then the process of nucleation,

$$Ca^{2+} + CO_3^{2-} \longrightarrow \underset{\text{ion pair}}{CaCO_3} \longrightarrow \underset{n \text{ is small}}{[CaCO_3]_n} \longrightarrow \underset{\text{crystal}}{CaCO_3},$$

may be helped in all steps by proteins. The suggestion is clear that such self-assembly may not need great supersaturation and may even be initiated almost at equilibrium. Just as this is approximately true for some crystals of $CaCO_3$ in living systems and of bone (the surface of bone is close to being in equilibrium with Ca^{2+}, H^+, HPO_4^{2-} in solution) so it may well be true that some protein/protein interactions also self-assemble with very small energy barriers. The problem of growth of structure is then that of the controlled supply of materials. The corresponding man-made equipment for growth is a chemostat in which supply is under feedback control of solution concentrations. The diagram in Fig. 7.42 of biological synthesis indicates by rate (R) and equilibrium (E) where rate controls may stop and where equilibrium control may be introduced. The complexity of biological syntheses requires that organisation to be based on management of each reaction pathway so that it is

Fig. 7.42 In all steps up to the formation of (folded) polymers the processes are energy-requiring and need enzyme catalysts. Self-assembly may only require correct transport, which is needed too in all other steps. As the number of large molecule components increases the fields around each changes and the steady-state shape changes with it. E, Equilibrium step; R, rate-determining step.

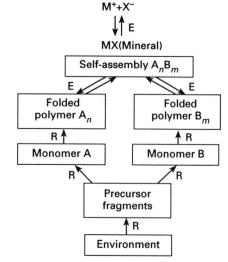

in communication with, say, at least two other pathways (see feedback control in Fig. 7.40) so that assembly of any unit is managed. In the figure we do not show the essential feedback to make a biological chemostat, but the control of, for example, bone formation, clearly has such an attribute. So far this description is of a stable structure and with little extension it can be used to understand how a shaped object can be built not just from repeating units but from large *different* molecules, proteins, see Section 5.6.4. Now such structures are solids, but we saw in Chapter 6 that liquids (and large dynamic living systems) could also have shapes.

7.18 Shape

A brief word is now required about shape before we analyse the problem in more detail (Chapter 11). 'Shape' is used as a description in preference to 'structure' when symmetry is ill defined and flexibility is permitted. The shape of any small semifluid system at equilibrium that is not controlled by a pattern as in a crystal must be due to tension, some mechanical constraint. Thus, as we have seen, the shape of a liquid surface is controlled by surface tension, (Section 6.2.2). The shape of a living system, obviously semifluid, must then be a reflection of tension generated by fibres acting on films of polymers. Now, shape is species-dependent so that the tension must be determined by controls closely related to those of all other species-dependent activities, especially metabolism. This dynamic metabolism constantly requires energy to renew fields, electrical and mechanical. The shape is then connected inevitably to the flow and homeostasis of an organism.

Now, in Chapter 6 we showed that in an equilibrated system of phases the removal of variation in concentration of components, given fixed physical conditions, generated not just a fixed phase system but also a fixed shape. The discussion hinged on surface area relationships to environmental conditions. If we are correct in saying that biological shape is fixed by kinetic constraints, then the only way in which variables can be removed (see below) is through feedback control related to the metabolic pathways. Shape is then not fixed in quite an absolute way but is very nearly so fixed by the enforced generation of flows (physical and chemical), which produce fields, including mechanical stress, within structure for controlling further flow. The distinction between a plant's not so well-controlled shape and an animal's very well controlled shape is then related to the distinction between the strengths of the metabolic homeostasis (see Table 7.25). We need next to see how energy enters these systems from some external energy source since synthesis and maintenance of structure (shape) are now seen to have a general demand for energy.

Table 7.25 Static and dynamic shapes

Static	Dynamic
Viruses	Muscles
Hair	Nerve termini
Shells	Bones
$SrSO_4$ skeletons	DNA
Cellulose walls	Vesicles

7.19 Energy, radiation fields and flow in temperature gradients

The simplest and most important source of energy that can be used to do work is a heat engine, which we can recognise by a temperature difference between

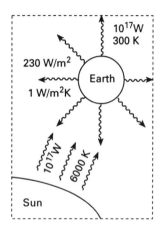

Fig. 7.43 The energy flows from the sun to the Earth and subsequent radiation and heat losses from Earth.

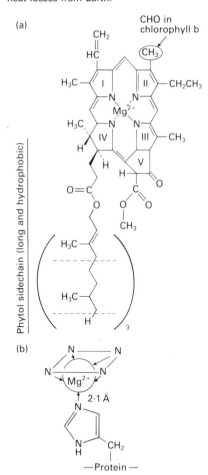

Fig. 7.44 (a) The formula of chlorophylls a and b and (b) their binding in proteins through a protein histidine to the magnesium atom of chlorophyll. Note the essential magnesium.

source and sink (Section 3.5). An example of a temperature difference between compartments is that of the sun relative to Earth (Section 3.7), but smaller temperature differences are around us everywhere. Thus, while matter is not necessarily exchanged, there must be radiative transfer since it is impossible to stop it. In certain 'fortunate' circumstances the radiative energy gain and loss may be balanced so that the temperature of the receiver is held in a steady state. The Earth's surface is some 30–50° hotter due to the presence of the sun and as a result of the steady-state balances shown in Fig. 7.43. Man's body temperature is about 37°C in any environment when properly clothed since the heat input from food digestion equals the heat losses, except in extreme conditions when other mechanisms, for example shivering and sweating, act to maintain the desirable temperature on which all of man's metabolism is based. We show later that the heat generated is controlled by hormonal signals to metabolism (see Chapter 13). The basic principle is the same as that of a feedback man-made thermostat. Hence, although compartments are quite different in flowing material content, in an organism we treat them as being in steady states of equal temperature, thermalised systems, so that we can use equilibrium free-energy considerations in the discussion of their relative capacity to do work. The requirement for control over temperature has already been described as necessary in a complex feedback control system in order to avoid variation in rate constants (Section 7.8.3). It is seen in both these cases that to maintain flow and temperature there must be added to all our previous considerations of kinetics an external energy supply and this supply must be controlled just as the material supply is controlled if homeostasis is to be secured.

7.19.1 Energy capture and control

We have already stated that radiation (light) as photons can be absorbed by atoms or molecules (Section 1.7). It is then essential for the receiving (molecular) system to have an absorbing unit. If the photons have enough energy (ultraviolet (UV) light) they may strike out electrons completely and turn a molecule into a radical ion. Lower energies (UV and visible light) can 'excite' electrons to higher energy levels and then either the molecule splits giving reactive free radicals, or it reacts with other molecules or it rearranges internally to give a different isomer. (The electrons may, of course, go back to lower energy levels and light is then emitted immediately (fluorescence) or delayed with minimal energy capture (phosphorescence).) Photons of even lower energy (near and middle infrared) can affect the vibrational states of molecules and at much lower energies (microwaves) affect the translational states (Table 7.26 and see Section 3.7). All these effects have been used functionally in natural processes, for example in biological or geochemical systems or in man's industry to trap energy. We will give some examples below, but the obvious one is that mentioned above of the atmosphere capturing energy that in turn gives long term stability to the flowing environment.

Some molecules are particularly sensitive to visible light; chlorophyll, for example, is one of them and has evolved naturally to be most effective in these respects (Fig. 7.44). The energy gained in the process (in which one electron is ejected from one molecule and then trapped by a different one) is used by plants to make carbohydrates and indirectly to oxidise water thus producing

Table 7.26 Photon energy and motion of molecules

Region	Energy (kcal mol^{-1})	Wavelength (nm)	Typical examples
Far ultraviolet	2860–140	10–200	Ionisation energies, chemical bonds, Lattice energies
Near ultraviolet	140–75	200–380	Chemical bonds, electronic transitions
Visible	75–35	380–780	d–d electronic transitions
Near infrared	35–10	780–3000	Semiconductor gaps, vibrational transitions
Middle infrared	10–1	3000–30 000	Vibrational transitions
Far infrared	1–0.1	30 000–300 000	Heavy atom vibrations
Microwave	0.1–3 × 10^{-5}	300 000–10^9	Rotational transitions and translational motion

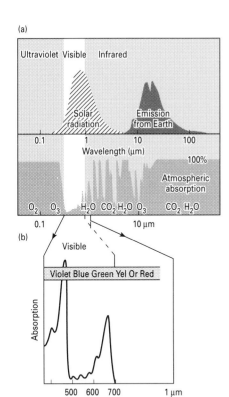

Fig. 7.45 (a) The radiation spectrum (see Fig. 7.43). (After P. A. Cox. *The elements on earth* (ref. in Chapter 8).) (b) The absorption spectrum of chlorophyll in the visible region; compare the sensitivity of the eye.

dioxygen. Life today depends on this process. Notice how a minor component is utilised to trap energy for the whole system by converting light into other forms of energy. (Note too that it requires magnesium.) Another light-sensitive natural process is the irradiation of dioxygen, O_2, by UV light in the stratosphere, forming a protective ozone layer. Without it advanced life on earth would be at risk since this ozone layer absorbs UV radiation of longer wavelength (giving back dioxygen) that might destroy, in particular, organic substances (see Table 7.26).

It is just good sense that man tries to use sunlight for many purposes and especially to produce energy, very much as nature does in driving photosynthesis. Man does it by constructing photocells where light (photons) forces charge separation, for example

$$h\nu + xy \longrightarrow x^+ + y^-$$

in such a way that x^+ and y^- are held separately in different compartments. In this way the energy of light is absorbed creating an energised state of separate x^+/y^- plus a field that affects all charged particles in the system. If sunlight is used for the purpose this sets limits to the energy of the photons and hence to the kind of charged species that can be obtained (Fig. 7.45). Nowadays man's photocells are built using suitably doped materials and semiconductors (see Chapter 2). We will not discuss this subject here in any detail but stress once more that both nature and man were forced by chance or purposefully (both leading to functional value) to select particular elements and combinations of elements for particular functions—magnesium in chlorophyll, silver salts in photographic film, selenium semiconductors for photocells, etc.—to make energy connections between regions at different temperatures through radiative (photon) emission, transfer and absorption.

Now, once light has been captured and a field gradient of any permanence has been achieved by constraining diffusion of y^- back to x^+ (in effect by holding different charges in separate compartments where we say that there is an established electric potential difference between x^+ and y^-), this separation can be used to do work. The system is only valuable, of course, if it is built with a barrier in space to prevent reverse diffusion, for example a membrane.

An extremely interesting process is the direct photodecomposition of water by sunlight (a real process, very inefficient in man's hands as yet but not so in biological systems)

$$2H_2O + (\text{sunlight}) \longrightarrow 2H_2 + O_2.$$

If this process could be made efficient we would obtain a practically endless and cheap source of fuel, converting light into thermal power, electricity and so on. The solution to this problem must be the use of catalysts that trap the light and activate the reactions (see Section 7.10.6), resulting initially in physical separation in space of charges. Promising results have been obtained, for example with finely produced titanium dioxide activated with platinum, as well as with other systems, but we are still far from a good, practical, efficient and cheap process. The charge separation, the energy of the sun, can be transformed into the energy of two highly reactive chemicals, H_2 and O_2 which, if mixed, go back (spontaneously) to H_2O (by sparking) with further release of energy that can be used to drive flow as is normally illustrated by a petrol engine (see Chapter 3).

The sequence of this or any equivalent process is:

(1) absorption of light in compartment I (inorganic or organic phase);

(2) local separation of charge (electrons from atoms), that is, storage of energy giving a capacity for work using (electrostatic) field energy within compartment I;

(3) flow of electrons (charge) to chemical reaction sites in two separate compartments II and III to give, for example, separated H^+ and OH^- (or $H^{\cdot}$ and $OH^{\cdot}$ which can give $H_2 + O_2$); flow is now generated in a structured field (note H^+ flow is equally useful);

(4) motion of separated electrolyte charges (H^+ and OH^-) under the (electrostatic) field to transduction sites;

(5) reaction of separate electrolytes, H^+ and OH^-, to give energised neutral chemicals; transfer of field energy to chemical bond energy.

Putting it in a different way,

(1)+(2) photocell gives *electronic* energy (excitation, diffusion, and then storage of energy in separate fixed x^+ and y^-);

(3)+(4) electronic energy gives *electrolytic energy* (free charges) and further spatial separation, that is, an electrolytic field;

(5): Electrolytic energy gives *chemical* bond energy by directed flow in the field

Once the energy capture reaches chemical bond energy we can manipulate energised chemicals almost as we wish (Fig. 7.46) since we know how to control the transfer and release of energy from chemicals, $CH_4 + O_2$ for example, by using connections between compartments and catalysts, all of which depend on selection of chemicals within compartments as described in parts B and C of this chapter. Industry and biology use such chemical energy in their syntheses by placing appropriate *catalysts* in special places to drive specific reactions. We deal with this problem again in Chapter 10 where we evaluate each compartment separately, once it has a fixed energy store, as if it were at local equilibrium. This is the basis of our discussion in Section 6.3.1.

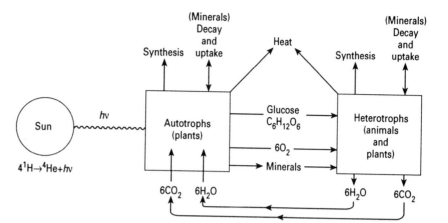

Fig. 7.46 The main routes of energy in the biosphere within chemicals.

7.19.2 Energy flow and shape

We have already distinguished between a plant and an animal in terms of the strength of homeostasis of metabolic activity (Section 7.18). Here we note additionally that an animal maintains soft material in a tensed state by a constant supply of energy, while a plant forms effectively much non-living structure, cellulose (wood), which once made maintains shape until it decays. A plant then uses energy to generate shape, but very little in its upkeep. The shape has qualities not unlike those of a mollusc shell. An animal uses energy to maintain tension, constantly renewing fields and flowing materials for resynthesis of structural parts (proteins and bone). To maintain the shape in harmony with activity requires a continuous supply of messages, so that nerves and brain and muscles and bone cannot be idle for long without deterioration. The energy cost of this constant activity is immense, but this is the nature of animal life—even to maintain structure it needs energy.

7.20 Thermodynamics and organisation

The essence of energy capture as described here and directed, for example, to synthesis, must not be thought to be against the laws of thermodynamics outlined in Chapter 3 and used to describe systems in Chapters 4–6. The above light-driven processes are accompanied by energy degradation (Section 3.7). In fact, if we view some activity such as that of living systems as driven by light, then what the activity does is to slow down the fastest transformation of light to heat in an energy exchanger. The steps are

$$\text{Light} + \text{material} \longrightarrow \text{Organisation} \longrightarrow \text{Material} + \text{heat}$$

so that, overall,

$$\text{Light (large quanta)} \longrightarrow \text{Heat (many more small quanta)}.$$

This is a transformation that is creating entropy. The hold-up within a steady state of material and energy called here *organisation* (not a state of very long life) is one described by *slow (even minimum) rate of entropy production*. (An equilibrium or a stationary state does not produce entropy since it is a condition of no change.) In contrast very fast change of light to heat occurs in the steady state of Earth's atmosphere.

7.21 Plans and information

There is a final step required if a kinetic scheme is to produce a particular observable *self-organised* dynamic (energised) construct such as an individual of a species. There has to be a means of generating a concerted relationship between all the activities based on energy and material supply within structural constraints and including the hidden systems of feedback control. The plan, a source of information connected to the development, is of the essence of an organism. We have seen how this might operate in Section 7.15 and had in mind there the DNA of cells. There is another form of planning that is not an unalterable instruction but a response to a possible environmental event. This planning is based on information and is processed by the animal nervous system, by the brain in its most refined form, into a memory. We shall not attempt to give any explanation of it as yet—see Chapter 13.

7.22 Thermodynamic efficiency and the ability to change state

We have now concluded our description of a balanced state of flow in a self-assembling system. There are two outstanding matters remaining, one of which we addressed in Chapter 3, thermodynamic efficiency, related to speed of a process, but the relative stability of a homeostatic state in response to environmental change is equally important. In Chapter 4 and again in Chapter 6 we discussed abrupt changes of physical or chemical states due to alterations of relative free energies of phases or components on change of conditions, temperature and composition. A strictly parallel problem must arise for a feedback steady state: it is only stable under particular conditions of energy and material input. A considerable problem is the maintenance or destabilisation of the environment for one flow system by *co-existing flows* that use the same materials and energy sources. Any one system could gain advantage either from greater speed or greater efficiency. This is looked upon by man as competition for resources, where competition depends on a combination of speed and efficiency, but we must also see that there is the possibility of a compromise combination of flow systems for mutual gain.

As explained in Chapter 3, all change that involves transfer of heat has an *absolute* limitation on the efficiency (thermodynamic) of utilising the energy of change to do work. (Here work is any activity involving mechanical machinery as explained in Section 3.6.) It was also shown that this absolute efficiency applies only to a system that is running reversibly—all change is infinitesimally small so that all parts are in equilibrium—implying that it is infinitely slow. This cannot be the condition of real systems where the rate of change is considerable. Their ability to do work is then further reduced from the *thermodynamic* efficiency (Section 3.6.2); hence we have to ask for the advantages of speed as opposed to efficiency. For example, maintenance of a steady state must depend on considerable rates of energy transfer to work and its survival can depend on speed of action so as to defeat destructive change. The system then requires selective catalysts. Lying behind such considerations is a subtle matter of element selection.

If we define efficiency of a system in terms of chemical products, that is simply in terms of growth, then very simple systems such as bacterial cells will be seen to be most efficient even when working quite quickly. Much more complex systems such as higher animals are in thermodynamic terms much less efficient but their more elaborate control mechanisms (energy dissipating without synthesis) give them great survival strength. Rather than competing it will be seen later that the two types of activity of higher and lower organisms can support one another since the first can supply chemicals while the second can supply protection. The natural selection of chemical elements in the two can be quite different.

The organised systems we have described are based on feedback maintenance of a steady state. The condition is sometimes called homeostatic. We shall use this word from time to time but it is unfortunate in that it appears to imply an inability to fluctuate. In effect it is a property of many feedback circuits that they are bound to oscillate around a mean, so that they are always searching. Moreover, they can be adaptable, switching to a new set of mean values of observables when the environment changes. The switch-over can be abrupt just as we have seen for a phase change of, say, ice $\rightleftarrows$ water at $0°C$ (see Chapter 3). Unfortunately, a small change can be such that the organisation collapses. (Through fluctuations a feedback circuit can go to a position of no return. A possibly helpful analogy is a switch from an ordered phase to a completely disordered one, for example CO_2 (crystal) $\rightarrow CO_2$ (gas), with a very small change in temperature.) The problem with the survival of a steady state is that effectively it is like a non-equilibrium ordered state; all its parts are co-operative. Dynamic co-operativity is at the heart of organisation; static co-operativity is at the heart of stationary states of matter. Around us we observe both, often in combination, that is, dynamic functioning within more or less static confines, all unstable with respect to their environment. Clearly, adaptability of organisation has some advantages but they can only be gained if any one condition is somewhat weak in organisational strength.

We conclude this chapter with a section summarising our attitude to organisation since we have now placed within kinetics the foundation of the most refined system of the natural selection of chemical elements—that for the *survival* of a self-reproducing organism. We have shown that it must be based on directed flow, that is, directed to the purposes of survival. It requires material for construction and energy to build chemicals but some of these chemicals must also generate plans for the building activity. The plans will be described in Chapters 9 to 13 and they are of two kinds—first a written message as in DNA and second a defined connectivity between signals and actions, messages and effector systems, as illustrated by the nervous system up to the brain and back to muscles. We have also hinted that there can be many different types of solution to organisational problems.

7.23 The trapping of energy and the evolution of organisation: a summary

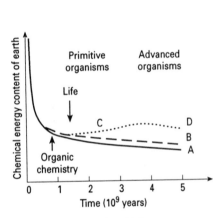

Fig. 7.47 The way in which energy becomes held up in unstable chemicals and then in organised systems. A, No organic chemistry; B, Abiotic organic chemistry; C→D, evolving life. (Not to scale.)

While we are interested—almost in passing—in the fundamental initial conditions, the big bang, which originated our present situation, we are in this book mostly concerned with the way the local Earth, especially its surface condition, has evolved geologically, biologically and through man's endeavours. Much of the development is controlled by kinetic factors. At first we want to separate the third aspect, man's endeavours, from the other two since the time-scale is different. Man had little effect on the Earth's surface until some 150–200 years ago. The Earth is 5×10^9 years old and biological systems may be 4×10^9 years old. We shall consider that from about 4×10^9 years ago the system of Earth and sun had settled sufficiently (in fact the planetary system is a crude organisation not a structure) so that we can treat the major part of the surface of the Earth as a system of local compartments under the effect of steady-state energy transfer from the Earth's core and from the sun. The energy capture was at first abiotic and, while some of it originated the creation of a steady state of organic chemicals, the major effect was the development of the air/sea/land surface (Fig. 7.47, line B). In the figure we see that line B continues to the present day. Line A represents the change in the absence of the sun. Line C then shows the development of early life and line CD the continuation of this early life after advanced forms appeared. For the most part, true organisation begins with life and is represented by a continuously increasing hold-up of the degradation of the sun's energy to heat.

While this diagram represents a general conclusion to the ever-heightening of the natural (useful) selection of the chemical elements and energy against the background of thermodynamics and kinetic controls, we have now to show in some depth the way in which these principles generated the observable chemical materials we see around us, especially in living systems. This is the subject matter of Part II of this book.

We begin Part II with a description of the Earth's formation in Chapter 8, and include the way in which the environment has evolved describing both equilibrated and non-equilibrated changes. This effectively is the story of the evolution of inorganic chemicals while cooling, reheating and cooling to form the Earth. Initial processes are at high temperature (Chapter 4). Through later energisation of these chemicals using sunlight at low temperature in the presence of water organic chemicals were generated as explained in Chapter 9. This is abiotic organic chemistry, though we use bioorganic chemistry to illustrate principles. It affected the environment to some degree. The coming of primitive life described in Chapters 10 and 11 gave a great impetus to organic and inorganic chemical diversity. This activity, which is not under thermodynamic control, might have led to a huge number of unrelated component chemicals, but in fact it has led to living organisation in which selective uptake and activity of chemical elements is ever more refined in species. There are new restrictions on variance, degrees of freedom, which we must understand (Table 7.27). The chemicals generated by life, amongst other factors, later changed the environment which created life and the environment and life have thus been forced to evolve together. This is the subject of Chapters 12 and 13. A further gross refinement of the use of chemicals in organisation has come about through the emergence of man.

Table 7.27 Variance of systems

At equilibrium	Steady state
Component composition and amounts: limited by transfer equilibria	Component composition and amounts: limited by uptake and output rates
Phase structure: limited by transfer equilibria	Compartment structure: limited by transfer in and out rates; structural units self-assembled
Temperature and pressure	Temperature and pressure
Fields	Fields: limited in large part by internal controls over rates (hidden systems) including physical and chemical barriers and plans* (DNA)

* Sets of instructions, for example a code, have to be given to existing machinery for the code to work, even though the code and the machine reproduce.

Man designs his own organisation (Chapter 14). We shall look on this activity as inevitably interacting with the environment and contributing to the cycling of material and the dissipation of energy (Chapter 15). Only in Chapter 16, the conclusion of the book, do we attempt to put Parts I and II together asking if the principles we have established in Part I and the practices in Part II can help us to see the possible future of the natural selection of the chemical elements. Throughout we have to concentrate on the reduction of variables, either in a static thermodynamically controlled system or in a steady state of flow (Table 7.27), so that the observed systems are not an infinite chaotic variety. It is our wish to use diagrams such as that of Fig. 7.47 to explain known patterns of change, such as that in Fig. 10.2, even of living systems.

We are now in a position to analyse the relationship between two statements.

(1) A condensed phase no matter how complex the component composition is a co-operative static unity.

(2) An organisation (or organism) no matter how complex its component and compartment composition is a co-operative dynamic unity.

To complete this part of the book we remind the reader of our first diagram, Fig. 1.1, which has now developed to Fig. 7.48 where there remain the fundamental problems of the nature of energy, matter and flow.

(In part of this chapter we benefited from assistance from Dr D. Edwards of Wadham College, Oxford.)

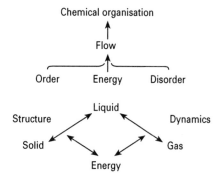

Fig. 7.48 A reminder of the central problems known in classical times (see Fig. 1.1 and Section 1.1), but now expressed in modern language leaving inherent unresolved problems of the evolution of the materials around us.

Further reading

The topics concerned in this chapter only in part are orthodox and there is no single appropriate reference that deals with them all.

Some parts were discussed in our previous book.

1. Fraústo da Silva, J. J. R. and Williams, R. J. P. (1991 (reprinted 1994). *The biological chemistry of the elements inorganic chemistry of life*, Oxford University Press, Oxford. See especially Part I, Chapters 2–7.

On catalysis, energy uptake and production and their control in biology the following reference provides excellent college-level reading:

2. Purves, W. K., Orians, G. H. and Heller, H. C. (1992). *Life—the science of biology*. Sinauer Associates, Sunderland (Massachusetts). Chapters 6, 7, 8.
3. Schuster, P. (1982). Irreversible thermodynamics—an overview. In Hoppe, W., Lochmann, W. H., Markl, H. and Ziegler, H. (eds.). *Biophysics*. Springer-Verlag, Berlin. Chapter 8.
4. Nicolis, G. and Prigogine, I. (1989). *Exploring complexity*. W. H. Freeman and Co, New York.
5. Prigogine, I. and Stengers, I. (1988). *Order out of chaos*, Heinemann.
6. Raines, R. T. and Hanse, D. E. (1988). An intuitive approach to steady-state kinetics. *J. Chem. Educ.* **65**, 757.

On neural networks the following paper is a good introduction to the subject and provides a comprehensive set of references:

7. Spining, M. T., Darsey, J. A., Sumpter, B. G. and Woid, D. W. (1994). Opening up the black box of artificial neural networks. *J. Chem. Educ.* **71**, 406–11.

Classical approaches to chemical kinetics are found in the following widely used textbooks (amongst several others).

8. Atkins, P. W. (1994). *Physical chemistry*, 5th edn. Oxford University Press, Oxford. Part 3.
9. Laidler, K. J. (1987). *Chemical kinetics*. Harper and Row, New York.
10. Connors, K. A. (1992). *Chemical kinetics in solution*. VCH, New York.

On biological kinetics the following texts provide ample information.

11. Roberts, D. B. (1977). *Enzyme kinetics*. Cambridge University Press, Cambridge.
12. Harrison, L. G. (1993). *Kinetic theory of living pattern*. Cambridge University Press, Cambridge.
13. *Structure and Bonding* **75**, 1–224 (1991). A series of articles on electron-transfer kinetics in biological systems.
14. Williams, R. J. P. (1988). Proton circuits in biological energy interconversions. *Annual Rev. Biophys. Biophys. Chem.* **17**, 71–97.

A thought-provoking text of a more general nature is:

15. Coveney, P. and Highfield, R. (1990). *The arrow of time*. W. H. Allen, London. See especially Chapters 6 and 7.
16. Weber, B. H., Depew, D. J. and Smith, J. D. (eds.) (1988). *Entropy, information and evolution*. The MIT Press, Cambridge, MA, USA.
17. Depew, D. J. and Weber, B. H. (1995). *Darwinism evolving*. The MIT Press, Cambridge, MA, USA.

Part II

The observed natural selection of chemical elements in both abiotic and biotic systems during their evolution

The evolution of inorganic chemicals on Earth

Come and I shall tell thee first of all the beginning of the sun and the sources from which have sprung all the things we now behold—the earth and the billowy sea, the damp vapour and the titan air that binds his circle fast round all things . . .

Empedocles of Agrigentum (*c.* 450 BC)

8.1 Introduction

Given the principles of chemistry, thermodynamics and kinetics outlined in the previous chapters, we can only follow the natural selection of the chemical elements at a sophisticated inorganic level by homing in on Earth. Earth itself may or may not be very similar to millions of other planets, but this we do not know. We know that it is quite different in inorganic chemistry from all other planets in the solar system and differs considerably from the moon. We can sense with some security too that only conditions of chemistry and physics, especially temperature, similar to those on Earth could have given rise, after cooling below 100°C at the surface, to the equally sophisticated organic chemistry that we describe in Chapter 9, and thus, with the pre-existing inorganic chemistry, could have generated life which, we shall see, illustrates the most refined chemical selection we know. Earth itself could therefore be a unique chemical laboratory formed 4.5×10^9 years ago and life, which probably appeared around 4.0 to 3.5×10^9 years ago, has then had a very long time to search out optimal element selection for survival. As we shall show, the steps that have occurred up to man exhibit an almost inevitable progression in the natural selection of the elements given the conditions around and on Earth. Moreover, given the time taken for man to appear, it could be that he too is unique as a self-conscious living organism. This slow development has depended on the approximate constancy of physical conditions over the whole period of 4.5×10^9 years. Earth has escaped from fluctuations that would be adverse for organic chemistry and life, yet fluctuations occur all the time in the universe. To what do we owe this circumstance? We must start from the origin of the elements in order to appreciate all the subsequent developments.

8.2 The formation of the elements in the universe

The Earth today has a gross composition that is related to that of the universe (Fig. 8.1), although several light elements, H, He, C and N, that occur in gaseous form, have very largely escaped from Earth and iron is more dominant especially at the centre. In this chapter we shall consider how this 'inorganic' Earth evolved. In essence, the major development corresponds to the evolution of inorganic chemicals effectively before the evolution of sophisticated organic chemicals began, before there was dioxygen in Earth's atmosphere, and long before man began his exploration of chemistry. The earliest event that we need to discuss following the big bang, now generally accepted to be the origin of the

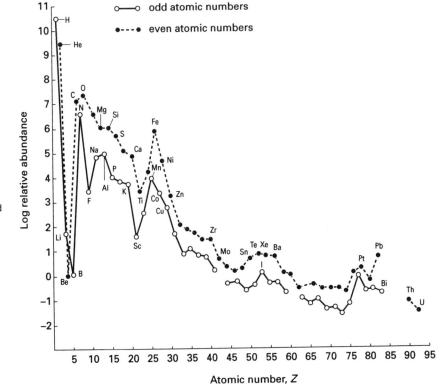

Fig. 8.1 Relative abundances of the elements in the universe (based on log (abundance of Si) = 6). Curve (●), even atomic numbers; curve (open circles), odd atomic numbers.

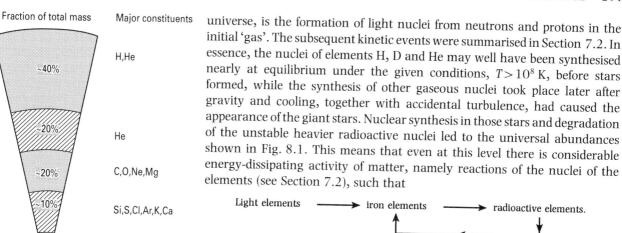

Fig. 8.2 Schematic diagram showing the 'shell' structure of a heavy star (around 25 solar masses) at the end of its evolution, just prior to a supernova explosion. The fraction of the total mass contained in each shell and the principal elements present are shown. (Of the 20 shown, 12 to 14 are essential for all life!)

universe, is the formation of light nuclei from neutrons and protons in the initial 'gas'. The subsequent kinetic events were summarised in Section 7.2. In essence, the nuclei of elements H, D and He may well have been synthesised nearly at equilibrium under the given conditions, $T > 10^8$ K, before stars formed, while the synthesis of other gaseous nuclei took place later after gravity and cooling, together with accidental turbulence, had caused the appearance of the giant stars. Nuclear synthesis in those stars and degradation of the unstable heavier radioactive nuclei led to the universal abundances shown in Fig. 8.1. This means that even at this level there is considerable energy-dissipating activity of matter, namely reactions of the nuclei of the elements (see Section 7.2), such that

Light elements $\longrightarrow$ iron elements $\longrightarrow$ radioactive elements.

Eventually, since there is a huge reserve of hydrogen in the universe, the radioactive heavy elements should come to a steady-state concentration while iron and related elements accumulate, but the universe is very far from that condition and expansion may reduce the numbers of giant stars so that development stops. Parts of the process have been recorded in the existing giant stars where some segregation of the elements has already occurred under gravitational forces (Fig. 8.2). As we shall see, gravity also plays a central role in the evolution of the Earth's mixture of inorganic chemicals. It is the relative *abundance* of elements formed from the nuclei more than 5×10^9 years ago and associated with the sun that has, in large part, decided the further natural selection of the elements on Earth.

8.3 The abundance of elements in the universe

The abundance of the elements is then a result of the greater or lesser ease of formation of their nuclei, that is, of the pathways and rates at which the nuclear particles, protons (p) and neutrons (n), have combined until the present day to give the different nuclei $[n(p)/m(n)]$. The pathway of synthesis (Fig. 7.3) to a large degree leaves in very low abundance three light nuclei, Li 3(p)3(n), Be 4(p)4(n) and B 5(p)5(n) (for reasons we need not discuss here), and then generates a series of abundant stable light nuclei—C, N, O and so on—that forms the basis of further synthesis. It is observed that certain later combinations of (p) and (n) also give very stable nuclei, so that during the process of building from $n(p) = 1$ towards $n(p)$ equal to about 100 some special nuclei accumulate. The numbers $n(n) + m(p)$ of particularly stable forms are 4(He), 12(C), 16(O), 24(Mg), 32(Si) and 56(Fe) (Fig. 8.1). (Notice again the peculiarity of the numbers 4, 4×3, 4×4, $4 \times 3 \times 2$, $4 \times 4 \times 2$ and $4 \times 7 \times 2$; the repeat of the number 4(He) is rather intriguing and suggests some kind of 'shell-like' structure of nuclei.) These stable nuclear forms mean that C, O, Mg, Si and Fe with H must always have had a dominant role to play in the universe, on Earth and in life. The peculiarity of the selection of atomic nuclei at temperatures below 10^6 K (see Fig. 7.1) is then that they owe stability to interactions between certain ratios of protons and neutrons. The so-called 'strong forces' acting can be described theoretically, but these explanations are, as stated in Chapter 1, beyond simple conceptual considerations. Already at iron, which has 26 protons and 30 neutrons and hence atomic weight 56, stability requires an excess of neutrons over protons. The excess increases until at element 92 there are 92 protons and over 140 neutrons. Such heavy elements turn out to be unstable and exist only through the flux of neutrons and protons that bombards the lighter elements (see Fig. 7.3). It is a characteristic of matter that, as it cools within a flux of radiation (here particles), some unstable systems result. This is a consequence of kinetic trapping (Chapter 7). The abundances represent the first step of natural selection, which will proceed in the pattern of Fig. 7.3 for unknowable lengths of time to come. If we ask why are there 92 'natural' elements then the answer is connected to fundamental constants in our universe for which we have no explanation.

There is little more of value that we can add, therefore, concerning elements in the universe, and our next step is to treat the elements in the context of the Earth in an inhomogeneous universe. Since Earth was formed by a particular perturbation of a particular star at a given time we must expect that the distribution of elements in it is to some degree fortuitous and not precisely the same as anywhere else in the universe. We have pointed out in Chapters 4–6 that often when we discuss such small compartments within a larger whole we are describing systems very far from universal equilibrium in both chemical and physical conditions but we may still usefully apply a local discussion of order/disorder equilibria. Thus we will turn to Earth asking about its chemical not its nuclear stability. It is worth stressing again that the nuclear numbers do govern abundances generally but *not* chemical stability in compounds; chemistry is governed by combinations of atoms that are controlled by the limitations upon how electrons fill space (and time) around nuclei (Chapters 1 and 2).

Now, in the above we have assumed that once nuclei are cooled they cannot react together. For example, we cannot carry out the reaction

$$2\,^{16}O \longrightarrow \,^{32}S$$

except at the very high temperatures of the stars. Chemistry (and life) is a protected Earth temperature subject! Little of what we study would exist at 1000°C, and no water solutions exist much above 100°C for long. The temperatures at which nuclear fusion or fission of light elements can occur are much higher, $> 10^6$ K. All of this tells us that much of the chemistry we study is confined to a very modest temperature range (see Fig. 8.3) where the compounds that have formed on cooling could be either thermodynamically or only kinetically stable (Sections 3.4.1 and 7.1). (In recent times, the last 50 years, man has challenged this limitation and has engaged on a minute scale in both fusion and fission of nuclei (see Chapter 1 and references 1 and 2 in 'Further reading').) Here thermodynamically stable means that the species in their observed local environment cannot be changed to other species unless energy is applied to them. Examples are the noble gases in our atmosphere and many rocks, for example, aluminosilicates. Kinetically stable means that the species to which we refer in their local environment can change spontaneously but very slowly giving out energy to their environment, for example the free iron of the Earth's core in the presence of dioxygen should give iron oxides but

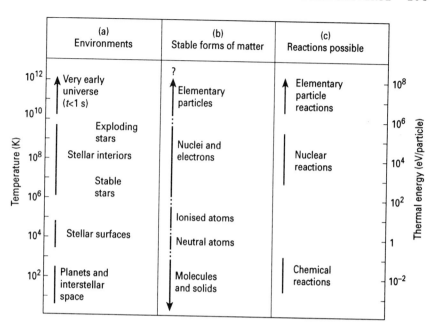

Fig. 8.3 Energy and temperature scales for chemical and nuclear processes. The scale on the left shows temperature, and that on the right indicates the average thermal energy for the particles present. Column (a) shows typical environments with different temperatures; (b) shows the stable forms of matter present; (c) indicates the types of reaction possible. (1.0 eV = 23.06 kcals mol^{-1}.)

because of kinetic barriers it does so only very slowly. These last two free elements exist on Earth only because, under the existing conditions today of temperature and pressure, there are compartmental barriers to their spontaneous change. All known living systems are composed of organic chemicals that are also only *kinetically* stable due to chemical trapping in our oxidising environment.

We do not usually recognise that the universe, as a whole, has kinetic stability only. As stated above, on an immense time-scale the universe is changing its elemental chemical composition very slowly from one based on the high-temperature starting element, hydrogen, towards one based on the most stable low-temperature element—iron. This means that all the time and in widely distant places compounds are being formed that remain unchanged for long periods and are effectively components (see Section 4.3.2). When later in the book we consider life forms as they are here and now on Earth, mainly organic compounds, we have to remember that there could be universes, and there certainly are bits of our universe, with such a different history of composition and temperature that the chemistry we associate with life could not exist there. All the planets in the solar system, except Earth, are in this class. But could there be some very different life system where element abundances and conditions are very different? What are the chemical limits of a self-assembling, reproducing organisation that we call living? We do not know yet; however, we do know that, unlike water, which is thermodynamically stable *on the surface of the Earth, the organic chemicals of life are only kinetically stable here.* We are not saying that this difference is the single fundamental feature of living systems (another is a temperature and a chemical composition that generated on Earth a large liquid water phase), but it is one fundamental feature, a controlled instability that is common to a limited set of compounds, largely organic compounds, in those life forms we know. This gives us one strong reason for devoting a separate chapter to the organic chemicals on Earth (see Chapter 9), that is, basically the peculiarity of the H, C, N, O distribution in compounds, and a further four chapters to an enquiry into Earth's biological chemistry (Chapters 10–13) where some 20 elements participate in steady, mainly aqueous, flowing states of remarkably controlled chemical selection. It will be shown that the essential *inorganic* elements of life are also in an unstable condition in living systems. Only towards its end will this chapter concern itself with such matters. Our immediate involvement is with the inorganic chemical form of Earth and its history, which has a closer connection to equilibria.

8.4 The initial formation of compounds and condensates

From nuclei and electrons atoms formed in stars at temperatures below 10^8 K (see Fig. 7.1). The dispersion of some of the material formed in the giant stars gave rise now and then to gaseous, ejected, plumes of atoms and then led to rapid cooling of some of this elementary atomic material. It followed that small molecules formed probably very close to equilibrium in the hotter regions ($< 3000°C$), depending on the relative affinities and abundance of elements (Table 8.1). Thus there was generated, approximately at equilibrium, *mixtures* of chemicals in small particles, liquids and solids with phase boundaries. Gravity next helped to form the 'cold' meteorites within which the low temperature controlled, or rather limited, the potential of chemicals for further reaction since some became hidden in the cores as solids while others remained exposed and, of course, the separate solids could not react with one another. (This was the beginning of the development of compartments (Section 6.3).) It is not unreasonable to consider much of this process as if it occurred at equilibrium.

Table 8.1 Development of the inorganic chemicals of the Earth in the solar system space

Element	Compounds
H	H_2 and H_2O, mostly lost (10^{-8})*
He	Lost
C	CO, (CO_2), mostly lost (10^{-4})*
N	N_2, $(NH_3$ and HCN?), mostly lost (10^{-4})*
O	MgO, SiO_2, Al_2O_3, CO, (CO_2)
Ca, Mg, Al, P, Si, Na, K, Ti, Cr, Mn	Oxides, silicates, $Ca_3(PO_4)_2$, in crust and mantle (and in meteorites)
S	SO_2, much lost, FeS
Cl	NaCl, HCl
Fe/Ni	Alloys based on Fe, since all oxygen was removed by C, Mg, Al, Si (also in certain meteorites)

* The approximate remaining fraction of elements is given in brackets.

8.4.1 The affinity of the elements for one another at different temperatures

Before we consider the way in which the Earth was formed we must appreciate that, as well as combining in different proportions (Chapter 1), elements have preferred affinities for one another at a fixed temperature (Chapter 2). (Their relative affinities, however, depend on temperature (Chapter 3).) A major concern of inorganic and indeed organic chemistry is the relative affinity of elements for oxygen since this is one of the three or four most abundant elements (Fig. 8.1) and combination with it gives the compounds of greatest stability for most elements (Section 3.4). Therefore we shall examine first the affinity and the change in relative affinity for oxygen of all elements with temperature, which can be done in a very convenient way using Ellingham diagrams such as that in Fig. 8.4 (and see Fig. 3.12). The affinity is a measure of the *equilibrium* partial pressure of oxygen above a mixture of the element and its oxide and the affinity scale can, therefore, be based on the logarithm of the oxygen partial pressure $\log p_{O_2}$, for example for the reactions

$$2M + O_2 \rightleftharpoons 2MO.$$

(A low partial pressure represents a high affinity (Section 3.5).) These affinities, while equilibrium conditions held, guided at least in part the development of the Earth. The slopes of the lines in Fig. 8.4 are related to the differences in entropy, mainly differences in the number of gaseous molecules, between the two sides of the above equation (see Section 3.2).

At high temperatures, say, $> 3000°C$, the affinity of oxygen for C (as CO) is greater than that for all other elements. This means that CO must have dominated many gas clouds from stars. At these high temperatures some H_2O would also have formed due to the abundance of hydrogen, but other oxides would have been unstable. On cooling to around $3000°C$ the elements of highest affinity for oxygen after carbon are Al, Ti, Si, Ca and Mg, and then, in sequence, (V, Cr), Mn, Fe, H, (Co, Ni), Zn and Cu; they should have formed oxides in this order. At still lower temperatures, below $1000°C$, the order of affinity changes but only somewhat for the metals, so that at room temperatures the affinity for oxygen follows the order

Ca, Mg, Al, Ti, Si, Na, K and then

(V, Cr), Mn, Zn, C, Fe, H, (Co, Ni), Cu, Ag, Au.

Carbon (as CO_2 or CO) now has a relatively low affinity for oxygen compared to the first group of elements, but remains above iron while hydrogen is still somewhat lower. To a first approximation these series do show which elements are to be found in meteorites or on Earth to this day as oxides or within oxyanions. The problem as to why further particular oxides did or did not form during this cooling is made more difficult since there may not have been enough oxygen for all elements to combine to give oxides. Again, kinetic limitations retained some oxygen with carbon. In any case, a few elements condensed before they could react with oxygen (from CO?), for example Fe. Some of the other elements and compounds that condensed at high temperature may have remained thermodynamically stable on cooling but others persisted in states kinetically protected through chemical or physical barriers, for example some NH_3 (Section 6.3.3). A first distinction must

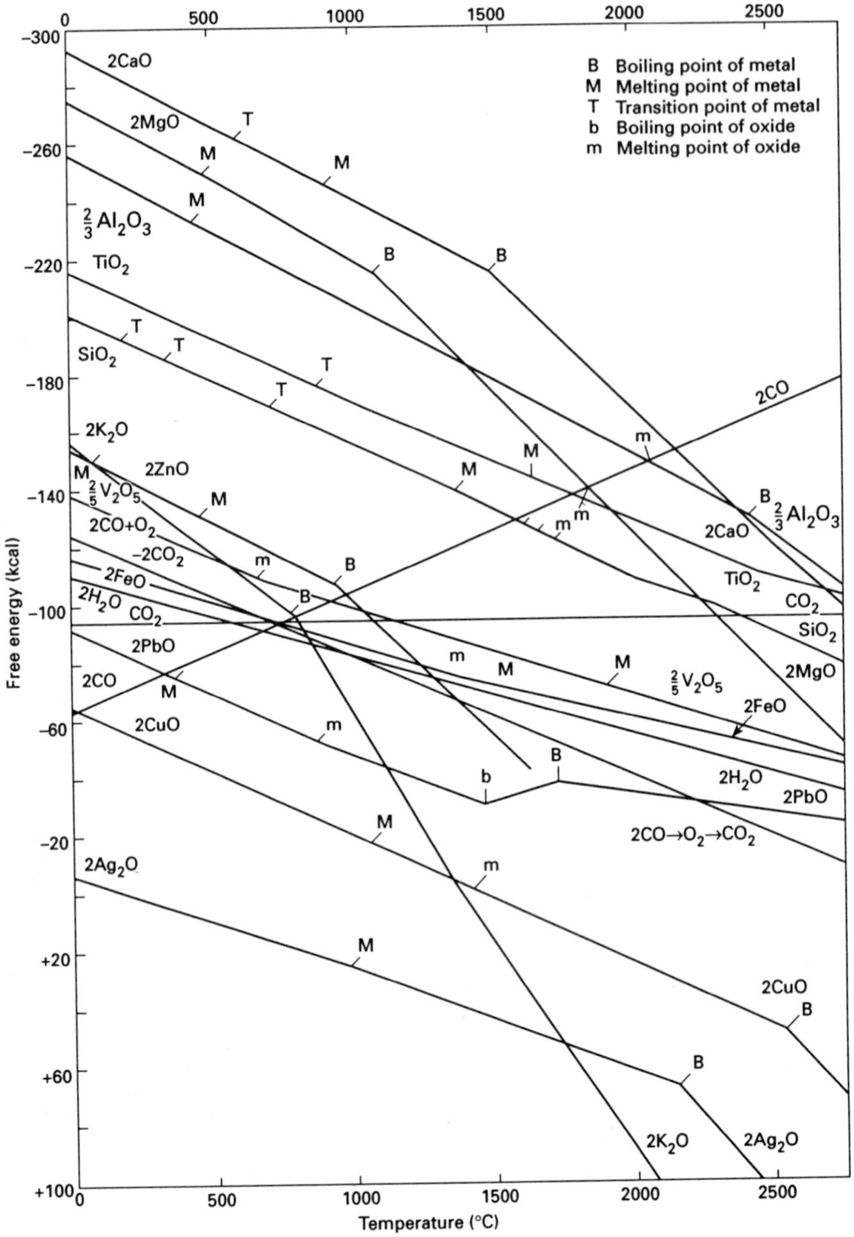

Fig. 8.4 The free energy of formation of oxides per gram mole of O_2 as a function of temperature (Ellingham diagram). The more negative the value of free energy, the more probable the formation of the corresponding compound (see Chapter 5). The slopes are dependent very largely on the entropy changes in the reactions (see Section 3.5), that is, the changes in the number of gas molecules in the reactions.

therefore be made between those abundant elements that formed compounds roughly in accord with equilibria and came out as solids, for example the stable basic often mixed oxides of Ca, Mg, Al, Ti, Si, all of which bind oxygen more strongly than iron; those that remained as gases such as CO, CO_2, H_2, N_2 also close to equilibrium; elements such as Fe, Au and Pt that remained free in the metallic state; and a few non-metals and metals that were trapped kinetically.

The different small solid particles that were formed first in space came to form the classes of meteorites of two major and one minor type (Table 8.2) that we still observe. We may suppose then that the condensed compounds formed at about 1500°C to give meteorites were not far from a physical/chemical

Table 8.2 Relative abundances in the solar nebula of elements critical to the formation of the Earth

Element	Relative abundance*
Gases	
H	250 000
C	10.0
N	3.0
Stone meteorites (1.0)†	
O	15.0
Mg	0.8
Al	0.08
Si	1.0
Sulphide (Fe) meteorites (0.1)†; iron meteorites (0.7)†	
S	0.4
Fe	0.6

* Since the dominant forms of H (H_2 and H_2O), C (CO and CO_2) and N (N_2 and NH_3) are gases, most of these elements, common in the universe, were lost upon the formation of the Earth. Only about one part in 20 000 of N and one part in 10 000 of C have been retained (relative to Si), yet they are central to life.

† Many meteorites are composites of oxide and iron alloy phases and there are meteorites that contain considerable amounts of carbon (see reference 2 in 'Further reading'). After each meteorite its relative proportion to stone meteorites is given in brackets.

equilibrium distribution, with much H_2, N_2, CO_2 and perhaps H_2O too, and the inert gases He, Ne and Ar very thinly dispersed around them. (This description still holds quite well if we allow for the formation on Earth of mixed oxides, for example silicates (see Section 2.11) and hydroxides as well as sulphides (see below).)

There is one peculiarity in the early formation of compounds—namely the appearance of much methane. If we assume that the pressure of hydrogen remained high at first then it may be that C—H bonds developed as the temperature dropped below 2000°C. This could only occur if the reactivity was still high enough so that the following reactions would take place (but see also Section 8.5):

$$[Ca, Mg, Al] + nCO \longrightarrow [C]_n + [Ca, Mg, Al \cdot O],$$
$$[C]_n + H_2 \longrightarrow C/H \text{ compounds}.$$

We consider the further chemistry of carbon compounds later, but an essential feature of the natural selection of chemical elements is the manner of formation of C—H and O—H bonds, which must have appeared through a combination of early reactions, perhaps at equilibrium. Somewhat later neither C—H nor O—H bonds were stable relative to combination with other elements, and therefore physical and chemical trapping must have prevented reactions such as

$$CH_4 \longrightarrow C + H_2 \uparrow,$$
$$Fe + H_2O \longrightarrow FeO + H_2 \uparrow.$$

A similar problem concerns the formation of NH_3 rather than N_2, and of H_2S rather than other sulphides. The only thermodynamically stable hydrides are those of a few non-metals, for example chlorine, but they are not stable in the presence of oxygen, that is, relative to the formation of water and chlorine.

8.5 The formation of Earth

As they cooled further from the conditions of relatively high temperature at which they had formed, the meteorites had too low a gravitational pull to hold the gases but gravity now played a large role in that it caused meteorites to coalesce. Upon forming a particular planet, namely the Earth, by this coalescence (accretion) of meteorites there was at first a somewhat random mixture of particles, non-equilibrated compartments, and now the large body held a considerable atmosphere although accumulated gases escaped continuously to some degree into space. Internal radioactivity reheated the Earth to around 1000–3000°C when the molten iron (melting point, 1540°C) droplets separated from the oxide (and sulphide) layers, again under the pull of gravity (Fig. 8.5). The iron, now at Earth's centre, has remained largely molten to this day, yet was unable to react with the gases that accumulated in planet Earth's atmosphere, that is gravity removed iron from CO_2 and H_2O for example, although they could react at temperatures such as 1000°C. Earth acquired and still has, then, four major *zones*—the two solid/liquid zones, metallic iron and mineral oxides (and some sulphides, see below), the sea; and

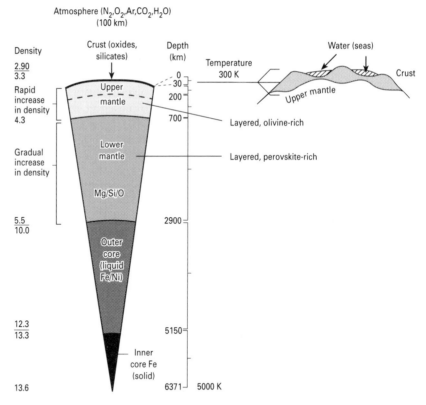

Fig. 8.5 A schematic section through the Earth. Notice that there is a huge source of energy in unbalanced chemical and physical (high-temperature) gradients of the Earth. This out-of-balance is due to the route of formation of the planet, that is, it is a kinetically controlled situation due to gravity. The out-of-balance shown in this figure has been increased by living systems via the production of dioxygen, O_2, and carbon (coal and oil) deposits. Some dramatic consequences of this physicochemical out-of-balance are earthquakes and volcanic eruptions and (possibly) life itself. See Fig. 4.14.

the atmosphere. Its total composition relative to the universe is given in Table 8.3. It is somewhat out-of-equilibrium in physical conditions and in chemistry, but this applies mostly to the uppermost levels since the core and the lower mantle are close to equilibrium given the abundance of the elements in them and the prevailing physical conditions. The upper region of the mantle is called the crust and will turn ot be of major concern for us (Fig. 8.5). This separation of compounds into major zones, which still exists today, is thus due partly to chemical equilibration, even though the temperature itself is still graded from core to surface, and partly due to gravity. Subsequent events,

Table 8.3 Ratio of logarithmic abundances on Earth relative to the universe*

Element	$\text{Log}\left\{\dfrac{[\text{Earth}]}{[\text{universe}]}\right\}$	Element	$\text{Log}\left\{\dfrac{[\text{Earth}]}{[\text{universe}]}\right\}$
H	−6.6	Si	0.0
He	−14.0	P	0.0
C	−4.0	S	−0.5
N	−6.0	Cl	−0.7
O	−0.8	Fe	+0.6
Na	0.0		
Mg	0.0		

* The logarithmic ratios are normalised to equal abundances in Earth and universe. Silicon is but 0.003% of the mass of the universe but 12.5% of the mass of Earth. (NB. log. 1 = 0.0.)

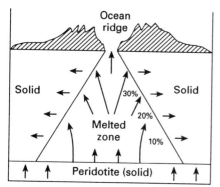

Fig. 8.6 Upwelling mantle melts to an extent that depends on the local temperature of the mantle. The percentages indicate the amount of periodotite that melts. Melting proceeds until the periodotite stops rising and starts flowing horizontally. The hotter the mantle, the deeper the melting begins. As a result, more of the mantle melts, creating a thicker crust.

activity of volcanoes and at crustal vents for example, have generated a vast number of small dispersed solid phases at the surface of the Earth that are not in equilibrium with one another or with their liquid and gaseous surrounds. The volcanoes and vents driven by the temperature gradient act against gravity to mix zones (Fig. 8.6). These are 'fluctuations', as defined above, which have given quite irregular patterns of elements locally (see below). We shall also have to consider the weathering of the crust, which, due to surface flow, also caused mixing of non-equilibrated particles.

In addition to the above some of the common metal elements that remained free from oxygen turn out to have a high affinity for sulphur, which itself does not have a very high affinity for oxygen. Hence, these metals formed a smaller sulphide zone within the upper mantle. The sulphides have a particular importance for the crust (see Section 8.5.1). Thus, five essentially non-exchanging compartments were formed in the crust: (1) magnesium–aluminium silicates, see Fig. 4.22 and 4.23; (2) gaseous oxides of carbon (plus inert nitrogen); (3) a smaller (iron) sulphide zone (until sulphur ran out); (4) metallic iron which largely remained in the centre; and (5) the water (sea) compartments. The distribution of the minority, metallic, electropositive elements (Fig. 8.7) was governed by their tendency: (1) to form particular compounds and possibly dissolve in the above as: (a) oxides, for example of Ti and Ca; (b) sulphides, for example of Zn, Cu and Cd; or (c) iron alloys, for example with Ni; or (2) their tendency to undergo reactions such as

$$M + Fe \cdot silicate = M \cdot silicate + Fe,$$
$$M + Fe \cdot sulphide = M \cdot sulphide + Fe.$$

As a result, elements more electropositive than iron (lithophiles) concentrated mostly in the silicate zone and those less electropositive (chalcophiles)

Fig. 8.7 Predicted condensation temperatures of the elements from a nebula with a total pressure of 10^{-4} bar. The symbols show the predominant solid phase formed and the temperature range of condensation. (From data in Ringwood (1978).)

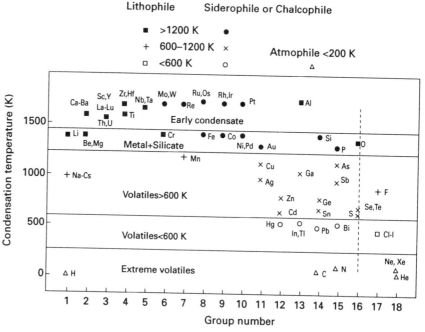

Table 8.4 Separation into analytical groups of elements*

Elements in analytical group

I	II	IIIA	IIIB	IVA	IVB	V
Cations						
Ag, Pb (W)	Hg, Cd Au, Ag Cu (Zn), As Sb, Sn Mo	Fe, Al Cr, (Si) (Be, Ti)	Mn, Co, Ni Zn	Ca	Sr, Ba	Na, K, Mg (Li, Rb, Cs)
Associated anions (pH)						
Cl^-, OH^- (0)	S^{2-} (Se^{2-}) (<1)	O^{2-} (OH^-) ($\sim$7)	S^{2-} (>7)	CO_3^{2-} ($\sim$7)	SO_4^{2-}	Cl^-, Br^-, I^-, NO_3^- soluble

* See also Table 5.11.

Table 8.5 Abundances of the elements in the Earth's crust by weight (after Gibson)

Relative abundance by weight

Element	%	Element	%
O	50	K	2.4
Si	26	Mg	1.9
Al	7.5	H	0.9
Fe	4.7	Ti	0.6
Ca	3.4	Cl	0.2
Na	2.6	P	0.1

Abundances expressed in g/(metric) ton*

<1 kg/ton	Mn, C, S, Ba, Cr, N, F, Zr, Zn, Ni, Sr, V
<100 g/ton	Cu, Y, W, Li, Rb, Hf, Ce, Pb, Th, Nd, Co, B
<10 g/ton	Mo, Br, Sn, Sc, Be, La, As, Ar, Ge
<1 g/ton	Se, Nb, Sb, U, Ta, Ga, In, Tl, Cd
~mg/ton	I, Pt metals, Ag, Bi, Hg, Te, Au, noble gases
~mg/1000 tons	Ra, Ac, Po

* The abundances of the remaining elements are conveniently expressed in g/(metric) ton (1 kg/ton = 0.1%) of crustal rock.

concentrated in the sulphide zone (see Fig. 8.18). (NB. This is close to the oxide/sulphide equilibrium exchange distribution (see Section 8.5.1).) The chalcophiles are mainly the metals of the so-called sulphide group of analytical chemistry and some softer non-metals (see Tables 8.4 and 5.10). (This fact tells us that the tendency of an element to appear as a sulphide or an oxide is much the same whether the precipitation of metals occurs from water solution or through competitive reactions of molten solids.) Elements still less electropositive than iron and that tend to form alloys with it (siderophiles), mostly nickel, concentrated in the metal phase. Small amounts of very electropositive elements, for example, gold and platinum metals, formed separate metal phases. Notice that this description closely follows the classifications of compounds of the elements given in Section 5.5.3, which were based on equilibria, and compare the distributions in analytical chemistry and in biological systems.

Elements that are gaseous or form gaseous non-metal compounds (atmophiles) (see Section 2.2) were to a large extent gradually lost to the cosmic space after the process of accretion since the gravitational attraction, even of large bodies (planets), was insufficient to retain them. This was the case for hydrogen and deuterium, for the light noble gases He and Ne, and, more slowly, for nitrogen and carbon compounds such as N_2, NH_3, CO, CO_2 and CH_4.

While these general divisions of zones and the chemicals in them are of great importance, it is most relevant for this book's purposes to refer to the very top layers of Earth since it is here that the greatest further development of the natural selection of the chemical elements occurred, that is, in the atmosphere, the seas and in part what we call the crust (Table 8.5). The crust constantly re-enters and mixes with the top zone of the mantle (so-called subduction) so that there has been progressive secondary change in it. The composition of the crust (Table 8.5) is a sum of a great number of non-equilibrated particles. Some relatively abundant elements are not readily available for reaction at the surface, of course.

There is one compartment that we have not described so far—the liquid hydrosphere. The first thing to say about it is that it is not easy to see how a water layer could have been formed given the low affinity of dihydrogen and dioxygen. The current view is that it resulted from the decomposition of hydrous silicate rocks in the remelting process, but another, perhaps as likely, possibility is that it came from the catalysed reactions of residual hydrogen gas with carbon dioxide, the form of carbon most prevalent at low temperatures (see Fig. 8.4) due to the reactions,

$$3H_2 + CO_2 \longrightarrow CH_4 + H_2O,$$

$$H_2 + CO_2 \longrightarrow CO + H_2O.$$

Simultaneously, these reactions could have generated methane, CH_4, and carbon monoxide, CO, which may well have contributed to the primordial atmosphere. Water, as ice, is found on many other planets and even in comets. This compound has the peculiarity that it forms relatively stable solid and liquid phases (and hydrates) below $100°C$ which, through the influence of gravity, have allowed it to remain continuously for 4.5×10^9 years on the surface of the low-temperature Earth in amounts that are strangely fortuitous for life (see below). As a liquid it presumably helps to control the temperature of the Earth today, especially at its surface. A different temperature balance on Earth, say $> 100°C$ (compare Venus), and the planet would have been dry long ago, while at a temperature of $-10°C$ the planet would have been covered in ice.

The detailed description of the composition of the aqueous phases on Earth is very complex and requires more analysis (see Section 8.6.2). Many elements have passed into the sea through weathering and have been deposited in the sea bed. The inorganic chemicals that remained in the sea gave rise to complex species with the organic chemicals in solution (in part from the atmosphere) and then to life. All the solutions are dilute (see Table 8.8).

8.5.1 The formation of sulphides

The sulphides on Earth may well have had a very great importance for the beginning of life and, although they are minor components even of the crust, they also have great importance in man's industry. We, therefore, devote a special paragraph to their selectivity. Figure 8.8 shows the energy balance of the reaction

$$MO + S \rightleftharpoons MS + \tfrac{1}{2}O_2.$$

It is clear that the formation of K, Rb, Cs and Ag sulphides is favoured relative to their oxides, and the same is true for the chlorides of these and other large ions such as Na, K, Ca, Sr, Ba, Tl, etc. Notice how size more than any other property of elements favours S or large non-metals such as Cl (see Fig. 8.8(a), (b)). The relative importance of size and electronegativity (Mulliken) or softness/hardness is discussed in Sections 2.4.1 and 5.5.3. Intriguingly, in the absence of enough oxygen to remove excess sulphur so that all the above reactions are driven to the left, that is, to oxides, hydrogen is also forced towards H_2S by large amounts of Al, Mg, Si, Ti and C. The other elements removed as sulphides include Cu and much of the Cd, Ni and probably Zn. As we shall see, the excess of oxygen in our present day atmosphere has produced a top layer that gives a false impression of the surface of early Earth where

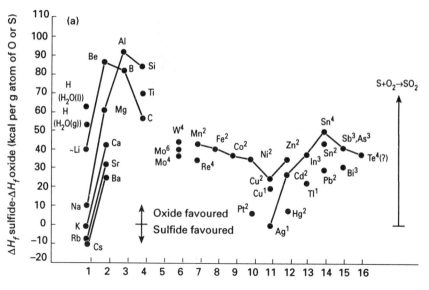

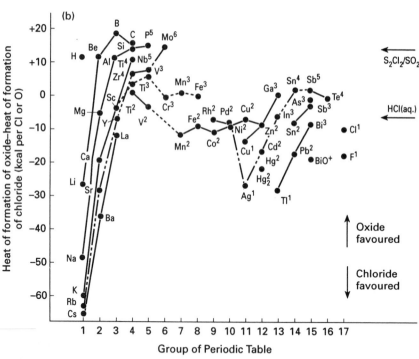

Fig. 8.8 (a) The difference in heats of formation of sulphides and oxides (per non-metal atom). The valence of the element is indicated where necessary. (b) The differences between the heats of formation of oxides and chlorides. Valence states are indicated.

sulphides, including H_2S, were present in considerable amounts. In Fig. 8.8 we also note the frequently observed relative stability order of all compounds MX: $Mg, Ca \ll Mn < Fe < Co < Ni < Cu > Zn$, which we have already noted in Section 5.5.3. In a competition in which there are limited amounts of chemicals it is the *relative* stability, not the absolute stability, that decides which partner any element will have. This principle is extremely important in all of chemistry based on thermodynamics. Thus, the above order also decides the *thermodynamic* preferences of metals in oxidation state II in solution and in biological systems. We then have the rough division

oxide chemistry		sulphide chemistry
Mg, Ca, Mn	Fe	Co, Ni, Cu
		Zn

that has dominated the Earth and living forms in agreement with equilibrium considerations. An overall picture of most of the elements in their most common combined forms on Earth is given in later sections.

Now, the Earth's surface has changed through its some 5×10^9 years, and at the same time the nature of living systems has altered continuously. In Table 8.6 we list the major eras and the organisms found in them (see also Fig. 12.1) so as to keep us in touch with the history of events following the formation of our planet. Note that it is only the Earth's surface, not its interior, that has evolved.

Table 8.6 Life's history on Earth

Years ago ($\times 10^6$)		Era
700	Major development of higher species	Phanerozoic
1000	20% dioxygen	↑
1500	Multicellular species; eukaryotes;	Proterozoic
2000	1% dioxygen	↓
2500		↑
3000		Archaean
3500	Prokaryotes; archaebacteria; life began	↓
4000		
4500	Formation of Earth	Hadean

In what follows we concentrate on the period after Earth had settled down to a low surface temperature. We must not miss the crucial events in the chemical selection in a relatively short period prior to this, that is, from the accident that produced the planets to their formation. In this period there was initially a high concentration of H and He and finally the distribution shown in Fig. 8.1. The actual chemical species that formed subsequently depended on the change of availability and the fall in temperature since trapping, both physical and chemical, regulated the possibility of reaction. Helium loss (Tables 8.3 and 8.7) would have been paralleled by H_2 loss at > 5000 K, where H_2O is not stable. Even at $100°C$ H_2O loss would have been very great. Carbon was retained as CO_3^{2-} but in the absence of O_2 would have been better retained as pure carbon, coal. If carbon had remained as CO the loss of carbon would have been as great as that of N_2. Thus the Earth's rate of cooling and its absolute temperature, its size (gravity field), as well as the chemical affinities (Chapter 2) and kinetics (Chapter 7) decided the early chemical combinations.

8.5.2 The primitive atmosphere

Given the complexity of the stages in the formation of the Earth as it cooled down to temperatures below $100°C$, we have to guess more or less the detailed

Table 8.7 Summary of data on the probable chemical composition of the atmosphere during stages 1, 2 and 3*

Stage 1 (early Earth)	Stage 2 ($\sim 2 \times 10^9$ years ago)	Stage 3 (Today)
Major components ($p > 10^{-2}$ atm)		
CO_2 (10 bar)		
N_2 (1 bar)	N_2	N_2
CH_4		O_2
CO		
Minor components ($10^{-2} < p > 10^{-6}$ atm)		
H_2 (?)	O_2 (?)	Argon
H_2O	H_2O	H_2O
H_2S	CO_2	CO_2 (10^{-3} bar)
NH_3	Argon	
Argon	(CO?)	
Trace components ($p < 10^{-6}$ atm)		
He	Ne	Ne
Ne	He	He
	CH_4	CH_4
	NH_3 (?)	CO
	SO_2	NO
O_2 (10^{-13} bar)	H_2S (?)	

* We are able to give a good account of stage 3 (Section 8.6.1) and a good estimate of stage 1, but the evolutionary period, stage 2, is hard to describe with any accuracy.

nature of the early atmosphere. It seems probable that in a first stage, at temperatures $< 100°C$, there was a high CO_2 pressure (10 bars†), some N_2 (1 bar), CH_4, H_2O and CO, and somewhat less H_2S, HCN and NH_3. It contained virtually no O_2 (perhaps 10^{-13} bar); see Tables 8.6 and 8.7. It is likely that, before water condensed, the element chlorine was present as HCl gas, since other elements do not have an affinity for chlorine higher than that of hydrogen, and so created an acid ocean. While H_2 was being lost at a fair rate there was always a reserve of reactive combined hydrogen in H_2S and in H_2O. The reaction of H_2S with metals in volcanic magma may well have generated H_2 and this activity could have been extremely intense before life began. (It is quite possible that the sea, when it formed, was somewhat acidic and almost black due to *catalytic* colloidal particles of reduced metal sulphides (see Section 8.5.1).) The compounds in the atmosphere were subjected to the effects of storms too, probably of unimaginable strengths. Small amounts of many C/H/N/O compounds were undoubtedly formed (Section 9.2) that built up through their kinetic stability at these low temperatures and these later dissolved in the sea and may have assisted life's beginnings.

We have to observe next that the formation of a non-equilibrium surface on Earth was concomitant with the creation of a planetary atmosphere compartment far from temperature equilibrium with the star (the sun) from which it came. Like the formation of stars, the formation of planets is another result of the action of gravitational forces which, in the case of the solar

† One bar is approximately equal to 1 atmosphere.

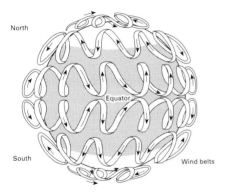

Fig. 8.9 Temperature differences around the world give rise to the winds. If the Earth did not turn, winds would tend to blow in a north–south direction. But the spinning of the Earth deflects them so that they blow east or west.

system, leave the planets relatively cold but bathed in the very high radiation field of the sun, which itself is trapped in its galaxy's gravity field. The radiation initiated further chemical reactions of the atmospheric gases, for example of $CO_2 + H_2O$, to give some organic compounds, and caused, *much later*, the formation of an ozone layer (see references in 'Further reading'). (The protection provided by the ozone is said to have allowed life to leave the sea and survive on land.)

The atmosphere is therefore constantly interacting with the surface of the Earth (including living systems) and with the sun, and suffers outpourings from inner zones. On the other hand the sun's energy falls unequally on Earth, so that the atmosphere has both a horizontal and a vertical temperature gradient, and a composition gradient. The fact that the atmosphere as a whole was and is far from equilibrium, is an unstable mixture and is highly energised means that it is always changing with time. It is a dynamic system and in flow due to these large energy-generating inhomogeneities (Fig. 8.9). The flow within the Earth's systems will be a topic of Chapter 15, but notice that the movements particularly of the air (and the water) on the Earth are constrained by the Earth's core, and their motions were initiated on formation of the Earth. Their impact on the solids causes erosion and so affects the Earth's surface. It is the dynamics of this complex system that has (fortunately) contributed to keep the Earth's average surface temperature close to $25\,^{\circ}C$ for 4×10^9 years although there have been fluctuations (but it cannot last for ever). The stability is in part related to the so-called greenhouse gas effect, see Section 5.8.

8.5.3 The nature of the early sea

The very earliest sea, as far as we can judge, must have been very different from that of today due to the peculiar initial conditions when water condensed. It is thought that initially this sea covered most of the mineral crust. It is quite possible that the rain that fell then was acid (Fig. 8.10); indeed, the original source of the chloride of the sea may well have been gaseous HCl since, as stated above, chlorine has such a low relative affinity for most elements other than H. In addition, there could have been some SO_2 and much CO_2 in the atmosphere which would also have contributed to acidity. The acid rain would have eroded surface and submerged minerals, some of which are basic, giving a less acidic solution. The best guess at the proto-ocean composition, estimated to be at pH 2 to 4, is that of Table 8.8. Very quickly, further erosion moved the pH up to around 5 and then, more slowly, to the neutral region where it has remained. During the rise in pH land masses separated from the sea due to fluctuations of the surface and its increase with the precipitation of carbonates as well as a result of a volcanic eruptions locally that gave sulphides (later to become sulphates). (The solubility products of such salts and their dependence on pH is described in Chapter 5.) The major precipitate was $CaCO_3$, removing eventually most of the calcium from the sea and most of the CO_2 from the atmosphere (Tables 8.7 and 8.8). This removal was continuous due to the constant transport of $CaCO_3$ from the equilibrium condition on the seabed to huge deposits on land. The sea quickly became effectively a saturated calcium salt solution, but maybe it had not changed dramatically before life was engaged in the carbon cycle (Fig. 15.10). The redox potential of the early sea,

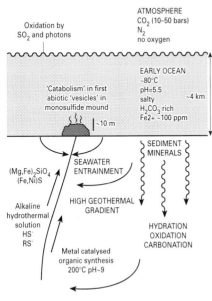

Fig. 8.10 A geologist's view of the interrelationship of the atmosphere, the sea and the sea bed when life began, 4×10^9 years ago. (After M. Russell.)

Table 8.8 The characteristics of the early ocean and of today

---

Proto-ocean (?)

pH = 2.0 (initial); $T = 80°C$

CO_2 and SO_2 not very soluble

HCl gives the acidity

Initially weak content of cations, but increasing to Ca^{2+}, 115 mM; Mg^{2+}, 95 mM; Na^+, 120 mM; K^+, 60 mM

Redox potential around -0.5 to 0.0 volts

Early ocean

pH = 8.0; $T = 55°C$

HCO_3^- (CO_2) high; SO_4^{2-} low; H_2S high

$Ca^{2+} \geqslant 10$ mM

Fe^{2+}, 1 mM; $Zn^{2+} \leqslant 10^{-10}$ M

Redox potential > 0.0 rising to < 0.4 volts

Late ocean (today)

pH = 8.0; $T = 25°C$

HCO_3^- (CO_2) high, and SO_4^{2-} (not H_2S) present

Average concentrations of cations are Ca^{2+}, 10 mM; Mg^{2+}, 105 mM; Na^+, 470 mM; K^+, 10 mM

Redox potential up to 0.80 volts at surface (O_2)

Fe^{3+}, 10^{-17}M; Cu^{2+}, etc., see Fig. 8.15

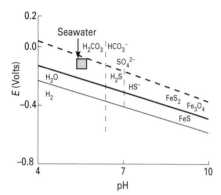

Fig. 8.11 A pH versus E diagram for some components of (early) sea water with special reference to sulphur/iron chemistry. (After M. Russell.)

shown in Fig. 8.11, was very reducing due to the presence of sulphides and hydrogen. The resulting trace elements and their speciation in the sea are given in Table 8.9.

The sea has never been just the result of this acidic rain and/or run-off from rivers of fresh water. Rivers are far from being saturated salt solutions and contain small particles as well as salts. The small particles settle at the river outlets to the oceans and develop into large delta lands, which together with wind-borne particles form new land masses. These soils, formed by erosion and gravity, can be very fertile, since they can contain the necessary inorganic elements for life, and maybe it was the development of sands and soils that allowed life to move from the sea. However, sands can also be extremely infertile as in the case of spreading deserts. The deserts are often very short of trace inorganic elements, apart from their shortage of water and organics, so that they may not sustain life easily. We return to a brief description of soils in Section 8.6.6.

Even today the elements found in the ocean are not in the same ratios as in the input from rivers. Partly this is due to selective biological uptake, but a major input of elements occurs through the volcanic eruptions and subsequent reactions in the hot cracks between continental shelves. It is now known that seawater has also always circulated from the edges to the centre of tectonic trenches where plumes of hot water arise (Fig. 8.10). The re-entering water plumes observed at the centre of trenches have been called 'black smokers' since they are black with particulate metal sulphides, especially of iron and manganese. They have high contents of H_2S, are very reducing and their low pH is very different from that of seawater. Thus, the precipitation of rain and the run-off from rivers are only two routes for element entry into the sea. The vents from the ocean floor are in many ways more interesting in that they create very strong and persistent local gradients of temperature, element concentration, reduction potential and pH. This is a very intriguing non-equilibrium (energised) mixture. If a zone of this composition happened to be trapped continuously in vesicles of some kind, that is, if it were enclosed by a membrane, then one would have here a possible precursor of a living system since it would have energy and chemicals contained in a semi-organised and catalytic system (see references 14 and 16 in 'Further reading'). Such a zone would be almost completely inorganic, including a vast variety of elements dominated by some 15–20 that are abundant and available. Is it coincidence that these very 15–20 elements are essential for life today?

The sea has always equilibrated with the atmosphere physically though not chemically. The formation of some organic compounds in the early atmosphere, for example, HCHO or HCN (also found in the galactic interstellar clouds), must immediately have led to an increasing concentration of reactive organic compounds in the sea. At the sea surface and especially at land margins light also generated more and more water-soluble organic molecules from CO_2 so that slowly the sea became a brew of organic molecules and inorganic colloids plus some soluble inorganic species. It is this mixture that advanced abiotic organic chemistry (Chapter 9) toward life (Chapter 10).

8.5.4 The early surface of the Earth

While we have been able to give a good impression of the whole Earth and of the early atmosphere, and a somewhat poorer one of the early seas and of the

Table 8.9 Some trace elements in the early sea*

Elements present

Fe^{2+}, Mn^{2+}, (Mo^{6+}), V^{4+}, (Ni^{2+}), W^{6+}, (Co^{2+}), Se as H_2Se

Elements largely absent

Cu^{2+}, Cd^{2+}, Zn^{2+}, Cr^{3+}, Ti^{3+}

* The assumption is that the $pH \geqslant 5$ and the amount of H_2S kept the sea as a reducing medium (see Fig. 8.11). The concentration of Mo^{6+} may have been lower than that of W^{6+} as Mo is precipitated as MoS_2 at low pH

layering of the deep inaccessible ocean regions of the Earth near ocean trenches, we will not have such an easy task with the rocks, soils and clays of the Earth's surface (Table 8.10) which have often emerged from the crust. There are two problems that complicate the situation. First, sudden eruptions with violent local increases of temperature force solids to dissolve into molten phases which, on cooling rapidly, do not come to equilibrium—all kinds of frozen solid solutions and mixtures are observed, containing many components. Second, erosion of material and subjection to high pressure on folding and unfolding under the surface at the edges of continental plates and re-emergence later left the peculiar sedimentary and metamorphic rocks. Scattered on the surface of the Earth there is therefore a huge variety of mixed mineral deposits (Table 8.10 and see Section 6.9). In the early periods sulphur would have had a considerable involvement in these surface reactions although at all times oxides dominated. It is even uncertain how the continents developed though they seem to have started as separate Gondwana and Laurasia lands, which split up in many ways over the geological periods (Fig. 8.12). The land masses drifted into different temperature zones and have thus suffered erosion and volcanic activity in diverse ways.

Unfortunately, therefore, when dealing with the natural selection of the chemicals of the Earth's surface, the convenient and simple approach based on chemical equilibria that we have used in Chapters 3–6 and for bulk considerations of Earth in this chapter is largely lost; it becomes only useful background knowledge. In part, the problems stem from the nature of the conditions in complex compartments such as these solutions and melts (Table 8.11). We must see the Earth's surface zones as a vast dynamic system in both a chemical and physical sense. Flow is seemingly random for the tectonic plates, but flow in the atmosphere and sea (see Chapter 15) is an effect of the field gradients, gravitational and radiation, in which Earth was generated and now sits. Fortunately for the story of the natural selection of

Table 8.10 The geological history of surface ore formation

Era (billion years before present)	Geological and chemical features	Major ores formed
Early Archaean (3.8–3.0)	Submarine trench formation; basic magma flows give *primary greenstones**	Fe, Ni, Cu sulphides, Au
Late Archaean (3.0–2.5)	Recycling of primary greenstone, hydrothermal processes	Cu, Zn hydrothermal sulphides
Early Proterozoic (2.5–1.7)	Uplifted crust erodes; O_2 produced by photosynthesis, oxidation of Fe^{2+}	Au deposits Banded Fe formations
Mid–late Proterozoic (1.7–0.7)	Thick continental crust forms Atmospheric O_2 increases; active sulphur redox chemistry	Ti, Cr oxides, Fe sulphide, Pt metals Co, Cu, U deposits
Phanerozoic (0.7–present)	Extensive crustal recycling	Hydrothermal Cu, Zn, Mo, Sn, Pb
	Tropical weathering conditions Secondary enrichment	Al, Fe resistates† Co, Ni, Cu minerals

* Greenstones—named after deposits in South Africa.
† Resistates—rocks resisting weathering.

(a)

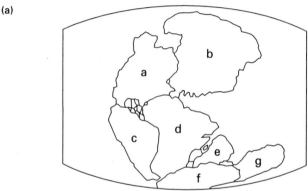

Fig. 8.12 The presumed form of the original continents: (a) $>200 \times 10^6$ years ago; (b) 120×10^6 years ago. Fragments of the original supercontinent are the continents we know today: a, North America; b, Asia; c, South America; d, Africa; e, India; f, Antarctica; and g, Australia.

(b)

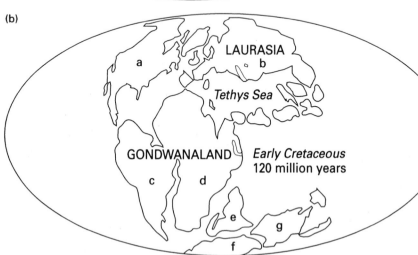

Table 8.11 Non–equilibrium inorganic phases on Earth

Phase	Nature of non-equilibrium
Many aqueous solutions	Fresh water brought in contact with land
Clay minerals	Eroded by weathering or activity of plants
Unoxidised rock	Product of volcanoes, ocean vents, etc.
Oceans	Circulation with life forms dominant in upper regions. Temperature switching
Air	Lack of catalysts for N_2/O_2 reaction; O_2 generally present
Frozen rock	Many examples, e.g. zeolites

elements, a much more dramatic and systematic development has occurred in the sea—the coming of life—but the sea remains a more easily appreciated environment than the sands and soils of Earth.

8.5.5 Some non-equilibrated inorganic compartments of interest: an aside

The most obvious of the non-equilibrium inorganic solid structures are the clays and zeolites. These structures are based on Al/Si/O/M lattices with open

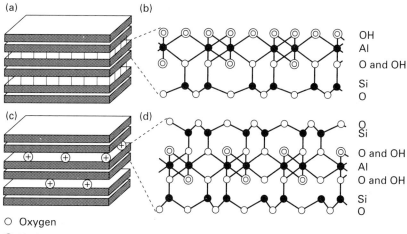

Fig. 8.13 Most clays are made up of stacks of silicate layers. In kaolinite (a) asymmetric layers are linked by hydrogen bonds. Each layer consists of a net of aluminium atoms and hydroxyl (OH) groups fused to a net of silicon and oxygen atoms (b). Other clays have symmetric layers, in which a silicon–oxygen net is fused on both sides to a metal–hydroxyl net; these layers are negatively charged and linked by positive ions (c). In illites (d) much of the negative charge arises from the substitution of aluminium for silicon atoms.

frames (Fig. 8.13 and Figs 7.18 and 2.28). They are not too far distant from the silicic acid $Si(OH)_{2n}O_{2-n}$ structures shown in Section 2.11. The zeolites (see Fig. 7.18) can be made in a variety of forms and, curiously, they have some features in common with proteins and polysaccharides, such as:

(1) preparation by energised condensation polymerisation;

(2) relatively ready hydrolysis;

(3) side chains that give acid/base catalysis;

(4) easy incorporation of metal ions for catalysis, now including reductive/oxidative reactions.

In one sense they are less complex than proteins in that they have fewer building blocks, just $Si(OH)_4$, but the blocks can be assembled in many non-linear ways and they are truly three-dimensional, not linear covalent systems (contrast proteins but also compare polysaccharides (Chapter 9)). Interestingly, there are theories of the origin of life that start from these silicate clays (see Section 8.8 and references 13 and 15 in 'Further reading'). A feature that they lack and that is of great importance in proteins is flexibility (see Fig. 3.15).

Much more dramatic than the development of inorganic chemicals *per se* has been the development of *both* inorganic and organic chemicals through life processes (see Chapters 10–13).

8.6 The later evolution of the Earth's chemicals

Now that we have described the major partitioning of the elements at the early period of the Earth's formation, we shall turn to a more detailed analysis of the atmosphere, the water and the crust on the Earth's surface today. In large part the change was generated by the large-scale production of dioxygen (by life) that converted Earth's surface into an oxidised rather than a reduced layer (Section 5.8.2).

8.6.1 The atmosphere today

Although the analysis of the present-day air was not easily accomplished, it was largely achieved before 1800. The air has two 'element' gases, N_2 and O_2, two 'compound' gases, CO_2 and H_2O, and some noble gases, particularly argon which is even more abundant than CO_2 (Tables 8.7 and 8.12). Thus, it has kept all the inorganic compounds necessary for the production of those organic compounds made from only H, C, N and O. We stress again that the air we know today is not the 'air' present when life started; the original 'air' contained virtually no O_2 (Table 8.7 and see Fig. 8.14). Again, today's air does

Table 8.12 Some chemically reactive gases of the air today*

Gas	Abundance (%)	Flux 10^6 tons	Extent of disequilibrium
Nitrogen	79	300	10^{10}
Oxygen	21	100 000	Taken as reference
Carbon dioxide	0.03	140 000	10^3
Methane	10^{-4}	500	'Infinite'
Nitrous oxide	10^{-5}	30	10^{13}
Ammonia	10^{-6}	300	'Infinite'
Dimethyl sulphide	10^{-8}	70	'Infinite'
Methyl chloride	10^{-7}	10	'Infinite'
Methyl iodide	10^{-10}	1	'Infinite'

* Abundance and flux (rate of annual flow through the atmosphere) of some of the gases of the air, and the extent to which this composition seems to violate the ordinary rules of equilibrium chemistry. (The word 'infinite' in this context means beyond the limits of computation.) Lovelock, J. (1979). *Gaia—a new look at life on Earth*. Oxford University Press, Oxford.)

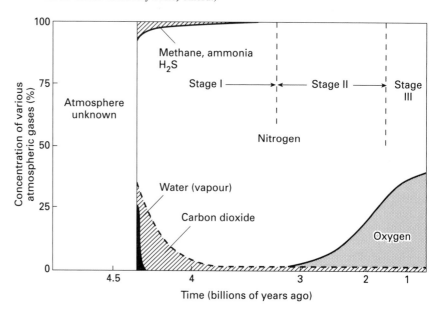

Fig. 8.14 Atmospheric composition, shown by the relative concentration of various gases, has been greatly influenced by life on Earth. Note that the total pressure has fallen greatly.

not have a balanced composition in a gravity field; for example, in the stratosphere it contains more ozone, O_3, produced by the action of the sun on O_2. Here is an example of vertical energisation; compare the crust of the Earth and the sea. Clearly, the composition of the air today does not relate to the composition of gases that should exist in equilibrium over the surface of the Earth (see Table 8.12). It is true that the water vapour pressure is not far from saturated, that is, humidity varies from 50 to 100 per cent, and this itself is the cause of water cycles. However, the nitrogen in the air should form some nitrogen oxides and the oxygen of the air should, as well as contributing some nitrogen oxides, more obviously consume all the carbon compounds on the surface, including all life. Even carbon dioxide is not very close to equilibrium with its solution in the sea or the formation of carbonates. Table 8.12 gives examples of all these out-of-balances in the atmosphere. We shall ask in Chapter 11 how this atmosphere arose from the original atmosphere and we shall find that it is due to life.

The situation is that the gases, especially H_2O, N_2 and CO_2, in the air are not very reactive at normal atmospheric temperatures, $\sim 25°C$. Even O_2, although a quite reactive substance, does not start to 'burn' organic matter easily in the temperature range $25–100°C$. In other words, air requires catalysts in order to react. Again H_2O, N_2 and CO_2 are stable molecules and to force them to become part of organic compounds, such as sugars, DNA, RNA and proteins, it is necessary to supply energy from the sun (as in biology) or from appropriate devices for chemical synthesis in a laboratory or factory and catalysts.

8.6.2 The nature of the sea today

The inorganic elements of the sea (Fig. 8.15) and of the lakes and rivers interact with the gases of the air. (Notice how thin is the layer of water at the surface of the Earth; seawater accounts for about 90 per cent of all water on the Earth, but oceans occupy only a tiny part of the planet. If the Earth is represented by a sphere with a diameter of 30 cm the oceans form only a

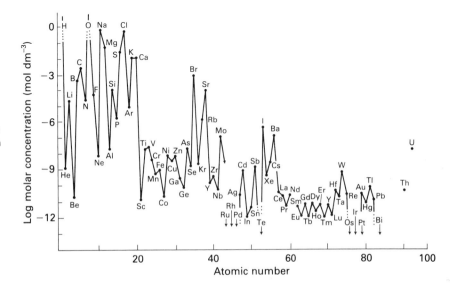

Fig. 8.15 The concentrations of elements in the *sea*. (After Cox, P. A., reference 2 in 'Further reading'.)

surface layer as thin as this sheet of paper and the largest abyss, ~ 11 km, would be no more than 3 mm deep!) Certain general features are worthy of note since, as we shall see, they affect element accumulation and combination in life.

The ease of extraction of elements into the surface of the sea from the land, that is, in the presence of dioxygen, is shown in Fig. 8.16. The removal of oxidised iron (Fe^{3+}) and other M^{3+} ions from seawater by precipitation as oxides/hydroxides contrasts with the relative availability of less abundant elements such as vanadium, molybdenum and selenium which are soluble in oxidised forms (the absolute values are given in Fig. 8.15). The situation at the bottom of the sea or of a lake can be very different since sulphide removes elements such as copper and to some extent zinc but does not remove Fe^{2+}. (It is believed that the conditions of deep static water are close to those of much earlier times when life began.) In these conditions, a column of water in the natural world has a steep gradient of elements (and other nutrients). The consequence is that different life forms appear in the different zones. The zones of the sea have a fairly fixed composition everywhere today, but over 4×10^9 years the changes have been very considerable (see Table 8.8).

Of course, there is also a vertical temperature gradient as well as one north and south from the equator. Such gradients, with that of pressure, adjust the solubility of salts; hence the composition in the sea is again variable. The sea and lakes are huge temperature-ordered structures horizontally, based on energy flow from the sun and the mantle, together with the fact that the seas also flow round the Earth in huge currents, such as the Gulf Stream. They are tidal too, whence there is a constant circulation of contents from zone to zone. All these features affect the chemistry in given localities in different ways, but the flows ensure that regions are of relatively constant analytical composition. Notwithstanding, the sea is never at equilibrium with the Earth, although some reactions are quite close to equilibrium. The composition given in Fig. 8.14 is then very much a general average. The organic chemical content of the sea is given in Table 8.13.

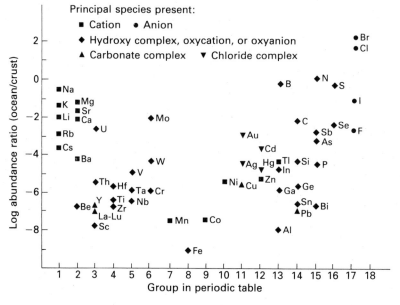

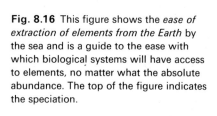

Fig. 8.16 This figure shows the *ease of extraction of elements from the Earth* by the sea and is a guide to the ease with which biological systems will have access to elements, no matter what the absolute abundance. The top of the figure indicates the speciation.

Table 8.13 Average concentrations of organics in seawater

Components	Concentration in seawater (μg carbon litre^{-1})
Free amino acids	20
Combined amino acids	50 (to 100?)
Free sugars	20
Fatty acids	10
Phenols	2
Sterols	0.2
Vitamins	0.006
Ketones	10
Aldehydes	5
Hydrocarbons	5
Urea	20
Uronic acids	18
Approximate total	340

There are also gross variations in the content of chemicals, vertically now, in the water column. In some cases this is due to the limited availability of light in deeper zones, but local increases may also result from the input of material from sediments at the bottom or from deep ocean vents while decreases may be due to removal by life at the surface. In particular, note that part of the deep ocean may be anaerobic as was the early ocean, that is, dioxygen-free, and may consequently have a high concentration of reduced chemicals.

8.6.3 The nature of freshwater today

The problem of freshwater aqueous solutions, especially those running over the land via rivers and lakes, is that there is a variation from the extremes of 'pure' water coming from rain or ice to soil water and then to water that is more closely related to seawater or even exceeds it in saltiness, for example the 'Dead Sea' (which is a lake). Although rain water is not free from dissolved compounds, it is nearly so. Once the water starts to flow over the land its composition changes due to its effects on rocks (erosion) and to biological contamination. There is, then, a local condition of freshwater, just as there is a local condition of the life in it, dependent on temperature, rates of flow and evaporation, and so on. We have to treat each system separately, but we can discern the principle that, to a good approximation, each system is constant under given conditions. This is attested to by the fact that year in and year out the flora and fauna over millions of years are seasonal and local—the cyclic flow of inorganic chemicals and life is never-ending. (The Chinese, see Chapter 1, were very aware of this and perhaps their realisation of an inability to change things caused their society to stagnate for a long time. It seems likely that the Western world today will reach the same conclusion by trial and error!) How should we look upon such chemical systems? Perhaps the simplest way forward is to note that, no matter what the source, the water in contact with land, though a dilute solution, becomes a solution much richer in certain elements than others. These are the common abundant elements that have reasonable solubility at pH = 6–8. They are listed in Table 8.14 (also see Fig. 8.15).

The difference between freshwater and the sea may appear to be of little concern to the overall purpose of this book—the natural selection of the chemical elements. However, it is possible that, although life started in a calcium carbonate saturated sea, it was its migration to the rivers that allowed development of bony phosphate animals and thus invasion of the land by animals following that by plants. It may well be that controlled use of Ca^{2+}, CO_3^{2-} and HPO_4^{2-} in different aqueous media has allowed man to emerge from life in water!

Finally, we must refer again to soil water which remains for long periods locked into clays (Section 8.6.6). This water is the source of nutrients for plants. Before we can discuss it thoroughly, however, we need to know the composition of soils (discussed in Section 8.6.6) after analysing the composition of the Earth's crust today and the trace element fractionation in rocks and soils of the crust (Sections 8.6.4 and 8.6.5), which greatly influences availability in any given area and era.

Table 8.14 Forms in which the main biological elements occur in aerated soil, water, rivers, lakes and sea (or in blood plasma)—simple species

Cations	Anions	Neutral species
NH_4^+, H_3O^+	HCO_3^-, CO_3^{2-}, NO_3^-	H_2O, $B(OH)_3$
Na^+, K^+	$H_2PO_4^-$, HPO_4^{2-}	CO_2, $SiO_2 \cdot nH_2O$
Mg^{2+}, Ca^{2+}	OH^-, F^-, Cl^-, Br^-, I^-, SO_4^{2-}	N_2, NH_3, O_2

8.6.4 The crust of Earth today

Since at low temperature—less than 100°C—most of the more abundant elements have a higher affinity for oxygen than for any other element (see Fig. 8.4) and since there is plenty of O_2 today, this element, held far from equilibrium, now totally dominates the composition of much of the top layers of Earth's solid surface (or crust) in metal silicates and carbonates where the metal is mainly Al, Mg, Ca and Fe or even some Na (silicate, granite, soils, clay); see Table 8.15 and Fig. 8.5) and H_2O. Oxygen represents 46.6 per cent of the calculated weight of the crust, 62.6 per cent of its atoms and 91.7 per cent of its volume (see Table 8.5). With oxygen in excess, locally at the surface new oxides have formed, such as those of iron and tin, and deposits of sulphates and nitrates have appeared. Much carbon, found as CO_2 in the atmosphere or dissolved in water, partly as H_2CO_3, is associated in carbonates

Table 8.15 Dominant distribution of the common elements in the Earth today

Oxides (silicates, carbonates phosphates)	Sulphides (selenides)	Elements soluble in water	Gases
(Be) Mg, Al, Ca, (Fe)	(Fe), Co, Ni, Cu, Zn	H(H_2O), Na, K, Cl	N_2, (CO_2), O_2
(B) C, Si, P	Cd, Hg, Mo, V	Br, I, (SO_4), Li	Noble gases, He, Ar

such as $CaCO_3$ and $MgCO_3$. Some of it is also found in coal and in the hydrocarbons of oil and natural gas. These are, like oxygen, products of life's chemistry (see Chapter 9), but they have formed new isolated compartments, as already stated, and they are not in equilibrium with O_2. This is a case of compartmental kinetic isolation. Finally, because surface oxygen is largely mopped up by light elements such as H, Al, Si, C, P, and now Fe, an excess of S will remain, particularly in deeper zones of water, and there several heavy metals that have lower affinity for O remain as partners of S, for example as Fe, Co, Ni, Cu and Zn sulphides, or occur as free metals or alloys.

We see, therefore, that there have been great changes in the atmosphere, the sea and the crust. It is these changes, we shall show, that have brought about the possibility of evolution in living organisms (Chapters 10–13), especially via the natural selection of the chemical elements in both inanimate and animate forms.

8.6.5 Trace element fractionation in the rocks and soils of the crust

The separations of the major or bulk elements in the Earth have now been described. Part of the bulk elements and the minor or trace elements tend to accumulate to some degree in all compartments as 'impurities' in, for example, major rocks or clays. This is a consequence of the partition equilibria discussed in Section 5.3. An example of the separate uptake of M^{2+} and M^{3+} ions into an aluminosilicate is shown in Fig. 8.17(a), (b). In Fig. 8.17 the ratio R is the observed concentration of a given element in the residual molten rock relative to that remaining after a given percentage has crystallised out. In other words, a downward curve shows fractionation of an element preferentially into the solid rock. We see that Ni^{2+} and Cr^{3+} are removed into a silicate or oxide rock leaving Ca^{2+} and Fe^{3+} behind in solution and therefore more available. (The reasons for this selective removal, crystal field effects, from melts or solutions, (see Sections 2.9.5, 2.9.6 and 5.5.3) are given in reference 12 in 'Further reading' and are based on equilibria.) The preference for a site of a given geometry in a *crystal* mineral is an indication of the relative difficulty of separating the element out of the mineral into a melt. This biased distribution against elements such as chromium and nickel especially in oxidising conditions by taking them into oxide lattices preferentially. The picture for sulphide ore formation is very different. To what degree has this influenced the very different uses of nickel in anaerobic as opposed to aerobic life (see Chapter 12)? Chromium has never been associated with sulphide.

More generally, there is, of course, the consideration of competition between sulphide and oxide for metals (Section 8.5.1). Similar considerations apply to the extraction of the element out of any crystal and into water. We stress that the major paritioning of the elements left traces of them in many different zones, for example both in rocks and the sea. These traces are essential to life.

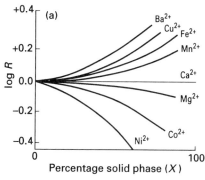

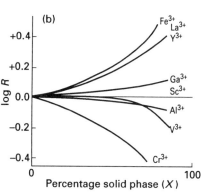

Fig. 8.17 (a) The uptake of divalent ions into a silicate. R is the ratio of the element in the liquid to that in the rock after X per cent crystallisation. (b) The uptake of trivalent ions into a silicate (R and X as in (a). (After Williams.)

8.6.6 The nature of soils and soil water

While we have often attempted to describe the composition of the Earth's surface in terms of the phase rule (Chapters 4 and 6) and aqueous solution

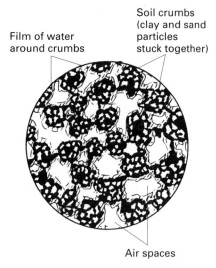

Film of water around crumbs

Soil crumbs (clay and sand particles stuck together)

Air spaces

Fig. 8.18 Magnified (diagrammatic) view of soil crumbs.

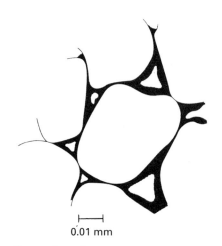

0.01 mm

Clay particles. Shaded areas represent water held by capillary action.

equilibria (Chapter 5), the possibility of doing so was lost once we came to the uppermost layer of the crust—soils. Soils are formed from sediments, but they are today grossly mixed with organic compounds. They are extremely heterogeneous and even one sample from any particular field contains many different compartments. The appearance of the soil is shown in Fig. 8.18. The size distribution of particles varies from gravel, with stones of around 1 cm and greater, through sand to clays with particle sizes around 0.001 cm or less (see marginal illustration). From place to place soil differs in grain size and composition, so that different life forms can benefit in different ways (compare deserts and forests). In many areas soils are deficient in particular elements valuable for plants and then animals.

The difficulty in describing soil is compounded by the presence of varying amounts of water. This soil water is different from that of the sea and from the fast run-off in lakes and rivers. In Table 8.16 we give the availability of elements in water from soils at very different pH values. The problems of describing speciation in such solutions are immense since there are present highly active surfaces of clay minerals, organic compounds such as humic acids, and a variety of anions and cations (see Section 5.6.2 for a discussion of the applicable conditional binding constants). The following points may be useful.

1. Lower pH values mean higher concentrations of free cations and anions (Table 8.16).

2. Lower pH values allow faster equilibration between soil and soil water.

3. Low pH will hydrolyse esters easily.

4. Oxidation is greater at higher pH.

The only realistic way to approach the quality of a given soil is through empirical study. A cycle of elements in soils, independent from biological input, is shown in Fig. 8.19 but a discussion of such cycles is reserved until Chapter 15.

8.6.7 Summary of element separation in Earth's compartments today

Looking at the element distribution in the light of the periodic table, (Fig. 8.20), we see that in the first row (top right) we have H, C, N, O in the atmosphere, in the hydrosphere we have mainly the extreme left- and right-hand members of this table, namely Na, K, Cl, S, Mg, (Ca) as well as H_2O, and in the geosphere we have the rest of the second row of the table, including Al, Si, P, some heavy metals and sulphur, but much sulphur has also gone with the central group of heavy (transition) elements into the lower layers of the Earth. These chemical associations of metals and non-metals are maintained throughout chemistry and this includes all of life, as we shall see; hence inorganic compounds in life are dominated by the same chemistry that dominates all 'cold' collections of atoms (in compounds) formed from giant stars, see the cover of this book. The distribution has not been without change, as we notice when we inspect in more detail.

It is obvious that the inorganic chemistry of the Earth's materials in the three physical states of matter is extremely complex locally. While we can use

Table 8.16 The availability of elements from soil waters of different pH*

Element	Availability of element ($\mu g \, l^{-1}$)	
	pH, 3	pH, 7.5
B	20	500
Cd	100	—
Mn	6000	500
Fe	2000	100
Co	1–5	1–5
Ni	1–5	1–5
Cu	750	50
Zn	7000	100
Mo		5
Se		1

*From Kabata-Pendias, A. and Pendias, H. (1992), reference 20 in 'Further reading'. Bulk elements such as Na, K, Mg, Ca are present in excess of $500 \, \mu g \, l^{-1}$ at pH = 7 and about equally so.

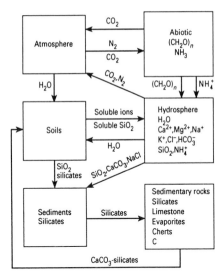

Fig. 8.19 Main abiotic exogenic or surface geochemical cycle. Little of this cycle has changed since Earth formed; O_2 is omitted. Note that sulphur and heavy elements are also excluded.

the discussions of physical equilibria, chemical exchange equilibria including redox potentials, and the effects of fields to give us a gross impression of the early Earth, the zones of greatest interest are the three on the surface—the crust, the aqueous zone and the atmosphere. There are important considerations of a quite different kind here. First, the application of equilibrium thermodynamics is not very helpful. The number of components is huge, due by origin to the elemental variables (90) but coming together now in non-equilibrated ways in all three physical states. In addition, none of these zones or even parts of them is homogeneous in temperature or pressure and each flows in turbulent patterns. Not surprisingly, almost every reasonable possible combination of elements is found locally. Again, the observed forms belong to different historical periods over which the general redox potential has shifted very markedly, Section 5.8.2. We are assisted in our thinking by the descriptions of matter given in Chapters 1–6 only to discover very great complexity in systems far from equilibrium; for example, much of the matter in the surface zones is in a trapped state and a full explanation of how this came about is unlikely to be obtained. This is especially true since the materials flow. (Explanations of other flows, for example, weather patterns, are notoriously prone to error.) However, on the Earth's surface and interactive with it life appeared. We must be aware that it has influenced all three zones and has become a fourth surface zone—the biosphere—with a very interesting chemical composition broken down within species, always having organised flow yet changing markedly with time.

The most remarkable influence of life upon the elements on Earth's surface has been the change in the range of redox potentials in the surface zone from -0.5 to 0.0 volts to -0.5 to $+0.8$ volts (Table 8.17). This change, brought about through the generation by life of dioxygen, has slowly caused changes in the sea, the air and the crust since about 3×10^9 years ago, see Fig. 8.14. The

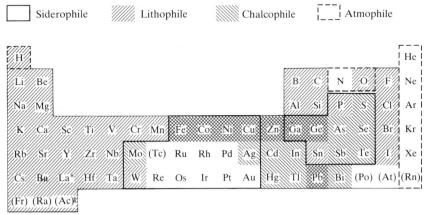

( ) Radioactive element of very low abundance

* Including lanthanides Ce–Lu

‡ Including actinides Th,U

Fig. 8.20 Geochemical classification of the elements. Lithophiles predominate as oxides, siderophiles as alloys. Atmophiles occur as gases while chalcophiles occur in the crust.

Table 8.17 Forms of elements and (availability)*

Element	Reducing environment	Oxidising environment
Iron	Fe(II) (high)	Fe(III) (low)
Copper	As sulphide (low)	Cu(II) (moderate)
Sulphur	HS^- (high)	SO_4^{2-} (high)
Molybdenum	$[MoO_nS_{4-n}]^{2-}$, MoS_2 (low)	MoO_4^{2-} (moderate)
Vanadium	V^{3+}, V(IV) sulphides	VO_4^{3-} (moderate)

* Note the switch in availability of iron relative to copper that must have had a profound effect on evolution as we will show later. Vanadium sulphides are quite soluble.

change was interactive with life, so that living forms and their elementary composition also changed. We describe all these changes in Chapter 12.

8.6.8 The elementary composition of living systems

Most of the elements and compounds commented upon so far are called inorganic and apparently are not associated with life. It may seem strange, therefore, to introduce living systems into the discussion of Earth's chemical composition, but we saw above that they have had a profound effect on the crust, the sea and the atmosphere. During the course of the early analysis, around AD 1850–1900, it was found that, seemingly, all the compounds obtained from living systems contained nothing but carbon, hydrogen, nitrogen and oxygen, plus a little sulphur and phosphorus. Increasingly, in the period 1850 to 1950, this led to a division in chemistry between the chemistry of carbon compounds formed with these elements, so-called *organic* chemistry, and the chemistry of the other elements, called *inorganic* chemistry. It was as if the essence of life had been discovered as a chemistry deeply based on carbon. (This, we believe, is as much a mistake as that which confused chemists generally about the nature of matter until about 1800.) It was based on undervalued water, always present in large excess over all other substances in living organisms, and on an inability to determine very small amounts of some elements in the presence of large amounts of others. The misconception has lasted almost to the present day. The refined and more discriminating use of analysis in the last 40 years (from about 1950) has shown that this view of life is not tenable and that all living systems are based on rather similar combinations of 'organic' and 'inorganic' elements (Fig. 8.21) although in

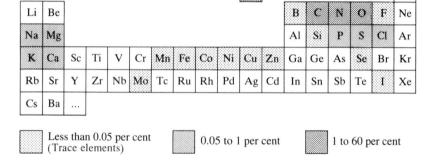

Fig. 8.21 The periodic table, showing the distribution of elements in the biosphere.

different amounts (see later chapters). The conclusion to be derived from these observations is straightforward: there is a required inorganic chemistry in life, and life must be included when discussing the evolution of inorganic chemicals on Earth, especially in the crust, the ocean and the atmosphere. We shall develop this theme after introducing organic chemistry in Chapter 9. There are also some intriguing speculative views on how life initially may have depended on *bulk* inorganic systems. Many elements are peculiarly associated with life, Table 8.18, and we shall look for the possible reasons in Chapters 10–13.

Table 8.18 Abundance (parts per billion; p.p.b.) of trace chemical elements in seawater and in some marine organisms (dry weight)

Chemical element	Total concentration in seawater (p.p.b.)	Maximum concentration in marine organisms (p.p.b.)
Ti	<1	10 000
V	1.5	280 000
Cr	0.3	1 400
Mn	0.03	4 100
Fe	0.05	86 000
Cu	0.3	7 500
Mo	10	6 400

8.6.9 Clays and sulphides: origins of early life?

Some of the most difficult inorganic systems to explain are the clay minerals, yet they are amongst the most important for life. Even before living organisms left the water, they began to bind strongly to sediments. At that time the sediments (mud) would be and have remained the major source of very many essential elements. Today the soils are still the source of water as well as of the trace elements, but not necessarily of H, C, N and O, and man's agriculture is totally dependent on this thin mineral layer. The soils are mainly composed of aluminium, magnesium and iron silicates, with many other elements present in small amounts. The soils retain organic and inorganic compounds by strong adsorption or even located in aqueous pockets or channels.

Some facets of soil chemistry do require further stress since, as mentioned above, it is believed by some that the surfaces of clay particles may have assisted very early life forms. A notable feature is the small particle size, the heterogeneity of phase structures in a given volume, and the fact that these clay structures are not only far from physical equilibrium but are also far from chemical equilibrium. Clearly now, the difficulty of reaching equilibrium in soils arises from the prevailing low temperature and the very slow rearrangement of silicate lattices even down to the loss of water. (The loss of water from them at high temperatures formed the basis of man's first industry—making pots and bricks (Chapter 14).) Some of the forms in which the clay structures appear are truly striking and life-like (Fig. 8.22). The long worm-like unit, shown in Fig. 8.22, which is called imogolite, is found in many clays. (The central hole is of diameter ~10 Å.) When we look at the

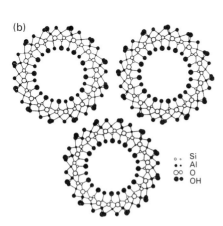

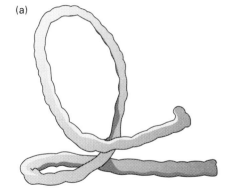

Fig. 8.22 Imogolite. An unusual clay structure. (a) Observed as a worm-like mineral in soil; (b) the atomic structure in cross-section.

○ Si
∴ Al
○○ O
●● OH

earliest forms of life based on cylinders (Chapter 9) we cannot help but wonder if these minerals assisted stabilisation of early organic forms. The tube-like structures grow and on fracture they multiply—their surfaces are also catalytic. The zeolites are related minerals with open structures (see Section 8.5.5). (Many of man's industrial processes use these frameworks (see Chapter 14.) Some sulphides seem to be able to form similar structures (see reference 9 in 'Further reading') and these have also been associated with life's beginnings. While all such schemes are speculative they inevitably lead to deep queries concerning the possible inorganic beginnings of life.

Other authors (see references 15 and 16 in 'Further reading') have stressed the potential roles of sulphides and not only in the formation of cylindrical tubes suitable for dividing space into the necessary compartments for accumulation of chemicals as occurs for silicates. Particularly there is the suggestion that sulphides could have been a primary chemical energy source and that, coupled to this energy, there would have been a natural catalytic surface for C/H/O/N reactions. The basic energy comes from the reaction

$$Fe^{2+} + HS^- \longrightarrow FeS \downarrow + H^+ \qquad \text{(proton gradient)}.$$
$$FeS + H_2S \longrightarrow FeS_2 + 2H^+$$

and

$$H_2S + CO_2 \longrightarrow [S] + C/H/O \text{ compounds} \qquad \text{(C capture)}$$
$$[S] + FeS \longrightarrow FeS_2, \qquad \text{pyrite (energy source)}.$$

(Pyrite is a peculiarly stable product, containing low-spin Fe^{2+}.)

This subject remains speculative and has only slight relevance to the present book which concerns the *observed* selection of the chemical elements in all systems, biotic and abiotic. We can never do more than guess at the origin of life.

8.7 Conclusion: Earth's evolution

The Earth is not just the result of a collapse of a combination of chemical elements to an equilibrium state. However, the description of equilibria in certain of its zones together with our somewhat detailed knowledge of the kinetics that influenced their formation led us to understand the reasons for the formation of an iron core and the silicate/sulphide mantle closely in accord with equilibria on cooling. The crust is much more difficult to understand from equilibrium considerations since it has changed considerably over 4.5×10^9 years and even before dioxygen evolved. This is also true of the contents of the sea, especially where sea and hot minerals meet at ocean vents. However, in this chapter we have been helped in our understanding of the crust and the sea by reference to the principles of element binding and the thermodynamics of large phases including both concentrated and dilute components, following Chapters 2–5. When we dealt with later formations of the crust due to sediments and clays (small phase volumes) the discussion became more difficult again and much of the chemistry was now found to be far from equilibrium although stable over very long periods, for example that of sands and soils. It certainly requires consideration of systems of limited size

(Chapter 6). The atmosphere and the organic chemistry of the biosphere are very different again. Fortunately, in the treatment of the atmosphere, which has changed greatly (see Table 8.7), we are still able to refer to a few elements or small molecules always found there such as O_2, N_2, CO_2, H_2O and Ar as we did in the early chapters, but, as we have seen, these elements are not nearly in equilibrium among themselves or with other parts of the Earth (Table 8.12). Organic chemicals present quite a new set of problems in that their formation represents change to complete dominance of kinetic control (Chapter 7) within many small phases, polymers (Chapter 6). It is these new features that have allowed a progression from abiotic selection of elements within organic chemistry (as will be discussed next in Chapter 9) to biological selection within organised microphases and including large molecules (as described in Chapters 6 and 10–13). Importantly the living organisms are not viable without some fifteen 'inorganic' elements. The importance of this present chapter to the description of the later development of organic and life's chemistry is that not only does it show the restrictions imposed by the abundance of the elements in the universe and hence on Earth but it also indicates that the level of their availability in the crust, the atmosphere or in waters is idiosyncratic perhaps to this planet and to particular periods of time. There has been a continuous adjustment of element availability with time. We shall see in Chapters 10–13 on living systems that these restrictions on and changes of availability are the major consideration in the evolution of living forms. Availability of elements will also be shown in Chapter 14 to have had a major, probably the most significant, impact on the development of man's industry and his culture. Since this industrial activity can impose itself even at the geological crust and sea level and therefore, like biological activity, affects availability, we look in Chapter 15 at present-day cycling of elements due to a combination of non-living, biological and man-introduced chemistry. There are two 'types' of chemicals always involved—those of 'inorganic' chemistry, largely examined in this chapter and Chapters 2–5, and those of 'organic' chemistry to be described in the next chapter.

Further reading

The following references provide extensive complementary information on the various topics discussed in this chapter, from the origin of the chemical elements to the formation of the solar system and of the Earth, its structure and dynamics, and the emergence of life based on a small number of elements available in the surroundings.

On the origin of the chemical elements

1. Kaufmann, W. J., III (1991). *Universe*. W. H. Freeman, New York. Chapters 7, 8, 22 and, in part 28, 29 and 'Afterword'.
2. Cox, P. A. (a) (1989) *The elements*. Oxford University Press, Oxford, Chapters 1, 3, 4, 5. (b) (1995) *The elements on Earth*. Oxford University Press, Oxford.
3. Mason, S. F. (1990). *Chemical evolution*. Oxford University Press, Oxford. Chapters 1–10 and 13.
4. *Scientific American* (1994). Special issue on the Universe and life on Earth. *Scientific American* **274**, 22–51.

5. Norman, E. B. (1994). Stellar alchemy—the origin of the chemical elements. *Journal of Chemical Education* **71**, 813–19.

6. Viola, V. E. Formation of the chemical elements and the evolution of the universe. *Journal of Chemical Education* **67**, 723–30 (1990) and see also *Journal of Chemical Education* **71**, 840–4 (1994).

7. Prantzos, N., Vangioni-Flan, E. and Casse, M. (eds.) (1994). *Origin and evolution of the elements.* Cambridge University Press, Cambridge. A series of papers reflecting our current knowledge of the subject.

On the structure and dynamics of the Earth

8. Press, J. and Siever, R. (1994). *Understanding earth.* W. H. Freeman, New York.

9. Brown, G. C., Hawkesworth, C. J. and Wilson, R. C. L. (eds.) (1992). *Understanding the Earth.* Cambridge University Press, Cambridge, Chapters 2, 3.

10. Mason, B. and Moore, C. B. (1982). *Principles of geochemistry* (4th edition). J. Wiley and Sons, New York. Chapters 1, 2, 3, 8, 9 and 10.

11. Henderson, P. (1986). *Inorganic geochemistry.* Pergamon Press, Oxford. Chapters 2, 4 and, in part, 10 and 11.

12. Phillips, C. S. G. and Williams, R. J. P. (1966). *Inorganic chemistry.* Oxford University Press, Oxford, Chapter 34.

On pre-biotic chemistry and the emergence of life

13. Cairns-Smith, A. G. (1982). *Genetic takeover and the mineral origins of life.* Cambridge University Press, Cambridge.

14. Edderfield, H. and Rudnicki, M. (1992). Iron fountains in the sea beds. *New Scientist* **134**, 31–35 (June).

15. Cairns-Smith, A. G., Hall, A. J. and Russel, M. J. (1992). Origins of life and evolution of the biosphere, **22**, 161–80, and see also Cairns-Smith, A. G. The first organisms. *Scientific American,* June 1985, pp. 74–85.

16. Wächtershäuser, G. (1990). Production of the first metabolic cycles. *Proceedings of the National Academy of the Sciences USA* **87**, 20–204.

17. Oró, J., Rewers, K. and Odom, D. (1992). Criteria for the emergence and evolution of life in the solar system. *Origins of Life,* **12**, 285–305.

On the elemental composition of the Earth, the sea and the atmosphere

18. Fraústo da Silva, J. J. R. and Williams, R. J. P. (1991). *The biological chemistry of the elements—inorganic chemistry of life.* Clarendon Press, Oxford. Chapter 1.

19. Adriano, D. C. (1986). *Trace elements in the terrestrial environment.* Springer-Verlag, New York.

20. Kabata-Pendias, A. and Pendias, H. (1992) *Trace elements in soils and plants* (2nd edn). CRC Press, Boca Raton, Florida.

21. Bruland, K. W. (1983). *Trace elements in sea water in chemical oceanography,* Vol. **8**, Chapter 45. Academic Press, London.

22. Wayne, R. P. (1991). *Chemistry of Atmospheres* (2nd edition). Clarendon Press, Oxford, Chapters 1, 4, 5, 9.

23. Butcher, S. S., Charlson, R. J., Orians, G. H. and Wolfe, V. (eds.) (1992). *Global biogeochemical cycles.* Academic Press, London, Chapter 2.

9

The evolution of organic compounds

In inorganic chemistry the radicals are simple; in organic chemistry the radicals are compounds—that is the sole difference. The laws of combination, the laws of reaction, are the same in the two branches of chemistry

> J. B. A. Dumas and Justus von Liebig:
> Memoir to the French Academy of Sciences
> *Annalen*, 1838, XXV, 3

9.1 Introduction

The designation 'organic', carrying its original sense of 'life' chemistry, has a broad meaning still centrally based on the element carbon in association with H, N and O, in both synthetic and degradative reactions. However, included under this heading today, with the chemistry of these four light elements, is some of the chemistry associated with heavier non-metals and that of some metals, when it overlaps with organometallic chemistry. Due to the diversity of organic chemistry, summarising it (especially in relation to life) is an awesome task, but it is necessary to do so in order to uncover some special principles of element association with carbon. It also leads forward to biological chemistry in the development of life, a combination of organic with much inorganic chemistry. Organic chemistry with inorganic chemistry is not simply linked to *biological chemistry* however, since this requires a much wider vision and an organisational ingredient which we leave until Chapters 10–13. In essence we have to show in this chapter the myriad ways of putting together several of each of the atoms C, H, N and O using single, double and triple bonds, as well as including S, P, Se and halogens in some compounds, the ways in which synthesis can be managed, and the ensuing properties of the resultant compounds. This would not be a problem peculiar to these elements but for the fact that, while the organic compounds of carbon are thermodynamically unstable (Fig. 9.1), they are kinetically stable in oxygen (air) and water for long periods, which gives them their particular importance in living systems and subsequently in some of man's industries. In effect, in many considerations of chemical equilibria each organic compound is operationally a component in that it does not exchange elements with other compounds (see Section 4.3.2). There must be equally as many thermodynamically unstable 'inorganic' compounds, but most of them are kinetically much less stable in the presence of air and water so that the diversity of unstable compounds that has been observed so far is much less and those known have been less studied. The instability of these inorganic compounds does not allow them to form many persistent, complex, assembled systems of molecules such as we see in living systems, at least on Earth. On the other hand, the number of thermodynamically stable inorganic compounds known is very large, while there are very few stable organic molecules. Carbon dioxide, CO_2, is the only thermodynamically stable 'organic' compound in the presence of O_2. Again CO_2 and CH_4 are the stable forms of carbon, except for carbon itself, in the presence of water and in the absence of free O_2, while in the presence of water and dihydrogen, at $pH = 7$, CH_4 is the only stable form of carbon (see Fig. 9.1). (NB. The data in Fig. 9.1 are related to the states H^+ 10^{-7} M and H_2 at unit pressure.) Let us see how the present situation, in which there is a great variety

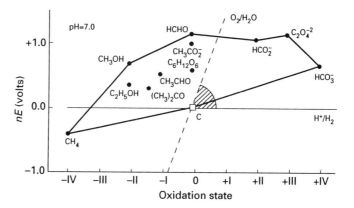

Fig. 9.1 The oxidation state diagram for carbon at pH = 7.0 and unit H_2 pressure against the H_2/H^+ electrode at pH = 7.0. In biological systems the in-cell cytoplasmic hydrogen carrier NADH has a redox potential against the H_2/H^+ standard potential close to that of H_2/H^+ at pH = 7 which is −0.42 volts, with an O_2/H_2O potential of +0.81 volts. Here the H_2/H^+ reference electrode at pH = 7 is given the value 0.0 volts for comparative purposes.

of organic compounds on the Earth, many of which are still largely associated with life, has developed.

9.2 The evolution of carbon chemistry: thermodynamics and kinetics

As discussed in Chapter 8, the initial reactions of carbon during the cooling of the universe were probably largely based on three features: (1) the stability of the simple oxides CO and CO_2 relative to one another and relative to the stability of the oxides of the elements with similar affinity for O_2; (2) the stability of carbon's hydride, CH_4; and (3) the stability of carbon itself (including graphite and fullerenes). It is apparent from Fig. 8.4 that at high temperatures, say, 3000°C to 5000°C, it is likely that much carbon and oxygen would have been in the form of CO and most of the residual oxygen would have been combined with Al, Si, Mg and Ca. As we have seen, the consequences of further cooling are dramatic since CO becomes relatively unstable thermodynamically, and the question arises as to the quantity that was left at temperatures as low as 1000°C. At this temperature CO remains in the gas phase while most of the oxides of Al, Si, Ti, Mg and Ca, now the elements of highest oxygen affinity, are condensed. Other elements, and particularly iron, at this low temperature have an affinity for O_2 higher than that of C (as CO or CO_2), but they had also condensed on Earth as elementary metals or in alloys and could not react readily with the oxides of carbon. Therefore, many such elements, metals, may well have failed to remove much of the oxygen from carbon just because of such phase separations, so that accidental kinetically stable mixtures of carbon gases, $CO + CO_2$, with large particles of oxide and metallic minerals formed eventually in planets around stars far from equilibrium at below 100°C. In this mixture there would also be much residual gaseous H_2. In such conditions catalysed reactions of the carbon oxides with H_2 on the inorganic surfaces could well have resulted in the considerable reduction of carbon to produce C—H compounds, even as far as methane and other gases, but also some heavier oils and small amounts of many organic compounds still seen in interstellar space (see Table 9.1). It is due to the kinetic stability of many of these compounds that they do not decompose rapidly. Our first concern is, as always in organic chemistry, with

Table 9.1 Molecules observed in interstellar space[*]

Number of atoms in molecule	Molecules			
2	H_2	CH^+	CH	OH
2	C_2	CN	CO	CO^+
2	NO	CS	SiO	SO
2	NS	SiS		
3	H_2O	C_2H	HCN	HNC
3	HCO^+	N_2H^+	H_2S	HCS^+
3	OCS	SO_2	$NaOH$	
4	NH_3	C_2H_2	H_2CO	$HNCO$
4	C_3N	H_2CS	$HNCS$	
5	CH_4	CH_2NH	CH_2CO	NH_2CN
5	$HCOOH$	C_4H	HC_3N	
6	CH_3OH	CH_3CN	NH_2CHO	CH_3SH
7	CH_3NH_2	CH_3C_2H	CH_3CHO	CH_2CHCN
7	HC_5N			
8	$HCOOCH_3$			
9	C_2H_5OH	$(CH_3)_2O$	C_2H_5CN	HC_7N
10	NH_2CH_2COOH			
11	HC_9N			
20–50	C_6H_{14} to $C_{16}H_{32}$			
60	C_{60} (fullerenes)			

[*] From Duley and Williams (1984). *Interstellar chemistry.* Academic Press, London.

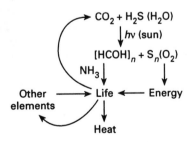

Fig. 9.2 x is the net thermodynamic energy needed for conversion of $H_2O +$ CO_2 to the unstable sugar $C_6H_{12}O_6$, but the pathway requires much greater energy since it is necessary to go via a one-carbon intermediate and at least two activation energy barriers ΔG_1^* and ΔG_2^* (transition states). Note. The final trapped state may be very stable kinetically due to large barriers to change, ΔG_3^* and ΔG_4^*.

The basic route for uptake of C, H, N, and O into living organisms using energy.

the pathways to such unstable products and their kinetic stability in trapped energised states (Fig. 9.2).

Examples of such pathways are:

$$CO + 3H_2 \longrightarrow CH_4 + H_2O,$$
$$CO + 2H_2 \longrightarrow CH_3OH,$$
$$CO + H_2 \longrightarrow HCHO \longrightarrow C_6H_{12}O_6.$$

(Notice that most larger molecular products are also not stable thermodynamically even to disproportionation; for example, glucose, $C_6H_{12}O_6$, goes with energy release to $3CO_2 + 3CH_4$ (see Fig. 9.1 and Section 9.10.2).) The early atmosphere of Earth also contained N_2, so that there were other catalysed reactions that gave products such as HCN and NH_3. The organic chemistry that followed (see Fig. 9.3), evolved on Earth and in interstellar space from a poorly equilibrated mixture of gases, H_2, CO, CO_2 (H_2CO_3), CH_4, NH_3 and HCN, which were now quite well retained by gravity, in the presence of liquid water, some H_2S (see Section 8.5.2) and a variety of catalytic solid surfaces. The catalysts were absolutely essential at the low temperatures of 300–350 K if further reactions were to occur in the absence of light (Section 7.19), but an input of energy from the sun or the interior of the Earth (see marginal illustration) was also required for the build-up of an *energised*, that is, even less equilibrated and therefore more unstable, system of relatively unreactive carbon chemicals. Such a series of compounds were produced and eventually gave rise to organic (and biological) chemistry, but it was only

(a) Types of reaction

Substitution

(includes condensation)

$CH_3OH+HCl \rightarrow CH_3Cl+H_2O$

Oxidation/reduction

$$CH_4 + O_2 \underset{\text{Reduction}}{\overset{\text{Oxidation}}{\rightleftharpoons}} CH_3OH + H_2O$$

Insertion

$CH_3CH_2\text{–}X+CO\rightarrow CH_3CH_2COX$

Addition

$CH_2{=}CH_2+Cl_2{\rightarrow}C_2H_4Cl_2$

Elimination

$CH_3CH_2OH{\rightarrow}C_2H_4{+}H_2O$

Fig. 9.3 (a) Simple classification of reactions showing (b) some intermediates (see Chapter 7) that are usually held by carriers or on catalysts as transients.

(b) Mechanisms of reaction

Heterolytic bond break
(Two-electron)

$$X{:}Y \rightarrow X{:}^- \quad + \quad Y^+$$

Intermediates

Nucleophile   Electrophile

Nucleophiles OH^-, Br^-, $:NH_3$, $H_2\ddot{O}:$

Electrophile Br^+, H^+, CH_3^+

Homolytic bond break
(One-electron)

$$X{:}Y \rightarrow X^\bullet + Y^\bullet$$

Intermediate
free radicals

Free radicals $H^\bullet$, $Cl^\bullet$, $RS^\bullet$

slowly with time that the sophistication of these compounds developed. All of organic and bioorganic chemistry is based to this day on synthesis using energy, starting from CH_4, CO_2, CO, NH_3 (N_2) and H_2O and employing catalysts. (Many of the Earth's stores of raw materials, in which we may include the Earth's available oxygen, oil, gas and coal, are, in fact, late biological products from CO and H_2O.) It is essential to see that the build-up of thermodynamically unstable molecules in relatively stable kinetic traps is the heart of organic chemistry (Fig. 9.2).

9.3 The basic reactions of organic chemistry

Since any required molecule of organic (or bioorganic) chemistry must be made from the stable precursor C, H, N, O small molecules, that is, from H_2O, CO, CO_2, CH_4, N_2, the progress to the required unstable molecules of greater molecular size has to be guided along a pathway. This is the essence of the science or art of organic synthesis and we can describe it in separate parts even though some of these are frequently combined. We shall consider the following:

(1) the relationship of H, C, N, O compounds to one another under classifications such as substitution, insertion, addition, elimination, and oxidation and reduction reactions (Fig. 9.3 and see Section 9.4);

(2) the stepwise mechanisms by which such changes of compounds can be

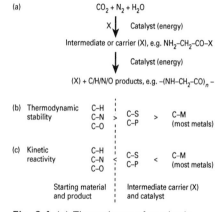

Fig. 9.4 (a) The pathway of synthesis depends on the sophisticated manipulation of weak bonds in intermediate carrier molecules and in the use of catalysts. The relative (b) thermodynamic and (c) kinetic stabilities of carbon —X bonds (Chapters 2 and 7) are summarised and the carbon bonds that are of greatest value in intermediate steps and in catalysis are indicated.

brought about (Fig. 9.4 and Section 9.8) by making and breaking bonds (Fig. 9.3);

(3) the intermediates that are most suitable to carry fragments of H, C, N, O units to one another (Section 9.9);

(4) the energy requirements for the steps (Section 9.10);

(5) the catalysts that are employed to overcome barriers (Section 9.16).

We shall introduce a much abbreviated version of organic and particularly bioorganic chemistry in this way starting with (1) a descriptive account of interrelationships between organic molecules, but we shall find it convenient to concentrate on aspects of acid–base reactions before oxidation/reduction reactions indicating the nature of *substitution* for the main part. It is one route essential to *synthesis*. A dominant concern will be condensation, removal of water, between groups such as —COOH and —NH$_2$, which is seen to be a substitution on C and N for OH and H respectively, that is,

$$—COOH + —NH_2 \longrightarrow —CO—NH— + H_2O,$$

but we shall see later that reduction of CO_2 and N_2 is an essential first step in living systems. Condensation includes the central problem of biology's and man's organic chemistry in water, that is, management of this solvent, water (parallel considerations apply in other solvents). We then turn to the transfer of fragments in intermediates and to the energisation of the condensation reactions before going on to redox reactions and catalysis. Other relationships of reactants and products will be mentioned from time to time. We must commence from the basic framework of carbon hydrogen chains.

9.4 Major small organic chemicals: abiotic organic chemistry

9.4.1 Saturated linear fatty chains (alkanes)

We shall introduce first some of the simplest organic products most likely to have evolved in quantity in the initial universal system of reducing conditions when there was no free dioxygen, namely, the hydrocarbons. They are the most kinetically stable organic molecules. With the formation of heavier oils, hydrocarbons and some fatty derivatives, all insoluble in water, there was created on Earth not only a new set of compounds but also a quite new set of out-of-equilibrium phases or compartments, that is, hydrophobic organic layers (lipids) as new solvents. We believe that it is this step that may well have made life possible and permitted the development of concentrated zones of many other organic chemicals, so helpful to further synthesis. It is worth noting that insolubility and gravity separates oils from water thus allowing large lipid phases to develop on top of or beneath the waters. The lipid molecules were protected by self-association in these phases. Again these oils, apart from the lightest fractions, are liquid in the same temperature range as that in which water is a liquid. It was a remarkably lucky result of chemistry and existing physical conditions (see Section 2.6.3) that allowed production of a new (to the universe) liquid phase, oil in a pre-existing solvent, water. There are several ways in which such oils have been made, for example

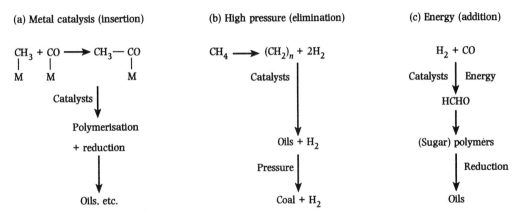

We need go no further with such speculation: these compounds were indeed formed on Earth in quantity some 4×10^9 years ago.

From what we have said, we start the discussion of organic chemicals from these oils (alkanes), which are essentially C/H combinations often in linear saturated arrays, for example

$$CH_3 - (CH_2)_n - CH_3.$$

Their physical properties are mentioned in Table 9.2 along with their common uses. Petrol has this formulation with n equal to about 6–8 (see Table 9.2). For

Table 9.2 Some normal hydrocarbons (alkanes)

Molecular formula	Name	Boiling point (°C)	Melting point (°C)
Natural gas			
CH_4	Methane	−161	−184
C_2H_6	Ethane	−88	−183
Bottled gas			
C_3H_8	Propane	−42	−188
C_4H_{10}	n-Butane	−0.5	−138
Gasoline (Petrol)			
C_6H_{14}	n-Hexane	69	−94
C_8H_{18}	n-Octane	126	−57
$C_{10}H_{22}$	n-Decane	174	−30
Kerosene			
$C_{11}H_{24}$	n-Undecane	194.5	−25.6
$C_{12}H_{26}$	n-Dodecane	214.5	−9.6
Gasoil (light)			
$C_{14}H_{30}$	n-Tetradecane	252.5	+5.5
$C_{16}H_{34}$	n-Hexadecane	287.5	18
Gasoil (heavy)			
$C_{18}H_{38}$	n-Octadecane	317	28
Biological lipids			
$C_{20}H_{42}$	n-Eicosane	334	36.7

NB. Note how liquid ranges increase as molecular weight increases.

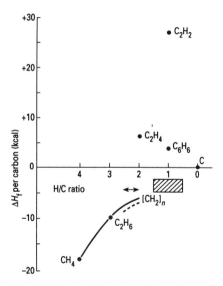

Fig. 9.5 The relative standard state heat of formation, ΔH_f, of C/H compounds per carbon atom. Solid line, linear chains; broken line, branched chains; double-sided arrow, cyclic XH_2 units; cross-hatched region, multiple rings (steroids).

CH₃ CH₃
  CH=CH

cis-butene

CH₃
  CH=CH
        CH₃

trans-butene

HO₂C   CO₂H
  CH=CH

Maleic acid

HO₂C
  CH=CH
        CO₂H

Fumaric acid

Cyclohexane (boat)
C_6H_{12}

Cyclohexane (chair)
C_6H_{12}

Fig. 9.6 The shapes of some organic molecules including *cis* and *trans* isomers.

CH₃
  CH—(CH₂)₃—CH
CH₃   CH₃
                 CH₃
CH₃

The steroid framework.

$n>8$ at 30°C the molecules self-assemble co-operatively as semi-ordered phases.

9.4.2 Unsaturated and non-linear C/H chains

If the ratio of H to C in oils (or in soaps) is lessened, we obtain —CH= links instead of (—CH₂)— links in compounds, for example as in ethylene and benzene. For obvious reasons these are called *unsaturated* rather than saturated compounds (see Table 9.3). The degree of unsaturation controls the properties of the fatty compounds (Table 9.3). Unsaturation also gives new structural possibilities to organic molecules since double bonds introduce stable geometric patterns, *cis* or *trans* (see Fig. 9.6). (Note how molecular shape is introducing itself here and even in saturated rings.)

In Fig. 9.5 we show the effect on the thermodynamic stability of decreasing the C/H ratio in compounds. The higher the ratio H/C, ultimately CH₄, the more stable the molecule. Branched chains are more stable than linear chains since more C atoms have a high H/C ratio. We can include those compounds that are based on saturated carbon–carbon networks, including alicyclic compounds, for example C_6H_{12}, and steroids (see marginal illustration) as well as those that contain double and triple bonds under 'unsaturated' compounds in the sense that the oxidation states of carbon in them are less negative than in simple saturated chains. This relationship can also be seen by noting their structural relationship to carbon in diamond as well as in graphite. It is generally true that saturated linear —[CH₂]ₙ— chains are more stable than chains containing double or triple bonds or than rings in the presence of H₂ or reducing agents giving potentials of about —0.5 volts at pH = 7.0. Energy is required to remove hydrogen from carbon.

Table 9.3 Some representative alkenes*

Molecular formula	Common name	Uses
CH₂=CH₂	Ethylene	Synthesis of polymers (plant hormone)
CH₃CH=CH₂	Propylene	Synthesis of polymers
CH₂=CH—CH=CH₂	Butadiene	Synthetic rubber
CH₂=CH—C(CH₃)=CH₂	Isoprene	Natural rubber, basis of terpenes, e.g. some vitamins, steroids
	Benzene	Solvent, raw material of aromatic chemistry
CH₃—	Toluene	Synthesis of TNT
	Naphtalene	Moth repeller, dyestuffs

* For long-chain unsaturated hydrocarbons see Table 9.4.

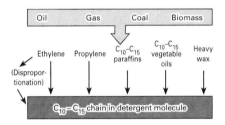

Fig. 9.7 From raw material to C_{10}–C_{15} backbones by refining. At the top are shown inputs giving rise to refined intermediates and leading to detergents.

High hydrogen pressure stabilises C—H bonds, but high external pressure at low pressure of H_2 will favour the loss of hydrogen and the production of heavy oils and ultimately coal. Note that disproportionation produces methane and carbon (coal),

$$C_nH_{n+m} \longrightarrow CH_4 + \text{carbon},$$

in all cases with energy loss so that the reaction goes readily if catalysed. Great skill is used in forcing formation of the most desirable C/H combinations, for example in oil refineries linked to the detergent industry (Fig. 9.7).

Many unsaturated systems are made by *elimination* or *oxidation* (Fig. 9.3) and their production is extremely important for the physical properties of fats. In particular, some polyunsaturated fats are distributed in the membranes of living organisms in a fascinating manner related to evolution (see Section 12.7) and were due to an historically late change in the mechanism of synthesis from elimination to oxidation, requiring some use of dioxygen. (The effect on physical properties is listed for acid derivatives in Table 9.4.) Chain branching also introduces important packing problems and alters stability of membranes as does an intermediate degree of branching (see Figs 9.8 and

Fig. 9.8 Micelle formation (below left) in which a lipid incorporates oil or fatty material. This is thought to be the transport mode for hydrophobic molecules in the animal bloodstream. A diagram of the bilipid layer membrane of a vesicle or a cell (below right) with (at (a) and (b)) a typical lipid, phosphatidylcholine. The ions surrounding and inside a *cell* are shown (below right) but the distribution is reversed in *vesicles* within biological cells. R,R′, fatty acyl chains. The cellular system has three compartments: inside water, outside water and membrane lipid.

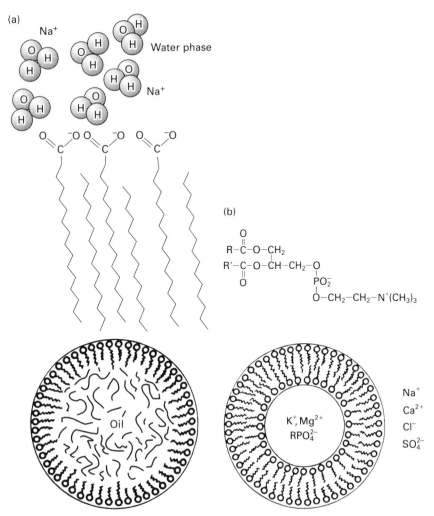

Table 9.4 Glycerol esters

$$
\begin{array}{c}
\text{CH}_2-\text{OH} \\
|\\
\text{CH}-\text{OH} \\
|\\
\text{CH}_2-\text{OH}
\end{array}
\quad + \quad 3\ \overset{\displaystyle O}{\underset{\displaystyle \parallel}{\text{RCOH}}}
\quad\longrightarrow\quad
\begin{array}{c}
\overset{\displaystyle O}{\overset{\displaystyle \parallel}{}} \\
\text{CH}_2-\text{O}-\text{CR} \\
|\quad\ \ O \\
|\quad\ \ \parallel \\
\text{CH}-\text{O}-\text{CR} \\
|\quad\ \ O \\
|\quad\ \ \parallel \\
\text{CH}_2-\text{O}-\text{CR}
\end{array}
\quad + \quad 3\ \text{H}_2\text{O}
$$

Glycerol   'Fatty Acid'   Fat (glycerol ester)

The R groups can be the same or different within the same fat. Furthermore, the R groups can be *unsaturated*, that is, the carbon chain of the fatty acid may contain one or more double bonds, and so is referred to as monounsaturated or polyunsaturated, respectively. Alternatively, the fat can be *saturated*, that is, all C atoms in the chain form σ bonds to other C atoms or to H atoms. Some common fatty acids are listed below.

			Melting point (°C)
Saturated acids			
Butyric	C_4	$CH_3CH_2CH_2COOH$	-5
Lauric	C_{12}	$CH_3(CH_2)_{10}COOH$	$+50$
Myristic	C_{14}	$CH_3(CH_2)_{12}COOH$	$+59$
Palmitic	C_{16}	$CH_3(CH_2)_{14}COOH$	$+64$
Stearic	C_{18}	$CH_3(CH_2)_{16}COOH$	$+69$
Unsaturated acids			
Oleic	C_{18}	$CH_3(CH_2)_7CH{=}CH(CH_2)_7COOH$	$+14$
Linolenic	C_{18}	$CH_3CH_2CH{=}CHCH_2CH{=}CHCH_2CH{=}$ $CH(CH_2)_7COOH$	-11
Arachidonic	C_{20}	$CH_3(CH_2)_4CH{=}CHCH_2CH{=}CHCH_2{-}CH{=}$ $CHCH_2CH{=}CH(CH_2)_3COOH$	-49

Whether the fatty acid is saturated or unsaturated, and the degree of unsaturation, is important to the food industry, in our diet and in the detergent industry as well as in biology.

9.9). The inclusion of specified allicyclic rings further alters the stereochemistry of the C—C chains and the way in which molecules can fill space. There is great interest then in the phase diagrams of mixtures of fats of all kinds (Sections 4.4.2 and 4.4.3). While double bonds are commonly found in linear molecules, they also allow the formation of ring structures of almost any extent. Both in biological systems and man's industry the use of such small unsaturated molecules in synthesis is of very great value (Table 9.3). Benzene, C_6H_6, is the first member of a planar series that extends all the way to graphite. All such unsaturated compounds are most unstable to reduction (Fig. 9.6). We shall have occasion to notice that unsaturation also increases reactivity of C/H frameworks (and C/H/N frameworks) so that the reactivity becomes more like that of S and P compounds. Thus, unsaturated organic compounds are useful in catalytic and carrier functions (Sections 9.9 and 9.10 and Chapter 10).

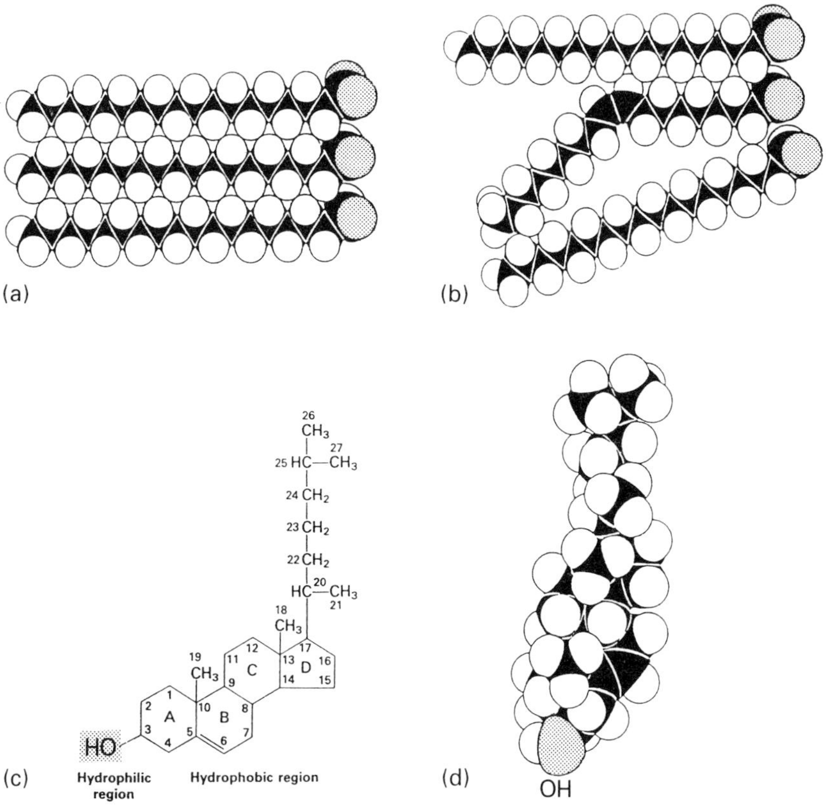

Fig. 9.9 (a) The highly ordered packing of fatty acid chains is disrupted by the presence of *cis* double bonds or by cholesterol. These space-filling models show the packing of (a) three molecules of stearate (C_{18}, saturated) and (b) a molecule of oleate (C_{18}, unsaturated) between two molecules of stearate. (c) Formula of cholesterol; (d) space-filling model of cholesterol.

9.4.3 Substitutional derivatives of C/H compounds

Carboxylic acids

Because of their particular importance in bioorganic chemistry and the development of life we treat carboxylic acids first. While still bound to metals, alkyl fragments, for example CH_3—CH_2—M, can react by insertion with CO to give new molecular acyl compounds, which with water give salts of long-chain 'fatty' acids (see Table 9.4) of the types that are related to soaps. An example of a substituted alkyl chain is

$$CH_3-(CH_2)_n-CO_2H.$$

The 'fatty' acids can be combined with multifunctional alcohols (see below) such as glycerols, formed by reduction of polyketones or aldehydes, to give esters (Table 9.4). While oils remain totally insoluble in water and form separate liquid phases (for example, petrol), soaps, for which $10 > n < 20$, form liquid films and membranes in the form of bubbles or vesicles, by constructing organic monolayers or bilayers with the carboxylate or glycerol heads of the molecules in contact with the water (Fig. 9.8). Such soap bubbles readily *disperse* in water as separate phases. In a mixture of oils, soap and water, some added chemicals go into one phase, water, and some go into the other, the oily soap. We shall find this to be most important in analysis, industry and particularly biology. Now, some of these bubbles, vesicles, have water inside

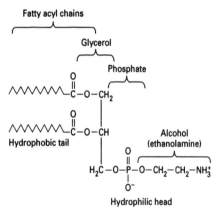

The structure of phosphatidylethanolamine, a typical phospholipid. The hydrophobic and hydrophilic parts of the molecule are the fatty chains and the glycerol phosphate, respectively.

them rather than air (Fig. 9.8). In fact, this step created on the surface of the Earth a second new division of phase space that separated outside and inside non-equilibrated *aqueous* phases (see Chapter 6), that is, a further new compartment, and so provided the opportunity of a new *localised equilibrium* chemistry in water (inside vesicles). (No life exists without this compartmental containment.) Incidentally, the vesicles generated a new phase 'shape', here just spheres, hollow or filled. The vesicles have many points of interest since the self-assembly of the hydrocarbon lipid chains, 'soaps', gives the anionic headgroups a locally energised unstable packing, especially on their inner surfaces (see below). This process is a further example of chemical self-assembly locally, distinguishing inside versus outside surfaces, and, unlike the separation of oil and water that depends partly on gravity, this process depends only on selective weak, co-operative, chemical forces and could well have given selective element (cation) binding (see Fig. 9.8(b)). Very importantly, these new liquid phases have surfaces that are commensurate with their bulk, so that surfaces of phases take on a particular relevance (see Section 6.2.3). Self-assembly in these cases does not meet an important kinetic transfer barrier; contrast mineral crystal formation (Section 7.17). As mentioned already the stability of the liquid phases is very sensitive to unsaturation in the chains and to the inclusion of cyclic ring compounds such as cholesterol.

Other non-metal substitutions

The next step in this much simplified description of organic chemistry is the general *substitution* of H bound to C by other elements in combination with hydrogen, for example, by NH_2, OH, SH or SeH, or even (much later in abiotic organic biological and chemical evolution) by halogens, for example Cl, Br or I. (Halogenation would have been very difficult some 4×10^9 years ago due to the reducing conditions.) The resulting compounds are called amines, alcohols, thiols and alkyl halides: some representative examples are

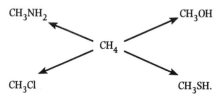

They can also be made by *addition*, for example of H_2O or NH_3 or HX across double bonds, or by *reduction* of, for example CO- or CN-containing compounds. (The reverse reaction to addition is called *elimination*, whereas exchange between groups, for example OH for Cl, is also called *substitution* (Fig. 9.3).) The introduction of more complex units, for example $-NO_2$, $-SO_3H$ or $-PO_3H_2$, so incorporating additional non-metals via the use of the oxygenated inorganic species, for example HNO_3, H_2SO_4, H_3PO_4, allows us to obtain nitrated, sulphated and phosphated compounds such as

$$CH_3ONO_2, \quad CH_3OSO_3H, \quad CH_3OPO_3H_2 \quad \text{(see Table 9.7)}.$$

These products usually arise from the reaction of fatty alcohols with inorganic acids by loss of water in *condensation* reactions, which prove to be extremely important for life and organic chemistry generally (see Section 9.5). Probably because phosphate was quite available at the earliest time in an anaerobic atmosphere, while nitrate and sulphate were reduced, phosphate compounds formed very early in living systems and were probably present in abiotic chemistry. An example of such a condensation reaction is

$$CH_3OH + HPO_4^{2-} \longrightarrow CH_3OPO_3^{2-} + H_2O.$$

Note again that none of these *esters* (as the products of reaction of acids and alcohols are called) are kinetically stable in water (and this includes the phosphoglycerate esters of membranes (see Fig. 9.8(b)), but they are not, in fact, very unstable either. Of course, unsaturation or chain branching can be introduced together with substituents. Once again this changes the properties and particularly affects self-assembly, for example of membranes.

Turning to the thermodynamic properties of these substituted molecules we can ask about the stability of, say, an alcohol to loss of water or an amine to loss of ammonia. To answer this question we consider *addition* reactions

$$CH_2{=}CH_2 + H_2O \longrightarrow CH_3CH_2OH, \qquad \Delta G° = -2 \text{ kcal},$$

$$CH_2{=}CH_2 + NH_3 \longrightarrow CH_3CH_2NH_2, \qquad \Delta G° = +15 \text{ kcal}.$$

which show that the reaction of removal of —NH$_2$ by —OH is favourable, that is,

$$CH_3CH_2NH_2 + H_2O \longrightarrow CH_3CH_2OH + NH_3, \qquad \Delta G° = -17 \text{ kcal}.$$

These observations gives us a feeling for the energy problems in the syntheses of organic molecules.

Alternatively, we can examine the reduction by hydrogen in reactions such as

$$CH_3CH_2OH + H_2 \longrightarrow CH_3CH_3 + H_2O, \qquad \Delta G° = -24 \text{ kcal},$$

$$CH_3CH_2NH_2 + H_2 \longrightarrow CH_3CH_3 + NH_3, \qquad \Delta G° = -7 \text{ kcal}.$$

Once again, the instability of substituted carbon compounds is clear and the problems of syntheses requiring subsequent protection are obvious.

A particular difficulty for the explanation of organic chemistry on early Earth arises from the fact that the introduction of double bonds to the heteroatoms as in ketones and aldehydes (Table 9.5) or imines also reduces stability further, so that they are all very unstable to reduction by hydrogen, but yet they are essential intermediates. Their lifetimes at relatively high temperatures (around the 100°C of early Earth) must have been short, especially in the presence of metal ion catalysts, which was inevitable on early Earth. The chemical and physical properties of all these simple compounds are described in elementary texts on organic chemistry (see 'Further reading').

As polarity and hydrogen-bonding ability are increased and/or charged groups are introduced, the organic compounds become more soluble in water and generally more reactive. The generation of polarity (see Section 2.5.1) can be achieved in several ways by introducing groups containing electronegative atoms. Simple examples of soluble neutral compounds are given by methyl alcohol, amino acids, glycerol, and so on. Solubility, of course, allows organic and biological chemistry in water (Chapter 5). By way of contrast, ethers, for

Table 9.5 Some representative aldehydes and ketones

Molecular formula	Common name	Uses
HCHO	Formaldehyde	Synthesis of polymers and dyestuffs
CH_3CHO	Acetaldehyde	Synthesis of acetic acid; fabrication of silver mirrors
$CH_3(CH_2)_5CHO$	Enantal	Synthesis of heptylic alcohol and enantic acid, used in perfumes·
$CH_2 = CH - CHO$	Acroleine	Raw material; very irritating (tears)
$CH_3·C = CH(CH_2)_2·C = CHCHO$ with CH_3 and CH_3	Citral	Artificial essences (lemon)
$CH_3 - C - CH_3$ with O (double bond)	Acetone	Solvent (e.g. for nitrocellulose to make gun powder and in making artificial silk)
$CH_3 - C - CH_2CH_3$ with O (double bond)	Metyl ethyl ketone	Solvent
$C_6H_5 - C - CH_3$ with O (double bond)	Acetophenone	Old hypnotic
	Muscone	Essence (musk)

example CH_3—O—CH_3 (see below), are relatively unreactive, of low polarity, and are not very soluble in water. (Some long-chain ethers generated the membranes of the very primitive archaebacteria.) The organic substituents —CO_2^- and —OPO_3^{2-} or —$O(RO)PO_2^-$ thus have a major role in biological chemistry, namely to give solubility in water *through charge*, for example in substrates, lipids, proteins, RNA and DNA.

We have now completed our brief description of the *relationship* between carbon compounds containing a variety of non-metal atoms. As we progress with these analyses the reader should keep in mind that almost every reaction has four dominant features: (1) it demands selection of elements; (2) it requires energy; (3) it needs a suitable solvent; and (4) it requires a pathway to the 'trapped' product (using a catalyst), which will have a reasonable lifetime, say, in water, only if it is stable to solvolysis (hydrolysis) (see Figs 9.4 and 9.5).

9.5 Shapes of organic molecules

We have already mentioned the shapes of organic molecules in Chapters 2 and 6 and, in passing, in Section 9.4.2. As well as the polarity, the shapes of organic molecules help to control their properties. For saturated compounds the basic feature is the tetrahedral arrangement of groups around any carbon (see Section 2.5.2). The corresponding molecules are not as easily packed as the flat ring of such unsaturated molecules as benzene. The introduction of aromatic rings into a compound can lead to stacking as in herring-bone packing (see DNA in Fig. 9.16) and once again assists self-assembly. Generally, the rules of shape around H, C, N and O atoms in organic compounds are simple (see Section 2.5.2), but these simple rules allow a fantastic range of complex structures and shapes especially in large polymeric molecules and their assemblies. We shall see this in the wealth of structures in biology (Chapters 10–13). Many of these shapes are dynamic, so that moving (mechanical) units as well as platforms or solid three-dimensional units can be made using organic chemicals, for example rubbers and plastics (Section 9.4). As mentioned earlier, the formation of condensed liquid phases and even of solids generates self-assembled macroshapes as in vesicles (Fig. 9.9). The properties of aliphatic systems are greatly influenced by unsaturation in chains or through cyclisation, and shape then influences the properties of membranes in biology (Fig. 9.9.).

9.5.1 Handedness within shape: optical activity

The handedness of certain organic molecules is illustrated in Figs 2.16 and 9.10 and is obviously related to the absence of symmetry in their shape. The problem of the origin of the particular chirality (optical activity) of some biological molecules, for example amino acids (L) and sugars (D), is as yet unsolved. What kind of evolutionary pressure has led some compounds to appear as just one optical isomer? Man's syntheses, in agreement with equilibrium thermodynamics (Section 3.3.1), yield only racemic mixtures (equal amounts of L- or D-forms) unless specifically designed to obtain one of them by some special catalytic or 'resolution' technique. Living systems, on the other hand, synthesise only one of the two although no logic demands such a choice. There could have existed two primitive systems, one based on one, the second on the other, but it is hard to see what kind of evolutionary pressure left just one (L) for amino acids and the other (D) for saccharides.

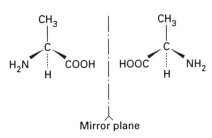

Fig. 9.10 An example of two optical isomers in L-alanine and D-alanine. Notice that the mirror images are not superimposable.

9.6 A general summary of organic chemical constructions

The reader may now take up an organic chemistry textbook and see how to put elements together in frameworks as in a three-dimensional jigsaw puzzle, using only C, H, N, O plus some S and P atoms, with particular combinations of single and multiple (usually double) bonds with well defined angles between bonds and any combining ratios. The ways are obviously infinite and in the last 150 years man has learned to handle this chemistry to make a bewildering variety of materials, foodstuffs, medicines, drugs, etc. (see Table 9.6), all of which have properties dependent on specific atoms, shapes and internal dynamics as well as upon their kinetic stability. While all of us know the power and success of this industry it does not yet challenge the variety of production of similar substances by living organisms, illustrated in Table 9.7. This gives us a reason to represent organic chemistry in the rest of this chapter largely by bioorganic chemistry† and at the same time to maintain a historical attitude to the natural selection of the chemical elements. Indeed, natural abiological and biological organic chemistry evolved first and are more than 4×10^9 years

Table 9.6 Some man-made organic chemicals

Application	Examples
Materials	Celluloid, Bakelite, Plexiglas, synthetic rubber, polyethylene, polystyrene, Teflon, PVC, silicone
Fibres	Nylon, Rayon, Orlon, Dacron, Acrilan, Terylene
Dyestuffs and pigments	Azodyes, phthalocyanins, quinacridones, triphenylene dyes
Drugs	Phenacetin, aspirin, veronal, salvarsan, sulphonamides, synthetic penicillins, anti-inflammatory steroids such as cortisone, tranquillisers such as diazepam (Valium), diuretics such as furosemide (Lasix)
Insecticides	DDT, lindane, dieldrin, aldrin, malathion, parathion, pyrethrins
Propellants	Fluorochlorocarbons (freons)
Sweeteners	Saccharin, cyclamate, aspartame, etc.
Detergents	Sodium *p*-dodecyl benzene sulphonate, polyoxyethylene (non-ionic)

Table 9.7 Some natural substances

Application	Examples
Materials	Cellulose, chitin, rubber
Fibres	Hair, silk, wool, cotton
Dyestuffs and pigments	Carotene, chlorophyll, astaxanthin (salmon, boiled shrimps, pink flamingo), anthocyanins (red, purple and blue colours of flowers and berries), melanine (tanning of skin and hair pigment)
Drugs, and hormones	Vitamins, hormones such as testosterone (male sex hormone) and oestradiol (female sex hormone), morphine, quinine, camphor, adrenaline, insulin, penicillin, tetracyclines, interferon
Scents	Musk, vanillin, geraniol, hyacinthin, ionone (violets) linaleol (oil of jasmine), carvone (oil of spearmint)
Sweeteners	Sucrose, fructose, glucose, maltose, etc.

† We distinguish *in vivo* studies, *biological* (organic) chemistry, from *organic* chemistry, *man's laboratory activity* generally, and from the study of 'organic' chemicals *extracted* from living systems, *bioorganic* chemistry.

older than man's organic chemistry. However, notice again that many of the problems of industrial organic, bioorganic and biological organic chemistry are the same: to synthesise *unstable* compounds from chosen starting materials (chemical elements eventually) by sucking in energy all the time, by using kinetic tricks particularly, and by the choice of reagents and catalysts so as to move intermediates along 'desired' pathways, that is, by using procedures so as to force chemicals to enter deep traps as in Figs 9.2 and 9.4. It is only lack of skill that restricts the success in the making of any organic molecule. We look first at a model of how biology manages syntheses, but we shall have to be aware constantly that most *biological* organic chemistry is carried out in water (notice that most of its compounds have oxygen atoms in them and are frequently charged), whereas man's organic chemistry is usually carried out in oil-like solvents. We must always note the presence of these solvents and consequent hydrolysis or solvation reactions, and later of the risk of redox reactions and their energetics, whenever we consider organic reactions in the presence of dioxygen or other oxidising agents.

If we now look back to Chapters 3–8, we see that there are no fundamental principles in the treatment of *small* organic molecules, which are different from those used to discuss inorganic salts. We refer to their physical and chemical properties using molecular language but in the discussion we always have in mind large numbers of identical individual molecules in a crystal or a solution that form a phase (or a compartment). In this way we used the phase rule for guidance on physical properties in Chapters 4 and 6, and we used statistical treatments of solutions in binding studies and reaction analysis in Chapters 5 and 7. There is then a common physicists' and chemists' approach to all chemicals, and organic chemistry differs only in that it is restricted to a small number of elements that have great diversity of combination. There is, therefore, no new problem that will limit discussion of molecules of up to several hundred atoms. However, when we treat very *much larger* molecules and especially in the context of biology we must be certain that this same descriptive language remains useful. In Section 6.8 we stated that when a folded molecule is large enough it can be treated as a co-operative 'phase' and that this 'phase' can self-assemble, not with identical molecules but with other very large molecules, 'phases', to give still larger bodies with idiosyncratically shaped 'structures', for example viruses and then living systems. At this point it is necessary to put in a cautionary word about the use of words in chemistry and their carry-over into biology. We are attempting in this book to understand the selection of the elements in materials and as we approach its most exciting development, that is, in biology, we must not assume without demonstration that we can treat biological (organic) chemistry in exactly the same way as we treat laboratory chemistry (see Section 1.1).

Before we engage in a further description of organic chemistry, therefore, the reader is reminded by the use of one example of the use of words in chemistry and biochemistry. The most confusing is the word *structure*. First, it refers to the *order* in *sequences* in which atoms are joined together and which restricts dynamics but only defines gross limits to the ways in which the atoms occupy space, for example CH_3CH_2OH. This primary 'structure' is the equivalent of a loose assembly of balls on a string or set of linked strings: it has some order and much configurational disorder or entropy but no sequence disorder. The importance of such dynamic physical properties for reaction is obvious in liquid (or gaseous) phases, while in solids it largely disappears—organic chemistry is generally molecular chemistry carried out in liquid solution. A quite different second meaning of *structure* is that of a rigid molecular construct such as a set of balls on a rigid rod or frame. Thus, we can say that individual *n*-hexane and benzene molecules have chemical 'structures' in the liquid state but the meaning is not quite the same: only benzene molecules have a *fixed shape*. In other words, *n*-hexane has more disorder or configurational entropy. Now both these liquid chemicals can also be put in a vessel when the liquid phase takes on a 'shape' in a container. It is *structured* in the sense that it is a phase restricted by a barrier, the vessel. This is a structure described by bulk phase surface properties, most obviously so in solids, and it can also be varied. However, if a sequence is long enough it may fold into a preferred *co-operative structure*, and this single molecular fold, described under 'secondary' and 'tertiary' *structure* (see Section 9.8.6), has all the attributes of a phase (Section 6.8). Here lies the confusion between structural transitions of small molecules, that is, their transitions of physical state, due to translational motion changes and transitions of large molecules introducing configurational disorder, both called melting. The word 'structure change' may thus mean different things. This is not a trivial point since long-range co-operativity is now the property of many large single molecules. Finally, liquids can flow in a field gradient so that their *momentum* is restricted. We shall call such 'flowing structure' *organisation* when the flow is contained on a directed path. It is equivalent to the 'structure' of the universe or of planetary systems. In man's chemistry this meaning of *structure*, in effect *organisation*, is applied to the directed flow of a gas or liquid such as benzene through a series of tubes as in separation equipment. As we shall see it is the advancement of structural organisation due to the properties of large molecules that is the highest form of natural selection of elements both in biology and in man's equipment. We must slowly turn our attention toward the properties of large molecules individually within the context of organic chemistry.

9.7 Basic practices of organic and biological organic chemistry

9.7.1 Introduction

When we describe either inorganic or organic chemistry in textbooks it is conventional to assume knowledge of very simple equipment, which then is not described and is not treated as part of the reaction system. Again, conditions are set by independent thermostats and pressure is equally controlled though often not stated. Thus, the choice of either non-connected (isolated) or connected systems is taken for granted. In fact, the equipment of organic chemistry is usually formed from inorganic materials in a 'frozen' state, glass or steel (Table 9.8). The situation in biology is quite different. All facets of a cell's existence including both those in its equipment, for example membranes, and its reactions are interdependent and will often be found to contain both organic and inorganic elements. Despite this problem, in this chapter we wish to give a view of some of the organic chemicals and their reactions within biological compartments as if the chemicals and reactions

Table 9.8 Biological and laboratory operations

	Laboratory	Biological
Containment	Glassware, metal vessels	Fatty acid membranes
Transfer	Filtration, distillation, extraction	Pumping through membrane channels
Solvents	Wide choice of organic solvents	Water or fatty acid membranes
Chemical intermediates	Use of many elements (>30) including metals	Use of some 15 available elements

could be isolated from their containers as in an organic chemical laboratory or factory. This is a usual biochemical or bioorganic chemical attitude but it is not a sound biological chemical approach. We start therefore by giving the briefest description of the containing system of a cell as if it were fixed before passing to some of its major reactions using the concepts described in Chapters 3 and 7, and finally, reverting to bioorganic chemistry, to the study of molecules isolated from living systems.

9.7.2. The apparatus of biological organic chemistry

The advantages of containment in compartments for reactions have been thoroughly analysed in Chapter 7 but are largely obvious. In essence, biological cells are analogous to vesicles in that they have a water solution inside as well as outside a membrane. The membranes of the cells are largely made from fatty acids related to soaps and detergents (in fact, much of man's soap is made from plant lipids or fats). Inside cell membranes is the aqueous cytoplasm and outside is the aqueous environment; hence we have minimally a three-compartment non-equilibrium system. If we find small enclosures inside large cells we still call them vesicles or sometimes *organelles*, but inside the vesicles or organelles the aqueous compartment is again different. Thus, biological organic chemistry is based on multiple non-equilibria in aqueous solutions and membrane phases, a multiplicity of compartments. Such compartmental restriction is extremely valuable in man's organic chemistry too, but, as stated above, is maintained by solvents, glassware, industrial plant equipment, etc. (Section 7.10.5). This structure is needed to maintain concentrations of reactants, of course, while keeping modes of transfer present which are to be used as necessary. Living systems have means of transfer just as man does, so that the operational modes are parallel. A way of looking at the gradual introduction of complexity and energy into chemicals is shown in Fig. 9.11. The huge difference between this organic chemistry and biological chemistry rests in the self-organising feedback flow in biological systems at all levels, which is yet to be described. We pass now to the isolated stepwise routes and the mechanisms by which synthesis is achieved within the apparatus described. The reader may wish to use an organic textbook to see parallel paths in man's syntheses and to observe the parallel use of compartments, transfer, catalysts and so on.

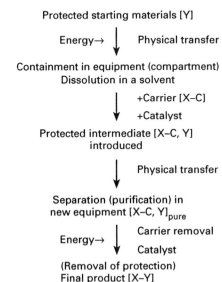

Fig. 9.11 The gradual introduction of complexity and energy into chemicals in biological or man's organic chemistry.

9.8 General introduction to bioorganic chemistry

Since the organic reactions in biological synthesis and metabolism have a close parallel with those in man's organic chemistry we can continue to follow the language and descriptions of organic chemistry (laboratory) texts, while we describe isolated (extracted) units from biology. This is often called bioorganic chemistry. All processes are then described in classes by reaction types.

Now, as stated, one of the most important acid–base reactions in chemistry is *substitution*, and in the case of bioorganic reactions the most important substitution is *condensation* (Table 9.9). In particular, the removal of water is the dominant substitution reaction of all leading to synthesis, for example, in the reaction

$$RCO{-}OH + NH{-}R' \longrightarrow RCO{-}NHR' + H_2O.$$

Let us consider the conditions necessary to bring about a designed synthesis, that is, by following Section 9.3:

1. The desired product must be considered so that the *appropriate supply of chemical elements*, starting materials, is made available (Section 9.8.1).

2. The *reaction vessel* must be designed so that the *conditions of reactions* (temperature, pressure, light, etc.) can be managed and the starting materials introduced appropriately (Section 9.17).

3. There must be a designed pathway with a succession of *intermediates* and *transfer operations*. Protection and/or carrier molecules may be necessary (Section 9.9).

4. The necessary *energy has to be introduced* (in every step probably) (Section 9.10).

5. *Catalysts* for individual steps will be required and need to be selected (Section 9.16).

Table 9.9 Some condensation products

Formula	Type
$CH_3COOC_2H_5$	Ester
Fatty acyl$-$O$-$CH$_2$ Fatty acyl$-$O$-$CH CH$_2$OPO$_3^{2-}$	Phospholipid (ester + phosphate ester)
$(C_2H_5)_2O$	Ether (see ring polysaccharides)
$^{2-}O_3POCH_2$ (sugar ring, OH, OH, OH)	Glucose-6-phosphate (see above) (see DNA, RNA)
$CH_3CO-NHR$	Amide (see polypeptides and proteins)

We shall now describe this approach for the condensation pathway leading to biological polymers. We choose to illustrate this with biopolymer chemistry since it is dominant in the refinement of the chemical selection of the elements in biology and introduces additional concepts. (Redox reactions will be described in Section 9.15.)

9.8.1 Condensation reactions in bioorganic chemistry

The central feature of biological organic molecules is the presence of polymers such as nucleic acids and proteins. We are, therefore, particularly interested in the synthesis of the large molecules from smaller ones, which requires participation of the element nitrogen, as well as carbon, hydrogen and oxygen.

Much of the simpler C, H, O organic chemistry has already been mentioned extending polymers as far as glycols, $-(CHOH)_n-$ (Table 9.4), and fats, $-(CH_2)_n-$ (Table 9.2). The properties of these compounds have also been described. The simplest organic chemistry inside the cell, which involves nitrogen as well as C, H and O, may be appreciated by starting from CO_2 of the air and ammonia, NH_3, which together give urea, a compound that is soluble in water, that is,

$$CO_2 + 2NH_3 \longrightarrow CO(NH_2)_2 + H_2O, \qquad \Delta G^\circ = +7 \text{ kcal}.$$

Urea was the first really 'organic' compound, that is, the first compound from a living source, to be characterised and synthesised by man (Wohler in 1832). This success truly represents a step in the evolution of our knowledge of chemicals but biological systems have used urea for some $4–3 \times 10^9$ years. The essence of the reaction is again the removal of water. (Since water is not present in dry organic phases, these phases help to allow the use of condensation reactions by man that will not proceed easily in water.)

Other compounds containing the CO group, comparable with urea, are the aldehydes and the ketones (see also Table 9.6),

$$\text{aldehydes } \begin{matrix} R \\ | \\ CO \\ | \\ H \end{matrix} \text{ and the ketones } \begin{matrix} R_1 \\ | \\ CO \\ | \\ R_2 \end{matrix} \text{ (see Table 9.6)}$$

which can also undergo addition and then condensation reactions with ammonia but can also react amongst themselves, for example in so-called aldol condensations. We may also devise other condensation reactions to give now an R—O—R link,

$$C_2H_5OH + C_2H_5OH \longrightarrow C_2H_5OC_2H_5 + H_2O.$$

$C_2H_5OC_2H_5$ is an ether (see Table 9.9), in this case diethyl ether, which is not very soluble in water (see above), and is used as an anaesthetic. The ether link is the basis of polysaccharide chemistry (Fig. 9.12), which involves polyalcohols in a multiplicity and variety of condensation reactions (see below). Many of these oxygen- and nitrogen-containing C/H compounds are polar and remain soluble or somewhat soluble in water, but, as we reduce the amount of oxygen and nitrogen in the molecules, which is the source of polarity, they dissolve increasingly in organic, lipid (membrane) phases.

Maltose (β-form)
(4-O-α-D-glucopyranosyl-β-D-glucopyranose)

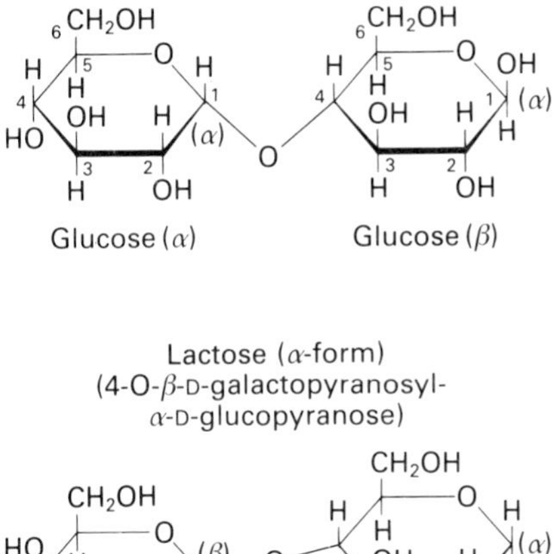

Glucose (α) Glucose (β)

Lactose (α-form)
(4-O-β-D-galactopyranosyl-α-D-glucopyranose)

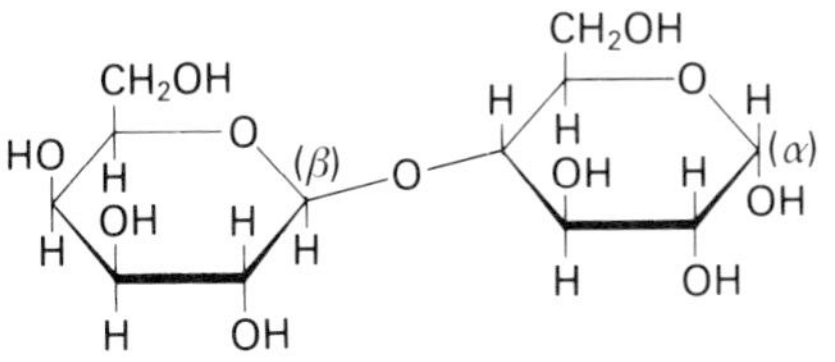

Galactose (β) Glucose (α)

Sucrose
(α-D-glucopyranosyl-β-D-fructofuranoside)

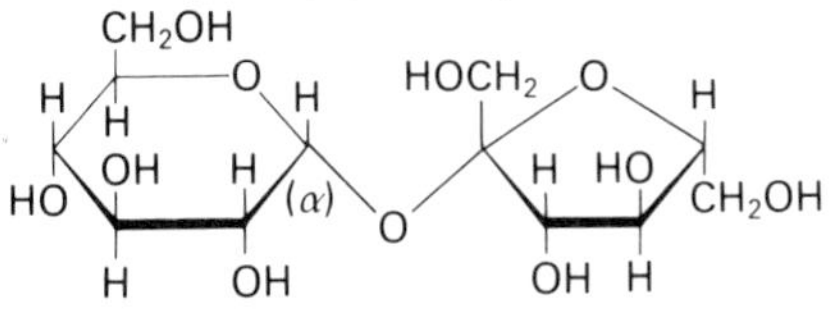

Glucose (α) Fructose (β)

Fig. 9.12 A simple saccharide, glucose in three disaccharides, showing an ether link to itself and two other monosaccharides.

Table 9.10 The natural protein amino acids

Name	3-letter symbol	1-letter symbol	Sidechain	Character	Metal binding
Aspartic acid	Asp	D	$HOOC-CH_2-CH{<}^{COO^-}_{NH_3^+}$	Acid (polar)	Mg, Ca
Glutamic acid	Glu	E	$HOOC-CH_2-CH_2-CH{<}^{COO^-}_{NH_3^+}$	Acid (polar)	Mg, Ca
Tyrosine	Tyr	Y	$H-O-\bigcirc-CH_2-CH{<}^{COO^-}_{NH_3^+}$	Neutral (non-polar)	Fe
Alanine	Ala	A	$CH_3-CH{<}^{COO^-}_{NH_3^+}$	Neutral (non-polar)	
Asparagine	Asn	N	$H_2N-CO-CH_2-CH{<}^{COO^-}_{NH_3^+}$	Neutral (polar)	
Cysteine	Cys	C	$HS-CH_2-CH{<}^{COO^-}_{NH_3^+}$	Neutral (non-polar)	Cu, Zn, Ni, Fe
Glutamine	Gln	Q	$H_2N-CO-CH_2-CH_2-CH{<}^{COO^-}_{NH_3^+}$	Neutral (polar)	
Serine	Ser	S	$HO-CH_2-CH{<}^{COO^-}_{NH_3^+}$	Neutral (polar)	
Threonine	Thr	T	$CH_3-\underset{OH}{CH}-CH{<}^{COO^-}_{NH_3^+}$	Neutral (polar)	Mg
Histidine	His	H	imidazole $-CH_2-CH{<}^{COO^-}_{NH_3^+}$	Basic (polar)	Cu, Zn, Mn, Fe, Ni
Arginine	Arg	R	$H_2N-\underset{\underset{+}{NH_2}}{C}-NH-CH_2-CH_2-CH_2-CH{<}^{COO^-}_{NH_3^+}$	Basic (polar)	
Lysine	Lys	K	$H_3\overset{+}{N}-CH_2-CH_2-CH_2-CH_2-CH{<}^{COO^-}_{NH_3^+}$	Basic (polar)	
Glycine	Gly	G	$H-CH{<}^{COO^-}_{NH_3^+}$	Non-polar hydrophobic	
Isoleucine	Ile	I	$CH_3-CH_2-\underset{CH_3}{CH}-CH{<}^{COO^-}_{NH_3^+}$	Non-polar hydrophobic	

Table 9.10 *continued*

Name	3-letter symbol	1-letter symbol	Sidechain	Character	Metal binding
Leucine	Leu	L		Non-polar hydrophobic	
Methionine	Met	M		Non-polar hydrophobic	Cu, Fe
Phenylalanine	Phe	F		Non-polar hydrophobic	
Proline	Pro	P		Non-polar hydrophobic	
Tryptophan	Trp	W		Non-polar hydrophobic	
Valine	Val	V		Non-polar hydrophobic	

9.8.2 Polyfunctional condensation

We turn now to polyfunctional organic compounds of the above kind in order to describe proteins, DNA and RNA, which again can be considered as derived from hydrocarbons replacing H bound to C by different groups. The most important case is that of the condensation polymerisation of amino acids, for example $H_2N—CH_2—COOH$, glycine, to give peptides. Amino acids themselves can be obtained in part by condensation reactions (from hydroxy acetic acid and ammonia) or by

$$HCHO + HCOOH + NH_3 \longrightarrow H_2NCH_2COOH + H_2O.$$

We shall describe their specific synthetic pathways in organisms in Chapter 10. The biological amino acids (Table 9.10) are *laevo* α-amino acids and hence they are related to α-hydroxyacids or to α-keto-acids, also difunctional organic compounds, through condensation reactions, such as

(The second step is a reduction that will be analysed later.)

The great problem in these reactions is, of course, how to make monomers selectively and then to make their condensation polymers, polypeptides, by removing water in water by energetically uphill reactions! We look at polymerisation itself and then return to pathways and energy input in Sections 9.9 and 9.10. (For pathways to the specific compounds required in biological systems the reader is referred to the next chapter and to books on biochemistry.)

9.8.3 Peptides and proteins

The steps of polymerisation of amino acids, Table 9.10, are of the kind

$$NH_2 \cdot CHR \cdot CO_2H + NH_2 \cdot CHR' \cdot CO_2H \longrightarrow$$

$$NH_2 \cdot CHR \cdot CONHCHR' \cdot CO_2H + H_2O.$$

When we carry out such condensation reactions n times we get a polypeptide of length n amino acids (Fig. 9.13) or a protein if n is large (see Fig. 9.14). In living organisms there are some 20–30 different groups R and R' in amino acids. We have a three-letter code for them, for convenience, just as we have a one- or two-letter code for elements (see Table 9.10). Putting elements together makes all materials around us; putting amino acids together makes all proteins, which with polysaccharides (see above) and nucleic acids (see below) are the main polymers of life. (We shall include fats $(-CH_2)_n$ as polymers although, strictly speaking, fats are not polymers.) Now, we know that only those molecules, large or small, that are polar will be soluble in water. We shall, therefore, need to look at the nature of R and R' in amino acids since these groups generate the specific nature of individual proteins via their folds, solubility and so on (see Table 9.10).

9.8.4 Nucleic acids

Let us now look at a different type of condensation reaction. If we take three urea molecules we can make a ring

Analogous condensation reactions give the nucleic acid bases of the central genetic materials DNA and RNA, deoxyribonucleic and ribonucleic acids, respectively. The five major bases of life are shown in Fig. 9.15 (also see the marginal illustration); such ring molecules are quite planar and quite soluble in water. When polymerised as indicated below (Figs 9.15 and 9.16) they are put together on a so-called DNA string (Fig. 9.15) where the bases are given a further single-letter code A, T, C, G, using bold type to avoid confusion with the nomenclature of atoms. We then have a 'language' or code that can be used to describe DNA (genes) corresponding to particular orderings of nucleic acid bases, for example,

$$A \cdot T \cdot C \cdot C \cdot G \cdot T \cdot A \cdot A \cdot T \cdot C \cdot G.$$

(Above) Two nucleosides; (below) two sugars of nucleic acids.

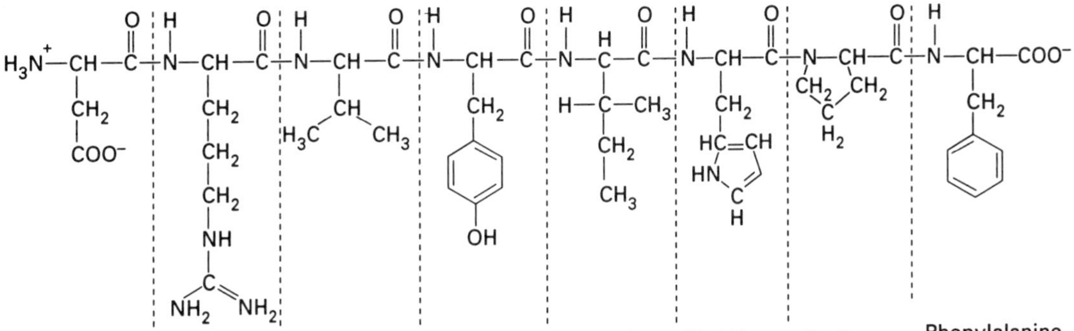

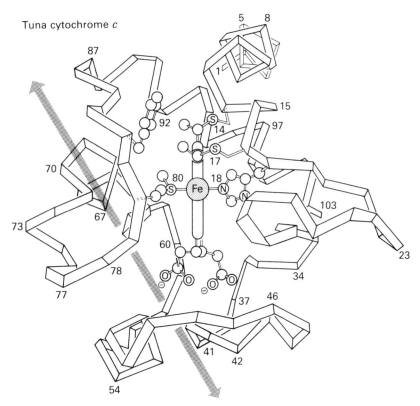

Aspartate (1)   Arginine (2)   Valine (3)   Tyrosine (4)   Isoleucine (5)   Histidine (6)   Proline (7)   Phenylalanine (8)

Fig. 9.13 Peptides, polypeptides and proteins are all made by linking together a selection of some 20 amino acids: the molecule shown is angiotensin II, an octapeptide (composed of eight amino-acid units). Angiotensin is a hormone with a role in controlling the blood pressure. Proteins may contain polypeptide chains hundreds or even thousands of amino acid units long. Examination of the chain above shows that some amino acids, (1) and (2), will prefer to be in water, while from (3) to (8) they will prefer to be in organic solvents. The sequence of amino acids is the primary structure.

Tuna cytochrome *c*

Fig. 9.14 A numbered sequence of amino acids in a protein, cytochrome *c*. Note how it folds around haem iron (shaded). The arrow indicates a possible path for a proton, see Section 7.5.3.

How is the DNA string held together with its bases? As we have seen above, while we can write condensation reactions of a carboxylic acid RCO_2H with an alcohol $R'OH$ as

$$RCO_2H + HOR' \longrightarrow RCO_2R + H_2O,$$

we can also write condensation reactions for phosphoric acid derivatives

$$RPO(OH)_2 + HOR' \longrightarrow RPO(OH)OR' + H_2O.$$

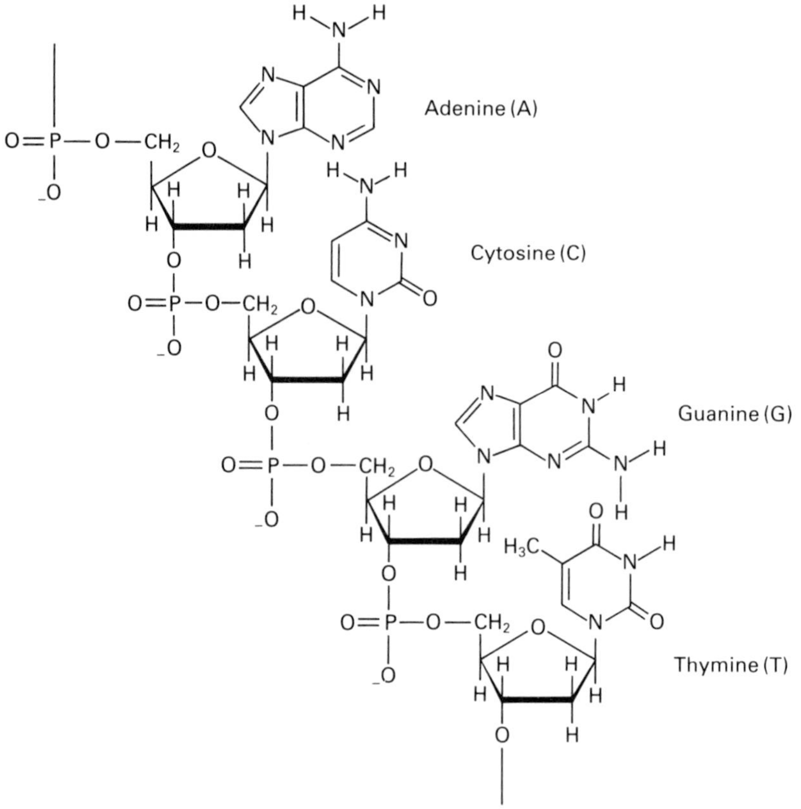

Fig. 9.15 A short string of single-strand DNA giving the formulae of four bases. In RNA thymine (**T**) is replaced by uracil (**U**) and deoxyribose is replaced by ribose (see marginal illustration).

Now, a glycol, $(CHOH)_n$, is just a multi-alcohol, and, in this sense, a sugar (see glucose which is a hexose, Fig. 9.12), is just a glycol. If we react a phosphoric acid with a glycol, for example, $CH_2OH \cdot CH_2OH$, we obtain

$$2PO(OH)_3 + CH_2OH \cdot CH_2OH \longrightarrow PO(OH)_2OCH_2 \cdot CH_2OPO(OH)_2 + 2H_2O.$$

The product of the reaction is still an acid, so that we can go on to make chains (polymers) of phosphate esters, for example,

$$PO(OH)_2OCH_2CH_2OPO(OCH_2CH_2OP)_nOCH_2CH_2OPO(OH) + nH_2O.$$
$$\qquad\qquad OH \qquad\qquad\quad OH \qquad\quad OH$$

Now, if we change each $—OCH_2CH_2O—$ to a saccharide (for example ribose; see Fig. 9.15) and we attach the bases **A**, **C**, **G**, **T** to it, we form what is called a *nucleoside*. In this way we have a string of nucleosides bridged by phosphate groups, forming repeating units called *nucleotides*, which constitute the genetic material, DNA. (Actually DNA is a double helical string, see Fig. 9.16.) The polymer is large and charged, hence it is a polyelectrolyte, that is, a salt soluble in water, so it cannot cross membranes. The charge is due to the fact that $(RO)_2PO(OH)$ is a strong acid. The same is true for RNA, but this polymer is a simpler string (Fig. 9.17), not a double helix except for short lengths of bases.

How this gene language arose and became translated into proteins and so became represented as living systems in species such as bacteria, plants and animals are central questions in biology. The *translation*, via an intermediate

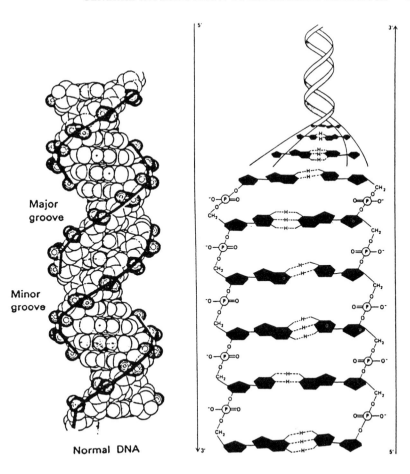

Fig. 9.16 Double-stranded DNA showing hydrogen bonding between **G–C** and **A–T** bases. Notice the grooves which are very useful for specific binding of proteins.

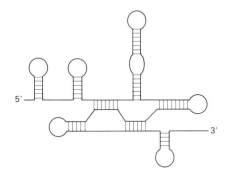

Fig. 9.17 The outline of alternating double helices and loops in a stretch of ribosomal RNA.

transcription to RNA (in which the sugar is ribose in place of deoxyribose in DNA and one of the bases, T, is replaced by uracil, U) is in the form of proteins, so that the A, T, C, G, language translates, using the amino acids of Table 9.10, into a protein language. That is, DNA makes RNA makes proteins makes the individual, or, as some biologists would say, 'the genotype gives (the foundation for) the phenotype'. (NB. Here there is not just the problem of faithful transcription and translation but a more difficult one as to how any specific sequence could have evolved (Section 11.3).)

Notice again that biological systems (or chemists) have to spend energy to drive all these polymerisation reactions by condensation, hence the polymers are all uphill in energy and are only kinetically stable, but as yet we have not described energy sources, suitable intermediates or catalysts for the processes.

One very special feature of the proteins and nucleic acids, the *major* polymers of biology, is that they are *linear*. It is a peculiarity of the kinetic stability of carbon that condensation polymerisation can be controlled so as to give polymers without cross-links.† Another feature is that, since the vast majority of these polymers (all DNA and RNA) carry charge and are salts, a

---

† There are many branched polysaccharides, for example starch, but they are not the major polymers of biology although they can play important biological functions, for example in cell recognition.

coun3ercharge will have to exist and will be critical. Finally, they have unique sequences of amino acids (proteins) or bases (RNA and DNA). The detailed problems of such syntheses will be analysed in Chapter 11.†

9.8.5 Weak bonds and the folded structures of polymers

So far in this outline of organic chemistry we have stressed the formation of strong covalent bonds between non-metal centres to form especially linear polymers in water. When we come to the larger linear molecules of proteins and nucleic acids, the way in which they fold is dictated by their sequences, generating the weak forces, especially hydrogen bonds and van der Waals forces between units of the polymer chain, see Section 2.6. Weak covalent bonds, —S—S—, may also form in proteins to stabilise a fold by cross-linking.

Two examples of the effects of the interactions are the folds of linear DNA (RNA) and of proteins. In Fig. 9.16 we show the way in which the bases of DNA come together to give the ideal double-stranded structure by matching hydrogen bonds. This is also a driving force for self-assembly. The whole unit, DNA (or RNA) remains charged and soluble in water (remember they are polyelectrolytes), and remains in large part linear and stretched out. There are regions down the centre of double-stranded DNA where the least polar parts of the structure come together so as to exclude water. We met this exclusion of water before when we described fats and the formation of membranes. Biological chemists have called the propensity of the less polar parts of molecules to get out of water as a 'hydrophobic force' (see Section 2.6). This is a second co-operative driving force of self-assembly. The general linearity of the structures of DNA and RNA is important since in living organisms they must be 'read' so as to be translated into particular proteins.

It will appear later that 'linear' double-stranded DNA does in fact bend back on itself which is only what could be expected for a polymer. The way in which DNA folds is guided by its interaction with Mg^{2+}, charged polyamines and proteins, as well as through special sequence runs. Specificity and control of expression are introduced in this way and, clearly through this effect, DNA sequences, genes, remote in a strand can be close in space. To a degree this lessens the notion of DNA as an all-dominant code. It needs partners.

The major driving force for self-assembly and folding in proteins is also H-bonding. In Fig. 9.18 we show the H-bonds in two protein secondary structure motifs, β-sheet and α-helix, and in Figs 9.14 and 9.19 the folded form of a tertiary structure of a protein. The folds require bends and these can be structured in the ways shown in Fig. 9.19, including what are called β-bends due to the H-bonds formed. The folds are also stabilised by hydrophobic interactions and by disulphide bonds.

In proteins it is the particular sequence, primary structure, of amino acids that allows the peptide bonds to form α-helical, β-sheet, bend or more open disordered structures. These secondary structures fold into tertiary forms. Where the sequence of R and R′ groups in amino acids is somewhat polar the protein, even when folded (Fig. 9.19), will remain more or less soluble in water, while, when it is very hydrophobic, the protein will enter the fatty acid

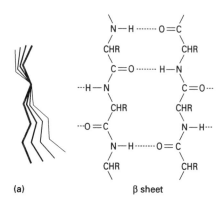

(a)   β sheet

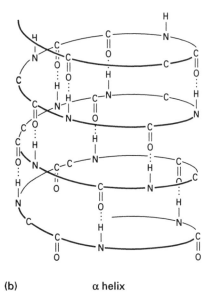

(b)   α helix

Fig. 9.18 Representations of the secondary structures of: (a) a β-sheet as in silk; and (b) an α-helix as in wool. The dashed lines represent hydrogen bonds between CO and NH and the C symbols represent the α-carbon atoms (the H and R groups attached to these are omitted, for clarity). The packaging of β-strands in the curved sheet is shown in (a).

† Referring back to the definition of components (Section 4.3.2), each sequence of monomers is a single component since it does not exchange monomers with any other sequence.

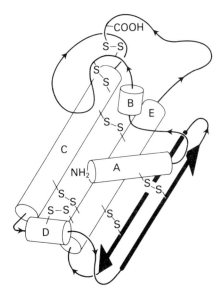

Fig. 9.19 An outline structure of a protein, here phospholipase A_2, showing α-helical runs of amino acids as cylinders, A to E, and antiparallel β-sheet runs as heavy black arrows. Disulphide cross-links are shown, and runs of no secondary structure appear as thin lines. The structure is relatively immobile and binds calcium in a constrained loop (see Chapter 10). (Reproduced with permission from Professor J. Drenth.)

membrane. Proteins then use their surfaces to give new self-assembled co-operative units within the compartments of the cell or to remain as free molecules in solution. Hydration of biological polymers is always in competition with intra- and intermolecular interactions. A further factor aiding association is, of course, charge matching of surfaces.

Polysaccharides are not necessarily linear and they perform a variety of functions such as C/H/O (fuel) storage units (starch and glycogen), and surface coverage of cells, and they also form a large part of the connective tissues between cells. These polymers, which are generally heavily hydrated and not permanently shaped, are described in greater detail in the references in 'Further reading', and again in Chapter 10. Finally, notice that all these biological polymers develop large surface areas with relatively few atoms and that their interiors are quite well packed (Fig. 9.20). It is the production of a diversity of surfaces (shapes) that is so important in evolution together with element selection limited to H, C, N, O, P and S, as seen in the diversity of bases, amino acids and sugars.

We have now shown in barest outline and by example how to make some major organic molecules and the major polymers of biology. Man's chemical skill allowed him to copy these procedures in the synthesis of polymers such as nylons, polyesters, polyurethanes, etc., but has also generated a great wealth of new organic chemistry. (As already stated, much of this synthetic chemistry is carried out in non-aqueous media where the condensation in, for example, polymerisation reactions can require removal of units other than H_2O.)

In contrast with those in organisms, most of these man-made polymers (see Table 9.7) are not synthesised on a template so they have polydispersed molecular weights—man's industrial equipment cannot yet match the sophistication of biological polymer synthesis. In all of this synthesis of polymers we must observe that the use of H, C, N and O, the light non-metals, is based on natural selection of these elements due to the very special kinetic (stability) and thermodynamic (instability) properties of their compounds. These properties differ from the properties of the heavier non-metals, Si, P and S, in that the heavier non-metals are all more reactive with water and oxygen (Figs 7.15 and 9.4). As we have seen silicon forms very thermodynamically

Fig. 9.20 (a) The outline fold of a protein flavodoxin of 138 amino acids. (b) An illustration of the packing of side chains in a plane some 5–10 Å below the top surface (after A. M. Lesk).

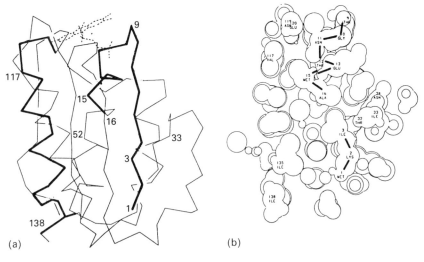

(a)

(b)

stable oxygen bonds in cross-linked compounds such as silicates (to which we have referred in Chapters 2 and 8), but in water very many of its other compounds, containing Si—H, Si—C, Si—N bonds, are of insufficient kinetic stability to be of great value. The natural selection of phosphorus and sulphur for their functional value in biology is described in Sections 9.8 and 9.9 where these elements are seen as the aids to light non-metal chemistry, especially in the building of polymers by transfer of units, C, H, N, O fragments. In this way, large-scale structure of a somewhat dynamic character has evolved. Many other elements are also essential and are involved in control and catalysis.

(The inclusion of phosphate in a polymer chain, especially an unfolded chain, is a sign of weakness to hydrolysis. This is advantageous to repair of DNA and to destruction of RNA, the turn-over of which is necessary in cells.)

9.8.6 The properties of folded polymers

One property of a linear polymer is that it can easily be read (see DNA above). However, linear random polymers when they come together form immense entanglements. Such polymers are useful in rubbers and, on cross-linking, are of great value in plastics. Linear polymers that crystallise lying side by side over considerable distances are valuable in structures. Such polymers are known in man's chemical industry and in biology, for example wood in celluloses and some proteins, for example keratins and collagens. The outstanding property that belongs to proteins uniquely is that they fold into three-dimensional shapes dependent upon the amino-acid sequence. There is no escape from the fact that it is the properties of these co-operative folds that generated life as we know it once it was possible to link synthesis of them to a code. A code itself is useless without a managed product that is functional. We shall refer time and again to the folds of polymers since these allow (see marginal note):

Polymer	Fold (Function)
DNA (RNA)	Linear (code)
Proteins	Three-dimensional (structure)
Polysaccharide	Random (p, T sensor)

(1) recognition between proteins due to their surfaces and hence the building of cells;

(2) recognition of small molecules, selective binding, acting in both catalysis by enzymes and communication generally even to the DNA code (here grooves in surfaces are particularly valuable);

(3) recognition of metal ions selectively, and acting both in enzymatic catalysis and in communication;

(4) response even to pressure, temperature and ionic strength when loosely folded (see polymer energy state distribution in Fig. 3.15).

The fold is not static but has localised mobility in enzymes and long-range mobility in mechanical communication devices. We will always be in the presence of proteins as we describe any biological system. It is very largely through the versatility of protein folds, based on different sequences, that evolution progressed, and it is undoubtedly through proteins that the 15–20 elements in all life came to be naturally selected and refined to the functional advantage of the organisation in a living system. In passing, it is the kinetic stability of the peptide bond that gives semipermanence to structures made from proteins, which are much more stable than esters such as DNA and RNA.

9.8.7 Summary of condensation reactions

The following features of the above reactions will not have escaped notice (see the introduction to Section 9.8):

Table 9.11 Essential features of organic chemistry

1. Selection of starting materials and reagents involved in synthesis associated with protecting and leaving groups
2. Choice of containing vessels
3. Choice of conditions (solvents, temperature, etc.)
4. Introduction of specified catalysts
5. Selection of separation and transfer methods

1. Containment in a compartment was essential—an apparatus was needed (Table 9.11).

2. Conditions must be controlled, for example temperature, pressure and solvent.

(NB. Each reaction pathway would be best carried out in an *isolated* reaction vessel, but this was not possible in the early biological cells nor was it in one sense desirable since no pathway can be allowed to be independent in a living organism (see Chapters 7 and 10–13).)

3. Each monomer is obtained from a few selected elements, here carbon from CO_2 and nitrogen from NH_3, HCN or N_2 reduction, while hydrogen comes from either H_2S or H_2O. These are the only redox steps needed; subsequently each polymer is made by condensation.

4. The monomer must be carried to the apparatus for polymer synthesis. In biology this is largely done by mobile compounds (coenzymes), while man has special reagents. Here S and P chemistry will dominate (Section 9.9).

5. The steps need an energy supply and a sink for water removed on condensation (Section 9.10).

6. Each synthesis needs to be guided and accelerated on a pathway by catalysts (Section 9.16).

The reader is asked to look at the parallel ways in which reactions are carried out in laboratory or industrial organic chemistry. Table 9.12 is inserted as a help in this effort. Condensation is stressed continuously here since it is the primary reaction of biology. In Section 9.9 there is an analysis of how units for synthesis are carried to correct places, and are recognised and incorporated using catalysts, and Section 9.10 considers how necessary energy is supplied. Throughout time evolution is of complexity of organic chemicals.

9.9 Intermediates: carriers of fragments: cofactors and coenzymes

For controlled synthesis and degradation, fragments of complex organic molecules must be handed along a chain of reactions. The carrier of an atom, ion or molecular fragment of any size must have the following properties:

(1) ease of movement between centres of reaction;

(2) relatively low stability compared with the desired product;

(3) relatively low kinetic barriers to transfer.

It follows from what has been said that the choice of carrier will be very dependent on the reaction path, the temperature, the pressure, the atmosphere (O_2, N_2, etc.) and the solvent system. The differences between organic chemistry in organic solvents and bioorganic chemistry in water are highlighted often by the use of organometallic chemistry in organic chemistry (see Section 9.16) and the use of heavy non-metal atoms in organic compounds in biochemistry. Table 9.12 lists some carriers in both systems—observe their parallel energetics and reactivity. Especially noticeable are the use of hetero-aromatic rings for H transfer, of RS— for carbon fragment transfer and of RO—PO$_3$— for oxygen-linked fragments (see Sections 7.7.3 and 7.7.4). The kinetic value of these units will be described after sections on their energisation, a necessary prerequisite so that the reactions of transfer proceed downhill. Thus the carrier unit should be of high energy and appropriate kinetic lability. It is frequently the first chemical to incorporate the smallest units in synthesis, for example H$^-$, CO$_2$, HPO$_4^{2-}$, NH$_3$, etc. In fact, control over many phosphate-based transfers in biology uses the very weak acid catalyst Mg^{2+}, while it is often the case that other carriers need no activation except strain induced by binding to selective centres on enzymes (proteins). When stronger catalysts are required, biology employs stronger Lewis acids, for example Zn^{2+} (while man may use Al^{3+}), or bases, for example RS$^-$ or R$_2$NH (while man may use alkoxides).

Table 9.12 Examples of differences between transfer reagents*

Reaction	Organic chemistry	Biological chemistry
Hydride transfer	NaBH$_4$, LiAlH$_4$	Pyridine nucleotide (NADH)
Electron transfer	Many metals	Flavin, Cu, Fe, Mn
O atom transfer	SeO$_2$	FeO, SeO
Dehydration	P$_2$O$_5$, CaCl$_2$, Na$_2$SO$_4$	Derivatives of pyrophosphate
Transfer of acyl	Acyl chloride	Acyl CoA
Transfer of —CH$_3$	Mg—CH$_3$, Al—CH$_3$	Vitamin B$_{12}$, methionyl CoA
	Organometallics	Pteridines
Transfer of —NH$_2$	NH$_3$	Glutamine

* Note the number and diversity of the elements listed.

9.10 Driving energy for biological reactions including condensation

9.10.1 Introduction

To drive condensation it is necessary to remove water selectively from chemicals in acid–base reactions controlled by catalysts using stored chemical energy. In man's chemistry a wide variety of methods can be used since there is open to him the use of organic solvents and solid isolated hygroscopic substances. The variety of solids that may be used for dehydration includes P$_2$O$_5$, silica gel, CaCl$_2$ and CuSO$_4$, but to be useful in solution, particularly in biology, they must not be insoluble in water and at the same time they must neither be of the most kinetically stable nor of the most reactive kind. Three major systems that satisfy these requisites are used to this day in biology: thioacyl, acylphosphate and pyrophosphate compounds. The formation of pyrophosphate itself depends on the previous production of acylphosphate in some cases (Section 9.10.2), and written here without ionisations

$$CH_3COOPO_3H_2 + H_3PO_4 \longrightarrow CH_3COOH + H_4P_2O_7.$$

To remove water in biology, the general idea then is to employ an energised, soluble, kinetically somewhat stable, dehydrating agent, and to this we add, for the reasons explained in Chapter 7, an acid catalyst such as H^+ and/or Mg^{2+}, a suitable basic catalyst, and so on. We shall next see how energy (thermodynamic instability) can be generated in the dehydrating agents, first by disproportionation of organic compounds.

9.10.2 Energy from chemicals: disproportionation

A different chemical reaction from condensation, which is important in organic chemistry, is

This is an *internal* oxidation of carbon 3 and a *reduction* of carbon 1; the reactions are internal and they are not treated as redox reactions. It can be called an internal disproportionation with elimination. More generally, we can write a series of disproportionation rearrangements involving carbon–bond breaking as

Notice how oxygen accumulates on carbon atoms different from those carrying hydrogen so that the thermodynamic instability is removed at the cost of energy loss. This is another way of stating (see Fig. 9.1) that this reaction is downhill and therefore gives energy. If we now include phosphate on the carbon atom being oxidised in the first reaction above, we start from a phosphate ester and finish with acylphosphate,

All the thermodynamic instability, that is, the drive to CH_3CO_2H from carbon sugar chemistry, is in this way collected and associated with the unstable intermediate carbon–phosphate bond, an acylphosphate. This acid anhydride can react with another phosphate without further energy change to give a pyrophosphate, P—P, which is the basis of energy retention in the glycolytic series of reactions in biology (Fig. 10.10). The scheme may be represented schematically by

$$2P + C_6H_{12}O_6 \longrightarrow 3CH_3CO_2H + P—P$$

where P = phosphate, or, using the carrier of phosphate, adenosine diphosphate,

Phosphate (P) + adenosine diphosphate (ADP) + glucose $\longrightarrow$ acetic acid + adenosine triphosphate (ATP).

We clearly see that this series of reactions requires the prior formation of sugars which contain the energy. They are formed by photosynthesis in plants and bacteria but may have been available from abiotic sources for early life. Their formation requires us to make a few remarks about initial energy sources in general.

9.10.3 Initial sources of energy

From the beginning of this chapter we have insisted that all organic molecules require energy in their synthesis from H_2O, CO_2, CH_4, N_2 or even carbon. The sources of this energy are ultimately the out-of-equilibrium of the zones of the Earth or the light from the sun. Whether we consider biological, laboratory or manufacturing chemistry we can see that it is possible to use these primary energy sources directly or to use secondary storage of energy in chemicals often from other life forms, for example, sugars, as described in Chapters 3–7. The forms energy takes are

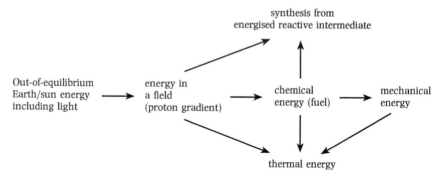

Of interest here are the necessary storage modes and the methods of transfer of the energy. As stressed in Chapter 6, energy can be stored against gravity (water reservoirs) or in electric charge distributions (condensers), that is, in fields, or in chemical fuels such as carbon compounds plus dioxygen gas (sugars, oil, gas and coal). The transmission of energy for man's organic synthesis uses chemical preparation of reactive intermediates from chosen chemicals energised originally in this way often from fuels from biological sources, for example gases (CH_4). In organisms the introduction of energised chemical bonds, P—O—P or C—S—CO—, involving heavy non-metal atoms is managed after carbon dioxide and water have been forced to form energised (CHOH) compounds, the sugars, which here are the main biological fuels. (This corresponds in large part to driving the reaction back from $CO_2 + H_2O$ to sugar (using light) before using the sugar to provide energy as above.) These two heavy atoms, P and S, are excellent for *carrying energy* and/or *energised*

fragments of organic compounds, as we saw in Section 9.8.7, since the kinetic stability of their bonds is intermediate between that of the much more stable C—N, C—H, C—O and C—C and that of the too kinetically reactive (to exchange) metal—N, metal—O, metal—H and metal—C bonds in water (Fig. 9.4 and see Section 7.6). In laboratory or industrial chemistry where *organic* solvents (not water) are used, these very different unstable metal–carbon transfer systems (intermediates) can be used to carry energised fragments. This instability is an essential feature of man's modern syntheses and may have been very important in primitive organisms. We described some features of this energisation in Chapters 6 and 7 and will return to it in Chapter 10. It is often the case that carrier, energisation and catalytic functions are incorporated in the same molecular unit for obvious reasons, that is, to avoid many step processes. Now in all these chemically driven reactions a *fuel* had to be and still has to be supplied and we describe its source from CH_4 or sugars using light energy in Chapter 10.

We turn again now to intermediate energised carriers showing why phosphorus and sulphur compounds are so useful in the further transfer of energised chemicals in bioorganic chemistry carried out in water. (The way in which biological cells make pumps (carriers) across cell membranes is described in Section 10.6.1. They are not used in bioorganic chemistry.)

9.11 Phosphorus bioorganic chemistry

In Chapter 2 we considered the chemistry of metal oxides and sulphides but left somewhat on one side the non-metal oxides and sulphides. At the beginning of this chapter we returned to the chemistry of carbon and nitrogen hydrides and oxides. We now turn to the 'organic' chemistry of phosphorus and sulphur in association with oxygen, hydrogen and carbon, and we also refer to a parallel chemistry of selenium.

Phosphorus is the simplest element to describe in its reactions with C, H and O. In essence, phosphorus oxide chemistry is totally dominant; there was little or no phosphine (P—C) chemistry until man made these compounds (see Chapter 14), and they are of trivial significance as far as this book is concerned (here we include P—H and P—M as well as P—C compounds). In fact, the affinity of phosphorus for oxygen is so great that at low temperatures we shall only be concerned with phosphate or its compounds (see Figs 5.14 and 9.21 which show the oxidation state diagram for phosphorus). The element, P, is therefore found in minerals in the form of phosphate salts, sometimes complex, and as phosphate ions, esters or anhydrides in biological organic chemistry. The nearest parallel is with the chemistry of silicon, but phosphorus tends to form simple anions or linear chains and not cross-linked structures to the same degree as silicon. Pyrophosphates are thermodynamically unstable, but kinetically they are reasonably long-lived in water. Condensed silicates (but not linear silicates) are stable in water at pH = 7 in the presence of a cation such as Mg^{2+}, but only monophosphate is thermodynamically stable in this solvent. Thus it is to pyrophosphate that biology turned for one relatively short-lived stored form of transferable energy. Note that there is no danger in its chemistry of a confusion between acid–base and redox reactions since

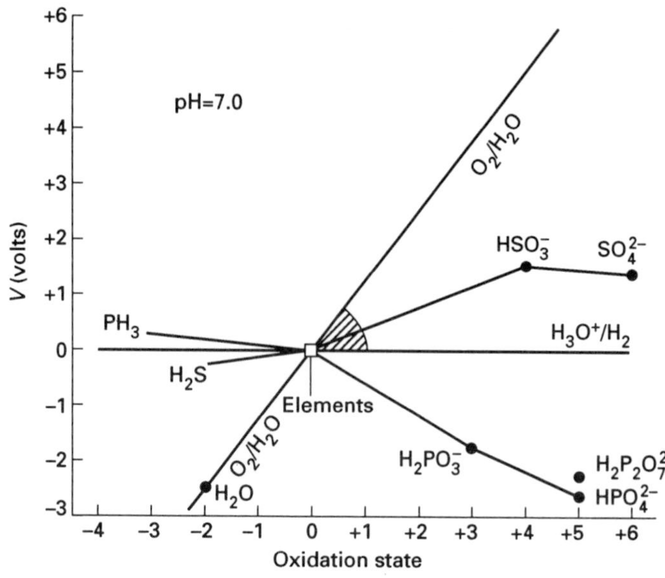

Fig. 9.21 Oxidation state diagram for S and P at pH = 7. Note that the reference potential is Pt · $H_2/10^{-7}M$ H^+, that is pH = 7.0, and not that of the standard H_2/H^+ one molar electrode.

phosphate is so difficult to reduce. This is then again a part of the natural selection of elements for their function.

In biochemistry pyrophosphate is bound to a carrier, adenosine monophosphate, and the compound is called ATP. ATP is the major condensation agent in biology together with other nucleotide triphosphates, NTP, and we note that nearly all biological polymer syntheses use these energised pyrophosphates, for example in the condensation reactions of amino acids, nucleotides and polysaccharides, linking H, C, N, O, S and P reactions. In effect bound pyrophosphate's (ATP) action as a soluble condensation agent is really just the last step in the action of P_2O_5 as a drying agent in man's organic chemistry

$$P_2O_5 + 2H_2O \longrightarrow P_2O_7^{4-} + 4H^+,$$

$$P_2O_7^{4-} + H_2O \longrightarrow 2PO_4^{3-} + 2H^+,$$

$$ATP^{4-} + H_2O \longrightarrow ADP^{3-} + PO_4^{3-} + 2H^+.$$

Thus ATP is a reagent that has stored chemical energy. We can write $-\Delta G_1$, its free energy capacity to drive reactions ($P = HPO_4^{2-}$), as

$$ATP + H_2O \longrightarrow ADP + P \qquad \Delta G_1 < 0.$$

Now any condensation reaction requires energy ($+\Delta G_2$)

$$X\text{—}OH + HY \longrightarrow X\text{—}Y + H_2O \qquad \Delta G_2 > 0.$$

Adding the two reactions together we have

$$X\text{—}OH + HY + ATP \longrightarrow X\text{—}Y + ADP + P \qquad \Delta G_3 = -(\Delta G_1 - \Delta G_2).$$

This is the way in which ATP drives much of the synthesis in cells but it must be protected from H_2O. Now since it carries energised phosphate it also acts as a phosphorylating agent and, clearly, whenever there is a —P—O—X bond it will be possible to transfer X while making —P—O—H.

9.12 Sulphur bioorganic chemistry

Sulphur centres can also be used for transfer of energised organic fragments in acid–base reactions. Sulphur as a reaction centre is, quite differently from phosphate, also extremely useful in oxidation–reductions (Section 9.15.2), and we need to note the general nature of all its reactions before we look at its value in the kinetic control of fragments by acid–base reactions.

The affinity of sulphur for hydrogen and carbon is low (higher for H than for C). (See Figs 5.14 and 9.21 for an oxidation diagram for sulphur.) It is higher for oxygen, but, as described in Sections 8.4.1 and 8.5.1, sulphur is not able to take oxygen from elements such as Al, Mg and even Fe. Thus, reduced sulphur chemistry could develop free from oxygen in the prebiotic era. (NB. There was also a considerable amount of H_2S in the early atmosphere.) On the other hand, sulphur has some considerable affinity for metals that were denied the opportunity to combine with oxygen, so metal sulphides are also common (see Chapter 8). The metal sulphides as M/S compounds also became central to biological chemistry (Section 10.5.3) for these reasons.

In this section we wish to explore the possibility of formation of C—S links in organic compounds. Table 9.13 shows that reactions such as

$$2CH_4 + S \longrightarrow (CH_3)_2S + H_2$$

have an adverse energy of some +30 kcal. The realm of C—S chemistry belongs, therefore, to energised kinetically stable substances only. One of the favourable possible reactions in abiotic chemistry to obtain S—C bonds is *addition*, for example

$$H_2S + CH_2{=}CH_2 \longrightarrow CH_3CH_2SH,$$

but we do not know if this reaction occurred very early on Earth. It is also possible (see Fig. 9.21) that some sulphite existed on Earth even when life started, but the present-day levels are due to the coming of dioxygen. Sulphide, if present, readily adds to aldehydes, such that

$$CH_3CHO + H_2SO_3 \longrightarrow \overset{\displaystyle SO_3H}{\underset{\displaystyle CH_3-CH-OH}{|}}$$

The resulting product is easy to reduce to the thiol CH_3CH_2SH.

Table 9.13 Bond energies for carbon and sulphur[*]

Bond	Bond energy (kcal mol^{-1})	Bond	Bond energy (kcal mol^{-1})
C—C	83	S—S (S$_8$)	54
C—H	98	S—H	87
C=O	191	S—O	(50)
S—C	~65	S=C	~137

[*] From Huheey, J. E., Keiter, E. A. and Keiter, R. L. (1993). *Inorganic chemistry—principles of structure and reactivity* (4th edn). Harper Collins College Publishers, New York. Notice that the balance of bond energies in theoretical reactions such as C—C + S—S → 2 S—C is unfavourable, i.e. C—S bonds are unstable relative to C—C and S—S.

Thiol esters could also have been made on metal sulphide surfaces by the reactions of CO and CH_4, that is,

$$MS + CH_3^-, CO \longrightarrow \overset{\overset{\displaystyle CO}{|}}{MS} + \overset{\overset{\displaystyle CH_3}{|}}{MS} \longrightarrow \overset{\overset{\displaystyle CH_3}{\underset{\displaystyle CO}{|}}}{\underset{|}{MS}} \, .$$

$$\overset{\overset{\displaystyle CH_3}{\underset{\displaystyle CO}{|}}}{\underset{|}{MS}} + SH^- \longrightarrow MS^- + CH_3COSH$$

The above sequence could have been a major abiotic pathway some 4×10^9 years ago. While we cannot be sure how thiols were formed we can now discuss their use in transfer reactions.

As pointed out by De Duve (see reference 4 in 'Further reading'), one possible starting point for the development of dehydrating agents in biology, that is, agents for driving the polymerising organic condensation reactions, was a thioacid, which, once formed, is highly energised, namely

$$R'COSH + H_2O \longrightarrow R'COOH + H_2S \text{ (driving energy, } \Delta G, <0) \quad (9.3)$$

The water molecule in step (9.3) can be derived from any condensation reaction rather than being water itself. Thus, thioacids can drive amide formation, that is,

$$R \cdot NH_2 + R'COSH \longrightarrow RNH \cdot COR' + H_2S \text{ (maintenance of energy)}$$

instead of $H_2O + R'COSH \longrightarrow R'COOH + H_2S$ (loss of energy as heat).

The reaction illustrates the general value of sulphur-based centres even in biology today, for example, in coenzyme A (Fig. 9.22). Sulphur compounds are also invaluable in redox reactions so that sulphur is functional in different ways (see Section 9.12). This, however, raises again the central problem of organic chemistry and biochemistry. It is possible to write down many reaction schemes involving one compound, which can equally well react with several others in different ways. The skill in operations such as those in man's synthetic chemistry is to make the reactants follow a 'chosen' path. Choice of path implies a whole series of considerations including the apparatus, coupling

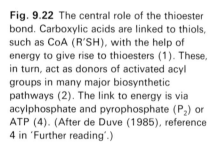

Fig. 9.22 The central role of the thioester bond. Carboxylic acids are linked to thiols, such as CoA (R'SH), with the help of energy to give rise to thioesters (1). These, in turn, act as donors of activated acyl groups in many major biosynthetic pathways (2). The link to energy is via acylphosphate and pyrophosphate (P_2) or ATP (4). (After de Duve (1985), reference 4 in 'Further reading'.)

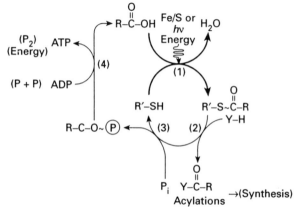

of reactions, conservation of energy as far as possible, transfer through space of compounds confined in special volumes, stereospecificity, positioning of attacking and leaving groups and, if we use catalysts, recognition of the components of reaction in desirable orientations (see Section 7.3). The first reaction systems must have been extremely simple, which is why we consider that the earliest reaction leading to the ancestors of today's carbon-based living organisms was the synthesis of fatty acids using metal compounds, sulphides and/or clays. Later, *enzyme* surfaces developed and the possibilities increased. (In passing the reader will have noticed the value of the selected chemistry of sulphur and phosphorus compounds. A summary is presented in Chapters 11 and 13.)

9.13 Relative thermodynamic and kinetic stability of organic molecules in water

We started this chapter with a description of organic compounds pointing to their instability relative to the elementary state of carbon, CH_4 and CO_2 (Fig. 9.1). We have now described a large number of organic compounds and we need to note next their *relative* stabilities in the absence of oxygen but in the presence of water. Here we mean unstable in the sense that their syntheses require energy. The order of increasing *thermodynamic stability* toward water is

$$\text{Anhydrides} < \text{thioacids} < \underset{\substack{\text{amides} \\ \text{(peptides)}}}{\text{amides}} < \underset{\substack{\text{phosphate esters} \\ \text{esters} \\ \text{(DNA)}}}{\text{esters}} < \underset{\substack{\text{ethers} \\ \text{(polysaccharides)}}}{\text{ethers}} < \underset{\substack{\text{hydrocarbons,} \\ \text{(lipids)}}}{\text{hydrocarbons,}}$$

$$\underset{RCO}{\overset{RCO}{\diagdown}}O \quad < \quad RCO{\cdot}SR \quad < \quad RCO{\cdot}NHR \quad < \quad RCO{\cdot}OR \quad < \quad R{\cdot}O{\cdot}R \quad < \quad CH_3CH_3 .$$

$$\left.\begin{array}{l} RCO{\cdot}OPO_3^{2-} \\[2em] (PO_3{\cdot}O{\cdot}PO_3)^{4-} \end{array}\right\} \quad < \quad ROPO_3^{2-}$$

There is also the question of the relative rate at which water decomposes them, that is of relative kinetic stability.

The order of *ease* of hydrolysis, that is the lack of kinetic stability, is

Anhydrides > thioacids > esters > peptides > ethers > hydrocarbons.

It follows that the more permanent building units from condensation reactions are fats (hydrocarbons), ethers (polysaccharides) and peptides (proteins), not esters (even such as DNA or RNA). Table 9.14 relates these trends to the bonds involved (Fig. 9.4). Of course, fats, simple *hydrocarbons*, are the most stable carbon compounds to hydrolysis. It also follows that the way to reach these kinetically stable states can only be through the more unstable compounds (see Fig. 9.2). We need to understand particular ways in which to reach relatively stable trapped states to appreciate how life commenced (and has continued), the maintenance of kinetic stability in the presence of thermodynamic instability and also how to transfer fragments in order to build required compounds and polymers. These are the subjects of Chapters 10–13.

Table 9.14 Selection of elements for kinetic stability

	C—H			
↑	C—C	C—N	C—O	C—F
Kinetic stability				
↑		(C—P)	C—S	C—Cl
			C—Se	C—Br
				C—I
		←Kinetic stability→		

Against this background it is easy to see when we look at the general nature of the energetics of organic chemistry pathways (Fig. 9.4) that the sequence of transfers in building must be

$$
\begin{array}{llll}
\text{Initial} & \text{energised} & \text{anydride} & \text{(esters)} \\
\text{condition} \longrightarrow & \text{fragment} \longrightarrow & \text{thioacid} \longrightarrow & \begin{array}{l}\text{peptides} \\ \text{ethers} \\ \text{lipids}\end{array} \\[1em]
\text{(Stable state)} \longrightarrow & & \text{(intermediates)} & \longrightarrow \text{(trapped states).}
\end{array}
$$

So far we have not referred directly to aromatic species. As stated in Section 9.4.2, they are more reactive than aliphatic compounds and can be so chosen that they are as valuable as anhydrides and thioacids in activated transfer. Examples include quinols, pyridines and pterins, many of which are used to transfer H^- and so are included under redox reagents (Fig. 9.23 and Section 9.15).

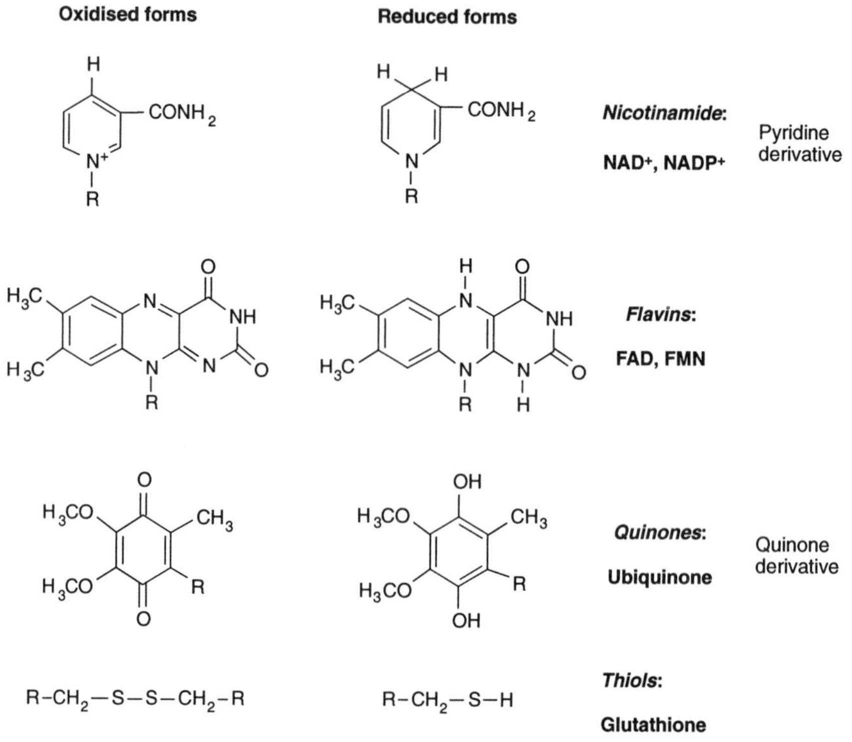

Fig. 9.23 Some organic redox reagents.

Oxidised forms	Reduced forms	
		Nicotinamide: NAD+, NADP+ — Pyridine derivative
		Flavins: FAD, FMN
		Quinones: Ubiquinone — Quinone derivative
R–CH$_2$–S–S–CH$_2$–R	R–CH$_2$–S–H	**Thiols:** Glutathione

9.14 Summary of the origins of acid–base organic chemistry

The way in which all the acid–base organic chemistry evolved in biological systems is quite unclear, but it has developed so that it is generally compatible with the original aqueous salt solutions (sea) of Na^+, K^+, Cl^- and the required acid–base catalysts that were probably restricted to Mg^{2+}, Mn^{2+}, Fe^{2+} (Fe^{3+}) and to some extent Zn^{2+}. The organic acid–base chemistry involved is that of exchange of groups between esters, ethers, amides and anhydrides. The control over reactions is somewhat different from that in modern organic chemistry which is largely confined to organic solvents. This means that reagents and modes of fragment transfer are made from different elements, but the principles of kinetic control are the same. It is the switches to different element uses that is central to evolution, natural and cultural, see Chapter 14.

Some of the biological chemistry took place in non-aqueous phases, lipids. These lipids, and indeed all the polymers described, generated very extensive surfaces which are such that reactions are often on surfaces more than in free solution, especially on the surfaces of enzymes. Many elements are required in addition to those that traditionally form part of organic compounds for carrier and catalyst functions. However, the imperative development was that of polymers and, in particular, that of proteins.

At this stage in this semihistorical description of the natural selection of combinations of the atoms C, H, N, O, P and S, dominant in biology, it is necessary to stress again that we are describing principles of organic chemistry and biochemistry rather than *in situ* biological chemistry—the *organised* turnover of these molecules in biology will be discussed later.

Now we have reached the end of our description of synthesis using acid–base chemistry, especially as concerns condensation reactions, including the manner in which these syntheses are energised. One factor of inestimable importance remains to be properly described. *At room temperature none of this chemistry will occur without catalysts.* The reader is asked to refer back to Sections 7.6–7.8 where we have already described the nature of acid–base catalysts as applied in organic chemical reactions (see also Section 9.16). It will be remembered that, overwhelmingly, they are inorganic ions, Lewis acids. Thus the organic chemistry we have described cannot be carried out just using the selected elements H, C, N, O, S, P and halogens, but requires elements such as Mg, Fe and Zn (in the laboratory Al, Sn, etc. can be used too). The sources of the elements are the minerals described in Chapter 8.

We pass on now to redox reactions in bioorganic chemistry where we will meet the same barriers to change which can only be overcome by using inorganic catalysts of a different kind. We return to the description of catalysts in Section 9.16.

9.15 Organic redox chemistry

(This section will be kept short since we will devote a considerable part of Chapters 10 and 12 to redox reactions in biological systems.) The reactions we have described so far are acid–base reactions in which the oxidation states of carbon, nitrogen or sulphur undergo no change. There is a further group of reactions, essential for life and in organic chemistry, namely redox reactions (Section 7.7.2) in which the oxidation states of elements are deliberately altered. The definition of oxidation state we shall use is that given in Section 5.7. The central atom here is carbon so that, since a C—H bond reduces the oxidation state by one and an oxygen atom in C=O increases it by two, CH_4 has a carbon oxidation state of $-IV$ and CO_2 has a state of $+IV$. Using this scheme we constructed Figs 9.1 and 9.5. Now, the oxidation state is also increased by halogen incorporation by one for each C—X and by two for sulphur incorporation C=S. The binding to metals, M, is treated as polar, and a methyl bound to a metal is written $M^+ \cdot CH_3^-$, for example when each M—C gives a reduction in oxidation state by one. All other combinations will be considered by convention to leave carbon oxidation states unchanged, that is, combinations of carbon with N, Si, P and Se. (No other combinations concern us.) In biological systems the redox chemistry of carbon is often linked to that of sulphur. We therefore include in Section 9.15.2 an account of sulphur redox chemistry and its connection to that of carbon while leaving the description of the redox chemistry of catalytic elements and much of the chemistry of dioxygen itself to later chapters. That discussion, which largely involves biological systems, re-introduces the redox chemistry of the metal elements and its evolution as outlined in Chapter 8 but now in the context of life. (NB. The oxidative chemistry is historically new ($1-2 \times 10^9$ years old) relative to reductive chemistry (4×10^9 years old).)

The obvious need for redox reactions in organic and biological chemistry in general is seen if we remember that the major source of carbon for life is CO_2, when reduction is needed, while the major source of carbon for man is a hydrocarbon, methane (CH_4), for example, when oxidation is required. These basic molecules are both thermodynamically and kinetically stable in many conditions, which means that energy is again needed to carry out reactions (for example using light in photosynthesis), intermediate carriers of redox fragments are necessary and special powerful catalysts are also essential. Carrier and catalytic elements that are quite different from those described for acid–base reactions are most useful; for example Fe, Co and Ni are redox catalysts and have an irreplaceable role even in early redox biological chemistry. They are still used in this mode today, and equally in man's industry. (Oxidation/reduction is also a major source of energy for organisms and man's industry as we stress in several later chapters (see Fig. 9.1).)

The energetics of making and breaking of bonds in organic molecules (with loss or gain of energy, Fig. 9.4) can then be divided into: (1) reactions that occur without the *relative* loss or gain of H or O when in Fig. 9.1 the oxidation state (horizontal axis) does not change but energy content alters on the vertical axis of the figure; this is the way in which we described rearrangements and condensation or addition of water in the previous sections; (2) redox reactions where either H or O are removed or added differentially, moving the oxidation state horizontally in Fig. 9.1. A number of reagents are listed in Table 9.15. Biological systems and man make use of a vast range of catalysts to bring about these changes.

The outstanding redox reactions are the removal or addition of a hydride (H^-) or of an oxygen atom (O). Both are two-electron reactions changing carbon oxidation state by -2 or $+2$. There is then a requirement for reagents to assist the introduction of H^- and O into carbon frameworks. Table 9.12 has already listed some of these but organic chemists have devised a vast range of hydride- and oxygen-carrying centres (especially using heavy non-metals and metals because of their advantageous exchange rates (see Section 9.15.2)). It is clear that carriers and catalysts are frequently based upon very similar reactive heavy elements or molecular constructs for obvious reasons. We shall not go further into these two-electron redox reactions here since they dominate much of biological chemistry together with one-electron redox

Table 9.15 Examples of redox reagents in biological and organic chemistry

Redox reagents in	
Organic chemistry	Biological chemistry
Reducing agents	
Li, B, Al hydrides; Na metal; Zn metal	Pyridine hydride (NADH); S, Se hydrides; flavins; Ni hydride (?)
Oxidising agents	
CrO_4^{2-}; MnO_4^-; SeO_2; Cl_2; Mn^{3+}; Fe^{3+}	FeO (IV); flavins; CuO; MoO_4^{2-}; Fe^{3+}

chemistry that we have described in Section 7.7.2 (also see Chapters 10–13). The catalysts for these reactions are largely metal ions, for example Fe^{2+} and Co^{2+}.

Great interest centres on those atoms and molecules that can catalyse or assist simultaneously the acid–base and redox reactions when in different states. In particular, we note the flexibility of hydrogen (H^+, $H^{\cdot}$, H^-), sulphur ($-S^-$ or SH, $-S-S-$) and also many metal ion centres (Fe^{2+}, Fe^{3+}). There is clearly a valuable range of reagents that can couple together reactions of the two kinds through their versatility. The power of the reagents as catalysts is in part related to their redox potential (Fig. 5.13) and in part to their ability to act as good attacking or leaving groups, that is, kinetic Lewis acid/base properties (see Tables 7.10 and 7.11). A central point is that C/N/O/H compounds are not of much value in either catalyses or as carriers of fragments for redox reactions relative to the value of heavier non-metals (S, Se) and metal ions.

We will not attempt to give extensive examples of the multitude of transformations that are possible by using redox reagents in chemistry—they are well documented and described in textbooks of organic chemistry. In the present section our main concern is to indicate that it is possibly the case that the chemistry of especially H and S (non-metals) and Fe (a metal) helped to generate organic syntheses even before the beginning of life and then in primitive life (Chapter 10) and finally that, with the coming of dioxygen, a great change occurred in all redox chemistry.

9.15.1 Redox reactions at low redox potential

The discussion in this section will consider the organic redox chemistry in the range from the H_2/H^+ reaction in water at pH $= 7.0$, -0.42 volts, to that at low H_2S concentration of the potential of the S_n/H_2S system which approaches 0.00 volts. In this system we treat sulphur, $-S-S-$, as an oxidising agent and hydrogen sulphide (H_2S) as a reducing agent (see also Fig. 9.23). We use this pair because they represent the situation in which it is believed that abiotic and biological organic C/H/N/O chemistry evolved originally (see Fig. 8.10). (We assume that metal sulphides or dihydrogen were the original sources of reducing equivalents.) The sulphur centres acted simultaneously as carriers of fragments involving acid–base exchanges and as redox catalysts (see Section 9.15.2). (All reactions were assisted by metals, especially iron. After the advent of a dixoygen atmosphere, the organic chemistry of life at pH 7.0 extended to the redox potential of $+0.8$ volts, with some extension to even higher values. An organic chemist can choose today a variety of more extreme reducing systems, for example metallic sodium Na^+/Na at -3.0 volts in acid solution, or oxidising systems, for example MnO_4^{2-}/Mn^{2+} at $+1.25$ volts in acid solution.)

Before proceeding we will remind ourselves of the basic redox chemistry of sulphur relative to that of H/C and O, already presented briefly (Section 9.12), since, as stated, this chemistry was central to early biology.

9.15.2 Organic redox chemistry and sulphur

The chemistry of sulphur has to be considered several times in this book for the following reasons:

1. As inorganic sulphide it dominates the availability of many metal elements, especially in the early period of the evolution of the Earth (Chapter 8).

2. In organic-sulphur chelating agents it generates a high selectivity for available metals (see Fig. 5.5); the complexes become excellent catalysts.

3. As $[S—S]_n$ in equilibrium with —SH it interacts and generates redox control in cells even today.

4. As oxidised R—S—S—R it forms stabilising cross-links in plastics, for example proteins and rubber tyres.

To these facts we add the following.

5. As RS^- and R_2S it forms energised thioesters and carriers of carbon fragments, for example —COCH$_3$ and —CH$_3$, as already described in Section 9.11.

6. As RS^- it is an excellent catalytic base.

7. Possibly as SO_2 it may have contributed to an initial atmosphere and sea. It certainly does contribute to acidity today in acid rain.

It is safe to assume that in the primitive Earth there was no equilibrium between water and the metal (iron) sulphide rocks as they were forced up by volcanic action. These rocks may well have been metal-element-rich and capable of giving H_2S on reaction with water, an acid–base reaction,

$$MS + H_2O \longrightarrow H_2S + MO$$

and H_2 in a redox reaction

$$M + H_2O \longrightarrow H_2 + MO.$$

The first reaction occurs today in volcanoes and sea vents (Section 8.5.1). The atmosphere produced is strongly reducing and as such can attack CO, CO_2 and N_2. This basic inorganic/organic chemistry would subsequently give all kinds of reduced carbon (and nitrogen) in energised fragments, that is,

$$(CO, CO_2) \longrightarrow (HCO) \text{ compounds,}$$

some of which may have included thiols, for example CH_3COSH, as mentioned already in Section 9.12. Thus very simple redox chemistry of sulphur, carbon and hydrogen could have been at the heart of abiotic organic chemistry forming new small molecules and, as mentioned above, these can react to give —S—C— fragments and larger molecules. Now these roles of sulphur (see marginal note) are in two-electron (H^-) redox and acid–base chemistry, but sulphur, like many metals, can also undergo one-electron reactions.

Sulphur in organisms

In cells	Outside cells
RSH/RS$^-$	R—S—S—R
RSCH$_3$	R—OSO$_3^-$
RS.COCH$_3$	(polysaccharides)

9.15.3 Free-radical reactions and polymerisations

A major important extension of organic chemistry, both biological and man-developed, has been the introduction of free-radical redox reactions, especially in polymerisation and cross-linking. This chemistry is due to addition or removal of one electron or one-electron agents, for example H˙ (see Section 7.7.4). No matter whether extreme oxidising or reducing radical intermediates are generated, there is the possibility of chain reactions so long as the radical does not suffer termination, that is

$$R^{\cdot} + R \longrightarrow R_2^{\cdot} + R \longrightarrow R_3^{\cdot} + nR \longrightarrow R_{n+3}^{\cdot}.$$

The realisation of such products of variable chain length (n not fixed) generated a large part of man's plastics industry. The possibilities can be extended by using diradicals to cross-link chains, for example, $^{\cdot}S$—$S^{\cdot}$ cross-linked polymers. An excellent example is given by its use in the rubber industry, especially in manufacturing motor car tyres, but the occurrence of such cross-links is very general in extracellular fluids of biology. This reaction, the polymerisation of sulphur, occurred naturally in pyrite, FeS_2, and many polysulphides before life began. In the very early stages of life the coupling of H_2S oxidation to S_n must have taken place by such radical reactions and may have initiated biological systems in their capture of hydrogen and to obtain energy. We see again how there was in this case a natural selection of elements for their functional significance and availability. Initiating agents for this polymerisation, not really catalysts, are one-electron reagents, for example Fe^{3+} and sulphide, or photochemical reactions. We shall see how this chemistry is exploited in biology in Chapters 10 and 12. Naturally, when dioxygen entered the atmosphere quite new radical chemistry became possible since dioxygen, a diradical itself, is able to generate a whole range of radicals by removal of $H^{\cdot}$ from molecules. The reactions require different catalysts of course. The catalyst has to be capable of one-electron reactions and metalloenzymes are frequently employed.

9.15.4 A note on selenium organic chemistry

In the context of sulphur chemistry we draw attention to the value of selenium which is the only further non-metal to appear in a coded amino acid, together with H, C, N, O and S (that is, not P and Si nor later non-metals, nor any halogen nor any metal). Selenium was available from early minerals as H_2Se and metal selenide colloids. It is a better transfer agent for hydrogen than sulphur in that it is more kinetically active. Perhaps this explains its use in primitive biology. In organic chemistry and many biochemical systems its reactivity in higher oxidation states is of greater importance. Thus, SeO_2 and R_2SeO are mild oxidants acting by two-electron, oxygen-atom transfer (see Fig. 5.14 and Section 12.4.1). This chemistry is found in later aerobic organisms.

9.15.5 Summary of redox organic chemistry

The diversity of organic reactions that are possible is very dependent on the oxidation–reduction potential that reagents allow the system to reach. Early in evolution, chemistry was restricted to hydride reactions and the use of water and CO_2 to bring C/H together with C/O chemistry. N/H chemistry was introduced through NH_3. In bioorganic chemistry these reactions are usually classified as *primary* metabolism and make use of special carriers of fragments and catalysts. A quite new situation arose with the advent of dioxygen in the atmosphere. Now C/O, H/O and N/O chemistry could be exploited. Much of it is treated in this book under bioenergetic topics. Bioenergetics changed remarkably as can be seen by referring to Fig. 9.21 where the involvement of the H_2O/O_2 couple produces large energies for living systems and man's engines. The diversity of new possible compounds includes the so-called *secondary* metabolites in which C—O bonds dominate. The possibilities can be seen by shuffling oxidation states at relatively high O/C ratios. New carriers and catalysts were required for these reactions and this subject forms a major part of Chapters 12 and 13. Here we take a brief look at some of the catalysts.

9.16 Nature's need for catalysts in organic chemistry

There is an essential feature of the chemistry that led to the production of sugars, amino acids and bases. As stated the elements carbon and oxygen were held in a reducing atmosphere so that C was present either as CH_4 or CO_2 (CO) while O was present either in H_2O or in these carbon compounds. Meanwhile N was present as N_2 or as NH_3, which is very soluble in water. Given that all these are kinetically stable molecules it is certain that the required known monomers of life—sugars, amino acids and nucleotides—could only appear if redox catalysts were present. The routes of synthesis are by redox reaction of carbon dioxide to formaldehyde or ketones. Formaldehyde polymerises readily and can easily disproportionate giving acids and keto acids. It is then that the constraints of the initial conditions for biological chemistry must be remembered. The biological catalysts and carriers for acid–base and redox reactions today are sometimes large complex coenzymes, but we must suppose that virtually all early acid–base and redox catalysis used simple compounds of the metal elements Mg, Mn, Fe, Co, Ni, W and Mo and the non-metals S and Se. Today redox chemistry is still activated by these elements. Zinc was also present in small amounts but, as we showed in Chapters 5 and 8, most metals not in this list were absent. Note especially the absence of copper. We now look at some organic/metal combinations to discover their involvement in such processes. Man's organic chemistry is in no way different and the use of catalysts not based on C/H/N/O chemicals is essential; however, the range of redox catalysts is much more widely based (Table 9.16) than that of acid–base catalysts.

9.16.1 Metal organic compounds in biological catalysts

In this chapter we have deliberately concentrated on non-metal (organic) chemistry especially in relation to biology. In Chapter 8 we concentrated on metal chemistry, especially in relation to minerals. The separation of the two in the development of prebiological and biological chemistry is, of course, artificial—since the Earth formed there has always been a vast association of metal and non-metal elements in oxides and sulphides, and there has also been a variety of metal/organic compounds in the sea. Then, when life began, it increased the variety of these associations based on natural selection in terms of functional value and availability. The link between the two kinds of chemistry is obvious too, in that, without the use of fire (organic fuel), man could not have made copper or iron. However, the full extent of the ways of putting metals together with *carbon* compounds could not be exploited in biology due to the very poor availability of two-thirds of the metal elements. It is in man's chemistry, which is the most recent development, that is, a twentieth-century evolution of chemistry, that the chemistry of metals with

Table 9.16 Ease of oxidation and choice of catalyst

Ease of oxidation		
Relatively easy	**Intermediate**	**Relatively difficult**
Compound		
R—CHO	R—OH	CH_4, C_6H_6
$\begin{matrix} R \\ R \end{matrix}$>CO, RI	C_6H_5—OH	R—Cl, R—F
$C_6H_4(OH)_2$	>C=C<	
Catalysts (man)		
Several transition metal ions/O_2	>SeO, >MoO	FeO(IV)
BH_4^-, AlH_4^-	Metals + H_2	Metals + H_2
Catalysts (biological)		
R—S—S—R	Cu oxidases, Mo oxidases	FeO(IV) (P-450)
R—SH		Ni enzymes (H_2)
NADH, flavins		Mo reductases (N_2)

organic compounds reaches its full expression. A most interesting probable early development of this kind of chemistry in biology is that of cyanide/metal ion chemistry leading to the porphyrins, but below we shall refer to other examples.

9.16.2 Organometallic and co-ordination compounds in organic chemistry

The incorporation of metal elements within the frameworks of non-metals C, H, N, O, S and P, can be done in two very different ways. In the case of water-soluble organic molecules, the presence of negatively charged centres or of centres that are strongly negative parts of dipoles gives rise to sites for or linkages to metal ions stable to hydrolysis and to attack by dioxygen. The resulting species are called co-ordination compounds and this type of chemistry is called co-ordination chemistry. The co-ordination compounds (complexes) are described in Chapter 5 under solution chemistry and again in Chapters 10 and 12 under biological chemistry. An example of a metal–protein complex is given in Fig. 9.24.

The second type of chemistry, in which there is a metal bound to an organic framework, is often based on metal–carbon (or metal–phosphorus) bonds and is called organometallic chemistry. The compounds are usually not water-soluble, and are open to hydrolysis and oxidation by air. Some examples are given in Fig. 9.25. These compounds are rare in biological systems (note vitamin B_{12}), but while some are used in man's catalysts most of them are just of academic interest at present. As in all chemistry a huge variety of combinations can be made, but the challenge is to uncover novelty of property whether immediately useful or not.

In neither co-ordination chemistry nor organometallic chemistry are the molecular constructs necessarily centred on one metal atom. There are cluster

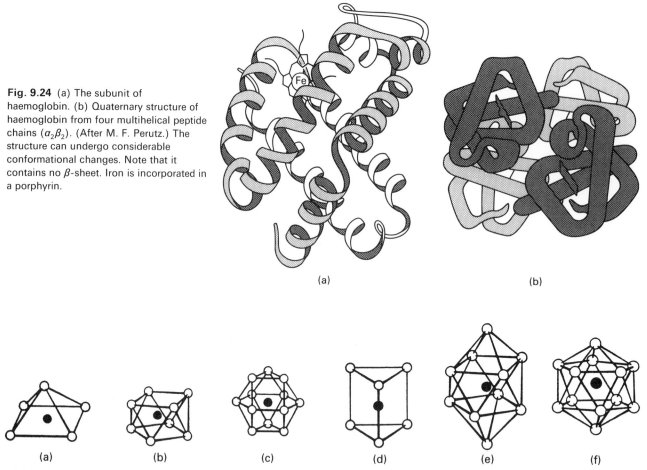

Fig. 9.24 (a) The subunit of haemoglobin. (b) Quaternary structure of haemoglobin from four multihelical peptide chains ($\alpha_2\beta_2$). (After M. F. Perutz.) The structure can undergo considerable conformational changes. Note that it contains no β-sheet. Iron is incorporated in a porphyrin.

(a) (b)

(a) (b) (c) (d) (e) (f)

Fig. 9.25 Examples of metal carbonyl clusters with interstitial atoms: (a) $Ru_5C(CO)_{15}$; (b) $[Co_8C(CO)_{18}]^{2-}$; (c) $[Rh_{12}H(CO)_{24}]^{4-}$; (d) $[Rh_6C(CO)_{15}]^{2-}$; (e) $Rh_{10}S(CO)_{22}]^{2-}$; (f) $[Rh_{12}Sb(CO)_{27}]^{3-}$. Carbonyls are not shown.

compounds with from two to very many metal atoms and the metal–metal bonding can be of very variable strength. Clusters of one kind or another also occur in biological compounds, but they are usually thought of as co-ordination complexes, for example Fe_4S_4 (see Section 5.6.1 and Fig. 10.8). This division between co-ordination chemistry and organometallic chemistry is, in some ways, arbitrary. For example, when phosphorus is the donor, not nitrogen, then the metal/organic chemistry is included in organometallic chemistry since the procedures, the use of non-aqueous solvents and the requirement for the absence of air, are similar. When sulphur is the donor then it is more usual to include the compounds within co-ordination chemistry together with those of oxygen and nitrogen donors, although M/S systems are often air-sensitive. It is extremely probable that this mix of inorganic and organic elements created the opportunity for the beginning of life, and, as we have seen, it was the deprivation of oxygen in the primitive atmosphere that allowed the very early extensive use of M/S, M/H and M/C chemicals with no sense of a category separation between them.

9.17 Organic chemicals in condensed phases

9.17.1 Introduction

We have now reached the end of this very brief outline of *molecular* organic chemistry in solution with particular reference to biochemicals and especially biological polymers. This is largely the conventional organic and bioorganic chemistry of textbooks (see 'Further reading'), but by itself it is not adequate to make a connection to biological systems. For this purpose three further steps are needed. First, the organic chemistry has to be knit together with very *selected inorganic chemistry*, which controls solution conditions and reaction pathway connections, apart from catalysis. Second, we have to take account of *the co-operative formation of condensed phases in addition to co-operative folding*. It is striking that much of inorganic mineral chemistry handles condensed systems such as silicates but conventional organic chemistry does not—it is mainly concerned with the properties and reactivity of small molecules. Yet, much of what is around us, including ourselves, is made from organic chemical condensates. Third, we need to analyse *the combination of compartments within flowing solutions*, which is the basis of *organised* biological chemistry (Chapters 10–13). We turn first to the assembly of condensed organic phases before moving to the understanding of living processes.

9.17.2 Self-assembly of organic molecules

The concept of self-assembly was described for simple crystals in Chapters 2, 3 and 8 and we commented on the self-assembly of oils and films and even lipid membranes at the beginning of this chapter. We note now that other large molecules, for example proteins, also self-assemble and do so in remarkably fixed patterns in biology with no simple repeating order except in very special structures such as viruses. The question arises as to whether the concepts of nucleation discussed in Section 7.17 apply to these cases. In that section it was stated that nucleation involved barriers due to surface instability such that some 50–100 units had to come together before self-assembly in a crystal could occur. The alternative is for self-assembly to proceed in thermo-dynamically downhill-directed single steps allowing readjustment. Perhaps as proteins are produced their assembly is in effect continuously downhill, first folding and then forming extensive somewhat mobile macrostructures, that is,

$$A + B \rightleftharpoons AB + C \rightleftharpoons ABC + D \rightleftharpoons ABCD \dots .$$

Thus diffusion controls (Chapter 7) would be favoured management tools of self-assembly, not activated processes (see Section 10.8.3). Proteins appear, however, to be positioned in space by transport, so there must also be an activated control over diffusion (Chapter 7C). The further questions we must ask, as we did in Chapters 4 and 6, are: as proteins assemble which are the degrees of freedom over physical states—liquid, solid, etc.—and how are shapes of assemblies controlled? We shall address these questions in the next section.

9.17.3 The phase rule and organic molecules

In Chapter 4 we pointed out that the phase rule allowed us to think in terms of the variables in a system as restricted by the number of independent components at equilibrium in local volumes, phases. When all elements present are in equilibrium at fixed temperature and pressure, the number of components is just the number of elements. When the amounts of each element are fixed, the resultant system of compounds is invariant. We used this approach in the discussion of the formation of the geosphere. Although not very satisfactory, this simple examination of elements allowed a considerable insight into the natural selection of the elements. The chemistry of the sea could be elaborated starting from the considerations of equilibria between elements or extremely simple compounds in dilute solution forming precipitates and complexes as discussed in Chapter 5. We were not overconcerned in much of Chapter 8, therefore, with small phases and surfaces since the phases were often quite large. Thus we were able to manage a good general description of the Earth from knowledge gained in Chapters 2–6, the main reason for the inadequacy of this description was the formation of dispersed condensed phases, small compartments in the crust, which has not allowed equilibrium at low temperatures to this day.

In organic chemistry, especially in its application to biological systems, there are two striking changes to such a description. The first is of a kinetic nature—no organic phase can be usefully described by reference to equilibria between the elements (H, C, N, O) or between simple compounds (H_2O, CO_2, N_2). The situation is that barriers to chemical reaction are so large that *each organic compound* becomes a separate component from the point of the phase rule, for example oil and water. Thus the variance in these systems is enormously increased. Second, the phases are small, especially biological vesicles or cells, so surface properties dominate. Prebiotic chemistry then produced at least five different valuable features allowing life to appear. The first three are:

(1) two new kinds of compartment (phases), the lipid vesicle membrane and the internal aqueous solution, which are thermodynamically unstable;

(2) a multitude of polymers from monomers based on H, C, N, O and P;

(3) additional local fields (electrostatic) that are important for small compartments.

Let us assume that the lipid phases self-assembled as vesicles. It is apparent that the proteins could be chosen so that they too will self-assemble either with the lipids or with themselves, see marginal illustration. Thus the variables would not be increased by this equilibrated association. We consider that this is very much a partial truth. Some protein assemblies in biology, for example in viruses (Chapter 6), are clearly of this kind but many depend on the initial selection of the compartment for a designated protein using energy to transfer it through the kinetic barriers generated by the membranes. Thus we need to be aware that transfer increases the number of possible variables by denying full equilibrium. As discussed in Chapter 7, this is physical barrier control. (The distribution, as in organic chemistry, requires energy). The physical barriers are observed to be shaped structures, see marginal illustration.

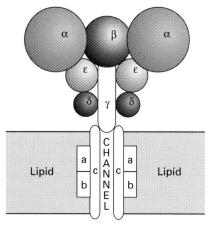

A Self-assembled proton pump from eight proteins and a membrane (see Section 10.6.2).

Now the shape of equipment is decided by functional need in organic chemistry but, remembering the discussion in Chapter 6, a non-biological shape represents, at equilibrium, a state based on the amounts of components. Once the composition and the amount of components were fixed the shape was invariant at equilibrium. This means that, if the amounts of components is fixed, the shape is fixed without a mould. We can only believe this to be so in biological systems by discovering the way in which composition can be fixed in a self-assembling system; this returns us to the problems of control over reaction systems in flow. We are stating, though, that control of composition in energised flow could control shape, but we must always be conscious of the fields that are simultaneously present.

The two further factors we observe to be necessary in prebiotic systems to develop such an 'organic' chemistry are:

(4) ways of distributing energy using gradients initiated by redox reactions across membranes so as to apply energy to movement of large as well as small molecules;

(5) considerable amounts of Mg^{2+}, Ca^{2+}, Na^+, K^+, Mn^{2+}, Fe^{2+}, Ni^{2+}, Cl^-, etc. needed to be selectively associated with lipids and polymers to form catalysts.

To be living, that is, to be able to develop and grow with precision, these activities clearly demand *organisation* in a self-assembling system to match that imposed on man's organic chemicals in the laboratory or in an industrial plant. (To reproduce needs, in addition, coded instruction.)

The observant reader will have noticed that almost all the concepts of Chapters 2–7 have been employed in an attempt to describe in Chapters 8 and 9 the nature of the inorganic and organic materials around us. Missing from this description is organisation, controlled flow to functional purpose and, in particular, the value of feedback, discussed in Section 7.13. The return to these concepts becomes almost immediate once we describe the next historical refinement of the natural selection of the chemical elements in biological systems (Chapters 10–13). When we come to Chapter 14 we shall ask about man's industrial activities and the presence or absence of parallel feedback control.

9.18 Abiotic organic chemical cycles

In this chapter we have described how abiotic organic chemicals arise by energisation of simple stable precursors. We have illustrated the discussion largely by reference to biochemicals but we have insisted that laboratory or manufacturing chemistry is very closely similar. The description contains the following cycle (compare Fig. 3.13 and note the discussion on thermodynamic efficiency in Section 3.6.2).

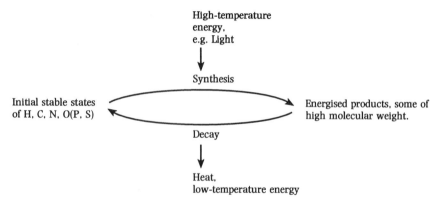

The route for energisation is not of necessity the same as that for decay but the final products are the starting materials, CO_2, N_2, H_2S, HPO_4^{2-} and H_2O. There is at the present time little knowledge as to how this cycle gave rise to living systems but we can sense that two features help to assist survival of the energised products in the above cycle. First, it was essential to increase the lifetime of products, which is the case when inherent *kinetic* stability increases with molecular weight even though thermodynamic stability decreases. Clearly, this is true if the energised product is a condensed phase object, for example a plastic, or wood, or a *folded* polymer, or a binding together of polymers. (In a sense the production of a folded polymer is protective and can be likened to formation of these (micro)phases.) The second assistance with survival of the energised product is its rate of production relative to that of decay, which can be managed by manipulation of catalysis. An example occurs in the production of C/H/O compounds, sugars as opposed to formaldehyde compounds (Fig. 9.2). In some cases transfer to a separate phase occurs as well; one example is the production of O_2 from liquid H_2O released to the atmosphere, no matter how it happened, and another is the formation of lipid phases.

9.19 Summary: the diversity of organic and biological chemistries

Underlying the evolution of organic chemistry is an increasing manipulative ability to transform combinations of H, C, N and O atoms (plus some hetero-atoms) into complicated frameworks with structural constraints and kinetic stability imposed on their dynamics. Its procedures allow isolation of individual organic chemical compounds. It is legitimate to treat such compounds as components of systems since they do not exchange elements. Thus, organic chemical manipulations introduce a vast new variability into the chemical components on Earth and increase the variance of the system. Their independence is kinetic in origin and will not operate at a high temperature. By treating them specifically in a carefully designed apparatus, in a chosen solvent, at a chosen temperature and pressure, with chosen reagents (also isolated as components) and with specific catalysts new chemical components can be made time and time again. All we have done in this chapter was to indicate some routes by which a few important products can be obtained. Organic and biological chemistry texts show the extreme diversity that can be achieved (see references 2 and 3 in 'Further reading'). We used as examples biological pathways to particular products employing catalysts, but in many respects the treatment was not different from the 'accidental' chemistry which belongs more generally in the environment (and is called inorganic), in which most of the elements are more reactive, where the general solvent is water and the sun is the source of energy. However, a striking feature of bioorganic chemistry is the use of co-operatively folded polymers, proteins.

Even though we used biochemicals to provide examples it is clear that *biological organisms* differ in their underlying principles from the organic chemistry described. Biological organisms produce simultaneously a *rich but limited variety* of organic (and inorganic) products *in fixed ratios and in fixed phases*. The independence associated with components has therefore been in some way lost from the chemicals, in order to produce an organised invariant living form. As we showed in Chapters 3 and 4, fixed ratios of compounds come about if, starting from some component chemicals, equilibrium is achieved. In this case we state that no new components are generated, that is, only interdependent chemical compounds are produced. If in the course of reaction to equilibrium new phases are formed, then the distribution in the phases is also fixed at fixed temperature and pressure and there is a loss in variance. We know, however, that biological systems are not at equilibrium and require continuous material and energy input. In this contrast lies the mystery of biology—the nature of its kinetic control. Chapter 7 provided us with a possible alternative explanation for the existence of interdependent chemical systems, with little or no possible variation, through feedback kinetic control. In our view the biological system can only achieve its high degree of invariance through such a multiplicity of feedback links. However, such a scheme must have a master information store that limits the number of operations in which it engages, an energy source and a communication network. The information store (DNA) is a governor of the living process, but cannot be the living process itself and must have been imposed upon an early somewhat more chaotic organic chemistry. The next few chapters will examine the nature of living systems in an attempt to discover the processes underlying life. Central to the discussion will be the postulate that it was only through the incorporation and use of the evolved inorganic chemistry of the sea (Chapter 8) with initially chaotically energised organic chemistry that life accidentally arose. It is especially the development of catalysts such as those shown in Table 9.17 that

Table 9.17 Catalysts in organic and biochemistry

| | Redox | |
Condensation	One-electron	Two-electron
Mg^{2+} (weak)	Fe^{2+}/Fe^{3+}	MoO
Zn^{2+}	Cu^+/Cu^{2+}	Ni—H
Al^{3+}	Co^{2+}/Co^{3+}	Co—Ch_3
RS^-	RS^-	RSeO
ROH	Haem	$\geqslant NR \rightarrow \overset{+}{\geqslant} N - R$
R—NH_2	Mn^{2+}/Mn^{3+}	MnO
Imidazole	Mn^{3+}/MnO^{2+}	

drives evolution, as it drives organic chemistry today. Thus changes in environmental chemicals with time (see Chapter 8) must be a driving force for biological evolution (Chapters 12, 13 and 16).

Further reading

Although the arrangement of the topics and the style of treatment in this chapter may seem unusual at first, the contents are orthodox and are discussed in an immense variety of texts which could be recommended to complement it. We indicate just a few titles (some simple and some more elaborated) but our choice is in no way limiting.

1. Hornby, M. and Peach, J. (1993). *Foundations of organic chemistry*. Oxford University Press, Oxford. This is one in a series of primers in the Oxford Chemistry Primer series (ed. S. G. Davies).
2. Roberts, J. D. and Caserio, M. C. (1977). *Basic principles of organic chemistry* (2nd edn). Addison Wesley. A book with a somewhat old-fashioned appearance but a clear logical development of the subject.
3. Stryer, L. (1995). *Biochemistry* (4th edn). W. H. Freeman, San Francisco. An excellent textbook.
4. De Duve, C. (1985). *Blue print for a cell*. Neil Patterson Publications, Burlington, North Carolina.
5. Nogrady, T. (1988). *Medicinal chemistry : a biochemical approach* (2nd edn). Oxford University Press, Oxford.
6. Streitweisser, A. and Heathcock, C. H. (1981). *An introduction to organic chemistry*. Macmillan, New York.
7. Pine, S. H., Hendrickson, J. B., Cram, D. J., and Hammond, G. S. (1980). *Organic chemistry* (4th edn). McGraw-Hill, New York.
8. Holum, J. R. (1994). *Fundamentals of organic chemistry* (4th edn). J. Wiley and Sons, New York (a simpler version, *Elements of general organic and biological chemistry* is also available, 4th edn 1995).
9. Volhardt, K. P. C. (1994). *Organic chemistry* (2nd edn). W. H. Freeman, New York. Up-to-date, visually appealing and informative.
10. Solomons, T. W. (1994). *Fundamentals of organic chemistry* (4th edn). J. Wiley and Sons, New York. A modern, widely used text.
11. Kenyon, D. H. and Steinman, G. (1969). *Biochemical predestination*. McGraw Hill, Inc. New York. An approach to the original synthesis of biological compounds in lines similar to those of the present chapter.

10

Early biological chemistry: the uptake and incorporation of elements in anaerobic organisms

We mean by 'possessing life' that a thing can nourish itself and grow and decay. . .

Aristotle (384–322 BC)

10.1 Introduction

We have described the basic principles of chemical binding at equilibrium in Chapters 2–6 after looking at the history of the development of ideas on the constitution of matter and energy in Chapter 1. This allowed us to consider the manner of the eventual appearance of the chemical elements as structured compounds and co-operative, condensed phases on the Earth. Simultaneously, in this process, energy was lost locally from the planet on cooling, but the pathways of energy loss often led also to a variety of chemical systems trapped out of equilibrium (Chapters 7–9). The processes are not difficult to follow. The facts that the materials in the crust of Earth were often out of equilibrium (Chapter 8) and that the Earth was far out of equilibrium with the sun then allowed new compounds, often organic, to be formed in solution or in the gas phase by the capture of energy. These compounds entered new (higher) energy traps, and acquired potential for further reactions, and so prebiotic organic chemistry developed (Chapter 9). In order to understand the underlying drive toward equilibrium, that is, toward the most stable state, we discussed in Chapters 3 and 6 the evolving forms of energy in the context of the evolving materials, both at equilibrium and in trapped states. We have also described the basic ideas of dynamic organisation of energised material as opposed to rigid structure in Chapter 1 and again in Chapter 7. It was necessary to clarify the nature of the development of all these aspects of materials, of energy and of kinetic traps and to include feedback controls over flow (organisation including structure) in Chapter 7 before turning to biological chemistry. Biological systems are self-organising, self-reproducing, energised, material traps in feedback flow, while they undergo evolution forced upon them in part by competition but equally by the necessary conditions for long-term survival, requiring protection in a changing environment (see below). In effect, they have become more and more longlasting and sophisticated chemical processes in kinetic traps (see Chapter 12). All biological progression involves a selection of chemical elements for 'use' to these ends and it is this fact that we shall follow in our further study of the natural selection of the functional capability of the elements. (Note immediately that element *function* is expressed now in dynamic organisation, flow, not just in static structure or shape.) Usefulness or functional value is defined as that property that assists survival in a hostile environment. The environment considered here is related to all the elements in the biological system since all these elements are in an energised condition.

The next two chapters then concern the nature of the early forms of life in order to develop the theme of continuous natural selection of the elements. We shall only be able to speculate about life's origins, and for its early manifestations we must concentrate on the very poor early biogeological records together with what look to be the simplest descendants alive today of the early forms of life. Since we believe that the earliest cells had no hard structures and formed in the absence of dioxygen, this forces us to describe the existing life chemistry of anaerobic unicellular organisms with but a single membrane (Fig. 10.1) as if they were the remnant of the earliest forms of life. There are great numbers of such cells living today, including many prokaryotic bacteria and archaebacteria†, all of which differ from one another only somewhat in chemical pathways of synthesis and degradation. All these organisms are based on proteins, DNA and a minimal group of 15 elements (Table 10.1). It is generally assumed that organisms of these kinds already existed at some time around $3.5–4 \times 10^9$ years ago when the atmosphere is known to have been almost totally free from dioxygen (see Table 8.7). It is also

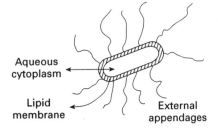

Aqueous cytoplasm

Lipid membrane

External appendages

Fig. 10.1 An outline of the most primitive kind of cell we can observe. There may well be an outer wall-like structure surrounding the membrane.

† The division of early species is not fully established. Here we shall use prokaryote to include all single-cell systems without a nuclear compartment (see Fig. 10.1), which often include so-called eubacteria and archaebacteria (Fig. 10.2). The eukaryotes complete the kingdoms of life in this figure but other authors give slightly different divisions, see reference 12.

Table 10.1 The absolute minimal element content of primitive life*

Element*	Source	Use
H	H_2S (air) HS^- (sea)	Organic molecules, energy capture
C	CH_4, CO or CO_2	Organic molecules
N	NH_3, HCN (sea)	Organic molecules
O	H_2O, CO, CO_2	Organic molecules
Na^+, K^+, Cl^-	Sea salts	Electrolyte balance, osmotic control
Ca^{2+}, Mg^{2+}	Sea salts	Structure stabilisation, weak acid catalyst (Mg^{2+})
$P(HPO_4^{2-})$	Sea salts	Organic molecules, energy transfer
S	H_2S (air) HS^- (sea)	Element transfer, energy metabolism
Fe	$Fe^{2+}/Fe^{3+}/S^{2-}$ (sea)	Catalysis

* The above 12 elements were inevitably incorporated into any vesicle formed in the sea in the period around $3–4 \times 10^9$ years ago. Others that were present in reasonable amounts but perhaps not incorporated of necessity initially were Al, Si, V, Mn, Co, Ni, Mo, W and perhaps Se, Br. The primitive organisms we know, such as archaebacteria, have approximately 20 elements.

quite possible that, at first, the organisms were not photosynthetic but formed in the deep dark trenches of the ocean, using material and energy available from there. There would then be two major liberating evolutionary steps: (1) the development of photosynthesis, which allowed cells to exist away from the energy and the food sources of the deep ocean trench; and (2) the progressive accumulation, due to photosynthesis, of dioxygen in the air, which allowed one part of life, symbiotic on photosynthetic life for reduced carbon, to become aerobic and independent from light as a direct source of energy, burning hydrocarbon fuel with dioxygen instead. The source of the hydrocarbon fuel was, of course, the product of metabolism of other organisms. The advent of dioxygen in the atmosphere is the simplest and clearest example of an element being energised into a long-lived kinetic trap as shown in Fig. 7.4. A corollary of its production is an increase of bound H in reduced CO_2 (all O_2 comes from $CO_2 + H_2O$) and we observe this hydrogen and carbon in life and in huge gas, oil and coal deposits.

Life was limited at first to the aqueous regions of the Earth, probably the sea; this was necessary to ensure easier transport of required materials in forms adequate for uptake and subsequent removal of rejected (waste) products. During the passage from anaerobic to light-dependent and then to fully aerobic organisms, with energy content increase (Fig. 10.2), two structural changes in cellular systems evolved: (1) internal membrane-enclosed compartments produced eukaryotic cells; and (2) cells came together not just in disorganised (or even relatively organised) colonies but in truly multicellular and differentiated cell structures. This then allowed multicellular organisms as we see them today to invade the land. Our task in this and the next chapter on simple primitive anaerobic cells is to give an account of the early prokaryote and archaebacterial cellular chemistry and its organisation up to the time of the generation of organised cellular systems which followed after the build-up of dioxygen in the atmosphere. (We shall include here some anaerobic

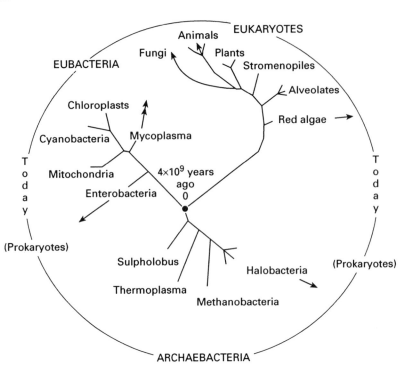

Fig. 10.2 The three kingdoms of life in a radial time sequence based on genetic analysis. The distinction between anaerobes and aerobes is not made here. All the kingdoms advanced with time but in very different ways. (See Woese, C. *New Scientist* 11 August 1990; and Sogin, M. (1993). *Science* **260**, 340.)

eukaryote chemistry.) Very much of this chemistry remains that of the cytoplasm of every cell to this day.

In this chapter we shall describe the essential *chemistry* of this anaerobic life, which, as stated, requires a minimal set of elements, at least 15 and perhaps 20 (see Fig. 10.3 and Table 10.1). Several of this set of elements, H, C, N and O, plus some S and P must be incorporated in compounds that are included in so-called primary metabolism leading to the synthesis of fats, sugars, proteins and polynucleotides (see Chapter 9). We shall examine the uptake and synthetic pathways for the incorporation of the light non-metal elements into these compounds first and will assume that there is but one compartment in which this essential chemistry evolved (Figs 10.1 and 10.4). The other essential elements are required to play roles described below. Chapter 11 is a description of the development of *dynamic feedback organisation* using the same elements in the pathways, initially also in one compartment. Organisation must have begun very early with: (1) basic *control* over primary metabolism, energy and uptake and incorporation of elements, which created a homeostatic internal composition; and (2) *regulation*, which we define as the management of the relative amounts of polymeric material, DNA, RNA and proteins, that were formed. This leads us on to consider the cell cycle as a steady reproductive mechanism, already present in very early cells, replacing abiotic haphazard accumulation and division. Following this, the earliest form of organisation, the ability arose to utilise space more effectively through the development of internal filaments and new compartments, vesicles and organelles. We do not know how or precisely when this advance led from prokaryotes to eukaryotes (Fig. 10.2). We deal with this topic at the end of Chapter 11 since such

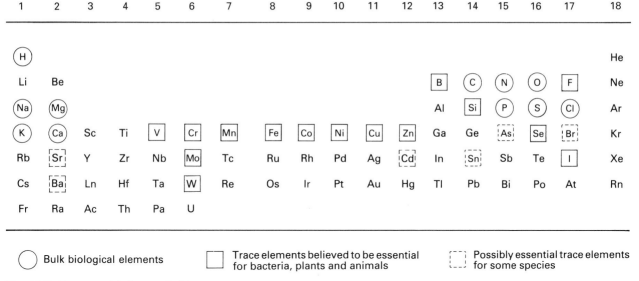

Fig. 10.3 The essential elements in life.

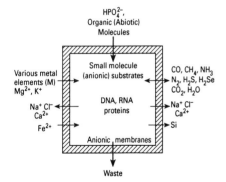

Fig. 10.4 The essential compartmental separation of elements found in all primitive cells.

structures may have evolved anaerobically without the new chemicals due to oxidation. Alternatively, this organisation developed only with a change in metabolism due to the drastic switch to aerobic chemistry resulting from the advent of O_2, which preceded the evolution of multicellular organisms. The most notable first change, as mentioned above, was the ability to use light as an energy source and we shall describe this change in this chapter. In turn, this led later to the production of dioxygen and a whole range of new chemicals (see Chapter 12). These chemicals undoubtedly affected the structures of membranes as well as metabolism. Thus, this change could have led to filament and vesicle formation but there is little evidence that such novel chemistry is required for these constructions.

The developing biological chemistry, through its waste and remnants, also must have altered the immediate aqueous and land surface as well as the atmosphere, and thus the 'mineral' crust of Earth itself slowly evolved (see Chapters 8 and 12). In the earliest phase of this development of life we assume that the environment in fact underwent little change.† It remained reducing, particularly the waters, with a plentiful energy supply due to the mixing of geological zones that were far from equilibrium, in the deep sea trenches for example. As described in Chapter 8, once the oceans reached neutral pH this restricted the availability of elements to a particular pattern. We shall now describe the first stage of this progression, that is, the basic primary chemistry of intracellular prokaryote life in an anaerobic and H_2S-dominated environment, stressing the roles of the 15–20 essential chemical elements.

---

† There may have been a change to neutral pH from very earliest prebiotic times when the ocean was probably acidic. This acidity could account, perhaps, for early use of tungsten rather than molybdenum in biology (MoS_2 is very insoluble at $pH = 5$), and sulphides and selenides generally would have been more soluble permitting greater use of selenium and elements such as nickel which are common in archaebacteria. Another change of possible consequence was the loss of CO_2 and its effect on the solubility of calcium.

10.2 The biological selection of major elements: general introduction

The dominant elements of both abiotic and biological chemistry are H and O in H_2O, water. It is water that has permitted biology to evolve, as we have stressed in several previous chapters: life must be a chemical process in a flowing liquid and water is and was the only liquid available. The next most important elements for organic chemistry in water are C and N which form many compounds with H and O. While some P and S are also incorporated, it is H, C, N and O that dominated prebiotic and biotic organic chemistry in water. We have already stressed the important features of this chemistry in Chapter 9. In effect these four elements were and are *abundant and available* and they have the following valuable properties in compounds that allowed biotic organic chemistry to evolve:

1. They form kinetically stable though thermodynamically unstable covalent combinations. This gives them a reasonably long lifetime in water.

2. Their synthetic chemistry is energy-requiring, but they trap energy easily, for example in sugars, and give a series of small monomer molecules. Their formation will be the subject of Section 10.5. Abiotic molecules, for example sugars, could have been one of life's first energy sources.

3. They readily form major classes of kinetically stable *linear* polymers, the synthesis of which is readily controllable by catalysts. We described these polymers in Chapter 9. The classes of polymers, separated by solubility in the two compartments of Fig. 10.4, are:

   (a) lipids, derived from fats $((CH_2)_n)$, which formed the containing phase, that is, membranes, since they are water-insoluble. In anaerobic conditions they are very stable. These membranes are to this day the basic containers of biology (see Figs 6.4 and 9.8) and some proteins partition into them.

   (b) polysaccharides, proteins and polynucleotides which are usually linear and water-soluble. (NB. Some polysaccharides, starch and cellulose are insoluble.) They can, in principle, then generate structures using unique orders of different monomers, though as yet in this book we have given no reason for any limitation on ordering monomers in a sequence. In biology there are some five bases in DNA and RNA, some 25 amino acids in proteins, and a great variety of monosaccharides. Their ordering and co-operative folding develop specific properties in the polymers (Section 6.8.1). These are the major final products of synthesis (Figs 9.13–9.20).

Non-metals in early organisms

H
C　N　O
(Si)　P　S　Cl
　　　　Se

The further co-operative association of polymers through van der Waals forces and H-bonds allowed self-assembly, bulk structure, so that space became occupied in selected ways (see Figs 10.4 and 10.5). Thus organic compounds, that is, combinations of the elements, H, C, N, O with some S and P, were selected for their ability to form kinetically stable units of great complexity, see marginal note. We are forced to conclude from the nature of life itself that even in prebiotic chemistry linear sequences of sugars, amino acids and nucleotides of considerable kinetic stability were formed and in considerable concentration. Their surfaces inevitably bound many elements, especially metal ions. Though relatively soft materials they readily bind one another to give shapes or frameworks assisted by these extra elements, for example Mg^{2+}, Fe^{2+}, Ca^{2+} and HPO_4^{2-}. Sufficient dynamics remains in some of the structures for long-range interaction as in elastic materials, so that frameworks and connecting mobile units (rods) can be built up to give the beginning of dynamic machinery (see Section 3.8). All of these features can be reproduced by man today with non-biological materials and in macro-size objects.

The manner in which this *chemistry* was brought about is described in Chapter 9 and need not be repeated here, but the restrictions on pressure (about 1 atm), temperature (270–370 K, 0–100°C) and solvent, H_2O, were very limiting for its development before life began. We cannot attempt to discuss in great depth the choices of materials that gave rise to life—for example how the selection of bases of DNA and RNA came to be related to the 25 amino acids so that the synthetic path of virtually all polymers evolved in the present synthetic order, that is,

$$DNA \longrightarrow RNA \longrightarrow proteins$$

transcription   translation

but it is our current belief that there had to have been a programme of gradual development of DNA to produce this present-day almost universal synthesis of sequences and we shall look into this briefly in Chapter 11. The main purpose of this chapter is to describe the particular way in which H, C, N and O are taken up, incorporated into monomers (Section 10.5) and then polymers (Section 10.8) within anaerobic cells, and the unavoidable association of this uptake and incorporation with that of some 15 other elements (Table 10.1).

We can do this systematically since analysis shows that, for the most part, particular elements are used functionally in the same way in all known organisms. We have stressed already that hydrogen, carbon, nitrogen and oxygen are the elements forming the basic polymers of life. They do not have outstanding functions in very simple forms, for example as CO_2 and NH_3, except for the case of H_2O. In addition to being part of water hydrogen plays a multiplicity of important roles and we use its known functions immediately as an illustration of the way in which we wish to describe the chemistry of other elements in addition to C, N, O in organisms.

Unlike carbon, nitrogen and oxygen, which form more than one covalent link and hence are the backbone of the monomers and polymers of biology (with phosphorus in the case of DNA and RNA), hydrogen forms but one bond and its covalent functional value is then quite different. First, it forms terminal groups of straight or side chains, $-CH_3$, $-NH_2$, $-OH$, in small molecules or polymers. The ability to form H bonds using the polar N—H or O—H and/or to use the non-polar characteristic of C—H to give hydrophobic association (Section 2.6) makes it possible for the polymers to fold. Second, hydrogen can exchange through three mechanisms as H^+, H^- and $H^{\cdot}$ (Section 2.7), which allows turnover of its chemicals but also allows it to function in other ways.

1. H^+ is a powerful acid, a very fast charge carrier and has a unique size. It is readily reduced at -0.5 volts at pH = 7.0. It binds to a large number of bases (Section 5.6.1), and this fact and its unique size allow biology to construct H^+ 'wires', that is conditions such that the proton can migrate

from one basic centre to another of equal or higher basicity (Fig. 7.1). Thus monitored flow of protons can be maintained between compartments of fixed but different pH, say, inside and outside.

2. H· (covalent H) is kinetically inert when held on carbon, but not so on a metal or some heavy non-metals or in coenzymes such as flavin. Thus it can be transferred and used in synthesis.

3. H⁻, which is often very reactive, is readily bound by metals or some non-metal centres. H⁻ is transported in biology by NAD (Section 9.13), and its incorporation or removal is assisted by catalysts.

The acid–base properties of H^+, $H^·$ and H^-, the redox properties of all three states, and the blocking nature of covalent C—H make hydrogen an ideal element in assisting energy and material capture within carbon compounds and in the subsequent chemical adjustment of fragments into organic compounds. The sets of compounds are then connected through series of catalysed steps and flow of material. We return to a description of much of this chemistry together with the small molecule chemistry of C, N and O in Section 10.5, but it is immediately obvious why hydrogen is a selected element in biology in the context of C, N and O chemistries.

10.3 Biological selection of minor elements

A simple and natural mistake to make in a first effort at an understanding of a chemical system is to assume that the *major* analytical components dominate the properties of the whole. When it was found, in the early nineteenth century, that, water apart, carbon, oxygen, hydrogen and nitrogen in various combined forms, together represented a large proportion of living material (Table 10.2), it was reasonable to suppose that other elements of the periodic table could be ignored. The prejudice has remained with us until very recently

Table 10.2 Average elemental composition of living material (plants and animals). Seawater is included for comparison

Element	Average composition (weight %)		
	Plants	Animals	Seawater
Oxygen	79.0	65.0	85.8
Hydrogen	10.0	10.0	10.7
Carbon	3.0	18.0	0.003
Nitrogen	0.3	3.0	5×10^{-5}
Calcium	0.1	2.0	0.05
Chlorine	0.07	0.2	2.07
Sulfur	0.01	0.3	0.05
Potassium	0.3	0.4	0.04
Sodium	0.03	0.2	1.9
Magnesium	0.08	0.05	0.15
Iron	0.02	0.004	1×10^{-6}
Manganese	0.12	1×10^{-5}	2×10^{-7}
Silicon	0.15	Low	0.0003

and organic (living) chemistry is often still supposed to be that of C, H, N and O, with some S and P. Only in the last 50 years have there been adequate analytical methods for the detection and quantitative evaluation of the amounts and roles of other elements. Today, the number of demonstrated essential elements in living systems is at least 20. In fact, a set of some 15–20 of these elements is required by *all* known living organisms (see Fig. 10.3). Given this knowledge, it pays us to inspect living systems against the background of the abundance and availability of the elements, especially since both changed with time as shown in Chapter 8. To what extent did this drive evolution? We may also ask why are so many of them vital for all organisms knowing in advance that the answer must be given in terms of their particular functional properties. They are obviously more fundamental than vitamins, special organic compounds, which lower living forms can synthesise. Of course, no living systems can make elements. We turn therefore to the chemistry of living systems to evaluate the functional value of the so-called minor *essential* elements, giving a few simple examples where such 'inorganic' elements play a fundamental role. Note again that it is the purpose of this book to demonstrate that elements in combination are selected by: (1) their abundance; (2) their combining potential controlling availability; (3) their potential to fall into stable kinetic traps; and finally (4) their *co-operative functional value*, fitness, in that time sequence. Man's activities continue this progression through *self-conscious* selection for functional value.

10.4 Examples of essential functions of inorganic elements

10.4.1 Osmotic pressure control

Any cell (Fig. 10.4), is an enclosed space holding many organic chemicals that are soluble in water; this space is exposed to a variety of external inorganic elements, especially ions, for example in the sea where, in all probability, life started. The membrane, composed of fatty acid molecules and proteins, is weak (Fig. 10.5) and yet it does not break even though there is a strong tendency for water to enter the cell to equalise osmotic pressure. Three

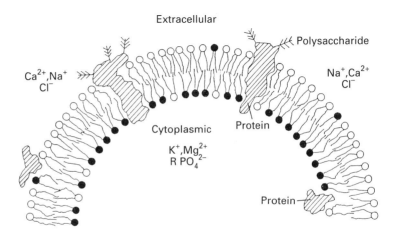

Fig. 10.5 A schematic representation of the assembly and distributional asymmetries of some biomembrane components. Lipid asymmetries are shown as a difference in the distribution of the polar head groups with the negatively charged lipids (filled circles) localised predominantly in the cytoplasmic leaflet (after D. Chapman). The lipids of archaebacteria differ from those of other bacteria which may also be an adaptation to conditions.

Extracellular

Polysaccharide

Ca^{2+}, Na^+
Cl^-

Na^+, Ca^{2+}
Cl^-

Protein

Cytoplasmic

K^+, Mg^{2+}
$R\,PO_4^{2-}$

Protein

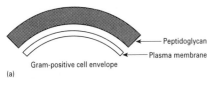

Gram-positive cell envelope

(a)

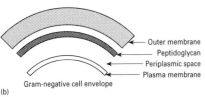

Gram-negative cell envelope

(b)

Fig. 10.6 Schematic diagrams showing the differences between the cell envelopes of (a) Gram-positive and (b) Gram-negative bacteria.

properties of cells maintain the cell volume and shape within the almost ion-impermeable membrane:

(1) a very strong *outer* wall, as in some bacteria and many plant cells (Fig. 10.6);

(2) a strong interconnected internal construction of proteins and filaments to which the membrane is attached, as in most animal and plant cells (Figs 7.29 and 7.30); much of this structure may be a late development;

(3) the continuous rejection of common salt, NaCl, by pumping (literally) the ions of the elements sodium and chlorine, Na^+ and Cl^-, out of the cells of *all* living systems, some of which have cells bathed in high salt, for example the sea (Fig. 6.13); uptake of some sodium is required by cells in very low salt for other reasons (see below).

Generally speaking, the cytoplasm of the vast majority of all living cells is comprised of $10\,mM\,Na^+$ and $100\,mM\,K^+$. Thus the energised balance of Na^+, K^+ and Cl^- is a fundamental feature of cell stability that is unrelated to the nature or presence of a DNA code. Today this distribution is interactive with the lipid components of the membrane making for inside/outside asymmetry (Fig. 10.5), and this asymmetry may have been present from the earliest times.

Before continuing, a very simple real example gives a striking illustration of the intimate relationship between a living cell and its inorganic/organic complement. Amongst the archaebacteria are those, called halobacteria, that live in 3–4 M NaCl solution, that is, in saturated common salt. The interior of the cell is 3–4 M in KCl which, if all of it were free, would be super-saturated. All the proteins and other cellular polymers of this species are especially evolved and are unstable in the salt conditions of around 0.1 M of normal cells! Here is a demonstration of the fact that the living chemistry we observe is a complicated interwoven chemistry of internal inorganic elements (even the simplest) and organic compounds that match the inorganic environment. The opposite case is that of organisms in freshwater when sodium and even more so potassium are taken up for the reasons given in the next paragraph.

10.4.2 Electrical neutrality

The interior of a cell is full of negatively charged organic species, for example DNA, RNA, fatty acid anions of membranes (Fig. 10.4), many substrates and other organic acid anions and inorganic anions such as chloride and phosphates. The *anionic* form of many organic molecules, which helps to keep them soluble, is an inevitable consequence of the general electronegativity of non-metals and it is then used to organise them, but this requires counterions to oppose repulsion (see below). The total concentration of these anions in cells must be quite high (say 10–100 millimolar) if association is to occur and if reaction rates are to be reasonably fast. The cell, if it is to be electrically stable, must have virtually complete compensation of this negative charge by equivalent positive charge, which has to be based on less electronegative (inorganic) elements, namely metals. Na^+ is very abundant, but living systems in high salt, for example the sea, cannot use freely the positive charge of this ion for the purpose since the entrance of Na^+ into the cells in quantity has to be prevented, as stated above, to control osmotic pressure and cell volume (a lot of water would have to enter as well). The cells accept instead the more dilute K^+ from the environment (while rejecting Na^+ in part) and, in order to keep the anionic charge as low as possible, they also reject most of the chloride, Cl^-. (Note that cells in freshwater pick up some Na^+ and some Cl^- while maintaining higher K^+, as if they had a memory of an earlier stage in evolution when they lived in high NaCl.) Putting together control of cell volume with control of charge generates an absolute requirement in all life forms for controlled levels of Na^+, K^+ and Cl^- (Table 10.2). In particular, the interaction of polyelectrolytes such as DNA and membranes with other molecules is also controlled by these ions and their corresponding salts (Fig. 10.5). The organic chemistry of life is 'designed' not around pure water as a solvent but around a fixed ionic strength aqueous solution generated by *specific* inorganic ions. In particular, DNA is sensitive to the substitution of Na^+ for K^+. K^+, not Na^+, is required in the telomere† structure of DNA and also by a number of enzymes. From the very earliest known periods, control over life in cells also required energy to ensure this balanced osmosis and approximate electrical neutrality and to maintain chemical structures (see Section 10.4.6). These seemingly trivial inorganic and physical chemical controls will later be seen to have evolved to form the bases of nerve and then brain primary message systems (Sections 13.4.1 and 13.8 and see also Section 6.4).

10.4.3 Cross-linking and precipitation

Now the problem of osmosis and electrical neutrality can be solved by careful control over Na^+ (pumped out), Cl^- (pumped out) and K^+ (allowed in), but there are two other common cations in the waters external to life—Mg^{2+} and Ca^{2+}. We have seen in our previous book that Ca^{2+} forms more insoluble salts than Mg^{2+} by connecting anions together (see ref. 1). Inside a cell this would be disastrous since the clumping together of anions, especially by calcium, would prevent reaction and reproduction of the anionic polymers; hence calcium, a 'poison', had to be very largely rejected even from primitive cells, while magnesium could be utilised, for instance as a mild catalyst in a fundamental connection with anion, especially phosphate, metabolism (see Section 9.11), and also could be used in stabilising RNA and DNA. Indeed, all cells reject calcium while accepting magnesium so that strong cross-linking inside cells is largely avoided. Outside cells cross-linking is desirable in order to form protective walls (Fig. 10.6), and calcium becomes more important than magnesium in external biological tissues, for example cell walls, and biominerals. (Later in evolution Ca^{2+} became the major external second messenger in information transmission (Chapter 12).) *All* known cells behave in this way, so we see that there are five simple inorganic ions around which all abiotic and biological systems evolved of necessity. In passing it is perhaps worth noting that the early reducing sea was probably low in sulphate, so that chloride would have been the only simple anion that it was necessary to avoid to a large degree.

Besides these five ions—Na^+, K^+, Cl^-, Ca^{2+} and Mg^{2+} (note that they are all in extreme groups of the periodic table and have either low or high electronegativity and are all available)—other inorganic elements are used in

---

† The telomere is the terminal part of DNA which controls copying.

essential fundamental steps of biological energy capture, information transfer and control, catalysis and development of shape (see Tables 7.9, 7.10, 7.12 and see Chapters 11 and 12). The following paragraphs introduce the reader to this variety of functions. As we go through these sections and indeed as we went through the above the reader is asked to keep in mind the question as to whether or not dissipative 'cell' systems could have arisen *before* coding. The system would not reproduce but could be increasing with time in kinetic stability by gaining control over internal conditions and so increasing the survival strength of the 'non-living' vesicle. It is selection for survival that concerns us here.

10.4.4 Electronic conduction

There is an inescapable need for life to build *polymers from carbon* compounds (on Earth no other element can substitute for carbon for this purpose). As pointed out in Chapters 7 and 9, all C/H/O compounds tend to disproportionate to CH_4 and CO_2; hence carbon was (primarily) only available as either CH_4 or CO_2 (Fig. 9.1). It was therefore essential to carry out partial oxidation of CH_4 or reduction of CO_2 to the stage H_2CO, since in this oxidation state carbon can be polymerised. In the oxidation or reduction reactions there is, therefore, a requirement for removal from or donation of the atom radical H to carbon atoms. This is achievable most easily in many compounds through the use of the separate movement of H as H^+ plus e, so that *electron transfer* (Section 7.5.1), (and *proton* transfer; see Section 7.5.3) in water is a necessary feature of biology. In fact, these redox reactions of CH_4 and CO_2 may well represent some of the earliest metabolic reactions to be incorporated in life (see Chapter 9). We shall see that these initial electron transfer steps of very primitive life have later become incorporated into very complicated series of electron transfer chains (see Section 7.5.1). For the moment we shall only observe the elements that are required to generate such chains, noting that they are needed to handle many fundamental redox reactions, not only of CH_4 and CO_2 but also of H_2, H_2O and N_2.

In Chapters 2 and 7 we stressed that the transition metals were best suited for giving electronic conduction in wires, for example Cu or Fe. However, as stated, if we space such metal ions a few angstroms apart in a matrix (up to 15 Å), an electron can still 'jump' (but slowly) from one metal atom to another utilising changes in their oxidation states. Indeed, it is the building of such inorganic metal atoms inside an organic matrix that has allowed evolution to develop many of the required electronic features seen in biology (Figs 7.9 and 10.7). Once again, putting selected inorganic atoms together with organic constructs generates the required property necessary for life, and we know of no living systems without such structures. Almost invariably, they have iron though, in principle, copper, manganese, cobalt, etc. could have been used also and are used here and there for their ability to undergo selective oxidation state changes and so connect particular electron transfer reactions together. Some of these reactions are best carried out by one of them only, for example manganese in the release of O_2 from water. We believe that iron was the element first used since it was and is the most abundant (Fig. 8.1) and was easily available on early Earth (Section 8.5.3). At least one element of this kind is essential for life. At some time in early evolution, judging from the

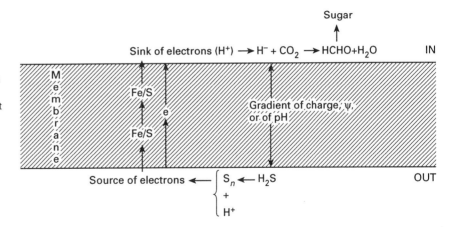

Fig. 10.7 A simplified description of electron flow and its connection to carbon dioxide reduction using a membrane. It is the presence of a transition metal (Fe) that is essential. In fact, the electron transfer usually meets the proton transfer pathway in the membrane (see 'Further reading').

archaebacteria, we know that several other metals (Mn, Co, Ni, V, Mo, W) also came to be used, more or less specifically, in different electron reaction paths.

In Fig. 10.8 we show some very simple Fe/S proteins that could have carried out one-electron reactions, perhaps before DNA was formed. (NB. we do not deny that some *organic molecules* could be and are now used as electron transfer centres but they require sophisticated synthesis, for example flavin and phaeophytins, and are unlikely to be extremely primitive.)

10.4.5 Catalysis

Just as organic chemistry in a laboratory requires a vast range of inorganic catalysts both for hydrolytic and redox purposes, so biological systems must utilise a variety of metal ions as catalysts in special proteins (enzymes) for the same purposes. In effect, it is not possible to visualise how purely organic catalysts, here proteins as enzymes, could have activated the small molecules, H_2O, N_2, CO_2, CH_4 and so on, which were the major H/C/N/O sources available in the primordial environment. Both main-group and transition metal elements must have been and still are essential in all forms of life as catalysts (Table 10.3), and a number of them must have been used not only in redox reactions but also in acid–base reactions. Which ones were used first we shall discuss later, but, by using several, selective action is acquired since each element has at least one unique potential value. Interesting examples are phosphate transfer which requires magnesium ions as a catalyst, the reduction of dinitrogen which requires Mo(or V) and Fe, RNA synthesis which uses zinc and DNA synthesis which needs iron or cobalt. Note again that some time later assistance was obtained from the use of special organic molecules, coenzymes, as catalysts, some of which also contain metal ions (see Table 10.6). We see then that the number of essential elements, apart from H, C, N and O, required by all life is already around ten.

Table 10.3 Early inorganic catalysts

Catalytic process	Element*
Acid–base	Mg^{2+}, Mn^{2+}, Fe^{2+} (Zn^{2+}) OH^-, SH^-, imidazole
Redox	Fe^{2+}/Fe^{3+} (haem), —SH/(S—S) Ni^+/Ni^{2+}(Ni^{3+}) (F-430) Co^{2+}/Co^{3+} (vitamin B_{12}) Mn^{2+}/Mn^{4+}

* Zn^{2+} was not very available in high sulphide conditions (Fig. 5.3). Archaebacteria use Fe, Co, Ni, Mn(?), Se, W and Mo. See also Chapters 7 and 9.

10.4.6 Energy capture—synthesis and gradients

Life as a process requires energy to move chemicals into kinetic traps of low thermodynamic stability but considerable lifetime. As we have seen in Chapter 9, the two major ways of so energising chemical systems, that is of increasing

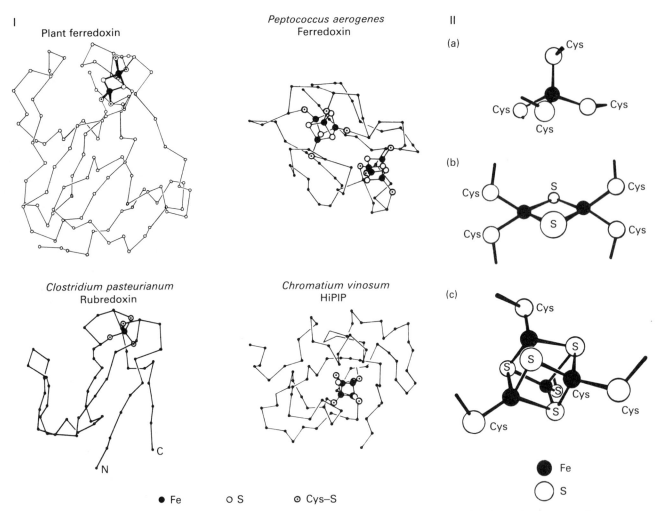

I
Plant ferredoxin

Peptococcus aerogenes
Ferredoxin

II
(a)

Cys
Cys
Cys
Cys

(b)

Cys
Cys
S
S
Cys
Cys

Clostridium pasteurianum
Rubredoxin

Chromatium vinosum
HiPIP

(c)

Cys
S
S
S
S
S
Cys
Cys
Cys

C
N

● Fe   ○ S   ◎ Cys–S

● Fe
○ S

Fig. 10.8 (I) The folds of some typical iron sulphur proteins. Note the β-sheet constructions. (After the work of many authors including L. Jensen, K. Fukuyama and R. Stroud.) (II) The co-ordination in iron/sulphur proteins: (a) rubredoxin; (b) Fe_2S_2 ferredoxins; (c) Fe_4S_4 ferredoxins. The Fe_3S_4 unit is like that of Fe_4S_4 with one Fe removed. ● Fe; ○ sulphur—those marked 'S' are sulphide, S^{2-}, while those marked 'cys' are cysteinate sulphur.

their potential energy, are: (1) putting atoms together in kinetically stable but thermodynamically unstable combinations, for example C with H *and* O (as in HCHO) rather than C with only H (as in CH_4) *or* C with only O (as in CO_2) (Fig. 9.1); (2) concentrating atoms (ions) in limited volumes (see Chapter 6). Those elements that combine 'covalently' are readily energised by the first procedure; the elements that form ionic solutions are best energised through concentration or field gradients. No matter where energy comes from, perhaps from the Earth's lack of equilibrium originally, or today mainly from light, disproportionation of chemicals, or burning of fuel, for example sugars, with dioxygen, it had to be used to create an electronic gradient to help incorporation of C or N with H and/or an electrolytic gradient to incorporate many elements that occur as ions.

We insert a short account of how this can be done to stress again the use of several elements.

The route of energy transduction for synthesis and for uptake is known to be

Energy $\longrightarrow$ charge separation e/H$^+$ $\begin{cases} \nearrow & \text{reduction by e of CO}_2 \\ \rightarrow & \text{production of ATP} \\ \searrow & \text{concentrated gradients} \\ & \text{of chemicals} \end{cases} \longrightarrow$ synthesis.

The production of the simplest of all inorganic species, e and H$^+$, that is, separated charges within and then across membranes, is the first chemical step. It is readily achieved by the absorption of light by a magnesium complex, chlorophyll (Fig. 7.44) as shown in Figs 10.7 and 10.9 and see Section 7.19.1 (as well as by other means). The charge gradient of protons is used either directly to produce other gradients by exchange, for example of Na$^+$, Cl$^-$ and Ca^{2+} ions, or to produce ATP (bound pyrophosphate) which can itself produce gradients by pumping of very many elements, free or in compounds, across

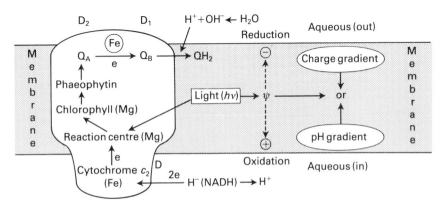

Fig. 10.9 The reactions of photosystem I. Light drives electrons and hence the separation of charge across a membrane to give an electrical potential, ψ. The separated charges can rest on metal ions or organic molecules. At the edges of the membrane the charges can migrate toward the aqueous phase to give a proton gradient and the chemical species generated can be used to drive organic reactions. Note the metals used. Q, quinone, an organic coenzyme used in redox reactions. NB. In real systems the proton at the top diffuses in the membrane to Q_B.

membranes. Reduction is achieved by recombination of e with H$^+$ at a particular chemical site, NADH, and is required for incorporation of CO$_2$, a process needing energy. Where the reduction uses H$_2$S it generates sulphur and, later, when it uses H$_2$O, it generates O$_2$ (Fig. 10.25). The capture of energy in this way uses iron in Fe$_n$/S$_n$ complexes for e transfer, then magnesium in chlorophyll, and later uses manganese in dioxygen release, and later still iron and copper in dioxygen activation. We draw attention to the fact that in these systems proton conduction in membranes is as dominant as electron conduction (see reference 1 in 'Further reading').

When we examine all the requirements as described in Sections 10.4.4–10.4.6, it is seen that in order that cells should exist it is necessary to use several different elements in different activities. While the uses of Na$^+$, Cl$^-$, K$^+$ and Mg^{2+} appear to be invariant, some eight transition metals, including Mn, Fe, Co, Ni, (Zn), Mo, (V) and (W), are used differently in early periods of evolution. Archaebacteria use Fe, Co, Ni, W, (Mo), Se, perhaps Mn and some Zn, apart from managing Na$^+$, K$^+$, Mg^{2+} and perhaps Ca^{2+} and, of course, H, C, N, O, S and P. It is very possible that life started with all these 15–20 elements from some precursor abiotic system that became coded in a reproductive manner (Section 9.8.4). All these elements were inevitably present and interactive with proteins of course, see marginal note.

(NB. In very primitive systems light may not have been used to generate the

Metal elements in early organisms

(Na) Mg
  K (Ca) ... V . Mn Fe Co Ni . (Zn)
          Mo
          W

( ) low levels in cells

e/H$^+$ gradients. As explained in Chapters 3–6, it is possible to use chemicals to drive electrical circuits and primitive life may have used the out-of-balance of the oxidation states of iron and sulfur in the ocean vents.)

10.4.7 Triggering of mechanical action

It is obvious that many biological activities require flexibility, while individual biological systems readily retain particular overall shapes. These activities must be timed (triggered) even in early anaerobes and include mechanical movements due to filamentous structures such as flagellae, as well as movements of DNA on the division of a cell. At the membrane level, endo- and exocytosis also require movement, but this feature probably came later. Even simpler activities, such as some transport, pumping or signalling systems, also need well-defined, timed, molecular *elastic* motions. Recovery is essential so that a steady state is maintained. The triggering of many mechanical activities is known to be a property of the switch of concentration of phosphorylated compounds (see Section 11.5.6). The energy for mechanical motion itself is supplied by $Mg^{2+} \cdot$ pyrophosphate (of ATP) hydrolysis (see Section 9.11). Relaxation is then brought about by metabolic regeneration of ATP to re-attain the initial resting state. The very earliest cells used ATP(Mg^{2+}) in these mechanical devices to bring about division. This is but one use of ATP, as already stated, but the essential role of phosphorus and magnesium must be stressed here. (Later we shall see that much of this activity becomes coupled with calcium functions, but, in the earliest cells, as already stated, the only certain essential feature involving calcium chemistry is that calcium should be largely rejected to avoid anion association in the cell when it could be used to cross-link the wall. (It may have been used similarly in a few internal structures.) This means, however, that a calcium pump was always essential.)

10.4.8 Summary of required elements

In these few paragraphs we see that there is an essential requirement in *all known living systems for an appropriate complement of inorganic as well as organic elements*. This links life, as we see it, to the evolution of the general chemistry on the surface of the Earth (element availability) and hence to the chemistry of the universe (element abundance). It is also true that there is a fundamental link to energy, stored in the reactive chemicals within the crust of the Earth or in the nuclear reactions of the sun. Life is a result of the uptake of both elements and this energy supply. Above all, however, we see that biology selects elements using energy along with functional value (see Fig. 10.30). Some elements were even rejected, for example sodium, calcium and chloride, and this also requires energy. Thus life can be considered as a fluctuation in the evolution of chemicals with time in the universe, in which the general unavoidable change toward de-energised states, cooling, is delayed by local systematic energisation, but with the additional particular feature of controlled flow and reproduction of all its chemical contents locally, that is to say, chemically developed organisation. This has led clearly to self-conscious use by man of chemicals as in present-day industrial chemistry, which is but a new way of utilising chemical elements in energised forms, that is, in kinetic traps, but now without controlled reproduction (Chapter 14). Much as in prebiotic times, the time taken for these progressions is not prescribed and possibly depends on adventitious events, such as changes in availability. We next turn to the more specific problems of the handling of individual elements, that is, their uptake and incorporation in 'primitive' cells. Later we must analyse the problem as to how given elements are taken up in closely defined amounts so that living species have an analytical identity. This is a matter of control and is described in Chapter 11.

10.5 Uptake and incorporation mechanisms of the major elements

The elements exist in external aqueous solution in very different forms and concentrations (Table 10.4 and see Table 8.14). Thus, in both uptake and incorporation they are associated with different physical/chemical routes. Table 10.5 illustrates the obvious facts that small covalent molecules of the light non-metals can be taken up easily and, because they must not be lost equally easily, they have to be incorporated through covalent bonding. H, C, N, O, S, Se and Si can pass through membranes as hydrides or oxides such as, CH_4, NH_3, H_2O, H_2S, H_2Se or H_2O, CO_2, NO (also N_2), SO_2 and $Si(OH)_4$. All must be covalently linked internally. Most other elements occur as cations, for example metal elements, or as anions, for example the non-metals Cl, Br, P, As and some forms of S. The anions referred to are HPO_4^{2-}, $HAsO_4^{2-}$, MoO_4^{3-} (included as a non-metal), Cl^-, Br^- and SO_3^{2-} or SO_4^{2-}. All such ions have to be forcibly moved across membranes in special channels by pumps that do not reverse. The elements become physically trapped inside and may or may not be covalently linked subsequently to the covalent assemblies of C, H, N, O. Once

Table 10.4 Available free ion concentrations in external aqueous solutions

Metal ion	Original conditions (molar)*	Aerobic conditions (molar)*
Na^+	$>10^{-1}$	$>10^{-1}$
K^+	$\sim10^{-2}$	$\sim10^{-2}$
Mg^{2+}	$\sim10^{-2}$	$\sim10^{-2}$
Ca^{2+}	$\sim10^{-3}$	$\sim10^{-3}$
Mn^{2+}	$\sim10^{-6}$	$\sim10^{-8}$
Fe	$\sim10^{-7}$ (Fe(II))	$\sim10^{-19}$ (Fe(III))
Co^{2+}	$\sim10^{-9}$	$\sim(10^{-9})$
Ni^{2+}	$<10^{-9}$	$<10^{-9}$
Cu	$<10^{-20}$ (very low), Cu(I)	$<10^{-10}$, Cu(II)
Zn^{2+}	$<10^{-12}$ (low)	$\sim10^{-8}$
Mo	$\sim10^{-9}$ (MoS_4^{2-}, $Mo(OH)_6$)	10^{-8} (MoO_4^{2-})
W	WS_4^{2-}	10^{-9} (WO_4^{2-})
H^+	pH low (5.5?)	pH 8.5 (sea)
O_2	$<10^{-6}$ atm	$\sim10^{-1}$ atm (21%)
CO_2	>10 atm	10^{-2} atm
N_2	$>?10$ atm (NH_3?)	~1 atm (78%)
H_2S	10^{-2}	Low [SO_4^{2-} (10^{-2} molar)]
HPO_4^{2-}	$<10^{-3}$	$<10^{-3}$ molar

* Except where other units are specified.

Table 10.5 Modes of uptake and incorporation of elements

Element	Major uptake mechanism	Major mode of incorporation
H, C, N, O, Si	Diffusion or pumping of larger molecules	Covalent binding
Na^+, K^+, Cl^-	Pumping	Free ion trapping
Mg^{2+}, Ca^{2+}	Pumping	Weak complexes
Transition metals	Pumping	Strong complexes
SO_4^{2-}, SeO_4^{2-}, MoO_4^{2-}	Pumping	Virtually covalent binding

the light non-metal elements are converted to more complex forms—usually water-soluble and often anionic—for example sugars, amino acids and so on, then reverse passage across membranes is again inhibited. When available from outside in food, these molecules too must be pumped into cells. Control is then gained over physically and chemically trapped forms for all elements of value. We now describe some specific cases.

10.5.1 Hydrogen and oxygen uptake and incorporation

Water passes readily through biological membranes and, therefore, a cell is always provided with a source of H and O. A further feature of water is its self-imposed dissociation to H^+ and OH^-, so that H and O are present in bound and reactive forms. In this respect, notice that the three chemicals H^+, H_2O and OH^- are always in relatively fast exchange. It is necessary for homeostasis of acid–base sensitive polymers then that the concentrations of H^+ and OH^- in cells are carefully managed at low values relative to a variety of possible external pH values and even though metabolism produces acids. As we have seen it is probable that the water on Earth was not formed initially at neutral pH. There are arguments that favour the view that the sea was initially acid due to solution of HCl from the atmosphere (Section 8.5.3), and it may be that early biological systems formed at this pH and had to remove H^+ from their cells. The pH rose rapidly, however, due to reaction with basic rocks, generating Ca^{2+}, Mg^{2+}, HCO_3^- ionic solutions, and has come to a value around pH = 8, which has been maintained in the sea. This steady-state pH is due to the equilibration of the carbon dioxide/carbonate reactions over the first billion years of the Earth (Chapter 8). However, there are still acid and alkaline solutions on some parts of Earth. No matter what the earliest conditions, life as we know it developed internally at neutral pH and today nearly all cells *maintain* in virtually all forms an *internal cytoplasmic pH of around 7 whatever the environment*. This requires a chemostat-like control over H^+ *uptake* and *rejection* which gives also a steady state for OH^- (see Chapters 7 and 11). Many features of life's chemicals in the cytoplasm depend on this pH close to 7, and this includes the activities of phosphates, many proteins and lipids. Clearly, the control over H and O incorporation in their active forms H^+ and OH^- can only be managed by the use of pumping. We shall describe the proton pump in Section 10.6.1.

The nature of the biological cell demands control over osmotic pressure

(Section 6.4.1) and at the same time an adequate intake of water is necessary for growth. Water in cells is kept under control by adjustment of the total ion plus molecule concentration, especially by adjusting K^+ levels (see Section 10.4.1). The control is very important in plant transpiration.

Covalent incorporation of H and O is tightly linked to C and N uptake as we describe next, but, we stress again, one early source of reactive hydrogen could well have been H_2S and even some H_2.

10.5.2 Uptake and incorporation of carbon and nitrogen: introduction

The initial, inevitably random reactions of organic chemical synthesis capturing energy in covalent bonds on the Earth's surface had to be based on small units of C, H, N, O, P and S. These reactions gradually developed into a sophisticated and 'purposeful' series of connected syntheses (and degradations) in living cell systems, especially of polymers built around carbon, nitrogen, oxygen and hydrogen. We do not know how this came about, but there are some central common features:

1. Carbon is readily available and easily diffuses through membranes as CO, CO_2 or CH_4. There is no uptake problem.

2. Carbon had to be incorporated initially from CO_2 or CH_4 into small molecules by reduction or oxidation (Fig. 9.1), and this is not easy either kinetically or thermodynamically. All C/H/O compounds will disproportionate to CH_4 and CO_2, so that degradation must be restricted, but as degradation, in particular, provides a source of energy, it may also be useful. Degradation and synthesis are always taking place at the same time in biology but by different routes. The systems are often *cyclic*, energy-requiring and energy-dissipating. They are irreversible processes in thermodynamic terms (Section 3.6).

3. Nitrogen had to be incorporated from N_2 or, at least, from NH_3. The first process is difficult since it involves a 6e reduction and has a large energy barrier.

4. Further energy had to be captured for the synthesis of C, H, O, N polymers from the small molecules.

5. Material had to be transferred from one pathway to another. (At first we consider only C/H/N/O/P/S transfers, but the statement is general (see Section 9.9).)

6. Energy, once captured, had also to be transferred from one pathway to another.

These seemingly competing needs for synthesis and degradation of the whole cellular system had to be brought into harmony and, clearly, material and energy had to be conserved.

We have described in Chapter 9, in general outline, the major classes of biological organic compounds and their relationships. Here we have to describe how they are integrated in a network of specific compounds and later we will seek to understand how this network is controlled and regulated so that the products reproduce the organism while it maintains itself (Chapter

11). This implies an analytical *homeostasis* in the ratio of H, C, N, O, P and S in the organic compounds of the cell while $pH(H^+/OH^-)$ and redox potential, e/H^+, are also maintained constant. We have, therefore, to go forward to a description of the particular pathways of biological H, C, N, O, P and S transfers. We stress that, of necessity, this is not a random set and much is common to all cells. We shall see later that we need additional homeostasis of perhaps as many as 10 'inorganic' elements to maintain the complete homeostasis of organic chemicals since the reactions of organic chemicals require catalysts as detailed above and must have a well controlled communication system and environment. We treat first the simplest pathway leading to and from sugars, $[HCHO]_n$.

10.5.3 C/H/O incorporation

The first problem is to describe how primitive biology incorporates C/H/O from the possible basic and initial sources of these elements: C from CO_2, perhaps as HCO_3^- (CO in some cases), and from CH_4; H from H_2O (H_2S), CH_4 or even from H_2; and O from H_2O or CO_2 (CO) but not at first from O_2. Of these molecules, CO, CH_4 and, to some degree, H_2O are difficult to activate to give intermediates while CO_2 and H_2S are more reactive. However, all of them enter a cell across membranes almost without hindrance. Initially, we may consider that reducing equivalents came from CH_4 or H_2S. The possible modes of incorporation of carbon are extremely likely to have involved metals in sulphides, for example Fe/S or Ni/S, and, since these same systems can provide, or act as a sink of, H^-, we can imagine the initial reactions as given in Sections 9.2 and 9.12, namely

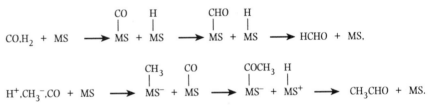

Here, C, H and O are simultaneously incorporated.

Once the carbon is in the form of carbonyl in ketones or aldehydes, further reaction is much easier (Section 9.16), and a range of somewhat larger molecules, sugars, as well as acids or alcohols, can be made by thermodynamically downhill reactions (Section 9.10) or by only somewhat energy-requiring paths, such as gluconeogenesis (Fig. 10.10). The formation of sugars must have been a very early and a very important step of incorporation of H, C and O. Sugars are quite stable, unlike formaldehyde, and are found in anaerobic systems and can be used to make polymers. They may well have been formed in abiotic reactions, but later, even in the earliest organisms, we know they could be synthesised and degraded. Certain sugars, pentoses (see Chapter 9), are also known to incorporate further CO_2 in living cells, so that the gradual build-up of carbon can be seen as it is today through the reverse of glycolysis or of the citric acid cycle (Fig. 10.11) or in the Calvin cycle of photosynthesis (Fig. 10.12), although this process was probably added last.

Once sugars have been made available (note that today energy from the sun is required to make it so), then the 'energy-containing' sugars can be degraded

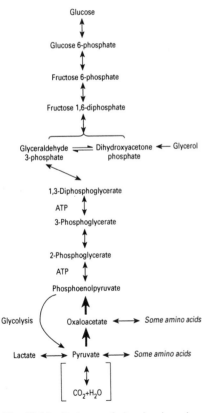

Fig. 10.10 Pathway of glycoloysis and gluconeogenesis. The distinctive reactions of gluconeogenesis are denoted by heavy arrows. The source of pyruvate is CO_2 and acetate (see Fig. 10.11).

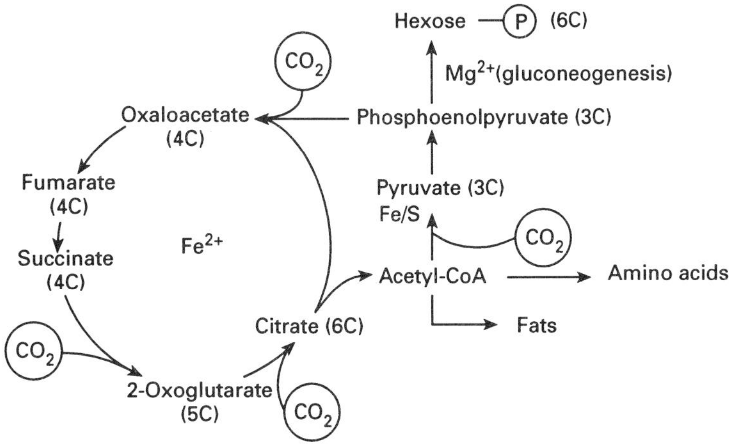

Fig. 10.11 The reductive citric acid cycle for autotrophic CO_2 fixation in *Thermoproteus neutrophilus*, an archaebacterium.

Fig. 10.12 (a) The basis of light-driven C, H, O incorporation; (b) A simple outline of the Calvin cycle for photoreduction of CO_2 and incorporation of C, H, O.

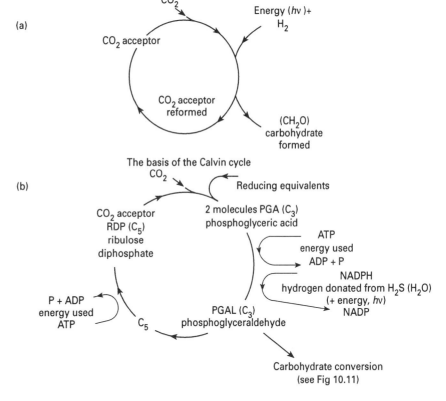

in part in the glycolytic pathway (Figs 10.10 and 10.11). This pathway, which is anaerobic, now uses the disproportionation of organic compounds (Section 9.10.2), such that

$$
\begin{array}{c} -\text{CH}-\text{CH}- \\ \quad | \qquad | \\ \quad \text{OH} \quad \text{OH} \end{array}
\longrightarrow
\begin{array}{c} | \\ -\text{CH}_2-\text{CO}(+\text{H}_2\text{O}) \end{array}
\longrightarrow
\text{CH}_3\text{COOH} + \text{energy,}
$$

and it then leads on to modes of incorporation of H, C, O into fats (Fig. 10.13) and amino acids (see below) and also makes higher 'energy-containing'

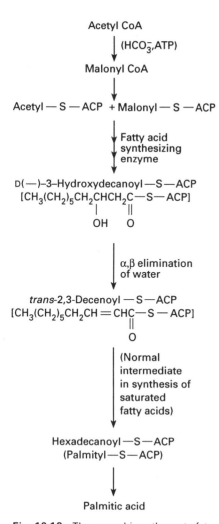

Fig. 10.13 The anaerobic pathway to fat synthesis in prokaryotic organisms. ACP, acyl carrier protein.

pyrophosphate in adenosine triphosphate (ATP). Several steps of carbon disproportionation are needed to make one pyrophosphate. The energy is required, as stated in Chapter 9, for the further syntheses of the biological polymers incorporating H, C, O with N, P and S (see Figs 10.10–10.13). Thus sugars are a basic driving energy source for the incorporation of other elements.

The links between the pathways of glycolysis and gluconeogenesis, fat synthesis (Fig. 10.13) and degradation, and the citric acid cycle are shown in Fig. 10.17. The question as to why these particular metabolic routes for incorporation are present can only be answered in terms of the overall stability (survival strength) of the formed organism, which is a way of saying that these routes and their products form a holistic co-operative system of functional chemicals.

While this basic C/H/O metabolism is usually almost identical in all organisms that we think of as being of recent origin, the earliest pathways may have been somewhat different. We see other possibilities in the archaebacteria, for example the pyroglycolytic pathway, some of which use metals such as cobalt and nickel more extensively (Fig. 10.14) and also use more primitive coenzymes, for example coenzyme M, for carrying carbon fragments. In some of these organisms, which may be very primitive, there is also some use of tungsten and selenium in hydrogenases (Fig. 10.14). The central use of Co, Ni and W has almost disappeared from more recent forms of life.

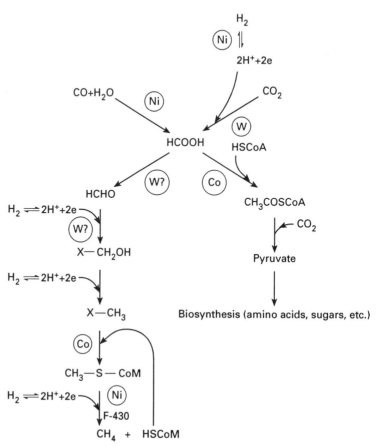

Fig. 10.14 C$_1$ metabolism in methanogens showing the possible involvement of especially Ni, Co and W in primitive life. X, Pterin-containing coenzyme (formaldehyde activity factor); HSCoA, coenzyme A; HSCoM, coenzyme M (HS—CH$_2$—CH$_2$—SO$_3$$^-$).

10.5.4 Nitrogen incorporation

Nitrogen is incorporated from NH_3 or N_2, or today via the oxides or oxyanions of nitrogen. There was and is no problem with the (primitive) incorporation of ammonia since it is a reactive molecule that condenses easily with aldehydes and ketones, but there are great difficulties in activating dinitrogen and even nitrate. Consequently, biological systems must again use the available catalysts and carriers of intermediates that the inorganic elements or coenzymes supply. Most N_2 fixation uses molybdenum and iron (Fig. 10.15), some uses vanadium and iron and a little is based on iron alone. (Surprisingly, today all nitrate activation depends on molybdenum as well.) The evolution of this molybdenum dependence for such an essential element as nitrogen is a major puzzle for the understanding of life's chemistry. Why molybdenum? Is it the 'fittest' available element? Or is its use new and did the use of vanadium pre-date it? Then we must ask why vanadium?

Once reduced to NH_3, nitrogen is readily incorporated into monomers by condensation reactions (see Section 9.8), using the carbon compounds from the citric acid cycle to give amino acids. In particular, glutamine and pyridoxal phosphate (see marginal illustration) became the effective coenzymes that are central to nitrogen distribution (Fig. 10.16). The development of this chemistry consists of further condensation reactions to give amides, as in peptides and proteins, and to give the nitrogen bases of DNA and RNA (Section 9.9.4). We shall not describe the pathways of synthesis of the nucleotide bases, but note that they are formed from a few amino acids, such as glycine, aspartate and glutamine. Energy is required as well as the ribose or deoxyribose (sugar) phosphate moieties for the RNA and DNA backbones, respectively.

Pyridoxal phosphate

Fig. 10.15 (a) An impression of the nature of nitrogenase with its different Fe/S clusters, its ATP-reaction site and (b) its active site for N_2 binding and reduction. The outline structure for both proteins is now known. The ATP acts on the first protein, which is in the form of a hinge, forcing a change on the Fe/S cluster so that electrons (and protons?) go to the FeMoco via P centres. Fe_7MoS_8 (homocitrate) has a cavity between two cubes. (The structure is taken with permission from Kim, J. and Rees, D.C. (1992). Structural models of the metal centres of nitrogenase. *Science* **257**, 1677–81.)

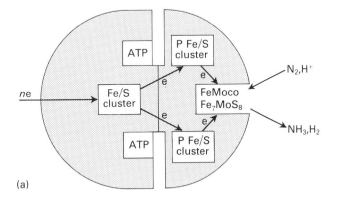

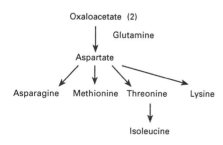

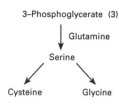

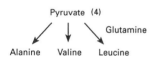

Fig. 10.16 Biosynthetic families of amino acid formation.

10.5.5 The uptake and incorporation of sulphur and selenium

The uptake of these elements presented no problem for cells originally since both were present as hydrides, H_2S and H_2Se, respectively, which cross membranes easily. The incorporation in carbon compounds by substitution and addition has been described in Section 9.2. The two elements are readily covalently attached, although energy is required for the reactions. Both Se and S are found in very primitive complex catalysts often with metal ions, for example Fe_nS_n, W/Se clusters. It could well be that such clusters were directly incorporated from inorganic sources some 3×10^9 years ago. Their functional use has been described in Chapter 9 and we have already stated that early biological systems used both elements in hydride and carbon fragment transfer as well as in catalysts.

We shall see in Chapter 11 that today uptake of S and Se and their incorporation have been made much more difficult since both elements exist largely as oxyanions, SO_4^{2-} and SeO_4^{2-}, which cannot pass through membranes and do not react directly with carbon compounds.

10.5.6 Interrelationships of C/H/O/N compound incorporation

While we have now described separately the production of several monomers of biological polymers there is also a necessary linkage between them to maintain a balance metabolism. The link from fats to sugars, for example, is via the intermediate pyruvate using the enzyme pyruvate dehydrogenase (PDC; Fig. 10.17). (Figure 10.17 also shows the link to the small amino acid, alanine). There is a very interesting feature of this primitive enzyme in that it is an Fe_4/S_4- and a lipoic acid-requiring protein complex. The essential character of metals and heavy non-metals in the catalyst and carrier functions of critical reactions must never be overlooked (Sections 9.9 and 9.16).

In linked reactions of the kind shown in Figs 10.11, 10.12 and 10.17 it is necessary to have carrier molecules in common for the fragments such as CH_3CO^- and to transfer energy (ATP). As explained in Chapter 9, where these fragments are to be used in relatively rapid synthesis they must be held by relatively weak bonds, that is by good leaving groups. R—CO-(-S—X) is, of

Fig. 10.17 The link between glycolysis, fatty acid synthesis and degradation, the citric acid cycle and the synthesis of amino acids. The centrality of the pyruvate dehydrogenase complex (PDC) to carbohydrate, fat and amino acid metabolism is clear. The nucleotide bases are made from the amino acids.

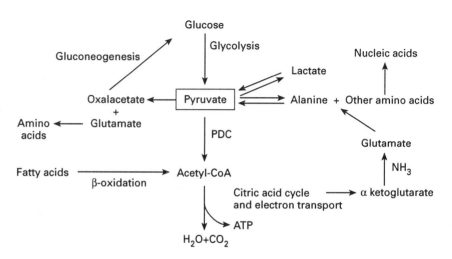

course, an excellent leaving system for transfer of R—CO—. Coenzyme A, Co-A, is one such group, —S—X, and is common to the pathways. That the pathways use the same mobile cofactors allows co-ordination of them all since they can be messengers between pathways, as well as carriers, allowing feedback control (see Section 7.8.2 and Chapter 11). Now many pathways require transfer of other identical units such as H^- and CH_3^-. In Table 10.6 we have given a list of some coenzymes (small molecules), each of which is used to transfer fragments in several pathways in synthesis or degradation giving the opportunity for co-ordination.

Table 10.6 Some activated carriers in metabolism

Carrier molecule	Group carried in activated form
(ATP)	Phosphoryl
NADH and NADPH	Electrons, H^-
FADH$_2$	Electrons, H^-
Coenzyme A (CoA)	Acyl
Lipoamide	Acyl
Thiamine pyrophosphate	Aldehyde
Biotin	CO_2
Tetrahydrofolate (THF)	One-carbon units
S-Adenosylmethionine	Methyl
Uridine diphosphate glucose (UDPG)	Glucose
Cytidine diphosphate diacylglcerol	Phosphatidate
Glutamine	—NH$_2$
Pyridoxal	—NH$_2$

These coenzymes may well have replaced earlier less selective metal ions on surfaces in prebiotic chemistry or early life. We know today from organometallic chemistry that particular metal complexes can readily transfer all these fragments. There is a strong suggestion that early life might have used such organometallic centres, which are inherently less selective, and that sophistication produced the present series of more complex recognisable centres based on larger units often containing the bases of nucleotides (see Table 10.6). Does this explain why nickel-based organisms are primitive?

10.5.7 The incorporation of phosphate

Our next problem is to show that, just as there is a network of pathways that are interwoven for H, C, N, O and S handling, so there are networks for the different modes of phosphate handling. We have to approach the problems associated with phosphate very differently from those for H, C, N, O, S or Se uptake since phosphate, being an anion, cannot enter a cell easily. The uptake of phosphate, like that of other anions and cations, requires pumps and they will be described in a general way later (Section 10.6). We shall therefore describe incorporation and in-cell pathways of phosphate transfer before describing uptake. As we saw in Section 9.10, phosphate itself is neither oxidised nor reduced but just transferred by forming both energised mono- and di-ester linkages and anhydrides through condensation reactions. In particular, the energy associated with pyrophosphate allowed phosphate to become

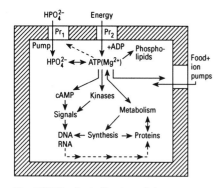

Fig. 10.18 An indication of the complexity of phosphate uptake and incorporation. Note the involvement of magnesium (see text).

Table 10.7 Standard free energy of hydrolysis of phosphate compounds*

	$\Delta G°$ (kcal mol^{-1})
Phosphoenolpyruvate	−14.8
1,3-Diphosphoglycerate	−11.8
Phosphocreatine	−10.3
Acetyl phosphate	−10.1
ATP	−7.3
Glucose 1-phosphate	−5.0
Fructose 6-phosphate	−3.8
Glucose 6-phosphate	−3.3
3-Phosphoglycerate	−2.4
Glycerol 3-phosphate	−2.2

* The direction of phosphate transfer in this table is downwards.

an energised carrier of many C, H, N, O fragments so that phosphate appears in many small coenzymes along with H, C, N and O with which it too can be transferred. It is also a part of the nucleotide monomer structure. A brief outline of some of its pathways is shown in Fig. 10.18 and the associated energetics are given in Table 10.7. In one form or another, phosphate is linked to virtually all the other small substrate pathways. It is no surprise, therefore, that phosphate compounds, like the other common carriers of H, C, N and O, appear again in our discussion (Chapter 11) in many in-cell message systems such as the nucleotide (N) derivatives cNMP, NMP, NDP and NTP as well as in substrate phosphates, protein phosphates and in membrane phospholipids. Phosphate compounds coordinate most pathways but also the energetics of these pathways.

The link from ATP levels extends to the concentrations of simple ions such as Na^+, K^+, Cl^- as has already been described (Section 10.4.2), but all these phosphate functions are only active if the level of $[Mg^{2+}]$ is correctly set. This functioning of Mg^{2+} as an essential catalyst in phosphate reactions leads us to the functional value of both. The use of second-period (of the periodic table) elements (Mg, P and S) in transfer and catalysis is striking (see Chapter 7B). The linking of different element activities based on sharing between pathways (Fig. 10.17) now begins to connect elements other than H, C, N, O into a defined pattern. The use of phosphate in free and bound forms as a central element in control and regulation is that it is indestructable in biological chemistry conveying energy and signals to all pathways, compare Ca^{2+} in later organisms.

10.5.8 Final stages of the incorporation of the major elements in polymers

The last step of incorporation of the major elements is the conversion of monomers to polymers, which then associate. The polymers are thermodynamically more unstable than their monomers due to the uphill character of chemical condensation and the loss of translational entropy. They are centres of strong retention of the elements C, H, N, O, S and P, however, since kinetically, especially because they fold, their decomposition is slow. Thus, folded polymers are excellent energy traps and so assist survival. In Sections 6.8–6.8.3 we showed that the assembly of the proteins generated further stabilisation as did assembly with DNA on double-helix formation. The proteins could then assist in the retention of other elements, for example metal ions (Section 10.7), while binding of these elements also gives the proteins (and nucleic acids) extra thermodynamic and kinetic stability.

10.5.9 Summary of major element incorporation

It is not quite justifiable to say that the primary pathways of metabolism of all cells, no matter how primitive (or advanced), are very much the same. However the similarities are sufficiently close so that we can use Fig. 10.19 to represent all of this primary metabolism, that is, basic uptake and incorporation. This figure is of immense importance in this book since it shows not just features of organic synthesis but stresses various features of the

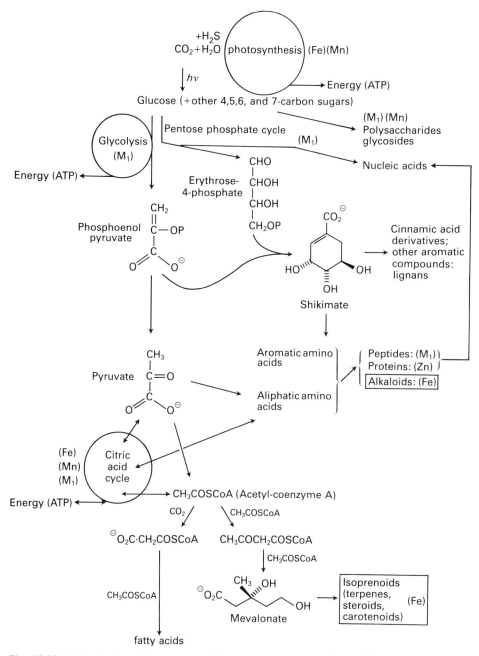

Fig. 10.19 An indication of the elaborate links between elements and metabolic paths that must be integrated in growth homeostasis. Note that both primary and secondary metabolism are involved. M_1 is Mg^{2+}, K^+, H^+. □ indicates products after the advent of dioxygen. (The figure is related to Fig. 10.17.)

organised self-assembling system that are found with very few variations in biology:

1. There are reaction pathways in units such as glycolysis, photosynthesis, fatty acid synthesis and the citric acid cycle. Their basic paths may be well separated in space. (Note that in primitive systems the citric acid cycle ran in reverse generating trapped carbon for synthesis.)

2. There is a network of energy production and distribution through ATP between the pathways.

3. There is a network of material distribution between the paths through common molecules, such as coenzymes, often based on S and P compounds (Tables 9.12 and 10.6), which carry H and C within the pathways as well.

4. In each pathway unit there are required catalysts. The catalysts are specific proteins with prescribed sequences for catalysis and recognition. Many of them have specific metal element requirements (see Figs 10.14 and 10.19).

5. The medium for transfer is either in water or along membranes, that is, on the membrane surface or inside it. The water inside each compartment, for example, the cytoplasm, is a medium of controlled and specific ionic strength and pH.

6. There must be flow, hence pumping.

7. It goes without saying that the whole is confined to a compartment, a cell, with all the advantages that this implies (Sections 6.3 and 7.11).

(Note that we do not yet refer to the role of control by small molecules or ions or of regulation by genetic material. We are concentrating on the uptake, incorporation and integration of elements and monomer substrates into (*connnected*) pathways or networks.)

While Fig. 10.19 shows *primary H, C, N, O metabolism* and while we have indicated the value of sulphur and phosphorus in the transfer of fragments and energy, the figure fails to show three other essential features of biological chemistry:

1. The uptake of the elements for the ionic medium in which reactions occur. As stated, the elements involved are Na^+, K^+, Cl^- and Mg^{2+} (and phosphate) (Section 10.6).

2. The uptake and incorporation of the catalysts that are essential for the reactions. Very many reactions require one or more of the elements V, Mn, Fe, Co, Ni, Mo and W, as ions even in primitive cells (Section 10.7).

3. The modes of management of the pathways. This is the topic of Chapter 11, where we shall see that different elements have different functional significance in communication.

We discuss the first two features in the next section.

10.6 The uptake and incorporation of electrolytes

10.6.1 Introduction to pumps

To incorporate cations and anions including phosphate into chemicals usually requires energy not as for N_2 or CO_2 fixation, but for uphill transport of an ion across a cell membrane. (As discussed before, the introduction of H, C, N, O, S and Se did not require an uptake mechanism but only incorporation steps since the forms of these elements that occur naturally diffuse through lipid membranes.) We therefore need to understand *electrolyte* uptake and rejection by pumps generally. How are ions such as phosphate, and K^+, and Mg^{2+}, taken into cells, and how are Na^+, Cl^- and Ca^{2+} rejected from cells?

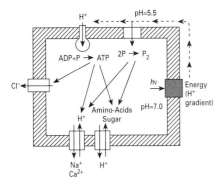

Fig. 10.20 Concentration of ions and food stuffs by a primitive cell. The membrane contains two types of proton pumps. The figure shows in principle the way in which it is possible to use H^+ gradients as the link of energy from redox reactions to ATP and how H^+ could then be the central ion in uptake exchange of ions and foodstuffs in a prokaryote.

There are two routes of uptake by pumping. The first uses *exchange* across the membrane from a primary ion gradient (Fig. 10.20). This gradient may well have been originally of protons produced from unstable inorganic compounds of the Earth's crust, possibly sulphides (Fig. 10.7), or later by light (Fig. 10.9 and see below). The second method utilises the chemical energy of either thioacids or pyrophosphate to pump ions across membranes (Figs 10.20 and 10.21). Both types of pumping are present in all known organisms today but, as stated, pyrophosphate needs primary pumping of phosphate and then energisation to make pyrophosphate, for example via glycolysis. Phosphate cannot be handled by covalent metal (sulphide) catalysts as for C, S and H, so it has to be processed via initial electrostatic binding and energy from gradients or glycolysis.

Let us look more closely at the nature of an ion gradient. Let us suppose that initial energy-utilising events generated an out-of-balance of ions, probably protons, across a membrane (Figs 10.7 and 10.9). The out-of-balance is both in concentration and in electrostatic potential field (Section 6.4.2). The available energy for work (see Chapters 3 and 6) is

$$\Delta G = \frac{RT}{nF} \ln \frac{c_{\text{out}}}{c_{\text{in}}} + zF(\psi).$$

where ΔG is the change in free energy and $zF\psi$ an expression in which ψ, the potential, is a variable (Section 6.4.2). While c_{out} is open to loss of energy to the whole environment by diffusion, c_{in} is contained. ψ is dependent only on the magnitude of the difference between internal and external charge.

The suggestion then is that the exchange of the ion gradient of protons is used first and foremost as the uptake method for the concentration of anions and cations and even neutral molecules such as sucrose in the cell (Fig.

Fig. 10.21 A hypothetical protein that couples an enzyme activity—ATP hydrolysis in this case—to ion transport across a membrane, thus acting as an ion pump. The design incorporates two β-sheet domains that together bind ATP in the closed conformation and hydrolyse it to ADP and phosphate, thus allowing transition to the open conformation. This transition is coupled to movements of the transmembrane α-helices that effect pumping of a cation, M^+, across the membrane. Reversal of the process where M^+ is H^+ gives an ATP synthetase. From Williams, R. J. P. (1994). *Current Biology*, **4**, 942–4.

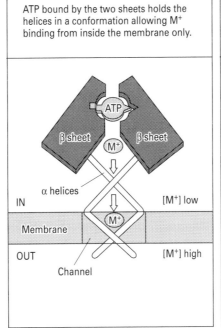

ATP bound by the two sheets holds the helices in a conformation allowing M^+ binding from inside the membrane only.

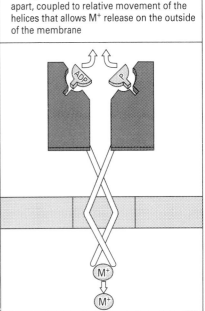

ATP hydrolysis allows the sheets to move apart, coupled to relative movement of the helices that allows M^+ release on the outside of the membrane

10.20). The way in which gradients of either charge or concentration can be utilised are described in biochemical texts. We give therefore just a brief description of proton pumps and the reversal of them, proton-driven ATP synthesis, as an example.

10.6.2 Proton and other ion pumps

The H^+ pump may, in fact, be the most primitive and most important of all the element pumps connected directly to energy capture. If sugar were available from abiotic reactions it would be converted to acids in the cell. Since many of the chemicals in a cell are extremely sensitive to pH, the removal of this excess acid was always necessary. The implication is that H^+ gradients became linked to exchange so as not to be wasted but also that these gradients were used to make ATP (Fig. 10.22). Subsequently, H^+ could also be driven from the cell by ATP. It is then possible to conceive that both H^+ gradients and ATP are linked to uptake or expulsion of every other element via exchange with H^+ across membranes or by ATP pumps (Fig. 10.21). We shall see that very basic processes, not only of the movements of ions K^+, Na^+, Mg^{2+}, Cl^- and Ca^{2+} and so on but also of simple compounds such as amino acids and sugars across membranes, became linked to either proton or ATP pumps or both. (In later multicellular eukaryotes, including plants but not animals, the H^+/ATPase (Fig. 10.22) generates *external and vesicular* pH = 5.5 with a cytoplasmic pH of 7.2. In animals the external pH is also close to 7.2 and it is an Na^+ gradient that is used in the uptake of food not H^+ gradients as in plants.)

Now it might appear that the ATPase proton pump and the ATP synthetase could not operate in the same cell and this must be true if there are no controls on their use. In Chapter 11 we show that such controls are present in biological cells, hence removing this dichotomy. In fact, it is known that various ATP pumps for ions can be reversed and that ATP synthetases can be driven by, for example, sodium ion gradients but under controls. Thus, there is a balanced state of proton flux and of phosphate potential [ATP]/[ADP][P], due to different controls.

A very large number of ions (and molecules) are pumped across cell membranes by these two means. Once in the cell they are prevented from leaving if the pumps are irreversible. The ions or molecules can also be retained in cells by incorporation in polymers or in complex ions to which we turn next.

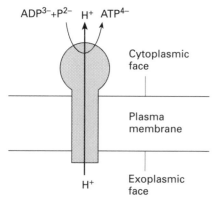

ADP^{3-}+P^{2-} H$^+$ ATP^{4-}

Cytoplasmic face

Plasma membrane

H$^+$ Exoplasmic face

Fig. 10.22 H^+/ATP synthetase pathway in membranes. The reversal is an H^+ ATPase, an H^+ pump using the energy of ATP, i.e. the reverse of Fig. 10.21.

10.7 Biological centres in proteins for metal ion incorporation

Once metal ions are inside cells they can bind selectively to the folded proteins. The protein frameworks can have donor atoms of oxygen (O), nitrogen (N) or sulphur (S) on their surfaces in particular geometric relationships. Some examples of such co-ordinating groups are $RCOO^-$, RNH_2 and RS^-, where the simplest R is $-CH_2-$. There is a naive way of looking at these protein reagents which is to say that RCO_2^- is modified CO_3^{2-} (carbonate), RNH_2 is modified ammonia and RS^- is modified sulphide (H_2S, HS— or S^{2-}). We then expect that the biological organic reagents (proteins) shown in Table 10.8 will be

Table 10.8 Chemical partners of metal ions in proteins

Probable ligands	Protein example
Mn ions (See Mg ions)	
Phosphate, imidazole	Lectins
Carboxylate	
Fe ions	
Porphyrin, imidazole, (H_2O)	Myoglobin, peroxidase
Porphyrin, imidazole, (R_2S)	Cytochrome *c*
Porphyrin, variety of ligands	Catalase, cytochrome P-450
Sulphur ligands	Rubredoxin, ferredoxin
Phenolates	Transferrin
Co ions	
Corrin, benzimidazole, carbanion of a sugar	B_{12} enzymes
Cu ions	
RS^-	Superoxide dismutase
Possibly $>N^-$ bases	'Blue' proteins, cupreins
Imidazole	
Zn ions	
Carboxylate	Carboxypeptidase
Imidazole	Metallothionein
$R-S^-$	Dehydrogenases
Imidazole	Carbonic anhydrase
Cd ions	
$R\!\!<^{\,S^-}_{\,S^-}$	Kidney proteins
	Metallothionein
Mg, Ca, (Mn) ions	
Carboxylate	ATPases
Phosphate, (imidazole)	Enolase
Serine —OH	Kinase
Threonine —OH	
Carbonyl	Calmodulin

selective in their reactions with metals, much as oxides, amines or sulphides have formed separately in the crust of the Earth (Section 8.5), or in simple inorganic analytical precipitations (Table 5.11), or complex ion reactions (Section 5.5.2). This prediction is borne out so that metals such as sodium, potassium, calcium and magnesium are combined with O ligands of proteins, if they are bound at all, while copper, zinc, iron and cadmium are combined with S and N ligands (Table 10.8). The nature of binding to N is intermediate; with Mn, for example, at least one nitrogen and several oxygens are coordinated.

Biological systems make proteins that are soluble either in water or organic solvents, here the membrane. Thus, a metal can be moved as a complex into a hydrophobic fatty acid membrane or left in the aqueous cytoplasm. Organic chemistry and thus biological organic chemistry increases vastly the sophistication of simple inorganic reagent chemistry in separation power using, for example, insertion of metal ions into chelates, but this sophistication requires energy and control due to the kinetic instability of the compounds used or formed. The implication then is that the same inorganic compounds in biology or associated with its organic chemicals may be only kinetically stable, for example Mg^{2+} in chlorophyll. All the modes of incorporation are described in detail in our previous book (reference 1 in the 'Further reading'). The complex ion chemistry involved is summarised in Chapter 5.

10.7.1 Incorporation of metals into cofactors

A remarkable feature of even the earliest organisms is the sophistication of the incorporation of metal ions not just in proteins as single ions or clusters but as specific complex ions of particular synthesised organic chelating agents. These chelating agents are known even in archaebacteria. The complexes (also listed in Table 10.8) act as cofactors or coenzymes. The synthesis of the ligands is a complicated multistep procedure as indeed is the subsequent introduction of the metal ion into them. Using enzymic methods the metals are literally chosen and inserted by means not yet fully understood (see Fig. 10.23). It is instructive to note that the earliest organisms, the archaebacteria, handle iron, cobalt, nickel (in porphyrin-like complexes, Fig. 10.23) and molybdenum and tungsten (in pterin-like complexes, see marginal illustration), but that the magnesium complexes (chlorophylls) appear in (later) algal systems using light. Perhaps most strikingly, the elements copper and zinc are not found in any such complexes except very rarely in pigments. Is this a consequence of the lack of availability of these elements in the sulphide-rich stages of evolution (see Fig. 5.3)? These metal-containing cofactors are in many ways like simple ions incorporated into proteins, some bind N-, some S-centres.

10.7.2 Elements in different oxidation states

We have pointed out in Chapter 2 and again in Chapters 5 and 7 that many elements exist in a variety of oxidation states. We can consider just Fe^{2+}/Fe^{3+} and $RS^-/R—S—S—R$ as examples of metals and non-metals that undergo redox reactions. This ability to switch oxidation state allows differential retention on either side of a membrane. Thus, iron can be brought into a cell as Fe^{3+} by a suitable carrier and then converted to Fe^{2+} by the more reducing

The formula of one molybdenum-containing cofactor. There are others with nucleotide phosphate extensions. In primitive life the same pterin binds to tungsten.

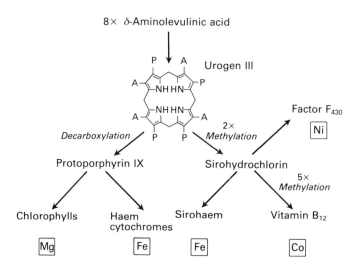

Fig. 10.23 The diverse ring chelates produced for different metal ions from one precursor.

potential of the inside of the cell, that is, the Fe^{2+} complex is unstable and releases free Fe^{2+} (Section 12.5.3). The Fe^{2+} ion can then be bound and reconverted to Fe^{3+} due to the differential strength of Fe^{3+} against Fe^{2+} binding in a special complex. Similarly, sulphur can be held as, say, —S—S— outside a cell but inside it is reduced to —SH. The redox potential of a metal system in particular depends on the coordination complex in which the metal ion is held (Section 5.8.1). A further example illustrates the point. It is likely that in primitive (and present-day) conditions cobalt would have been taken up as Co^{2+}, which is rather labile, but in the cell it is incorporated as Co^{3+} inside the complex molecule vitamin B_{12} where the binding constant of Co^{3+} is $\geqslant 10^{20}$ times higher. Exchange from such a situation is extremely slow so that retention is effectively irreversible. (This irreversibility applies to the incorporation of Fe in haem, of Mg in chlorophyll and of Ni in factor F_{430}.) Of course, the control exerted over the metal redox states by the proteins is one way in which catalytic function can be enhanced.

10.8 Summary of early element uptake and incorporation and of developing functional use

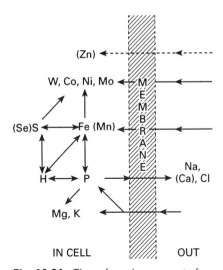

Fig. 10.24 The enforced movement of some elements into and out of primitive cells. The use of zinc in these anaerobes is uncertain. The arrows indicate coupled movements into cells.

In Table 10.9 we list the modes of incorporation of elements of interest in biology. The modes of retention are very different due to the strength of binding, its character, and the kinetics of exchange described in Chapters 2, 5 and 7. These elements all have interwoven functions as shown in Fig. 10.24 and as illustrated for a typical cell in Fig. 10.25. Now it would be quite wrong to leave the impression that, once initial modes of uptake and incorporation have been achieved, there is no further development possible. Evolution drives the element incorporation to greater and greater survival strength which implies that the elements in cells become more and more highly refined functionally. The way in which this is managed for non-metals such as H, C,

Table 10.9 Modes of element incorporation

Mode*	Element (compound) concerned†
Free diffusion followed by covalent incorporation	CO_2, (O_2), H_2O, H_2S, N_2
Membrane-restricted diffusion followed by trapping behind membranes after pumping	Na^+, K^+, Cl^-, Mg^{2+}, Ca^{2+}, Mn^{2+}, Br^-, I^-
As for (b) followed by covalent incorporation	(SO_4^{2-}), HPO_4^{2-}, (MoO_4^{2-}), $B(OH)_4$, (HVO_4^{2-})
Carrier molecule diffusion before co-ordination compound formation	Fe, Co, Ni, (Cu), (Zn), Mn, (Mo), W

* Elements can be included or excluded by these modes, so that Na^+, Ca^{2+} and Cl^- are pumped out as ions while poisonous metals such as Cd, Hg, and so on, can be pumped out in compounds.
† Elements in brackets above were not available (in the form shown) in very early periods, say 3–4×10^9 years ago, i.e. in high sulphide and pH < 7.

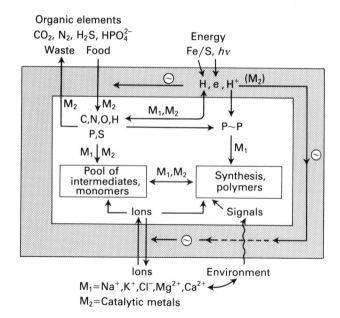

Fig. 10.25 A schematic picture of the roles of different elements in a primitive cell and a low redox potential environment. O_2 and SO_4^{2-} were introduced later.

N, O, P and S is by mutational variation of sequences or by chemical modification of amino acids or nucleic acids. These changes affect the structure and dynamics of the polymers. Thus they increase the fitness of the organism. Simultaneously, this chemical refinement of structure and dynamics occurs for the other elements incorporated but now there are two ways forward. The first depends upon utilising the mutations in proteins to improve the fitness of the properties of the metal-ion-binding site, while the second depends upon refinement of the metal element content of the cell both as to which ions are incorporated and in what concentration. (There is yet a third refinement, which is to uncover more effective use of the available energy from the environment, which requires the evolution of use of both non-metals and metals within space. The final refinement is in organisation which is a matter of the co-operative functional use of the selected 15–20 elements.)

10.8.1 Catalysis and metal ions

The combination of trace elements with polymers progressed from thermo-dynamic selection, based on the binding constants of complexes such as those of Table 5.8, to more sophisticated forms of incorporation (see above and our previous book). The reason is that the metal ions are not just incorporated for structural reasons. The metal ions are, of course, catalytic, and the development of catalysts has been crucial for the syntheses of organic molecules, although the syntheses must be done while recognising that the destruction of the energised molecules is the most probable outcome unless trapping is associated with the product. Thus, catalysts by themselves are not sufficient and they must be coupled to energy-driven processes for synthesis. (We may hazard the guess that, initially, the incorporation of trace elements into organic chemicals was always a risk but that it was eventually coupled to useful products for polymer synthesis. Overall greater survival was the known outcome.) In fact two elementary catalysts initiate extremely important reactions in all organisms: (1) Fe/S clusters of membranes allow electron transfer to generate energy (Fig. 10.25); (2) Mg^{2+} (or Zn^{2+}) catalyses phosphate reactions, that is, all ATP reactions. The fact that they were then held in proteins allowed specificity of reactions to develop in enzymes. Specificity in redox catalysis is particularly assisted by management of the redox potential of the catalytic atoms, related to how different oxidation states are bound in proteins (Section 5.8.1). Thus, diversity of incorporation became of increasing value.

10.8.2 Fitness of individual sites: the entatic state

In this chapter we have been concerned to show that every biological organism uses and probably has always used some 15–20 chemical elements since this is the way in which it could best survive, that is, become efficient in using absorbed energy to make compounds in kinetic traps. We shall go on to consider the different ways in which efficiency can be described for all organisms, including higher organisms in Chapter 12. It is not necessary to develop the same degree of sophisticated analysis in order to see that even very primitive organisms evolved not only the common and effectively the best use of the 15–20 readily *available* elements but that the individual elements were put to use in very refined (optimal) ways functionally. The Darwinian expression is that '*fitness*' for a task evolved. Although in the end we must see fitness as a property of an organism in an ecosystem, we have pointed out in our previous book that the use of an element for a function can be refined (made fitter) by:

(1) capturing the element in a suitable combination with other elements (this is far from just a thermodynamic association);

(2) ensuring that the structure and dynamics of the site are optimal, as in the so-called entatic state, see marginal illustration, which uses the geometry of the site and its dynamics to optimise function (Fig. 10.26);

(3) placing the unit carrying the site in as advantageous a position as is possible.

These developments largely depend on the synthesis of suitable proteins, and placing them in suitable compartments.

10.8.3 The incorporation of proteins into compartments

Protein synthesis itself has been described in Chapter 9 as an energised condensation reaction of amino acids. In a primitive cell and perhaps in prebiotic structures, proteins dominated organic chemistry. Their importance lies largely in their folds. Unlike nucleic acids, polysaccharides or fatty acids they generate idiosyncratic co-operative shapes and dynamics so that they can be utilised in many functions both in binding elements, for example in catalysts, and in transferring them, for example in pumps (Fig. 10.21). They are also the basis of structure having sufficient rigidity and yet flexibility to allow development. The usefulness of the proteins is greatly increased by placing them appropriately in space where they can dominate the last stage of the incorporation of the elements in organisms—spatial fitness. For this

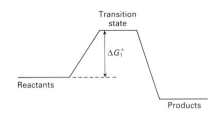

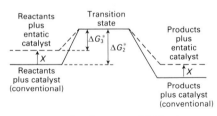

Top: conventional pattern of ΔG_1^* of reaction. Bottom: pattern of reduction of ΔG_1^* to ΔG_2^* by non-activated catalyst. ΔG_3^* is the barrier for an activated catalyst (see Section 7.7).

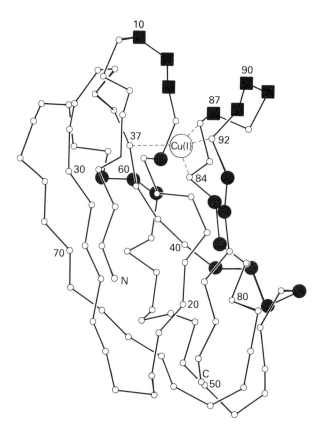

Fig. 10.26 The β-barrel fold of the copper protein, plastocyanin, which is a rigid structure designed to fit Cu(I) but not Cu(II) that becomes entatic in the rigid site. Note the difference from the protein mechanical rack of haemoglobin which is made from helices that can move and share strain with the metal ion. The filled squares and circles show binding zones for donor or acceptor proteins in self-assembly.

purpose the synthesis of proteins is often carried out on a membrane when the protein can be exported as it is synthesised or placed in the membrane as a pump. For primitive cells this allows use of any space between the outer wall and the inner membrane (Fig. 10.27) often called a periplasmic space, for positioning of any element.

The positioning of proteins can also be around the cell (Fig. 10.5) since different regions have different curvature. Curvature is related to surface energy which is minimised by particular arrangements of particles on a surface (Section 6.2.2 and Fig. 7.26). Thus, a cell's activities can be distributed

Fig. 10.27 Prokaryotic proteins destined for locations other than the cytosol are synthesised by ribosomes bound to a membrane. A signal sequence on the nascent chain directs the ribosome to the plasma membrane and enables the protein to be translocated across this membrane.

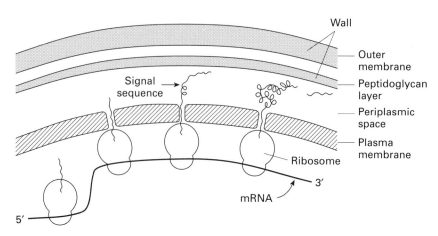

around the cell, allowing, for example, exit and entrance systems for ions to generate fields and currents. Such fields can then locate other objects around the cell. These arrangements of proteins (and lipids) may well be self-assembling, generating molecular machines (see Section 6.8).

Interest in proteins extends therefore to their surfaces, since through them proteins self-assemble now together to higher aggregates and with lipids, DNA or RNA. These aggregations are largely driven by *downhill* processes, although proteins are directed to the different compartments of a primitive cell—the cytoplasm, the membranes, or the outside and periplasmic spaces. We shall return to this problem but we have already seen in Sections 7.16 and 7.17 that thermodynamic co-operative binding leads to fats self-assembling through hydrophobic forces and to ions crystallising as salts through electrostatic forces and that a combination of forces leads to ions binding to organic molecules. Bringing all the same binding forces into play between the surfaces of proteins generates a very great selective potential for folding and then combination. This is seen in the association of protein catalysts with substrates and is surely responsible for much protein self-assembly (see Sections 6.8.1–6.8.3). Since there is unlikely to be a major kinetic barrier to the association, the requirement is only for a sufficient concentration to overcome dilution. These associations may remain open to relatively rapid dissociation, so that a cell can adjust its function.

We would like to suppose that we have now given adequate insight into the more permanent covalent uptake and incorporation of C, H, N and O molecules and of the somewhat less kinetically stable S and P compounds, and of the uptake of all ions. At all periods of time these processes evolved. The remaining question is how the energy resources were developed.

10.9 Primitive development of energy sources

The possible original sources of energy were:

(1) inorganic chemical gradients of materials such as those between different sulphides, for example of iron, but we cannot be sure of this;

(2) organic molecules such as sugars resulting from abiotic synthesis.

The subsequent advance made by biological systems early in evolution was to turn to the sun as an energy source (see Sections 6.6 and 7.19 and Fig. 10.28). Given that light could be captured, it became possible to generate sulphur from sulphide and then dioxygen from water. Dioxygen would have immediately been a dangerous chemical but for the fact that it was expelled to the atmosphere. Later it became a source of new energy and chemicals. The metabolism that developed using dioxygen is the subject of Chapter 12. Here we are stressing the upgrading of energy capture for the increasing survival strength of the kinetic traps of a cell. (NB. Copper in Fig. 10.28 is not essential but became used later in the photosystem.)

The picture of energisation of the cells is that, in evolution, the organisation increases as the energy content increases so that energy input from light gives a huge impetus to development (Fig. 10.29). We shall describe the physical

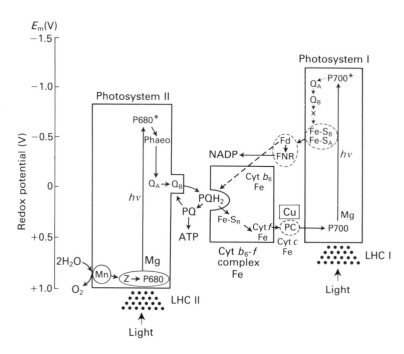

Fig. 10.28 The scheme for the energising of electron potential in two stages uses light. Water is effectively turned into O_2 and $2H_2$ or, in equivalent electrode potential terms, into an electrochemical cell of 1.2 volts. All arrows except those labelled hv show electron flow. PS, Photosystem; Q and PQ, quinones; PC, plastocyanin (Cu); P, photopigment; FD, ferredoxin; fp, flavoprotein; LHC, light harvesting complex.

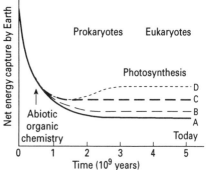

Fig. 10.29 The upgrading of energy by anaerobes. A, The fall in energy content of original Earth; B, the capture of energy in abiotic mainly organic chemicals; C, the capture of energy by anaerobic life from chemicals (A + B); D, the capture of energy by anaerobic life with light capture.

advances in the next chapter but note immediately that the use of light demanded new elements in new places. In fact, manganese became central to evolution since its chemistry allowed cells to generate dioxygen from water on being oxidised to the oxidation state +IV in MnO_2. Thus, photosynthetic organisms (plants) are generally higher in manganese.

With these developments in energy sources came structural advance both in membranes and in internal vesicles and filaments—the beginnings of the eukaryotic cell, which we will describe in the last sections of the next chapter. Cells began to contain within themselves specialised vesicles and higher organisation was then demanded (see Chapter 11). We do not know the order of events, but at about this time the significance of some elements changed. Examples are the uses of Ca^{2+} and Cl^-; both are needed in photosynthetic apparatus and became valuable in message systems affecting many mechanical events inside the cell. We know that such roles for Ca^{2+} are absent in prokaryotes. However, it was the production of dioxygen that eventually changed the course of evolution by changing the availability and then use of many elements. These changes will be described in Chapters 12 and 13.

10.10 Summary of uptake and incorporation of elements in anaerobic organisms

The intention of this chapter is to show the advance from prebiotic to primitive biotic and then to more sophisticated anaerobic cell chemistry. We have observed that biological systems are an advanced form of trapping energised states. Being energised they constantly degrade and must be replaced. Although we know today that this is managed under the influence of coding polymers, DNA, we cannot start our description of living systems from such a sophisticated molecule. From a consideration of survival we were forced to see

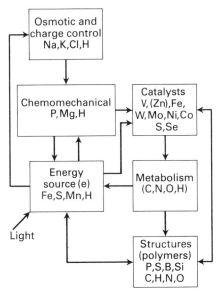

Fig. 10.30 The interrelationship of the functions of the elements is given here to stress the intimate connection between them and energy. This diagram is only an aid to thinking about the ways in which energy and metabolism have to be linked, frequently using inorganic chemical functions, in primitive systems. Ca and Cu are deliberately omitted until Chapter 12.

the advantages of compartmental separation, the formation of a cell membrane, and then of limiting chemistry to particular compounds that are kinetically stable. The effective route was stabilisation by binding and constraint by catalysts to particular pathways for element uptake and incorporation. Thus, we looked at what we take to be primitive cells and asked, given the abundance and initial availability limitations upon the chemical elements (Chapter 8), what were the minimum requirements for survival including replacement but not yet copying. It was readily shown that some 15–20 elements are absolutely essential. The immediate concern was then to discuss their uptake and incorporation, illustrating meanwhile the roles they played. We can look upon this activity as an increasing energy upgrading of chemicals (Fig. 10.29).

This chapter has therefore begun to reveal the special nature of biological chemistry in that it has shown the selective even specific uptake and incorporation of those 15–20 elements in a nearly constant ratio for a given species even in the most primitive cells in such a way as to generate ever-increasing fitness, that is, *functional value*. This is a different selection of the chemical elements from that seen in simple thermodynamic selection (in fact, it goes contrary to this selection) and is partly based on incidental kinetic trap selection as seen in prebiotic chemistry. It is kinetic trap selection that is manipulated to gain the survival strength of co-operative organisation (Fig. 10.30). The biological compounds must not be seen just as independent components optimised in use but should be viewed as having mutual dependence upon one another where this mutual dependence is a selection pressure through kinetic protection demanding replacement or reproduction (synthesis). Of necessity, there are 'chosen' metabolic pathways to provide energy and intermediates for the 'chosen' polymers just as there are 'chosen' elements. Moreover, it is clear that each and everyone of these chosen pathways is in communication with each and every other one (a necessary condition of balanced stability). Thus, while energy capture, uptake, incorporation and associated syntheses are all required, they must all be managed to produce a kinetically stable whole organisation as is observed in living organisms. Such stability in a self-assembling system requires not just uptake and incorporation (synthesis) but needs to flow and to have an informed communication system, that is, it must have an information centre (a plan) and a network of messengers. The parallel with human society is not exact, but the relationship is close as we shall see in Chapter 14. In the next chapter we look for the nature of the information centre and the communication network in primitive cells, that already demand energy, material and machinery in order to create organisations. These central controls vastly increase kinetic stability of flow.

Further reading

The most comprehensive general reference for this chapter is our previous book.

1. Fraústo da Silva, J. J. R. and Williams, R. J. P. (1991). *The biological chemistry of the elements. The inorganic chemistry of life* (3rd printing, 1994). Oxford University Press, Oxford.

The aspects related to element incorporation and metabolic pathways are covered in all good biochemistry textbooks. Some classical texts were recommended in Chapter 9; the following are simpler but much used text books.

2. Conn, E. E. and Stumpf, P. K. (1976). *Outlines of biochemistry* (4th edn). J. Wiley & Sons, New York.
3. Metzler, D. E. (1977). *Biochemistry—the chemical reactions of living cells.* Academic Press, New York.

The primitive forms of life and the fossil records are described in three recent books.

4. Danson, M. J., Hough, D. W. and Lunt, G. G. (ed.) (1992). *The archaebacteria: biochemistry and biotechnology.* Biochemical Society, Portland Press, Colchester.
5. Balows, A., Trüper, H. G., Dworkin, M., Harder, W. and Schleifer, K.-H. (ed.) (1992). *The prokaryotes* (2nd edn). Springer-Verlag, Heidelberg.
6. Benton, M. J. (ed.) (1993). *The fossil record* (2nd edn). Chapman & Hall, London. This book includes records of the earliest 'bacterial' remnants found in mineral deposits.

The following papers relate somewhat indirectly to the topics discussed in this chapter but complement information and provide interesting reading.

7. Oró, J., Rewers, K. and Odom, D. (1982). Criteria for the emergence and evolution of life in the solar system. *Origins of Life* **12**, 285–305.
8. Wächtershäuser, G. (1990). Evolution of the first metabolic cycle. *Proceedings of the National Academy of Sciences, USA* **82**, 200–4.
9. Hemming, A. and Blotevogel, K. H. (1985). A new pathway for CO_2 fixation in methanogenic bacteria. *Trends in Biochemical Sciences* **10**, 198–200.
10. Woese, C. R. (1982). Bacterial evolution. *Microbiological Reviews* **51**, 221–71.
11. Joyce, G. F. (1989). RNA evolution and the origins of life. *Nature* **338**, 217–24.
12. Margulis, L. and Schwartz, K. V. (1988). *Five kingdoms.* W. H. Freeman, New York.
13. Levett, P. N. (ed.) (1991). *Anaerobic bacteria.* IRL Press, Oxford University Press, Oxford.
14. McClendon, J. B. (1976). Elemental abundances as a factor in the supply of mineral nutrient elements. *Journal of Molecular Evolution*, **8**, 175–95.
15. Silver, S. and Misra, T. (1988). Bacterial element requirements. *Annual Reviews of Microbiology*, **42**, 717–43.

Note added in proof

Details of the involvement of hydrogenation reactions in primitive cells are becoming known. The structure of the nickel–iron hydrogenase of *Desulfovibrio gigas* has been solved illustrating a special (entatic) state of nickel. (Volbeda, A., Charon, M.-H., Piras, C., Hatchiklan, E. C., Frey, M. and Fontecilla-Camps, J. C. (1995). *Nature*, **373**, 580–7.) The outline organisation of the complex I which oxidises NADH with quinones has shown that effectively all electron transfer is peripheral to the membrane making proton diffusion the major current carrier in the membrane. The importance of the proton diffusion in photoreaction centres and in particle IV, cytochrome oxidase will be described later (Chapter 12). See Williams, R. J. P. (1961). The functions of chains of catalysts. *Journal of Theoretical Biology* **1**, 1–13.

11

Early cellular organisation in anaerobes

All things by immortal power
Near or far
Hiddenly
To each other linked are
>Francis Thompson *The mistress of vision* (1859–1907)

11.1 Introduction

Chapter 10 described which elements, compounds and reaction sequences have been selected in order to generate a living system. This was but a descriptive analysis of connected but uncoordinated rates of uptake and reactions in pathways, although it did also show the need for energy coupled to element incorporation. We stressed that the very fact that the chemistry was energised made it more probable that degradation would occur rather than synthesis. Since survival must depend on driving synthesis more efficiently than degradation, a living system cannot just push energy into all syntheses randomly or into reversible catalysed reactions. In order to bias effective over random synthesis and over degradation a considerable degree of *organisation* is necessary, where organisation implies flow of material and energy in managed dissipative systems (Fig. 11.1). We know that life is only identifiable and has superior survival character because it has organised its own production of a self-protected set of molecules in a cell and this must be examined next. Now, an additional feature that we can show has survival value is that life self-reproduces. Reproduction is copying and in biological chemistry implies coded activity. We cannot solve the initial problems of the origin of the whole of this efficient organisation and reproduction in any very convincing way today but we can imagine some possibilities for its beginnings. What we shall do in this chapter, therefore, is to have a brief look at one or two of the possible ways in which life may have started. Assuming that a coded activity is required, we begin by considering how it and required organisation could have arisen, but very shortly we shall change course to look at how an already coded life is organised in the simplest anaerobic cells known today. In this way the essence of the early organisation in life, the beginnings of the highest form of natural selection of elements, as observed in the previous chapter, will be seen. It is a pious hope that clarification of the nature of early life may help to show us the way life started in a general but not in a detailed chemical sense. Only at the end of the chapter and again in Chapters 12 and 13 do we go on to consider why, once life as a reproducing organisation had appeared, it would be forced to evolve to more and more advanced coded life forms being driven in large measure by opportunities provided by environmental change.

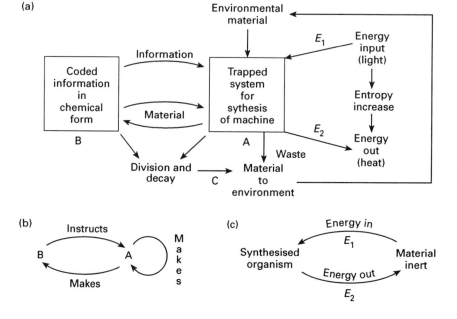

Fig. 11.1 To generate a living system (a) it is necessary to have sources of energy and material, a coded programme (B) and a machine (A). The essence of the machine is that it provides routes to synthesis (as irreversible reactions) since synthesis has to absorb energy, E_1, and decay generates it, E_2, see (c). (b) The code and the machine evolve together.

11.2 Co-operative stability in dissipative systems

In the previous chapter we have also seen how an energised state can be built up; an essential feature of the process is protection from dilution and degradation (Table 11.1). The first step needed was therefore a limit on 'cell' volume. Thus all activity is seen as taking place within a diffusion-limited space, the self-assembled cell or primitive vesicle, to restrict entropy (Fig. 11.2). Chemical protection could then be achieved by a flow of materials into kinetically stable polymers within this vesicle (Fig. 11.2). We shall consider next how such a set of polymers could have developed into life.

As stressed in Chapter 10, through a multitude of rather weak interactions sets of polymers that stabilise one another through *co-operative* interactions can be selectively built. There is no need for a repetitive pattern as in NaCl crystals. It is the matching of surfaces of, say, proteins (for structure) and polynucleotides such as RNA/DNA (for coding later) separately or together, that assists the survival of both (see Table 11.1 and Figs 11.3 and 11.4). This self-assembly is one required feature for survival. There are further factors which will assist the survival of the two types of molecules that are non-random and that were bound to be uncovered. We have observed that both sets of polymers are made from the same elements, H, C, N and O, but not in the same ratio, as well as from some different elements P (RNA, DNA) and S (proteins) (Table 11.2). If effective cooperative stabilisation of different polymers was to result, the initial material resources in the form of H_2O, NH_3 and CO_2 and similar small molecules needed for synthesis of both kinds of polymer had to be shared in a *controlled way*. Both syntheses also required energy (Fig. 11.2), and the chemical intermediate for carrying energy to both became ATP or some other nucleotide triphosphate. Thus sharing of energy in a *managed way* also had to arise (Fig. 11.5). Finally, the amounts of several

Table 11.1 Factors assisting survival

---

1. *Kinetically stable bonds*
   Fats > amino acids and sugars > esters
   Peptides and polysaccharides > polynucleotides (RNA/DNA)
2. *Folding* strength of polymers
   Proteins > polysaccharides or RNA/DNA
3. *Cross-linking*, covalently or by metal ions,
   Proteins > polysaccharides or RNA/DNA
4. *Association* into larger *assemblies*. Self-assembly.
   Fats ≥ proteins > polysaccharides or RNA/DNA followed by co-operative association of different polymer types
5. *Membrane confinement* to assist (4)
6. Maintenance of *electrolyte content* to stabilise the vesicle of (5) against osmotic and electrical breakdown
7. Development of *catalysts* including heavy metals to assist production of monomers and polymers of (5)
8. Trapping of *energy* within the vesicle of (5)

Later events

9. *Copying*: self-reproduction using coded molecules
10. *Feedback* control and regulation of production and concentration of all chemical components and of physical conditions

---

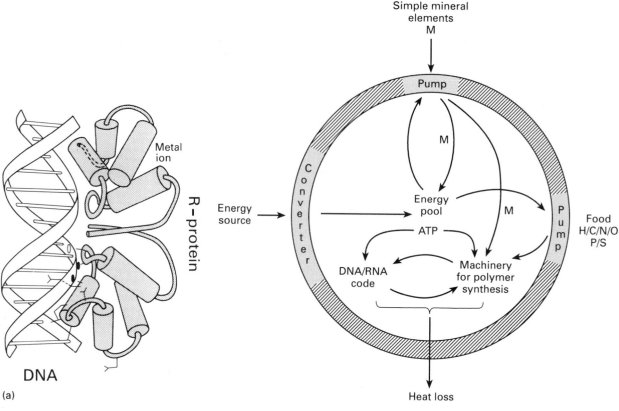

Fig. 11.2 An idealised cell requiring elements for metabolism, that is, food H/C/N/O/P/S, simple minerals, M, and energy within an enclosed space.

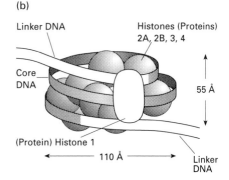

(a)

(b)

Fig. 11.3 (a) The way in which a helical set of two protein segments could fit into the grooves of DNA to stabilise both. Note the value of mobile helices. Several metal-binding regulatory proteins may be of this kind. The structure is from the works of B.W. Matthews and T.A. Steitz. (b) Schematic diagram of a nucleosome. The DNA double helix (shown as a band) is wound around an octamer of histone proteins (two molecules each of 2A, 2B, 3, and 4, shown as spheres). The unit is quite stable. Histone 1 binds nucleosomes together from outside.

Table 11.2 Elements required for biopolymer construction

Polymer*	Elements
$CH_3(CH_2)_n COOH$ (fats)	H C O (N) (P)
$-NH-CH-CO-$ (proteins) $\quad\quad\mid$ $\quad\quad R$	H C O N (S) (Se)
RNA/DNA (polynucleotides)	H C O N P
$(HCHO)_n$ (polysaccharides)	H C O (N)

* The ratios of H:C:O:N are different in all four polymers and the amounts of the polymers are very different in different species.

proteins, not just one particular protein, required to protect a given set of nucleotides, RNA/DNA, are not necessarily in simple ratios. (Several proteins are required since RNA/DNA has a huge surface with a variety of exposed nucleotide sequences while a single protein folds and by itself can provide only a few specified bases in a sequence of DNA with protection.) In essence (today in higher organisms), protein spheres are placed inside (histones) or outside (transcription factors) helical DNA (Fig. 11.3). It follows that the rate of

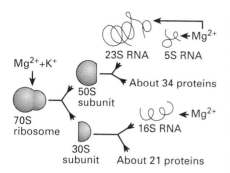

Fig. 11.4 The ribosome is a self-assembling unit for protein synthesis. Ribosomes can be dissociated into about 55 proteins and three RNA molecules. The symbol S is a size unit for RNA. Note the need for Mg^{2+} and K^+ to further increase the stability.

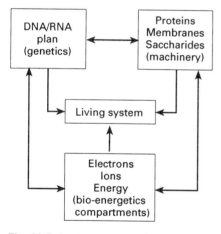

Fig. 11.5 A scheme for the interaction between required features of a biological steady state.

production of all kinds of polymers had to be commensurate, that is regulated. There is no escape from the conclusion that effective co-operative protection for survival needs co-operative syntheses as well as co-operative binding. This requires information so as to control the amounts of different materials taken up and produced and their mutually stabilising sequences. Thus crossed feedback control of all syntheses is needed for optimal protection, that is, optimal survival. (There is no suggestion yet that one set of polymers (DNA, RNA) *codes* for the other).

As stated, this set of two (or three) polymers is only one form of protection for survival. Another is for a quite different set of protein sequences to bind in or on to lipids to stabilise the vesicle membrane (Fig. 10.5). The same arguments apply: strong co-operative stabilisation will demand a special set of lipids to make membranes and proteins since again the elements C, H and O and energy (ATP) must be shared in synthesis when the amounts synthesised must be co-ordinated. Feedback control of the syntheses of both is required for the stability of the whole membrane. The discussion also applies to the formation of polysaccharides.

It is now reasonable to assume that the binding of proteins to lipids and to DNA/RNA will stabilise all three polymers in particles within or on the membrane (see Fig. 11.9). It is also reasonable that the internal solution should be adjusted to assist survival by using the membrane proteins and energy in pumps to provide basic selected materials, needed not just for synthesis but for polymer and cell stability, for example using K^+ more than Na^+ (Fig. 11.6). Finally, once energy is available, it is necessary to have appropriate amounts of catalysts, some 5–10 additional elements to promote synthesis in measured amounts. Thus, there could be brought about a rough and ready cooperative production of an association of polymers using some 15–20 elements in a variety of molecules and in solutions all in fixed amounts (see Chapter 10). The heightened survival strength implies that this system would appear more frequently than any other. All that is lacking now, compared with a real simple cell, is a specific directing code. An extremely important point to stress, given the above conditions for efficiency of activity to generate survival and without invoking a code at first, is that the best conditions demand an analytical content of elements in balance. This balance will characterise an organism that grows in a specific physical and chemical pattern. The balance, homeostasis, is of the 15–20 elements and may or may not vary somewhat with external conditions but it will define a species just as shape and DNA do.

As yet the proteins and all other polymers we have described are not synthesised in a given sequential manner from monomers produced in controlled regulated amounts. It is only the mutual binding that determines selective stabilisation of sequences. Again, the degree of sharing of resources, elements and energy is only generally protective in its resultant organisation, and nothing in this scheme prevents lots of different RNA/DNA molecules and proteins with no protection from being built and then hydrolysed. Such a system is wasteful and clearly has a lower survival value than one in which only polymers of useful sequence for stability are produced. It is implied that a coded relationship between sets of syntheses is an improved form of protection. Thus, instead of producing proteins independently from the production of RNA/DNA, it is better to produce coded protective RNA/DNA on useful coded

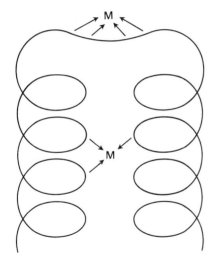

Fig. 11.6 Two binding modes for metal ions that can stabilise bonds or tertiary folds of polymers, both DNA(RNA) and proteins.

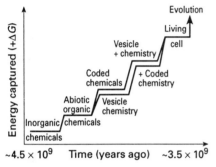

Fig. 11.7 The energy captured by life is shown as a development from abiotic capture of energy, proceeding by two possible alternative routes. The captured energy is in many forms—chemical bonds, concentration gradients, electrochemical gradients, and so on (see Chapters 3 and 6).

proteins or, putting things as we observe them today, to produce coded protective useful proteins on coded RNA/DNA so that energy waste in synthesis is eliminated. (In fact, the DNA wraps around proteins in nucleosomes (Fig. 11.3(a).) As stated, we do not know the steps that have brought about this state of affairs, which must have an informed crossed feedback system.

The system for survival we have described meets the demands of logic; it would not reproduce itself but could be repeated accidentally time and time again since it represents production in a co-operative *stable trap* (Chapter 7). It is not the system we observe today, which is semi-conservative copying of a polymer of one kind, that is, *DNA*, coupled to reproduction of proteins. Now, the process of reproduction demands that, as well as reproducing a code, a cell should divide and take half its complement of machinery, proteins and selected solutions into each daughter cell. For example, polymerase must be synthesised and inherited, K^+ not Na^+, and Mg^{2+} not Ca^{2+} must dominate the cytoplasm, and energy must be captured. Thus, while each RNA/DNA strand has to double (at least) to maintain the system, in addition, for synthesis to continue, the RNA/DNA *after doubling* has to carry with it certain essential features that are duplicated, including an ability to *synthesise* as well as to copy. The whole cell machinery must be doubled, including the confining membrane, ionic solutions, and 15–20 chemical elements.

We require therefore: (1) *a copying capability* in addition to (2) a synthesis of small molecules, monomers, with a *feedback control* from a mechanism for uptake of elements; (3) a *feedback regulatory* action linked to the copying, transcription and translation machinery for production of DNA, RNA and proteins in correct amounts; and (4) an energy capture system. We shall describe the four features in this sequence before we give an outline of cell division. We are following either one or the other of two possible paths to life from inorganic chemicals (Fig. 11.7).

(It is salutary to look back at Chapters 4–7 to see how we are beginning to analyse restrictions of variance (degrees of freedom) of the type of component present and its concentration due in part to equilibrium binding and phase formation but overwhelmingly to co-operative kinetic feedback limitations in some way related to a code.)

11.3 The beginning of molecular copying

If some special co-ordinated set of polymers of given sequence survived longer than others and was equally easy to produce then copying of this special set was an advance for the survival of (pre)biotic polymers without random generation of sequences. Moreover, copying could well enhance the speed of stable polymer production. Copying depends on self-recognition during synthesis which is the basis of crystal and liquid crystal formation demanding co-operativity. Clearly, unfolded sequences of linear polymers only can be copied.

The simplest copying process is that in which the monomer units of one

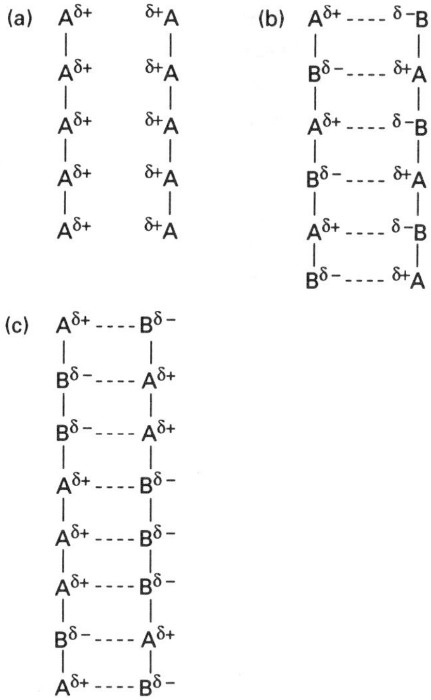

(a)
```
  A^{δ+}      ^{δ+}A
  |           |
  A^{δ+}      ^{δ+}A
  |           |
  A^{δ+}      ^{δ+}A
  |           |
  A^{δ+}      ^{δ+}A
  |           |
  A^{δ+}      ^{δ+}A
```

(b)
```
  A^{δ+} ---- ^{δ-}B
  |           |
  B^{δ-} ---- ^{δ+}A
  |           |
  A^{δ+} ---- ^{δ-}B
  |           |
  B^{δ-} ---- ^{δ+}A
  |           |
  A^{δ+} ---- ^{δ-}B
  |           |
  B^{δ-} ---- ^{δ+}A
```

(c)
```
  A^{δ+} ---- B^{δ-}
  |           |
  B^{δ-} ---- A^{δ+}
  |           |
  B^{δ-} ---- A^{δ+}
  |           |
  A^{δ+} ---- B^{δ-}
  |           |
  A^{δ+} ---- B^{δ-}
  |           |
  A^{δ+} ---- B^{δ-}
  |           |
  B^{δ-} ---- A^{δ+}
  |           |
  A^{δ+} ---- B^{δ-}
```

Matching sequences: (a) of a single monomer; (b) of two different monomers in a repeating pattern; and (c) of two different monomers in an irregular pattern.

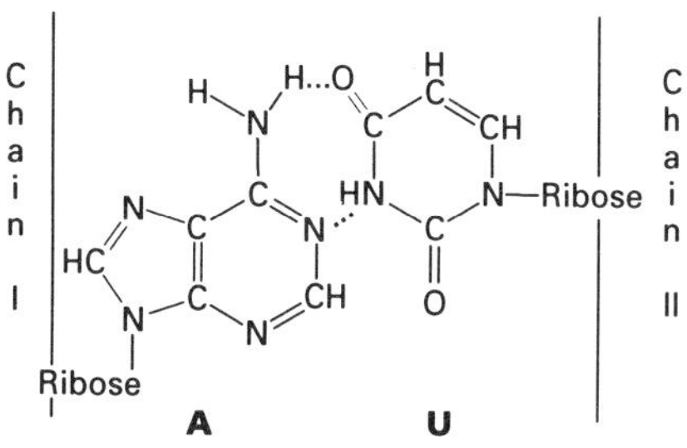

Fig. 11.8 The H-bond pairing of **A**—**U** in an RNA double helix.

strand A_n bind to those of another, A—A recognition, while building. When we remind ourselves of the nature of chemical interactions through electrostatics whether this be charge–charge or dipole–dipole interaction, including H bonds (Sections 2.6.1 and 2.9), we note that a stronger interaction develops between pairs of A—B units of opposed polarity than in pairing of A—A or B—B units, especially if A and B monomers are required to be soluble and bind in water, that is by the opposed polarity of donors and acceptors (Sections 2.9 and 5.2). Thus one of the most likely recognition schemes is between two polymer strands $[AB]_n$ using H bonds, see marginal illustration. This is possible for all three major classes of biological polymers, but only proteins (and polysaccharides) have general monomer binding through non-sequence-specific, backbone interactions and they fold. Now, we know that today the polymer engaged in reproductive recognition is DNA, virtually unfolded, where AB interaction is based on side-chain differences and this applies to all known cells, so it is an obvious temptation to attempt a description of the beginnings of copying in life starting with an $[AB]_n$ nucleotide polymer. We can follow this line of discussion, since, in principle, it might not be seen to matter which $[AB]_n$ system, that is, of any kind of monomers in linear polymers we choose, in order to understand the principles of copying. In fact, the tendency in recent discussion has been to choose RNA rather than DNA itself since RNA has a simpler sugar, ribose, is able to form more stable single chains, and can fold itself partially to give some catalytic activity aiding self-synthesis. In principle, it is possible to imagine therefore an RNA-only life—no DNA. (NB. We are deliberately putting protein and membranes to one side for the moment.) We must add one further complication. The sequence $[AB]_n$ is unique. To introduce variety we can consider sequences $A_x B_y$ in any order of A and B, for example, A·BB·AAA·BA which would be copied as B·AA·BBB·AB. only to give, after dissociation and the next round of copying, A·BB·AAA·BA. Thus, a given strand of *fixed* structure carries *information* about its parent and whatever activity is in this *information-carrying molecule* is exactly reproduced after two rounds of polymer synthesis whether it be structure, dynamics or catalysis to daughter units. Once such reproductive activity started in a vesicle we cannot call it anything but living.

11.3.1 Coded polymers: RNA life

In the case of RNA we can believe that the simplest reproducible polymer would result from a linear strand polymer with units **A** and **U** of the ribonucleotide type (Fig. 11.8), since **A** and **U** recognised one another A↔U but not self, not A↔A or U↔U. Thus two RNA strands of two bases could recognise one another in faithful reproduction. Moreover, it is known that RNA polymers (in the presence of Mg^{2+} ions) are catalytic, and are often called ribozymes, so that the presence of one strand could help the bond-making (or -breaking) process essential for synthesis. Of course, if the direction is to be toward synthesis not degradation then there needs to be more than a catalyst; there has to be energy capture to make **A** and **U** units, which have to be concentrated, and to carry out polymerisation. Reverse reactions have to be prevented. Again, since all known life is cellular, we now assume that the RNA replication is in a vesicle. The simplest additional assumption is that

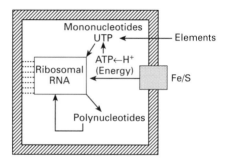

Fig. 11.9 Proposed RNA-life. The membrane was made by the production of oils (Section 9.3). The monomers were made from basic available forms of elements, H_2O, CO_2, NH_3 and HPO_4^{2-}. The source of energy was, for example, an energised Fe/S particle trapped on the membrane. The system reproduces when the duplex polymer, say $[\mathbf{A \cdot U \cdot A \cdot U}]_2$, is released as two single strands only for each to be trapped in a new synthesising vesicle. The improbability of such a scheme emerging is very great, yet some such event happened. Note that it is not necessary to have a simple $[\mathbf{A \cdot U}]_n$ unit and a variety of units of any ordering of $\mathbf{A}$ and $\mathbf{U}$ is reproducible with fidelity on second copying (Fig. 11.8).

association with certain membrane surfaces inside a vesicle or on a trapped mineral particle will *help* RNA synthesis (see Fig. 11.9). Synthesis could be helped too through the energy-trapping activity, for example, of some particle such as an iron sulphide also stuck to the membrane (Fig. 11.9), giving through intermediate proton gradients (see Section 10.4.6) the required nucleotide triphosphates, UTP and ATP. This organisation would be considered to be living and needs but a persistent flow of energy and elements (Fig. 11.5). (Note that this presupposes production of lipids before reproduction and proteins in the lipid will have been needed to hold the iron sulphide units. However abiotic lipids could have been scavenged at first.)

Now this line of conjecture is common today largely because we need to imagine some organised simple system in order to start experiments on energised assembly. It is clear that there are many problems even with this, the simplest, proposal. The difficulties increase many-fold as soon as we introduce proteins, that is, polymers from a second series of monomers known to be made in a particular quantitative way in all life forms ever observed. We know that today proteins are synthesised from RNA within a machinery, ribosomes, but, while the process is open to formal description (Fig. 11.10) its origin is not explained. Yet proteins, not RNA, are of the essence of the only *working* biological machinery we know. Thus, even though it may be the case that the selection of RNA as a self-reproducing polymer represents an interesting possible starting place for thinking about life, the next step, the selection of proteins by an RNA life, is of much greater sophistication and yields a set of polymers, proteins coded by RNA/DNA, of much greater complexity and infinitely greater functional value. As stated, proteins fold extremely selectively. The survival of biological systems as we see them was and is undoubtedly a problem of protein generation, whether or not it was so initially.

In the following section we shall not concern ourselves with which came first, proteins or RNA, or whether they developed together as is probable (see Section 11.1), but with a known scheme for syntheses of them based on

Fig. 11.10 Diagram of the protein synthesis apparatus (see Fig. 11.4).

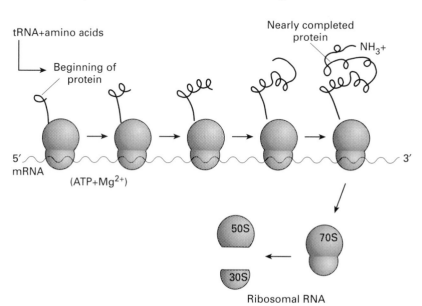

copying, *translation*. Later we must see how to control the amounts of the polymers synthesised.

11.3.2 Syntheses of matching sequences of different polymers

When we look at the problem of matching production of polymer sequences of different monomers, here the production of proteins of some 25 different amino acids with the information in sequences of four (no longer two) different nucleotide bases on ribosomes, it is easy to see mathematically that the presence of four bases allows a code of three bases at a time, a codon, to produce the more than ample 64 possible orderings. There is enough information then to code for the known numbers of amino acids plus stop and start readings to fix protein chain length (Table 11.3). The physical chemical explanation is more difficult since there is no obvious chemical *binding* connection between *side-chains* of amino acids and the *triplets of bases*. Once again we are forced to use imagination which applies to any consideration of a coded RNA/DNA:amino acid relationship. A simple representation of the problem uses two alphabets, say, Roman AB for two such bases, for example AB, and Greek $\alpha\beta\gamma$ for three amino acids. We state that in some way different A/B triplets can bind α, β or γ, as separate molecular carriers, for example ABA·γ, ABB·β and BBA·α. Then we assume that through base pairing a ribonucleotide sequence such as BAB·BAA·AAB· can only lead to formation of a γ—β—α peptide. Of course, any ordering of the three triplets will occur unless at the same time ABA plus ABB plus BBA separately can be bound to a preformed opposite strand of correct sequence and on which AB matches BA, etc. (see Fig. 11.10). This is the transfer-of-information system found in biology and is called *translation* of one

Table 11.3 The genetic code. The 64 triplet codons are listed in the 5′ → 3′ direction in which they are read. The three termination (term) codons are given

	U		C		A		G	
U	UUU	Phe	UCU	Ser	UAU	Tyr	UGU	Cys
	UUC	Phe	UCC	Ser	UAC	Tyr	UGC	Cys
	UUA	Leu	UCA	Ser	UAA	Term	UGA	Term
	UUG	Leu	UCG	Ser	UAG	Term	UGG	Trp
C	CUU	Leu	CCU	Pro	CAC	His	CGU	Arg
	CUC	Leu	CCC	Pro	CAC	His	CGG	Arg
	CUA	Leu	CCA	Pro	CAA	Gln	CGA	Arg
	CUG	Leu	CCG	Pro	CAG	Gln	CGG	Arg
A	AUU	Ile	ACU	Thr	AAU	Asn	AGU	Ser
	AUC	Ile	ACC	Thr	AAC	Asn	AGC	Ser
	AUA	Ile	ACA	Thr	AAA	Lys	AGA	Arg
	AUG	Met	ACG	Thr	AAG	Lys	AGG	Arg
G	GUU	Val	GCU	Ala	GAU	Asp	GGU	Gly
	GUC	Val	GCC	Ala	GAC	Asp	GGC	Gly
	GUA	Val	GCA	Ala	GAA	Glu	GGA	Gly
	GUG	Val	GCG	Ala	GAG	Glu	GGG	Gly

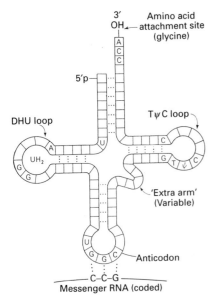

Fig. 11.11 Features of tRNA molecule for glycine (see Fig. 11.10). For a full explanation of tRNA functions see references 3 and 7.

polymer kind AB to another, $\alpha\beta\gamma$. In textbooks of biochemistry the details of relationships of this kind are described using the real cases of transfer RNA (Fig. 11.11 and see Fig. 9.17), not a simple triplet of bases, to carry amino acids. Such a system could only arise if it had survival value of the kind: proteins $\alpha\beta\gamma$ produced by nucleotide polymers AB are protective mutually and not very vulnerable to hydrolysis. (It is generally true today that the stable RNA and DNA are bound by proteins as in ribosomes Figs 11.3 and 11.4.) Of course, the programmed synthesis needs energy and catalysts, yet it must not run backwards to monomers. We are being forced to invent more and more conditions in order to generate a code which is translated. (A coded sequence is inherently improbable since it is ordered, see Chapter 3.)

While this account is a descriptive possibility of life's beginnings we have noted only one way of initiating the known system which is to state that in some sense selected RNA (even small fragments) and selected peptides are mutually protective, that is, allow survival to be enhanced in specific ways through mutual recognition (and synthesis). Of course, we know that short RNA(DNA) sequences do bind selected peptides. Therefore it is tempting to presume that we should be searching for a simplified scheme for the build-up of a coded relationship in which we take a limited number of bases in one polymer and a limited number of amino acids in another and ask if the code itself reveals the interactive initial protective nucleotide sequence/peptide sequence system, that is through an observed mutual stabilisation of proteins and RNA (see Figs 9.17 and 11.4).

There is here an applied significant favourable interaction between the particular bases and the particular amino acids above a decipherable coded relationship between the two. We can look for this by investigating the translation code and how it might have begun.

11.3.3 The beginnings of translation

Here again there is not sufficient knowledge, but a possible conjecture is that the bases C and G came before the extended code of C, G, A, U. There is today a coded relation between these two bases and the proposed earliest (and usually still the most common) smaller amino acids, glycine (GG), and alanine (GC), together with proline (CC) and the basic amino acid, arginine (CG) (Table 11.3). Now, these thoughts on development are speculative; however, Eigen, who postulated this scheme, states (see reference 9, in 'Further reading') 'such assertions [that is, these particular correlations] are nothing but plausible'. Another interesting feature of these four amino acids is that three of them are *known* to be associated with β-sheet and β-turn protein structures (see Chapter 9). Such structures are the most stable and useful in catalysts. All the primitive Fe_n/S_n proteins of energy capture devices are β-sheets (see Figs 10.8 and 11.9). Of course, positively charged arginine peptides will bind to phosphate of RNA and DNA, and hence there could well be mutual stabilisation.

The codes of nucleic-acid bases related to the larger amino acids are *also* given in Table 11.3. These larger amino acids, especially leucine, lysine and glutamine, are particularly correlated with the helical structures of proteins. It is now known that helices are associated with mechanical devices more than with basic catalysis in biology (Section 7.5 and see Fraústo da Silva and

Williams, reference 6 in 'Further reading'). There is good reason to suppose that mechanical devices would be a later development; they are more sophisticated than just protection (binding) or catalytic structures.

Now, however, our discussion is running into great difficulty—where did these bases and amino acids come from? Today they are made by protein-catalysed reactions and energy sources. The suggestion that has been advanced that they were generated accidentally by electric discharge and light is somewhat unsatisfying. We shall not discuss the nature of the origin of the code further since, although it is an intriguing problem, it is one where ideas greatly exceed the evidence needed to prove them. Incidentally, however, some intriguing issues have been raised concerning the selective use of elements within the polymers of life.

11.3.4 The selected elements for the coded polymers

Before we leave the nature of the code itself, that is, the RNA(DNA)–protein relationship we should see that it is connected to only a few selected elements in organic molecules. The coded amino acids are a limited set formed from a particular group of readily available (in primitive times) non-metals, and these non-metals also form the bases and saccharide units of DNA and RNA, namely,

$$
\begin{array}{c|ccc|c}
 & \mathbf{H} & & & \\
\hline
\mathbf{B} & \mathbf{C} & \mathbf{N} & \mathbf{O} & \mathrm{F} \\
 & \mathrm{Si} & \mathrm{P} & \mathbf{S} & \mathrm{Cl} \\
 & \mathrm{Ge} & \mathrm{As} & \mathbf{Se} & \mathrm{Br} \;.
\end{array}
$$

These are most of the common elements that readily form *covalent* bonds in molecules of some quite high polarity (Section 2.5) and all these bonds are of considerable kinetic stability. This remarkable grouping excludes the halogens probably because the early reducing environment could not generate the redox chemistry necessary to introduce F, Cl, Br or even I into carbon compounds (Section 12.4.1 and 12.5.5). Again, B, Si, Ge and P are not introduced into amino acids since the necessary reductive power to make C—B, C—P, C—Si, C—Ge or C—As bonds was not available. Other modes of incorporation, for example as esters, are of lower kinetic stability. Incorporating metal elements is, of course, a much more complex problem since these elements exchange quite rapidly from complexes of all kinds in water. Clearly, their inclusion in *coded* small or large molecules is difficult.

A particularly intriguing feature is the coding for seleno amino acids. Early in evolution selenium was present as selenide (H_2Se) and therefore could be introduced in an (unstable) C—Se—H bond much as sulphur was introduced (Section 10.5.5). The fact that a code exists for Se-amino acids means that coding is involved for all possible kinetically stable prebiotic C—X covalent element combinations where X was available. There is a hint here that amino acids (peptides) and bases (polynucleotides) grew up together in mutual recognition in a chance environment—the reducing atmosphere of early Earth. It was not absolutely necessary to code for selenium—it happened to be there and so formed a covalent compound that could be used to some advantage for survival, for example in catalysis (Section 9.15.4 and 12.4.1). The peculiarity of the position of the uncoded phosphorus, always as phosphate, not bound in a C—P bond, will not have gone unnoticed. C—P itself is difficult to form and very unstable. Phosphate exchange from covalent ester incorporation is, however, more rapid than that of the above elements in C—X bonds, and it therefore has special value in *exchange*, not coding for structure. In effect, it is introduced as a chemical modification of monomers and polymers as is glycosylation. Phosphate's presence in DNA and RNA means that there is an advantage to their instability as well as to their linearity.

In an oxidising atmosphere, such as that of today, we have to wonder whether we (man) could not adapt the nucleotide code to introduce for example C—Br-containing amino acids into proteins. Biology today can make C—Cl, C—Br and C—I links. We conclude that the present code is but one example of using naturally selected chemical elements for information purposes, that is, to form a control language for instructing the synthesis of polymers, under particular environmental conditions. It is not the only possible use of the chemical elements for a language but it is the case that the only non-metal elements with accessible chemistry in anaerobic conditions were used as governed by the environment existing some 4×10^9 years ago. (A quite different and more difficult question is the choice of amino acids, for example why tryptophan?)

11.4 Control and regulation: introduction

In essence, what we discover as we stumble forward looking for explanations of the earliest conditions for the production of life is that any life that we know of is very complex in organisation. We saw in the last chapter by analysis and by reasoning why life needed polymers based on specified reaction steps, energy and some 15–20 elements. Moreover, we described advances in survival above if there was *co-operative* stabilisation, self-organisation, based on a code. In this chapter we shall see next that, in order to evolve a living system, that is, to give it greater survival strength, we need to *correlate*, *organise* the metabolism using flow in and out of the cell and to and from information-carrying polymers, RNA/DNA. Everyday experience shows that any organisation such as this that has parts distantly connected also needs management which is only achieved by messengers carrying instructions from one coordinating centre (synthesising unit) to another. Given the lack of firm knowledge as to how such complexity grew even up to the state of the simplest primitive cell, we are forced to abandon the analysis of the possible beginnings of life and start a somewhat different enquiry from the simplest known cell asking directly about the messengers within management, which we shall divide into *control* and *regulation*. On the one hand, there is the *co-operative regulation* of the production of RNA and proteins, that is, production of polymers in correct amounts based on DNA but there is also the *co-operative control* of the uptake and incorporation of elements, of energy, and of metabolism, degradation and synthesis of covalent molecules (discussed in Chapter 10) to be considered. Each of these activities can be broken down into a flow diagram as in Figs 11.1 and 11.2. They occur in different parts of space (Fig. 11.2). Thus we are re-introduced to the problem that worried Chinese thinkers particularly, the nature of sustained controlled flow. We can even trace its origins, that is, of seemingly organised momentum, to very close to the beginning of the universe, followed by general further outward motion with turbulence (Chapter 1). In Chapter 7 we showed that, in any such out-of-equilibrium system, flow will continue to occur and is controllable by fields, initially macroscopic and gravitational, produced by turbulence. However in a *microscopic* system 'fields' refers dominantly to ordered molecular structure carrying charge, that is electrostatic fields, produced by condensed systems of chemicals, rather than the gravitational fields of the astrophysical world. Tackling the nature of life must be primarily a problem in the evolution of managed flow in electrostatic fields utilising the selected chemical elements in self-organised activities, but to be maintained life must also capture energy, of course, see marginal note.

This chapter has so far described co-operatively bound but very different *molecular* polymers of biological systems, which help to bring about such organisation (Sections 11.1–11.3). In Sections 11.5 and 11.6 we shall describe the *chemical control of reaction pathways of elements and small molecules* using intracellular message systems and ways of ensuring that energy and chemical elements are distributed in appropriate proportions for the needs of integrated cellular syntheses and degradations. We then look at *chemical regulation of the polymer syntheses* and the messenger systems there involved (Sections 11.8 and 11.9). Thirdly, we shall describe the cell cycle (Fig. 11.1 and Section 11.11) and differentiation. Eventually, all management has to be totally integrated. All 15–20 essential elements have their interconnected roles to play in these activities (Section 11.10).

Later sections of this chapter (see Sections 11.7–11.12) describe the *physical organisation* of primitive prokaryote cells, going forward with evolutionary advance through changes in membranes, filaments and vesicles. These changes allowed different single-cell shapes to evolve. Throughout this analysis only anaerobic conditions will be described. Later developments were associated with the codevelopment of new chemistry, including the use of captured light and then dioxygen production by simple cells. In turn, the utilisation of these new energy and chemical resources led to further metabolic developments, but we shall leave the complexity of all these later developments, especially those within advanced multicellular organisms, until Chapters 12 and 13. In an intermediate period this novel chemistry (of life) allowed a new energisation of the atmosphere and of the surface of Earth (Section 8.6) which then back-interacted with the evolving forms of life (Fig. 11.5). There is then an evolving environment as well as evolving life forms and their combination will be described in its essential principles at the end of this chapter since the processes began before dioxygen came into the atmosphere in large quantities. In a local sense, parts of the surface of the Earth and of life became intermingled, a statement which reflects the ideas of Gaia (see reference 6 in 'Further reading' and Chapter 15). Since the early life we are discussing was presumably in the sea we shall not be concerned with any geophysical changes of land masses in this chapter, see Section 8.5.4.

In a straightforward manner this discussion will lead us toward the organisation that man is now imposing on the Earth's chemicals. There is no

Expansion of Universe

Big-Bang

Outward ↓ Flow

Gravity Field

Turbulent ↓ Flow

Stars (Galaxies)

Electrostatic ↓ Fields

Atoms and Molecules

↓

Condensates

↓

Flow in Liquids
(Life)

new principle governing his activities but just a new application of the natural selection of the chemical elements in relation to organised functional value (Chapter 14). At the microscopic level this selection reaches its height in biology but at the macroscopic level in man's industry.

We now go back to the principles of Chapter 7 where we introduced feedback controls on physical flow and on reaction sequences.

11.5 Feedback control of element and small molecule concentrations

11.5.1 The cell as a chemostat: feedback to pumps

First we need to show how cells can maintain rather fixed concentration conditions of elements as ions. In essence, there are two mechanisms to consider. The first is a process well known in man's pH chemostats in which a concentration of an element, here H^+ (pH), is monitored by a device that switches on supply if H^+ is deficient and switches it off if H^+ is in excess. To manage such a system cells must have pumps (inlets) and channels (outlets) controlled by gates, which, given that the proteins in them can act as mechanical devices controlled by chemical concentrations (or sometimes electrolytic potentials), can be given a simple pictorial presentation (Fig. 11.12 and see Section 10.6). Energy is of course, required. Initially in evolution a major requirement was to stop too much Na^+, Cl^- and Ca^{2+} as well as H^+ entering the cell, and then to stop too much outward pumping when critical low levels had been achieved (Fig. 11.12(a)). The concentrations developed are then fixed by the feedback from the element itself to the pump in a specific way for each element. This applies to H^+, Na^+, K^+, Mg^{2+}, Ca^{2+}, Cl^- and HPO_4^{2-}, etc.

We can readily increase the sophistication of the control network for those elements that are present in less readily available amounts outside cells by the second mechanism which uses synthesised carriers (Fig. 11.12(b)). This is achieved by the synthesis of chemicals that capture the required element. A detailed specific *well-proven* example is given later using the iron scavenger chelates called siderophores (Section 12.5.1). Although they evolved after the coming of dioxygen, their essence is common to a large number of uptake systems that from very early times had:

(1) a synthesised recognition compound, Y, for a desired element or compound, for example, a scavenger, a chelating agent or protein, which could be linked to a pump in a membrane (Fig. 11.12(b));

(2) an energy source necessary to drive this chemical in or out of the cell connected to the pump and channel, respectively (Fig. 11.12), which are also attached to recognition devices; the energy source could be a field gradient, a concentration gradient or an energised chemical system;

(3) a switch-on/switch-off control, which in a feedback system is part of the pump or acts at the pump, so that deficiencies or excesses of the desired element are recognised and prevented.

While (2) and (3) are just like the pumps for ions already described above, the scavenger itself involves additional syntheses and hence regulation of

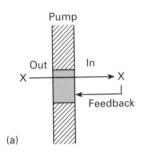

(a)

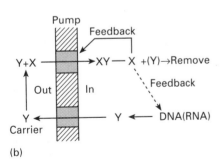

(b)

Fig. 11.12 (a) An ion pump with feedback from the internal ion, X, concentration. (b) A pump for an ion, X, plus a carrier, Y, which requires synthesis. There is now a need for feedback, high X, and feedforward, low X, to the DNA (or RNA) for carrier synthesis as well as feedback on the pump.

additional enzymes. (Of course, the numbers of pumps and channels can also be regulated at the level of their synthesis (see later).) It is easy to see that self-control of each ion could now be at the *pump activity* level, as in Fig. 11.12(a), or it could be at the level of *synthesis* and *destruction* of the carrier synthesising proteins (Fig. 11.12(b)) or at both. A suitable combination of these two in *feedback* arrays will see to it that the cell works with adaptable concentration of a given chemical scavenger and all its associated apparatus, that is, the synthesis and destruction of the proteins for making the scavenger, pumping and switch systems. This has an optimal value for one element so as to give efficient running of a particular cell, commensurate with the interaction with other parts of the cell machinery. A particular cell requires optimal values for all of the essential 15–20 elements (Fig. 10.3) so they must become linked.

Element, simple ion, concentration gradients can be linked energetically to one another in many ways. Two simple ways are as follows:

1. The membrane has a protein that exchanges ions, for example Na^+ for H^+, and this exchange has a gating mechanism attached to it dependent on both $[Na^+]$ and $[H^+]$.

2. One ion, for example H^+, closes the pumping mechanism of another.

Given these cross-connections between ion flows, the cell becomes a cross-connected feedback network of all the simple ions as described in Section 7.13. It is possible to extend all the pumps and exchanges to include molecules.

These mechanisms are general to almost all elements as ions or in compounds and require energy that is usually supplied in the form of ATP so that, while there are specific mechanisms for the uptake of each element and control over these mechanisms, there is also a general link to the energy status of the cell. In this way all the elements can have their own particular balance related to ATP, that is, the element phosphorus, produced by metabolic pathways. In essence there is a cross-connected feedback which generates kinetic cooperativity. Our next problem is then the control over such metabolism.

11.5.2 Feedback in metabolic pathways of the elements hydrogen, carbon and oxygen

A good example of feedback control in metabolism of the most common elements, H, C and O, is provided by the glycolytic pathway (see Figs 10.11 and 11.13). In the latter figure we include the connections of the pathway to some feedback agents, that is, the major products of glycolysis, ATP and pyruvate. If there is excess of either then it pays the cell to store energy and elements C, H and O in glycogen, that is, to reverse the process of glycolysis through gluconeogenesis which is achieved by not quite the same pathway as glycolysis (Fig. 10.11). Thus ATP and pyruvate act back on early steps in glycolysis, for example, at phosphofructokinase, to slow down the enzyme of an early substrate in the pathway. Such end-product feedback is very common in metabolism. A general scheme is given in Fig. 7.21. However, another feedback agent acting on glycolysis is citrate which belongs to a second pathway, the citric acid cycle (Fig. 11.13 and see Fig. 10.12). Citric acid also has a feedback interaction with the fatty acid synthesis network. Thus all the C/H/O pathways are controlled in a linked manner (Fig. 11.13).

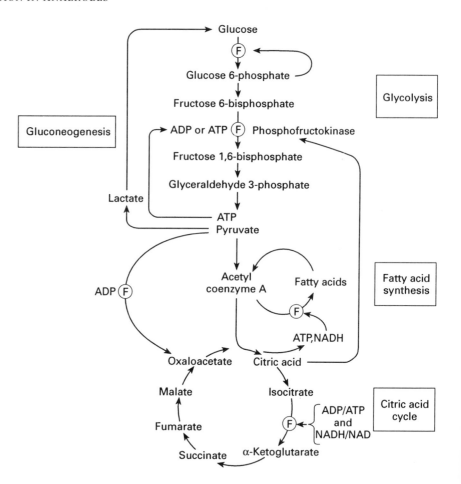

Fig. 11.13 Feedback control sites, F, of glycolysis, fatty acid synthesis and the citric acid cycle. Steps most affected can be acted upon in a negative or positive manner. Note that feedback controls glycolysis and the citric acid cycle at crucial early steps in the pathways; this increases the efficiency of the pathways and prevents excessive build-up of intermediates. Note also that the compounds, inhibiting or activating enzymes, are often the energy-carrying or chemical-carrying compounds themselves—ATP, ADP, NAD$^+$, NADH, that is, coenzymes.

11.5.3 Feedback control of the incorporation of the element nitrogen, and of amino acid synthesis

The syntheses of amino acids and their internal feedback controls are shown in Fig. 11.14. We need to note that the synthesis of amino acids has already been linked to the amination of small C/H/O fragments from glycolysis and the citric acid cycle and that these two pathways produce the energy, as ATP, for the amino acid synthesis as well as the glutamic acid required to carry nitrogen as glutamine (Section 10.5.4). The production of amino acids is therefore geared to the production of fats and sugars. The concentrations of amino acids have to be controlled individually so that excesses of particular ones do not build up. The feedback control in Fig. 11.14 shows that, while some amino acids block all syntheses, for example, threonine and lysine, others only block their own synthesis, for example, methionine. (In passing note that glutamine synthetase requires Mg^{2+} or Mn^{2+}.)

11.5.4 Feedback control of nucleotides: incorporation into polymers

We first draw attention to the fact that the small molecules used in the code of RNA and DNA are the same molecules as the triphosphates that are used in

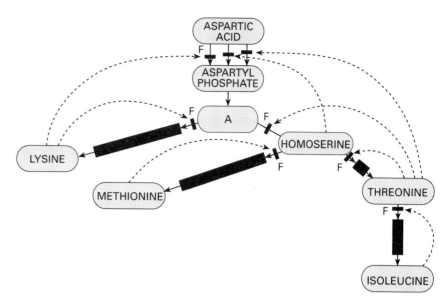

Fig. 11.14 Feedback control governs synthesis of four amino acids—lysine, methionine, threonine and isoleucine—from aspartic acid in the bacterium *Escherichia coli*. All four amino acids and one intermediate product act as allosteric inhibitors, either at the first step of the synthesis or at critical branch points. Long black rectangles, series of enzymes; short black rectangles labelled 'F', feedback points. A is a branching intermediate.

energy transfer and as major parts of the coenzymes of acid–base metabolism. Thus, the concentrations of five of the carriers for energy and material transfer (ions and substrates) and controls of the metabolism are made from *the very same* molecules, NTP, where N is one of the five bases, as are monomers of DNA (RNA). We know that not only A, but also C, T, G and U are so used in feedback control in different pathways to give flexibility to signalling (Table 11.4). The interdependence is extremely valuable and totally desirable for obvious organisational reasons giving stability to a cell (see the diagrams of cross-talk in feedback circuits in Section 7.13). (In effect, DNA (RNA) cannot be made if there is an inadequate supply of coenzymes or nucleotides of course.)

The syntheses of RNA and DNA nucleotides have to be controlled at different rates since the required amounts of the DNA nucleotides are different from those of the RNA nucleotides. Nucleotides designed for DNA also have to be modified from those for RNA by conversion of the ribose in them to deoxyribose. The enzyme for this reduction is ribonucleotide reductase (Fig. 11.15). To achieve synthesis of balanced amounts of different bases in DNA deoxyribonucleotide synthesis is controlled by feedback from the deoxyribose nucleotides (Table 11.4), while the RNA synthesis is controlled by feedback from the ribonucleotides (Table 11.4). Both syntheses require energy in the

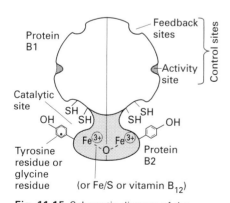

Fig. 11.15 Schematic diagram of the location of the catalytic and regulatory sites of *E. coli* ribonucleotide reductase. (Adapted from Thelander, L. and Reichard, P. (1979). Reduction of ribonucleotides. *Ann. Rev. Biochem.* **48**, 133–58.)

Table 11.4 Some feedback controls of enzymes for nucleotide synthesis

Enzyme	Feedback inhibitor
Aspartate transcarbamylase (RNA)	CTP
Cytidine triphosphate synthetase (RNA)	CTP
Ribonucleotide reductase (DNA)*	Several free nucleotide triphosphates

* The most primitive ribonucleotide reductase uses vitamin B_{12}, cobalt, or an Fe/S cofactor, while the most recent enzymes use an iron or manganese cofactors that only appeared after dioxygen was introduced (Chapter 12). All the enzymes are controlled in a similar manner (Fig. 11.15).

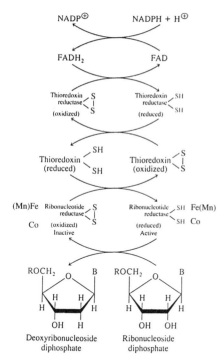

Fig. 11.16 Summary of the steps in the formation of deoxyribonucleotides. Electron transport from NADPH to ribonucleoside diphosphate in *E. coli* involves a succession of redox reactions that interconvert sulphhydryl groups and disulphide bonds. R represents the diphosphate group and B represents the nitrogenous base of the nucleotides.

form of NTP. There is also a link to redox control through thioredoxin (Figs 11.16 and 11.24) and Fe/S proteins. We begin to see how the supply of monomers is controlled by the synthesis of precursors and their concentration which is linked to the supply of elements such as Fe.

11.5.5 Different pathways for degradation and synthesis in metabolism

It is a feature of the above sets of reactions that degradation with energy production and synthesis with energy use are on somewhat different metabolic pathways (Fig. 11.13). The transport of H in redox degradation uses NADH, while in redox synthesis it uses NADPH. The fact that different pathways (enzymes) and different transfer and control agents (coenzymes) are often used for degradation and synthesis means, of course, that equilibration of materials and energy carriers is prevented but trapping for synthesis is controlled in a very central metabolic pool. Thus we write reactions not as in catalysed equilibrium balance,

$$\text{Energy} + \text{substrates} \left\{ \begin{array}{c} \text{Degradation} \\ \xrightarrow{\hspace{1.5cm}} \\ \text{Enzyme (M)} \\ \xleftarrow{\hspace{1.5cm}} \\ \text{Synthesis} \end{array} \right\} \text{products} + \text{energy},$$

but as a controlled network,

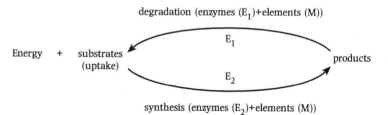

It is feedback control from both substrates, specific to particular paths, and energy (ATP usually), general to all paths, that switches off one or the other in the second (cyclic) scheme, while in the first scheme any feedback affects both sides of the reaction equally. The advantages of the second (cyclic) scheme are then self-evident in that circumstances decide between: (1) degradation for energy; and (2) storage of energy and material (synthesis).

This concludes the description of the co-operative controls over the incorporation of H/C/N/O as elements by feedback of the products of the pathways. A diagrammatic summary of the basic primary metabolic processes in anaerobic cells (but, in fact, in all cells we know) is given in Fig. 11.17. In this description we have constantly referred to energy control, which introduces the fifth element, phosphorus, as phosphate. In the next section we consider phosphate pathways and thereafter the controls exerted by some 10–15 other elements. A different aspect of control emerges in that, while the function of all the molecules we have described so far lies in reaction sequences of metabolism and while now and then one or another of these molecules acts in a control capacity as well, when we turn to phosphate (and other elements)

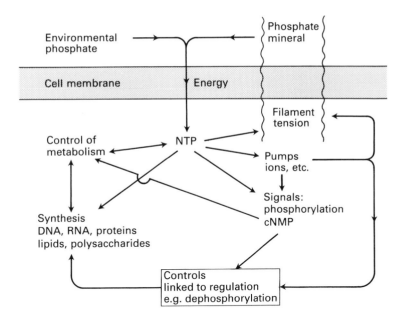

Fig. 11.17 The central connection of phosphate to nucleotide triphosphates and then to control and regulation.

the function of the molecule is more frequently as a messenger of control and it is not necessarily involved in the pathway chemically, and certainly not through change of oxidation state.

11.5.6 Phosphate metabolism and its feedback

Figures 11.17 and 10.18 showed the major pathways of phosphate from its initial entry. The levels of ATP have been observed to be held constant in particular steady states of cells. The steady state of flow demands an adequate fixed source of energy (light or food), which is then generally related to the so-called phosphate potential (see Table 10.7). As mentioned before this is defined by the position of balance of [ATP] relative to [ADP] and [P], that is,

$$\text{Phosphate potential} = \frac{[\text{ATP}]}{[\text{ADP}][\text{P}]}$$

This potential closely reflects the adequacy of all energy supplies to the cell. If the ratio [ATP]/[ADP] [P] is high then the cell progresses toward division but if the ratio is low the cell rests. Thus ATP, ADP and P must act in feedback loops,

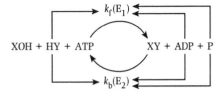

where ATP, ADP, and P bind to the enzymes E_1 and E_2 to control rates of the reaction, k, of substrates XOH, HY and XY, but they also control the rates of many other reactions (Table 11.5 and Fig. 11.13). The control obviously connects to the pH through the acid–base equilibrium since the pK_a of phosphates is around 6.8 (Section 5.6.1). Thus, homeostasis of ATP and H^+ are linked (see Section 10.6.2), and are ultimately linked to all acid–base

Table 11.5 Known feedback in pathways

Pathway	Feedback	
	General (P)	Selective
Glycolysis (C,H,O)	Mg^{2+} ATP*	Citrate, pyruvate
Citric acid cycle (C,H,O)	Mg^{2+} ATP*	Fe^{2+}, citrate, NADH
Nucleotide synthesis (C,H,N,O)	Mg^{2+} ATP*	Nucleotides (Fe^{2+}?)
Amino acid synthesis (C,H,N,O)	Mg^{2+} ATP*	Amino acids

* The control is usually the phosphate potential, that is, [ATP]/[ADP][P], which increases directly as the supply of energy increases until it self-controls and is linked to [H^+].

homeostasis and, hence, through condensation reactions may well control all polymer synthesis and energy transfer in biology, including pumping (see Fig. 11.17).

(The *fitness* of phosphate in all these chemical reactions and message systems is clear. It is the best element for carrying condensation energy, for signalling energy status to catalytic systems and for the regulation of catalysts due to its easy hydrolysis kinetics (Section 9.13) and its lack of redox biochemistry. For every element in biology there should be *a fit function* (Section 10.4), and we trust this is already obvious in the polymers formed from H, C, O and N.)

In this discussion we are treating phosphate biochemistry as part of general *linked* control, not just of metabolism, and four other factors strengthen this approach:

1. ATP and ADP are largely MgATP and MgADP so that the presence of Mg^{2+} is then part of the internal message and control systems over acid–base phosphate metabolism in cells. We cannot discuss phosphate potential control without discussing Mg^{2+} concentration.

2. In the presence of equimolar Mg^{2+}, ATP and ADP + P can bind protons to some degree. The result is that there is a network of messengers between pathways based on three elements Mg^{2+}, H^+ and P, and two compounds, ADP and ATP. The separation of Mg^{2+} and H^+ from P messages is seen, however, in the use of a further messenger cyclic AMP (cAMP), which does not bind H^+ until pH = 1 (Fig. 10.18) and does not bind Mg^{2+}. (The other nucleotides are used similarly in controls of some other pathways.)

3. ATP is part of a complex system of messages in cells through its ability to make not only cAMP but also protein-P which together act in many enzyme controls and are also major separate factors controlling the synthesis of proteins at the DNA level, and therefore in regulation. (They will be shown later to be in large part dependent on calcium input to eukaryote cells for the initiation of a response to external events (Fig. 13.7), though this is not so in prokaryotes.)

4. To complete the feedback control ATP itself is used in pumps to control the levels of Mg^{2+} and H^+ in cells as described in Section 10.6.1.

The complexity of the message systems to general acid–base metabolism based on the elements H^+, P and Mg^{2+} cannot be overstressed and involves a whole range of enzymes such as: (1) kinases, which catalyse protein phosphorylation, substrate phosphorylation and so on; (2) phosphatases that reverse the phosphorylation; (3) cyclases that produce cAMP from ATP; and (4) esterases that hydrolyse cAMP to AMP; furthermore, (5) ATP is used to produce all the other nucleotide (N) triphosphates and thus the corresponding network of control compounds, NTP, NDP, NMP, cNMP for other pathways. The suggestion is that H^+, Mg^{2+} and HPO_4^{2-} were always at the very heart of cell control through acid–base reactions, see marginal note.

11.5.7 Additional elements in controls over pathways

The description of phosphate messenger systems introduces the general point that control of metabolism rests in part with C/H/N/O compounds but, because these compounds are inadequate in other functions demanded by a cell, it is to other elements (in compounds) that we must look for additional vital activities, such as energy transfer, osmotic and electrolytic balance and catalysis. All of them must be controlled and involved in control to maintain homeostasis. While describing phosphate controls we were forced to introduce Mg^{2+}. Referring back to Sections 9.12 and 10.5.5, we see that, when describing oxidation–reduction and additional carrier functions, sulphur is equally essential and in Section 11.6 redox balance will be re-introduced. We stressed in Section 11.1 that, from the very initiation of living processes, catalysts must have been present. Here metal ions and heavy non-metals are essential and have always been used but they too must be held in controlled concentration. The control of uptake is at the level of the feedback pumps, but, once inside a cell, the ions, for example Fe^{2+}, act in a vast range of reactions. Thus many catalyses can be linked to a common pool of several free metal ions by exchange, for example

$$\text{Any } Mg^{2+} \text{ enzyme} \rightleftarrows Mg^{2+} + \text{apoenzyme (inactive)}$$

or

$$\text{Any } Fe^{2+} \text{ enzyme} \rightleftarrows Fe^{2+} + \text{apoenzyme (inactive)}.$$

The conclusion is simple: any catalytic and exchanging element can act as a control point in the overall cell metabolism, and this is more or less a necessity if reaction pathways are to be coupled effectively to the *rates* of total material throughput. It is then of interest to see that Fe/S and Fe/O clusters act in enzymes as critical as archaebacterial pyruvate dehydrogenase (Section 10.11) and ribonucleotide reductase. In different organisms Co, Fe or Mn act in this particular enzyme but a metal is always essential. A final example which may also be primitive is that there are phosphatases concerned with transcription factors which require Mn, Fe and Zn, for example the calcineurins. (In passing we must not ignore the maintenance of the concentrations of the simplest elements as ions, Na^+, K^+ and Cl^-, which are required for osmotic and electrolyte control (Sections 11.5.1 and 10.4.2).)

Elements in acid–base control

C, N, O, H (slow)
substrates

$\downarrow\uparrow$

(fast) H, P (intermediate)

$\downarrow\uparrow$

(Zn) Mg, K . (fast)

Na, Ca, Cl   out

11.6 Redox potential controls

We have described the need for redox state changes of carbon compounds within the glycolytic, fatty acid and citric acid pathways. The transfer of hydride by two-electron reagents in anaerobic cells was based on the NADH/NAD coenzyme balance. Now the redox potential of this couple is low and as a transfer agent it does not equilibrate with the transfer of redox equivalents within one-electron reactions. Here the potential inside the cell appears to be set by the reactions of —SH carrying molecules, where there is a connection between the Fe^{2+}/Fe^{3+} reactions and the 2 —SH/—S—S— redox couples, see marginal note, but not between the metal couple and NADH/NAD. Thus the redox potential is not precisely fixed in a cell since not all couples equilibrate with the same carriers of electrons. In fact, this is essential since the redox potential difference between the outside environment and the inside of a cell gives a major source of energy for anaerobic cells, which cannot absorb light. Thus we have controls of some pathways by NAD/NADH, etc as described, but of others by iron and/or —SH/—S—S— systems, in thioredoxin for example, such as in the synthesis of DNA (Fig. 11.16).

While redox control is linked almost inevitably to pH in organic chemical reactions, the pH itself is uniform in any one compartment of a primitive cell except extremely locally where there may be membrane invaginations or fast reactions. However, in more advanced cells of many compartments each compartment has its own pH and redox potential. Note how H connects redox and acid–base systems.

11.6.1 Summary of acid–base and redox control

We may now give some general conclusions concerning controls. In order to gain adequate control over small molecules and ions the following are necessary:

1. The substrates of a given path should back-interact within the path.

2. The fragment carriers between paths should feedback to all paths that utilise the common elements carried by the fragment. The carriers in general are mobile coenzymes, for example NADH, NADPH, CoA.

3. The components of catalysts should be mobile and give a general interrelated control over many paths. For example H^+, Fe^{2+} and Mg^{2+} exchange can be utilised to control acid–base reactions, while Fe^{2+} or thiolates control activity between redox pathways.

4. The cell is always under osmotic and electrolytic control using pumps.

5. Energy should be made available to all pathways of synthesis and uptake.

6. Energy-utilising pathways should be isolated from the reverse reactions that generate energy. Thus it is preferable to have two sets of carriers—one in synthesis, the other in degradation. We observe in degradation ATP and NADH, while in synthesis we find GTP and NADPH. Some pathways use ATP, some GTP, some UTP and so on, but for initial energy they all ultimately rely on ATP.

7. It may pay to separate paths by compartmental isolation so that some molecules are tightly constrained in space to one set of paths. In primitive cells inside and outside cell reaction paths and paths within membranes could be separated.

8. The pH and the redox potential should be poised, see marginal note.

Table 11.6 Examples of elements used in early controls*

Element	Control (mode of use)
H	NADH (NADPH), mobile coenzymes
e/H^+	Thiolate $\rightleftharpoons$ disulphide (thioredoxin)
C	CoA (acetyl is the C-fragment), mobile coenzyme
N	Glutamine
P	Very many NTP, cNMP, P, NDP, NMP
Mg^{2+}	Intimately involved with P (exchange)
H^+ (pH)	Intimately involved with P, S and proteins
Fe^{2+} (Fe_n/S_n)	Free Fe^{2+} in enzymes (exchange); redox processes
S	Used with Fe in Fe/S proteins; redox processes
Mn^{2+}	Free Mn^{2+} in enzymes (exchange)
K^+, Na^+, Cl^-	Free ions acting on mechanical stress systems (H_2O levels or osmotic pressure)
Fe(haem)	Control in slow exchange

* The relationship of the same elements to regulation is described in Section 11.8.

We summarise the extensive network of control elements in Table 11.6. Each element has at least a dual capacity in that it has the direct functional significance referred to in Section 10.4 but it also has an interactive function as a part of the cross-linked feedback circuitry. It is this cross-linking of control circuitry that makes the organisation within biological organisms so difficult to compare with that of man's computers (see Section 7.13). Although the central principles are common within control theory, the control in biology is linked to a protective system for the *survival strength of a growing organism* that requires 15–20 elements to be held effectively constant in a great variety of components. While the components are in a thermodynamic sense independent (as they must be by definition), they are no longer independent in this flow system containing a feedback network controlling their physical movement (pumps) and their chemical transformation. We return to this point at the end of the chapter but we now see that the analytical content as well as the code, DNA, describe the cell although today neither alone can provide an adequate description. DNA must have grown up in an internal environment already largely present in a cell. It codes *qualitatively* for proteins but the *quantitative* reading of it belongs to a previous demand for *survival fitness* linked to the metabolism of 15–20 elements in a network, dependent on energy.

11.7 The control of shape: mechanical controls

It is clear that, just as the chemistry of a cell, down to its elemental analysis and its DNA, is a characteristic of a species, so is its shape. Maintained shape is a correlative of physical homeostasis, which parallels chemical homeostasis and equally requires maintenance and control. The question which arises here, as it did in Chapter 6 on crystal shape, is the degree to which shape is controlled by chemical content and external fields. In some primitive cells the shape is *apparently* managed by the construction of a relatively undistortable, that is, semi-rigid, outer wall formed by peptidyl cross-linked polymers, for example in Gram-negative bacteria (Fig. 10.6). These walls do not communicate in an obvious manner directly to the inside of the cell but appear to act merely as barriers. The inner membrane then presses toward this outer wall and maintains a positive osmotic pressure upon it. It is this (turgor) pressure that is controlled so that pressure becomes related to the uptake of osmoregulating ions or molecules. (A not very different system appears to exist in plants, which also have strong (cellulose) walls, and even to some degree in shellfish which have a fixed external mineral shell (Chapter 13).) Perhaps one may compare a balloon (Section 6.3.4).

By itself this simple account is unsatisfactory, however, since the cell grows during the cell cycle and the wall is extended with it. During growth there is an accumulation of small molecules and proteins so that pressure is always being applied to the wall. There then has to be a control device that synthesises wall materials according to the magnitude of this internal pressure. All the components of a cell then grow together following an established pattern so long as chemicals are produced in fixed ratios. This feature is common to very many organisms. We then conclude that shape is not independent of composition and environmental conditions, which is exactly the conclusion we reached in Chapter 6 concerning the shape of crystals but there we referred to thermodynamic limitations. Here we are saying that the shape is a consequence of linked feedback controls on synthesis both of internal and wall material.

There is then the question of the origin of a shape other than that of a sphere for any cell. A vesicle formed like a soap bubble is spherical (see Section 6.2.2). The fact that no known cell no matter how primitive is spherical correlates with the fact that cells are growing towards division. The most primitive known shape of a cell, a closed cylinder, is clearly selected to favour controlled division and to avoid the random break-up of a swelling spherical, balloon-like cell. We can only assume at present that the appearance of a cylindrical cell depends on the way in which the components of the membrane give rise to a 'phase' separation in which some compartments favour high curvature and others favour a flat surface. As is conventional, we have to state that this functionally valuable shape arose through trial and error. Survival was then improved with structured division which is precisely timed.

Once the shape was no longer controlled by the outer wall but by permanent filaments running both under the flexible membrane and perpendicular to it and to some central nuclear axis (Fig. 7.29), shape could develop according to new rules since tension could be distributed anisotropically and locally by application of energy to the elastic filaments. Shape was then subject to modes of energy supply to filaments which themselves were controlled. It appears that the main mode of energy supply has always been through ATP, thus connecting shape to acid–base reactions. Now, the necessary placing of ATPases along filaments for this catalytic activity could be used to manipulate shape, provided a signalling system was evolved to indicate a 'desirable' shape change once a rigid wall was abandoned. A well-known example is pseudopod formation when an amoeba cell with a soft outer membrane attaches itself to a surface. The first shape change is at the point of contact since the membrane curvature changes (Section 6.2.2). This causes a change in the properties of the membrane and in itself will relay a message to the nearby filaments so that the external spatial anisotropy of the environment is reflected in the internal shape anisotropy of the cell. The change in shape acts as a mechanical switch or could cause a field switch if ions are redistributed along surfaces. The further response needs ATP to amplify it. The change in cell contact could also be used to open channels in the membrane locally so that external chemicals could enter and trigger local ATPases through proteins and then go on to trigger metabolic changes. Thus shape comes to have a relationship to the local phosphate potential [ATP]/[ADP][P] and thence to the Na^+, K^+, Cl^-, etc. ionic components of turgor pressure, and to various metabolic activities. In principle, cell colonies as well as single cells can derive shape from such a set of activities, and this process would be aided once connections between cells evolved (see Section 12.10). We mention a well-known later example of such flexibility of cells: red blood cells pass through capillaries smaller than the size of an isolated cell by adjusting their cylindrical radius using energy to contract anisotropically when a contact signal enters the cell. (Note we do not know when cells evolved to be flexible but in this paragraph shape (a physical property) is related to *environment* much as are chemical properties and needs messages from cell *outer surfaces*.)

11.7.1 Fields, flow and shape

The construction of anisotropic shape immediately allows construction of interactive fields since the curvature changes can carry charge differences. Once shape evolved, channels and pumps could also be put in different places since proteins can move about within membranes and seek minimum energy opposite curvature. Pumps and channels could then see to it that inorganic ions of selected elements would circulate. The ends of filamentous cells have currents associated with them (Fig. 11.18) and these current patterns as well as the differentiated cell surface could be recognised, thus allowing colonies of cells to build up, but now beginning to use *external* solutions.

We have frequently stated that the generation of new compartments has an

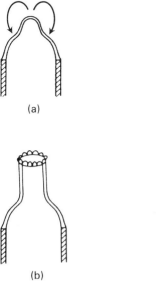
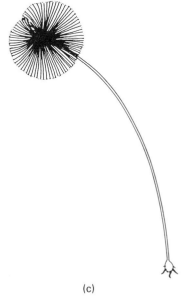

Fig. 11.18 (a) The growth currents around the tip of an acetabularia cell which are part of the relationship between cell morphogenesis and element functions. The currents are believed to be part of the guidance system that leads to the final shape by attracting certain proteins to locate in particular regions leading to (b) and then (c).

M^+ M^+

(a)

(b)

(c)

advantage in evolution, but so has the anisotropic distribution of activities along membranes and filaments (Section 7.11). Step by step we are building an immensely complex *organisation* associated with co-operative *structured* parts but also with co-operative flow of information, energy and material (Fig. 11.2). During the increase in complexity, element selectivity must have increased in accord with the capability of the element to act in a new function. Given feedback controls, this leads to a greater and greater refinement of the analytical concentration of elements in cells, so that all the time complexity of organisation is increasing interactively with the natural selection of chemical elements.

What is being proposed here is that after an initial cylindrical cell developed an early evolutionary advance was the internal structuring of anaerobic cellular systems through filamentous proteins so that cells had particular shapes. The shapes were consistent with positioning of a variety of activities in regions of different curvature. New activities included:

(1) motility with a 'motor' at one end or at the edges of the cell (Fig. 11.19);

(2) sensing with devices often at the opposite end from the motor. (Motility with sensing allows searching for nutrients.);

(3) disposition of feeding (entry) and waste (exit) systems;

(4) energy capture by localised ion circuits, possibly in vesicles.

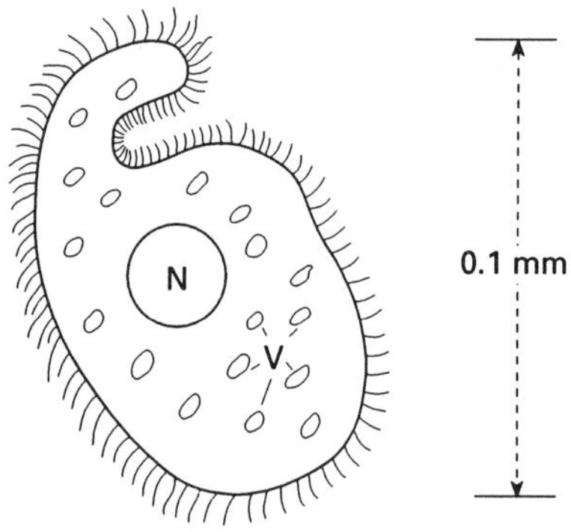

These developments bring advantages for survival of the chemical system but reduce the replication rate due to increased complexity and inevitably increasing size. Hence, it becomes advantageous to reduce complications, for example by not attempting certain syntheses basic to all life. Instead, the more complex systems became dependent on other simpler organisms for basic chemicals. By allowing very primitive organisms alone to carry out the very primitive metabolic activities, more sophisticated organisms had to feed upon them so devoting their lifestyle to more and more sophisticated actions including searching. Life's progression depends deeply on such division of responsibility making higher living organisms dependent on very primitive systems while providing protection for the same lower organisms, say in symbiosis. Only the most primitive organisms can survive independently. Symbiosis in ecosystems had to become a major part of the natural selection of chemical elements if complexity was to develop (see Chapter 13). Automatically, as shape, separation of function and symbiosis developed, there arose a requirement for more sophisticated communication (controls). (Notice that finally complexity cannot develop for isolated cells but is the resultant of an ecosystem. We shall see this throughout evolution. Complex organisms evolved with bacteria which themselves evolved (but not far).) One obvious early possible development with these changes was the use of the Na^+, Cl^-, and Ca^{2+} gradients (see Fig. 11.32), in messages from the outside.

Fig. 11.19 An anaerobic ciliated protozoa with a nucleus (N) and many small vesicles (V). The protozoa can digest bacteria and they are able to swim in a directed way. The nucleus came later (Section 11.12).

11.7.2 Early mineralisation

The greatest problem in the description of the earliest organisms is the absence of deposits of minerals that we can with certainty state to be of biological origin. The minerals that might be expected are amorphous silica, calcium carbonate and iron oxides. There are deposits of such minerals in various parts of the world which are claimed as evidence for life billions of years ago. Generally speaking, however, anaerobes (primitive life forms) do not have mineralised shells. We shall return to mineralisation in Section 12.12 where we shall discuss the huge increase in mineralisation after 7 or 8 hundred million years ago.† (See references 18 and 19 in 'Further reading'.)

11.8 Regulation: introduction

The description of the feedback mechanisms so far outlined did not include those that limit the concentrations of the polymers, that is proteins, RNA and even DNA, within regulation. Now it is DNA that is today the centre of all management of polymers since it is the masterplan. A reproducible plan gives fidelity to reproduction but also gives sophistication to the protein and RNA balances in cells since only certain sequences are allowed to be made. The reproduction of DNA itself is controlled by the cell cycle, but we must see that, for much of the time in a cell's activity, it is the steady *levels* of RNA and proteins produced from a fixed DNA that are of interest. The protein levels required will depend on the material available for synthesis, of course, but ultimately these *levels* are maintained by the cell's regulation. Production of any protein (or RNA) must also have a feedback system so as to prevent deficient or excessive concentration. In effect, a part of DNA becomes hidden by messenger-protein binding or is revealed by protein dissociation from it (Fig. 11.20). This can be achieved either by using a concentration of a specific

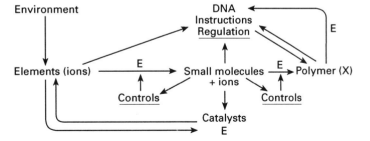

Fig. 11.20 The polymers (X) acting on DNA are transcription factors activated frequently by small molecules or ions with a wide variety of time constants. E is an enzyme. Instructions are passed to a polymer (X) or directly to DNA.

protein, which itself is reversibly bound to DNA, as an on/off signal (see Fig. 11.3(a)) or by a set of on/off signals from the metabolic or uptake events to a DNA-bound signal protein. The latter signals can be due to an up/down change of an ion concentration, such as that of H^+, HPO_4^{2-} (covalently linked) or Fe^{2+}, or of variations of concentrations of complex units such as haem, glutathione and so on, or even of some important substrate, for example galactose. Such signal proteins are called *transcription factors*. There is both acid–base- and redox-related regulation (see Table 11.6) and both are generally slower than control.

† Our lack of knowledge of the ocean water in which the earliest life appeared leaves us with deep problems in the description of especially early calcium biochemistry. If the ocean was acidic then $CaCO_3$ would have been soluble at pH = 5 although its solubility would have been depressed if at the same time CO_2 was of 10^3 times higher pressure than it is today. It may be that changes in Ca^{2+} ions in water profoundly affected early evolution. Again at pH = 5 molybdenum is insoluble in the presence of H_2S but vanadium and tungsten may have been sufficiently available to have been preferred.

The general scheme is illustrated first for the case of some proteins related to the acid–base reactions of phosphate metabolism. The choice of phosphate is again due to the simplicity of its chemistry and its general value in biology. It may well be the case that phosphate has always been, and still is, the dominant signalling element in regulation. This regulation is known to be present in both prokaryotes and eukaryotes, and to be similar in them in many chemical respects, and to be related to energy (ATP).

11.8.1 Phosphate regulation of genes

Phosphate is the only form of phosphorus of interest in biology since it shows no redox chemistry in biological aqueous solutions, although it has many combined forms. Immediately, this lack of redox chemistry makes phosphate regulation easier to discuss than that due to carbon, nitrogen or sulphur compounds. It has been found that a vast range of protein and enzyme production is limited by phosphorylation-dependent regulation, and the obvious conclusion is that this is part of the need to relate cell activity to energy, that is, the phosphate potential [ATP]/[ADP][P], and to the concentration of the nucleotide bases, for example ATP. The following scheme illustrates the process:

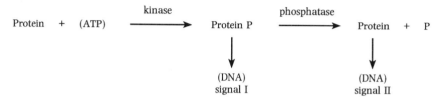

where P is simple inorganic phosphate. The protein in either the phosphorylated or in the dephosphorylated condition may bind to and 'instruct' the DNA, causing synthesis or stopping it, and the resting state may contain more or less of the phosphorylated protein (Fig. 11.21). This regulation relates directly to the metabolic and uptake *control* systems making and using ATP discussed above, and it has selectivity opposite different reaction paths which use different kinases and different phosphatases. The phosphorus regulation system is complex since it requires expression of two extra proteins, a kinase and/or a phosphatase, apart from energy capture machinery to make ATP. These proteins are *enzymes* and therefore are themselves open to control related to particular metabolic paths (Fig. 11.21).

Before commenting further on such systems we look at a second regulatory role of phosphate in cyclic esters, for example

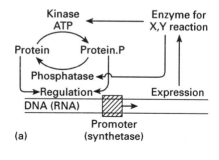

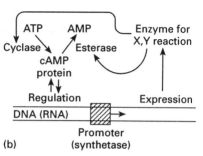

Fig. 11.21 The two modes at which ATP levels can regulate expression: (a) via phosphorylation and (b) cAMP, both of which are linked to a message from X, Y.

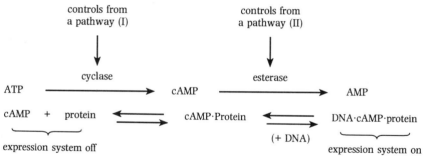

The binding of cAMP to a protein is a simple equilibrium but the system is again complex in that it requires production of a different set of proteins, cyclases and esterases, and both are influenced by metabolic controls. There are now two ways of handling the binding of phosphates in phosphorus regulation. The two are quite separate in the sense that protein-bound P and cAMP do not exchange P rapidly, that is, they are separate components, and are produced by different enzymes under selective controls, but their feedback links to ATP will also assure that they are in communication. In both cases it is necessary to have both catalysed activation and deactivation of phosphate compounds to reach a steady state. Why should there be two ways of allowing the element P to talk to DNA, thus controlling expression of RNA and through it of proteins?

First note that energised P is linked to a wide range of activities via nucleotide triphosphates, NTP (ATP, GTP, CTP, UTP and so on), and to DNA and RNA themselves as well as to many lipids and coenzymes. The level of ATP itself is specifically connected to protein kinases and a vast number of metabolic responses, while GTP, CTP and UTP are connected to separate systems of protein synthesis, glycoside synthesis and so on. ATP is also linked to many pumps for Na^+, K^+ Ca^{2+}, Mg^{2+} (Ni^{2+}), H^+ and so forth. It is therefore the [ATP]/[ADP][P] level that controls a whole range of activity and later it became the first form of energy transduction from photo or oxidative phosphorylation. It is not surprising then that it provides regulation of genes for the production of the catalysts involved with many pathways. (There is no point in producing enzymes if the elements in free or combined forms are not available in the cytoplasm. Hence regulation must be complex in its connection to the whole cell content.) That it does so in two ways

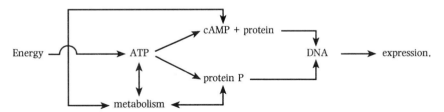

allows an extra degree of variety since P binding and structure in Protein P is H^+ dependent while cAMP binding is not. Both are Mg^{2+}-dependent for their formation via MgATP. Through the above-mentioned equilibrium binding all cAMP-dependent systems respond simultaneously to cAMP concentration changes with relatively fixed rates; while the phosphorylation mechanism is specific to particular transcription factors, a protein kinase has a specific substrate.

The main feature of this phosphate regulation is that syntheses of groups of proteins are under the management of transcription factors which themselves are proteins which are also controlled by specific small molecules. The activation of the protein synthesis then takes place under independent but connected links that are related to different sources of material, food. Since different food materials are in different pathways we can see that where the H/C/O food source is a sugar one set of enzymes is needed, while where it is an amino acid or a fat quite another set is required. The proteins need to be inducible when needed or repressed when not required. One scheme is:

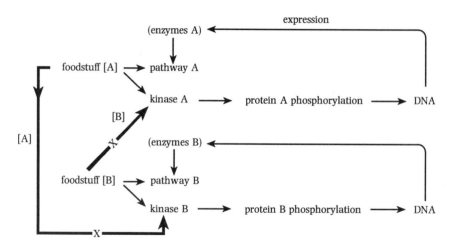

Anaerobic acid–base regulation

C, N, O, H
substrates
↓↑
(fast) H, P (slower)
↓↑
Mg, (K) (fast)

Here X represents repression by one foodstuff (or its product) over the pathway of another. In place of kinase B we could place a system for production of signalling. Thus, pathways have regulatory management through production of catalysts as well as feedback control. We see that a cell that can change its expression is inducible, that is, it can differentiate, so as to perform different tasks. This flexibility of DNA expression is of extreme value, and it will allow one cell to generate a variety of cell type activities in primitive life, and of cell types in multicellular life from the same DNA. These differentiated cell types will be composed of different elements, but we may not conclude that this variety is continuously variable as we shall see in Chapter 16. Now, it is not just phosphate amongst elements that regulates expression, much though phosphate is dominant for good chemical reasons, see marginal note.

11.8.2 Other examples of regulation by non-metal elements

The derivatives of phosphate provide one example of the way in which non-metal covalent reactions can be used in regulation. Other examples of the use of non-metal elements are given in Table 11.6. Carbon, like phosphorus, can be used in more than one way in different processes—methylation, glycosylation and acetylation of DNA—while carbon and nitrogen can be used in the attachment of N-acetyl glucosamine. This brings us to a general point concerning covalent regulation of the expression of RNA and then proteins.

There are three ways in which regulation messages can be managed chemically:

(1) direct attack on DNA or RNA bases (methylation for example);

(2) removal or introduction of a covering molecule on DNA and which affected reading, by reversible ion-binding;

(3) modification of a protein transcription factor by covalent binding of a small unit, for example, phosphate or an ion with which it is in moderately slow exchange.

These different regulation modes have very different kinetics of exchange of the groups introducing the message. Thus methylation/demethylation is slow, of the order of days, phosphorylation/dephosphorylation of a protein factor followed by association with or dissociation from DNA may take hours, but ion exchange can be much faster. The distinction between processes based on time is similar to that for control. There are then ways of adjusting the activity of a cell on a short-term basis or ways of long-term modification (differentiation). In the limit, differentiation could be of longer term than cell turnover so that particular cells reproduced differentiated cells.

Non-metals and some metals can also regulate in more complex processes. In the case of a C/H/O or a C/H/N/O compound it binds highly selectively to a transcription factor switching on or off a gene or usually a series of genes that are the production units of proteins (enzymes) which are themselves the catalysts for the pathway of transformation or uptake of the signalling molecule. The *lac* (lactose) series of genes, that is an operon, is of this kind (Fig. 11.22) and haem, *cyt*, genes may also be of a similar nature.

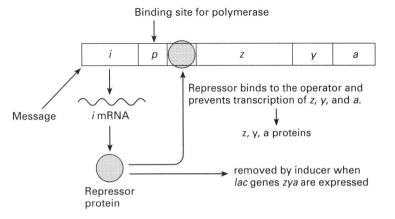

Fig. 11.22 Diagram of the lactose operon. As in the case of Fe uptake, the expressed proteins z, y, a belong to uptake and catalytic systems for lactose. Note *operon* is the name of a co-ordinated unit of gene expression.

11.8.3 Regulation by metal cations

The simplest form of regulation by cations is that in which a specific protein for binding the cation, a transcription factor, binds to the DNA in the metal-bound form but differently or not at all in the metal-free form. This is a simple case of complex ion equilibria regulating expression. A well known example (Fig. 11.23) is

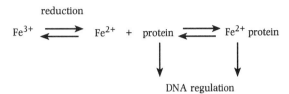

where the free protein instructs the DNA to make scavengers for Fe^{3+} (Figs 11.23 and 12.18), while the Fe^{2+} protein instructs the DNA to make enzymes for general metabolism.

It is not yet clear how many of the simple ions work directly in this fashion in primitive cells, that is, on proteins that bind DNA, but at least Mn^{2+} and Fe^{2+} are so involved while H^+ and Mg^{2+} may well be or may affect the phosphorylation steady state instead so that they generate changes in DNA-

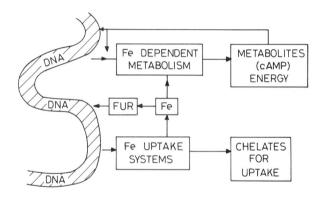

Fig. 11.23 The regulation of iron uptake by iron concentration. FUR is the *Ferric Uptake Regulatory* protein which binds differentially in the iron-bound and iron-free states to DNA. (This particular example is from an aerobic organism but the principle is general.) See Fig. 11.12.

(a)

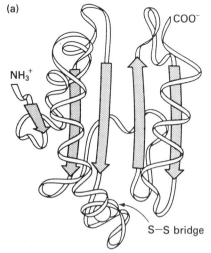

Thioredoxin $\overset{S}{\underset{S}{|}}$ +NADPH+H$^+$ ⟶

thioredoxin $\overset{SH}{\underset{SH}{<}}$ +NADP$^+$

(b)

HO$_2$C–CH$_2$–NH–CO–CH–NH–CO–CH$_2$–CH$_2$–CH–CO$_2$H
$\qquad\qquad\qquad\qquad\qquad\qquad\qquad\qquad$ NH$_2$

Fig. 11.24 (a) The structure and reaction of thioredoxin. (After A. Holmgren and C.-I. Branden.) (b) Glutathione.

Anaerobic redox regulation

C, N, O, H
substrates
↓↑
e, H (fast)
↓↑
Fe, S (slower)

binding proteins through their influence on phosphorylation rather than by binding to such proteins directly. Note that H$^+$, Mg^{2+}, Na$^+$, K$^+$ and Cl$^-$ all bind rapidly in ionic exchange, while the other elements are usually retained for long periods. Later in evolution Zn^{2+} and Cu$^+$ act in the same way (Chapter 13).

11.8.4 Redox regulation of protein production: sulphur and iron

The major forms of sulphur in biology are related to reduced H$_2$S (S^{2-} and RS$^-$), or oxidised —S—S— (sulphur bridges) forms, and later in aerobes to SO$_4^{2-}$. The main primitive regulation of DNA by redox reactions via sulphur chemistry that does not involve a second element uses the reaction

$$2RS^- \underset{\text{reductase}}{\rightleftharpoons} R\text{—}S\text{—}S\text{—}R,$$

where either the reduced or oxidised form may bind to DNA. The major signalling system of this kind uses the protein thioredoxin (Fig. 11.24(a)) or possibly the small molecule glutathione (Fig. 11.24(b)), which binds to a protein transcription factor. A quite separate sulphur system is associated with metal ions. Thus, it is known that the combination of iron with sulphide to give Fe/S clusters (Fig. 10.8) that can dissociate from proteins, is used in regulation as well as free Fe^{2+}, thus also connecting the two major redox centres in cells—Fe and S. The exact nature of this regulation is not known but involves dissociation/association of one or other of the units Fe$_4$S$_4$, Fe^{2+} and S^{2-} with a protein or even a redox change of state of any one of them within a protein. In fact this Fe/S, Fe and S regulation may belong to extremely primitive life that has persisted to this day in modified form, see marginal note.

In this outline of regulation we have tried to show that there is a network of management cross-related through feedback through proteins to the levels of effectively all the elements (and hence components) in the cell. Expression is dependent on the environment giving rise to differentiation.

11.8.5 Control plus regulation—an example: nitrogen fixation

Before turning to its management, the fundamental importance of the nitrogenase enzyme for present-day life must be stressed. Nitrogen is acquired from ammonia, going through intermediates such as glutamate, for the synthesis of all proteins and all polynucleotides (see Fig. 10.16). Nitrogen (N) is as essential as C, H and O for polymer construction—even for the headgroups of lipids and the side-chains of many polysaccharides. Thus it might well be expected that it is controlled and regulated by virtually every element in the cell. It is an anaerobic incorporation process.

We can summarise the conjoint nature of control at the level of metabolism and of regulation of the gene by considering the fixation of nitrogen, for example in the bacteria *Klebsiella pneumoniae*. The metabolic pathway using nitrogenase is shown in Fig. 11.25 giving, from molecular nitrogen and reducing equivalents, the product ammonia. The reducing equivalents come from pyruvate, which is central to carbon metabolism through several pathways (Section 11.5.2). The production of ammonia is in itself of little value unless it is quickly taken up into carbon compounds. In fact, NH_3 is a poison at quite low levels. The controls on the nitrogenase enzyme cannot be over the freely diffusing molecule N_2, but can be through the supply of electrons by restricting diffusion from the reductase, or through the level of ATP which promotes reaction, while ADP inhibits it. Feedback control is also established by the carrier of the element N ($-NH_2$) glutamine (an amino acid of proteins). Thus the control rests in effect in: (1) the phosphate potential, as is usual; energy is required to drive electrons to the N_2 site at a low potential: (2) the carriers of $-NH_2$ and C/H units. Both energy control and substrate control are used, so that, in the absence of ATP or in the absence of excess substrate or reducing equivalents, electrons do not leave the iron protein in the step before the iron/molybdenum protein (Fig. 11.25). The substrate level control is also via the carrier, coenzyme A, the absence of which can block electron transfer from the flavoprotein (Fig. 11.25). Free coenzyme A concentration again reflects the

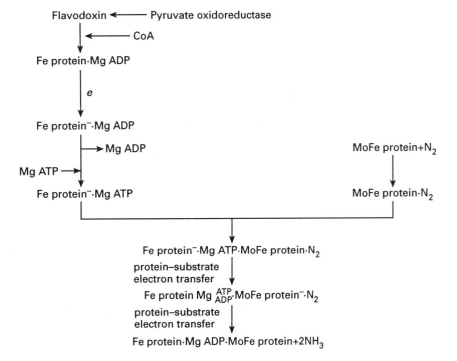

Fig. 11.25 The pathway of electrons and the use of energy ATP in the synthesis of ammonia from dinitrogen.

levels of free carbon/hydrogen substrates (see glycolysis cycles (Fig. 11.13)). Thus in the *control* of nitrogen fixation there are very wide links to energy through a phosphate carrier, ATP (plus Mg^{2+} and H^+), to carbon and to hydrogen (reducing equivalents) via a coenzyme carbon-fragment carrier, CoA, and to nitrogen metabolism via glutamine. Finally on the control side there is also the uptake of the metals Fe and Mo, needed for synthesis of the enzyme. Both uptakes are in turn linked to ATP levels via cAMP. The iron level is also critical in that it feeds forward to energy production via many reaction paths which produce ATP. Thus at least eight elements *control* N_2 fixation.

The *feedback* network of *regulation* of nitrogenase is shown in Fig. 11.26. Starting from the genes on the left we see that the gene products from *F* of the major reaction system for nitrogen conversion to ammonia (and then to glutamine (Gln)) are regulated by a negative feedback for the whole gene series through the genes *A* and *L* (genes are written in italics). Protein transcription factors bind glutamine. There is also a positive ATP regulation that manages a phosphorylating protein acting at *L*. These genes regulate the synthesis of all the proteins and enzymes of nitrogenase itself by preventing the polymerisation of RNA of the nitrogen-fixing genes. Some of these genes relate to only one enzyme as shown in Fig. 11.26, but others relate to a series of enzymes, for example, those required in the synthesis of the FeMoCo cofactor.

This extremely complicated set of controls and regulations ensures that the elements C/H (of substrates), N (of N_2, glutamine, nucleotides (and proteins)), P (of ATP and hence of energy), O (for energy in dioxygen) and linked to protective devices) and several metal ion levels are in co-operative feedback communication. It is the feedback that helps to decide the nature of the organism. For example, the feedback in nitrogenases in *Azobacter* extends to a complex communication network between Mo, V and Fe since it can use three different nitrogenases, but they are switched on or off according to the supply

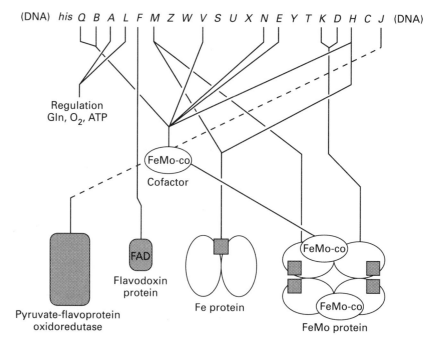

Fig. 11.26 *Nif* genes required for nitrogen fixation as arranged in *K. pneumoniae* and their respective gene products. The regulation by product, glutamine (Gln), (dioxygen) and ATP (energy) is shown. This genetic structure for O_2 became fully necessary only after its advent. All genes except *A* and *L* generate RNA for the production of proteins directly or for the production of a synthesis protein unit for FeMo-co.

level of the different metals. This example shows the extreme complexity of one vital (to all life) metabolic step ($N_2 \rightarrow NH_3$) in a *simple* bacterium. As other kinds of complexity developed, for example in plants and animals, survival fitness led to abandonment of this basic nitrogen metabolism altogether and to a reliance on symbiosis.

11.9 The different pathways of polymer synthesis and degradation

Clearly, the pathway of degradation of polymers of all kinds is not energy-requiring since the products are more stable than the reactants. Thus, the enzymes of hydrolysis, phosphatases, proteases, saccharases and so on, do not use nucleotide triphosphates or diphosphates but the synthesis of these polymers requires energy-giving intermediates in polymerases. The forward and backward paths are then

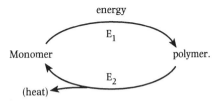

The enzymes, E_1 and E_2, are quite distinct and their syntheses are regulated by opposite excesses of substrate and product. In glycolysis energy is obtained but the synthesis of glucose and then of glycogen, a polymer, requires energy. The different pathways and controls are here placed at the beginning of the reverse path using pyruvate that is (see Fig. 11.13),

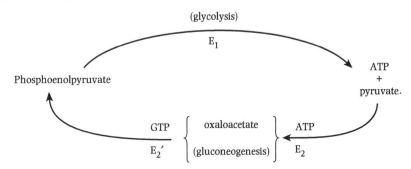

11.9.1 Cross-talk in regulation

The simple description of regulation given above does not account for the complex behaviour observed even in the cells that we are calling primitive, that is, anaerobic cells with a single internal compartment. It is clear that distant parts of DNA can be affected by the binding of a regulatory factor at a particular place so we have to assume that DNA bends back on itself in selected ways. Of course, this removes the idea that DNA is a simple linear message. Again many genes appear to be under multisignal regulation when there has to be some kind of cooperativity along the DNA which is not consistent with a simple linear run of information. Finally, reading of a gene message requires special reading enzymes, DNA transcriptase and RNA translatory machines. These machines are open to control so that message reading is complicated. The RNA-reading machine requires energy but it usually needs zinc. We shall observe time and time again that zinc plays a central role in biology, but in the most primitive systems it would appear to be entirely secondary to iron. Iron dominates very much of early biology acting directly as a cation, in Fe/S complexes and in haem as a messenger in both control and regulation.

11.10 The integration of element chemistry in control and regulation: summary

It is essential that the reader now recognises that the description of the natural selection of chemical elements described in this chapter has taken a large step forward from that in Chapters 8–10. In those chapters on abiotic chemistry and on biotic pathways we could only describe the selective distribution in different 'phases' (compartments) and the combination of elements in compounds either under equilibrium conditions or in kinetic traps. The arrival in kinetic traps could have been either through loss of energy, as indeed was the case for arrival on the way toward an equilibrium condition, *or* through *energisation* into the trap. (Remember that each element or molecule that does not exchange elements is treated as a separate component in a trap independent from all others.) Moreover, we were not concerned with the long-time turnover of these components. A large variance in chemical components and compartments was the result—almost chaos. The step forward in this chapter is the consideration that in a living system, that is, in biological chemistry, there is co-operative *kinetic interrelationship* between all the compartments, elements and compounds (components) in any one cell. Once the self-assembled DNA or code had been defined then the conditioned state of the cell was almost precisely reproducible, but only if given the above regulations and controls. It is the case, as shown in Fig. 11.27, that all the

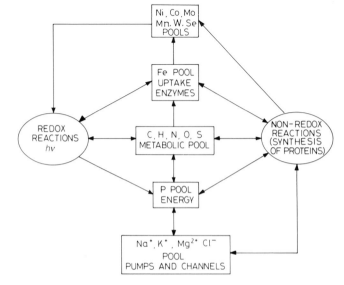

Fig. 11.27 A simple view of the necessary interaction between a minimum of 15 elements to maintain a primitive cell.

different activities of such a cell are tied to one another, which relates to the energy and material uptake and loss required to establish this steady state in the cell. There has evolved an active, dynamic, condition of flow (of energy and material) that can be *apparently* stationary (or, as we shall see, in a continuous series of changes in progression) and that is virtually chemically and physically unique much as one would describe an equilibrium phase of many components and of a very narrow composition range. Thus, the advance described in this chapter is to a fixed *dynamic co-operativity* (of activity) in the natural selection of elements.

Now, long-term survival of an energised condition is only achievable if it can be reproduced. We shall now analyse such survival in some further detail.

11.11 The cell cycle: introduction

As far as the simplest cells are concerned, development is reproduction, but, of course, division had to be timed not just to reproduce DNA but to generate a total cell complement that would allow two daughter cells to survive. We have described the regulatory and control systems that could lead to a balanced ever-increasing complement of all that is required and that, at its simplest level, can be analytically reduced to fixed relative amounts of 15–20 elements in a great variety of compounds and as free ions. (A cell can also differentiate, that is express particular genes under particular circumstances, as we have described.) We turn to the cell cycle as seen today which must allow the analytical content of a cell to build before DNA reproduction and before division (Fig. 11.28). Obviously, a new control or regulation, which is a *timed sequence* of events and cannot be reversed is required.

Prokaryote cells grow and divide when they have an adequate supply of nutrients. In other words, their life cycle is dependent on a surplus of chemical elements and energy and nothing more. In the absence of such conditions the cells remain in a resting phase in which there is use of stored chemicals to maintain a low residual steady-state activity. Deprivation of some of the 15–20 critical elements, for example Mg or P, over long periods of time, even at rest, leads to cell degradation and death.

The cell cycle, which is common to all cells to a good approximation, is

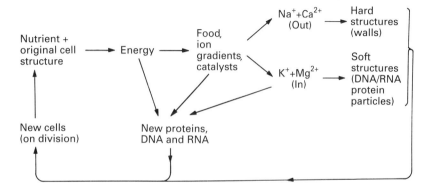

Fig. 11.28 A diagram showing the partitioning of selected elements inside and outside a cell, all of which are essential for the cell's stability.

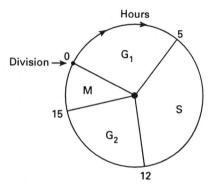

Fig. 11.29 The cell cycle G_1, basic proteins duplicated (4–5 hours); S, synthesis of nucleotides and DNA synthesis started (7–10 hours); G_2, machinery of division made, DNA synthesis finished (3–5 hours); M, mitosis to division (1–2 hours).

shown in Fig. 11.29. There is an initial resting state: leaving this resting state toward division, the cell first passes through a preparative stage in which monomers, some polymers and catalysts (the machinery for metabolism) are built up. This is called the G_1 condition. There follows the synthesis stage, S, in which DNA is replicated. The next stage, G_2, builds the machinery for division, which itself is a separate stage, M (mitosis). The stages are distinct and a cell can be brought to a stop at different stages using different drugs indicating that different management structures are operative during each different phase.

For such a process as cell duplication to be reproducible strong regulation and control must exist. In large part, the proteins synthesised at a given stage are under the regulation of the special signalling proteins, cyclins, which have timed periods of existence. Thus we can consider that these cyclins regulate expression of other proteins by binding to DNA either directly or co-operatively with other proteins. As we stressed above, the cell cycle goes to the G_1 condition only in the presence of adequate element and energy supplies. It is considered that a dominant signal is again the phosphate potential of a cell

$$\text{Phosphate potential} = \frac{[\text{ATP}]}{[\text{ADP}][\text{P}]},$$

which enables a kinase to phosphorylate specific regulatory proteins, here the so-called cyclins, and which will ensure the controlled uptake and incorporation of 15–20 elements. Phosphatases later remove the phosphate of course.

At some stage, and due no doubt in part to the accumulation of small molecule intermediates and new proteins, it is readily conceivable that these intermediates, again assisted by the appropriate phosphate potential, switch off the above kinases using phosphatases, and generate a new protein kinase, that is, a new phosphorylation system that can selectively drive the cell into and through the S phase. In fact, we can only see one way in which to drive timed sequences of cyclic DNA activity, namely, by timed sequences of production and then destruction of regulatory proteins, finishing with the mechanical actions of spindle formation and division, that is mitosis.

In the description of biological activity, here the cell cycle, we wish to stress that, even though we can analyse events as being under the influence of specific switches (here cyclins, kinases and phosphatases, which geneticists stress), there is an underlying need—the dependence on the phosphate potential. This potential is itself directly dependent on H^+ and Mg^{2+} ion concentrations and indirectly dependent on the elements that assist in transduction of external energy to ATP, that is, it requires such elements as Mn, Fe and so on. There is no adequate description of a cell cycle that does not include the functioning of these elements or that does not include the need for their controlled uptake and incorporation (Chapter 10) to allow division to create two daughter cells.

11.12 Developmental evolution of anaerobic cells

We have now concluded the description of simple anaerobic cells with no internal vesicles and no permanent filamentous structures, which is the earliest form of life we know of. The description is based on existing organisms, but today these organisms must be chemically far advanced relative to the original form of life even though we have used them as models. We have observed that in no way could anything resembling such living systems exist without the 15–20 chemical elements of Table 11.7. The next step forward we shall take is to go from the description of one-compartment single cells to cells in which there are several compartments with a connecting internal set of filaments and to cells that depend on other cells for a complex lifestyle. We remain concerned with strictly anaerobic organisms. Their limitations appear

Table 11.7 Functional value of the elements*

Osmotic controls: information (electrical) transfer and store	Na^+, K^+, Cl^-, H^+, Mg^{2+}, Ca^{2+}, HPO_4^{2-}
Chemical–mechanical transmission	Mg^{2+}, Ca^{2+}, HPO_4^{2-}
Acid–base catalysis	Non-metals and divalent and trivalent ions, e.g. (Zn^{2+}), Fe^{3+}, H^+, N, S
Redox catalysis	Transition metal ions and some non-metals, e.g. Fe, Mn, Co, Mo, W, Se, S
Structural role (excluding the organic polymers)	Si, B, P, S, Ca, Mg, (Zn), (Ba), (Sr)
Chemical energy transmission and storage	P, S, (C)

* The functional value of the elements arises partly from the chemistry and partly from the energy input to that chemistry. The energy input can be to the gradient of the free concentration of an element, to a chemical bond of given kinetic stability or to the synthesis of a polymer in a compartment that acts as a trap.

to be that no truly multicellular systems could arise, but there could be and was great diversity of species in that some organisms developed so as to use light while remaining *small* prokaryotes (100 μm) and even to generate dioxygen as a biproduct, a poison for all strict anaerobes. It must also have happened that, early in evolution, a membrane developed that specialised in the transduction of light energy to chemical energy. An example is seen in the presence of thylakoids within chloroplasts (Fig. 11.30), for example in cyanobacteria. There are clearly two branches of evolution, one using light and the other not so doing, apart from the archaebacteria (Fig. 10.1). (We observe immediately that no archaebacteria use the chloroplast light-collecting system and that no *large* anaerobes are photosensitive except through compartments such as chloroplasts. Why is this so?) A quite separate evolutionary development led to large single-cell organisms which are eukaryotes. In essence, they enclose the nucleus and some other vesicle systems by membranes separate from the cytoplasmic membrane (Fig. 11.31). Some eukaryotes are anaerobic and will be described here. All aerobic systems are described in Chapters 12 and 13.

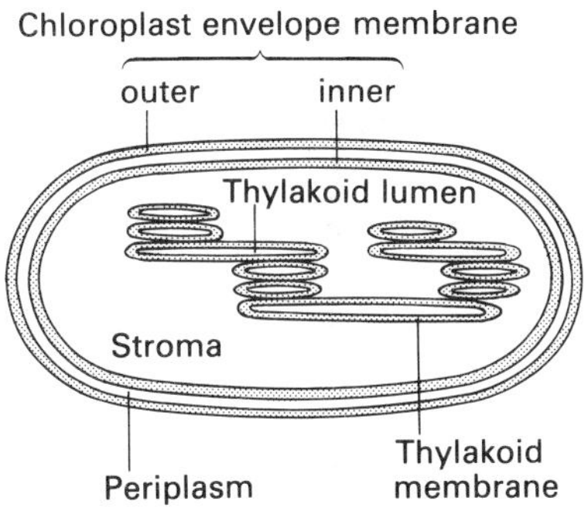

Fig. 11.30 Diagram of chloroplast, 5–10 μm long.

Table 11.8 Compartments possible in single anaerobic cells

	Membrane limited
Nucleus	Yes
Ribosome	No
Vesicles	Yes
Thylakoid	Yes
Vacuoles	Yes
Centrosome	No

11.12.1 Intracellular vesicles and their membranes

The change from a prokaryote to a eukaryote structure (Fig. 11.31) has an uncertain dating in evolution, but it may have taken place before the advent of dioxygen. In effect, a change in the internal structure such that nuclear events were separated from those in the cytoplasm is advantageous especially as the chromosomal structure increased in length and complexity. Clearly, the large nucleus must be protected. Examination of the nuclear membranes does not as

Fig. 11.31 (a) A schematic diagram of a eukaryote. (b) Typical prokaryotic and eukaryotic cells, based on electron microscopy. Not every prokaryote or eukaryote has every feature shown here. NB. The mitochondria were not added until the dioxygen level was considerable, that is, until aerobic conditions. (From *Five Kingdoms* by Margulis and Schwartz (1988). Copyright © W. H. Freeman and Company, New York. Reproduced with permission.)

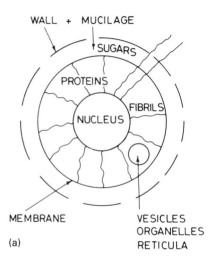

(a)

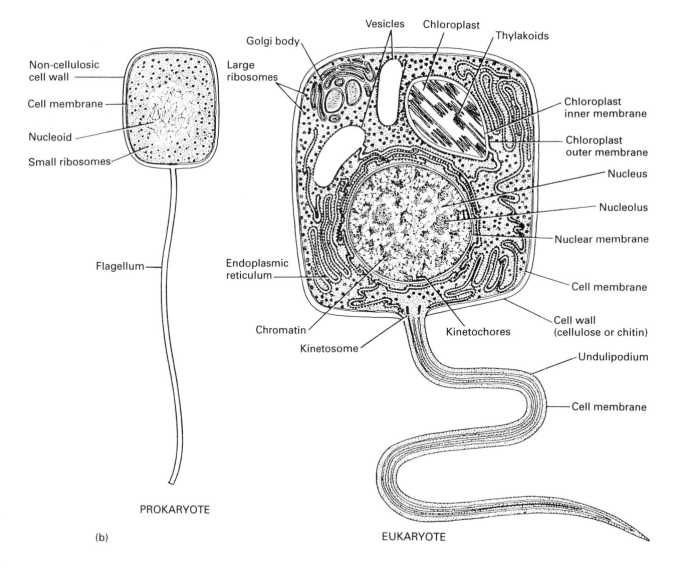

(b)

yet indicate any special chemical feature to associate it with a change in metabolism and this development is entirely consistent with the primary metabolic pathways. Similarly, the internal network of filaments, including tubulins, actinomyosins and microfilaments, is composed of proteins that are not modified by dioxygen reactions. Furthermore, these proteins, or ones closely related to them, were required for cell division in prokaryotes and for motility. Thus it may well have been that the subtle nature of the cell division process in prokaryotes developed into a family of eukaryotes without any major change in chemical pathways but based just on the advantages of dividing space. It was this development that allowed: (1) manipulation of the outer membrane separated from cell division; and (2) stabilisation and location of internal vesicles including the nuclear membrane. There is also little evidence that any novel di-oxygen dependent lipid was required for the making of the *vesicle* membranes.

Finally, there evolved a group of cytoplasmic membrane-associated proteins, which attach themselves to some vesicles and to filaments. There are, for example, the proteins related to the spectrin of the red blood cell (Fig. 7.29). Once again, they are proteins requiring only anaerobic metabolism for synthesis. Thus, the initial state of the eukaryote cell seemingly required no change in the chemical elements† used but only in the way in which they formed compounds and hence generated new compartments while the cell grew considerably in size (10–100 μm; see Figs 6.25 and 11.31).

(One piece of evidence that could count against this discussion is that the outer membranes of extant eukaryotes usually (perhaps always) contain alicyclic compounds that are partially oxidised, for example cholesterol. This molecule (marginal illustration) is synthesised by a path that requires dioxygen at a low partial pressure. The question we pose is, 'Did eukaryotes appear before cholesterol came into their membranes?' We shall *assume* that they did but that the membranes remained primitive without cholesterol and without arachidonic acid, both of which are synthesised using dioxygen (see Chapter 12).)

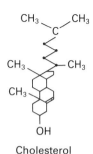

Cholesterol

11.12.2 Contents of internal vesicles

Once vesicles evolved, a major step forward providing new reaction compartments and boundaries, each aqueous compartment in them was kinetically distinct as was each vesicle membrane itself so that elements and compounds in eukaryote cells are a mosaic of chemical content (Fig. 11.31). The character of the vesicle membranes is much as it is for the cytoplasmic membrane in so far as there are zones differentiated by their curvature, but they do not have internal filaments. This curvature localises activities and hence must cause gradients of many kinds around and within all vesicles. A spherical vesicle is excluded from this categorisation. Complexity is increased by the filaments in the cell, which locate not only the vesicles but have built into them new zonings of the cell through the facts that they have directional properties linking them to the vesicles' membranes in different local spatial

† It is noteworthy that the one element which grows immensely in importance from prokaryotes to the most primitive eukaryotes (even anaerobic) is calcium. Now the calcium concentrations changed with the evolving ocean and the falling level of CO_2 in the atmosphere. Did these changes allow organisms to develop vesicles and contractile filaments?

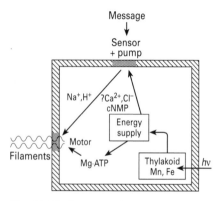

New element controls (vesicular)

HCNO
substrates

$\downarrow\uparrow$

H, P

$\uparrow\downarrow$

?   Mg, K   ?

Na, Cl, Ca
(Vesicle)

The release of Na, Cl or Ca from vesicles could act to control the cytoplasm.

Message
$\downarrow$
Sensor
+ pump

Na⁺,H⁺ ?Ca²⁺,Cl⁻
cNMP

Energy
supply

Filaments

Motor

Mg·ATP

Thylakoid
Mn, Fe

$h\nu$

Fig. 11.32 The most primitive connection between an external event (message) and mechanical motor reaction using energy. Compare the reactions of plants and animals in Chapter 13. Note the possible involvement of Ca^{2+}.

regions and that they too carry activities. In such complex self-assembled systems, which we and others have described in detail elsewhere, there can be no *simple* overriding chemical principles. The cell generates a multitude of compartments each of which is zoned in itself. This zoning was dramatically increased by the capture of organelles, chloroplasts, which in their original form were photosynthetic bacteria.

If these vesicles resemble for the most part the vesicles found in more recently evolved plant and animal cells, then their element content was and is more like that of the aqueous external solution, the sea, than of the cytoplasm. They contain Ca^{2+} and possibly are at a lower pH. Thus they could be used like today's lysosomes to degrade complex molecules to obtain food. Again these compartments could be employed, as we shall see in Chapter 13, to trigger locally internal events (by release of calcium) after external excitation, see marginal note. The triggering could then be connected to external filaments via motors (Fig. 11.32).

To summarise, what has been achieved by the compartmentalisation and zoning inside single cells is an evolution of chemical element and compound selection in zones. The nucleus was placed in a compartment free from protein synthesis, from highly acidic conditions, and from high calcium concentration while neighbouring compartments contain these features. The cell increased in size and in lifetime. It became able to ingest or digest smaller prokaryotes and, we believe, even to symbiotically incorporate them, retaining in part their functions. There developed 'cells' within cells, organelles such as chloroplasts (and later mitochondria) fitted into position by filaments (Fig. 11.31). The use of light to give energy via free-radical chemistry was therefore safely placed in a new compartment, organelle, the chloroplast. But, nevertheless, this is not without sacrifice of two kinds: (1) the increased period of time taken to reproduce; (2) the loss of flexibility and of differentiation in response to the environment. The net gains are a part of long-term physicochemical fitness in the natural selection of chemical elements but there is a problem looming all the while in this development. The cell became very complex and its control and regulation became much more complicated and time consuming. As in all organisms the cell began to need delegation of control and regulation so that parts are partially independent while contributing to the whole, which is the case for the chloroplast and the mitochondrion in that they retained much of their own DNA. There was another step possible which altogether abandoned complete self-reliance—symbiosis, which we consider next.

11.13 Symbiosis and differentiation

The gradual development of complexity and the sharing of tasks in different compartments could be extended from internal vesicles and organelles to external associations. In these associations chemical dependence on a second organism evolved. The usual situation, seen in Fig. 11.33, is for a large complex cell to be associated with smaller simpler cells, often bacteria. The function of the smaller cell is to synthesise very basic chemicals, for example, NH_3 or amino acids from N_2, while getting in return certain carbon compounds. Such association, symbiosis, requires a signalling system so that

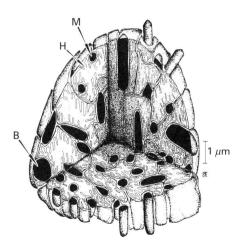

Fig. 11.33 A tripartite symbiotic consortium between methanogenic bacteria (M), eubacteria (B) and hydrogenosomes (H) recently discovered as a permanent intracellular complex in an anaerobic ciliate (*Cyclidium porcatum*). The whole complex is about 1/100th of a millimetre from top to bottom.

the several organisms, at least three in Fig. 11.33, can operate efficiently. In this way the beginnings of cell–cell communication can be considered to have been developed in anaerobes. Was calcium signalling introduced?

An alternative is for cells to form loose sponges of differentiated cells when, for example, outer cells take over the role of protection while inner cells digest and supply chemicals. The cells can dedifferentiate since the differences in genetic expression in any cell, for protection or digestion, are only under temporary regulation.

The symbiotic or the differentiation routes to increased cell survival are made by distributing chemical elements differently in different cells. Such complexity in organisation could have developed in stromalite-forming organisms and is certainly seen in sponges and fungi. We are approaching true multicellular life through cell–cell and even species–species co-operative dependence. What is lacking, of course, is a sufficiently permanent disposition of cells relative to one another in space.

11.14 Summary

Chapters 10 and 11 have now to be seen in the context of the developing theme of this book, the natural selection of the chemical elements, as it has developed historically in life. While the production of the elements in their relative abundances, next as molecules, and then in co-operative condensed material on planets (Earth is the one we know best) is a remarkable progression of this selection, the introduction of life on Earth is at quite another, higher level of sophistication introducing *organised* capture of energy and material and not following the retarded drift to order/disorder balance, that is, to equilibrium, for some local systems. We have presented possible origins of such organisation and also of primitive living systems to the best of our ability. The strength of this organisation then is its survival value under the constraints of feedback kinetics. For kinetic survival there are seen to be very general requirements:

(1) containment in a cell;

(2) an adequate supply of the required 15–20 chemical elements in aqueous solution;

(3) an adequate source of energy;

(4) a coded control and a regulatory molecular unit;

(5) machinery including a reproductive mechanism.

These and these features alone dominate a living cell. All uptake and incorporation (pathways) came to be prescribed and, through the feedback mechanisms, the chemicals in them are not independent variables. We shall explore this deduction in the next chapters but an obvious implication is that if we analyse the DNA it has to read, related to cell activity, the element composition. No matter what the product of the gene or the gene-regulated proteins, we can add them up in an elemental analysis. DNA and element composition are conditionally correlated; one does not dominate the other.

We are stressing that a coded molecule such as DNA has no value unless the code can be read in a controlled way so as to express the cell contents *in toto*. *DNA is not an independent substance and its sequential structure is no more than a device for generating and reproducing what had already been discovered about co-operative functioning* (Section 11.2). It needs a reading machine, a synthesis machine, a confined space, a controlled supply of energy and a controlled uptake and incorporation of elements in correct proportions.

Since we have tied all the element concentrations together with all the levels of compounds made from them, even DNA, we have reached the point where we can see that, by using feedback systems, a cell is extremely restricted in its chemical degrees of freedom, variance, in a manner that parallels the constraints imposed by phase rule equilibria (Chapters 4 and 6).

Given a beginning, such a system will evolve not mainly in competition or by competition but by the simple fact that it only exists because it survives the ravages of time—it is unstable and evolves unwittingly to survive longer. The use of all the machinery down to the level of all the elements then improved in fitness along with function. (This includes physical shape which, as we have shown, is not an independent feature.) The mechanism of improvement is random selection based on mutation internally but it is driven also by environmental possibility and change (Chapter 8) which select the fittest. We do not know of a direct way for the environment to affect the organism but indirectly it can change the probability of mutation related to that part of DNA which is most affected by the change of environment. It then becomes critical to follow environmental change and life's evolution. All these advances are thermodynamically uphill and require ever-increasing energy (Fig. 11.34).

As the evolution progressed it ran into the problem of complexity—refinement of activity for survival. The first escape was compartmentalisation but this introduces more difficult communication control. A second escape is incorporation, a third is symbiosis and a fourth is multicellular growth but the best solution is to divest to others some functions, obviously the simplest since these are the common needs, allowing co-operative ecological growth. This is ecological organisation. It is the evolution of co-operative organisation that we shall follow in the next chapters while we observe the growing sophistication of the natural selection of the chemical elements.

A final paragraph is necessary referring to Fig. 10.2 which shows

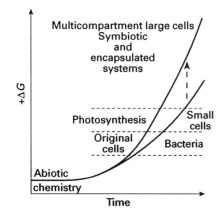

Fig. 11.34 The increase in absorbed energy per cell. Fast reproductive rates could be maintained for advanced bacteria but not for the larger cells and certainly not for eukaryotes.

evolutionary trees. The trees *all* grow with time and diversify, and all have branches in different environments; thus all increase in species membership with the evolving environmental diversity (see Chapters 12 and 13). This is a matter of the survival of the fittest in particular chemical situations. Diversity amongst bacteria is almost infinite since the non-living chemical environment is of huge diversity itself (Chapters 8 and 9). To add to the multitude of living systems complexity increased, but eventually or perhaps even very soon increased complexity became dependent on simplicity (bacteria). Complexity will again be shown to evolve with environmental differences. By their very nature the complex systems will inevitably develop in a minority of forms provided that simple systems handle the basic element chemistry for them. Evolution of complexity, a minor living feature numerically, is sustained by conservative simplicity in the use of the natural selection of chemical elements.

Further reading

Three excellent textbooks, beautifully illustrated, very informative and recently re-edited provide excellent complementary reading in specific topics:

1. Villee, C. A., Solomon, E. P., Martin, C. G., Martin, D. W., Berg, L. R. and Davies, D. W. (1989). *Biology* (2nd edn). Saunders College Publishing, Philadelphia.
2. Purves, W. K., Orians, G. H. and Craig Heller, H. (1995). *Life—the science of biology* (4th edn). Sinauer Associates/W. H. Freeman, New York.
3. Lodish, H. D., Baltimore, D., Zipursky, L., Berk, A. and Matsudaira, P. (1995). *Molecular cell biology* (3rd edn). Scientific American Books/W. H. Freeman, New York (at more advanced level).

The following references contain useful discussion of cell organisation.

4. *Scientific American* (1994). **271**, 52–98.
5. Margulis, L. and Schwartz, K. V. (1988). *Five kingdoms*. Freeman & Co., New York.
6. Fraústo da Silva, J. J. R. and Williams, R. J. P. (1991). *The biological chemistry of the elements*. Oxford University Press, Oxford.
7. Stryer, L. (1995). *Biochemistry* (4th edn). W. H. Freeman, New York.
8. Kauffman, S. A. (1993). *The origins of order*. Oxford University Press, New York.

Other useful references:

9. Eigen, M. (1982). Stages of emerging life. *Journal of Molecular Biology* **19**, 47–61.
10. Hemsley, A. R., Collinson, M. E., Kavach, W. L., Vincent, B. and Williams, T. (1994). The role of self-assembly in biological systems. *Philosophical Transactions of the Royal Society, Series B (London)* **345**, 1–136.
11. D'Arcy Thompson (1966, reprinted 1988). *On growth and form*. Cambridge University Press, Cambridge.
12. Peacocke, A. R. (1983). *The physical chemistry of biological organisation*. Clarendon Press, Oxford.
13. Harold, F. M. (1990). Morphogenesis in micro-organisms. *Microbiological Reviews* **54**, 381–431.
14. Hawksworth, D. L. (ed.) (1994). Biodiversity: measurement and estimation. *Philosophical Transactions of the Royal Society, Series B (London)* **345**, 1–136.
15. Danson, M. J., Hough, D. W. and Lunt, G. G. (ed.) (1991). *The archaebacteria*, Biochemical Society Symposium no. 58. Portland Press, London.

16. Levett, P. N. (ed.) (1991). *Anaerobic bacteria*. IRL Press at Oxford University Press, Oxford.
17. Gordon, K. H. J. (1995). Were RNA replication and translation coupled in the RNA world. *Journal of Theoretical Biology* **173**, 179–94.
18. Mann, S., Webb, J. and Williams, R. J. P. (ed.) (1989). *Biomineralisation*. VCH, Weinheim, Germany, especially Chapters 1, 2, 9 and 10.
19. Suga, S. and Nakahara, H. (ed.) (1991). *Mechanisms and phylogeny of mineralisation in biological systems*. Springer-Verlag, Tokyo.

The following references give various approaches to the origin of life.

20. Day, W. (1981). *Genesis on planet Earth*. Shiva Publishing Ltd, Nantwich, Cheshire.
21. Oparin, A. I. (1957). *The origin of life on the Earth* (3rd edn). Oliver and Boyd, Edinburgh.
22. Bernal, J. D. (1967). *The origin of life*. Weidenfeld and Nicolai, London.
23. Calvin, M. (1969). *Chemical evolution*. Clarendon Press, Oxford.
24. Miller, S. L. and Orgel, L. E. (1974). *The origins of life*. Prentice-Hall Inc., New Jersey.
25. Cairns-Smith, A. G. (1985). *Seven clues to the origin of life*. Cambridge University Press, Cambridge.
26. Scott, A. (1986). *The creation of life*. Basil Blackwell Inc., Oxford.
27. Shapiro, R. (1986). *Origins—a skeptic's guide to the creation of life on Earth*. Heinemann, New York.
28. Gesteland, R. F. and Atkins, J. F. (ed.) (1993). *The RNA world*. Cold Spring Harbor Laboratory Press, New York.

12

The structure and chemistry of organisms after the advent of dioxygen

Evolution is a change from an indefinite incoherent homogeneity to a definite coherent heterogeneity

Herbert Spencer
1820–1903 *First principles*, Chapter 16, §138

12.1 Introduction

In Chapter 10 we described the selective uptake of the chemical elements in early life to the degree to which this was possible given the poor fossil and geochemical evidence. We depended for our description largely upon present-day descendants of primitive life. As well as looking at the uptake of the elements we were concerned with the way in which they were incorporated in organisms since this involved other selection processes. We showed that many elements, between 15 and 20, were always essential for this anaerobic life and that those used for any function were chosen to a large degree by abundance, availability and inherent selective chemical properties as observed in isolated atoms or in very small molecules (Chapter 10). These properties were then refined so that the fitness of elements for particular functions was enhanced. There are two obvious parts to such refinement within self-assembly—a physical or compartmental selection and a chemical combination selection. While some of the selective interactions in local volumes could be achieved using the drive to thermodynamic equilibrium as described in Chapters 2–6, it was also necessary for many other processes to utilise energy in uptake and synthesis, as analysed in Chapter 7. Kinetic management was therefore required over many pumps and pathways, that is control systems, and over syntheses leading to the production of all the polymers including DNA, that is by regulation. We found that only through co-operative feedback communication in complicated networks could we see a way in which survival strength could have been achieved and increased in cells in all three features, element uptake and incorporation, metabolic pathways, and polymer production relative to random prebiotic chemistry (as illustrated in Chapter 11). The cells described were anaerobic and presumed to be primitive, although we extended our description to photosynthetic organisms and to eukaryote structures. In this chapter we shall show how the physics and chemistry of biological systems has developed since some 2 to 1×10^9 years ago simultaneously with a considerable but slow change in the environment, namely in the availability of elements and energy (Fig. 12.1). In particular, we shall be concerned with the effects of the *slow introduction of dioxygen* into the atmosphere (Fig. 12.2). Any such change must be seen to have been damaging for the original primitive life systems and could only have become fortuitously favourable as the living system adjusted (by chance?) to generate new protection, new pathways of metabolism and new organisation. There can be no guarantee, therefore, of favourable development from any such newly evolved chemical, which may

Fig.12.1 Biospheric, lithospheric and atmospheric evolution on the primitive Earth.

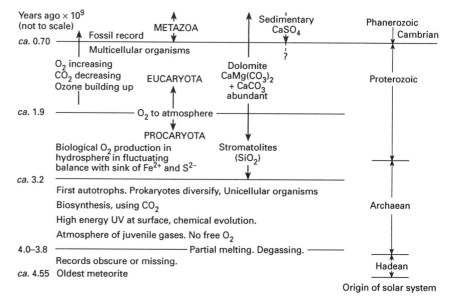

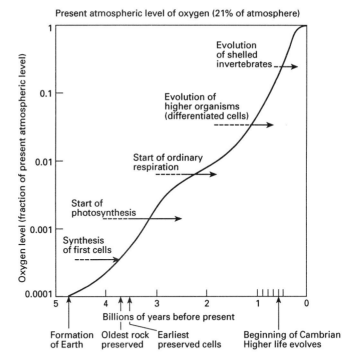

Fig.12.2 One hypothesis of the evolution of oxygen in the atmosphere in relation to the origin of life and the evolution of higher organisms. (From *Earth* (4th edn) by Press and Sievers. Copyright © 1986 W. H. Freeman and Company, with permission.)

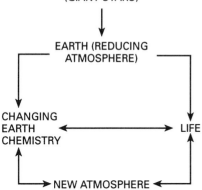

Fig.12.3 The locked-in interaction of changes of the chemistry of the atmosphere, the Earth and life.

be a poison, as O_2 was, since there is little and very slow feedback control over environmental change. Life and its evolution today are surely still dependent on chance mutations in DNA, but they have always been also locked into the physical and chemical nature of the evolving apparently abiotic environment (Fig. 12.3).

The environment itself is not a passive partner soaking up waste products of life such as calcium salts, carbon dioxide and dioxygen (allowing that waste can be re-used), but it also has imposed upon it from internal and external sources various changing abiotic activities (see Chapter 8) as well as biological actions (Fig. 12.1). The activities include physical variations in the energy from the sun, losses of chemicals from the atmosphere, changes of chemicals due to volcanic activity, continental drift and changes in the rate of rotation of Earth around the sun and of the rotation of the Earth on its own axis. None of these activities has a fixed, timeless, character. Together they impose on Earth and on biological systems not only long-term changes but obvious short cycles of such features as pressure, temperature, light intensity, etc., which are known to cause irregular ice ages and yearly patterns of the weather. These cycles of physical conditions have inevitable chemical counterparts, causing changes in the composition of land, sea and air due to solubility and volatilisation variations, and involving all the chemicals that were or came to be involved in life. Thus, any consideration of even such factors as continental drift (see Fig. 8.12), which is on a time-scale comparable to that of evolution, must show that, in part, the physicochemical environment helps to drive biological change. We know too that in evolution there have been major extinctions of species as well as development on a broad front (see Fig. 12.6).

On the surface of the Earth the chemical activities, measured in terms of rate of change of turnover of chemicals in cycles, are, in fact, on a somewhat equal scale for the two systems, one of which we call living (animate) and the other of which we call dead (inanimate). If we ignore the distinction between living and dead and treat the whole as just chemistry then we shall see that there is, on the surface of Earth at any one period, one intertwined system of cycling chemicals (Fig. 12.3 and see Chapter 15). This cycling must, however, be progressive on a long time-scale since neither system is at equilibrium, as we saw in Chapter 8. Just as life through DNA is subject to random insult so the Earth's chemical systems can suffer random impact, but there is an inevitable overall direction of life towards higher organisation, which provides greater survival strength so long as more energy is put in to move it away from equilibrium (see Chapter 11) while Earth's chemical systems tend toward equilibrium (Chapters 3–6 and 8). The system of living things increases fitness, a chemical kinetic stability of flowing material; the system of inanimate things increases in entropy, a measure of thermodynamic stability. To some degree the two systems are in a competition that only the second can win finally. When discussing vast time-scales we must be aware that today, although there is a balance of chemicals from various sources (inputs) and sinks (exits) (see the cycles of Chapter 15), this does not preclude a change in the atmosphere and the sea due to a quite sudden increased or decreased activity of the mantle toward equilibrium. Chemically, a sustained increase of Fe/S output, for example from volcanoes, would lower the O_2 pressure and increase acidity. Changes in life would be very considerable and it could even revert to the anaerobic state described in Chapter 10. It is by looking at the effects of the introduction of dioxygen that we can appreciate such possible events. While we have stressed the chemical dependence, the physical dependence on temperature, T, and pressure, p, is always determinant too. In fact, just as in the discussion of the phase rule, and thermodynamic equilibrium generally, we found that there were constraints based on 90 (elements) + 2 (T and p) variables (Chapters 3–6), so there are the same constraints in the case of living systems. Although fixed amounts of available elements, of equilibria and of fixed imposed fields do not limit biological systems in a strict fashion, these systems are constrained almost equally now by the limitations of feedback control and regulation, chemical kinetics (related to the code of DNA), necessary for survival and acting on the preferred elements (Chapter 11). We shall come back to the nature of the constraints only after we have reviewed the nature of aerobic biological conditions. We attempt a fuller description of constraints in Chapter 16.

In this chapter, therefore, we shall be concerned mainly with the changing patterns of chemicals (and physical conditions) associated with life's evolution. We shall look at what has happened in stages before we ask why it has happened. First, we outline the observed physical changes in appearance of living systems, starting some 1×10^9 years ago and 3.5×10^9 years after the formation of Earth (Fig. 12.4), following relatively good geological records. It was by this period that dioxygen increased so as to approach present-day levels, more exactly some 700 million years ago (Fig. 12.2). All classifications of biological species were based originally on the physical characteristics of the

Fig.12.4 (a) The evolutionary pathways of known animal lineages showing the common ancestors of each branch. (Adapted from Barrow, J. (1991), *Theories of everything*.) (b) Some common evolved forms of plants. The numbers in parentheses are $\times 10^6$ years ago.

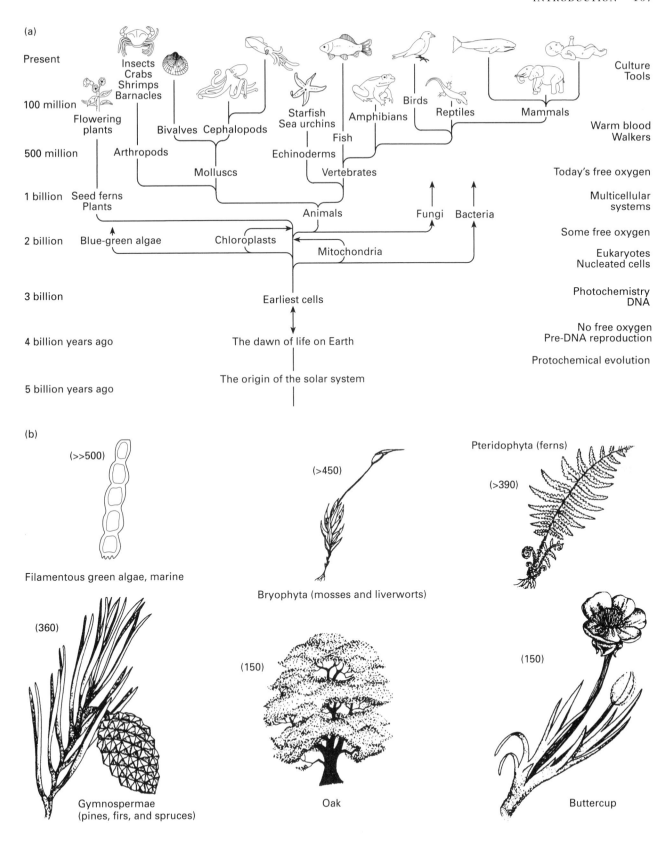

(a)

Present

100 million

500 million

1 billion

2 billion

3 billion

4 billion years ago

5 billion years ago

Flowering plants

Insects
Crabs
Shrimps
Barnacles

Bivalves Cephalopods

Arthropods

Seed ferns
Plants

Molluscs

Starfish
Sea urchins

Echinoderms

Amphibians

Fish

Birds

Reptiles

Vertebrates

Mammals

Culture
Tools

Warm blood
Walkers

Today's free oxygen

Multicellular
systems

Some free oxygen

Eukaryotes
Nucleated cells

Photochemistry
DNA

No free oxygen
Pre-DNA reproduction

Protochemical evolution

Blue-green algae Chloroplasts

Animals

Fungi Bacteria

Mitochondria

Earliest cells

The dawn of life on Earth

The origin of the solar system

(b)

(>>500)

Filamentous green algae, marine

(>450)

Bryophyta (mosses and liverworts)

Pteridophyta (ferns)

(>390)

(360)

Gymnospermae
(pines, firs, and spruces)

(150)

Oak

(150)

Buttercup

Table 12.1 The geological time-scale of life after the advent of dioxygen*

Era	Period	Epoch	Distinctive features	Years before present
Cenozoic	Quarternary	Recent	Modern humans	11 000
		Pleistocene	Early humans	1 700 000
	Tertiary	Pliocene	Large carnivores	5 000 000
		Miocene	First abundant grazing animals	23 000 000
		Oligocene	Large running mammals	38 000 000
		Eocene	Many modern types of mammals	54 000 000
		Palaeocene	First placental mammals	65 000 000
Mesozoic	Cretaceous		First flowering plants; extinction of dinousaurs and ammonites at end of period	135 000 000
	Jurassic		First birds and mammals; dinosaurs and ammonites abundant	192 000 000
	Triassic		First dinosaurs; abundant cycads and conifers	223 000 000
Palaeozoic	Permian		Extinction of many kinds of marine animals, including trilobites	280 000 000
	Carboniferous	Pennsylvanian	Great coal-forming forests; conifers; first reptiles	321 000 000
		Mississippian	Sharks and amphibians abundant; large primitive trees and ferns	345 000 000
	Devonian		First amphibians and ammonites; fishes abundant	405 000 000
	Silurian		First terrestrial plants and animals	
	Ordovician		First fishes; invertebrates dominant	495 000 000
	Cambrian		First abundant record of marine life; trilobites dominant, followed by massive extinction at end of period	570 000 000
	Precambrian		Fossils extremely rare, consisting of primitive aquatic plants	700 000 000

* Source: Brown, J. H, and Gibson, A. C. (1983). *Biogeography.* Mosby, St Louis; see also Ricklefs, R. E. (1993). *The economy of nature—a textbook in basic ecology* (3rd edn). W. H. Freeman & Co, New York.

organisms that have appeared since that time (Fig. 12.4 and Table 12.1). The geological records also date some chemical changes, either due to biological action or not, and our second task is to give the known and probable resultant changes in the evolution of the availability of inorganic elements on the Earth's surface during and after the advent of the dioxygen atmosphere (see Section 5.8.2 and Chapter 8). It was the chance-operated re-selection of the uptake of elements into biological systems, so as to trap this potential for new chemistry in life, as well as the randomly operated increase in organisation through DNA mutations that acted to improve survival under the new conditions (see Chapter 13). These changes extended the possibility of creating new organic molecules, that is, of forcing new constructions *upon* the combinations of H, C, N, O, P and S (Chapter 9). Such changes could and did generate new compartmental organisation up to the level of the evolution of organs and ecosystems of mutual dependence again as observed from the geological record (Section 12.2.1).

While we examine the changes that give rise to multicellular organisms we must remember that unicellular organisms adapted too. Above all, bacteria, the dominant species on Earth, adjusted their metabolism to the new conditions. There developed also a vast range of other unicellular species that are either obligatory aerobes or can switch lifestyle from anaerobic to aerobic

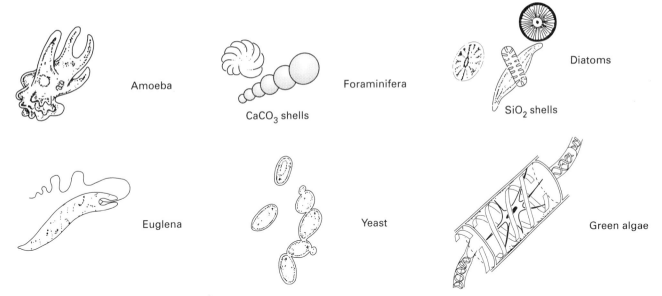

Amoeba

Foraminifera

CaCO$_3$ shells

Diatoms

SiO$_2$ shells

Euglena

Yeast

Green algae

Fig.12.5 Some aerobic unicellular organisms.

(Fig. 12.5). Their adaptability is a part of their survival strength. They too utilised the new availability of chemical elements and struggled to maintain concentrations of those essential elements that became harder to obtain.

After giving an account of the new inorganic and organic chemistry of life we return to the nature of the very complex self-assembling organisms of today. What has to be explained is not just evolved physical form but which chemical selection could generate a particular 'fit' organism, the reason for increased length of cycle of living organisms through birth, development and growth, morphogenesis and reproduction, to death, and why so many species exist. What do we mean in fact by the 'fitness' to survive of a living system in chemical terms when so many different species exist together? Much of this discussion is deferred to Chapters 13 and 16.

As we explained in Chapter 8, the vast effect of a switch to an oxygenated atmosphere was due to the nature of the original Earth which was decided by a relative deficiency of oxygen compared to the sum of all other available elements so that not all the elements could become oxides, that is, they could not attain their most favoured combined chemical status and, therefore, they remained reduced. Once the Earth cooled there was a further barrier to the possibility of elements becoming oxides: the Earth's temperature was and is too low for reactions of many elements with dioxygen to proceed at a measurable rate (see Chapter 7). Thus it could happen and, in fact, did happen that, when the cool surface of Earth captured energy and generated dioxygen through life, this element was in a new kinetic trap in a special compartment, our atmosphere (Fig. 12.3), and kinetic conditions were established in which dioxygen only slowly changed the state of the sea and the Earth's crust. Before the final build-up of free dioxygen could happen, elements (in compounds or free) that have a considerable affinity for oxygen, for example iron and sulphur, were forced to change their chemistry on Earth's *surface* by reaction with it until they were effectively fully converted to oxides; for example

Non-metals in aerobic organisms

H
C N O (F)
Si P S Cl
Se Br
I

Contrast Section 10.2.

sulphide went to SO_4^{2-}, and Fe^{2+} went to Fe^{3+}. This major change had knock-on effects on other elements both outside and in organisms (Table 12.3 and see marginal note). We consider that it is this combination of physical conditions and chemical reactions due to the production of dioxygen and ozone, (note that the ozone layer is also protective against UV radiation) that has directed the evolution of life in the last 1×10^9 years (Fig. 12.4). Of course, the appearance of dioxygen meant that some chemical elements had been deprived of it by the additional, newly possible, energy-driven changes of living systems. The best examples are the oxides of both carbon and hydrogen, $CO_2 + H_2O$, which have formed a separately energised, high ($+\Delta H$) state, seen in living systems and as oil, gas and coal. In the opposite direction, ammonia was oxidised and became dinitrogen. These are further examples of trapped kinetically stable chemicals in the presence of dioxygen. We would be foolish to believe that it would be impossible for a different evolution of chemistry balanced between life cycles and non-biotic activity to occur again since this surface system is not in equilibrium. As we know full well today, through discussion of the greenhouse effect, the temperature at the surface of Earth is a variable in this physical–chemical steady state as are its chemicals, such as CO_2 and CH_4. Section 16.10 examines some possibilities, but a sudden violent fluctuation of a variety of kinds can never be excluded. Fluctuations have often driven the major developments of chemical selection in the universe, for example the formation of the Earth!

12.2 Developing organisation

As explained in Chapter 10, before the coming of dioxygen there could well have been the development of anaerobic organisms with complex structures. They used energy and carbon sources in a relatively simple way (Chapter 10). Filamentous constructs that could retain vesicles or other membrane-limited compartments in single cells could have been built somewhat later. They could also have developed a flagellum motor. Again colonial and symbiotic life and some differential cellular activity must have been possible (Section 11.3), and we have good reason to believe that the main source of energy was already sunlight (Section 10.4.6). However, the geological record does not allow us to do more than to speculate about much of this. One such speculation suggests that prokaryotes developed in a multitude of forms at first, all with basic axial symmetry (Fig. 10.1). They had an ability to bind to small particles and hence build mats of inorganic particles stuck together by organic material. The principal depositions would be of the saturated compounds in the sea, $Si(OH)_4$ and $CaCO_3$. To this day we can find evidence of these mats in the stromalites, the earliest of which date apparently to before 3×10^9 years ago (Fig. 12.1). Further development could have come about by trial and error association of different anaerobic forms either internally or externally (Fig. 11.33). All of this discussion is very difficult to substantiate and to date due to the lack of proper geological records. In order to carry on a discussion based on evidence we turn in this chapter to the known progression of life from the coming of considerable levels of dioxygen, that is, from 1.0×10^9 years ago, where the geological record is becoming more and more secure. We have to assume, with the present state of knowledge, that internal compartmentalised systems already existed, that is, the age of the eukaryotes had begun long before this.

12.2.1 The geological record of evolution

Figures 12.6 and 12.7 give a more quantitative impression of the geological records of plant and animal development from 500×10^6 years ago which is to be seen against the presumed picture of earlier times (Table 12.1 and compare Fig. 12.4). The record is also based on species likenesses in physical appearance and on molecular sequence identities within DNA, RNA and proteins. It has been found that some organisms in groups of species come and go so that continuity is broken: evolution is not a simple progression. Our ultimate aim is to give this observed evolution a physicochemical explanation in terms of selected elements. It is probably true that all this development required some kind of random selection at the gene level, but such a process is of little chemical interest. Interest arises when we concentrate upon the result and treat it as if it had been designed to meet circumstances—that is, as if it was driven in a particular way. Survival of the fittest should then lead to the best designed physical or chemical system opposite the coexisting environment. In other words, what principles or practices *of value* in the use of elements was random selection forced to discover as the environment changed? We define value in terms of *survival*, which has a number of collateral components, including the ability to have a long kinetically stable life, to reproduce, to forage for food, to protect from predators, and so on. This is called the 'fitness' of an organism. Particular 'fitnesses' in chemistry are not necessarily compatible with different environments, and we must treat developments of biological systems within a variety of constraints. (The geological record may well be analysed in more detail in future so as to show not only what shapes

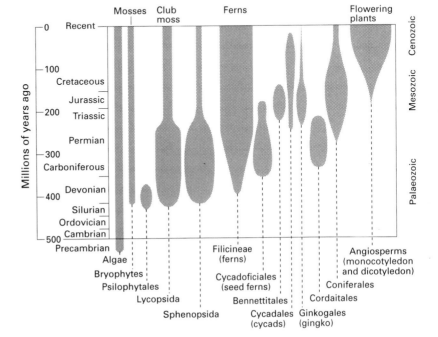

Fig.12.6 The history of the plant kingdom. The width of the shaded area gives the magnitude of the population. At the left *periods* are shown, and on the right the *eras*.

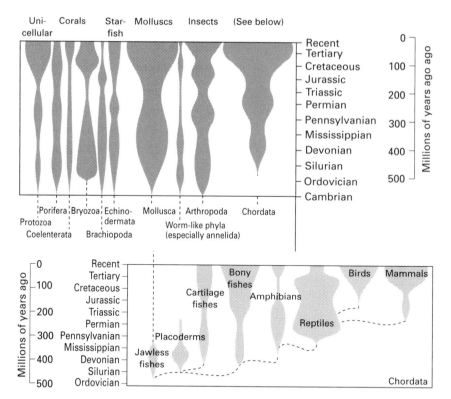

Fig.12.7 The history of the animal kingdom. The width of the shaded area gives the magnitude of the population. The invertebrates are shown above.

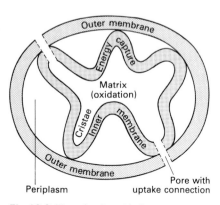

Fig.12.8 The mitochondria have two aqueous and two membrane compartments. Many of their proteins are made outside the cell cytoplasm and then are placed in different compartments of the organelle. The membranes of the organelle have shape, and different proteins are also positioned in lateral organisation. The distribution of charge is also differential both across and along membranes. These organelles are thought to have been generated through capture of primitive aerobic bacteria into advanced (anaerobic) eukaryotes.

and organisation developed and could be maintained, but also which trace elements were used including the particular selection of each element for functional use.) Of course, we must remember that the nature of evolution is such that the bulk of primary metabolism and synthesis in the cytoplasm as seen in anaerobic *cells* had to be maintained. Life built and still builds on what it had, or has, and there are no new basic primary metabolic foundations in all the systems we know—there is but one basic form of life's metabolism in the cytoplasm of cells, that of anaerobic life, which we described in Chapters 10 and 11. (There are minor variations, of course.)

12.2.2 Development of intracellular compartments in aerobic cells

Now it is not only the case that the organisms we are describing are *expanded* physical cells of the same kind as in strict anaerobes, see Chapter 11. Three other changes have to be noted: (1) the introduction of new compartments within single cells (Figs 11.3 and 12.8); (2) differentiation of cells; and (3) organisation of cells into organs within a whole organism. All three represent compartmental development associated with a complexity of inter- and extracellular physical and chemical networks. As discussed in Chapter 10 (and compare Chapter 7), it is clear that increasing the number of compartments allows a gain in overall functional capability. Particular acts of chemical and physical sophistication are best carried on in isolated zones so that different reaction paths are not confused, and then equilibria are established by the diffusion of all the chemicals involved within one vessel. It is restriction of

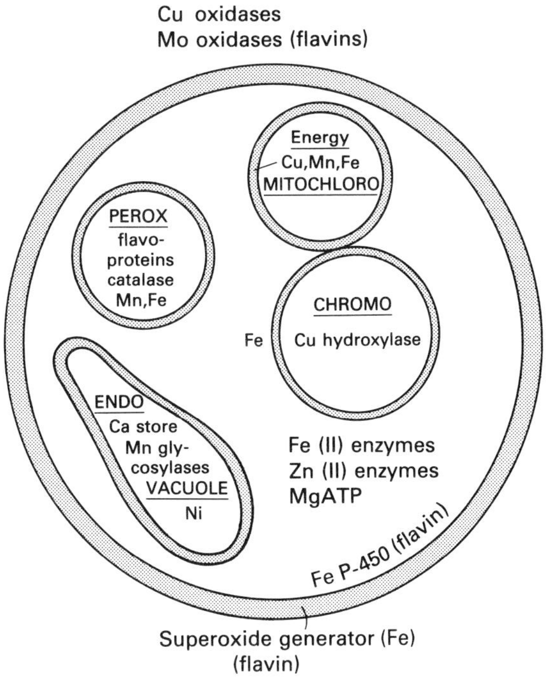

Fig.12.9 A schematic indication of some of the different membrane-separated compartments in an advanced cell. PEROX is a peroxisome; MITOCHLORO is either a mitochondrion or a chloroplast; CHROMO is a vesicle of, say, the adrenal granule; ENDO is a reticulum, for example the endoplasmic reticulum. Other compartments are lysosomes, vacuoles, calcisomes, and so on. Localised metal concentrations are shown.

mixing, apparent increase in degrees of freedom, that is necessary if development is to occur even if new restrictions of other kinetic kinds are introduced, for example restrictions associated with much lower reproduction rate and complexity of communication.

Figure 12.9 illustrates the increasing number of *intra*cellular vesicular compartments and the introduction of two new organelles, the chloroplast and the mitochondrion. Both organelles were very probably evolved from prokaryotes trapped inside initially anaerobic cells. They allow the (dangerous) production and use of molecular dioxygen as an energy source to be placed separately and safely away from much of the anaerobic metabolism. Each of the compartments has new chemistry and a novel distinctive chemical element composition, so we must look again at element availability, localised uptake and incorporation, now in the presence of dioxygen. This new chemistry also appears in the membranes of aerobic bacteria.

Some of the other compartments may have developed within anaerobes but their complexity increased with time in aerobes. We can illustrate this by reference to such compartments as the Golgi apparatus and the recticula formed from it (Fig. 12.10). The Golgi helps to make small vesicles in cells and to distribute proteins that have been modified, sulphated and glycosylated, before they are exported. Protein export is a major activity of multicellular organisms as is the production of secretory granules (see Chapter 13). A second illustration is the production of peroxisomes (for example see Fig. 12.11), which help to remove dangerous chemicals such as H_2O_2 but also serve as zones of special dioxygen-based metabolism and of protective reactions.

During these developments cells took on new shaped structures so that the proteins and vesicles from the Golgi went to special places in the cell presumably secured for them by recognition at fixed positions on filaments.

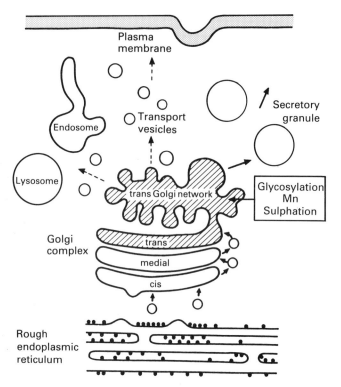

Fig.12.10 A representation of the Golgi apparatus, which is supplied with proteins by ribosomes of the endoplasmic reticulum for processing and then transfer to specific functional compartments.

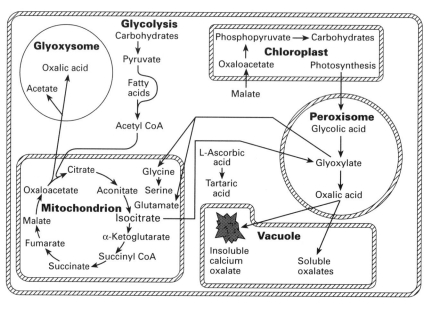

Fig.12.11 Summary of metabolic pathways in plants. Major vesicular pathways are shown.

12.2.3 The chemical problems of evolution

We have described the basic principle behind living organisms as one of the survival of a co-operative chemical reaction system with feedback within a chemical and physical trap, needing energy and material, Chapters 7 and 10. Evolution can be no more or less than increasing physicochemical complexity in them. Looking at the evolutionary trees of primitive and advanced organisms as in Fig. 12.4, we see two obvious features. First, the increase in complexity is not linear but always branching and, second, branches once created generally survive. Thus, increasing complexity exists together with increasing diversity.† The chemical question is what simultaneously stimulates complexity and generates diversity? It is absolutely clear that random mutations in DNA alone do not explain the observations of the evolutionary tree even though they record it. Anaerobes did not develop diversity in the way aerobes have done: archaebacteria have not generated very much complexity. The view taken here will be that the pattern of evolution is driven by the advantage taken of the changing availability of the chemical elements in the sequence

Changing element availability ⟶ protective reaction ⟶ useful system
          (poisons)   ('poison' acts as   (complexity
                                              messenger)   of use of
                                                                      'poison')

Diversity evolved with complexity but the development of complexity gives rise to problems as well as advances. The main problems concern management; the more complex the system becomes, the stricter the management necessary to maintain it in a feedback balance of growth (see Section 7.1 and Chapter 13). Within evolution the restrictions on cells of complex organisms became stricter so that the chemical system became more able to give rise to differentiated cells, different organs, etc., but less able to allow de-differentiation, that is, to be adaptable. All plant cells can revert to a precursor stem cell and regenerate whole plants, and reptiles can regrow limbs, while higher animals can only develop and regrowth is minimal in most organs. De-differentiated growth within a tightly managed system is even cancerous. (We may consider that the higher the degree of co-operativity the narrower the composition range of stability as in a phase diagram, see Chapter 4.)

Given this approach to evolution, our task is to show how the change in element availability with time allowed new chemical components (initially poisons) to be produced. Then we consider how these poisons can be countered so as to generate protection to increase survival. The final stage is to modify the protective devices, in diversification to more complex organisms, so that survival is increased once more. As we shall see, the reason why such diversity leads to a coexistence of species is that complexity brings with it for one species a dependence on the simplicity of others as the best method for the whole to survive. Very simple systems are extremely strong survivors, for example bacteria, but they cannot occupy, though they may share, the same niches as plants and animals.

As an example of diversity and dependence we can refer to one of the earliest divisions of living organisms which separated those that could use light as an energy source from those that could not. The cell systems that could absorb light were able to maintain existing energy status with minimum scavenging and hence could evolve a sedentary lifestyle—the plants. The cell systems without light-capture devices had to scavenge to survive and the only way

† The diversity shown in evolutionary trees has to be seen against the background of the extremely large number of simple (prokaryotes) that have evolved as well as the small number of organisms of great complexity.

forward was to evolve motors and sensing devices, which became the muscles and nervous system of animals, that is, they had to increase in complexity. It is obvious that two such systems could exist independently, but equally clear that the second could come to depend on the first for food (elements) and thus increase its survival strength. Now, light is not the only resource and we can imagine a variety of different chemical dependencies, one of which is the dependence on dioxygen, that is, aerobes versus anaerobes, within different chemical niches. Thus dependence again developed: there are, for example, molybdenum-dependent nitrogen-fixing bacteria that are essential for plants, which themselves provide the bacteria with reduced carbon and so on. There can also be quantitative differences involving dependence on more of one chemical element and less of another in rather similar species, but which may come to depend upon one another and so can also drive evolution into more and more diversity. This theme will be developed in Chapters 13 and 16. Here we must concentrate on the major changes of chemistry throughout multicellular systems with the advent of dioxygen.

12.3 The introduction of new metabolism

It might be thought that the very fact that the use of dioxygen became coupled to novel organic chemistry contradicts the above statement that primary internal cytoplasmic metabolism did not change. In fact, it does not. The DNA/RNA/protein code is little changed; all cell constructions are built from lipids, proteins and polysaccharides via the same cytoplasmic reactions with a redox potential around 0.0 volts, and the pH is still 7.2. The basic primary chemistry is that of primitive prokaryotes. Thus, it is mainly to the outside of the cytoplasm that we must turn for new metabolism.

'Outside the cytoplasm' here refers to two very different regions. The first is the inside of vesicles and the second the outside of the cell. We group them together since:

(1) neither region is maintained at a low redox potential;

(2) both regions are higher in Ca^{2+} and H^+ since the pumps on the cytoplasmic surface point away from it, that is, to the outside or into vesicles;

(3) in neither zone is protein synthesis possible.

As mentioned above many vesicles carry out new oxidative reactions (Fig. 12.11), and this allows production of oxidised aromatics, adrenaline, for example, and peptides or even proteins for export. The handling of sulphate (an oxidised product from sulphide) is managed in the Golgi vesicles (Fig. 12.10). In the new acidic vesicles hydrolysis occurs in the lysosomes, and they and the vacuoles of plants have some digestive functions at pH = 5.5. The digestive or oxidative activity associated with vesicles is also a large part of protection so that strong chemical attack in the cytoplasm is avoided.

The new metabolism outside cells also includes much digestion without energy conservation. However, it includes a great variety of new assembly and protective devices as well. The most primitive is the development of an extracellular matrix that is oxidatively cross-linked, for example collagen and chitin, Section 12.10. More sophisticated external processes are those of wound-healing (blood-clotting in animals and browning in plants which are dependent on oxidative reactions) and immune responses.

Energy transduction at high redox potential is also removed from the cytoplasm and put safely into organelles. The major new reaction associated with dioxygen in mitochondria is a reversal of the reductive citric acid cycle to give energy but it is still used in many syntheses as well as in degradation (Fig. 12.12). The product, ATP, now made in large amounts (for export to the cytoplasm) remains the normal energy-carrying chemical. In plants the additional production of oxygen from light takes place in a second organelle, the chloroplast (Fig. 11.30), where ATP is also made. Thus the metabolism of the vesicles and the organelles supplies: (1) the inside of the cell with some chemicals and much energy to be used there in primary metabolism in a reductive environment; and (2) the outside of the cells with other chemicals to

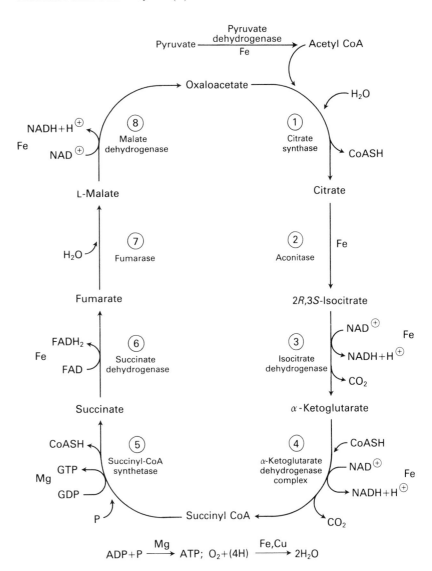

Fig.12.12 The *oxidative* citric acid cycle. Note the use of coenzymes. The NADH and FADH$_2$ are oxidised to give ATP, see Fig. 12.16. CoA is coenzyme A.

be used in an oxidative environment. Much of the oxidative metabolism is included under the heading *secondary metabolism*. The elements used in it are qualitatively and quantitatively different from those used in primary metabolism. We turn directly to the chemical elements that were made more available in the presence of dioxygen and that gradually came to be used.

12.4 Changing element availability with rising dioxygen partial pressure

The slow development of a dioxygen atmosphere had a gradual effect on many elements on the surface of the Earth and in the sea (Chapter 8). To a first approximation, we may consider that there was a gradual rise in redox potential of the surface of the sea as the partial pressure of O_2 rose with the pH constant at 7.5 or thereabouts. The initial redox potential in the reducing atmosphere some 4.5×10^9 years ago could have been well below 0.0 volts. If we take this value as a starting point and then introduce dioxygen up to present-day levels then the potential rose slowly and today, near the sea's surface, must be around that of the dioxygen/water couple of $+0.8$ volts at pH $= 7$ (Fig. 5.12). The change was slow and remains far from complete since the reducing components of the early surface and sea formed a large redox buffer, especially through iron sulphides. Since these still remain at the bottom of stagnant water a vast new out-of-equilibrium set of compartments, niches, was generated vertically in the sea. Considering the non-metals first, we see from Figs. 5.14, 5.20, 5.21 and 12.13 that the elements had to change to higher redox states in the *timed order* $H_2 \rightarrow H_2O$, $NH_3 \rightarrow N_2$, $H_2S \rightarrow SO_4^{2-}$, $H_2Se \rightarrow SeO_4^{2-}$, $I^- \rightarrow I_2$, other halides $X^- \rightarrow X_2$ or XO^-, nitrogen $\rightarrow NO_2^-$ and NO_3^-. In fact, the last two changes cannot occur more than fractionally even today since O_2 is not a powerful enough oxidant and there are kinetic limitations to the chemistry. Dioxygen itself could only appear in quantity after most of the surface H_2, NH_3, H_2S, H_2Se, Fe^{2+} and so on had been oxidised.

Amongst the metals (Fig. 12.13) it is noticeable that the order of change due to dioxygen advent is largely due to the difference in binding strength and solubility of sulphides relative to oxides in the salts, such as

$$CuS\downarrow + 3/2O_2 \longrightarrow CuO\uparrow + SO_2$$

where $\downarrow$ represents more insoluble and $\uparrow$ represents more soluble (see Fig. 5.3). Zinc, cadmium and copper, especially, became more available in that order. They are the more abundant rather 'b'-class elements (see Fig. 2.23 and Table 5.12). At the same time, iron became of greatly reduced availability.

$$4FeS \longrightarrow 4FeO\uparrow + O_2 \longrightarrow 2Fe_2O_3\downarrow.$$

Other changes of soluble metal anions took place, such as

$$MoS_4^{2-}\uparrow + 6O_2 \longrightarrow MoO_4^{2-}\uparrow + 4SO_2\uparrow$$

and some similar reaction of vanadium, which did not affect availability but required adjustment of uptake. The loss of both sulphide and iron put the whole of Fe/S chemistry in jeopardy, but this chemistry was and still remains one intracellular 'bedrock' of life! There had to be new uses, but old uses had to be protected in cells and separated compartments have a huge advantage

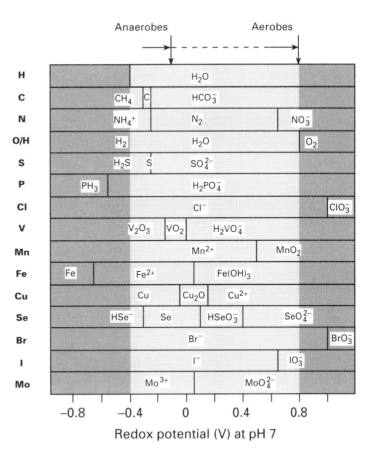

Fig.12.13 The redox states of various elements covered by the redox range of H_2/H_2O to O_2/H_2O (after P. A. Cox). Note the presence of sulphide affects the range for copper very strongly in a positive sense.

when such diversity of chemistry is required. The changes in metal elements and compounds related to evolution is shown in Fig. 12.14, and they too occurred sequentially, see also Fig. 5.20.

12.4.1 The chemical and biological reactions to dioxygen increase with time

The first response by organisms generally had to be protection *against* newly evolved dioxygen and must have used bound H, as in NADH, converting O_2 accidentally to O_2^- and H_2O_2, and then these two back to water. The cytoplasm, therefore, had to and still has to have added protection from these two reduction products of O_2. If we follow this description then we must look for the beginnings of a defence against the presence of O_2, not only simple oxidoreductases (to eliminate O_2), but also simple superoxide dismutases, SOD (to remove O_2^-), and simple peroxidases and catalases (to remove H_2O_2). It is very difficult to provide definite evidence for the development of enzymes such as those listed in Table 12.2. However, in the sulphate-using prokaryotes, at one time thought to be strict primitive anaerobes but which can survive very low O_2 pressures, all three protective enzymes, and no other 'uses' for these, have been found. We shall therefore examine these prokaryote organisms in a little detail.

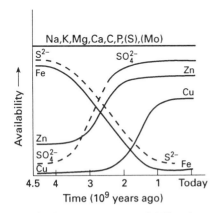

Fig.12.14 The changing availability of some elements with time and as redox potentials rose due to dioxygen pressure increases. Note that it is solubility as well as standard redox potential that controls availability. NB. $NH_3 \rightarrow N_2$ as $S^{2-} \rightarrow SO_4^{2-}$.

Table 12.2 The evolution of protection against dioxygen

	Prokaryotes* (early)	Eukaryotes
SOD	Mn, Fe	Cu, Zn
Peroxidases	Haem	Haem, Se
Catalases	Haem, Mn	Haem
Oxidases	Haem	Cu, Fe

* Including sulphate bacteria. Copper, zinc, and Se/O chemistry belong to later periods of evolution than iron or manganese chemistry.

Sulphate users may well have evolved very early as O_2 first oxidised sulphide (Fig. 12.13). The sulphate bacteria had an advantage in that they got (and get) energy from the reduction of sulphate even though it is a poor way to utilise the oxidative power of dioxygen. (Only 4 kcal mol^{-1} of energy is extractable, but sulphate must have been the first relatively plentiful newly available source of energy in water, ultimately derived from dioxygen (Fig. 5.21).) The reactions required new enzymes, catalysts, that is, new elements or old elements used differently, to manage sulphate. For example, these bacteria appear to have switched the molybdenum pterin cofactor known to be used in aldehyde reactions in primitive organisms (Fig. 10.14 and Section 10.7) to the $SO_4^{2-} \rightarrow SO_3^{2-}$ reaction. This molybdenum cofactor has a (new) balanced S/O coordination sphere (alternatively perhaps earlier tungsten was used)

with the metal in oxidation state VI, which became an intermediate in the transformation of SO_4^{2-} to SO_3^{2-}. (Figure 5.13 shows that the redox potentials from soluble Mo(III) to MoO_4^{2-} are all low and close in energy, making oxomolybdenum (oxotungsten?) chemistry both available and valuable in the earliest evolutionary changes of redox potential, for example, of sulphide to sulphite and then to sulphate.)

The known retention of nickel hydrogenases in these sulphate organisms means that hydrogen metabolism was not yet lost at the somewhat higher potentials that had arisen. NADH from hydrogen could then be used wastefully to remove O_2 as H_2O_2 or O_2^{-}. The known development in sulphate bacteria of peroxidases or catalases to remove H_2O_2 based on *haem* is consistent with the low redox potential of haem in its early proteins. The removal of superoxide in sulphate bacteria is due to an iron superoxide dismutase, which is again primitive, that is, it is found in many prokaryotes but not in eukaryotes. In this context 'anaerobic' sulphate bacteria do not appear to contain copper, the metal ion used in superoxide dismutases of eukaryotes. In the environment of sulphate bacteria the O_2 pressure is too low to generate a sufficient redox potential to liberate much copper or even zinc from sulphides.

A parallel exists between the development of sulphate-reducing bacteria and bacteria able to handle N_2. The demand for nitrogen is very great and, as dioxygen removed reduced forms of nitrogen, especially NH_3 as N_2, there was not only a thermodynamic difficulty but also a large kinetic barrier to reducing N_2 back to ammonia in cells. These barriers required highly active

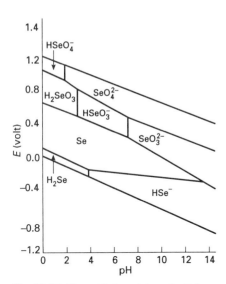

Fig. 12.15 The oxidation state potentials as a function of pH for selenium (see Section 5.7). $C_{Se} = 10^{-6}$ M.

molybdenum and Fe/S catalysts to generate intermediates (as indeed does SO_3^{2-} reduction; see above). Molybdenum and Fe/S centres have the correct range of redox potentials to be naturally available and to be used as catalysts in the N_2 reactions of the nitrogenase. The Fe/S centres are sensitive to O_2 and great protection of the nitrogenase from dioxygen became necessary. It is provided by the accompanying release of hydrogen by these enzymes. The process, $N_2 \rightarrow NH_3$ is extremely costly in energy, ATP (Sections 10.5.4 and 11.8.5), but more energy was slowly being made available from dioxygen production. (The nitrogenase reactions are limited to this day to bacteria protected from dioxygen and may well predate the arrival of multicellular organisms at higher O_2 tension by two billion years (Fig. 12.2). Nitrate, nitrite and the handling of oxides of nitrogen have become a further possible source of nitrogen now for aerobic bacteria, but, if we follow the above timing of redox changes, Fig. 5.20, these oxides of nitrogen and their enzymes must have been late arrivals. As we shall see they are largely dependent on copper.)

The next stage in the rise of redox potential changed the role of selenium from its use in hydrogen or reducing systems generally to its use in protective reactions for the removal of increasing amounts of *organic* peroxides, which are not easily removed by haem peroxidases but which must have become an ever-increasing menace to cells. The selenium redox potential diagram is shown in Fig. 12.15. The dominant species at pH = 7 changes from HSe^- to $HSeO_3^-$ as the oxidising potential is increased toward +0.4 volts. Simultaneously, there arose the possibility of converting I^- to I_2 and hence iodinated phenols such as thyroxine could be made. These iodinated phenols are to this day used as protective poisons by many organisms. Strikingly, de-iodination of the iodophenols takes place through a selenium enzyme, suggesting that the changed uses of the redox reactions of iodine and selenium took place at a similar time in evolution. Later, in animals we find that iodination is used in synthesis of a messenger, thyroxine, by haem peroxidases. Some primitive iodination (later general halogenation) seems to be dependent on vanadium, a metal which also has a low range of redox potentials.

When all the surface NH_3, H_2S, sulphide and selenide were removed by dioxygen, the final stages in redox potential upgrading of non-metals produced small quantities of the poisonous oxides of nitrogen and the halogens, for example NO, NO_2 and ClO^- (Fig. 12.13). They were initially removed by reduction using haem enzymes and then developed for use in bacteria and then presumably in higher organisms, often involving copper. The possibilities are

New chemical $\rightarrow$ poison $\rightarrow$ detoxification $\rightarrow$ signal (uses) $\rightarrow$ simple incorporation (N)
(protection) in system

energy source (ATP) from NO_2^- .

As stated above, use is made of nitrate and nitrite in denitrifying bacteria as sources of energy and elements. The same oxides of nitrogen have been converted into signalling systems in animals. Thus, if we follow the rising redox potentials of non-metals, we must propose that evolution developed using first S_n/S^{2-} then to using SO_4^{2-}/S_n then to using SeO_4^{2-} and IO^- and, finally, to using O_2 and N/O in that order in bacteria. Mitochondria are believed to be incorporated forms of nitrobacteria which were converted to O_2 metabolism only!!

The corresponding *changes in availability* of metals with the gradually increasing redox potential of surface waters are shown in Table 12.3, see marginal note. Note that the metals concerned are present only in traces and that they alter catalysis and signalling, not bulk chemical features. In stages with the rise of redox potential, when most of the H_2S had been oxidised, there must have been first the simultaneous loss of Fe/S complexes from the sea, precipitation of $Fe(OH)_3$, and the gain of zinc from the oxidation of ZnS. The handling of such a new poisonous metal as zinc (a threat to iron systems since it binds to the same centres) could not be by the same method (metabolism) as that for removal of non-metals and had to use changes in chelation and pumping out of zinc ions or its chelates. Thus, probably at first zinc was removed by the protein metallothionein and much of it was placed in vesicles. This may happen still in some algae. Slowly, zinc was then made into a more and more useful hydrolytic catalyst, especially in vesicles away from iron. The useful dependence on zinc in eukaryotes has become of immense consequence as it developed from some early prokaryote use in digestive and synthetic enzymes to employment in a vast variety of eukaryote enzymes and then in cell–cell signalling of several kinds (Fig. 13.28). The binding of zinc in signalling systems is less strong than in (primitive) enzymes, since exchange is required, and arose as higher free zinc concentration developed later in evolution.

The changed competition outside cells between sulfide and oxide centres for metals was strongly biased by the production of O_2, but inside cells and especially in cytoplasmic cellular fluids some but much less directly changed competition between metal elements could be allowed. As stated, oxidation potentials there remained quite low due to redox buffering. The equations that illustrate the changed level of competition for sulphur ligands can be written as

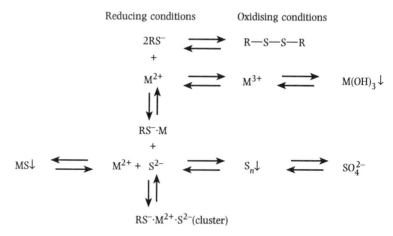

There are here several interacting effects that led to new association of some metals and loss of association of others with both thiolates and sulphide. Nickel sulphide complexes virtually disappeared in eukaryotes while zinc thiolates appeared. Iron was and is now distributed in Fe^{3+}—O—Fe^{3+} centres as well as in remaining iron/sulphur cluster proteins but simple iron thiolates disappeared.

Increasing redox potential next removed more and more sulphide and finally allowed copper sulphide to dissolve in sizeable amounts to give

Metals in aerobic organisms

Na Mg
K Ca...V.Mn Fe (Co)(Ni) Cu Zn
Mo

Contrast Section 10.4.6.

Table 12.3 Major element availability changes with the advent of dioxygen*

Non-metals

$CO_2\uparrow$	$H_2\downarrow$ $N/O\uparrow$	$O_2\uparrow$ $S/O\uparrow$ $Se/O\uparrow$	$Cl—\uparrow$ $Br—\uparrow$ $I—\uparrow$	

Metals

$V/O\uparrow$	—	—	$Fe/O\downarrow$	—	$Cu\uparrow$	$Zn\uparrow$
—	$Mo/O\uparrow$					

* $\uparrow$indicates increases in availability. The formulation $N/O\uparrow$ indicates an increased presence of nitrogen oxides while $Cl—\uparrow$ indicates an increase in covalently bound chlorine. (CH_4 and NH_3 disappeared.)

considerable free copper (see redox potential diagram, Fig. 12.13). Probably a substantial rise in copper did not occur, however, until between 2 and 1×10^9 years ago (see the E_h/pH diagram in Fig. 5.18). Copper was then a very considerable new poison for all biological species.

The arrival of copper must have been a direct special threat to the already developed zinc systems since copper binds much more strongly than zinc to thiolates. However, we observe today that the cysteine zinc-finger thiolates do not bind copper(I) competitively in the presence of metallothionein. Thus, it is extremely likely that most cell systems survived either by sequestration of copper using various special thiolates such as the metallothioneins or by rejecting the excess of copper to vesicles or outside (or by both actions). It is found that very little copper remains in the cytoplasm. A method of rejection of copper by an ATP-using pump, which is similar to that for rejecting calcium, is now known. As with all other newly introduced poisons, there then evolved uses of copper, outside the cell and there copper came into its own in oxidases (see below). (Iron, which could have been used in principle in these reactions, had been removed as $Fe(OH)_3$.) Copper slowly became the major redox element away from the internal reducing media of cell cytoplasm. (Note that in eukaryotes the sole use of copper in the *cytoplasm* is as copper superoxide dismutase.) Thus, we find copper associated with the external metabolism of nitrogen oxides and with various message activities in vesicles as well as in most of the *major oxidative enzymes for cross-linking extracellular matrices*.

It is fascinating to reflect that the changes in dioxygen partial pressure gradually forced association of metals with thiolates from one end to the other along the Irving–Williams series. Metals changed in availability from before the loss of sulphide (H_2S) in the order for M/thiolate binding,

$$Fe^{2+} \text{ before } Zn^{2+} \text{ before } Cu^{2+}(Cu^+),$$

so that residual organic thiolates became bound increasingly in the complexing strength,

$$Fe^{2+} \text{ less than } Zn^{2+} \text{ less than } Cu^{2+}(Cu^+).$$

The exploitation of conditional binding constants (Section 5.6.2) here is quite remarkable. The use of pumping, local distribution of proteins and control of redox potential then manipulated these thermodynamic factors so that activities were placed in compartments for functional use. This is a good example of a combination of thermodynamic and kinetic factors in an overall co-operative flow.

12.4.2 Coevolution of the use of new non-metal compounds and metal elements

Our thesis is that the increasing effectiveness of the application of energy will lead to an increasingly locked-in homeostasis of an improving life form in a fixed environment. In the above we have concentrated on the consequential fact that each new chemical coming into this environment must be a poison for the existing, well-adapted organism. The next step is for the organism to be 'aware' of the poison and then to convert it into an innocuous form. This requires synthesis of a reacting protective chemical, and it is a plausible assumption that this is best achieved by using the poison as the messenger of its own presence and then causing a reaction to it. The final step is to develop value for survival from the newly produced material. Protection from metal poisons can be secured by complexation and subsequent rejection by pumping, while protection from organic molecules can only be secured by metabolism. Now, as we have explained, rapid metabolism is only achievable using metal ions. Since the particular metal and non-metal poisons (or non-metal deficiencies such as $NH_3 \rightarrow N_2$) are generated in a timed redox sequence, we can readily conclude that this sequence would inevitably lead to the new metal that simultaneously became available being used as the new catalyst for the new non-metal removal by metabolism. An example, see above, is the use of selenium in the removal of iodinated organic molecules. The thesis is that the new reaction was first used for protection and then in new pathways and it is obviously best carried out away from the pre-existing primary reductive metabolism since it is largely oxidative. This objective is achieved automatically by pumping the new metal (and its metabolic activity) outside the cell or into a vesicle. Such compartments have great advantages in protecting the primary metabolism going on in the cytoplasm while generating new products. The use of copper provides the best example.

Before leaving this look at overall strategy we must realise that a possible twist in the development is to use an existing metal or coenzyme catalytic centre in a new protein. Effectively, this combination generates a new metal catalyst. The example of the switch of molybdenum chemistry in sulphate and dinitrogen from aldehyde metabolism is paralleled by that of iron and haem iron to dioxygen metabolism. Here there is no increase in availability (in the case of iron there is even a loss), but there is an increase of functional fitness.

The combination of new non-metal and new metal chemistry is summarised in Table 12.4, but there was not just a new potential for chemical reactions but also a new source of energy—dioxygen itself—which we examine next.

12.4.3 Energy transduction from dioxygen

The development of an environment with a possible redox potential as high as that of O_2 (at a partial pressure of 0.2 atmospheres), gave rise to the possible

Table 12.4 Poisons (new chemicals) after the advent of dioxygen

Non-metal	metal
N_2	$Mo(O)$, $V(O)$
SO_4^{2-}	$Mo(O)$
RI	$Se(O)$
Rising O_2 (initial)	$Fe(haem)$, $Se(O)$
Rising O_2^- (initial)	$Fe(O)$, $Mn(O)$
Rising H_2O_2 (initial)	$Fe(haem)$, Mn,
Oxidised organics, e.g.	$Fe(haem)$ in cells,
phenols and sterols	Zn proteins
High O_2 (late)	Cu^{2+}, Fe haem
High O_2^- (late)	Cu^{2+}, Fe haem
Further oxidised organics	Cu^{2+} (outside cells)
Oxides of nitrogen	Cu^{2+}
Oxides of Cl, Br, I	$V(O_2)$, $Fe(haem)$

use by some organisms of reducing compounds in the reaction

$$(CHOH)_n + nO_2 \longrightarrow nCO_2 + nH_2O + energy.$$

The energy capture possible here is very large and is the basis of much of modern life. Today this reaction is at the heart of the oxidation of fats and sugars in the citric acid cycle for production of ATP, the main energy transporter (Fig. 12.16). However, such a development was slow since the O_2-atmosphere built up only gradually. We have seen one intermediate stage where O_2 first converted H_2S to SO_4^{2-}, when SO_4^{2-} was then usable as an oxidising agent for ATP production. With free O_2 at a somewhat higher level, dioxygen could be used directly with haem iron catalysts at a potential around $+0.2$ volts and, finally, as it rose to its present partial pressure and a redox potential $+0.8$ volts, it could be used with copper, which was only available at high O_2 pressure. The result is an expansion of the energy generating S^{2-}/S_n with the H_2/S_n systems successively to the H_2/SO_4^{2-} system at very low pressure O_2 (with haem enzymes) and then H_2/low pressure O_2 (with further haem enzymes) to the final situation of H_2/high pressure O_2 (with Cu, haem enzymes) where H_2 represents here any oxidisable hydrogen in organic molecules as well as the free gas.

Instructively, the handling of dioxygen by cells in order to produce energy may therefore have passed through the following stages:

1. O_2 coupled to a dihaem oxidase which is a simple development from the H_2S/S_n system at the level of quinol oxidase. The energy capture efficiency remained low. Do not forget that H_2S disappeared as O_2 appeared.

2. A similar oxidase to (1) but using copper at the active site; the energy capture efficiency was still low.

3. A high potential cytochrome oxidase using three atoms of copper, bound zinc and two haems. The energy capture efficiency is much greater.

The sequence, detailed in biochemical textbooks, follows adaptation to increasing partial pressure of O_2 concomitant with increasing presence of copper and zinc and lower iron availability. (Note that copper is always toward the outer membrane surface in all these reaction systems.)

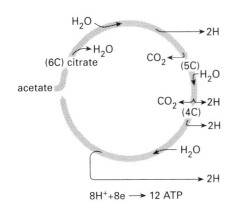

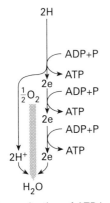

Fig.12.16 The production of ATP in steps in the citric acid cycle by coupling to electron and proton gradients (see also Figs 10.7 and 10.23). The 2H oxidised in each step is used as $2H^+$ in the gradients coupled to ATP-synthetase.

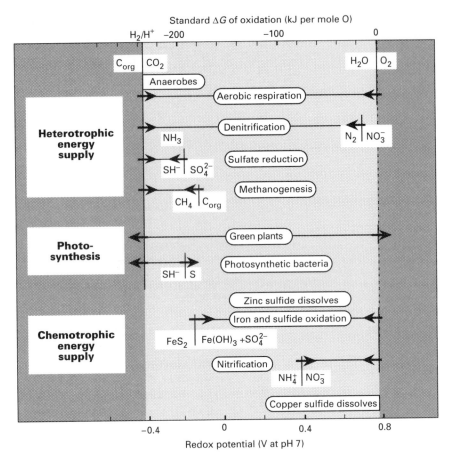

Fig.12.17 The redox potential ranges of certain organisms and simple reactions allowing concentration changes of 10^{10}. (From Cox, P. A. (1995). *The elements on earth*, Oxford University Press, Oxford).

In conclusion, vast changes in metabolism and energy sources using new elements or old elements in new forms became possible—a summary is given in Table 12.5 and Fig. 12.17.

12.5 New uptake and incorporation of elements

The changing character of the environment forced upon the evolving organisms not only the possibility of new energy capture, new pathways of synthesis and new signalling modes, but also quite new difficulties in the selective handling of both previously and newly available elements in many compartments (see Fig. 12.21). The problem concerned a whole variety of metals and non-metals that bind competitively (see Sections 5.5.3 and 5.6.5) and extended to a range of new organic compounds. We can only illustrate major effects and we shall look particularly at the effect on the metabolism and the organic products of iron, copper and zinc.

At the same time, changes of ligands for incorporation were forced upon metals by oxidation. Thus, after the coming of dioxygen, the changes of metal coordination are those shown in Table 12.5. These alterations are due to a combination of changes in metal and in organic compound availabilities and

Table 12.5 Switches in metal co-ordination in evolution

Ligand	Original, M	Present-day, M	Example
Thiolates	Fe^{3+}, Ni^{2+}	$Zn^{2+}/Cu^{2+}/Cd^{2+}$	Rubredoxin† Alcohol dehydrogenase Zinc fingers Metallothionein
$-CO_2^-/N$	Mn^{2+}/Mn^{3+}	Fe^{3+}, Zn^{2+}, Ni^{2+}	SOD* FUR
Porphyrin	Fe*, Ni, Co	Mg^{2+}, Fe, (Co)	Chlorophylls* Porphyrins* Ribonucleotide
O^{2-}	Mn $\diagup^{O}\diagdown$ Mn	Fe $\diagup^{O}\diagdown$ Fe	reductase
Tyrosine	—?	Fe	Catalase
Dithiolate	W(?)	Mo	Aldehyde reductase
$-CH(CO_2^-)_2$	—	Ca	Osteocalcin

* Archaebacteria did not incorporate Mg in chlorin to make chlorophyll. They made nickel and cobalt ring chelates, F-430 and vitamin B_{12}, but little iron porphyrin.
† Rubredoxins are only present in prokaryotes, zinc fingers in eukaryotes only.

must be viewed against a background knowledge of conditional stability constants (Section 5.6.2).

To illustrate the problems we describe some features of the distribution of iron first in aerobic bacteria and then in higher animals. In the following description we need to illustrate control and regulation together. The handling of elements such as iron is obviously critical to the whole homeostasis of aerobic cells, as it was to anaerobic cells.

12.5.1 The new handling of iron uptake

When dioxygen built up above a given low level it removed much iron from the environment. However, fixed levels of iron remained essential for many cellular processes and even for the use of dioxygen. We need to show how a new set of feedback controls and regulatory systems allowed iron levels to be maintained. At the same time, as stated, some products of dioxygen reduction, such as superoxide, are extremely poisonous so that, simultaneously, a defence against poisons had to be developed and iron was obviously valuable in this and other new functions all under feedback *regulation* related, of necessity, to the primitive primary metabolism inherited from anaerobes.

The regulated uptake of iron has been studied in a number of aerobic bacteria, for example *E. coli*. The uptake is governed by the synthesis of small chelating agents produced by oxidative metabolism and then rejected into the environment (Fig. 12.18). These agents, called siderophores, bind iron and then re-enter the cell through series of proteins integrated in the cell membranes. On entering the cell the iron binds to a protein, FUR,† as well as being generally distributed into the proteins that the cell requires, which is part and parcel of control. The iron-bound FUR binds to DNA and stops production of siderophores. At this stage iron-dependent metabolism is re-established, for example in the citric acid cycle. The feedback dependence is quite clear:

† FUR is an abbreviation of *f*erric *u*ptake *r*egulatory protein.

(a)

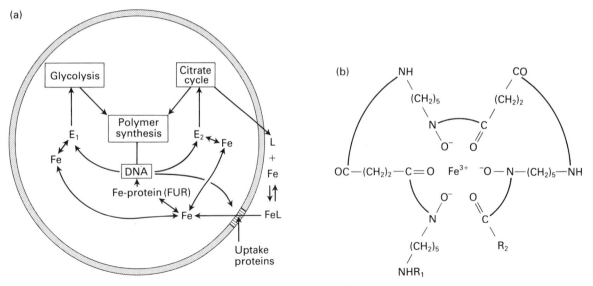

(b)

Fig.12.18 The scavenger system for iron in bacteria. A siderophore scavenger chelate, L, is shown in (b).

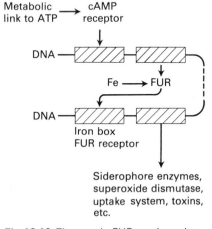

Fig.12.19 The protein FUR regulates the iron-box gene so as to produce the range of proteins shown. FUR itself is produced under the control of a gene regulated by cAMP. Thus phosphate and iron are connected through the genes as well as in metabolism, see also Fig. 11.23.

expression of proteins by DNA (via RNA) is iron-dependent. The proteins involved and which are switched on by the *absence* of iron are the enzymes for the production of the chelates and the proteins of the uptake mechanisms for the iron chelates in the outer membranes. It may well be that FUR requires zinc to be able to bind to DNA; thus iron and zinc are linked together at the DNA level in aerobic prokaryotes.

The iron-binding protein FUR is also regulatory for the production of other enzymes protecting the cell from dioxygen-derived poisons including super-oxide dismutase. Thus, the supply of iron for use in oxidative metabolism is linked to the production of enzymes for protection against accidental generation of poisons.

Similar mechanisms exist for the uptake of copper and manganese, for example, but environmental chelating agents are not so essential in bacteria since these metals became more available than iron.

12.5.2 The linking of new iron uptake to phosphate: controls and regulation

The next step to consider is the production of FUR itself. FUR is produced under the influence of a cAMP-dependent binding protein that also binds to DNA (Fig. 12.19). Thus, a cell that is deficient in cAMP cannot pick up iron. The nature of cAMP as a cell regulator is well known (Chapter 11) and provides a link back to the anaerobic systems central to all cellular purposes in that the concentration of cAMP is connected to the ATP concentration in the cell, which is a measure of its energy status, that is

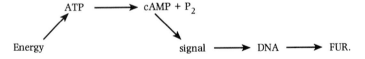

Thus the uptake of iron, which is linked deeply inside the mechanism of oxidative phosphorylation, that is, to energy use in ATP production, is feedback-controlled by the product of this production via cAMP. The role of ATP is absolutely crucial to condensation reactions, see Chapter 9, and iron is almost equally essential for redox reactions in aerobes. To maintain homeostasis the feedback connection between Fe and phosphate is and always has been essential. It illustrates the requirement for cross-linked feedback to give homeostatic stability.

Now we know that iron levels are very probably connected to the uptake mechanisms of bacteria for many other elements via Fe/S regulatory proteins that bind to DNA and RNA. Amongst them are those for both Mo (via *nif* genes) and Ni. Iron uptake is also linked to cobalt (vitamin B_{12}) uptake through the general system of proteins produced in the membrane for metal siderophore uptake (Fig. 12.18).

In many bacterial cells it is advantageous if the cell can switch from aerobic to anaerobic conditions. The switch requires a dioxygen-dependent regulatory protein that binds DNA differently in the presence and absence of O_2. This is also an Fe/S-regulated process in many such facultative anaerobes.

12.5.3 Iron uptake in higher organisms and its connections

In multicellular higher animals and plants the system for iron uptake by cells has been upgraded since transport between cells is now necessary. In plants the changes are small. The roots are able to exudate siderophores to take in iron and this iron is then passed on to an internal iron carrier, citrate. The stem and leaf cells have receptors for iron citrate and iron is stored in the protein vesicle ferritin. The nature of the feedback control is unknown but it may be rather like that in bacteria, that is, directly at the level of DNA transcription of RNA.

In higher animals the carrier of iron in circulating fluids is a protein, transferrin, for which there are receptors on the cell surfaces. The iron is passed to a store in ferritin from which it is distributed internally (Fig. 12.20). The entire system is under *regulated translation* at the RNA level by an iron sulphur protein, which, like aconitase in the citric acid cycle, can undergo a reaction of the type

$$Fe_4S_4 \rightleftharpoons Fe_3S_4 + Fe^{2+}.$$

The uptake of iron is linked, therefore, not just to controls and regulations of those processes that require iron but is connected to virtually all cellular processes in aerobic bacteria and in the cytoplasm of plant and animal cells. We shall see this shortly when considering the relationship of iron to much of secondary metabolism and then to communication networks between cells (Chapter 13).

12.5.4 The uptake, distribution and incorporation of other elements

While we have described iron uptake and its regulation and control, similar systems have evolved for other metals (see Fig. 13.2). Typical cases are those of

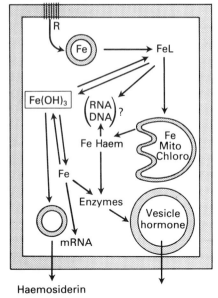

Fe(transferrin)

Haemosiderin

Fig.12.20 A generalised scheme of iron uptake for multicellular animals. The differences from Fig. 12.18 arise in compartmentalisation, absent in prokaryotes, in the changed nature of chemicals, for example ferritin, and in the pathways of control via RNA. Some controls go from haem iron to DNA, but the *fur* iron box is missing. L is a ligand and Mito and Chloro refers to organelles. The RNA regulation is of transferrin as well as of ferritin $Fe(OH)_3$ and of the receptor, R.

of copper and zinc. Here carrier proteins in the circulatory systems of animals, albumins, are known and pumps at least for copper have been described. Storage of the metals takes place in a variety of metallothioneins inside cells. The connection with DNA in a regulatory role for several such proteins is known and the regulation is probably linked to several other factors that bind DNA. We describe these features in Chapter 13. The selective incorporation of these transition metals into proteins follows the prescriptions given in Chapters 5 and 10 and in our previous book (see reference 1 in 'Further reading').

The distribution of non-metals is also a problem in multicellular systems. The most studied example is that of dioxygen itself, carried by haemoglobin, haemocyanin and haemerythrin. The last two appear to be relatively clumsy systems dropped later in evolution, although persisting in those organisms that evolved around 500 million years ago. The biochemistry of these three helical allosteric proteins is described in standard texts. Of the other non-metals, C/H/O/N circulate as glucose or other small water-soluble molecules, phosphorus circulates as phosphate and sulphur as sulphate and both have carrier proteins while some complex molecules such as lipid-soluble hormones, for example iodine in thyroxine, are carried in fat globules. It is these molecules which must now be recognised by receptors and pumped into cells using H^+ or Na^+ gradients or ATP energy. Inside the cells the functions of the elements remain generally similar to those in primitive cells, but some details have changed.

12.5.5 Changes in the uses of some non-metals

With the emergence of a dioxygen atmosphere the basic metabolism of phosphate as seen in anaerobes remained unaltered but one new feature came into play. A series of sugar phosphates, inositol phosphates, became bound to a particular fatty acid in the membrane—arachidonic acid, a product of dioxygen oxidation. The inositol tri- and tetraphosphates and perhaps others became major signalling devices in cells especially when coupled to calcium (Section 13.3.2).

Sulphate, absent from *strict* anaerobes, became of value in the sulphated polysaccharides which allowed a very open mesh of polymers to be created around a cell facilitating necessary rapid diffusion to and from it. The special feature of sulphate is that it gives very strong acid esters, RSO_4^- which do not bind to ions or form H bonds easily.

The oxidation of nitrogen generated nitrate whence it was possible to gain access to a new nitrogen source. The denitrifying bacteria can handle all the oxides of nitrogen, but nitrogen remains incorporated as NH_3 only. The new chemistry did give opportunity to create new messengers, in this case nitric oxide, NO. Note the close association of nitrogen oxide metabolism with copper and molybdenum enzymes outside cells of bacteria.

Finally and apart from carbon metabolism, we need to look at the halogens. Once again, the presence of oxygen allowed them to become oxidised and then incorporated. In particular, iodine has taken on a major role in the hormone thyroxine, but in many organisms in the sea there is a rich variety of bromine and chlorine compounds (see references). (The development of this oxidative chemistry reaches its height, however, only in the hands of man.)

12.6 The distribution of elements in compartments

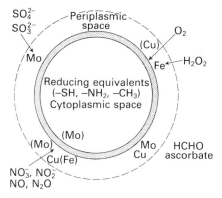

Fig.12.21 The distribution of elements in the periplasmic compartment of aerobic bacteria.

If we look first at the development of aerobic bacteria, the simplest cells, we find that they have changed little internally (Table 10.1). As stated, primary metabolism had to remain intact. On the other hand, new elements and possible metabolites surrounded these organisms and new sources of energy became available. To utilise these compounds and their energy content and at the same time to avoid confusion of metabolism in the cytoplasm the obvious advance was to develop compartments (Chapter 7). Figure 12.21 is an idealised description of a Gram-negative bacterium with an extra compartment, the periplasmic space. The compartments available are

$$\text{Outside} \parallel \text{periplasm} \parallel \text{membrane} \parallel \text{cytoplasm.}$$

Figure 12.21 shows how the new metabolic processes, oxidative phosphorylation and handling of oxides of S and N, are managed in the membrane or in the periplasm. The dangerous oxidative products that could enter the cell, O_2^- and H_2O_2, are removed in the cytoplasm by Fe(Mn) enzymes. It is particularly noticeable that virtually all copper, the new oxidative metal catalyst, is used in the outer membrane or in the periplasm. A new use has also been found for *oxomolybdenum* in handling SO_4^{2-} and NO_3^- often outside cells (Table 12.6).

Table 12.6 Substrates of metalloenzymes in periplasm

Copper enzyme	Molybdenum enzyme	Haem iron enzyme
NO	NO_3^-	e (cytochrome *c*)
N_2O	SO_4^{2-}	
O_2	Dimethylsulphoxide	
Cytochrome oxidase	Aldehyde oxidase	Cytochrome oxidase (O_2)
(outer face of membrane)		(outer face of membrane)
(NO_2^-)		NO_2^-

12.6.1 Element distribution in eukaryote cells

Inspection of eukaryote cells shows that the compartmental complexity is so great that, at present, many analytical uncertainties remain. However, we can give an outline of the distribution of the elements. Figure 12.11 shows a plant cell and Fig. 12.22 a typical animal cell but do remember that each differentiated cell is different in element distribution. We see that, just as in bacteria, the metabolism and elements in the cytoplasm are only slightly modified while the organelle and vesicle chemistry, but especially the intercellular circulating fluids (not the true outside), are quite different. The distinctive features are that the extracellular zone is truly an oxidising medium and operates using, for example, high redox potential copper proteins, but the cytoplasm of cells is effectively anaerobic.

The storage of elements also became important, and we have had occasion to note amongst others:

• calcium—in many vesicles and biominerals;

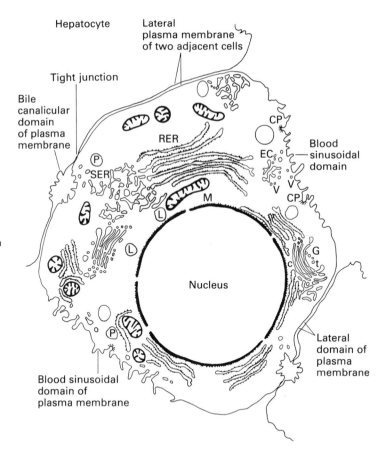

Fig.12.22 Section through a hepatocyte illustrating the various organelles and membrane systems. EC, endocytic compartment; G, Golgi apparatus (c=*cis* face; t=*trans* face); L, lysosome; M, mitochondrion; P, peroxisome; RER, rough endoplasmic reticulum; SER, smooth endoplasmic reticulum; V, vesicle; CP, coated pit.

- iron—in ferritin in the cytoplasm;
- phosphate—in bone and some pellets;
- zinc—in metallothionein clusters in the cytoplasm and in vesicles as Zn^{2+};
- copper—in metallothionein clusters, as Cu^+;
- manganese—in vesicles and vacuoles.

These stores can act as buffers giving a thermodynamic control over homeostasis. The stores can be located in selected cells; for example, the liver is heavily loaded with iron and some parts of the pancreas, the brain and the reproductive tract are high in zinc.

12.7 New small organic compounds

We cannot describe in any detail the changes in biological organic chemicals with the coming of dioxygen since there is such a huge variety of so-called secondary metabolites often produced by iron enzymes (Table 12.7). They extend over the following types of bioorganic chemicals:

1. Membranes (Table 12.8). The synthesis of cholesterol, present in eukaryote and higher organism membranes, extended the membrane flexibility and

Table 12.7 Production of new organic compounds

Catalytic element	New compound or reaction
Mn	O_2, H_2O_2, glycosylation?
	Lignin
Fe	Hydroxylation of sterols, phenols
	Cytoplasmic oxidation in animals
	General oxidation in plants
	Chitin, NO, thyroxine
Co	None
Ni	None
Cu	Oxidative cross-linking
	Oxidation of peptides
	Oxidation (extracellular) of amines, ascorbate, sugars, etc.
Zn	Peptides from proteins
	Hydrolysis
Mo	Nitrite from nitrate
	Sulphite from sulphate

Table 12.8 Lipid (membrane) components (fats) *

Prokaryotes Archaebacteria	Bacteria (cyanobacteria)	Eukaryotes (O_2)
Saturated	Saturated	Saturated
Some unsaturated	Some unsaturated	Some unsaturated
Ethers in membranes	Polyunsaturated†	Polyunsaturated†
	Sterols†	Sterols
		Later sterol hormones†
		IP$_3$? Arachidonate?
	Haem	Prostaglandins†
	(Chlorophyll)	Halogen compounds

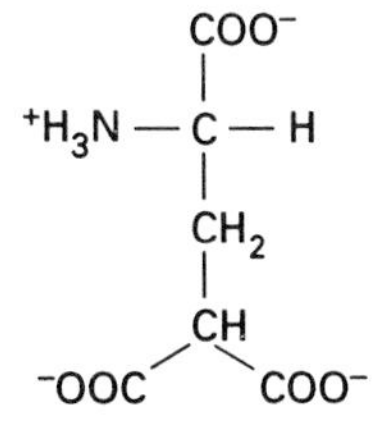

γ-**Carboxyglutamate**

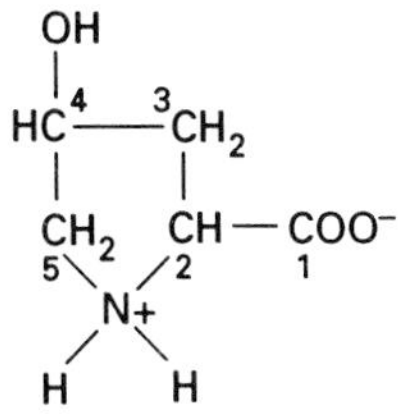

4-Hydroxyproline (Hyp)

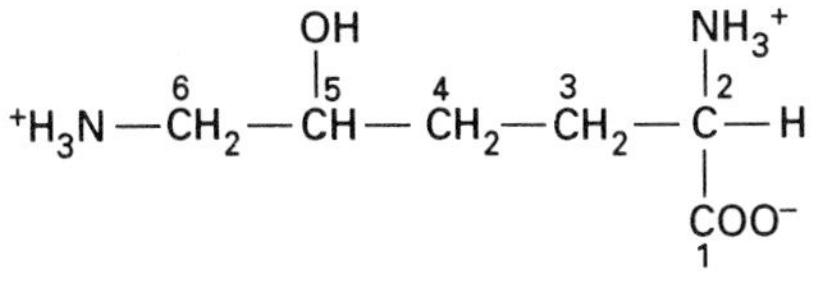

5-Hydroxylysine (Hyl)

* Source: Mead, J. F., Alfin-Slater, R. B., Houston, D. H. and Popjak, G. (1986). *Lipids: chemistry, biochemistry and nutrition.* Plenum Press, New York.
† Require dioxygen in desaturases (flavin + Fe_n/S_n) or oxygenases (Fe or haem).

its temperature dependence related to 'phase' stability, so allowing a greater variety of membrane functions. The effect on 'phase' formation is described in Chapter 4. The production of arachidonic acid allowed phosphate signalling to advance.

2. Proteins (Tables 12.9 and 12.10). The modification by hydroxylation and oxidation of amino acids such as proline, lysine, arginine, glutamic acid, tyrosine and tryptophan (see marginal illustration) allowed new protein structures to be built especially outside cells. It also created new messenger systems. The link to biomineralisation is of great consequence. New series of proteins arose opposite particular elements. Novel peptides were also made.

3. Polysaccharides. The sulphation of polysaccharides generated an open-mesh polymer network around cells, which is essential so as to allow access to the cell.

4. Plants produced special cross-linked polymers, lignins.

5. Oxidation of cholesterol and arachidonic acid and other related hydrocarbons has led to a range of messengers and hormones, such as the prostaglandins.

6. Halogenation of phenols in particular has created both hormones and poisons (and note antibiotics).

7. A vast range of new colouring and tanning agents were produced as well as such compounds as alkaloids, of little known value for survival, which came about through oxidases.

All these activities are described in books under *secondary* metabolism. Many of the reactions are also found in unicellular organisms, even bacteria of today. However, very many more are absent from bacteria and to go much further with discussion of the changes in metabolism we need to note the value of the new chemicals to bacteria and then to eukaryotic cells, with several compartments, and finally to multicellular organisms. A glance at the above list shows that many new uses of organic chemicals involve extracellular matrices or as messengers in multicellular systems. We deal with the first group in this chapter while reserving the discussion of messengers and hormones in signalling to Chapter 13.

12.8 New proteins and enzymes

It is apparent from the above that series of new proteins were required for uptake, incorporation, storage, structures (Table 12.9) and enzyme activities (Table 12.10) as well as for control and regulation. It is not possible yet to put the syntheses of these proteins into a rational system but their presence could not be allowed to cause confusion of the anaerobic primary metabolism and yet must dove-tail with it. The way in which the proteins arose has to be a matter of unravelling the DNA changes. We see this as a process of mutation but it would be nice to find that the exploitation of mutation was not general to all the DNA in a search for progress. Thus, the modification by random selection for copper function, for example, might well be due to an increased susceptibility to local DNA mutation related to particular proteins for copper binding (Fig. 13.27). The best targets for exploration were the thiol-rich proteins since they were already involved with iron and zinc. (Some new enzymes which appeared are listed in Tables 12.13 and 12.14.) The simplest mutational change could have been to modify the proteins for zinc, with which copper would interfere, so that the modified metallothioneins would control copper function. In fact, there are copper proteins of the metallothionein kind, including one known as ACE-1 which binds to DNA. There is also a thiol-dependent copper pump that has evolved from a calcium pump, though the zinc pump has not been found yet. In parallel, the evolved sites for zinc, zinc fingers, may be distant relatives in binding sites to the rubredoxins—iron proteins. Their co-ordination chemistry is very similar.

A more taxing problem is the slow development of structural proteins in cells. There is the basic suggestion that early filamentous proteins were actin-like but that subsequently the tubulins evolved. The latter are extremely

Table 12.9 Some new non-enzymic proteins of particular elements

Element	Protein
Mn	Concanavalin
Fe	Haemerythrin
	Ferritin, transferrin, haemoglobin, myoglobin
Co	None
Ni	None
Cu	Metallothionein, Cu-pump, haemocyanin
Ca	γ-Glutamic acid proteins in extracellular matrix calmodulin and many related proteins, annexins, etc.
Na/K	Na^+/K^+ ATPases pump (animals (vertebrates?) only)
Zn	Metallothionein, zinc fingers
P	IP_3 (synthesis and receptor), phosphoproteins of shell and bone

Table 12.10 Some new specific metal-ion catalyses

Small molecule reactant	Metal ion	Examples
N_2	Mo(Fe)	Nitrogenase
NO_3^-	Mo	Nitrate reductase
SO_4^{2-}	Mo	Sulphate reductase
$O_2 \rightarrow H_2O$, NO, N_2O	Fe	Cytochrome oxidase
Oxygen insertion (high redox potential)	Fe	Cytochrome oxidase
SO_3^{2-}, NO_2^-	Fe	Reductase
$H_2O \rightarrow O_2$	Mn	Oxygen-generating system of plants
H_2O_2/Cl^-, Br^-, I^-	Fe(Se)(V)	Catalase, peroxidase
H_2O/urea, CH_3CO^-	Ni	Urease

important proteins since they give rise to a structure upon which vesicles can move around, but their production is not known to depend on O_2.

12.9 Metabolite and protein distribution

Just as the complexity of eukaryotic metabolism could have led to confusion of function at the levels of elements so that to avoid this problem elements were pumped into different compartments, so it happened with metabolites and proteins. In particular, we find oxidation reactions in mitochondria or

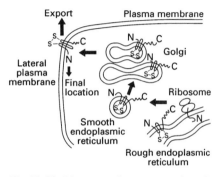

Fig.12.23 Diagrammatic representation of the topography and probable intracellular trafficking route of a protein.

peroxisomes, and in smaller vesicles many organic chemicals are prepared for export (see Chapter 13). Table 12.11 gives examples of the association of metals with organic molecules in different parts of cell space.

Protein distribution is much more complex in higher cells than in prokaryotes (Section 10.8.3). The proteins now have many destinations to a variety of compartments and are not sorted solely by signal sequences for crossing membranes. There is an additional pre-treatment of proteins for delivery to specific vesicles in the Golgi apparatus (Figs 12.10 and 12.23). Here a protein is modified, for example by glycosylation and sulphation, and passed through several steps before being located in a vesicle or a membrane. The vesicles themselves have selected destinations guided by the filament construction, including the supply of proteins to the outside of the cell, so that a cell is a complex organisation with proteins distributed in space. The sequence information in DNA becomes, therefore, not just the protein fold but also more and more strongly the protein location in space. Those proteins directed to the membrane play a large part in deciding cell–cell organisation as well.

We turn next to the special case of the production of proteins and other polymers for the extracellular matrix.

Table 12.11 Compartments and reactions in aerobic eukaryotes

Compartment	Reactions	Elements
Cytoplasm	Synthesis of DNA, RNA proteins	Mg, Zn, K
	Glycolysis	Ca, Mg
Mitochondria	Oxidative phosphorylation Citric acid cycle	Fe, (Cu), (Ca), (Mg)
Chloroplasts	Photophosphorylation	Mg (chlorin), Fe(Cu)
	Carbon assimilation	Mg
Sarcoplasmic reticulum	None(?)	(Ca)
Golgi apparatus	Glycosylation Sulphation	Mn
Endoplasmic reticulum	$2RS^- \rightarrow R—S—S—R$	(?) (Ca)
Vacuoles	Urea hydrolysis	Ni, (Mn)
Peroxisomes	Oxidative degradation	Fe, Mn
Circulating fluids	Hydrolysis	Zn, Ca, (Na)
	Oxidations	Cu
Vesicles	Peptide oxidation	Cu
	Adrenaline synthesis	Fe, Cu

12.10 Putting together cells with filaments: multicellular systems

The development of vesicles and filaments in cells and, more clearly, flagellae outside cells may belong to the anaerobic world (Section 11.12). It is even likely that some differentiated cells existed in early colonies of such organisms. There must have been loose fibrous structures holding the whole of a colony together using calcium. We shall describe the biomineralisation of these structures shortly. Further development toward multicellular organisms depended on the production of connective tissue of a more permanent nature (Fig. 12.24). Today we find these connective tissues in various forms of cross-

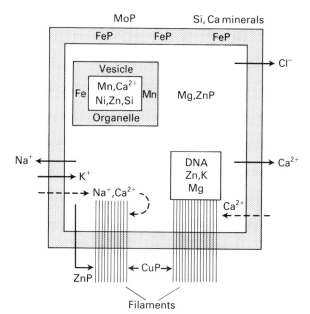

Fig.12.24 The distribution of elements in a eukaryote cell that is part of a multicellular system. P, protein. Compare Figs 10.24 and 10.25.

Table 12.12 Cross-linking processes

Reaction	Catalyst
Hydroxylation of proline	Iron proline hydroxylase
Hydroxylation of lysine	Iron lysine hydroxylase
Oxidative coupling of collagen	Copper lysine oxidase
Cross-linking of chitin	Copper tyrosinase
Synthesis of lignin	Manganese and iron peroxidases
Incorporation of calcium	Calcium
Disulphide bridges	(?)

Table 12.13 Metal enzymes of the external matrix

Enzyme	Metal	Activity
		Hydrolysis of
Elastase	Zn	Elastins
Collagenase	Zn	Collagens
Stromelysin	Zn	Proteoglycans
Gelatinase	Zn	Gelatins
		Oxidation
Lysine oxidase	Cu	Cross-links collagen and elastin
Laccase	Cu	Cross-links via phenols
Phenol oxidase	Cu	Cross-links via phenols
Chitinase	Mn/Fe	Destroys chitin

linked plastic materials (Tables 12.12 and 12.13). Such materials are formed in both plants and animals of all kinds. The cross-linking is mainly due to oxidative, free-radical coupling since coupling of polymers by condensation is not achievable outside cells where there is no dehydrating agent such as ATP or $RS \cdot COCH_3$. The fact that copper enzymes are used (Table 12.13) dates this development at a time when considerable amounts of copper had accumulated. We have already seen that this could only have happened after the dioxygen pressure rose considerably. This may explain the huge surge in evolution starting some time after 1×10^9 years ago (Figs 12.1 and 12.4).

Now, in multicellular development cells must also divide and grow and must not be hindered by too permanent an extracellular system (Fig. 12.24). In effect, the demands of growth and firm structure are contradictory. Biological systems solved the problem by a repeated make/break/remake succession of reactions that depend on copper oxidases for cross-linking and zinc proteases for cutting to make space (Table 12.13). (The increasing value of zinc is of great interest, and not only in hydrolysis, since it could only be that, as dioxygen increased in partial pressure, this metal, which increased in waters, was used in cross-linking the polymers of tyrosine-containing chitin, in the connective tissue of worms, for example, after these chitin substances were produced by copper enzymes. In vesicles zinc proteases could also act in hydrolysis.)

12.11 The redox metabolism outside cells

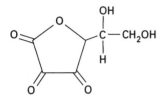

Ascorbic acid

Ascorbate

Dehydroascorbic acid

The creation of a large external liquid pool outside cells in the extracellular flowing matrix generated quite new problems of redox metabolism in this environment. We shall see that the mineral balances are managed by organs developed in animals. pH could always be looked after by CO_2/HCO_3^- exchange across cell membranes and by other buffering anions, but the use of redox reactions in a controlled manner is a much more difficult problem. Clearly, the real potential cannot be allowed to be that of O_2/H_2O system, 0.8 volts, since this potential would destroy many organic compounds needed to make extracellular matrices. Now, a plant maintains itself in a reducing chemical environment through the use of light in photosynthesis, and has a large excess of carbon at the $(CHOH)_n$ oxidation state level, for example as cellulose. Interactive with the cellulose structure, oxidative enzymes are required for cross-linking the matrix and so on, but the potential must never be allowed to increase greatly even locally. Plants developed a mechanism for limiting the effective oxidation potential outside and to some extent inside cells in vesicles to about +0.2 volts using a new chemical, ascorbic acid, (marginal illustration), largely with haem peroxidases as the oxidative enzyme.

The aerobic animal world has no way in which to maintain the low potentials, of course, except through the absorption of reduced substrates from plants or bacteria. The ultimate source is again the photochemical reduction of CO_2 and water to the carbon oxidation state level of —CHOH—. This constant consumption of plant reducing material allows the cytoplasm of all animal cells to be maintained effectively in a moderate reducing external medium.

The question must then arise as to the manipulation of the redox buffering of the animal extracellular fluids where we know that —SH groups are largely required to be converted to —S—S—, that iron is transported as Fe^{3+} and that copper is used at potentials above +0.2 volts in many enzymes, for example for cross-linking collagen. It would appear that two mechanisms have developed using circulating sugars such as glucose and the vitamin (for man) ascorbic acid. In animals both the oxidases for these substrates, which maintain a relatively low extracellular redox potential, are based on copper (Table 12.14). The enzymes responsible for maintaining a modest homeostatic redox potential *in vesicles* also use ascorbate with new copper oxidases (Table 12.14). Thus, the use of reducing sugars by animals appears to be a more recent innovation than that of the earliest uses in plants. Plants often use haem in circumstances where animals use copper in enzymes; for an example consider hormonal oxidative systems, Chapter 13. Plants came first.

Table 12.14 Ascorbic acid and copper

Enzyme using ascorbate	Comment
Ascorbic acid oxidase	Extracellular redox buffer (Cu)
Lysyl hydroxylase	Collagen synthesis and cross-
Proline 2-oxoglutamate dioxygenase	linking in which last step requires extracellular copper (Cu)
Dopamine β-mono-oxygenase	Vesicular enzymes (Cu)

NB. Glucose oxidase is a copper enzyme.

12.12 Biomineralisation: introduction

One major continuously refined homeostasis of elements outside cells is that of the mineral elements, Na^+, K^+, Cl^-, Mg^{2+}, Ca^{2+} amongst others. This topic is covered in Chapter 13 while here we refer specifically to the elements whose external homeostasis is related to biomineralisation. (See references 18 and 19 in Chapter 11.) Biomineralisation belongs to the extracellular compartment. The minerals that biology can handle are limited by the availability of elements in aqueous solutions. The earliest known biominerals are $CaCO_3$ and $SiO_n(OH)_{4-2n}$, both of which are close to their solubility limits in the sea. The fossil record indicates that possibly the calcium phosphates that appeared later were produced after life entered freshwater, in which calcium concentrations are likely to be $< 10^{-4}$ M. The precipitation with which we are concerned must therefore be controlled by the organism since the concentration of phosphate is greatly elevated in organisms (2×10^{-3} M) with respect to the sea (10^{-5} M) where calcium is 10^{-2} M. Bony fish perhaps re-entered the sea from freshwater having evolved from worm-like species (Fig. 12.25).

Plants on land do not form calcium carbonates or phosphates, but only oxalates and silica. The plant maintains a low extracellular pH of 5.5, so that the salts of weak acids are of high solubility. Insects rarely precipitate any salts, except for calcium citrate in the early stages of development of some spiders. The origin of insects from non-flying soft-bodied worms (annelids) differs from the origin of birds (from bony fish) and their calcium metabolism is quite different. In some sense, an insect is more closely related to a plant and can use similar methods in oxidative reactions. The structure of connective tissue in insects is overwhelmingly based on chitin, not collagen. The difference in evolution extends to the formation of organic teeth in insects and worms/chitons, that are cross-linked by zinc, as compared with those of animals (apatite, calcium phosphate).

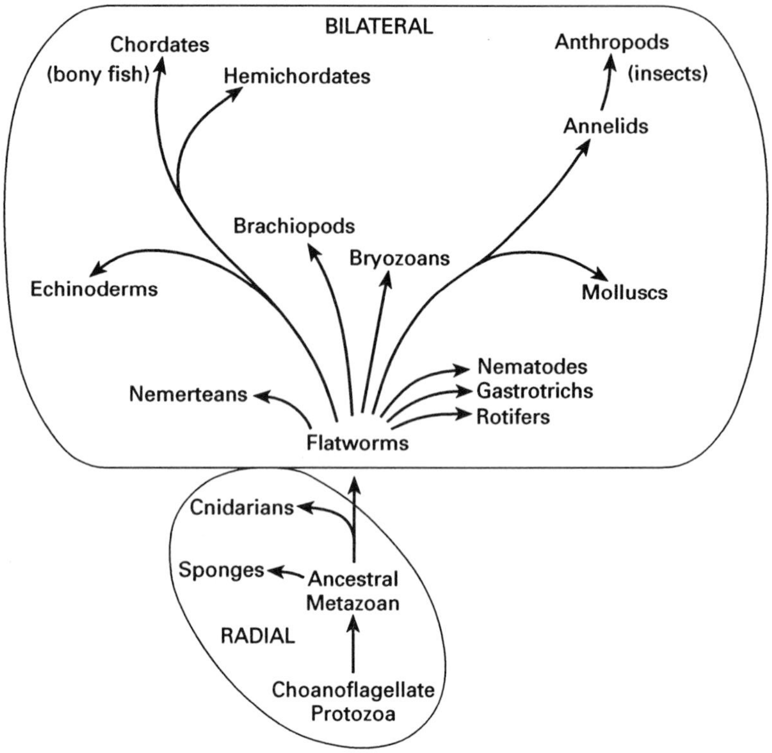

Fig.12.25 Distribution of symmetry within the animal kingdom and the development from ancestral worms to modern animals. After Barnes, R. D. (1987). *Invertebrate zoology.* Saunders College Publishing, Philadelphia. Reproduced with permission.

12.12.1 Biominerals: examples

The mineralisation of biological systems changed from the haphazard to the functionally regulated as cells developed elaborate structure and then multicellular organisms appeared (Table 12.15). The first example of disordered building was the stromalite, mainly built from amorphous silica. The next example we take here is the silicification of structured, perhaps totally

Table 12.15 The main inorganic solids in biological systems

Cation	Anion	Formula	Crystal	Occurrence	Function
Calcium	Carbonate	$CaCO_3$	Calcite	Widespread in	Exoskeleton
			Aragonite	animals and plants	Gravity, Ca store
			Vaterite		Eye lens
	Phosphate	$Ca_{10}(PO_4)_6(OH)_2$	Hydroxyapatite	Shells, some bacteria, bones and teeth	Skeletal, Ca store (piezoelectric)
	Oxalate	$Ca(COO)_2 \cdot H_2O$	Whewellite	Insect eggs	Deterrent
				Vertebrate stones	Cytoskeleton
		$Ca(COO)_2 \cdot 2H_2O$	Weddellite	Abundant in plants	Ca store
	Sulphate	$CaSO_4 \cdot 2H_2O$	Gypsum	Coelenterate statocysts	Gravity
				Plants	S store, Ca store
	Silicate			Phytoliths	
Iron	Oxide	Fe_3O_4	Magnetite	Bacteria—chitons' teeth	Magnetic device
		$FeO(OH)$	Ferritin	Widespread	Iron store
Silicon	Oxide	SiO_2	Amorphous (opaline)	Sponges, protozoa abundant in plants	Skeletal, deterrent
Magnesium	Carbonate	$MgCO_3$	Magnesite	Reef corals	Skeletal

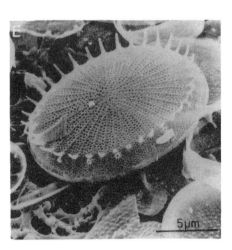

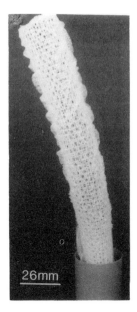

Fig.12.26 Siliceous diatom: gel-like silica membrane bounded.

Fig.12.27 Siliceous sponge: organisation on a macroscopic level.

Fig.12.28 An example of a radiolarian.

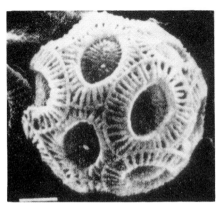

Fig.12.29 The structure of the complete shell of a coccolithiophorid, *Emiliania huxleyi* (scale bar, 1 μm).

anaerobic cells that have a network of internal filaments as well as a network of supporting filaments (spectrins) under the cell membrane. In principle, the decoration of the filaments could be the reason for the fantastic shapes and forms of the diatoms (Fig. 12.26). A small degree of differentiation could well have led on to the functionally useful structures of silica in advanced sponges (Fig. 12.27). The rigidity of these structures suggests underlying cross-linking of extracellular organic polymers, which may have required oxidative reactions using dioxygen. It is known that spectrin, the protein that underlies the membrane of many cells, evolved close to the time of the switch to multicellular organisms.

Silicas are not the only materials that could have been structured in this way and, in fact, most external biominerals may have been so structured. An example is the calcium carbonate of corals and radiolaria (Fig. 12.28). Sophistication could come about by the deliberate build-up of units of the external structures within vesicles before their deposition outside cells. No doubt such materials as the $CaCO_3$ of coccoliths (Fig. 12.29) were also built like this and the individual units may not have needed cross-linked polymers. The systematic putting together of units did not yet provide a firm continuous external structure for organisms, but this can be solved in two ways as follows. Internally, the units can be based on a fixed central point or can grow under a membrane as does the $SrSO_4$ of acantharia. (Note the use of sulphate minerals often in waters of very low O_2 tension.) The use of external scaffolds depends on chitins to give rigidity as is well known in seashells. Here the shell is a true extracellular mineralisation. Chitins are very difficult to break down and generally they use minerals to build units that outlive the lifetime of the organism, for example, mollusc shells (Fig. 12.30) and insect coats.

There is still another possibility—that of developing biominerals inside organisms, for example in bony fish and man. The material used is invariably

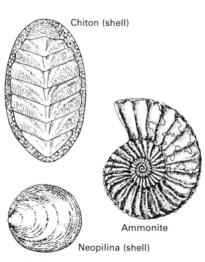

Chiton (shell)

Ammonite

Neopilina (shell)

Fig.12.30 The forms of some monovalve shellfish showing different variations on conical shape and illustrating at the right the logarithmic spiral. Here the shells are dominant in the appearance of the animal, but the control over shape rests internally and grows two-dimensionally.

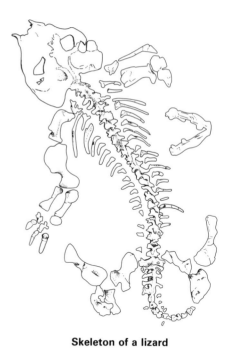

Skeleton of a lizard

Fig.12.31 Skeleton of a lizard, which grows in all directions.

$Ca_2(OH)PO_4$, bone, since it grows in the same way as an organic matrix (Fig. 12.31) outside the cell, but inside the organism.

When we look at the shapes of minerals and compare then with biological shapes we have to remember the use of both co-operative thermodynamic binding as described especially in Chapter 2 and co-operative feedback kinetic flow as described in Chapter 7.

12.13 Summary

This chapter has shown that the advent of a relatively high pressure of dioxygen in the atmosphere has generated concomitantly the development of new physical and chemical characteristics of organisms. Undoubtedly, the first effects of dioxygen were the gradual release of new elements, lowered amounts of other elements and formation of new forms of a third group of elements in the environment, possibly in the order:

(1) decreasing amounts of H_2 and reduced forms of nitrogen and carbon in the atmosphere and a corresponding *relative* increase in N_2 and CO_2 while much carbon and hydrogen were buried as coal, oils and gases;

(2) lowering of H_2S levels and replacement by SO_4^{2-};

(3) lowering of selenide levels and replacement by SeO_4^{2-};

(4) change of form of molybdenum from MoS_4^{2-} to MoO_4^{2-};

(5) loss of Fe^{2+} and precipitation of $Fe(OH)_3$; gain of zinc from oxidation of ZnS;

(6) introduction of iodination;

(7) introduction of copper in solution from oxidation of CuS;

(8) introduction of oxides of nitrogen (and of lighter halogens, but not F).

The changes took place in a sequential manner as the oxygen tension slowly increased.

Note that little change took place in the levels of Na^+, K^+, Mg^{2+}, Ca^{2+}, Mn^{2+}, $Si(OH)_4$, HPO_4^{2-} and Cl^-. There were smaller changes in nickel and cobalt levels as well as increases of 'poisonous metals' from oxidised sulphides, for example Cd^{2+}, Pb^{2+}, Hg^{2+}, and Sn^{2+}.

The changes in the inorganic chemistry allowed developments of organic compounds inside cells, especially in vesicles and membranes but generally redox processes in the cytoplasm remained unchanged. Cell structures altered through production of new proteins, but it is uncertain exactly when some of the proteins shown in Table 12.9 appeared.

The big change came in cell–cell organisation with the evolution of *cross-linked* connective tissue of various kinds (Table 12.12), dependent on both oxidative and hydrolytic chemistry, that is, on new enzymes of iron, copper and zinc especially. Simultaneously, biological mineral element homeostasis and mineralisation increased in sophistication, Tables 12.15 and 12.16. A network of differentiated cells was generated that evolved into organs, all organised in space.

We should also notice that within the evolution of many species some elements became of very different value, see Fig. 12.32, within the branches of the evolutionary tree and that some processes were also lost (Table 12.17). Why had this to be so?

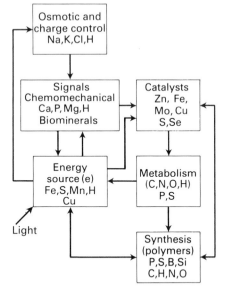

Fig.12.32 The interrelationship of the functions of the elements is given here to stress the intimate connection between them and energy. This diagram is only an aid to thinking about the ways in which energy and metabolism and their feedback have to be linked frequently using inorganic chemical functions. Note that Cu and Zn are added to Fig. 10.30.

Table 12.16 Biomineralisation: combinations of elements

Anion	SO_4^{2-}	CO_3^{2-}	HPO_4^{2-}	OH^-, O^{2-}	S^{2-}
Cation	Sr, Ba	Ca, (Mg)	Ca (Fe)	Fe	Fe, Cd

NB. Carbonates, phosphates and sulphides evolved before sulphates and oxides.

Table 12.17 Sophisticated activity lost (L) or gained (G) in animals

Activity	Organism	Special element or compound
Essential amino acid synthesis (L)	Animals	None known
Degradation of cellulose (L)	Animals	None known
Nervous activity, brain (G)	Animals not plants	Na/K
Shell and bone formation (G)	Animals not plants	$CaCO_3$, $Ca_{10}(PO_4)_6(OH)_2$

Missing from this account is the development of a communication system to integrate the activities, which is essential in any organisation. The next chapter describes the newly devised external messengers and the modified internal messenger systems of multicellular organisms. In this chapter and the next it becomes apparent that the environment and evolution move together, linked as they must be by the very nature of the way in which complexity can evolve in kinetic schemes. Since gross changes are needed in vesicle and extracellular chemistry, where the new reactions occur, but not in the cytoplasm we are forced to ask do changes in the environment impose differential rates of random mutation within DNA? Such a possibility has enormous advantages for the response time of evolution. We are also forced to see too that complexity carries the disadvantages of slow reproduction, poor adaptability and risks inherent in complex systems. The only way forward is and was to abandon self-reliance. Have we learnt this lesson?

Further reading

The main reference for this chapter is our previous book in which details of each section are given.

1. Fraústo da Silva, J. J. R. and Williams, R. J. P. (1991). *The biological chemistry of the elements* (3rd printing (1994) with corrections and additions), Oxford University Press, Oxford.

Other useful references to specific topics are:

On the consequences of the increase of dioxygen in the atmosphere:

2. Knoll, A. H. (1991). End of the Proterozoic eon. *Scientific American*, October issue, 42–9.
3. Williams, R. J. P. and Fraústo da Silva, J. J. R. (ed.) (1978). High redox potential chemicals in biological systems. In *New-trends in bio-inorganic chemistry*. Academic Press, New York.

On symbiosis

4. Margulis, L. (1981). *Symbiosis in cell evolution*. W. H. Freeman, San Francisco.
5. Margulis, L. and Sagan, S. (1987). *Microcosmos—four billion years of evolution from our microbial ancestors*. Allen and Unwin, London.
6. Sapp, J. (1994). *Evolution by association—a history of symbiosis*. Oxford University Press, Oxford. (A historical account with many references.)

On halogenated compounds

7. Neidleman, S. N. and Geigert, J. (1986). *Biohalogenation*. Ellis Horwood Ltd, Chichester.
8. Gribble, G. W. (1994). Natural organo-halogens—many more than you think. *Journal of Chemical Education* **71**, 907–11.

Note added in proof

The structure of cytochrome oxidase has been solved and shows that all electron-transfering centres, copper and haem iron are away from the cytoplasm in the membrane. The longest conduction path is of protons not of electrons in the membrane (Iwata, S., Ostermeier, C., Ludwig, B. and Michel, H. (1995). *Nature* August issue). See also reference 1 of Chapter 13.

13

Organisation in advanced organisms

All that is limited by form, semblance, sound, colour
Is called an object.
Among them all, man alone
Is more than an object.
Though, like objects, he has form and substance,
He is not limited to form. He is more;
He can attain to formlessness†
Chuang-Tzu (Fourth century BC): Texts (xix.2) Wholeness—Adapted
Thomas Merton

13.1 Introduction

In Chapter 12 we have shown that the development of organisms from 3 to 1×10^9 years ago caused an increasing dioxygen partial pressure and gradually changed the availability of many elements on Earth. Many

† Formlessness arises through the self-conscious imagination.

organisms themselves then altered and/or adjusted their uptake and incorporation mechanisms for elements and so became analytically different in a changed environment (Fig. 13.1). These changes in element use led eventually to a massive development of life in the form of multicellular aerobic organisms. The fact that many of these organisms could build hard mineral structures has left us with a remarkably good fossil record about them but only dating from less than 1×10^9 years ago (Figs 12.6 and 12.7). The evolutionary development of the chemistry was matched by changes in the shape and function of single cells and also by the appearance within multicellular shaped organisms of a diversity of differentiated cells (Fig. 13.1). Internally and externally, cells increased the complexity of their compartments, and even new oxygen-utilising organelles, mitochondria, were incorporated. Later, within multicellular differentiated systems, similar cells came together to form interactive large organs. Continuously, the survival of complicated species increased, but these higher organisms also became increasingly reliant on lower organisms thus generating ecosystems of *co-operative* biological chemistry.

In this chapter we examine the organisation that is necessary to maintain these new cellular and organism structures and chemical patterns against the background of certain characteristics of earlier life that had to be maintained, that is, the primary metabolism of primitive anaerobes. As stated in Chapter 12, the coding schemes of the gene polymers and the polymers themselves,

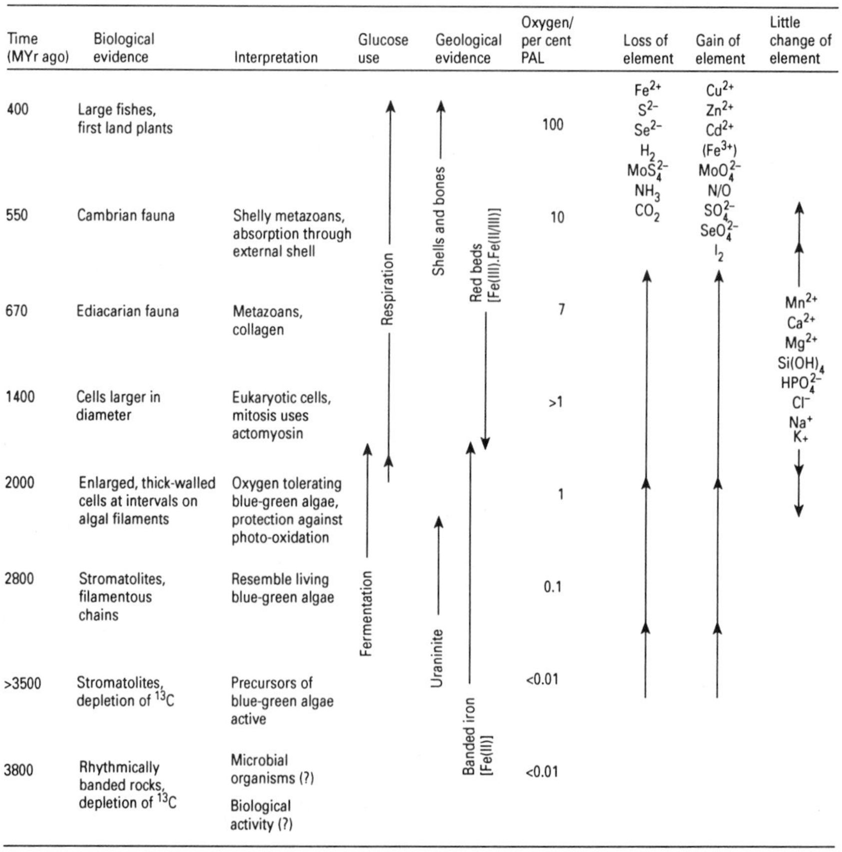

Fig. 13.1 Evolution of the biosphere and the atmosphere—a summary of the biological and geological evidence that suggests how oxygen levels in the atmosphere may have progressed towards their present-day values and how the availability of chemical elements changed. PAL=present atmosphere levels. (Based on Cloud, P. (1983), *Sci. Am.* **249** (3), 132.)

Time (MYr ago)	Biological evidence	Interpretation	Glucose use	Geological evidence	Oxygen/per cent PAL	Loss of element	Gain of element	Little change of element
400	Large fishes, first land plants		Respiration	Shells and bones	100	Fe^{2+} S^{2-} Se^{2-} H_2 MoS_4^{2-} NH_3 CO_2	Cu^{2+} Zn^{2+} Cd^{2+} (Fe^{3+}) MoO_4^{2-} N/O SO_4^{2-} SeO_4^{2-} I_2	
550	Cambrian fauna	Shelly metazoans, absorption through external shell		Red beds [Fe(II).Fe(II/III)]	10			
670	Ediacarian fauna	Metazoans, collagen			7			Mn^{2+} Ca^{2+} Mg^{2+} $Si(OH)_4$ HPO_4^{2-} Cl^- Na^+ K^+
1400	Cells larger in diameter	Eukaryotic cells, mitosis uses actomyosin			>1			
2000	Enlarged, thick-walled cells at intervals on algal filaments	Oxygen tolerating blue-green algae, protection against photo-oxidation		Uraninite	1			
2800	Stromatolites, filamentous chains	Resemble living blue-green algae	Fermentation		0.1			
>3500	Stromatolites, depletion of ^{13}C	Precursors of blue-green algae active		Banded iron [Fe(II)]	<0.01			
3800	Rhythmically banded rocks, depletion of ^{13}C	Microbial organisms (?) / Biological activity (?)			<0.01			

DNA and RNA, remained unaltered, and the major enzyme-catalysed pathways to carbohydrates, fats, amino acids, nucleotides and coenzymes were also kept (but see also Section 13.10). There were, however, changes in many specific products amongst small molecules, proteins, fats and carbohydrates due to new secondary metabolism, but very much of this is related to vesicle and extracellular chemistry. We have also shown in Chapter 12 that several such new proteins and polysaccharides were needed to make the extracellular matrices in particular and to allow cells of quite novel shapes to evolve within these matrices, for example nerve cells. A major task outlined in Chapter 12 was therefore to describe the dependence of the changes in organic chemicals upon the cell complement of the chemical elements—both those found in early life forms, but often now in different (new) compounds, and those newly available from the changed environment. Just as we insisted in Chapter 11, which described anaerobic organisation, that its chemical pathways (analysed in Chapter 10) had to be integrated, so here we must insist that integration is even more necessary in and between the complex compartments of higher organisms if we are to explain the constancy of growth patterns within multicellular species. The task of describing integration is more difficult not just because the whole multicellular construction is more complex but also because it is a cobbling together of very much of the pre-existing cytoplasmic activity of anaerobes and the new activities generated by dioxygen. Of course, this means that as communication evolved it had to integrate with the pre-existing communication network of the anaerobes. A considerable part of this management will be shown to be dependent on inorganic elements.

Elements will be taken in turn at first to show their new functions and also their connections to those developing organic chemicals that became part of management. Once the dependence on individual element has been described, we can extend our illustration to interelement dependence. There are two parts to the management as described in Chapter 11: *control* of pathways, by catalysts, and *regulation* of the synthesis of proteins including that of the catalysts. We remind readers of the distinction between them in the next section.

13.1.1 Control and regulation: a reminder

Control acts at the level of metabolism and one part of it is concerned with the use of proteins including catalysts but not with their productions. Some of these controls act on the well-recognised *internal* cellular pathways of energy and material metabolism, for example glycolysis, the citric acid cycle and so on, while new ones will involve especially hormones and transmitters which go between cells. A second part of control acts on the use of energy in contractile systems, movement and changes of cell shape. Some of this control needs to be faster than and separate from the controlled use of catalysts and other proteins. As described in Chapter 7, all these controls must remain based on feedback circuits.

Regulation acts at the level of the gene and is cytoplasmic. As stated already, it was little altered from that in anaerobes by the development of multicellular organisms at least in terms of major principles. It now as then depends on the expression of proteins, some of them novel, regulated by the states of DNA

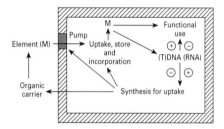

Fig. 13.2 The switch on (+) and off (−) between normal functional use of an element (M), which is in adequate supply, to the synthesis of systems for uptake when the element is in short supply. Element homeostasis must be maintained at all times. T is the transcription protein.

(RNA) binding proteins. In a particular example it was shown in Chapter 12 that the uptake of iron in aerobic *Escherichia coli* had undergone enforced changes from that in the anaerobic prokaryotes, but the basic theme was the linking of an on/off switch of synthesis of proteins for iron transport and uptake systems that controlled the level of iron in the cell. When iron levels fell, the synthesis of the scavenging system was switched on, while, when adequate iron was present, the DNA was informed, normal metabolic activity took place and synthesis of iron transport systems was reduced. Figure 13.2 (compare Fig. 12.18) shows the network which has a parallel for many elements. It is the production of proteins for carrier systems, and for pumps and stores that ensures the appropriate level of elements in a cell. There must be then a mutual feedback between element levels, the syntheses of protein pumps and carriers, and metabolic activity (Fig. 13.2). (The connection between the DNA and the Fe^{2+} concentration in some bacteria is the transcription protein, FUR.)

For higher organisms a major new problem is the regulation of production of extracellular connective tissue in relation to the production of intracellular materials of cells. It is tempting to suppose immediately that, since these extracellular structures are relatively recent in origin, their production will be strongly associated with the newly available elements, for example zinc and copper. This was found to be the case in Chapter 12. However, there are many message systems using a variety of elements (and compounds) that contribute to the integration of such external and internal activities. It is the connection between the uses of elements and the communication network as a whole that we seek to elucidate in this chapter. Therefore we treat control and regulation together noting that while some differentiated cells respond rapidly to controls there are many slow regulatory responses. Since the elements used in distinct activities are different we remind the reader of some major distinctions between elements that are important for the functioning of communication networks.

13.2 The elements in message transmission

In Chapter 2 we stressed that chemical elements were very different mainly in their three characteristics: (1) size; (2) electronegativity; and (3) variety of valence states. These characteristics, which are properties of atoms or ions, caused them to bind selectively with a *thermodynamic* constant to organic centres, giving complexes as described in Chapter 5. The control over the *kinetic* selectivity of a complex (Table 13.1) was related to the same factors in Chapter 7. We pointed out that, when a protein matrix was used as a binding agent, it could firstly distinguish ions (atoms) of different sizes even where the binding constant was small for all ions (atoms). Thus the structure of a protein could be such that even with very low binding strength it could act as a discriminating filter or channel. In biological systems there are two cations in large excess over all others, namely K^+ and Na^+, which bind very poorly; they are very different from one another in size. K^+ can then be allowed to flow into cells through protein channels not permeable to Na^+ (Section 7.5.3). In many respects protein channels for a weakly binding anion such as chloride have the

Table 13.1 Time-scales of mobility or exchange

Less than a second	Less than a day	Greater than time of destruction of binding protein
Na^+, K^+, Mg^{2+}, Ca^{2+}, H^+ Cl^-, HPO_4^{2-}, SO_4^{2-}	Some coenzymes	Chlorophyll Some Cu^{2+} and Zn^{2+} proteins
Many substrates	Fe^{2+}, Mn^{2+} Some Cu^{2+}, Zn^{2+}	Some Fe-enzymes, e.g. haem-binding
NTP, NDP, NMP Mobile coenzymes NADH, Q		

same selectivity characteristics. Such cations and anions are associated with the fastest cell messages since they diffuse without hindrance.

Next, using moderately strong binding centres, proteins can be designed to recognise selectively divalent ions such as the large Ca^{2+} ion and the small Mg^{2+} ion and to bind them differentially even in the presence of excess K^+ and Na^+ and low concentrations of other not very available ions such as Mn^{2+}. Ca^{2+} and Mg^{2+} are associated with moderate rates of secondary message transmission. Finally, strong binding of more electronegative ions is possible to the N/S-donor centres of proteins (see Section 5.5). In Chapter 5 we also outlined other electronic factors that contribute to their selectivity. The cations in this last group are only able to exchange on long time-scales of the order of hours. The ions concerned, mainly Fe^{2+}, Zn^{2+} and Cu^{2+}, are the slowest metal ion messengers, but remember that some of them transfer electrons.

The conclusion is simple: biological systems can distinguish between even very similar elements and so can use those that are available to great advantage in diverse capacities increasing their fitness of use. We need to illustrate how these different ion capacities (see marginal note) and the kinetics of their binding processes have been developed in the communication systems of aerobic organisms particularly of those that are multicellular.

There is a corresponding set of binding constants and exchange rates for organic messenger molecules (Table 13.2), but here exchange is not usually just a reflection of off-rate and binding constants. The input and removal of simple ions to a cell is by channels and pumps but that of organic molecules is often by synthesis and degradation using enzymes (Chapters 7 and 9). Both sets of messengers can be managed with a great variety of rate constants. Now that the basic principles have been re-introduced we can look at the eukaryotic cell communication systems at first within single cells.

ELEMENT FUNCTIONS

(a) *ACIDS AND BASES ONLY*

Na Mg . Si P . Cl
K   Ca
    Zn

(b) Also *REDOX ACTIVE*

        H
    C N O
        S
        Se I
V . Mn Fe Co Ni Cu
 . Mo .

13.3 Single eukaryotic cells: control and regulation

After the advent of dioxygen it is easier to consider first the new management in an idealised eukaryotic *single* cell before going on to that in multicellular systems, since we have already described the basic features of the changes from the prokaryote to the eukaryote cell. The eukaryotic cell has numerous internal compartments as well as in many cases an inner and an outer

(a)

Acetylcholine

$$CH_3CO\cdot O—CH_2—CH_2—N^+(CH_3)_3$$

Adrenaline

$$CH_3\overset{+}{N}H_2—CH_2—CH \overset{\displaystyle —OH}{\underset{\displaystyle \ \ OH}{\diagdown}}$$
OH

(b)

Cyclic AMP

Inositol 1,4,5 triphosphate (IP$_3$)

(a) Messengers between cells; (b) messengers inside cells.

Table 13.2 Some new organic messengers*

Messenger‡	Release	Removal	Time constant† (s)
Acetyl choline	From vesicle	Hydrolysis	Fast ($<10^{-3}$)
Adrenaline	From vesicle	Oxidation	Moderate (<10)
Inositol triphosphate	From membrane	Hydrolysis	Moderate (>1)
Nitric oxide	From arginine	Reaction?	Slow ($>10^2$)?
Peptides	From vesicle	Hydrolysis	Slow ($>10^2$)
Steroids	From fatty bodies	Oxidation	Very slow ($\geqslant 10^3$)

* NB. Note that phosphates are used in signalling mostly in cells and oxidised organic molecules between cells.
† Release from receptor.
‡ See marginal illustration for some formulae.

encompassing membrane (Fig. 13.3). All these compartments have to grow and multiply in harmony. Therefore, just as is the case for the prokaryotes, it is necessary for all compartments to be in connection with one another so as to maintain the cell as a whole in a homeostatic (co-operative growth) condition. Clearly, since different activities are now in different organelles and vesicles, a messenger system is required that goes between these compartments and maintains homeostasis. The messenger system needs to communicate not only to material distribution but also to the distribution of energy and tension, the last so that the relative positions of vesicles are managed. There is much suggestive evidence today that an early and major evolutionary development of management was the introduction of an extensive series of connections between all these compartments using many of the same signals across internal membranes as were used within the prokaryote cytoplasm. The new membranes of the organelles and vesicles have feedback pumps for some

Fig. 13.3 The corresponding figure to Fig. 10.24 for a generalised eukaryote cell without external controls being shown. Notice the nucleus, vesicles and organelles. Not included are the membranous internal filaments which determine shape. The whole remains in a growth homeostatic state over long periods. Some 15–20 elements are involved. The internal M$_1$ signals connect the vesicle metabolisms and the environment.

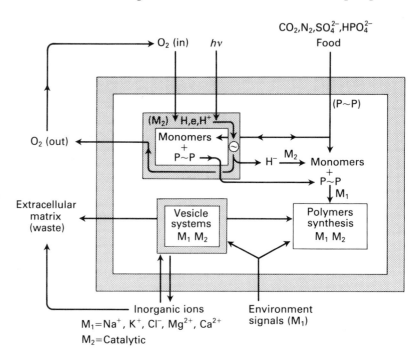

organic metabolic pathway substrates as well as for ions so that this relationship between pathways in and between different organelles was and is maintained. In particular, the citric acid cycle is now in the mitochondria and has to share C/H/O substrates with glycolysis, gluconeogenesis and fatty acid synthesis and degradation in the cytoplasm (Fig. 12.11), and it has to give energy, in the form of ATP, to the cell. Communication is maintained in part as before through coenzyme exchange, for example of NAD/NADH and ATP/ADP, now across the mitochondrial membrane. The metabolites and ions that also communicate in the balanced steady-state condition, for example H^+ and Mg^{2+}, do so as before but now through mitochondrial membrane exchangers as well as through outer membrane and vesicle pumps. This does not prevent such compartments from being held in different steady states, of course. Some vesicles and organelles are held at a pH some two units lower than the cytoplasmic value, and element levels in different compart-ments may be very different. The selective uptake of ions such as Fe^{2+} into mitochondria is also required since porphyrin synthesis takes place there. What has also happened is that many of the internal solutions of vesicles have an ionic milieu more like that of external solutions, low in K^+ and high in Ca^{2+}, for example. Much of this change relates to compartmental separation of activities with new attendant gain in control but no real chemical pathway novelty. New features did emerge, however, in that (1) there is extensive internal cross-talk using new systems of organic and inorganic messengers; (2) there is new communication with the external solutions; and (3) some vesicles contained new oxidative reaction pathways. We begin with those systems that came to use calcium. Remember that we are dealing with the effects of a messenger in a single cell at first.

13.3.1 The extended use of calcium as a messenger

Due to the placing of calcium pumps or exchangers, the levels of calcium are quite high in all vesicles and somewhat higher in organelles relative to that in the cytoplasm, where it is $< 10^{-7}$ M. The dehydrogenases of mitochondria and the dioxygen production enzymes of chloroplasts depend on these somewhat elevated calcium levels. Thus, free calcium became part of the homeostatic signalling between compartments of cells. It also became a major, if not the major, communicating ion between the outside and inside of a cell using both Ca^{2+} in the environment and in vesicles such as the endoplasmic reticulum. It is difficult to know when and in what way the calcium signal from outside the cell evolved, but it could have been quite early in the recognition of external chemicals by single, even anaerobic, eukaryotic cells perhaps allowing them to take appropriate action using mechanical motors, flagellae, see Section 11.12.2.

The calcium signalling from the outside to the inside of cells is today based on the following sequential steps (Fig. 13.4). An incident, a stimulus, at the surface of the cell, for example contact with a foreign body or chemical, causes calcium inlet channels to open. The upsurge of calcium in the cell can be local, for example due to a receptor protein (Fig. 13.5), or delocalised over the whole cell by complete depolarisation of the cell membrane potential, for example of a muscle cell. In either event the calcium entering the cell is recognised by one or

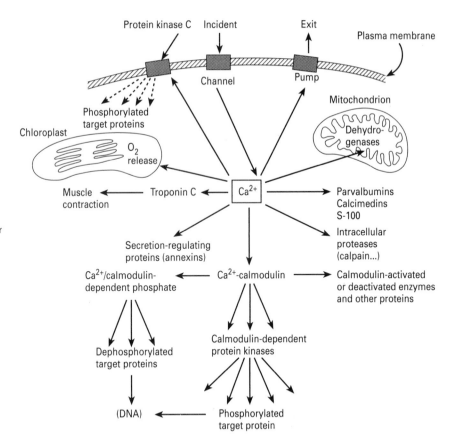

Fig. 13.4 Condensed overview of the interaction of Ca^{2+} with intracellular proteins (after S. Forsen). The extracellular systems are equally important.

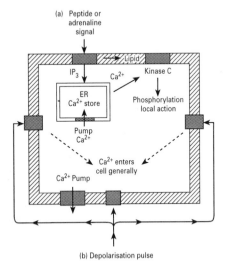

Fig. 13.5 The two kinds of calcium signal entering a cell due to: (a) a localised receptor for transmitters or hormones; and (b) a Na^+/K^+ depolarising pulse which is delocalised. ER, endoplasmic reticulum.

another or many of the newly evolved calcium receptor proteins (Table 13.3). The proteins undergo conformational changes and, according to the connection they make in the selected cell, trigger off many responses, for example:

(1) contractile activity—troponins (fast);

(2) metabolic changes—calmodulins (moderately fast);

(3) vesicle discharge—annexins (intermediate speed);

(4) relay of message to DNA for protein synthesis, here through calmodulins, S-100 or other proteins connected to kinases for phosphorylation of regulatory proteins or to protein phosphate phosphatases also leading to regulation (slow) (Fig. 13.4).

The exact way in which a given incidental stimulus triggers a particular response is described in many texts (see 'Further reading'), and one basic receptor protein reaction is shown in Fig. 13.6.

Since this input of calcium would jeopardise the cell or force it to change if sustained, this ion is rapidly pumped out again to re-establish the resting homeostatic level (Fig. 13.4). (The implication is that, since channel and pump activities differ in different cells, the basal level of Ca^{2+} concentration as well as the response to calcium are not exactly the same in different differentiated cells of one organism just as they are different in resting and growing states of aerobic bacteria.) The particular calcium pump proteins may

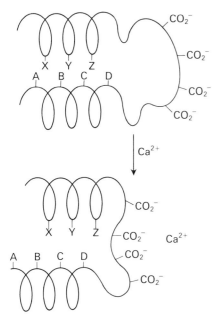

Fig. 13.6 A schematic presentation of the way in which Ca^{2+} stimulates a change in conformation of a protein that carries the above helix–bend–helix structure, called an EF-hand.

Table 13.3 Classes of calcium-binding intracellular proteins

Protein	Ca-binding sites	Probable function
Troponin	Classical EF-hands†	Fast triggering of muscle contraction
Calmodulin	Classical EF-hands	Fast triggering of many events
Parvalbumin	Classical EF-hands	Buffer*, relaxation factor*
Calcineurin	Calmodulin-like	Protein phosphatase, two calmodulins?
Calpain	Calmodulin-like	Protease
Calbindin	Some classical, some non-classical EF-hands	Slower buffering and transport
S-100 (many proteins)	Some classical, some non-classical EF-hands	Buffer*, transport factor*, connecting linker
Annexins	Phospholipase A_2 hands (not EF-hands)	Organisation of cellular constructs* (ion channel*)
Phospholipase A_2	Not EF-hands	Lipase

* Possible function, demonstrated in model or particular systems rather than *in vivo*.
† See Fig. 13.6.

or may not be of recent evolution and, in principle, all such activity could have evolved in a single eukaryote cell, for example, yeast.

As stated, we must see how this new set of external to internal signals is co-ordinated with the pre-existing internal cell signals inherited from the anaerobic era. This leads us back to consideration of phosphate metabolism and signalling, which is, as stated in Chapter 11, the major primitive communication system and had to be carried over (with modification) into advanced cells. Calcium had to link to kinases and phosphatases.

13.3.2 Phosphate metabolic and signalling changes

It is important to see that no change in homeostasis or triggering of activities in a cell could be allowed to occur through calcium signals without leaving existing phosphate signalling virtually intact, since phosphate is placed at the centre of controls and regulation through the NTP and cNMP reactions in the primary metabolism of all cells (Fig. 11.17). To this end the new features are that many of the nucleotide triphosphate (NTP)-dependent (kinases) and cyclic nucleotide monophosphate (cNMP)-dependent steps, primarily of ATP (adenine nucleotides), are only somewhat modified, switched on or off by calcium concentration changes (Fig. 13.7), while calcium entry is linked back to only a few (new) internal phosphate-dependent signals. Many of these calcium phosphate-dependent processes act as quite fast enzymic controls, but many act on slow transcription factors right down to the gene level, that is, on differentiation by regulation. Calcium is not involved as far as we know in direct action (through proteins) at the gene level (Fig. 13.4). The entangled network of calcium and phosphate control and regulatory activity is not yet fully understood, but it is clear that the new calcium signalling in eukaryotes (outside⇌inside) is complementary to and interwoven and co-operative with the phosphate signalling (inside only) inherited from prokaryotes.

An additional signalling system that is connected to both phosphate and

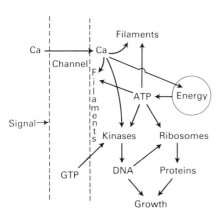

Fig. 13.7 The intimate linking of calcium (and phosphate-based compounds) to various internal cell compartments, including those for energy supply, that is, mitochondria and chloroplasts; mechanical stress, that is, filaments; and synthesis of DNA, RNA and proteins. Figure 13.16 shows the further connections to extracellular calcium in eukaryotes and multicellular organisms.

calcium and that came into being during the build-up of dioxygen uses a membrane-bound inositol triphosphate, IP_3, see marginal illustration, p. 510. To understand the events we must refer to Fig. 13.5 again. An external stimulus, for example a hormone, triggers *locally* the release of IP_3 from the membrane and it in turn causes *local* release of calcium from the *internal* vesicles. This calcium backdiffuses to activate, with or without other lipid agencies in the membrane, for example arachidonic acid, a variety of kinases for protein phosphorylation. The phosphorylation systems, which are often stimulated by the newly introduced hormones (see Sections 13.4.3 and 13.4.4), may also involve additional series of kinases, the so-called tyrosine kinases. These phosphorylated proteins go on to cause changes in protein expression at the gene level. Thus the phosphate signalling is extended but not fundamentally changed. As a further example we mention the new set of phosphate-dependent signals from outside the cell to phosphorylation systems, kinases called G-proteins (Fig. 13.8). These proteins, which are in the membrane, specifically utilise guanine, G, nucleotides and are activated by certain other hormones. Just as phosphorylation can generate transcription factors so too can removal of phosphate bound to proteins. In all this analysis we see that control of small molecules and regulation are hard to separate and are differently adjusted in material and rate by the same messengers, here calcium and phosphate esters, in different cell types, e.g. in a muscle and in an oocyte depending on the cell surface receptors.

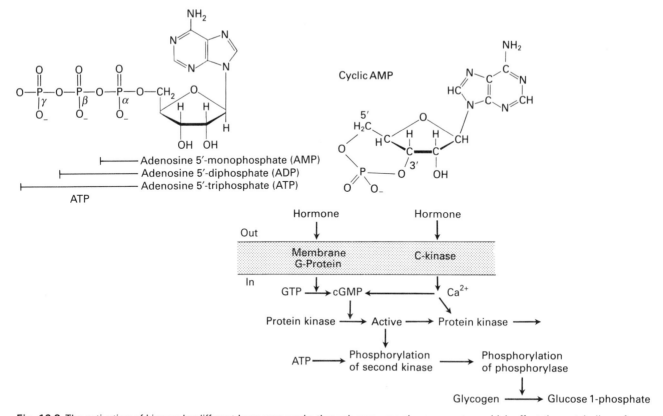

Fig. 13.8 The activation of kinases by different hormones works through many membrane receptors which affect the metabolism of NTP and c-NMP; lower diagram and Fig. 13.4. The transformations of NTP are general, given only for ATP above.

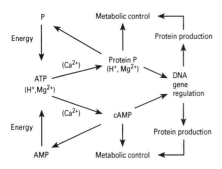

Fig. 13.9 The intimate linking of the four elements, H (H$^+$), Mg^{2+}, Ca^{2+} and P (phosphate), in a eukaryote. It is easy to see how feedback controls maintain a homeostatic resting-level of all if reference is made to Chapter 12.

We begin to build up a picture of the inside of an advanced eukaryotic cell, in particular, of its homeostasis and signalling (Fig. 13.9) based on the fast exchange of Mg^{2+} and H$^+$ (always communicating to phosphates), of the somewhat different exchange characteristics of Ca^{2+} (faster on but slower off due to binding) and of the relatively slow covalent bond chemistry of phosphate, now related not only to internal cell metabolism but also to external events impacting upon the membrane. However, we have not shown the relationship of these new phosphate- and calcium-based systems to dioxygen effects and the development of aerobes. To do this we must consider the stimuli that affect the eukaryote cell and here we have to refer to it within a multicellular system. The source of any stimulus is now a neighbouring cell that sends out organic messenger molecules. (We will consider the relationship to external inorganic ion messengers, for example calcium, again later.) The organic messengers that trigger (Table 13.4) are, for example: (1) various

Table 13.4 Examples of the production of new signalling modes

Signalling molecule	Mode of production
Peptides	Hydrolysis from proteins; export from vesicle store (zinc enzymes)
Oxidised peptides	Oxidation of terminal glycine (copper enzymes)
Oxidised aromatics	Export from vesicle store (copper enzymes)
Oxidised steroids	Membranal oxidation (haem enzymes)
Iodinated aromatics	Stored in lipid micelles (new iron (haem) enzymes)

small molecules made from dioxygen reactions; (2) peptides controlled by zinc enzymes; and (3) peptides oxidised by copper proteins (see Chapter 12). Many of these messengers are hormones. (Remember that zinc and copper are the two major metal ions that became more available after the advent of dioxygen (see Figs 12.14 and 13.1).) It is signals (2) and (3) that relate to the IP$_3$ and G-protein system in particular. In addition, the IP$_3$ binding in the membrane is largely if not exclusively to arachidonic acid lipids. The arachidonic acid is itself a messenger in the membrane (Fig. 13.5) and leads on to the synthesis of other hormone-like substances, the prostaglandins, via further oxidation. All these chemicals, arachidonic acid, related prostaglandins, and many aromatics and peptides for signalling plus the new levels of the metal ions copper and zinc, resulted from dioxygen-based changes in metabolism or in catalysts. Virtually all exhibit strong binding and slow exchange, Tables 13.1 and 13.2. We illustrate the complexity of this internal signalling in Fig. 13.10. We need

Fig. 13.10 A much simplified description of the joint roles in signalling carried on by new inorganic ion systems, Zn, Cu and Fe, and new organic transmitters and hormones. R$_1$ and R$_2$ are receptors. Cell A here controls cell B.

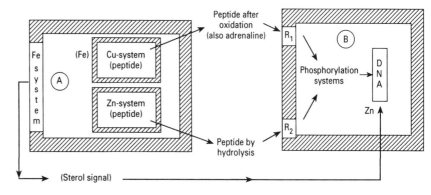

Cytoplasmic membrane

Lipid	Vesicle lipid	

Protein		Spectrin

Tubulin

Tubulin

Actin

Vesicle

Actin

Fig. 13.11 A schematic representation of the relationship between the cytoplasmic membrane, an internal membrane (vesicle), and the protein network. One series of proteins, for example spectrins, lies under the outer membrane, bound to it, while another series, here called tubulins but including a variety of filamentous structures, actins, criss-crosses the cell and holds vesicles in place. The vesicle membrane may form part of the outer membrane reversibly. One cell can speak to another through the release of messengers such as angiotensin from vesicles (see also use of nitric oxide).

to observe later the synthetic paths for the organic messengers and the placing of metals required for the synthesis in the vesicles that release these signal molecules. Again all the *internal* responses could have developed in single eukaryote cells, although we have referred to hormone responses from neighbour cells. Before we refer again to the organic messengers there is another aspect of phosphate and calcium homeostasis and signalling that we need to describe.

13.3.3 Messages to mechanical devices

Before we leave the internal messenger systems that are related to calcium and phosphate (and that could have developed in a single cell) we need to notice that calcium acts as a fast messenger for mechanical activity too since together with phosphate it controls tension in cell filaments. Here we illustrate its effects in both single cells and multicellular organisms. Thus, the shape of an organism or cell, its ability to release vesicles (Fig. 13.11), its ability to move, and the dynamic variation of its shape, all depend on the intricate receptor systems for the second messenger, calcium, and on energy released from ATP (Fig. 13.7). The receptors are a series of proteins related to calmodulin, called troponins, gelsolins and annexins (Table 13.3). On binding to Ca^{2+} the receptors cause tension to build in several filamentous proteins and can even cause a phase change from sol to gel of the cytoplasm. We have shown previously that somewhat parallel *multicellular* shape-building in animals was connected to dioxygen chemistry through the cross-linking of collagen and it is this collagen that connects to the internal contractile filaments (Sections 12.3 and 12.10). Calcium and ATP levels, therefore, have a role in maintaining the steady tension of many linked cells including slow posture (shape) muscles, as well as causing fast transient changes in tension of fast muscles (Fig. 13.12) and maintaining shapes of single cells. (NB. Plant cells are much more confined in shape by cellulose walls.)

Fig. 13.12 (a) Diagrammatic representation of the frog sartorius muscle. In fact, several cells (b) coalesce to form the muscle units. The cells contain (c) fibrous proteins which form working filaments (d).

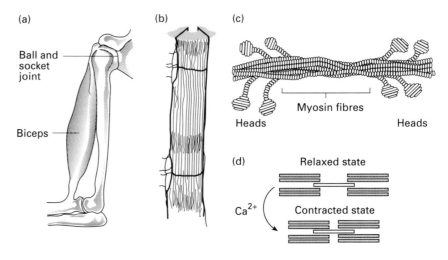

(a)

Ball and socket joint

Biceps

(b)

(c)

Myosin fibres

Heads

Heads

(d)

Ca^{2+}

Relaxed state

Contracted state

13.4 Multicellular organisms

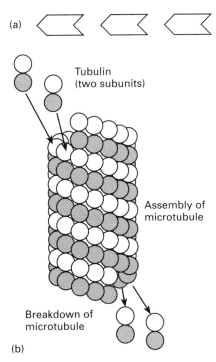

Fig. 13.13 (a) Tubulin units are shaped to come together in long strands, which wind around as in (b). (b) The strands formed in (a) pack further into helical bundles in a tube-like form.

We shall now turn to additional developments of some of these calcium-dependent releases that are found in multicellular organisms and that are stimulated by electrolytic messages (Fig. 13.5). We have discussed two communication modes to single cells: physical (mechanical and electrolytic), generally fast, and chemical (transmitters and hormones) which often are slower but, before describing these connections in a *multicellular* organism, we show that cell shapes had to evolve so as to take full advantage of very fast 'inorganic' ionic message transmission in such complex organisms by making cell–cell contacts at long distance.

13.4.1 Maintained cell shape: single nerve cells

A major feature of diversification of cells in the eukaryotes was their development of *shape*. We have shown that shape at *thermodynamic equilibrium* is a reflection of the components in the system and external fields (Sections 6.2.2 and 7.18) and that shape is again not a variable in a *kinetic steady state* (Chapter 7) but reflects the thermodynamic influence of components plus steady-state flow, tension and energy supplied, say, from the effective phosphate potential. Shapes of cells were modified especially by the evolution of a new cytoplasmic protein, tubulin (Fig. 13.13). These proteins make relatively rigid thin tubes, microtubules, and so support long-range narrow membrane protuberances from a cell body as seen today in nerve cells (Fig. 13.14). Transport of vesicles is permitted along these filaments to the synapse. Cell extension and growth is connected to pseudopod activity near the nerve cell synapse through tension development of other filaments, probably of the actin/myosin kind as in muscle. Now these long tubular cells provide a survival advantage to a complex multicellular body in that they allow directed

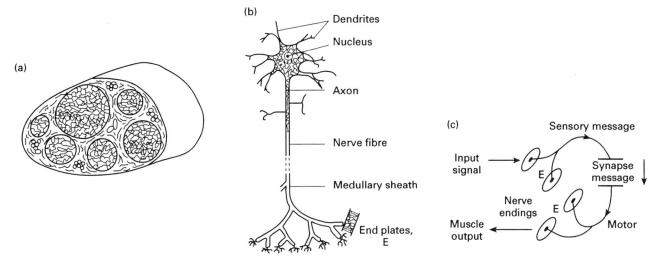

Fig. 13.14 (a) Cross-section of a nerve trunk with individual nerve fibres grouped into six separate groups. (b) A nerve cell or neuron. (c) Sensory and motor nerves connected via a synapse.

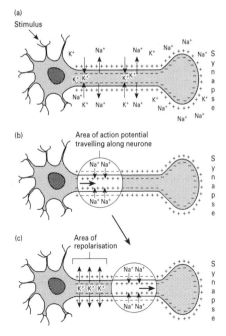

(a)
Stimulus

(b)
Area of action potential
travelling along neurone

(c)
Area of
repolarisation

Fig. 13.15 Transmission of an impulse along an axon. (a) The dendrites (or cell body) of a neurone are stimulated sufficiently to depolarise the membrane to firing level. The axon is shown still in the resting state and has a resting potential. (b) and (c) An impulse is transmitted as a wave of depolarisation that travels down the axon. At the region of depolarisation, sodium ions diffuse into the cell. As the impulse progresses along the axon, repolarisation occurs quickly behind it. (After Villee *et al.*, reference 1 in 'Further reading'.)

and faster *specific* communication between cells at some distance from one another than can be obtained by the undirected diffusion of chemicals from one cell to another. The cell protuberances are close parallels in fact to the wires of an electronic device (Figs 7.33 and 7.35). The fastest possible mode of information transfer along the nerve protuberances is electrolytic depolarisation (Fig. 13.15). This possibility of a communication device, different from osmotic and electroneutrality control (Sections 6.4.1 and 6.4.2), required a new set of controls over the current carriers Na^+ and K^+. The controls are based on a new ATPase pump (in animals and hence sometime after dioxygen levels rose) which we have described in Fig. 6.16 and Section 10.6. The pump associated with this gradient generates a cell membrane potential, and a running discharge of this potential became the fast message system. The device has the great advantage that Na^+ and K^+ bind so weakly that, on change of concentration, they do not communicate with metabolism. The mode of charge carriage is better therefore than an H^+/Na^+ transfer much developed in prokaryotes, since H^+ by binding connects to metabolism. The Na^+/K^+ pump is only found in animals that have these long nerve cell connections, that is not in plants. It has a grave disadvantage in that it can only communicate electrolytically along a single cell tube and requires a connecting mode to an adjacent cell at a synapse (Fig. 13.14). (Here and there, where adjacent cells have a gap linking them (a gap junction) or where contact is sufficiently tight, an electrical message can jump between the cells.)

Thus the Na^+/K^+ electrolytic message, once it reaches the nerve cell terminus, must activate a second message system to cause chemical transformation in a second cell. The activation is of organic transmitters released from inside the cell terminals but it requires the Na^+/K^+ message to be transformed to a Ca^{2+} message at the releasing terminal since only Ca^{2+} can bring about the required conformational changes to induce organic chemical release (Fig. 13.16). The organic messengers are of selected kinds to diversify the messages, for example in nervous tissue including the brain (Table 13.5). The messages are normally relaxed quickly, since more persistent messages to a nerve cell cause the calcium/phosphate system to introduce regulated cell growth (memory?) or differentiation and change of shape, see Section 13.8.

It follows from the above account that communication between differentiated cells in a multicellular organism must be generated to maintain a steady-state relationship between their shapes and their activities and to allow them to be open to excitation by varying the dose of stimulant to which one set of cells is exposed by another, for example, due to an external event. This must

Fig. 13.16 A schematic representation of the Na^+/K^+ depolarisation message interacting successively with a calcium second message and an organic transmitter, A, released from vesicles. A, Acetylcholine or γ-aminobutyric acid (GABA).

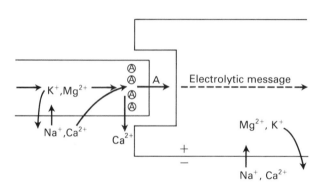

Table 13.5 Signal compounds between cells in the brain

Fast signals

Excitatory (+)/inhibitory (−) amino acids
Glutamate (+)
Glycine (−)
GABA (γ-aminobutyric acid) (−)

Monoamines

Noradrenaline
Dopamine
Serotonin
Acetycholine

Inorganic signals

Nitric oxide
Carbon monoxide?
Na^+, K^+, Ca^{2+}, (Zn^{2+})

Slow signals

Neuropeptides (large number >50)
Substance P
Cholecystokinin
Corticotrophin-releasing factor

(a)

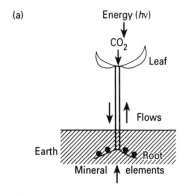

(b)

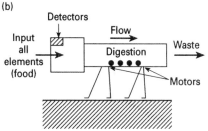

Fig. 13.17 (a) The basic plant structure receives energy from the sun and carbon from the air. All other elements (15–20) come from the Earth as inorganic minerals or from bacteria ($N_2 \rightarrow NH_3$) shown as filled circles. (b) The basic animal structure. Motors are connected to arms, legs, wings, fins, etc. The animal acquires 15–20 elements and energy from food, and requires a multitude of assisting bacteria.

apply to all cells within an organism as well as to nerves and muscles. The signals exhibit a great variety of timing and it seems very probable that an animal must receive a continuous base level of messages to maintain its steady state as well as different pulses to generate heightened or altered response. Very likely the very chemicals initially rejected by single eukaryote cells (and used as poisons to protect them from other cells) became these external constant chemical messengers once cells came together in multicellular organisms. Examples are the uses of sodium and calcium ions discussed above, but many of the organic hormones synthesised by oxidation may also have been produced initially as protective poisons.

In order to facilitate the relationship between the activities of differentiated cells in organs it was necessary to evolve further functional shapes in which organs were positioned to advantage; for example epithelial surface cells are flat.

13.4.2 The functional shapes of multicellular organisms and communication

The complexities of the structure of multicellular organism development cannot be overstressed. It is not only local communication that is required but also long-range transfer of messages and material. For this reason the development of plants had to be quite different from that of animals. The plant has its major source of ionic materials in the soil or water, but its source of energy, light and other materials, such as small covalent molecules CO_2 and O_2, is above these condensed phases in the air so that the two uptake mechanisms had to be at opposite ends of the plants (Fig. 13.17(a)). Flow is required and is very often achieved by simple osmotic pumping in a permanent structure. Very fast reaction is not necessary so the messages are conveyed in the streaming liquids.† (A plant can die back and then rejuvenate.)) For an animal the way to get energy and materials is very different, namely by consuming other organisms. The searching for this food requires movement of the organism with no particular overall body design except for a cylindrical structure for input and waste. Worms, for example, are just tubes (Fig. 13.17(b)), and so is the human gut. The essence is a filter feeder that requires a muscular pump—the heart—to help move absorbed food around the body and several other organs to help in distribution, assisted by the liver and kidney. The animal must move about and search all the time and may well need to move quickly since it is both a predator and a prey. Therefore it needs electrolytic (nerve) *fast* messages to connect to fast muscles. The improvement in muscle or ability to move has been relatively trivial in hundreds of millions of years, but the variety of searching devices has led to the brain as a centralised nerve co-ordinator of all activity and it has developed rapidly. Its activity is continuous and structures cannot be allowed to die back without loss; new connections are constantly formed. A standard book (ref. 17 in 'Further reading') on the evolution of animal shape shows the continuous development of sensors, brain, nerve, and muscle with connections to feeding over some 10^9 years. Notice that much of plant tissue is inert polysaccharide, cellulose, supporting rigid tubes, while much of animal tissue is flexible muscle protein to assist various movements. The structures of the extracellular matrices are as described in Chapters 9 and 12. The two very different developments in plants and animals required somewhat different natural selection of the elements as we see in Fig. 13.18. Lying behind all this new activity and communication there is now the additional need to manage the chemical element content and the proteins of extracellular systems as well since, effectively, organisms are carrying within themselves the environment for most of their cells.

(There are excellent books describing the physical developments described above but no comparable chemical accounts to parallel the one we now give.)

13.4.3 Extracellular fluids and multicellular communication

As the new cell–cell co-ordination of animals uses long-range electric communication dependent on nerve-to-nerve and nerve-to-muscle contact, it is no longer just the in-cell balances that matter. The extracellular fluids of animals have to be of very narrowly fixed ionic composition if nerve transmission and muscle action are to be of great sensitivity since these message systems depend on ion gradients across membranes. The control of body fluids is also required for osmotic pressure and electrical balance and for the basal homeostasis of in-cell metabolic activity. (The animal carries its environment with it.) It is then an absolute requirement that uptake and rejection of the simple ions H^+, Na^+, K^+, Cl^-, HPO_4^{2-}, Ca^{2+} and Mg^{2+} should be precisely controlled within the whole body. This whole body control is essential to maintain other activities as well, such as the level of glucose sent to cells to supply energy everywhere. To produce the necessary flow the multicellular systems developed special organs, for example the non-stop beating heart. In the case of mineral element management the organ for controlling rejection is the kidney, but to achieve homeostasis it has to be

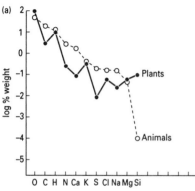

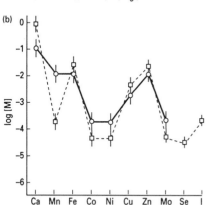

Fig. 13.18 (a) Concentration (logarithm of average weight percentage) of major chemical elements in plants and animals (including mineral deposits). (Data from Calvin, M. (1969). *Chemical evolution.* Clarendon Press, Oxford.) (b) Estimates of the total trace elements in plants (circles) and animals (squares) including precipitates. [M], per cent dry weight.

† There is a slow message system in parts of plants based on calcium/potassium potentials.

linked to the uptake of mineral elements from the digestive tract. Both systems need to co-operate and hence communicate centrally. They do not do this through the fast nerve system (in contrast to muscles) since there is no requirement for fast response, but through chemical messengers that regulate the proteins controlling ion concentrations in all cells and that manage circulating fluids. The messengers are distributed through these circulating fluids from centrally placed glands, and the glands themselves respond to the central nerve system (Fig. 13.19 and Table 13.6). The circulating chemical messengers for regulation of the levels of proteins in cells for the uptake and rejection of minerals are certain corticosteroid hormones (Table 13.6). It is their task to communicate to cells that manage mineral intake and rejection, and they do so by acting on DNA-binding proteins in the appropriate epithelial cells. These protein receptors for the hormones are DNA-transcription factors called zinc fingers. (NB. Zinc fingers are not found in prokaryotes.) The transcription factors therefore regulate the slow but steady production of the proteins (see Fig. 13.10) that manage mineral element uptake and rejection. Clearly, they alone do not control all aspects of mineral element handling since this was essential earlier in the cytoplasm of anaerobic prokaryotes, where the signalling was already linked to phosphate (kinase/NTP and cNMP signals; Chapter 11). Undoubtedly, the mineral handling of all cells depends now on both phosphate and steroid signals to DNA in cells of aerobic multicellular organisms. Notice how the first set of signals we described, often peptides, acted on the outside membrane of the cell, and then could only reach the DNA by activating phosphate signals, using Ca^{2+}, while many of the hormones,

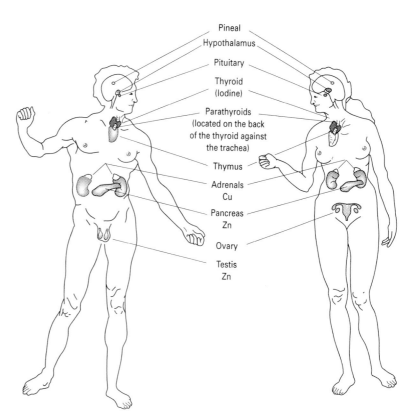

Fig. 13.19 The endocrine glands of man with some specially required elements indicated. (Adapted from Purves et al., reference 2 in 'Further reading'.) Note that organs such as liver are rich in other elements, for example, iron.

Table 13.6 Some human hormones

Secreting tissue (*) or gland	Hormone (*)	Chemical nature	Targets	Important properties or actions
Pancreas	Somatostatin (Zn)	Peptide	Digestive tract; other cells of the pancreas	Inhibits insulin and glucagon release; also decreases secretion, motility, and absorption in the digestive tract
Adrenal medulla	Adrenaline, noradrenaline (Cu, Fe)	Modified amino acids	Heart, blood vessels, liver, fat cells	Stimulate 'fight-or-flight' reactions: increase heart rate, redistribute blood to muscles, raise blood sugar
Adrenal cortex	Glucocorticoids (cortisol) (Fe, Zn)	Steroids	Muscle, immune system, various other tissues	Mediate response to stress; reduce metabolism of glucose, increase metabolism of proteins and fats; reduce inflammation and immune responses
	Mineralocorticoids (Fe, Zn) (aldosterone)	Steroids	Kidneys	Stimulates excretion of potassium ions and reabsorption of sodium ions
Stomach lining	Gastrin (Zn)	Peptide	Stomach	Promotes digestion of food by stimulating release of digestive juices; stimulates stomach movements that mix food and digestive juices
Lining of small intestine	Secretin (Zn)	Peptide	Pancreas	Stimulates secretion of bicarbonate solution by ducts of pancreas
	Cholecystokinin (Zn)	Peptide	Pancreas, liver and gall bladder	Stimulates secretion of digestive enzymes by pancreas and other digestive juices from liver; stimulates contractions of gall bladder and ducts
	Enterogastrone (Zn)	Polypeptide	Stomach	Inhibits digestive activities in the stomach
Pineal	Melatonin (Cu)	Modified amino acid	Hypothalamus	Involved in biological rhythms
Ovaries	Oestrogens (Fe, Zn)	Steroids	Breasts, uterus, other tissues	Stimulate development and maintenance of female characteristics and sexual behaviour
	Progesterone (Fe, Zn)	Steroid	Uterus	Sustains pregnancy; helps to maintain secondary female sexual characteristics
Testes	Androgens (Fe, Zn)	Steroids	Various tissues	Stimulate development and maintenance of male sexual behaviour and secondary male sexual characteristics; stimulate spermatogenesis
Most cells	Prostaglandins (Fe)	Modified fatty acids	Various tissues	Many diverse actions
Thyroid	Thyroxin (Fe, Se, Zn)	Iodinated aromatic	Mitochondria?	Stimulates metabolic rate

* Associated metal element.
Adapted from Purves *et al*. reference (2).

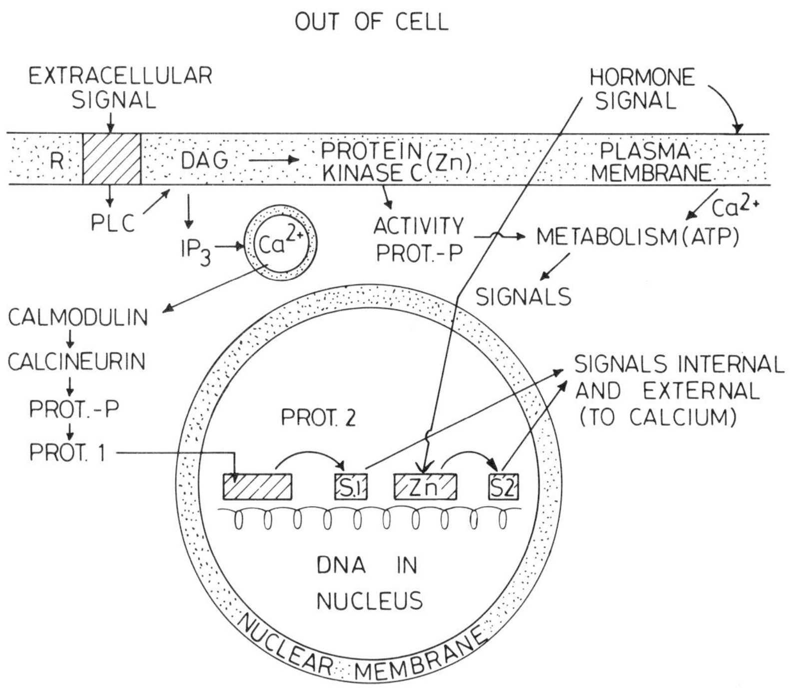

OUT OF CELL

Fig. 13.20 The different signals between a membrane using release of modified cytoplasmic internal phosphate signals (linked to prokaryote systems) and the totally new systems of eukaryotes linked to oxidative sterol metabolism and zinc receptors. DAG, 1,2 diacylglycerol; PLC, phospholipase C.

especially steroids that we describe now act directly on DNA-binding proteins since they pass through membranes (Fig. 13.20).

The use of hormones and transmitters in slow signalling between cells and especially zinc-protein transmission to DNA is new to the era of aerobic eukaryote and multicellular organisms which is concomitant with the rise of dioxygen. The above example of mineral control extends to all aspects of growth and differentiation so that a multicellular organism has a variety of new chemicals and proteins for both *control and regulation* including those for growth and morphogenesis. The connections from glands to organs to activities are illustrated in Table 13.6. The connecting chemistry lies, as stated, in the production of new organic chemicals (hormones and transmitters) with very high binding affinities and new proteins for passing messages to DNA, both of which will be seen to be related to the changed availability of the chemical elements (Table 13.7). Returning to the cases of hormone synthesis, the link is sometimes directly to oxidative secondary metabolism using Fe and Cu catalysts, enzymes, with O_2 and sometimes to new hydrolytic enzymes using Zn. In the case of transcription (receptor) proteins there is already the example of a direct link to DNA via zinc fingers just described and used by a vast variety of hormones, for example steroids (Table 13.6), that are soluble in organic solvents, that is messengers that diffuse through membranes and would have been of little use except as poisons to single cells. A second example is the new use of cell surface protein receptors for peptides or adrenaline which stimulate IP_3 release (Fig. 13.5) also already described. (Note again that the release of IP_3 is from arachidonic acid, itself a product of oxidative metabolism.)

An observation concerning the above description is that it is clearly necessary, in animals especially, to make more and more connections between

Table 13.7 Messengers connected to elements

Element	Compartment	Organic link (speed)
Iron	Internal	Adrenaline (intermediate)
Haem iron	Internal/external*	Sterols (slow) Plant hormones (slow)
Zinc	Internal (vesicles)/external*	Sterols (slow) Peptides (slow)
Na/K	Internal/external*	ACh, Ca (fast)
Calcium	(Internal) vesicles/external*	Na/K, GABA (fast) Adrenaline, etc.
Copper	External (vesicles)	Adrenaline (intermediate) Peptides (slow)

* Different in different organisms. Plants have more external iron systems. ACh, acetylcholine; GABA, γ-aminobutyric acid.

activities in order to maintain the whole continuous homeostasis, and the homeostasis itself has to be more and more finely balanced. The balance is not just a matter of maintaining homeostasis of chemicals both inside and outside cells but of holding constant the number of cells of given types in organs while all cells have a much shorter lifetime than the organisms. The extracellular matrix carries both repair mechanisms (blood-clotting for example) and protective systems (immune cells). We can but indicate here that that they are stimulated by growth hormones and many other message systems frequently linked to circulating calcium, see references. In this book we wish to stress only the selective use of certain chemical elements. All such systems return to us to the great problems of increased cross-connected complexity: it introduces some advantageous survival features but its also introduces risk (Section 7.22). Indications of the problems of out-of-tune balances of minerals and/or proteins are given in Table 13.8, and cancer (and possibly Alzheimer's disease) is probably the greatest danger when phosphorylation of proteins is mismanaged. The homeostasis of elements does not just apply to the simplest cations and anions. It becomes necessary to have tight control over all essential 15–20 elements since each one engages in cross-talk to an ever-increasing number of activities as the multicellular life gets more complex. It also became of increasing value to control temperature since all kinetic rate

Table 13.8 The activity product of plasma calcium and phosphate in various pathological states

Condition	Ionised concentration (mM)		Concentration product*
	$[Ca^{2+}]$	$[HPO_4^{2-}]$	$[Ca^{2+}] [HPO_4^{2-}] \times 10^7$
Normal	1.30	0.81	1.05
Hypoparathyroidism	0.90	1.63	1.47
	0.50	2.56	1.28
Hyperparathyroidism	2.20	0.57	1.25
	2.20	0.40	0.88
Vitamin D deficiency	0.95	0.57	0.54

* The danger here is a precipitation of or lack of bone formation. *Very* fine homeostasis is required.

constants should be held at constant relative values if homeostasis is to be sensitively managed. Warm-blooded thermostated animals were a major evolutionary advance. Before we discuss the problems of such complexity† we must look in some further detail at some of the elements other than calcium and phosphate that came to be used in novel ways and in networks.

13.4.4 Zinc and its organic messenger network

As described in Chapter 12, due to the oxidation of sulphides zinc became a readily available cation perhaps 2×10^9 years ago. While it was probably used as a rare but effective trace element in earlier primitive anaerobes, including prokaryotes, it became as common as iron in animals and quite common in plants after the advent of dioxygen (Fig. 13.1). Its early role was undoubtedly in some syntheses (for example of RNA) and in Lewis acid catalysis of some degradations especially as a part of digestion. Its use as a Lewis acid was greatly extended both as a generator and a receiver of messages in aerobic eukaryotes. We have already mentioned the very extensively used zinc fingers as receivers in the transcription factors of eukaryotes from yeasts to man. It could be that 10–20 per cent of all eukaryotic transcription factors are zinc-dependent. We have seen that the messenger molecules received are virtually all membrane-soluble, hydrophobic, organic compounds derived from dioxygen activation and that they are frequently related to growth or differentiation. Their response must, of course, dovetail with that of the underlying messengers for primary metabolism, that is, the phosphate compounds mentioned before (Fig. 13.21) and to mineral, K^+, Na^+, Cl^-, Mg^{2+} and Ca^{2+} uptake.

We turn next to zinc as a generator of messages, also mentioned earlier. The involvement of zinc in enzymes has been described in Chapters 5, 7 and 12 (see 'Further reading'). The new features of this activity in multicellular organisms are that zinc enzymes are employed both in the production of peptide hormones (from proteins in vesicles) which act at the surfaces of cells as water-soluble messengers and as the catalytic centre that removes these peptides subsequently so that a short-time burst of peptide hormone can be released against a basal level. Figure 13.22 shows some cellular zinc connections.

In summary, the Lewis acid functions of zinc are extremely widespread and

Fig. 13.21 The extensive connections of zinc to multicellular controls and regulations. Zinc links to sterol-like hormones, to peptide hormones, to the external matrix and to a vast list of enzyme and control activities in cells. It links to DNA via metallothioneins and zinc fingers and to RNA via transcriptases. Is Zn a hormone?

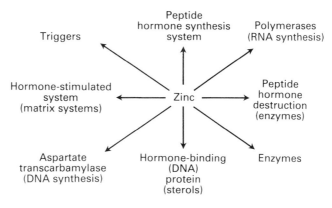

† A further novelty is the control over membrane fluidity due to the incorporation of cholesterol, also a product of dioxygen metabolism (see phase diagram, Fig. 4.24).

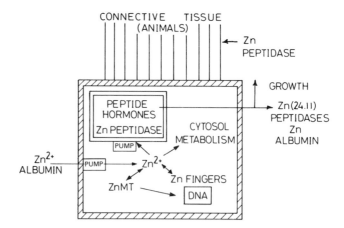

Fig. 13.22 The relationship of zinc activities to cellular structure. MT, metallothionein.

to some degree it has taken over the central role from free iron observed in the prokaryotes both as a Lewis acid and as a producer of small molecules (Section 11.8.3). This undoubtedly reflects the changed availability from a sulphide-rich medium in which $Fe^{2+} > Zn^{2+}$ to an oxygen-rich environment, where $Zn^{2+} \gg Fe^{3+}$. This change may also have arisen because Fe^{2+} and Fe^{3+} are dangerous redox catalysts and need greater sequestration in aerobic higher cells to avoid DNA damage leading to mutations and possibly cancer.

In a variety of sites, therefore, zinc became involved in many parts of the (slow) cell/cell signalling system but its role also increased in digestion and degradation. We have observed its new degradative role in conjunction with the new proteins of the extracellular matrix in Chapter 12, which is clearly a modification of digestion, in this case controlled self-digestion! It is also known that large amounts of zinc are in 'mossy fibres' of the brain and in the male reproductive tract. It is very hard to exaggerate the role of zinc in eukaryotes and it has been described as the dominant inorganic 'hormone', just as calcium is the dominant 'inorganic' second messenger (Table 13.9). Free zinc had to be controlled therefore in order to became an internal cell control and regulator in homeostasis. (Remember that iron and haem still maintain central roles as does phosphate.)

As already stated, in cells free zinc itself is probably controlled by a group of proteins, the metallothioneins, which are synthesised in response to zinc levels and may be transcription factors as well. These new proteins are first found in photosensitive, oxygen-producing prokaryotes, for example in cyanobacteria

Table 13.9 Element connections to zinc fingers*

Effector system	Element connected (M)
Glucocorticoids (sterol)	Na, K levels (kidney)
Vitamin D (sterol)	Ca transport proteins, osteocalcin and even metallothionein (Zn), alkaline phosphatase (P)
Thyroxine	Iodine metabolism (Fe), (I), Ca–ATPase (Ca), connective tissue/Ca
Retinoic acid	Zinc dehydrogenases (Zn)
Sex hormones	Not known
Haem	Synthesis requires zinc

and perhaps they are self-regulated, that is, by zinc metallothionein itself, a co-operative buffer protein. It is an essential feature that the zinc concentrations in the different systems of Fig. 13.22 communicate among themselves to maintain homeostasis and growth, and in turn the zinc level must be in homeostatic balance with all the other 15–20 elements of an organism. (It is not known if the free zinc concentration in the cytoplasm of eukaryotes is higher than in prokaryotes. It may be that zinc is very tightly held in enzymes and more loosely held in vesicles, in metallothionein and in zinc fingers.)

13.4.5. Summary of acid–base messengers

It is useful to summarise at this stage and to make a pattern of co-operative connections between non-redox elements and messages, see marginal note. We have described the inorganic Lewis acid connections of Fig. 13.20. Now, as stated, such a system can only work if the levels of free Na^+, K^+, Ca^{2+} and Zn^{2+} are accurately fixed during growth for otherwise messages cannot be of managed intensity. Not only is homeostasis managed through controlled uptake and release to and from the inside of cells (Section 13.4), but for several elements it is also kept close to thermodynamic balance locally outside by precipitates, for example, shells and bones (Figs 13.1 and 13.23). This is in fact the best way to fix an invariant concentration but the local precipitation requires an extracellular protein network. Some of the production of essential proteins for bone and shell formation is therefore linked of necessity to the same set of steroids and zinc fingers described for internal mineral level controls. (The less flexible plant external constructs are linked often to the precipitation of opal silica and this is relatively ill-controlled.†)

We need to see next how oxidative secondary metabolism is managed, that is, the production of hormones using dioxygen, which connect to zinc fingers, to calcium release and to phosphate metabolism. The oxidative reactions depend on transition metal ions, and are more often regulatory.

13.4.6 Redox communications: free iron and the new messenger networks

In this section we exclude haem iron but include all other forms of iron that can exchange from bound forms, for example Fe^{2+} and Fe_nS_n clusters. The free iron then connects to most of the simple iron enzymes in cells since most exchange iron but over relatively long periods of time, say hours. The iron is stored in ferritin, a vesicle-contained $Fe(OH)_3$ precipitate in the cell (see Section 6.8.3) that, in principle, controls free iron levels in the cytoplasm very precisely. Given the homeostatic mechanism mentioned in Section 10.7, the free iron levels now regulate the synthesis of a large number of required Fe- and Fe_nS_n-containing enzymes, which are central to primitive and modern cell metabolism and to iron uptake and storage as discussed in Chapter 10. The new activity with which we are concerned here is the production of some of the organic messengers, for example those stored in vesicles such as adrenaline or

Modern acid/base connections

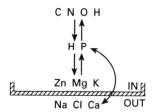

See Sections 11.5.7 and 11.8.1.

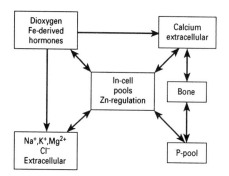

Fig. 13.23 The link between in-cell production of regulatory and control systems and the extracellular matrix extends to control over bone, which as a precipitate feeds back a control over the concentrations of Ca^{2+} and HPO_4^{2-}.

---

† Silica deposition is beautifully well-controlled in diatoms and it is thought that $Si(OH)_4$ acts as a messenger to DNA in these organisms. It is the dominant mineral in plants but the controls are unknown.

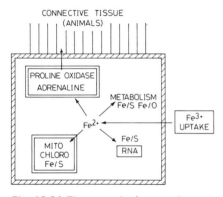

Fig. 13.24 The network of connections from the in-cell iron pool. The connections are better known in aerobic bacteria than in higher cells.

the prostaglandins. (We shall describe in more detail under copper the adrenaline signal produced from the vesicles of the adrenal gland (see Section 13.4.8).) The placing of the relevant new oxygen-using iron enzymes is often in the membrane, sometimes in the outer membrane of vesicles, or sometimes in the endoplasmic reticula. A list of the messenger molecules produced in this way is given in Table 13.10. (There is also the link described in Section 12.9 to the production of connective tissue in the Golgi vesicles, for example via collagen proline hydroxylase.) The whole activity is summarised in Fig. 13.24. The messenger molecules are destroyed by oxidative attack, often by other iron enzymes. We note that even cholesterol and arachidonic acid are products of such iron-catalysed syntheses using dioxygen to attack terpenes. Thus, through the newly available dioxygen, iron is linked to a great variety of new messenger activities, but very few if any of the iron enzymes are in the extracellular fluids of animals (contrast copper enzymes; Section 13.4.8).

Table 13.10 Secondary metabolite production catalysed by non-haem iron proteins

Reaction*	Substrates/products	Location
Hydroxylation	Prostaglandins, adrenaline	Vesicle/reticulum membranes
Dioxygenation	Aromatic molecules	Extracellular (soil bacteria)
Proline hydroxylation	Proline in proteins	Golgi (?) for connective tissue synthesis
Superoxide dismutation	O_2^-	Organelle (or cytoplasm of prokaryotes)
Lipid oxidation	Arachidonic acid	Membranes
Scavenger synthesis (Fe)	Siderophores	Prokaryote exterior
Penicillin cyclation	Penicillin	Prokaryote exterior

* NB. Many of the reactions are in bacteria, and haem-enzymes may carry out the same reactions in multicellular organisms which need protection from free iron.

13.4.7 Haem iron and the new networks

While it is sometimes the case that the oxidative enzymes are based on simple iron ions, in other cases they are based on haem iron. The importance of haem iron is ever-increasing in evolution, especially in multicellular plants and animals, while the simpler iron enzymes are more often found in aerobic bacteria. Examples of the latter are penicillin synthetase and some sterol oxidases. The maintenance of the Fe/S proteins in all cells is a remnant of prokaryote anaerobic life and often haem systems are additions to that life in the cytoplasm. Note that haem iron is a less dangerous oxidising centre than free iron due to restricted exchange of the metal or of haem. Haem iron in animals and plants has, in fact, a more important role than free iron in the network of new intercellular messages (Fig. 13.5). Let us take the animal cell first, noting the connection with primitive cells. As stated earlier it is quite possible that haem was the first protective catalytic centre in prokaryotes to be used in relation to the products of peroxide and dioxygen reactions since we know that H_2O_2 and its products must have been poisons initially and available

Thyroxine (T$_4$).

**Indoleacetic acid
(an auxin)**

Zeatin (a cytokinin)

**Gibberellic acid
(a gibberellin)**

Ethylene

$$CH_2 = CH_2$$

Abscisic acid

Examples of the five classes of plant hormones requiring Fe for synthesis or removal.

Table 13.11 Long-term messengers associated with haem iron

Messenger	Synthesis
NO	From arginine due to haem
Thyroxine (I)	From phenols due to haem
Sterols	From cholesterol due to Fe or haem
Plant hormones*	Various, often due to haem iron

* See marginal illustration.

low-potential haem enzymes were used to destroy them. Presumably, later the haem proteins were used to signal the presence of such oxygen-generated poisons as H_2O_2 and NO. (CO also generates signals, but it was present very early in the evolution of the Earth, and many CO-binding haem enzymes are found in primitive and advanced prokaryotes.) Haem-proteins are also used in animals in signalling, but now the production of the signal molecule, NO, from arginine and its recognition are both dependent on haem. A parallel use of haem-proteins, familiar as the carriers of O_2 in haemoglobin of course, is in signalling of O_2 levels to DNA. Both these small molecules (O_2 and NO) are extremely important in intercellular signalling. Still later in evolution we consider that dioxygen reactions catalysed by haem proteins became the common route to many oxidised organic molecules, for example sterol messengers. Interestingly, the destruction of sterol hormone messengers seems to be very dependent on the same kind of enzymes, P-450 oxidases, as is their synthesis (Table 13.11). It appears that the major long-term role of the sterol-related hormones is in growth and cell differentiation, that is of course a pre-requisite of truly organised multicellular systems, that have to develop from single cells.

Now, while in animals *peroxide* (H_2O_2) enzymes, as opposed to dioxygen-utilising haem enzymes, seem to be confined largely to defence (with the very notable exception of the synthesis of thyroxine (see marginal illustration on p. 528) in vesicles), in plants both the *synthesis* and destruction of a great range of organic molecules, including hormones, is dependent on haem enzymes using H_2O_2. Many of the compounds are protective poisons reflecting again the pattern:

New chemical leads to *external poison* and requires *internal protection* (destruction) which leads to *use of the 'poison' as a messenger* which leads further to *internal use* as an incorporated unit.

It may be that alkaloids of unknown function in plants represent compounds at an intermediate stage of evolution. Moreover, especially in plants, haem-enzymes (peroxidases) act outside cells and are linked to the levels of *growth* hormones, the production of the cross-links of lignins (Fig. 13.25), and perhaps the destruction of both of these series of organic molecules. Haem-enzymes are not common outside cells in animals.

The different development of plants and animals in terms of the natural selection of chemical elements is illustrated here. All through this chapter we have seen that the organisation of a plant is more primitive than that of an animal and it can be considered that the handling of messenger systems in the

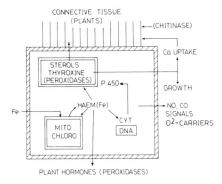

Fig. 13.25 The multiple new roles of haem-enzymes in eukaryotes after dioxygen became available. The use of high-potential haem proteins is the most striking novelty. Undoubtedly haem had a role in reductive chemistry in primitive cells.

plant world is chemically less sophisticated and more open to risk which the animal has to avoid. Here the risk in using H_2O_2 is especially clear.

In this book we cannot go into further details about the vast number of organic molecules and their feedback control produced by secondary metabolism using haem proteins for catalysed production and destruction (see 'Further reading'). A final point of considerable importance is, however, that the control of the synthesis of haem itself is not in the cytoplasm but in the mitochondria. Thus, advanced cells have given up porphyrin synthesis in their cytoplasm. Haem itself also became a regulating factor like free iron, not only in mitochondria but for the cell DNA, and must clearly be produced in homeostatic relationship with its own proteins. There are various regulating proteins, called CYT (cytochrome production factor), that bind DNA and act as transcription factors for the production of enzymes such as cytochrome P-450, catalase and so on.

13.4.8 Copper and new messengers

Copper became available even later than zinc due to the insolubility of its sulphide (Section 5.5.3). This metal is today linked to the production of two messenger systems, adrenaline and, more generally, amidated peptides in animals (Fig. 13.26). It is Cu^+ with which we are concerned, not Cu^{2+}. It is also linked to the oxidative removal of amines, that is a feedback connection, and to ascorbate oxidation. Adrenaline functions in short-term activity, generating a state of preparedness in muscles. As an amine it is also oxidatively removed by copper enzymes.

The production of one of the amidated peptides, melanin, and of similar hormones for the control of coloration and cross-linking of connective tissue illustrates an intriguing possibility since melanin is itself directed at the activation of certain cells called melanocytes, which employ the *copper* enzymes tyrosinase or phenol oxidase. It would be curious if the synthesis of a hormone, a messenger organic molecule, in one cell type could be frequently related to the activity of the same hormone in a different receptor cell and to its removal there, both cells using the same new metal system in the three activities, here copper. The peptide syntheses and hydrolyses dependent on zinc and the sterol production and destruction using haem iron may be other examples (Table 13.12). (Compare also the co-operative use of simple elements such as selenium and iodine (Section 12.4.2).†) It is interesting that all this activity of copper oxidases was probably more critical later in evolution than that of the free iron and haem oxidases. The activities are all associated with oxidases in vesicles or in external fluids, especially in animals. Thus, copper is the centre of the animal, not plant, *external* oxidases, while iron is the centre of *internal* oxidases. The availability of copper is the necessary progenitor of connective tissue. The pattern of use appears to follow the evolution of the environment in a remarkably close way: iron availability fell as that of the (poison) copper rose.

The peculiar production by copper enzymes of amidated peptides is best

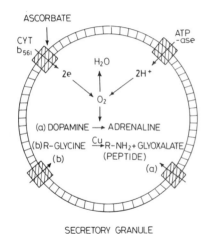

ASCORBATE

CYT b$_{561}$

ATP-ase

H$_2$O

2e 2H$^+$

O$_2$

(a) DOPAMINE → ADRENALINE

(b) R–GLYCINE $\xrightarrow{Cu}$ R–NH$_2$ + GLYOXALATE
(b) (PEPTIDE)
(a)

SECRETORY GRANULE

Fig. 13.26 An illustration of a secretory granule (vesicle) containing copper oxidative enzymes. Note the extra pumps required and the use of high-potential ascorbate as a redox control.

† If it is the case that the introduction of two new poisons occurred at roughly the same time in evolution, then the fact that they are then *mutually* assimilated eventually as useful agents (through genetic change) could suggest that the poisons themselves *direct* the most probable mutations. This is a localised Darwinian mechanism for the natural selection of elements.

Table 13.12 The relationship of metals to production and destruction of a slow signal

Signal molecule	Production	Sensor	Removal
NO	Haem enzyme	Haem enzyme	(?)
Sterols	Cytochrome P-450	Zn-protein/DNA	Cytochrome P-450
Adrenaline	Cu/Fe enzymes	Cell surface (Ca)	Copper oxidase
Oxidised peptides (melanin)	Cu enzymes	Cell surface (Ca)	Zn enzymes?
Peptides	Zn enzymes	Cell surface (Ca)	Zn enzymes
Thyroxine	Haem peroxidase	Zn-protein/DNA	Se enzyme
Plant hormones	Haem peroxidases	(?)/DNA	Haem peroxidases

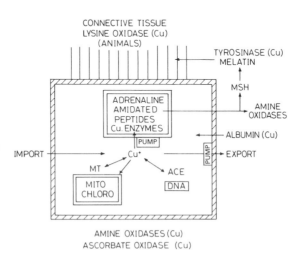

Fig. 13.27 The multiple roles of copper, all in vesicles or in extracellular fluids. MSH, melanocyte-stimulating hormone; MT, metallothionein; ACE is a transcription factor.

related to the diversification of a system of peptide messengers already related to zinc (Section 13.4.4). These, like other peptides, are probably removed by zinc enzymes. Once again, we are reminded of the joint action of zinc and copper as seen in the regulation of their storage by metallothioneins and in their joint link to connective tissue (compare Figs 13.22 and 13.27). As stated in Section 12.4.1, it may well have been that the metallothioneins were first produced to manage zinc levels but that this control spread to protect against the newest poisons—copper (and cadmium).

13.4.9 Other elements in messenger systems

As stated in the previous chapter, the advent of dioxygen did not just alter the chemistry of metals. New messengers were also developed from new oxidative non-metal chemistry, for example iodine in thyroxine, albeit with the use of metals in synthesis and even as receptors. Examples are given in Table 13.6. An interesting example of the use of non-metals in catalysis is in the metabolism of thyroxine, which requires haem iron peroxidase for its production in a vesicle but is destroyed by a *selenium* enzyme and recognised by a zinc finger DNA-binding protein. Note the changed ability with time (Fig. 5.20) to oxidise halides thus introducing organohalogen chemistry, later to be much valued by man. Man's industrial chemistry provides the latest step in the evolution of high oxidation states of metals and non-metals.

MODERN REDOX CONNECTIONS

CNOH

$\updownarrow$

H e S

$\updownarrow$

Mn Fe (Co)(Ni) Cu
Mo Se

See Sections 11.6 and 11.8.3.

The remaining metal elements involved in catalysis of the production of organic molecules after the advent of dioxygen are manganese and molybdenum. Manganese is directly involved in the production of dioxygen itself and it may well be a messenger in plants as well as in animals. It is involved in the metabolism of lignin but no other simple messenger system is known to be related to it. There is much manganese in plants and its role may well be greatly underestimated as yet. The role of molybdenum in message systems appears to be very limited though it is connected to xanthine (and aldehyde redox reactions primitively) and to sulphate, dinitrogen and nitrate metabolism, more recently.

The catalytic trace elements cobalt and nickel do not appear to be used in new oxidative reactions much though they are still used in *reductive* reactions in anaerobes. It seems as if the production even of nucleic acids for DNA is today more dependent on oxidative reactions than on reductive actions of cobalt or of Fe/S enzymes (Section 11.5.4), see marginal note. Is evolution in complicated organisms removing nickel and cobalt from positions of importance? Are they only more useful than Mn, Fe, Cu and Zn in the reductive formation of M—C and M—H bonds in anaerobic conditions?

13.5 Interactions between and complexity of messenger systems

We have now seen that any one element is connected to a very large number of messenger activities. In a sense we can look upon each element as if it were an electron in a circuit. The circuit has feedback as in Fig. 7.34, so that there is never a sustained excessive or deficient current. Each such circuit has grown in complexity from its simple and rather flexible use inside prokaryote cells to more delicate functional value in multiple compartmental uses in multicellular organisms all of which are co-operatively cross-connected. Multicellular organisms grow slowly so that long-term sustained signalling is required as well as fast switching. The imposition in multicellular organisms of much greater physical and chemical complexity and sensitivity of control and regulation demanded essential further refinements of the system, such as refinements of temperature control and of ionic strength to reduce fluctuations. Now, in these higher organisms as in lower organisms each element circuit has to be linked to all others so that there are many interactive carriers of information and we have to extend these links also to new and old organic (C, H, N, O, P, S) messengers. We shall examine the integration of these in a schematic fashion before we return to the problems that this complexity introduces.

13.5.1 The integration of element messages in networks

We are deliberately emphasising in this book the selective use of chemical elements. The examples given above of new activities arising upon the advent of dioxygen, giving rise to external message systems and linked to the production of extracellular matrices, were associated mainly and separately with Ca^{2+}, Fe^{2+}, Zn^{2+} and Cu^+. We must see next how these activities associated with individual elements are connected together. In an organised body individual message systems have to be co-ordinated, and a minimum need is for the connection of each system to *two* others so as to make an integrated network. We choose to illustrate the cross-connection of sequential messenger systems using first a figure which is very largely schematic (Fig. 13.28).

In this figure we have selected some of the features of the descriptions of individual element activities given in Section 13.4, but we now attempt to relate two cells to one another. One obvious connection is between the uses of copper and zinc. Thus, both copper and zinc are involved in the synthesis of peptide hormones as well as in the metabolic destruction of some of these same hormones. Copper and zinc are also deeply involved in the formation and breakdown of extracellular matrices, especially in animals, and both are buffered by metallothioneins. (Zinc is further involved in DNA expression through sterol hormones and zinc fingers.) Table 13.13 gives some indication

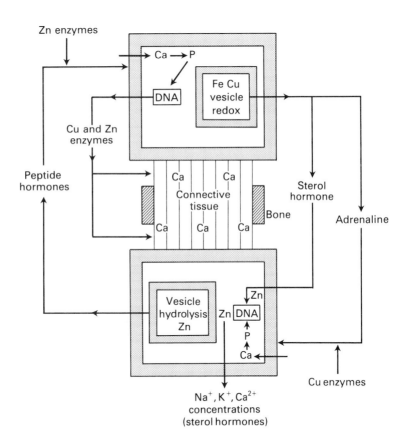

Fig. 13.28 A schematic diagram illustrating the connections between cells based on inorganic and organic chemicals but stressing the links between different elements.

Table 13.13 Cu/Zn-dependent long-term signals

Hormone	Production	Receptor	Destruction
Adrenaline	Cu/Fe oxidation	G-Protein, Ca, P	Cu oxidation
Amidated peptide	Cu oxidation	G-Protein, Ca, P	Zn hydrolysis
Other peptides	Zn hydrolysis	G-Protein, Ca, P	Zn hydrolysis
Sterols, thyroxine	Fe oxidation	Zn-fingers	Fe oxidation
Sodium/potassium	Sterols, pumps	Zn-finger depolarisation	ATPase pump

of the total involvement of these two metals but it is immediately apparent that further elaboration involving other elements is necessary in order to appreciate the network of messages and activities between cells. The roles of calcium, especially in the stabilisation of connective tissue, and of iron in the production of hormones such as adrenaline and steroids are clear and related to copper or zinc. In addition, the effects of adrenaline messages are upon calcium levels in cells and the effects of many steroids are on the zinc-finger-controlled production of proteins to manage calcium uptake (amongst other mineral elements). As in any other network, there is an interwoven set of connections that leads from one part of a circuit, here involving one element, to another, using a second element, only to return to the first in some other form involving a third and fourth element at intermediate steps. It is only through such a system of feedforward and feedback connections that a long-term holistic co-operative homeostasis of elements or other outputs can be maintained either in a cell or in a computerised operation.

A recently discovered connectivity between elements in different cells lies in the use of nitric oxide, NO, as a signal controlling slow muscle relaxation. The NO is produced by a haem (Fe)-dependent enzyme which is controlled by calcium levels. The receptor for NO in a neighbour cell is another haem protein which activates a GTP kinase (a phosphate (P)) system. This is an example of linked *control* between cells dependent on several cross-connected element messengers, but we must also see that *regulation* of proteins in different cells has to be cross-connected via messengers too so that cell–cell growth is in harmony. A specific, further new example is the effect of certain selected peptide messages from one cell to a transcription factor in another. The peptide causes Ca^{2+} entry into the receptor cell where Ca^{2+} activates a phosphatase, calcineurin, which in turn removes phosphate from a latent transcription factor and so activates it. Now calcineurin contains both zinc and iron (Fe^{2+}) and hence is only optimally active when these elements are all present in homeostatic amounts. Here four elements with widely different time constants as messengers are interlinked.

This book has as its central purpose the natural selection of the chemical elements yet within evolution we must see that the selection is in part governed by the ability of the organism to respond through the synthesis of new proteins such as calcineurin. How novelty of proteins is introduced is not known for certain. However just as we see many elements in control and regulation through communication between cells so we must see that because the proteins are the basis of dynamic structure within the molecular machinery of cells there is the need to keep protein synthesis in tune between cells and cellular organs of fixed cell content, both of which require messages of a much more sustained lifetime.

If we have described correctly the effect of introducing dioxygen into life's chemicals with its effects on the availability of other elements then the transformation is that together they generated multicellular organisms of differentiated cells. There are demonstrated required parts to the transformation, which so far relate to messenger networks, including:

(1) new structures of slow turnover;

(2) control and regulatory elements and molecules with large binding constants and slow exchange kinetics;

(3) fast responses in some cells which did not affect the long-term homeostasis;

(4) cumulative, repeated, fast responses in further cell-types which generated long-term memory (growth in a particular organ, the brain).

13.5.2 Total integration of element functions

The integration we have discussed so far in this chapter is that of the messenger systems within and between cells. They are seen to be understood at least in principle. We need next to appreciate how this integration is linked to the functions examined in Chapter 12. These were:

(1) the controlled uptake of some 20 elements;

(2) the production and repair of organic cellular structures, that is, differentiated cells (in organs within larger organisms);

(3) the use of these elements to establish structures in the form of mineralised connective tissue;

(4) the ability to develop surface fields that constrain activity and presumably allow morphological development;

(5) the multiplication of cells through the directives of the central organisation of messages, in some way linked to the whole DNA;

(6) the reproduction (sexual) by special cells;

(7) the appropriate distribution of energy as well as of materials between the chemical pathways.

Much of this linkage must be the focus of future analysis. While its complications are already very clear in textbooks of biochemistry and physiology, in this book we are stressing an underlying problem—the way in which individual chemical elements are naturally selected for functional value within the whole organism—and our next step is to give a brief view of the uptake of different elements within organs rather than within cells and to discuss how such activity can be maintained in a whole organism (Table 13.14). An organ such as the liver has several types of cells in an organised unit based on cell–cell homeostasis and, as discussed, has a shape related to that homeostasis. The picture that we wish to develop is one in which different organs have different element compositions and yet are externally connected to one another and to a final overall management system. It is worth stressing that organs can be extracted and, when given appropriate energy and material, they survive over very long periods.

13.5.3 Organs and their elements

Once we begin to describe large systems of differentiated cells, we must refer to their organisation within organs. It is then necessary for messages and responses to be due to organs rather than to single cells, hence organs must have a network of signals both internally and to other organs (Fig. 13.29). In the context of this book, it is the management of the element composition of the different organs and messages to and from them that concern us next. It is clear even to the eye that the liver of animals is heavily loaded with iron in the

Table 13.14 Examples of the relationships of organs to hormones and elements

Organ	Message hormone (transmitter)*	Receiver
Nerve (fast)	Na^+, K^+, Ca^{2+}, ACh, GABA, etc.; catecholamines	Nerve and/or muscle
Adrenal gland: medulla (intermediate); cortex (slow)	Catecholamine (Cu); glucocorticoids (sterols) (Fe)	Nerve and muscle; many cells: control of protein synthesis; Zn-receptor; bone Ca^{2+}/PO_4^{3-}
Pituitary (slow)	Many peptides (Zn); thyrotropin; ACTH; growth hormones	Thyroid and adrenal glands (general in cells)
Pancreas (slow)	Insulin (Zn); glucagon	Liver (many organs); glucose distribution
Thyroid (slow)	Thyroxine (I); calcitonin	Zinc finger: membrane

* ACh, acetylcholine; GABA, γ-aminobutyric acid; ACTH, adrenocorticotrophic hormone.

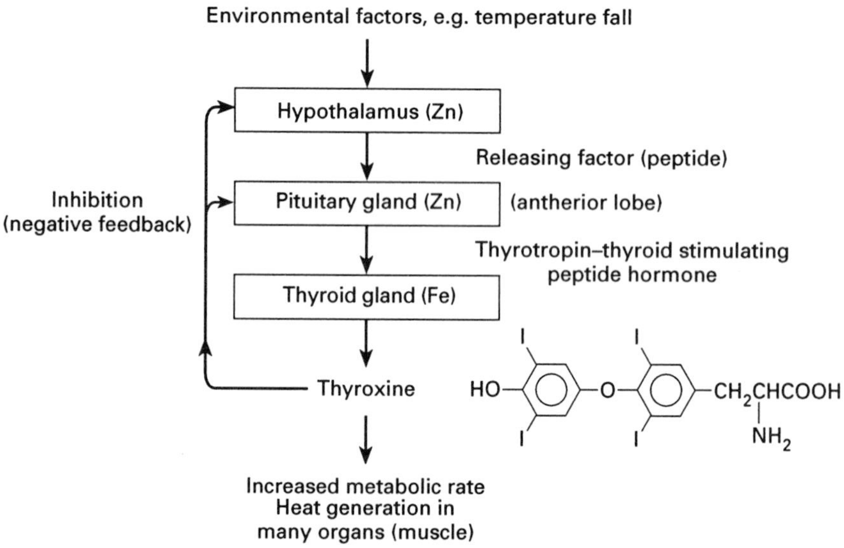

Fig. 13.29 An illustration of feedback at the level of organs in an organism. Note the use of inorganic elements, Fe, Zn and I, as described earlier, but which are differentially concentrated in different organs.

form of haem iron. It is also the source of much vitamin B_{12} (cobalt), which is required for haem synthesis. Quite generally, differentiation into organs implies differentiation of elements since different functions belong to different organs and these functions require different elements.

The system of organs therefore had to have distribution modes for elements within the flowing liquids of the whole body. In these circumstances, synthesis of carrier proteins for selected elements and specific membrane recognition devices for these proteins were required. One such protein is transferrin, the iron carrier, and liver cells in particular have many transferrin receptors for it. The circulating proteins for Cu, Zn and vitamin B_{12} (Co) are also well known as well as those for special organic molecules such as sterols and thyroid hormones. Thus we need a map of element transfer within the whole body that takes into account quantitatively the separate reception of elements by organs.

Without going into detail, the development of organs in spatial patterns in multicellular organisms is seen to involve ever greater complexity than that exhibited in simple cell–cell communication even at the level of chemical element distribution employing a network of signals to maintain regional homeostasis. Yet all this construction hides one problem—how could such an organisation arise?

13.6 The problem of development

Here we meet an even greater problem for our understanding. While the final form of an advanced organism is readily described in textbooks of physiology and the message and distribution systems between organs are analysed in textbooks on biological chemistry, all such organisms develop from a single cell (Fig. 13.30). Up to now we have treated the adult body of an organism as being in a homeostatic state of chemicals, and it is very probably correct to assume that this homeostasis decides the final morphology of an organism so that the morphology of an organisation is fixed for a given chemical flow just as

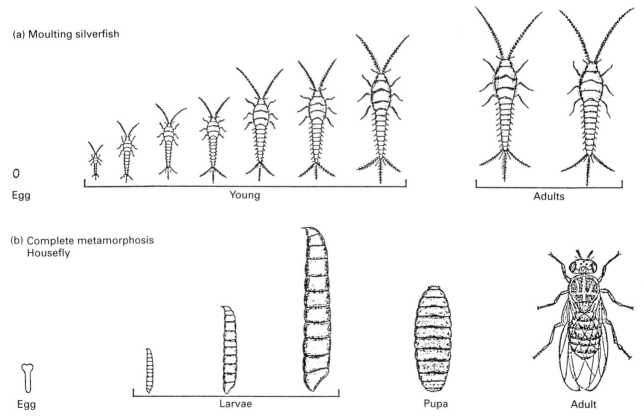

Fig. 13.30 The development of insects from a single-cell eggs: (a) by growth; (b) by growth and metamorphosis to a *fixed* shape. (See *Scientific American*, July 1994.)

the morphology of a crystal was fixed thermodynamically for a given composition of a chemical system (see Chapter 6). The problem acutely present in this description, however, is the origin of the flowing chemical network starting from the initial single cell, the egg. This requires a different set of controls based on *continuous modification in time of the homeostasis*, branching as symmetry is lost (Fig. 13.31) and local organs develop. We turn then to the further complexity of multicellular life—development with diversification of cell function. This can only mean that expression of DNA is constrained in time with the differential distribution of different elements in special cells in organs in defined spatial patterns. The natural selection of chemical elements here evolves with time in an individual organism.

13.6.1 Adaptability, differentiation and morphogenesis

The ability of a cell containing a particular code to behave in different ways has developed from the process of adaptability in bacteria to reversible differentiation in cell colonies (and even in certain types of cells in plants and animals) and has reached a final condition in which a large cell population made from a single cell has divided into groups of cells (in organs) with each group *permanently*, that is, irreversibly, acting in one role within an integrated whole animal organism such as man (Table 13.15). This most sophisticated condition, reached in higher animals, implies that DNA expression can be effectively regulated in organs after development periods of considerable length.

A very similar description can be applied to the shape of organisms. At the simplest extreme of multicellular life there is little more than a loose

Fig. 13.31 The growth pattern of animal cells from fertilisation. While even cell A has polarity, the distinction between different cells begins at the 8–16 cell step (D–E), and then cells become of increasingly divergent shape and function. This is the period of morphogenesis. Organisms only become truly homeostatic, it is suggested, with the appearance of a fetus with a close resemblance to the adult form. H stresses the divergence of cells as the inside becomes totally distinct from the outside of the organism, G, before infolding occurs. The infolding may be related to the surface currents as surfaces change during packing.

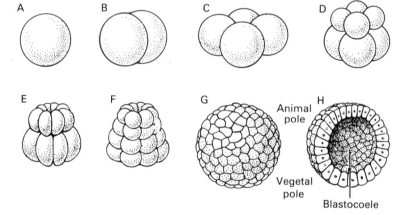

Table 13.15 The growth of organisation

Unifunctional single cell	Earliest cells, bacteria
Adaptable single cell	Modern prokaryotes and eukaryotes: bacteria/yeasts
Colonies of adaptable cells	Sponges, slime moulds
Colonies of differentiated but revertible cells in organs	Plants, worms (reptiles)
Organised organs of little adaptability or reversibility	Higher animals

association of adaptable differentiated cells, as in a sponge, and shape is poorly defined—cells can even be interchanged in space and function. More shape (using the word loosely) can be ascribed to organisms such as the jellyfish, while the soft bodies of cockles and mussels are contained within a formed shell (Fig. 12.30) holding some shaped organs. In these animals much replacement of tissue is possible, for example in jellyfish. Parallel examples are observed in plant life where we observe only fixed patterns but not fixed overall shape. Morphology is not well defined and cells can differentiate and de-differentiate even to the degree that virtually all cells can switch function to grow into a plant. The development of message systems, temperature control and so on is limited and within a much looser construct than in a higher animal. As stated before, the separate tissues of plants retain a plasticity such that loss of one organ can lead to a change of body plan. In contrast, in higher animals morphological control is permanently determined after a relatively long period of development. Development may be continuous from fertilisation to birth, to the adult stage, or may undergo relatively sudden switches such as that from the form of a caterpillar to that of a butterfly, metamorphosis. It is noticeable that the ability to restructure or to regrow lost features of form or shape decreases to a greater and greater degree as complexity increases. We wish to consider how this is related to tighter control of element composition.

13.6.2 The problem of the development of an element network

Once again we may refer to the growth of crystals. In the steps before a final form appears, say, in the growth of NaCl crystals there is a period of 'nucleation' in which shape is flexible (Section 7.17) and the atoms re-arrange themselves since the overall energies internally do not at this stage govern local packing due to the magnitude and variety of the surface energies. Thus, the particles flow within the nucleation system. Only after a critical size does the shape become fixed for the rest of growth. In the crystallisation of a non-stoichiometric compound composition varies as the crystal grows, i.e. with crystal size, even with constant solution conditions. We maintain that there is a close parallel between this process and the development of a multicellular system (Fig. 13.31).

This leads to the suggestion that, during the development period, element concentrations must change. A very striking fact related to this account of morphogenesis is that chemical elements in the wrong concentration in animals are teratogenic. An example is lithium. Again, deficiency of iodine, needed for the thyroid hormone, causes defects in growth patterns, and the addition of thyroid hormone, identified with the element iodine, can cause morphological restructuring. (A precisely similar observation applies to the effects of impurities on crystal morphology. We refer the reader back to Chapter 6 where we consider impurities and surface energies.)

13.7 Conclusions concerning the element content of organisms

As far as this book is concerned, we are now reaching a very elementary conclusion. All the patterns of living systems we observe are based on flow of

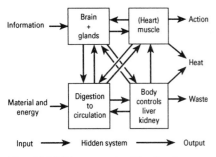

Fig. 13.32 The network of feedback through both nerve and chemical messages gives the hidden system a steady state and an ability to respond intermittently (see Section 7.14). Material transport is also required in the hidden system.

energy and material under co-operative feedback and feedforward management. The supply of energy can come in different forms but is transformed quickly into selected chemical potentials. It is the description of the particular chemicals and their potentials that is proving so difficult, yet underlying this description is a simple analytical statement in terms of the concentration of some 20 chemical elements. *The chemical element composition must be a complete description of a species at all stages of development and within all its separate organs* (Table 13.16). The simplicity of this fact is hidden by the complex intake systems of animals, but it is obvious in bacterial or plant life. No matter what living system we describe we can use the diagram of Fig. 13.32 where rejection is kept in relation to uptake so that the hidden (living) system (Section 7.13), here the organism or organ, has a balanced flow of elements or has successive changes of such flow. This balance also ensures a fixed shape for an organism or organ at all stages. We reduce the nature of species to the level of the way in which a code can utilise energy and *inherited* machinery to form an organism or organ from about twenty environmental chemical elements in fixed ratios. During development the ratios will change but again in a fixed pattern and as organs form they will have the chemical elements in fixed ratios. This fixing of the ratios fixes the shape and activity of the whole and of the organs in the whole.

Table 13.16 The ionic composition, during development, of human serum and muscles and of human and pig brain

| | Human serum composition at age (years) | | | | Ion | Ionic composition (meq kg^{-1}) of human skeletal muscle (heart muscle*) at age | | | | |
| | | | | | | Prenatal (weeks) | | After birth | | |
	5–12	12–18	20–30	70–90		13–14	20–22	Newborn	4–7 months	Adult
Na$^+$ (meq l^{-1})	142.0	142.0	142.0	142.0	Na$^+$	101	91 (46)	60 (64)	50 (60)	36 (58)
K$^+$ (meq l^{-1})	4.9	4.9	4.8	4.8	K$^+$	56	58 (81)	58 (54)	90 (50)	92 (66)
Ca^{2+} (mg/100 ml)	9.8	9.8	10.0	10.3	Mg^{2+}	12	11	14 (11)	20 (11)	17 (13)
Mg^{2+} (mg/100 ml)	2.0	2.0	2.0	2.0	Ca^{2+}	6	7	4 (7)	3 (8)	3
P (mg/100 ml)	4.8	4.3	4.0	3.5	Cl$^-$	76	66 (41)	43 (45)	35 (50)	22 (47)
					P	37	40 (50)	47 (47)	64 (50)	59 (49)

Electrolyte composition (meq kg^{-1})

| | Human brain at age | | | | | Pig brain at age | | | | |
| | Prenatal age (weeks) | | | | | Prenatal | After birth | | | |
Electrolyte	13–14	20–22	Newborn	Adult	Senescent	46 days	Newborn	3 weeks	4 weeks	Adult
Na$^+$	97.5	91.7	80.9	55.2	Rises	82.8	59.9	64.1	61.4	60.7
K$^+$	49.6	52.0	58.2	84.6	Falls	56.7	85.8	74.6	75.9	76.0
Cl$^-$	72.1	72.6	66.1	40.5	Rises	62.3	46.8	48.9	45.2	41.1
Mg^{2+}		8.4	7.9	11.4	Unknown	9.5	10.6	9.2	8.0	12.2
Ca^{2+}		4.9	4.8	4.0	Rises					
P	57.0	52.2	54.0	109.0	Falls	57.0	82.5	82.0	92.5	125.0

* Heart muscle matures early and, in keeping with being a rather slow muscle, has low ionic ratios, K/Na, Mg/Ca and P/Cl.
Data from Comar, C. L. and Bronner, F. (1960). *Mineral metabolism*. Academic Press, New York, see reference 19.

Pantothenate

Riboflavin
(Vitamin B$_2$)

Nicotinate
(Niacin)

Pyridoxine
(A form of Vitamin B$_6$)

Structures of some water-soluble vitamins.

Table 13.18 Basic set of 20 amino acids

Non-essential	Essential
Alanine	Arginine
Asparagine	Histidine
Aspartate	Isoleucine
Cysteine	Leucine
Glutamate	Lysine
Glutamine	Methionine
Glycine	Phenylalanine
Proline	Threonine
Serine	Tryptophan
Tyrosine	Valine

Table 13.17 Dependences for required compounds and elements

Requirement	Bacteria	Plants	Animals*
(N,C)amino acids	No	N only	Yes
Coenzymes	No	No	Yes
Fats†	No	No	Yes
Sterols	(Absent)	No	No?
Fe	No	No	Yes
Co	No	No	Yes
Mo	No	No	Yes
P	No	No	Yes
ATP (energy)	No	Yes (chloroplasts)	Yes (mitochondria)

* Ultimately the only direct intake of elements by animals is of H_2O.
† Animals cannot synthesise a range of unsaturated fats including linoleate and arachidonate (essential for signalling). They cannot convert fats to glucose.

We have come a long way from the considerations of single prokaryote cells to ever-increasing complexity in multicellular systems with organs. However, in one sense little has changed. Some twenty chemical elements are assembled under direction from an inherited plan plus machinery. The plan and the machine are in feedback, feedforward relationship so that life is a combination of them both. It was the primitive machinery, perhaps before there was a code but including the basic containing membrane, which recognised the simple materials. It developed into the earliest cells which, through controls, could lie dormant or almost so in the absence of food. However, this style of life is in complete contrast with the lifestyle of a complex animal organism that needs a constant supply of material (food) (Table 13.17) and energy to maintain the communicating cells during development and in which, thereafter, the flows remain in balance. (Plant life is intermediate, only requiring nitrogen fixation, Table 13.17, but being able to lie dormant for long periods.) This constant demand means that the complex animal systems, which are free-roving, could improve the chemical traps they were in by improved scavenging and protection based on ever better sensory devices. The vicious circle of such improvements in sensing is that they require ever more complexity and refinement of internal equipment and an ever larger energy input. The complexity also requires long growth/reproduction cycles. It has inherent risks, since a mistake in the network or growth can damage the whole organism before reproduction is possible (Section 7.22). A way of reducing the risk to the organism is to somehow reduce the number of functions it has to perform. For a roving animal the way forward that was discovered was to discard very primitive metabolic syntheses that are essential to all organisms from bacteria up, thereafter relying on scavenging products from lower organisms by digestion. The dependences which then developed are described under *essential* amino acids (Table 13.18), fats and vitamins (Table 13.19). Since pathways have associated chemical element requirements, species became different in trace element composition. Some of the dependences are particularly striking. Although a roving animal had no way of getting an adequate intake of energy or materials without consuming vast amounts of prepared energy and material from other forms of life, the curious feature is the abandonment of production of many (but not all) of the simplest units, for

α-Tocopherol
(Vitamin E)

Retinol
(Vitamin A)

Vitamin K₁

Calciferol
(Vitamin D₂)

Structures of some fat-soluble vitamins.

Table 13.19 Vitamins† required by man

Letter designation	Synonym	Function
Fat-soluble		
A	Retinol	Enters into photochemical reaction in rods in retina of eye: morphogenesis factor?
D	Calciferol	Calcification and hardening of bone, absorption of calcium and phosphorus
E	Tocopherol	Poorly understood; an anti-oxidant
K	Phylloquinone	Required for synthesis of certain blood-clotting factors; mechanism unknown
Water-soluble		
B_1	Thiamine	Required as coenzyme for decarboxylation and certain other processes in cellular metabolism
B_2	Riboflavin	Forms flavin coenzymes (FAD, etc.), required as hydrogen carriers in cell respiration
PP	Nicotinic acid	Forms coenzymes NAD and NADP, required as hydrogen acceptors in cell respiration
B_5	Pantothenic acid	Forms coenzyme A, which activates certain carboxylic acids in cellular metabolism
B_6	Pyridoxine	Forms coenzyme required for synthesis of amino acids from carbohydrate intermediates
B_{12}	Cobalamin	Required for formation of haem
M or Bc	Folic acid	Similar to B_{12}
C	Ascorbic acid	Required for function of intercellular material; biochemical role as antioxidant
H	Biotin	Required as coenzyme for carboxylation reactions in cellular metabolism

†See marginal illustrations for structures on pp. 541 and 542.

example some amino acids, fats, and coenzymes (vitamins) on top of an inability to obtain for itself any of about 15 mineral elements (see Tables 13.17 and 13.20). Even the synthesis of haem has been left to mitochondria. Clearly today we are at an intermediate stage of development of such dependences since the common chemicals of all life include *all* amino acids, *all* nucleotides, *most* fats and so on, yet many syntheses have not been lost. Man, through organic chemistry, has gone some way along the line with his industry to providing a backup system for himself so that he will remove dependence on other life forms, but as we shall see this lacks a fundamental feedback connection to life's chemistry. Before turning to man's industry, Chapter 14, there is a need to look more closely at the organisation that has already been forced upon higher forms by the intense complexity of *coexistence* with more lowly organisation in other cellular systems. (It is said that 90 per cent of cells in a human are not based on human DNA but on the genes of coexisting bacteria.) To handle the coexistence the control system of the higher species has developed new internal organisational power. In individual species the highest level of control is now in the central nervous system to which we turn finally. It is using this system that man can apparently evolve chemical independence through his industry (see Chapter 14)—he has become self-conscious of his own chemistry.

Table 13.20 Elements taken into by plants and then absorbed by animals

Element	Source	Absorbed form
Non-mineral elements		
Carbon (C)	Atmosphere	CO_2
Oxygen (O)	Atmosphere	CO_2
Hydrogen (H)	Soil	H_2O
Nitrogen (N)	Soil	NH_4^+ and NO_3^-
Nitrogen (N)	Bacteria	RNH_2
Mineral nutrients		
Macronutrients		
Sodium (Na)	Soil	Na^+
Phosphorus (P)	Soil	$H_2PO_4^-$
Potassium (K)	Soil	K^+
Sulphur (S)	Soil	SO_4^{2-}
Calcium (Ca)	Soil	Ca^{2+}
Magnesium (Mg)	Soil	Mg^{2+}
Silicon (Si)	Soil	$Si(OH)_4$
Micronutrients		
Iron (Fe)	Soil	Fe^{3+} complex
Chlorine (Cl)	Soil	Cl^-
Manganese (Mn)	Soil	Mn^{2+}
Boron (B)*	Soil	$H_2BO_3^-$ and HBO_3^{2-}
Nickel (Ni)	Soil	Ni^{2+} complex
Cobalt (Co)	Soil	Co^{2+} complex
Zinc (Zn)	Soil	Zn^{2+} complex
Copper (Cu)	Soil	Cu^{2+} complex
Molybdenum (Mo)	Soil	MoO_4^{3-} complex

* Not essential for animals, see reference (2).

13.8 The brain: a phenotypical organisation centre

Whereas we see in all cells an organisation centre in the DNA, the complex organism demands *integration of activity* of the whole developed animal (phenotypic control) to a greater and greater degree as dynamic complexity increases. The problem is one of connecting sensors of the environment to activity of internal organs. It is useless before conception, of course, and therefore is largely a chemical development after birth. Here the first feature of importance in this book's approach to evolution via selection of chemical elements is to follow the elements in brain with time (Table 13.16). Brain cells develop in element content and, as expected, in shape as these are not independent factors (Section 7.18). Thus the brain becomes a quite new three dimensional morphological field, an element map of experience, made by electrolytic connections and growth, and ready for recall. Its information content is *individual* to each member of a species and not closely related to the linear information in DNA. The control circuit grows as the animal explores and by back interaction with senses then activated. In other words 'the plan' in the hidden system of Section 7.21 is no longer just the inherited DNA and its machinery but has additionally an environmentally conditioned character which is not inherited. The individual is not predestined to behave rigidly as the DNA dictates except in so far as such features as body-shape and, organ function are concerned. The gradual development of idiosyncratic behaviour as the nervous system of the brain has evolved is very clear in any examination of animal species. Before we see the consequences of this development we need to note the special chemical nature of the brain which is protected by a special spinal fluid even of slightly different electrolyte composition from the rest of the body.

We have already noted that most organs grow with fixed shape and chemical element content. One particular and curious feature of the brain is that its cells grow with varying shape whence we are forced to conclude that the growth proteins that arise in changing amounts in brain cells are indicative of local chemical changes and that these changes have come about through the influence of external factors upon sensing systems. The brain is then an ever changing chemical morphogenic pattern, not due to DNA but to experience.

The early development of a central nervous system as a progressively evolving co-ordinating centre, a brain, was driven by the advantage gained from a memory which allowed the individual animal to position itself or to adjust its state of awareness according to the environment. The important step from the point of view of this book is that this awareness slowly developed further into *an ability to adjust the environment* to suit the needs of the animal, such as protection. An early example is that fish and birds build nests. This management of physical surrounds during the growth of the organism is very different from the growth of a shell which is a part of the organism and appears to be wholly related to DNA instructions. The use of this ability to plan *physical* features, to build a nest for the future, has become heightened in man to the programmed use of almost all the *chemical* elements, that is to an expanded natural selection on a rational basis. It must be a peculiarly changed nature of the brain that has allowed this ultimate control. We turn to this most intriguing part of evolution in Chapter 14 where we describe the way in which man's chemical insight has been used for more than 10 000 years in a highly empirical way but has only become fully rationalised, as described in Part I of this book in the last 150 years. By becoming self-conscious of his chemical nature, and not just of his physical surrounds, man has become able to direct the future purposeful selection of chemical elements as well as that of his physical surrounds. While we describe this novel evolution in the next chapter, we must be aware first of the nature of man in relationship with the most recent developments of organisation in which many organisms come together to generate an ecosystem.

13.9 Cellular dependencies: genetic management towards an ecosystem

The development of cells must have followed an evolutionary path (Table 13.21) such as

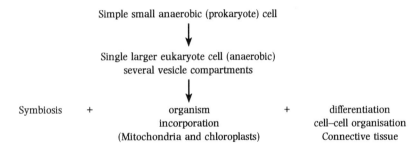

At each step there is delegation of central function to local function to gain the advantages of compartmentalised organisation rather than simple central control. The increase in complexity is reflected in part by the increased size of the DNA (Table 13.22). We are concerned again to show the part played in these changes by the chemical elements, changes that were generated by dioxygen and its effects on the environment.

Table 13.21 A simplified evolution of integrated organisation

---

Early prokaryote

Anaerobic metabolism in which developing homeostasis is based on several chemical element concentrations. Cell has no shape? Division accidental. Many ion pumps but only potassium channels. No morphogenic pattern. No control of environment.

Later prokaryote (after dioxygen)

Both anaerobic and aerobic metabolism in which homeostasis is based on several chemical element concentrations. Cell has outer constructions, even walls and a periplasmic space, i.e. external morphology. Division controlled, i.e. morphogenesis. Many ion pumps but only potassium channels. Environment monitored and utilised via external enzymes. New systems of catalysts based on Cu, Mo, Se and so on.

Eukaryote, unicellular

Homeostasis now stricter and based on element concentrations and morphology but external and internal based on organisation due to filaments and vesicles. Cell has characteristic shape and phases of growth and morphogenesis. Many pumps and many channels: sensing currents around shaped exterior, e.g. Na^+, H^+ and Ca^{2+}. Vesicle rejection to give poisons and mineral protection.

Eukaryote, multicellular

Strict homeostasis as for unicellular eukaryote but extended to external fluids and cell/cell communication. Cells and cell/cell interaction give shape, etc. Many pumps and channels connected to circulating fluids. Nervous systems of currents around and between cells. Vesicle rejection gives hormones and transmitters to circulating system. Environment of cell controlled in element composition. Development of external filaments controlling through tension external chemical potential. Long periods of morphogenesis followed by even longer periods of homeostatic growth and occasional metamorphosis. Temperature control.

In Section 11.12.2 we indicated how compartments within cells took up different elements and so developed functional idiosyncratic activities locally. With organisation this distribution became based on differences in cell types based on one type of DNA. Finally, and in contrast, symbiosis and organism incorporation introduce multiple DNA-types into the stabilisation of co-operative individual organisms in ecosystems. Individual men are in fact 'ecosystems', containing many bacteria, if the word is used loosely.

Table 13.22 The DNA content of various cells

| Organism | Size of DNA genome | | Maximum number of proteins encoded[*] |
	Number of base pairs	Total length (mm)	
Prokaryotic			
Escherichia coli (bacterium)	4.0×10^6	1.36	3.3×10^3
Eukaryotic			
Saccharomyces cerevisiae (yeast)	1.35×10^7	4.60	1.125×10^4
Drosophila melanogaster (insect)	1.65×10^8	56	1.375×10^5
Homo sapiens (human)	2.9×10^9	990	2.42×10^6
Zea mays (corn)	5.0×10^9	1710	4.0×10^6

* Assuming 1200 base pairs per protein. Size is doubled for diploid cells.

13.10 Summary: complexity, management and survival stability

The complexity of any system increases survival but introduces risks of its own. Multiplying refinement has limitations and management of the operations within a system under one central control may well become less effective than a looser arrangement of separated functions under different control centres. A rugged supply-and-demand compromise control where supply-and-demand activities are largely independent in different organisations, but either or both can fail unless they moderate activity in a matching feedback/feed forward way, has survival strength that in practice exceeds that of production based on *ab initio* introduction of everything from simple elements upwards in one place. Not only does this appear to be true in the macroeconomic activities of man, but it seems to be what has happened in evolution in order to assist survival. As multicellular organisms or even eukaryotic single cell activity developed, it would appear that the survival strength of a multiplicity of different species doing different things yet supplying one another (Table 13.17) exceeded the survival strength in many environments of those species that attempted to develop complexity using only basic chemical elements and energy directly from the environment. This is not to say that primitive bacterial cells cannot survive by simply replicating, since they have done so, but they are very limited: they require an aqueous medium and are not very good at scavenging material or light. Neither can they occupy many of the niches that plants and animals fill. Forms of association or symbiosis are then the most successful strategy since the bacteria, single prokaryote cells, have a permitted flexibility with a fundamental simplicity such that they can provide basic chemicals for all forms of life. Perhaps the most surprising result of symbiosis is the presence of mitochondria and chloroplasts in other living single or multicellular eukaryote systems. These are the major energy capture devices, neither of which are fully coded in the gene of the parent organism, but we have already noted their relative thermodynamic efficiency (Section 7.22). Photosynthetic organisms such as plants and aerobic animals then have small incorporated power stations with independent controls and different DNA. The content of chemical elements in such units also differs from that of the cytoplasm. (In passing we observe that man's power stations are also well separated from his other activities.)

A different symbiosis is seen in the nitrogen-fixing bacteria often associated with plant roots. Here the bacteria take carbon compounds from the plant and hence energy in exchange for nitrogen compounds.

The increase in evolution of symbiosis is already seen in the need of lowly eukaryote organisms such as single cell coccoliths for a vitamin, folic acid, that is a part of a coenzyme. Here the food, prokaryotes, for the coccolith, supplies a required compound that was not easy to synthesise. We could continue to explore the requirements of more and more sophisticated organisms, Table 13.21, but the case of man, the most sophisticated animal, gives a very easy appreciation of ever-increasing dependence on other species with ever-increasing complexity. *In fact, evolution proceeds by co-operation as well as by competition. Survival strength, as we now see it, is of a community of species.*

Tables 13.18 and 13.19 list two major requirements—amino acids and vitamins (mainly parts of coenzymes)—but to this list we must now add initial uptake of all elements (Table 13.20) except for H and O in H_2O, a supply of energy to make ATP, and the demand for a balanced diet of fats. This means that man has almost totally sacrificed chemical and energy independence for sophistication while searching for external sources of them both. The fact that in man so many pathways are either not available or, in the case of mitochondria, are associated with a symbiotic DNA, means that an animal such as man has a very abnormal balance of the chemical elements even when compared with plants, Fig. 13.18, which carry out most simple syntheses for themselves.

13.10.1 The ecosystem and man

The conclusion we have reached is that multicellular development was bound to increase in complexity as newly available elements were incorporated but could only do so by coexistence with simpler forms. Complexity is eventually self-defeating and the escape from this dilemma is only possible within an ecosystem of the simple and the complex. The system is based on physical conditions generated 4.5×10^9 years ago on Earth and on chemical availabilities under reducing conditions that could and did change in one direction—towards physical complexity within limitations. Only later advanced life was based on chemical oxidation, thus introducing new chemical risk. A new twist to this evolution was the appearance of self-conscious man. By inspection and empiricism he has discovered the limitations imposed by Earth's restrictions on life's condition and also how these limitations can be overcome. The opportunities are now re-opened to build organisation based on the 90 elements of the periodic table in a wide variety of oxidation states since all can be made available. Thus, man is creating quite new physical and chemical traps utilising temperatures ranging from 0 K to $> 10^6$ K. Into what complexity can this power lead? In the next chapter we will look at the development of man's industry and then in Chapter 15 at the ways in which elements cycle in the present-day world. There is, however, a central problem. In effect, it appears as if man is attempting to escape from the ecosystem in which he was born and to create a self-sufficiency but we must be aware of the chemical sophistication required to achieve this end and of the risks to which man exposes himself in his new endeavours (see

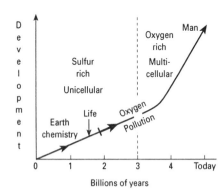

Fig. 13.33 The figure shows the development of complexity but this does not coincide with survival strength when met by environmental change. Such survival strength belongs to bacteria. (NB. New Earth chemistry is interactive with evolution.) Compare this figure with Figs 14.13 and 16.12.

Table 13.23 Functional value of the elements

Information (electrical) transfer and storage	Mechanical transmission	Acid–Base catalysis	Redox catalysis	Structural role (exluding the organic polymers)	Chemical energy transmission and storage
Na^+, K^+, Cl^-, H^+, Mg^{2+}, Ca^{2+}, HPO_4^{2-}	Mg^{2+}, Ca^{2+}, HPO_4^{2-}	Non-metals and divalent and trivalent ions, e.g. H^+, (N) (S) Zn^{2+}, Fe^{3+}, Mg^{2+}	Transition metal ions and some non-metals, e.g. Cu, Fe, Mn, Co, Mo, Se, S	Si, B, P, S, Ca, Mg, (Zn)	P, S, (C)

NB. The functional value of the elements arises partly from the chemistry and partly from the energy input to that chemistry. The energy input can be to the gradient of the free concentration of an element, to a chemical bond of given kinetic stability, or to the synthesis of a polymer in a compartment that acts as a trap.

Fig. 13.33). The experiment has been done before by the biological systems that introduced dioxygen and the vast range of new chemicals this generated. Any such development by man must be viewed against the background of the curious nature of living species—they show already a highly (natural) selected functional specificity in their chemical element composition (Table 13.23). We return to the implications of their chemical nature and that of man's industrial development in Chapter 16.

Further reading

Three excellent textbooks with much general information.

1. Villee, C. A., Solomon, E. P., Martin, C. G., Martin, D. W., Berg, L. R., and Davies, D. W. (1989). *Biology*, 2nd edn. Saunders College Publishing, Philadelphia.
2. Purves, W. K., Orians, G. H., and Craig-Heller, H. (1992). *Life—the science of biology*, 4th edn. Sinauer Associates, Sunderland, Massachusetts.
3. Lodish, H. D., Baltimore, D., Berk, A., Zipurski, S. L., Matsudaira, P. and Darnell, J. (1995). *Molecular cell biology*, 3rd edn. Scientific American Books, W. H. Freeman & Co., New York.

Other recommended references

4. Fraústo da Silva, J. J. R. and Williams, R. J. P. (1991). *The biological chemistry of the elements—the inorganic chemistry of life*. Oxford University Press, Oxford (1994, 3rd printing with additions.)
5. Ricklefs, R. E. (1993). *The economy of nature—a textbook in basic ecology*, 3rd edn. W. H. Freeman & Co., New York.
6. Wilson, E. O. (1992). *The diversity of life*. Allen Lane, The Penguin Press, London.
7. Mo, M.-W. and Fox, S. (ed.) (1988). *Evolutionary processes and metaphors*. Wiley, Chichester.
8. Romer, A. S. (1968). *The procession of life*. Weidenfeld and Nicolson, London.
9. Margulis, L. (1981). *Symbiosis in cell evolution*. Freeman, San Francisco.
10. Sapp, J. (1993). *Evolution by association*. Oxford University Press, Oxford.
11. Hopper, A. E. and Hart, N. H. (1985). *The foundations of animal development*, 2nd edn. Oxford University Press, Oxford.
12. May, R. (1987). Chaos and dynamics, Chapter 5, in *Dynamical chaos* (ed. M. V. Berry, I. C. Percival and N. O. Weiss). Princeton University Press, Princeton.

13. Depew, D. J. and Weber, B. R. (1995). *Darwinian evolving.* MIT Press, Cambridge, Massachusetts.

Books on signalling and regulation

14. Goldbeter, A. (ed.) (1989). Cell to cell signalling. Academic Press, New York.
15. *Calcium and the cell* (1986). Ciba Foundation Symposium no. 122. Wiley, New York.
16. Hennecke, H. (1990). Regulation by metal–protein complexes. *Molecular Microbiology*, **4**, 1621–8.
17. Thomson, K. S. (1988). *Morphogenesis and evolution.* Oxford University Press, Oxford.
18. Simons, P. (1992). *The action plant.* Blackwells, Oxford.
19. Williams, R. J. P. (1970). Cation distributions and the energy status of cells. *Bioenergetics*, **1**, 215–25.

14

Man's selection of the chemical elements

The Kingly Man . . . had he all the world's power he would not hold it as his own; if he conquered everything he would not take it to himself. His glory is in knowing that all things come together in One and life and death are equal.

Chuang Tzu—Texts: *The kingly man* (Fourth century BC)

14.1 Introduction

Before we begin to look at the remarkable changes in the natural selection of the chemical elements brought about by man we need to see the purpose of his activity, which is largely feeding, protection and comfort, which includes pleasure. Man is a relatively long-living animal of about 100 years lifespan. Once he had become sufficiently successful as a reproducing species (genotypic isolation through strong internal protection) expanding over the Earth to a considerable degree, his survival strength evolved rapidly through phenotypic protection (Table 14.1). Protection in biological systems involves the internal chemical systems and external physical and chemical armoury, but clearly the internal systems are not readily altered, although man has begun to approach

Table 14.1 The nature of phenotypic activity as related to protection

Activity	Protection from
Agriculture	Starvation
Medicine	Lower organisms
Manufacture of weapons	Other animals and other men
Manufacture of materials	Inclement conditions through clothing and housing
Mining, forestry, etc.	Lack of materials, lack of fuel
Organisation (through numbers, division of activities and regulation)	All of the above
Communication (use of signals, language, writing and electromagnetic waves)	Lack of information to control organisation

such alterations (Section 14.8). It need not be thought that any special development, for example self-conscious reflection, that man possesses was required to start him building up many external forms of phenotypic protection any more than that these were required for initial basic internal protection—both protection forms exist in lower animal species. Thus, the internal immune system has a long evolution in even lower animals and parallel protection exists in plants. As far as external protection is concerned, where genotypic expression and phenotypic development may become confused, it is enough to remember that some molluscs live in rock holes, silk worms build cocoons, and termites, birds and fish produce mud houses and mould clays with straw (nests), to give just a few examples. Furthermore, a central feature of the internal and external systems of advanced animals is group communication, which strengthens their organisation and hence protection. An obvious case is that of insect communities. For a long time communication amongst animals has been based on smell, sight, gesture and, of course, sound. The change that came about in man's general phenotypic development, and as far as we know is absent in other animals, was twofold. First, he began to know about himself using especially language, a totally new communication system, and, second, he began to develop through experiment a 'scientific' knowledge of his surroundings. Self-knowledge has constantly led to some degree of care for others as well as to a bewildered search for a meaning or purpose of life, no matter whether it took the form of a passive mystical belief in Great Powers (Gods), who are purported to be unlike man yet concerned about him, or the belief in underlying abstract systems of perfect beauty, as expressed by Plato or by mathematicians, in which man might glory in contemplation. The second knowledge (scientific) of his surroundings has built up an empirical (science-based) activity that has vastly exceeded that of other animals, if they have any, since it incorporates considerable underlying systematic strength, which now relates not only to the controlled use of the environment but also to the management of the behaviour of man himself in a society. This scientific knowledge is not passive, instinct-based, as is knowledge in other animals (or, at least, it seems to be), since it involves rationalisation and constant experiment which extends to the examination of life itself. This chapter is not about the whole of this experimental approach but about one part of it—the understanding and development of man's chemical systems—while looking at life in this context. We shall not attempt any

Table 14.2 History of element usage; the use of natural materials (e.g. wood, stone) without processing is excluded*

Application	Era of first major usage of element or its compounds			
	Prehistoric (before 2500 BC)	Pre-industrial (2500 BC–AD 1750)	Industrial (AD 1750–1940)	High-technology (after AD 1940)
Metals: vessels, tools, coins, weapons, etc.	Cu, Au	Fe, Zn, Ag, Sn, Pb	Al, Ni, Mo, W	Zr, Nb
Metals: construction, transport	—	—	Al, Cr, Mn, Fe	Be, Mg, Ti
Fuels and explosives	C	N, S	H	U, Pu, Th, H
Glass, ceramics, refractories	Na, Al, Si, Ca (clays)	Pb	Mg, Zr, Th	Li, B, La–Lu
Pigments and dyeing	—	Al, Fe, Co, Cu, Cd, Hg, Pb	Ni, Zn, As, Se	Ti
Pharmaceutical	C, N, O, H (plants)	S, As, Sb, Hg	Bi, Br, I, Ra	Li, Pt
Fertilisers and pesticides	—	—	N, K, P, Cl, As, S, B, Br, Hg, Cu	Sn
Industrial chemicals and catalysts	—	—	C, N, F, Na, S, Cl, K, Hg, Pt	Ar, Rh, Ba, La–Lu, Re
Electrical and electronics	—	—	Fe, Cu, In, Pb, W	Si, Ga, Ge, As, Li, Cd, Ni, Se, Ta, Ir
Household goods and chemicals	—	—	C, N, Na, Cl	B, P, Br, Sn, C, H, N, O, F

*After Cox (1995), reference 2 in 'Further reading'. Note each column adds to the previous ones.

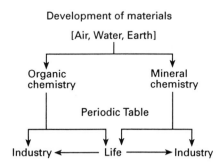

Development of materials

[Air, Water, Earth]

Organic chemistry — Mineral chemistry

Periodic Table

Industry ← Life → Industry

Fig. 14.1 The earliest 'accidental' developments were of basic 'elements' recognised in classical times, that is, of air, water and earth. The later development by living systems has two parallel arms, by nature and by man (industry). It is almost as if nature recognised the value of many of the elements of the periodic table before man did since elements from at least 15 groups of the table are used by nature (Fig. 14.2).

discussion of mystical or abstract beliefs and will concentrate on the analysis of the natural (purposeful) selection of chemical elements by man based on the experimental approach, to strengthen his survival with as little self-effort as possible (see Table 14.2).

As stated before, by 'natural selection' we imply the way in which chemical elements are combined (in mineral or biological materials but now including the products of man's activities; see Fig. 14.1) driven by: (1) the tendency to form ordered systems (spatial order), and (2) the tendency to form organised systems, persistent through flowing. The formation of ordered systems is seen in nuclei, in atoms and in chemical compounds, whether in the mineral or the 'organic' biological world. The formation of organised systems is largely seen in the biological world, but we must not forget the dynamic structures of nebulae and planets, and the physicochemical cycles on the Earth's surface (Chapters 8 and 15). In its essentials man does little extra other than to increase ordered structures and organisation through activity external to himself which is the essence of development of control over a larger and larger volume. This endeavour has gone on in nature for at least 3×10^9 years. The new feature that man brings to this natural selection is that he introduces a reasoned approach to both order and organisation over huge distances and structures based on empirical knowledge, and in so doing he extends the previously possible biological chemistry. Our first task is to recognise the way in which this rational semiempirical activity has developed, initially in a rough and ready way, and then especially through the knowledge of the chemistry embodied in the periodic table (Fig. 14.2). At the same time man has begun to understand the nature of biological systems so that he can now use chemistry to adjust internal genetics in the continuation of his control over other living systems through breeding. Later we shall turn to the risks man runs if he does not recognise that in all evolutionary steps the introduction of new chemicals

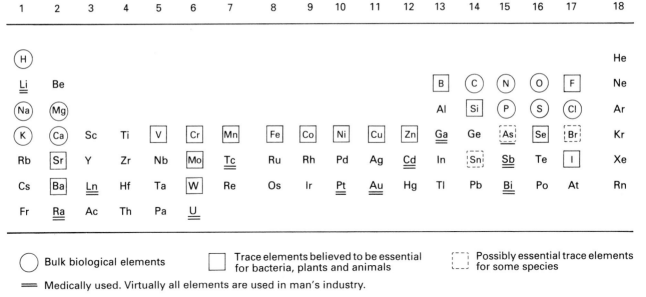

Fig. 14.2 Distribution in the periodic table of the elements essential for life, used by man or for medical use.

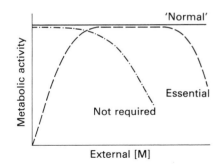

Fig. 14.3 Diagram of the effect of external concentration of M, [M], on metabolism. ('Normal' corresponds to optimal activity.) Essential elements can be toxic at high concentrations; 'not required' elements can be poisons at low concentration, but may be therapeutic.

is an addition to, even an imposition on, a pre-existing homeostatic biological multicellular ecosystem that cannot adapt genotypically except over millions of years. Thus, there arise the general problems of so-called pollution and ecosystem change, which will exist for every element that has its availability altered by man. (NB. Unicellular systems are very adaptable.)

It is important therefore to recognise again that any impact of a new chemical within internal biological evolutionary development proceeds via the route

$$\text{Poison} \longrightarrow \text{protection} \longrightarrow \text{use.}$$

Man's danger is that he has not (yet?) the power to manipulate his own genetics so that he can go through such development quickly. Thus, he becomes exposed in the sense that he is developing external *use* of elements without prior internal *protection* and before he has sensed what is to some degree a *poison* (see Fig. 14.3).

An earlier example in nature was the use of photosynthesis to generate dioxygen by anaerobes which excluded the anaerobes themselves from the top surfaces of Earth. We return to the potential risks after describing the chemical nature of man himself and the chemistry of his industry.

14.2 The internal chemistry of man

The internal biological chemistry of man is not very different from that of other higher animals, particularly mammals, except in minor details derived from his increased heterotrophic character. Man needs many elements and compounds of course (see Fig. 14.2), but he obtains them largely in adequate combined form from his food, plants and other animals, not so much from the air he breathes or the water he ingests. The question for him is not just the kind of elements that he requires but the amounts and forms (chemical species) in which some elements must be provided. As we stressed earlier, his complexity as a higher animal leaves him requiring elements, vitamins and essential fats and amino acids, as well as suitable sources of H, C, N, O. He cannot use cellulose for example, which is the most plentiful C, H, O, compound. His dependence on other life forms is complex through these constraints (Table 14.3). Hence, he may suffer from unbalanced supply, lack or excess, which

Table 14.3 Some elements required by higher and (symbiotic) lower species

Higher organism	Lower organism (symbiotic)	Special elements required in the symbiosis
Plants	Bacteria	Mo, V, Fe (N_2)
Ruminants	Archaebacteria (Plants)	Ni, Co (C/H compounds)
Man, animals	Bacteria (Plants)	All, but special compounds of C, H, N, O, e.g. vitamins and amino acids, Co(B_{12}), etc.
Animal and plant cells	Mitochondria	Fe, Cu (O_2)
Plant cells only	Chloroplasts	Fe, Mn, Cu ($h\nu$, O_2)

can cause deficiency or toxicity problems, respectively. Furthermore, deficiency can result from natural unavailability (primary deficiency) or from interference from other elements or compounds in the absorption process which may decrease availability (secondary deficiency). Examples of both kinds are widely known: goitre resulting from lack of iodine; iron and zinc deficiency resulting from excess of phytic acid in the food (flour), etc. Table 14.4 gives many other examples of metabolic perturbations or diseases resulting from deficiency or excess of elements. Knowing such facts, man corrects his balance of elements and that of his plants and animals by diet and fertiliser additives.

Excess uptake of even the essential elements may lead to manifestations of toxicity; this is also true, of course, for the other, non-essential elements, except that in this case the limits for toxicity are usually much lower— generally they are poisons and can cause death at relatively low levels depending on the kind of ingestion (through the circulatory, respiratory or digestive systems) or on their chemical form. Aluminium, for example, is practically non-toxic when absorbed through the digestive system but quite toxic when absorbed through the circulatory system (as in the old blood dialysis machines); barium ion is moderately toxic at high concentration but barium sulphate can safely be ingested (in effect, it is used as a contrast agent in X-ray medical diagnosis of diseases of the gastrointestinal tract). These observations can be summarised as in Fig. 14.3 which shows that for the essential elements there is an optimal range of supply related to the needs; the non-essential elements always have toxicity problems associated with their ingestion and some can be lethal at relatively low levels.

The question for man, therefore, is how to absorb the required amount of the essential elements in adequate form and avoid the non-essential toxic elements and compounds while expanding industry. The absorption takes place

Table 14.4 Examples of metabolic disturbances and diseases in humans and animals caused by deficiency or excess of trace biological elements

Element	Effect of deficiency	Effect of excess	Observations
F	Increased incidence of dental caries	Fluorosis (teeth)	Favours structural resistance of teeth
Si	Growth depression: anomalies in bone and cartilage	Silicosis (lungs)	Possible cross-linking of proteins
V	Growth depression; failure of reproduction	Unknown except at high doses	Possible effect on Na^+, K^+-ATPase; vanadate competes with phosphate
Cr	Impaired glucose tolerance; elevated serum lipids; corneal opacity	Toxic as CrO_4^{2-} (through reduction to Cr^{3+}); lung cancer; contact dermatitis mostly in men	Potentiates insulin (?); template for a component of glucose tolerance factor (?)
Mn	Skeletal and cartilage defects; depressed reproductive male function	Manganism (locura manganica)—psychiatric disorders (memory, speech, hallucinations)	Role in the synthesis of chondroitin sulphate; glycosylation
Fe	Anaemia; general weakness	Haemochromatosis	Component of haemoglobin and enzymes of respiratory chain
Co	Anaemia; anorexia; growth depression; white liver disease (in sheep)	Heart failure; hypothyroidism	Component of vitamin B_{12}; involvement in synthesis of haemoglobin
Ni	Growth depression; impaired reproduction; prenatal mortality (in animals)	Lung cancer; contact dermatitis, mostly in women	May replace other transition metals in their sites
Cu	Anaemia; ataxia; defective melanine production and keratinisation	Liver necrosis, e.g. in Wilson's disease; hypertension	Component of oxidative enzymes involved in haem synthesis; cross-links elastin and collagen
Zn	Anorexia; growth depression; sexual immaturity; hypogonadism; skin lesions; hyperkeratosis; depression of immune response; acrodermatitis enteropathica; teratogenic effects	Relatively non-toxic except at high doses	Component of many hydrolytic enzymes; component of zinc fingers—affects gene expression
As	Impairment of growth and reproduction (animals)	Often poisonous	Blocks sulphdryl groups of enzymes
Se	Endemic cardiomyopathy (Kesham disease) and osteoarthropathy (Kashin's Beck disease); nutritional muscular dystrophy (ruminants); exudative diathesis (poultry)	Selenosis—hair and nails loss; blind-staggers and chronic alkali disease (grazing-cattle)	Component of glutathione peroxidase
Mo	Growth depression; defective keratinisation; hyperoxipurinaemia	Anaemia; persistent dysentery (grazing animals); gout-like syndrome (USSR)	Interference with Cu^{2+} absorption
I	Goitre; cretinism	Goitre; thyrotoxicosis	Component of thyroid hormones

through the gut, but we will not deal with such problems except to say that the percentage absorption may vary considerably so that the requirements of the ingestation are usually much higher than the actual needs (see Table 14.5).

From what has been said earlier in this book we can easily understand the reason for the particular optimal ranges of uptake of the different elements and compounds. In order to ensure co-ordinated metabolism, co-operative organisation, so that homeostasis, a balanced utilisation of the elements, is maintained, products must be made in the correct interrelated amounts at the right time. Feedback control mechanisms help to keep homeostasis, but they

Table 14.5 Essential elements in humans*

Element	pH 7 form	Serum concentration	Human amount	Daily allowance
Na	Na^+	140 mM	70 g	1–2 g
K	K^+	4 mM	130 g	2–5 g
Mg	Mg^{2+}	0.8 mM	22 g	0.3 g
Ca	Ca^{2+}	2.4 mM	1100 g	0.8 g
Mn	Mn^{2+}	10 nM	12 mg	3 mg
Fe	$Fe(OH)_3\downarrow$	17 μM	4 g	10–20 mg
Co	Co^{2+}	2 nM	1 mg	3 μg vitamin B_{12}
Cu	Cu^{2+}	17 μM	80 mg	3 mg
Zn	Zn^{2+}	14 μM	2.3 g	15 mg
Cr	$Cr(OH)_2^+$	3 nM	6 mg	0.1 mg
Mo	MoO_4^{2-}	6 nM	5 mg	0.2 mg
Cl	Cl^-	104 mM	80 g	2–4 g
Si	$Si(OH)_4$	0.1 mM	18 g	>20 mg
P	HPO_4^{2-}	1.1 mM	600 g	1 g
S	SO_4^{2-}	24 mM	120 g	0.7 g Met†
Se	$HSeO_3^-$	1 μM	5 mg	0.1 mg
F	F^-	2 μM	2.5 g	2 mg
I	I^-	0.4 μM	30 mg	0.15 mg

* In addition to H, C, N, and O.
† Essential amino acid methionine.

can only be effective if the supply of 'raw materials' is not affected for too long a period. It is not only the supply of these materials that needs to be controlled: the rates of the metabolic reactions are affected to different extents by the temperature, hence this variable must also be controlled internally and kept constant within a narrow range or else co-ordination is lost. Man's complexity is very great and hence the feedback system cannot tolerate changes of temperature (see Section 7.8.3). The best and most critical example is the brain, which, like an advanced computer, is very carefully thermostated.

We can derive the conclusion that man's activity must be and has always been primarily directed to the satisfaction of two basic needs: (1) feeding, which has meant gathering, hunting and fishing, and then physical development of husbandry and agriculture; and (2) physical protection, especially through housing and clothing. Protection against disease came next, reflected first in the search for drugs from natural sources, then in the development of the so-called iatrochemistry (mostly chemistry of inorganic drugs) of the sixteenth century and, finally, of organic chemistry. The progression to higher and higher sophistication today utilises inorganic elements in drugs (see Table 14.6).

Curiously, or perhaps not, two other objectives different from feeding and overall protection, which are not common to other animals, also directed man's activity from ancient times: comfort and pleasure, as seen in the early development of pigments for painting and dyeing, jewels, parchment and then paper. The boundless pursuit of these objectives gave us our culture. Gradually, man realised that it was the development of chemistry external to himself that could make possible all other advances, including very much of his culture. With these objectives in mind he has evolved an industry in order to get all kinds of elements into suitable physical and chemical states for all his

Table 14.6 Medical uses of unusual elements in inorganic drugs

Element	Medical use
Lithium	Hyperactivity drug
Boron	Neutron capture therapy
Fluorine	Tooth protection
Aluminium	Antiacid
Chlorine	In several antibiotics
Copper	Anti-inflammatory
Zinc	Wound healing
Platinum	Anticancer
Gold	Arthritis; anti-inflammatory
Bismuth	Drugs for gastric diseases
Arsenic	Antisyphilis (now obsolete)
Antimony	Antischistosomiasis
Barium	X-ray diagnosis (gastrointestinal tract)
Mercury	Antiseptics, diuretics (now obsolete)
Selenium	Antiseborrhoeic
Tin	Antiboil and anti-acne drugs
Gadolinium ⎫ Manganese ⎭	Magnetic resonance imaging

purposes. What was previously done automatically by living organisms has become an extremely self-conscious pursuit, perhaps more of pleasure than of survival. Do these objectives conflict?

14.3 The evolution of man's industrial inorganic chemistry

The first great step forward taken by man was the evolution of inorganic chemistry. Man had, of course, no knowledge of the theory of chemistry until some 200 years ago, yet he started to use procedures that bring about deliberate chemical change at least 30 000 years ago. He became then a practising empirical chemist. The first major discovery probably was the use of fire (energy) to drive water out of clays. While the heat of the sun was of some value for this purpose, the discovery of fire allowed controlled higher temperatures to be used to make pots, bricks, tiles and so on. It is no wonder the Greeks and the Chinese gave fire, water and earth such priority in their schemes (Figs 14.1, 1.1 and 1.5). The basic reaction, which occurs in many geological zones, especially hotter climes, is the enforced removal of water, according to the scheme

$$\geq Si-OH + HO-Si\leq \longrightarrow \geq Si-O-Si\leq + H_2O.$$

The reaction (see Section 9.8) is called a condensation reaction and, if repeated with many molecules of $Si(OH)_4$, condensation polymerisation gives SiO_2 in which every Si atom is linked via oxygen to four other atoms of Si. The resulting mineral is quartz (Fig. 2.18). If the reaction is restricted by incorporating Al, Mg and perhaps Na, K and many other elements, the minerals produced from Si—OH groups on condensation can have one-, two- or three-dimensional chains of Si—O—Si units and are like most of the rocks on Earth (see Fig. 2.27). In clay minerals, by way of contrast, many of the OH groups remain uncombined so that small clay particles on heating can be joined together as shown above. In effect, this is a step from clay to rocks. The basic reaction here, producing all kinds of polymers in one, two or three dimensions, is exactly the same as that which is at the basis of life's evolution—condensation polymerisation in one dimension produces DNA, RNA, proteins, polysaccharides (branched) and even many intermediates in the synthesis of fats. The same reaction is also seen in biological mineralisation using $Si(OH)_4$ (see Chapter 13). Knowledge of how to make bricks allowed man to build large protective structures external to himself matching mollusc external structures in this sense, for example in shells.

It took thousands of years for man to understand this chemistry but further practical use of inorganic chemistry came long before understanding. The empirical study of minerals led from simple pots to more sophisticated ceramics, glasses and porcelains, as well as to mineral pigments, for example PbO and so on. (The pigments, greatly developed, are still the basic materials of the art of painting.) Man also learned to be more sophisticated in his chemistry; for example, instead of heating the material that had a high affinity for water at low temperature he discovered a drying agent at room temperature, so that he needed only to mix and make a water slurry of hydroxylated powders with that agent, an anhydrous compound, when new condensed solids formed. Many calcium salts (lime) are drying agents and can be mixed in slurries with small clay particles,

$$nCaO + 2nSi(OH)_4 \longrightarrow \left[Ca(OH)_2 \cdot (-\overset{|}{\underset{|}{Si}}-O-\overset{|}{\underset{|}{Si}}-OH) \right]_n.$$

when calcium silicates form an amorphous lattice which sets.† The materials obtained are mortars, plasters and cements from which concrete is derived. Today these condensation reactions are the basis of massive industries which on the top surface of Earth are as large as nature's activity. Looking around at today's buildings and roads give us an idea of the dimensions of such activity. (Notice that this chemical procedure mimics the use of pyrophosphate as a drying agent in biosystems (Section 9.11) and also the use of P_2O_5 in organic chemistry.)

Later, man's development of inorganic chemistry through the use of fire came to have a new direction; more than 10 000 years ago man discovered that by heating some minerals in a wood fire he could obtain metals. To judge from available evidence of goldware and jewels he knew earlier of native gold from natural sources, but he soon learned how to work it with fire since it melts

---

† The product, an example of which is the so-called Portland cement, has a composition that corresponds to Ca_2SiO_4 (26 per cent), Ca_3SiO_5 (51 per cent), $Ca_3Al_2O_6$ (11 per cent) and $Ca_4Al_2Fe_2O_{10}$ (12 per cent), but the setting process involves a complex series of reactions, not quite well understood, in which extensive chains or rings are formed.

$$(-\overset{O}{\underset{O}{Si}}-O-\overset{O}{\underset{O}{Si}}-O)_n$$

easily. However, the real chemical discovery was that certain minerals gave him metallic *copper*. The basic chemistry used carbon from wood, that is, charcoal, such that

$$C + CuO \longrightarrow CO + Cu.$$

Turning to Fig. 8.4, the Ellingham diagram, we can see that the oxide of any element can be turned into its metal at a temperature where the line for carbon lies above the line for that metallic element. For copper the temperature is lower than that of a wood fire; hence the metal separates easily. The reaction that man found here is, of course, a reduction, which usually corresponds to the release of oxygen from one element and its uptake by another or, more generally, to the removal of an element from combination by taking a second element as the removal agent.

By heating and reducing arsenic, tin or zinc minerals with copper minerals (all of which lose oxygen easily to carbon or lose sulphur to the air as SO_2) man then made alloys—bronzes and brasses. (*Note that all the mineral clays and alloys he made and used initially have no stoichiometry.*†) With these he made coins and vessels as well as powerful weapons. Of course, metals are easily shaped at high temperature but are hard at low temperature. The bronze age lasted thousands of years before man moved on by increasing the temperature of his fires. This led to the production of iron and, eventually, a vast range of iron/carbon alloys—steels—which are much stronger than bronzes. Man then became an aggressive as well as a well-protected animal. (Compare these processes with Earth's formation, Chapter 8.)

Other metals can also be used separately or in alloys, for example, Sn, Pb, and so on. Some of these materials allowed man to govern the flow of water in pipes, others to build bridges and all manner of constructions. Some of the more recent uses of metals and alloys are included in Table 14.7 and Fig. 14.4, which also shows the annual consumption of the most used elements. Thus, the discovery of fire and its use in chemistry allowed human societies to develop and protect themselves. Only when man purified and analysed all these materials did rational arguments about chemistry start to play a part in

Table 14.7 Some of man's uses of metal and alloys*

Use	Metals
Utensils	Sn, Cu, Zn, Fe
Building frameworks	Fe (steel), alloys (e.g. Mg/Al)
Pipes and containers	Pb
Wires (conductors)	Cu, Fe
Ornaments	Au, Ag, Pt
Catalysts	Cu, Pt, Mo, Fe, Ni
Atomic energy	U, Pu, B

* While the big leap in biological evolution was due to O_2 (oxidation) which increased the availability of some elements and their oxidation states in compounds (Chapter 12), the big leap for man was in the reduction of chemicals to provide a source of all *elements*.

† No wonder that stoichiometry only appeared important after 1800 but since then it has been unduly stressed due to the detailed study of molecules before that of solids (see Chapter 2).

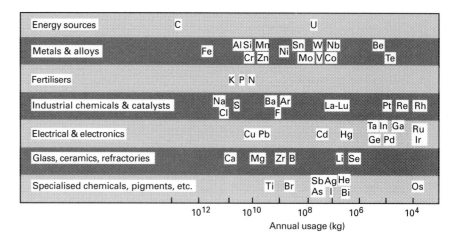

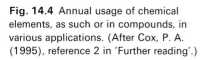

Fig. 14.4 Annual usage of chemical elements, as such or in compounds, in various applications. (After Cox, P. A. (1995), reference 2 in 'Further reading'.)

civilisation's development, that is, after 1800 and, especially, after 1850. The value of these discoveries to the arts lies in the materials used which have changed progressively with chemical discovery and knowledge of physics.

Coming very close to today man has now explored the use of a great variety of inorganic materials *which he makes* based increasingly on an *understanding* of processing. These materials include light alloys, for example (Al/Mg), very heavy metal elements, for example U and Pu, and non-metals such as Si. Reference to Fig. 14.7 shows that their extraction comes about through the application of even higher temperatures to get from oxides to the elements. Clearly, there is a slow but steady progression over 10 000 years in man's ability to develop local hot spots, that is, factories, for his reactions. The novelty in this natural selection of chemical elements is that man found that pure elements could be forced into chemical and physical traps, and that many elements have functional values that biology had failed to find (Fig. 14.4).

14.3.1 Inorganic agricultural developments

Before man discovered the periodic table he had begun to fertilise his fields to increase crop yields. At first he used organic manures and substances such as seaweeds. By around 1700 he had learnt to apply phosphates (bone meal) and then, by employing acids, to get phosphate from natural rock minerals:

$$Ca_3(PO_4)_2 + H_2SO_4 = Ca_2(HPO_4)_2 + CaSO_4.$$
$$\text{(insoluble)} \qquad \text{(soluble)}$$

During the nineteenth century he also used lime to make the soil less acidic. As he began to understand inorganic chemistry related to life he introduced more chemical elements. In this century, the big steps forward have been to apply nitrogen fixed as ammonia, which required the development of a high-energy catalysed industrial process—the Bayer synthesis of NH_3 from N_2 and H_2,

$$N_2 + 3H_2 \longrightarrow 2NH_3,$$

and potassium as potassium phosphate. In the last 20 years or so, trace elements have also been introduced into fertilisers (or in feed for animals, Table 14.8). Slowly, as man has learned that 15 to 20 elements are needed for all life

Table 14.8 Some elementary substances used in agriculture

Substance	Value
NH_4^+, NO_3^-, salts (N)	Nitrogen supply
Phosphates (P)	Phosphate supply
Potassium salts (K)	Potassium supply
Borates (B)	Boron supply
Lime $(Ca(OH)_2)$	pH control (calcium supply)
Cobalt licks (Co)	Bacterial synthesis of vitamin B_{12}
Selenate (Se)	To prevent disease in man and animals
Lanthanides (Ln)	Improve crop yield

(see Fig. 14.2), he has applied this knowledge even though medical and biological science studies concentrated largely on the non-metals alone. Thus, he used fertilisers with little understanding of plant requirements but from observed effects and consequent knowledge of need. Man also discovered very early in his self-conscious history that he needed to replace salt, NaCl, losses. Salt then became of great commercial value. This is but one example of a 'medical' use of inorganic elements.

14.3.2 Catalysis and inorganic materials

The preparation by man of some inorganic materials in quite unusual states (for nature) has provided materials that proved to be highly active as catalysts for the transformation of both inorganic and organic substances. It is the surfaces of the materials that are active and so these heterogeneous catalysts resemble enzymes. An example of a catalytic oxide, a zeolite, is shown in Fig. 14.5(a), and a use of it is shown in Fig. 14.5(b). An example of a catalytic metal in a solid support of silica gel is shown in Fig. 14.6. Many other extremely important transformations are brought about by solid state catalysts. They are often able to withstand high temperature and pressure, so that man has again opened up the way to chemistry not known on the Earth before. Remember that the stress here is on the selected use of chemical elements for particular purposes, as is the case in biological catalysts. However, man's range of elements is about 90, whereas that of biological systems is only about 20. We return to another source of catalysts in Section 14.5.

14.3.3 Man's development of energy sources using chemical elements

The use of sources of energy has a complex history (Fig. 14.7), which, like the use of chemicals, could only be put on a rational basis after the development in the knowledge and understanding of the questions discussed in Chapters 1–7. In effect, this meant that, without understanding of 'fire' (energy) (Fig. 1.1), see Chapter 3, man could only progress clumsily in his development of industry. He managed over several thousand years to drive temperatures of ovens up to some $1200\,°C$ using wood and charcoal, but, as understanding of

(a)

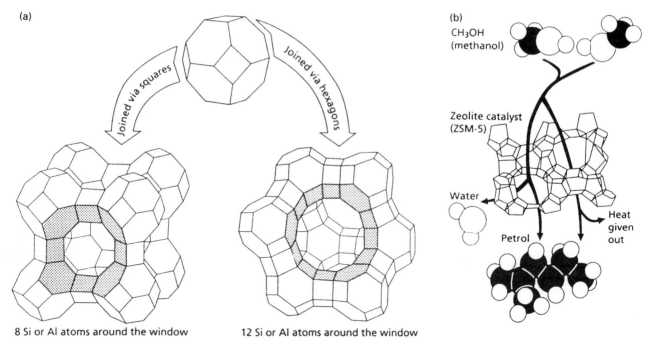

8 Si or Al atoms around the window   12 Si or Al atoms around the window

(b)

CH₃OH
(methanol)

Zeolite catalyst
(ZSM-5)

Water

Heat
given
out

Petrol

Fig. 14.5 An example of a zeolite showing (a) the distribution of silicon and aluminium atoms around the window, and (b) a typical reaction catalysed by this zeolite—the conversion of methanol into petrol. (After Schriver *et al.* (1994), reference 3 in 'Further reading'.)

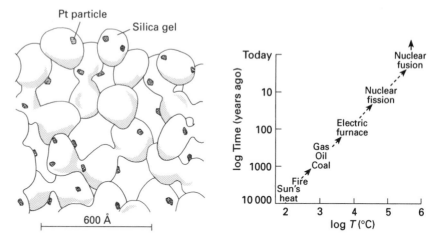

Fig. 14.6 Schematic diagram of metal particles (platinum) supported on silica gel. (After Schriver *et al.* (1994), reference 3 in 'Further reading'.)

Fig. 14.7 The evolution with time of the range of temperatures achieved on Earth using different sources of energy.

chemistry and physics progressed, man could consider ways of raising temperatures to around $3000\,°C$ (electric furnaces) and then to $10^6\,°C$ in atomic fusion reactors (Fig. 14.8). While he failed to grasp material possibilities earlier, he also took a very long time to realise the energy sources needed, which were around him. He discovered wind power (wind mills) and water power (water mills) thousands of years ago, but the development of

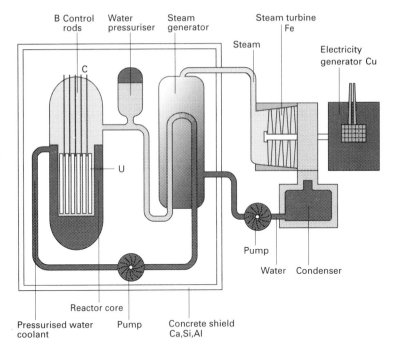

Fig. 14.8 Schematic diagram of an atomic fission reactor. Note the use of particular chemical elements in different parts of the reactor.

power from the burning of wood and some oils to that of coal and gas arose only in the last three centuries. Conversion to electricity belongs to the last two centuries and atomic power to the last 50 years. None of these developments could occur without particular substances not only consumed in the generation of energy but also in the manufacture of the appropriate materials for power stations and wirings for power distribution. The last 200 years have, in fact, seen a spectacular 'chemical age' in the development of required inorganic materials for functional use through the energy industry. The main energy sources are still largely fossil fuels (Figs 14.4 and 14.9). There is a huge coal, oil and gas repository in the Earth due to the production of another huge repository of dioxygen in the atmosphere—an endothermic reaction driven by the sun's radiation energy and mediated by particular life forms—green plants

Fig. 14.9 Percentages of various sources of energy used in the United States from 1850 to 1990. (Data from US Energy Information Agency (1991). From *Understanding earth* by Press and Sievers. Copyright © by W. H. Freeman and Company. Used with permission.) Note the increased use of coal to meet the needs of the industrial revolution at the expense of wood and other traditional energy sources, and its progressive relative decrease in favour of oil, natural gas and nuclear energy. In absolute terms the use of wood has not diminished.

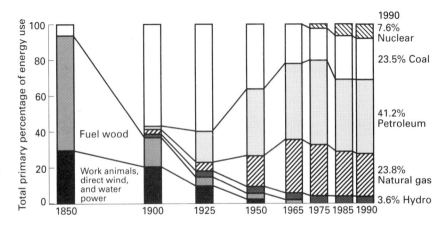

and photosynthetic bacteria, that is,

$$2H_2O + CO_2 \xrightarrow[\text{sun}]{\text{life}} CH_4 + 2O_2.$$

The fuel is subsequently used in transport machines, to drive reactions and to maintain physical conditions of temperature (heating) and illumination (lighting), as clearly seen everywhere around us by reversing the above equation so that

$$CH_4 + O_2 \longrightarrow CO_2 + H_2O + \text{energy}.$$

The modern style of life and, we might even say, modern culture and civilisation, depend critically on this 'fossil' energy. (The use of atomic power will be essential by 2050 if present day lifestyle is to be maintained.) But this was not the sole contribution of the chemical elements; the inorganic products have many uses and, although it is not usually recognised, they too have made much of modern 'culture' possible. We dedicate a short section to this topic.

14.3.4 Culture and inorganic materials

In the introduction to this chapter, we mentioned that inorganic pigments allowed early pictorial art. The advanced use of such materials and, more recently, an understanding of colour and how to produce it, form the basis of the paint industry (inorganic and organic) and, for example, of colour television. Paint for art work, especially in glass, enamel and ceramics, requires particular elements, and their manipulation is now a well developed science. The ability to print books or reproduce documents, for example on photocopiers, depends also on specific materials, for example lead and selenium. These, however, are not the only areas where man's created chemicals can assist the imagination to give rise to reflection and pleasure. The development of musical instruments, for example, is due in large part to the uncovering of the chemistry of metals, for example the brass section of an orchestra or the strings of a piano. Instrument builders have always been empirical physical scientists, but it was the chemists' task to supply materials for the equipment which allowed culture to develop from the primitive drum and cave paintings to the modern audiovideo equipment, tapes, disks, etc. so much linked to today's culture. At both extremes selected chemical elements play an irreplaceable role.

Before concluding this section let the chemist be proud of his profession. His initiatives, first in almost prehistoric ages, put man on a civilising path. There would be few musical instruments, no paints, no books, no radio, no television and hence few ways of revealing to one another our inner thoughts and feelings had there been no chemistry. The chemist's work can now be rationalised and this rationalisation will lead on to many new forms of art. He must be careful, however—living systems are a part of an even earlier chemistry and they cannot change internally at all easily, thus they can be damaged by new chemistry. We return to this problem later since we must now look at the discovery of new properties of materials that allowed further large-scale developments in different (or complementary) areas.

14.4 The development of industrial organic materials

Air

H₂N
        N₂
        \
         C = O   O₂
        /
H₂N   CO₂

Urea   H₂O

Organic (life) chemistry became the rearrangement of the atoms of the air!

Previous to his development of chemistry,† man did not distinguish organic and inorganic substances in a clear way. He obtained both materials from plants, animals or the mineral world, and all were based on recipes or 'elixirs' in which organic and inorganic substances were mixed unwittingly. The synthesis of urea (Wohler, 1826; see marginal illustration), a compound known from life and containing nothing but H, C, N and O led to the separation of the two branches of chemistry which increased from 1850 to 1950 in an unfortunate manner.

The synthesis of organic materials by man is, therefore, a more recent and much more subtle development than that of inorganic materials. His basic materials today are gas and oil. The initial process is cracking to obtain suitable units of hydrocarbons. These compounds are progressively oxidised so as to make alcohols, ketones, acids and then a great range of chemicals (see Chapter 9). Some of the most important are polymers in the synthesis of which it is often necessary to remove water, reaction (14.1), but without excess heat which would have led to the reaction with dioxygen, (14.2), at least in air. The following example illustrates the differences,

$$\text{Desirable:} \quad R\text{—}COOH + H_2N\text{—}R' \longrightarrow R\text{—}CO\text{—}NHR' + H_2O, \qquad (14.1)$$

Undesirable (for example):
$$R\text{—}COOH + H_2N\text{—}R' + 3O_2 \longrightarrow ROH + R'COOH + CO_2 + H_2O + NO_2. (14.2)$$

The methods devised to remove water parallel those of biology, but man quickly discovered that a preferred route was to work in non-aqueous solvents, using hydrophobic liquids to obtain by extraction a huge variety of H, C, N, O, S combinations from plants and animals and then the same and many other combinations by synthesis.

There were three particularly interesting developments in this organic chemistry. The first was, in effect, the enlargement of protection, provided in a limited way by the internal biological immune system, through the use of therapeutic substances, drugs (medicines), while the second was the enlargement of external protection, clothes and useful construction materials, which followed the discovery of plastics. In both of these developments the earliest uses did not employ synthesis. Instead, in the first case, extracts from plants, animals or even bacteria were used and in the second case vegetable fibres, fur from animals, wool, silk, etc. were the choices. It is only really in the last 100 years that man has deviated from the path of utilising natural organic chemicals from biology. A third field in which important progress has been made is that of the agrochemicals—insecticides, pesticides and herbicides, particularly chloro, phosphorus and carbamate derivatives (see Section 9.4.3). Naturally, these are extremely relevant for the yield and quality of crops. As explained in Chapter 9, it is only since 1830 that man has begun to appreciate how organic compounds could be made, but, just as use of inorganic chemicals has allowed development of man's material world and

† We may say that the first use of organic chemistry was the destructive oxidation of wood and oil. This use extended to coal and coke raising the temperature of fire slowly to 1200°C. Figure 9.1 shows the energy gain at the cost of generating CO_2.

then his cultural expression, so has the use of organic chemicals in multiple and varied ways. The full magnitude of these developments was described in Chapter 9. As stated above, man now makes, from organic chemicals, plastics, dyes, clothing, coatings, insecticides, pesticides, herbicides, as well as drugs and a variety of other products both for protection and for cultural purposes.

14.5 Organometallic and complex ion chemistry: homogeneous catalysts

Over the period 1950–80 it was increasingly realised that, as well as the combinations of H, C, O, N, S, P and halogens of traditional organic chemistry and the combinations of the 70 metal elements amongst themselves or with the non-metals O, S and halogens (inorganic chemistry), it was possible to produce, quite easily, compounds of the elements C, N, P, H, plus any ratio of metals and semimetals in isolation or in clusters of any size. This part of chemistry is now called organometallic chemistry. It is nothing more than a coming together of traditional inorganic and organic chemistry carried out in the style of organic chemistry. What was required for the synthesis of such compounds was the initiative and the skill to use the methods of kinetic control, known already in principle, to produce quite new extensive series of compounds. (There is no doubt that the extension along similar lines will allow the boundary between metals and alloys and non-conducting organic solids to be crossed time and again.) As stated before, the evolution of biological chemistry itself almost 4×10^9 years ago may well have been based on parts of this field of organometallic chemistry before it developed traditional aqueous complex ion chemistry, since biological chemistry could have started in an inert atmosphere with activated metal oxide or metal sulphide surfaces plus CO_2, CH_4, CO, HCN and so on (Chapters 9 and 10). Man had to start his synthetic chemistry in an atmosphere of dioxygen, and so organic metallic chemistry, which generally has more reactive compounds than organic chemistry, was for a long time hidden from him. Thus, it was the ability to isolate systems in protected compartments that allowed new chemistry to be developed.

In the course of their studies, workers in the field of organometallics have synthesised a huge variety of compounds based on single metal atoms and on clusters. Some examples are given in Table 5.16 and Fig. 9.25. The units are so large in some cases that they can be likened to small parts of continuous metallic solids. In these compounds, which today are often stabilised by phosphines (inherently unstable in air) for convenience of synthesis, there can be bound H_2, CO, N_2, —CH_3, CO_2, NO, CN, and, in fact, a great variety of 1, 2, 3, 4, etc. carbon atom fragments. All these small molecules and fragments are reactive while bound to metal atoms and many compounds are active catalysts (see Fig. 14.10). It is this synthetic chemistry that makes possible the development of inorganic catalysts comparable to solid metal and metal sulphides (where sulphide replaces phosphine) in the initial organic (organometallic) chemistry on Earth, which predated life. This synthetic chemistry also needs energy capture, but it is entirely possible that this could have been generated from the rearrangement of unstable mineral surfaces or from chemicals created at high temperatures but stabilised on the sudden cooling of Earth.

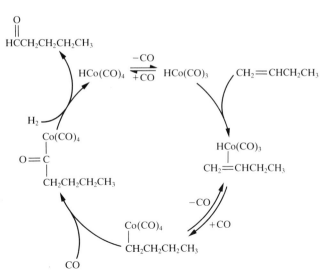

Fig. 14.10 A cyclic industrial process—the hydroformylation of 1-butene to pentanal—in which catalytic compounds of cobalt (cobalt carbonyl complexes) are used. (After Schriver *et al.* (1994), reference 3 in 'Further reading'.)

Organometallic chemistry is closely linked to complex ion chemistry (Chapter 5) and both relate to the major highly active enzyme catalysts (Chapters 9 and 10). Man, with great ingenuity, is now not only doing those very reactions that biology manages using similar reagents and solvents but has increased the range greatly using new intermediates and new catalysts in chosen solvents and under chosen conditions. These reagents may be made from elements in very low abundance, but since catalysts are used in very small amounts the energy cost of preparation is not large compared to the tonnage of upgraded chemicals obtained. Thus, we find Pt, Ir and Rh used in the synthesis of hydrides, much as we observed Mo to be used as a catalyst in biology. The major structures of man or biology remain based on the abundant lighter elements of the first three rows of the periodic table, but man himself is not restricted in his choices, which may extend to all rows of this table and are open to an almost infinite variety of combinations for different purposes. Slight variations in such combinations can provide remarkably different products; see in Fig. 14.11 the curious effect of some quite related Zr catalysts.

Figure 14.12 shows an organometallic cluster compound. Here a kinetically stable array of metals, P, C and H is put together. There was no possibility of the creation of this molecule except by design, that is, self-conscious thought. Unlike previous synthesis involving ideas from known materials, these syntheses have created substances not available before, at least on Earth. In effect, there is no limit to the way in which all the elements can be combined to give a variety of useful materials now required in large amounts.

The development of organometallic and complex ion chemistry illustrates again a general point concerning chemistry. It has become obvious that the combination of elements that was generated on Earth before life began is a limited set. It is limited by the abundances and availability of elements, by their combining strengths as well as by the kinetics of chemical change. As we have seen, life generated quite new sets of compounds using C, H, O, N, P and S extensively, but also combined forms of some 15–20 other elements. Many of these compounds are only kinetically stable and their syntheses require energy input. Limitations initially were the reducing atmosphere, the presence of

Isotactic polypropylene

Syndiotactic polypropylene

Fig. 14.11 Examples of the effect of the structure of closely related zirconium catalysts on the type of polymers, polypropylenes, formed.

Atactic polypropylene

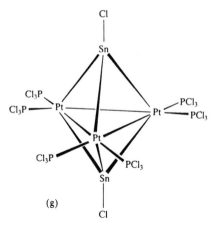

Fig. 14.12 An organometallic cluster compound with Pt—Pt and Pt—Sn bonds and P and Cl donor ligands.

water and the inability to overcome certain thermodynamic and/or kinetic barriers. Some of these barriers were removed with the advent of an oxygen atmosphere. Man, in the last 200 years, has changed all this. He can use any atmosphere, any oxidising or reduced conditions, any temperature, any pressure and any solvent he wishes to manipulate conditions of combination of now over 100 elements, several of which he himself has made. The huge range of possibilities was first opened by inorganic and organic chemists in the synthesis of small molecules which were before unknown on Earth. We stress that much of this innovation and indeed of that in biological systems owes itself to the introduction of new catalysts often using novel elements. Examples of these are the multiplicity of hydrides and many relatively small organic compounds. By 1900 the range of syntheses was rapidly increasing into those of alloys and ceramics and then large polymers. The end of the present century has already seen the development of methods for the synthesis of biological molecules of all kinds, even up to large pieces of DNA and RNA. As mentioned above, perhaps the best illustration of the novelty of these syntheses is in the combination of organic and inorganic techniques. Finally, man is able to manipulate or mimic biology's synthetic procedures—genetic engineering—and sooner or later he may be able to produce a novel life form. A major coincidental point, which must not be missed in all this innovation, is the introduction into the environment of many new relatively kinetically stable compounds, for example organic fluorides, and the risks they generate (Chapter 16).

14.6 Compartments, catalysts and energy

The organic and biological chemistry described in Chapters 9–13 consisted essentially in driving chemicals into energised kinetic traps and using them in

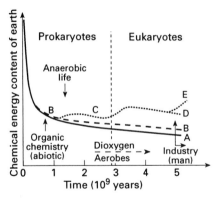

Fig. 14.13 The evolution with time of the chemical energy content of Earth due to the effect of organised energy capture (by living systems and due to man's actions; see Figs 7.47, 10.29 and 16.12). Decline of the chemical energy content of the Earth: A, in the absence of organic chemical trapping; B, in the presence of abiotic organic chemistry; C to D, in the presence of organised energy capture (living systems) with (E) atomic energy or better energy capture from the sun.

these traps (see Chapter 7). The activities in man's industry are no more than an extension of these procedures (Fig. 14.13 and compare Figs 7.47 and 10.29). A material is driven into a kinetic trap by introducing energy and, especially where this is done at high temperature, by isolating and allowing rapid cooling after reaction to keep the product in the trap. Examples are found in the production of metals used in building construction, machines and communication systems. The metallic materials are frequently made from thermodynamically stable oxides, yet in the metallic form in which they are obtained they are thermodynamically unstable in the presence of oxygen. Their preservation in that form depends on their physical state and upon surface barriers to reaction with O_2. Of course, there are still other ways of carrying out reactions that involve energy.

Extremely important, however, is the control of the rate of chemical transformation through the use of catalysts and isolation in controlled conditions, that is, in compartments. As described in Chapter 10, this is also seen in biological systems which mainly require for the incorporation of energy into compounds the isolation of material in compartments and the location there of catalysts. Thus, reaction is part of a flow diagram such as,

$$\text{Separate reactants} \longrightarrow \underset{\text{(in a compartment)}}{\text{catalysed reaction}} \longrightarrow \text{product} \qquad (14.3)$$

where energy can be introduced, especially in the provision of reactants. Man follows this procedure especially in many factories, working at modest temperatures. Observe here the skill in the design of compartments, connections and controls as mentioned in Chapter 7C.

At this stage of the examination of man's input to the use of chemical elements we have noted some extremely important features that have extended the natural selection of the chemical elements.

1. Man uses all the chemical elements, both naturally occurring and (a few) synthetic. To a large degree he has overcome the problem of availability, though in the long term this could come to haunt him (see Table 14.9).

Table 14.9 Some examples of element availability altered by man

Element	Change	Cause of change
Lithium	+ (Very small)	Medical applications
Boron	+ (Considerable)	Intensive agriculture
Carbon: carbon dioxide	+ (Large)	Coal and oil burning
Carbon: methane	+ (Large)	Intensive cattle farming
Nitrogen: nitrate	+ (Large)	Excess fertiliser
Oxygen	(+)(−)(Large)	Depletion of ozone layer
Fluorine	+ (Considerable)	Drinking water additive, control of tooth decay, refrigerant gases
Aluminium	+ (Considerable)	Sewage treatment, acid rain
Phosphorus	+ (Considerable)	Excess fertiliser and water softening
Sulphur	+ (Considerable)	Fossil fuel burning in the power industry (acid rain)
Chlorine	+ (Considerable)	Organic compounds in drugs, pesticides, polymers
Many metal elements	+ (Small)	Mining and industrial uses (low solubility)

2. Man has discovered energy sources that allow him to use quite different preparative methods, for example high and low temperature and pressure.

3. Man has evolved compartments of novel kinds, new catalysts, and new connecting systems for flow.

4. In effect, man has uncovered many new ways of putting elements together for functional use.

Now, man is part of biology and biology not only evolved new materials but it developed *organisation*. Man has also had to use organisation (outside biology) so as to extend the strengths of his novel constructs. This is the subject of the following section, but before we deal with it we have to analyse briefly the transfer of material, which with the transfer of instructions is an essential feature of organisation.

14.6.1 Transfer of material

Just as in biology, it is necessary that in a system of compartments with activity there are modes of transport to and from them of many kinds of material and of energy. Man slowly developed modes for moving materials and fuel, not only between compartments at close distance, laboratories, but also for bringing materials and fuel from long distances to the organised compartments (Table 14.10). He started, as one might expect, with the traditional pre-existing

Table 14.10 Development of transport and communication

		Transport modes	Communication modes
To	1750	Animals, ships	Animals, mail (man) and fire
	1850	Rail	Telegraph
	1900	Road	Telephone
	1950	Aeroplanes	Wireless radio, television, radar
	2000	Superfast planes, superfast trains, manned-satellites	Fax, laser beams, electronic mail, optical fibres (light pipes)

modes of transportation—animals on land and boats on the sea—but he now has available a network of transport systems using engines and pipes of various kinds on land, in the sea and in the air.

Our concern in this book is with the chemical elements involved in such systems today, noting particularly the accelerated rate of change of use of such elements—from iron and steel, copper and zinc to light magnesium and aluminium alloys, to organic plastics, to composite materials—which have made the transport systems possible.

14.7 Organisation and control in chemical change

Slowly, and almost unconscious of the fact at first, man has developed *organisation* into his chemical activities. He has built more and more compartments and other connecting constructions not only for multiple-batch single-compartment activity but also for flow processes in which compartments are effectively located in stages of successive reactions. The most evolved form of these syntheses is the continuous *cyclic* process in which the output of one stage is *linked by feedback* with the input of the following stage and the desired product is withdrawn at some particular step. These are parallel to many biological processes (see Figs 10.11 and 10.12), and several examples are of practical use today (see Figs 14.10). The next step needed in this development is, of course, total feedback (kinetic) control, so that the amount of product required regulates the input rate (Figs 7.38 and 7.40). This could be done by man himself, but eventually it requires a quite new development leading now to robotics and computer-driven processing (Table 14.11). Of course, prior to that, man had to develop the basic hardware for communication systems. They have, as their basis, the flow of electrons in conductors and semiconductors originally made from copper and today from germanium and silicon and newer materials such as gallium arsenide. Electromagnetic fields now replace the mechanical devices of biological systems and electronic circuits replace chemical transmitters. Development has been extremely fast and will undoubtedly continue (Fig. 14.14), and all the time we wish the reader to have in mind the employment (evolution) of new materials based on the functional value of elements.

The two aspects of kinetic control that are thoroughly understood by man, compartmental or physical control and chemical reaction control by transfer or message, now come together not only in biology but in an industrial plant.

Table 14.11 Evolution of methods of process control

	Method
Up to 1850	Manual
1900	Mechanical
1940	Electrical
1960	Electronics
>1980	Robotics, computer-driven

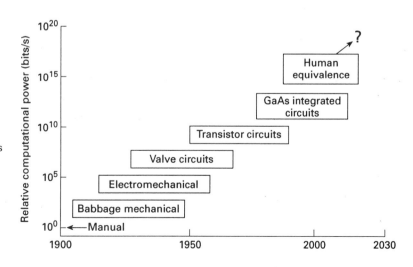

Fig. 14.14 The evolution with time of computer power represented by the evolution of computer technology from mechanical machines to todays' high-speed electronic processors. Also shown is the equivalent measure of unaided human calculating power.

The whole of the operation is ideally under one central computer (the nerve centre or brain), but there are subsidiary control centres at points in the system. It is of the essence of these large organised factories that monitoring takes place at every stage of both physical and chemical conditions and these are fedback from sensors to the control systems. The control system then forwards messages so that situations are adjusted according to a preset programme. The development is exactly parallel with that which we consider to occur even in lower biological forms (Chapters 10–13), and man's advance over 200 years has been to learn very quickly how to construct organised systems based on chemical materials selected for their functional value both as regards products and also as regards construction. There remains however an important distinction between biological and manufacturing processes. In biological systems the cycling of elements is nearly complete today (though it was not 2×10^9 years ago). Manufacturing processes do not cycle their products but sooner or later they will have to. Given today's huge activity, man has to evaluate carefully the corresponding environmental impact of his products, which now gives reasons for concern. He organises his own society, of course, and has the possibility to moderate trends but, most importantly, he has to begin to see himself within the context of the biology of the planet Earth in order to avoid the risk of misunderstanding his chemical nature and thus run into some disaster or another (see Section 14.9 and Chapter 16). We have learned at the ends of Chapters 11 and 13 that man is not an isolated organism in chemical terms but a product of the cooperative activity of many living forms. He must look after many if not all of them. Again slowly but surely his industry becomes an ecosystem! (Of course he cannot look after the deep zones of Earth which through sudden eruption could destroy advanced life.)

14.8 Genetic manipulation: a new industry?

Now that man knows that DNA is the master plan, the genetic material, and that its composition is defined in sequences of four bases, he can consider what to do with this knowledge. Before he even considers manipulating himself (except for medical purposes), there are two possibilities for his industry. The first is in the control of organisms other than himself. Given the control and regulation complexity of complex forms his best approach could well be to use bacteria. In principle, they could make many organic materials. The second is to attempt to devise a new code altogether. Life, as we know, is based on one kind of primary metabolism which arose in reducing conditions. Given an oxidising environment what would have been the initial range of monomers and polymers and the code that developed from them? What could a new biological system do? Any change impacts on the existing ecosystem of course.

14.9 The parallel and the orthogonal activities of man and biology

We stressed in our study of biological systems the evolving use of the chemical elements in relation to the structuring of space, that is, the ever-increasing division into compartments, the management within that space of localised specific activity and the enhancement of the value of an element in its selected site (see especially Chapter 13). The overall functional usefulness, which we have called *fitness*, then depended on the ability to apply energy so as to manage synthesis and the essential flow of material and energy within the organisation, which also required the use of the elements (usually in compounds) in messenger systems. All this activity was possible through appropriate element selection. In this chapter we have seen that the whole of man's effort is based on the same fundamental activity, survival, even if today it is confused by his demand for pleasure. The parallel between the two systems can be seen in the diagram

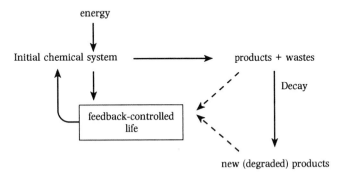

The major changes in the last 150 years have been in the new chemical systems brought into service by the increase in the use of energy (in new ways), which then required machinery of a complex variety using new electronic feedback circuits in order that the whole would be organised.

Now we have seen that the stability of a biological species is built on the strength of the feedback control processes but that the organisation grew more and more complex as it came to include more and more species in stable ecosystems that used a selection of elements with great sophistication (Chapter 13). Man is part of this ecosystem but is his industry? We need to look more closely at what man does. As in the case of a biological ecosystem we must ask: is man's society stable? What influences have new degraded products on life?

Undoubtedly, man produces waste and Tables 14.12 and 14.13 list some of his products which generate it. Many of the industrial products have lifetimes that exceed that of man himself, so that before decay returns material to the initial state a large bank of 'intermediate waste' is produced. However, in this respect there is little immediate distinction between this intermediate waste and the previous production of huge biological deposits of oil, coal, natural gas or of biological minerals such as calcium carbonates and silicates, and of gases such as O_2, CO_2 and CH_4, which 'polluted' the Earth by thus adjusting its chemical balances. There is a distinction, however, in that the elements man *selects and uses and then wastes* are not the same as those used and wasted by biology. The obvious 'threat' is that these 'new' elements can have the same effect upon existing life (see Fig. 14.13) as did the introduction of dioxygen some 3 to 1×10^9 years ago. How do the 'newly' selected elements affect the ecosystem and its potential development (Fig. 14.15)?

Again man uses energy but he produces it not from the sun but from other resources, some from biological materials and some of mineral origins, and

Table 14.12 The top 20 metals: world production*

Metal	World production (tonnes year^{-1})
Iron	7.16×10^8
Sodium (salt)	1.68×10^8
(carbonate)	0.26×10^8
Calcium (lime)	1.12×10^8
Potassium (salts)	51×10^6
Aluminium	15×10^6
Copper	$> 7 \times 10^6$
Chromium (chromite)	9.6×10^6
Barium (sulphate)	6.0×10^6
Zinc	4.9×10^6
Manganese	4.85×10^6
Lead	4.1×10^6
Titanium (TiO_2)	3×10^6
Zirconium (zircon)	7×10^5
Nickel	3.25×10^5
Tin	1.65×10^5
Strontium	1.37×10^5
Molybdenum	8×10^4
Tungsten	4.5×10^4
Uranium	3.5×10^4

* Data from Emsley, J. (1991), *The elements* 2nd edn. Oxford University Press, Oxford.

Table 14.13 The top 20 synthetic chemicals

Synthetic chemical	Rank*	Catalytic process
Sulphuric acid	1	SO_2 oxidation, heterogeneous
Ethylene	2	Cracking, heterogeneous
Lime	3	
Ammonia	4	$N_2 + H_2$; heterogeneous
Sodium hydroxide	5	
Chlorine	6	Electrocatalysis, heterogeneous
Phosphoric acid	7	
Propylene†	8	Cracking, heterogeneous
Sodium carbonate	9	
1,2-Dichloroethane	10	$C_2H_4 + Cl_2$; homogeneous
Nitric acid	11	$NH_3 + O_2$; heterogeneous
Urea	12	‡
Ammonium nitrate	13	‡
Benzene	14	Petroleum refining; heterogeneous
Ethylbenzene	15	Alkylation of benzene; homogeneous
Carbon dioxide	16	
Vinyl chloride	17	Chlorination of C_2H_4; heterogeneous
Styrene	18	Dehydrogenation of ethylbenzene; heterogeneous
Terephthalic acid	19	Oxidation of *p*-xylene; homogeneous
Methanol	20	$CO + H_2$; heterogeneous

* Based on tonnage, from *Chemical and Engineering News* survey of US industrial chemicals (1989).
† Primarily used to make polypropylene in a catalytic polymerisation.
‡ Synthesis based on starting materials produced by a catalytic process.

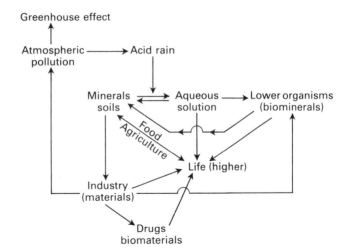

Fig. 14.15 A generalised scheme of the way in which man's industry and his environment interact with living systems.

with increasing intensity. Is the way in which man uses energy a threat to the temperature balance of the Earth's surface, which is partly based on the present ecosystem?

Man also uses chemicals to combat other species which he regards as 'unfriendly' and to aid those he believes to be 'friendly'. In doing so does he create long-term problems? For example, in his treatments of sewage (using chlorine compounds), in treatments of soil (using agrochemicals containing covalent chlorine and fluorine) and with his various protective chemicals including those in agriculture (using copper) and for preventing corrosion (using tin) man is *attacking living systems*. At the same time he is seriously considering introducing new species.

While facing these questions, man must believe that he can manage all these problems taking limited risks. There may also be a new way forward which is to adjust biological rather than chemical systems to our own ends but there is no certain knowledge that this will be possible. We will return to these problems in Chapter 16, after we have looked in Chapter 15 at the total balances (in cycles) of the whole chemical activity on the Earth's surface—abiotic, biological and man-devised.

Further reading

The following references are largely complementary to the purpose of this chapter: the purposeful selection by man of the chemical elements.

1. Fraústo da Silva, J. J. R. and Williams, R. J. P. (1991). *The biological chemistry of the elements—the inorganic chemistry of life.* Oxford University Press, Oxford (3rd printing, 1994). Chapter 22 and see references in this chapter.
2. Cox, P. A. (1995). *The elements on Earth*, Oxford University Press, Oxford.
3. Schriver, D. F., Atkins, P. W., and Langford, C. H. (1994). *Inorganic chemistry* (2nd edn). Oxford University Press, Oxford. Chapters 10, 16, 17 and 19.
4. Cardwell, D. (1994). *The Fontana history of technology.* Fontana Press, London.
5. Press, F. and Siever, R. (1986). *Earth* (4th edn). W.H. Freeman & Co, New York.

6. Press, F. and Siever, R. (1994). *Understanding Earth*. W.H. Freeman & Co, New York.
7. Wayne, R. P. (1993). *Chemistry of atmospheres* (2nd edn). Oxford University Press, Oxford. Chapters 1, 2, 4 and 5.

On environmental problems at the layman level.

8. Lovelock, J. E. (1979). *Gaia, a new look at life on Earth*. Oxford University Press, Oxford.
9. Lovelock, J. E. (1988). *The ages of Gaia*. Oxford University Press, Oxford.
10. Gribbin, J. (1988). *The hole in the sky*. Bantam Press, London.
11. Gribbin, J. (1990). *Hothouse Earth—the greenhouse effect and Gaia*. Bantam Press, London.

See also references in Chapters 1, 2, 11 and

12. Mains, G. (1976). *The oxygen revolution*. David and Charles, Newton Abbot, Devon.

Note added in proof. The difficulties which man may face in his use of chemicals to bias competition (and co-operation) between living species is illustrated by the problems modern medical and agricultural practice now face through mutational changes in lower organisms. These changes are forced on an ecosystem by man's chemical attack. See Weatherall, D. (1995). *Science and the quiet art*. Oxford University Press, Oxford.

15

Element cycles and their evolution

Between Heaven and Earth nothing goes away that does not return . . .
I Ching (*Book of Changes*, Appendix II)
(transl. J. Legge)

15.1 Introduction

In this book we are examining the natural selection of the elements. We have seen in Chapter 8 the way in which the reactions of atomic nuclei in the stars are slowly, in a fixed sequence, changing the abundances of the elements almost uniformly in the universe. There is a kinetic limitation on these processes and no final equilibrium or steady state has been reached. This first step of physical natural selection is based on this nuclear kinetic and thermodynamic stability. In Chapter 8 we also saw how the natural selection of elements, once formed, at first followed affinity for one another, given their abundances, to give compounds and then condensed phases in a second step.

The initial combination of matter was then fixed largely by thermodynamic physicochemical selection (Chapters 2 and 3). However, after rapid cooling and when planets formed, phase separations occurred due to fields of gravity and in these cold bodies some compounds were produced in the presence of others with which they should react but they could not because the temperature was too low. It was then necessary to define many chemicals as components in compartments, not in equilibrated phase systems, in an operational manner. Physical restriction in phases and kinetic chemical control of selection, generating components, in this third step (Chapters 4–7), had taken over from thermodynamic chemical selection. At the same time, these inorganic compounds were now and then subjected to volcanic action (Figs 8.6 and 15.2) as well as weathering, which forced energised compartmental and chemical separation upon them. Some inorganic and small organic compounds on the surfaces of planets were also made to react and to be elevated in energy, giving new components, through the action of solar radiation. Of course, the new compounds produced were unstable and decayed (slowly) since they were energised. There began, therefore, a constant cycle of prebiotic activation and deactivation of matter on the surface of our planet—the Earth (Fig. 15.1, cycling stage I). In this cycling, new element selection in kinetic traps was possible and, at some stage, copying began as a special kinetic trap. This development generated an organic self-assembling system for the natural selection, self-selection, of elements utilising external energy sources and, eventually, a living system. Like all other excited and self-assembling conditions, life must decay and cycle (Fig. 15.1) but, as we saw in Chapters 10–12, it also became self-developing in complexity—a Darwinian natural selection that included further element selection. In this development, the highest form of the natural selection of elements, there were two cycling stages (Fig. 15.1); the first, cycling stage II, up to the coming of dioxygen, was unicellular and relatively simple (Chapter 10), while the second, cycling stage III, was predominantly multicellular and much more sophisticated at the level of cell–cell interaction (Chapter 12). Now, we cannot understand the development of stage III, without seeing that, once stage II developed a large capacity, it became interactive with the cycle of the inorganic/organic 'dead' chemistry of the surface of the Earth, stage I, and forced development on it. Due to its ability to select certain elements, generating dioxygen in particular, the early living systems as a whole then *readjusted* and *energised* inadvertently new combinations of the elements on much of the Earth's surface. We have seen in Chapter 12 the way in which this drifting steady state of the cycles of elements in life built up ever new cycles of the environment so generating the conditions for even greater furthering of life's evolution. It is probable that the whole system has only approached a rough steady state on the surface in the last few hundred million years. It is the very fact that the early prokaryotes with their energised chemistry initiated a new energised chemistry over which they had no control that generated the surface chemistry of the Earth which, in turn, effectively forced evolution to move onwards in stage II of this Darwinian natural selection of the elements toward the present cyclic steady state (see Fig. 15.2 (a)). Since every compound so formed is unstable they all must decay. Concurrent with the biological cycles is the close-to-cyclical behaviour of the unstable gaseous and liquid surface of Earth due to the out-of-equilibrium between the Earth and the sun, that is, the movement of air and water. To this

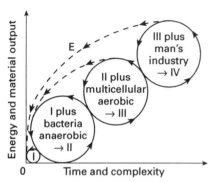

Fig. 15.1 At each stage (I to III) of chemical evolution there has been an approximate steady state for some period (say, 1×10^9 years) which is followed by a transition to a new approximate steady state. The steady states are based on higher levels of complexity. The transitions are due to a change in chemical element availability. (The origin 0 is an equilibrium state that, by definition, has no complexity.) The fourth cycle does not exist as yet since we are in a transition period. At present man's activity may even be driving chemical systems back toward equilibrium by tapping reserves (see process, E). I, the first cycle, is abiotic.

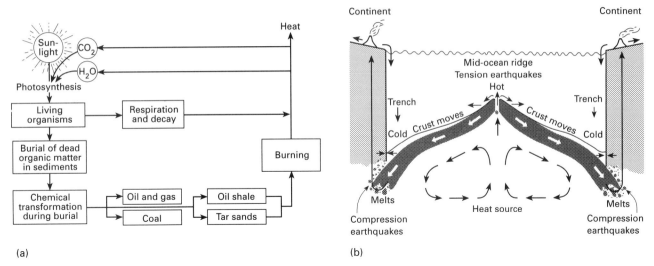

(a)   (b)

Fig. 15.2 (a) Photosynthesis produces organic matter that is buried and transformed, and so becomes a fossilised product of photosynthesis—a fossil fuel. The fossil fuel has been very slowly recycled but increasingly rapidly due to man. (b) An illustration of the partially cyclic chemistry of the Earth's crust. Older crust is laterally moved to trenches where it is recycled by descending convective limbs. (From *Earth* (4th edn) by Press and Sievers. Original source: US Geological Society and NASA.)

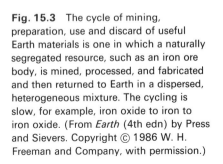

Fig. 15.3 The cycle of mining, preparation, use and discard of useful Earth materials is one in which a naturally segregated resource, such as an iron ore body, is mined, processed, and fabricated and then returned to Earth in a dispersed, heterogeneous mixture. The cycling is slow, for example, iron oxide to iron to iron oxide. (From *Earth* (4th edn) by Press and Sievers. Copyright © 1986 W. H. Freeman and Company, with permission.)

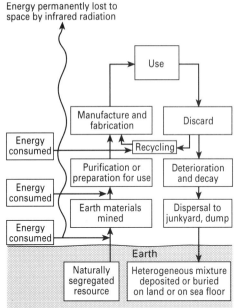

cycling is further added a partially cyclic chemical input from the non-equilibrated layers of the Earth's upper zones (Fig. 15.2 (b)). This implies that most compounds or mixtures on the surface of the Earth, such as the air itself, the sedimentary rocks and the waters generally, must be interwoven with the biological cycles to a large degree. The evolution of chemical change now continues in a final stage, IV, as described in Chapter 14, with the ascent of man (Fig. 15.1). With man the natural selection of elements takes a peculiar twist—self-conscious purposeful functional use (Fig. 15.3)—but it must become interactive with the pre-existing earlier forms of life as well as with abiotic chemistry and its cycles, and eventually has to form a stable energised cycling of matter on the Earth. It is this fourth stage that is now causing some worry since the nature of the new cycle is not clear. We are especially uncertain of its time constants.

In this chapter, we will consider two kinds of cycling at the present time, so that we include man's inputs. Initially we need to discuss the *physical* cycling of the gaseous, liquid and solid phases (compartments) of the Earth, that is, the atmosphere (air), the hydrosphere (water) and the lithosphere (rock or earth), see Sections 15.3–15.5; then we shall turn to the way in which different chemicals move in these and other cycles, Sections 15.6–15.12. Of course, the physical cycles are interwoven with the cycles of different elements. (We shall indicate where there is a non-cyclic change present.) First, however, we must consider the nature of all cycles and the forces which drive them.

15.2 The nature of 'cycles'

Truly complete cycles involve no change of input and overall output except of forms of energy. As we have seen, although there is no change in any such full cycle of physical or chemical systems, there must be overall entropy increase (Chapter 3). If there is to be no overall change of the material in a cycle and no change of the physical conditions, temperature and pressure, internally, then the entropic drive is the increase of quanta externally. For example,

$$\text{One quantum of light} \longrightarrow n \text{ quanta of heat.}$$

The drive on the cycle is illustrated in Fig. 15.2(a). The sun appears to be an infinite resource and indeed it is on a billion years scale. When describing cycles we often ignore such energy losses or entropy gains.

A second (possible) *energy*-driven so-called cycle due to non-equilibration within Earth is, to a degree, incomplete in material and/or in change of physical conditions of material—it is not a fully cyclical processes (Fig. 15.2(b)). Although, in principle, the high temperature of the inner part of the Earth can be used to drive various geological movements and biological changes while heating the colder air, in fact the interior of the Earth changes in physical (and chemical) properties. A massive uprising could not return to the present balance. Another type of so-called cycle of material has an input from an exhaustible but huge concentration reserve to a second exhaustible but huge system. For example, the oxygen of H_2O cycles through various mainly biological processes to O_2, dioxygen, but some of this gas is lost to the universe all the time. Over short periods of time the cycle is, like all other cycles, in a steady state through which material moves with input equal to output plus losses. Now, while the cycles that only have energy conversion as a net input to output will last as long as the sun, 5×10^9 more years (hence the Earth has a thermostated temperature), the material cycles can be upset by novel inputs. Thus, today's dioxygen cycle did not exist 3 to 2×10^9 years ago and probably reached a steady state only some 1×10^9 years back (Fig. 15.1). Variations in cycling are also to be expected if there are new material or energy inputs caused, for example, by changes in chemical activities on Earth due to man. The essence of stability of a steady state cycle is permanence of input and output.

Given these background notions we can now examine the physical cycles of the solid, liquid and gaseous phases of the Earth, and those cycles of the individual elements, before we turn to their origin and finally their total interaction (Sections 15.13–15.15).

15.3 The cycles of the rocks

The flow of material on the Earth's surface is due to two very different processes: (1) the upwelling of molten rock to form part of the crust (Figs 8.6 and 15.2(b)), and (2) the weathering of the rocks which forms new sedimentary deposits in the sea† (Figs 15.4 and 15.5). Due to the motion of

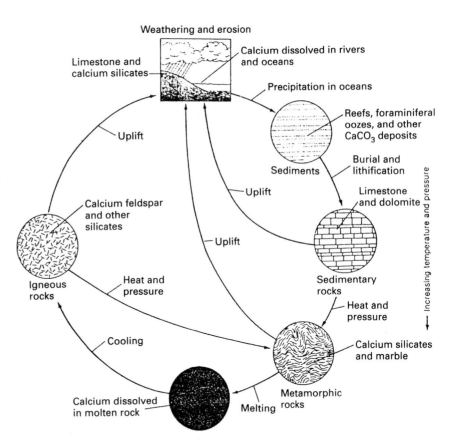

Fig. 15.4 The rock cycle. Rocks are weathered to form sediment, which is then buried. After deep burial the rocks undergo metamorphosis or melting, or both. Later they are deformed and uplifted into mountain chains, only to be weathered again and recycled. Some injection of rock from the upper mantle is irreversible, that is, non-cyclic. (From *Earth* (4th edn) by Press and Sievers. Copyright © 1986 W. H. Freeman and Company, with permission.)

† Meteorites (and cosmic dust) fall continuously on the earth; although only a few of them are found—most fall on the seas and oceans, polar regions, deserts, forests and other uninhabited regions—it is believed that their weight may reach 2000 tonnes every year. Once in the earth they are degraded by weathering and erosion processes and their constituent atoms are incorporated in the soils, are absorbed by the plants and then enter the food chain reaching the higher animals, even man. At the same time, both the living species and the earth give back to the atmosphere many substances, mostly gases and dust.

In this way, the universe around us is tightly linked to the earth and to the living forms that populate it, so that one is correctly entitled to speak of a continuous exchange of matter between our planet and the cosmic space.

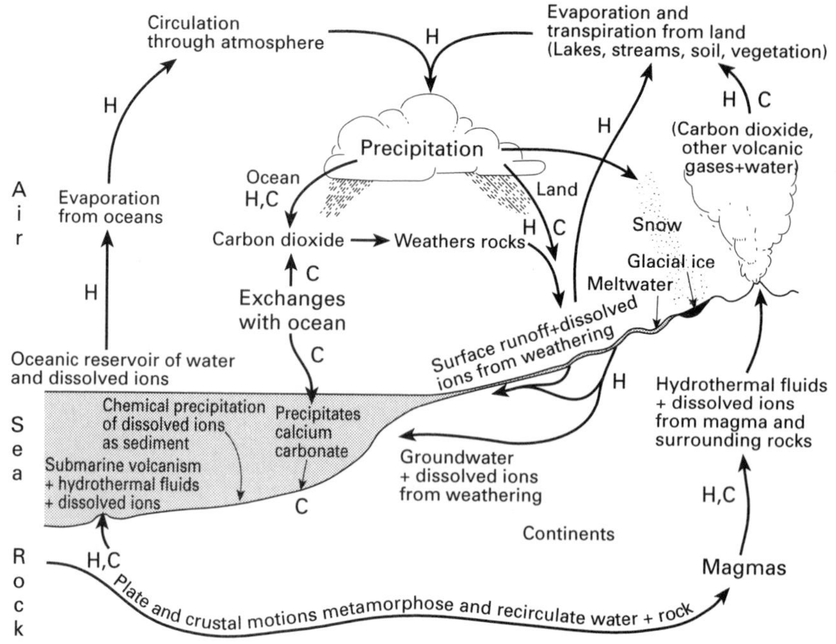

Fig. 15.5 Overall hydrological cycle. Routes of carbon dioxide are shown by the letter, C; water by the letter, H. (From *Earth* (4th edn) by Press and Sievers. Copyright © 1986 W. H. Freeman and Company, with permission.)

continents, the sedimentary rock can be submerged beneath other rock and re-emerge at some distant date only to be eroded again. This erosion gives back very many elements to the sea, so that the sea is constantly replenished (see Figs 8.19 and 15.5), and any losses due to precipitation are replaced. Notice that the energy for the cycle comes partly from the core of the Earth and partly from the sun (and the moon) which cause movement of air and water. The cycles and irreversible processes described here have both a physical and a chemical character since the movements of the rocks occur both due to transport of large particles, under gravity for example, and due to solution or reaction between molten or liquid phases. Huge deserts are produced by air-borne particles, huge delta-lands by the rivers and massive ocean shelves by glaciers, but none of them are permanent. The time scale of the cycle is on the order of $\geq 10^8$ years. Man is not involved in this cycling except through mining, which affects the very top surface locally.

15.4 The cycle of water

One way in which the elements hydrogen and oxygen are cycled is through their physical motion as water. The cycle has two parts: the part most commonly recognised is transport by evaporation followed by precipitation (rain) and then flow back to the source (Fig. 15.5).

This process is not a simple and independent cycle since it interacts with erosion (hence the cycling of rocks is not independent either). Moreover, the pattern of flow evolves since oceans and river courses change. Of course, the cycle uses mainly the sun's energy. We have, however, to include a second part since there are cycles within the sea itself, both along the surface and deep into the ocean, that are extremely important for the mixing of dissolved or suspended chemicals, for movement of material in estuaries, etc. The sea also cycles, as does some freshwater, deep into the rocks. From the sea, the water enters the edges of deep ocean trenches (Fig. 15.2(b)), and re-emerges later through 'springs' in the trenches. Of course, the physical cycling of H_2O also carries with it the cycling of other elements that may be lost to the rocks or taken from them. We shall look at the time scale of individual elements later, but deep circulation of the oceans is of the order of 10^2 to 10^3 years, while circulation through the ocean trenches is orders of magnitude longer. Man affects the water circulation in a very minor way at present. (There is a fear, however, that by changing greenhouse gases sea (and air) flows may be upset.)

15.5 The cycle of air

The third physical cycle is that of the air which moves both along the surface of the Earth and in a vertical column (Fig. 15.6). Very little is lost, but we shall see that the different components of the air, while remaining in quite closely fixed proportions, undergo quite different chemical cycles, in particular, those of H, C, N and O. The time of atmospheric physical cycles can be quite short, but a year is required for complete horizontal cycling at its slowest. Vertical mixing appears to be very slow, of the order of years, since the ozone layer does not mix with lower layers. Destruction of the ozone could, therefore, take hundreds of years to repair. The mixing of air and water is not included, but is shown separately in Fig. 15.5.

The first three flows we have described are those of the three phases: air (gas), water (liquid) and earth (solids). These movements were noted by early thinkers, both Chinese and Greek (see Fig. 15.7), who have incorporated ideas

Fig. 15.7 The continuous flow of rivers and waterfalls that so struck the Chinese (Section 1.2). Chinese painting by Ma Yuan-Koung (Ming period).

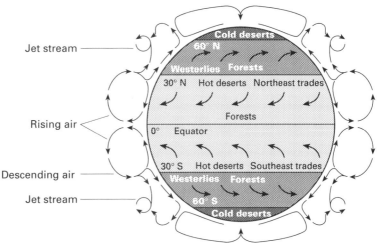

Fig. 15.6 Circulation of Earth's atmosphere shown as we would see it if we could stand outside Earth and observe vertical movements like those indicated by the continuous arrows and surface winds like those shown by the broken arrows (see also Fig. 8.9). (After Purves, Orians and Heller, reference 2 in Chapter 11.)

of continuous cyclic flow into their philosophy, but, of course, they knew nothing of the chemical elements that concern us in this book.

We shall now dissect the ways in which these different elements flow. As stated before, this involves ignoring the distinction between 'living' chemicals, 'dead' chemicals, and man's chemical industry. On the surface of the Earth all 90 naturally occurring elements of the periodic table are in long-term flow. It is very important for man to understand the steady states of the flow of the elements in different zones, especially when he adds unusual chemicals, for example chlorofluorocarbons or substantial amounts of common chemicals, such as CO_2, to the environment. Indeed, much activity on the Earth's surface is strictly cyclical and is either much faster or as fast as life's evolution, which means that changes in either drive the other (see Fig. 15.1).

15.6 The cycle of oxygen

We take this example of an element cycle first since it reveals the complexity of the interaction between elements. In Chapters 2 and 8 we have shown that on the surface of the Earth most, if not all, elements are found in large part in oxides. The exchange of this oxygen in the elemental cycle occurs in all three phases of matter, that is, in solids, liquids and gases. If we start from dioxygen of the air then it cycles with the ozone layer, a process that does not involve a second element (Fig. 15.8). Note that this dioxygen comes largely, but not totally, from the photosynthetic reactions of living substances and its primary origin is H_2O. The dioxygen back-reacts with the organic chemicals of life (Chapters 9 and 11) giving H_2O, CO_2 and so on. Thus the oxygen, hydrogen and carbon cycles intermingle. The dioxygen also reacts with sulphides from many sources, with nitrogen in the air and with many minerals, and most of the reactions are cyclical. Of course, we must not omit man's industry since

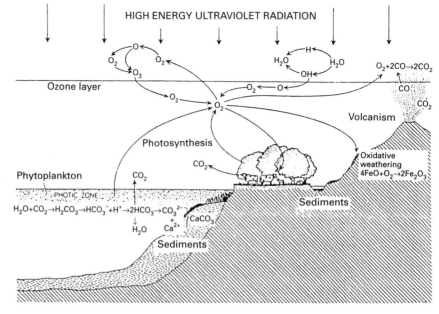

Fig. 15.8 The oxygen cycle. Oxygen exists in the atmosphere mostly as O_2 (21 per cent volume), H_2O, O_3 and CO_2, nitrogen oxides, SO_2, etc. In the lithosphere it exists in combined forms—oxides, silicates, carbonates, sulphates, etc., which do not exchange quickly as well as in organic compounds, for example, humic acids. Photosynthesis and respiration in plants are in approximate balance and far outweigh other inputs and outputs. The cycle turns over in approximately 2000 years. (After Cloud, P. and Cibor, A. (1970). *The biosphere*. Scientific American Books.)

dioxygen is a component of his major fuels, the 'fossil' fuels, and is changing the cycle. Indeed, the dioxygen of the air can also be considered as a 'fossil' fuel. It is not our purpose to describe in detail the dioxygen balances; we only indicate its cyclic complexity. Many different chemical cycles have evolved with living systems (see Chapters 9 and 11), and thus the oxygen cycle is now a mixture of prebiotic, biotic and man's chemistry, which are totally interactive. This, like all other cycles, is not strictly reversible since some dioxygen is lost through escape to outer space and some through going into thermodynamically stable traps. The time scale of the dioxygen (oxygen) cycle is compartmental from years to millions of years.

15.7 The cycle of hydrogen

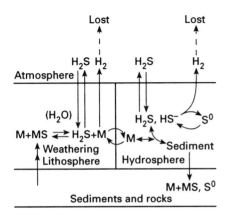

Fig. 15.9 A proposed cycle of prebiotic hydrogen with sulphur. Dominant species are H_2S and HS^-, as well as sulphides. Note the loss of H_2 to outer space.

Even though the oxygen cycle is complex, the hydrogen cycle would be very simple but for the presence of life. There was, for many millions of years, little hydrogen present except as H_2O and H_2S when the hydrogen cycle was linked just to that of water, for example in condensation and hydrolysis reactions, and to the cycle of prebiotic sulphur, for example in redox reactions (Fig. 15.9). As life developed gradually, this cycle became more complex and started to involve chemicals such as NH_3 and CH_4. Very few other hydrogen compounds exist today except in life. With the increase of dioxygen in the environment much of the H_2S cycle has been restricted by oxidation and this is true also for NH_3 and CH_4. It is the case too that little free H_2 is involved now in the cycling presumably because H_2 would not cycle, being continuously lost from the planet to some degree. This must also be true for the very light noble (and inert) gas helium, which is lost continuously from rocks, while the heavier noble gases, argon, neon, krypton, xenon and radon (produced continuously from ^{226}Ra) remain as a fixed part of the atmosphere and are involved in the physical cycle of air. Clearly a large part of the hydrogen cycle is of short time-span.

15.8 The cycle of carbon

The global scale of the carbon cycle is indicated in Fig. 15.10. It is again essential to see the complexity and the coupling of movement of this element, mainly with H and O. Note that the entry to a cycle or exit from a cycle can be blocked by a phase separation, especially where the phase is hidden in deeper rock, as happens with carbonates but also with the hydrocarbon gases, fuels and coal. (Another example of phase separation is that of the iron of the core of the Earth.) Sooner or later the light materials come back to the surface, whether or not man assists this process, and then re-enter the cycle. These observations force us to remember that the cycling is dependent upon the time-scale of our considerations. We can forget some reserves for as long as we do not start to expose them. If we use them, for example carbon, then the steady-state levels of the cycle change, for example CO_2 increases while C and C/H compounds are reduced.

The carbon cycle and that of oxygen are locked into the energy that reaches

Fig. 15.10 The carbon cycle.
Atmospheric CO_2 is mainly assimilated by
green plants, algae and autotrophic
microorganisms in soils, particularly
methane-forming bacteria. The absorbed
CO_2 is converted into organic compounds,
especially sugars, and part of these are
used for growth whereas the other part is
used in combustion reactions with O_2 for
energy. The resulting CO_2 formed in these
reactions is given back to the atmosphere.
In the hydrosphere, a significant part of
CO_2 combines with calcium to form
calcium carbonate and much of this is
produced by sea animals to be used in
shells. The rate of turnover of carbon in
this cycle is estimated to be of the order of
33 000 years/cycle, but man's activities
(industry) are increasing the level of CO_2
in the atmosphere, which was 290 p.p.m.
in 1940 and is now about 355 p.p.m.,
contributing strongly to the so-called
'greenhouse effect'. The same is true for
CH_4, which is also a 'greenhouse gas'. The
amounts in reserves are in billion tonnes
and the transfer rates in billion tonnes per
year. Fossil fuel burning releases
additionally 5 billion tons per year. After
Wayne, R. P. (1993). *Chemistry of
atmospheres*. Oxford University Press,
Oxford.

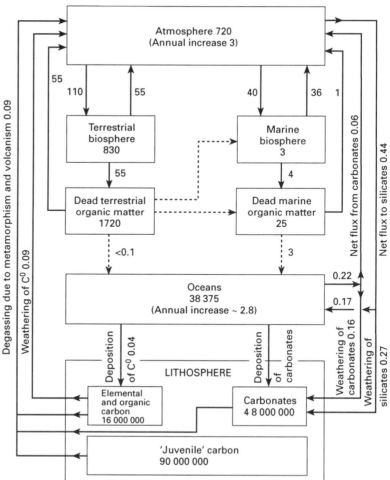

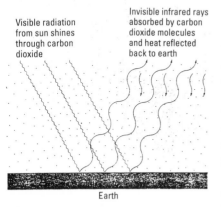

Fig. 15.11 The greenhouse effect. Just
as the glass of a greenhouse transmits light
rays but holds in heat, the carbon dioxide
of the atmosphere transmits visible
radiation from the sun but absorbs and
reflects back to Earth the infrared rays from
the surface. Water vapour, methane, alkyl
halides, and polyatomic molecules in
general are also 'greenhouse' compounds.

the Earth's surface. This energy is partially reflected and is trapped in the
atmosphere opposing the cooling tendency and providing a thermostatic
control of temperature due to what is called the greenhouse effect (see Fig.
15.11). Here we see a link between a chemical chemostat and a physical
thermostat. Changes in CO_2 change the temperature of the cycle which
changes many other components since kinetics depend on temperature. It is
difficult to compute exactly what new input of chemicals will lead to what
novelty in these cycles, but they will change!

The carbon cycle already existed in prebiotic times, but was extremely
simple involving mainly the cycling of CO_2 to and from the atmosphere to
minerals such as carbonates. There are, however, suggestions that some CO_2
would even then undergo photoreduction to aldehydes and then to methane
(see Section 10.5.3). The methane could be reoxidised to CO_2 but not by
dioxygen at that time. The point to be made here is that the cycling of carbon
on the surface of Earth was utterly changed by life, mainly primitive life,
altered again by the advent of dioxygen, and is now being altered once more by
man, through CO_2 emissions and forest depletion, and through CH_4 from
intensive farming (Chapter 14), with knock-on effects to all other cycles.

15.9 The cycle of nitrogen

The nitrogen cycle today is shown in Fig. 15.12. As with the hydrogen cycle, it is important to remember the chemistry of nitrogen. This element gave rise to very few prebiotic compounds which were stable, except perhaps NH_3. It is today prominent as N_2 in the atmosphere or in the products due to living systems, biological or man-made; the main examples are N—H or N—O compounds, for example, NH_3 and NO or NO_3^-. The lack of major chemical reservoirs of nitrogen in the waters or rocks makes its cycle very sensitive to change. As mentioned above, the coming of dioxygen removed HCN and NH_3 to a large degree. In fact, we note that the cycling of nitrogen has been greatly accelerated (catalysed) by living systems. This is a relevant observation since there are always two features of a cycle—the nature of the component chemicals in it and the speed of the steps. Catalysing a given step and not others alters, of course, the form in which the element is stored towards steps after the catalysed one. The cycling time of nitrogen is on the time scale of hundreds of years. We need to know the *residence times* of all other elements in

Fig. 15.12 The nitrogen cycle. The largest reservoir of N_2 is the atmosphere (79 per cent volume), which also contains a small proportion of nitrogen oxides. Nitrogen-fixing organisms (bacteria, free-living or symbiotically associated with the roots of some plants) convert N_2 into NH_3 and organic compounds containing nitrogen such as amino acids, nucleotides, etc. Industrial fixation consumes a considerable part of atmospheric N_2 and, as a result, there may be a negative balance between the total nitrogen fixed and that given back by denitrifying bacteria in the soils and aquatic media, but it is necessary to take also into account the contribution from volcanic phenomena, which probably balances inputs and outputs. The nitrogen cycle takes some 3000 years. Quantities as in Fig. 15.10. After Wayne, R. P. (1993). *Chemistry of atmospheres*. Oxford University Press, Oxford.

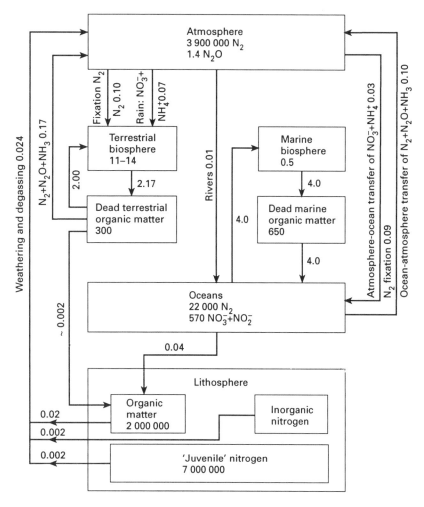

various parts of the cycle which tell us the proportion of an element in a given state (see Figs 15.10 and 15.12).

15.10 The cycles of sulphur and phosphorus

The last two elements we shall consider in some detail are sulphur and phosphorus. In this way we cover all the most common elements of life: H, C, N, O, P and S. The sulphur cycle is shown in Fig. 15.13 (compare with Fig. 15.9); it is energised by the interior of the Earth and comes into a new balance extremely slowly when the situation undergoes changes. Once again, we have to be aware of the effects of life and of the coming of dioxygen on this cycle. In particular, there is the novel production of SO_2 in this cycle which causes the phenomenon known as acid rain, for example. (Sulphur chemistry, like oxygen and iron chemistries, is not truly cyclical since in the long term the FeS will remove all the O_2 and all H_2 will be lost.)

The phosphorus cycle, shown in Fig. 15.14, is another vulnerable cycle since phosphates are rather insoluble and phase separation may occur. In some ways it is preferable to present two cycles for this element: one (Fig. 15.14(b)), of the rocks (land-based), is long term, and the other (Fig. 15.14(a)), of aqueous solution (water-based), is short term. Man affects the water-based cycle. Notice that the phosphate cycle is quite different from all the others so far described in that no redox chemistry is involved; thus, the coming of dioxygen did not affect it directly, much though life does.

Fig. 15.13 The sulphur cycle. Sulphur exists in the atmosphere mainly in the forms SH_2 and SO_2 (or SO_3); in the hydrosphere the major forms in aerobic media are the sulphate ion, SO_4^{2-}, and organic sulphur compounds, for example, dimethyl sulphide $(CH_3)_2S$ (DMS) produced by certain algae. In media deprived of oxygen (for example, ocean vents) sulphur appears as H_2S, SH^-, SO_3^{2-} and $S_2O_3^{2-}$ as well as metal sulphides, some of which are soluble. In the lithosphere there are several sulphur-containing minerals, mostly sulphates and sulphides as well as the native element. The major source of SO_2 is volcanism and man's industrial activity, burning of fossil fuels, since sulphur (from biological material) is always a component of these.

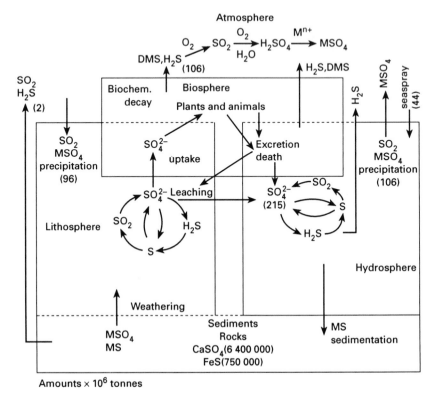

Amounts $\times 10^6$ tonnes

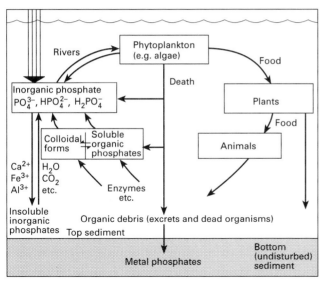

(a) Water-based phosphate cycle

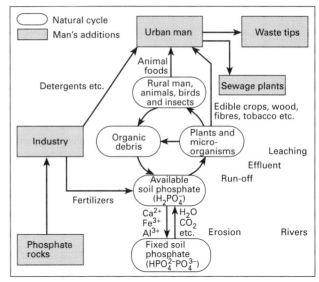

(b) Land-based phosphate cycle

Fig. 15.14 Phosphorus cycle. Whereas the cycles of most non-metals, for example C, O, N, S, involve oxidation–reduction reactions, that of phosphorus operates without them. The main form of phosphorus in the earth is as phosphate ion or its salts, particularly those of calcium. In the biosphere, the phosphate esters, especially the nucleotides, the polynucleotides and the phospholipids, are the dominant phosphorus species. Man's industrial activity is again important in this cycle since phosphate is mined and then added to the soils as fertiliser to promote the development of useful plants, but it may be leached to the rivers and the sea by rain fall and then cause uncontrolled growth of other species that may ultimately endanger animal life, for example of fish, due to the excess consumption of oxygen.

15.11 The cycles of light elements and metal ions

The cycles shown in Figs 15.8–15.14 are for a number of non-metals—H, N, O, C, S and P. These cycles could occur, in principle, without the involvement of other elements. In practice they then would go exceedingly slowly. In real circumstances the steps are catalysed by organisms and so become faster. In Chapters 7, 10 and 11 we described the catalysts for the individual cycles. It is then clear that there are some apparently very strange features. First, no nitrogen could cycle rapidly without molybdenum, either as N_2 or NO_3^-. Equally, no sulphur could cycle through sulphate without molybdenum. The critical role of this element in both cases requires explanation. Again oxygen would not cycle rapidly without Mn and very little hydrogen as H_2 would have cycled in the past without Ni. Furthermore, iron is also required as a partner in every case. In addition, the phosphate cycle is associated with magnesium. Thus, these cycles, which in part are biological but in part are geochemical, are associated with a group of metal elements. We see here several relevant examples of the close relationship that exists amongst the chemical elements which has in biological chemistry its deepest meaning. Close inspection of each cycle shows that it is linked to one or more other cycles. *Minor element changes, through catalysis, are as important as direct major element changes.* Thus, with very little structure, the whole surface regions of the Earth—atmosphere, waters and rocks—are for a given period in a long-term steady state, physically and chemically, but drifting slowly to new steady states. In a very real sense, the whole is a network of interacting cycles—it has apparent organisation, but one thing is certain: it can only grow in complexity up to a limit before it returns towards stationary traps or equilibrium as the energy transfer fails.

15.12 The cycling of other elements

All other elements cycle on the immediate surface of the Earth in relatively short periods of years (Table 15.1). The major cycles of concern are of those elements used by living systems. However, in each case we must ask about the involvement of man today. In several of the above cycles (Figs 15.8–15.14) we have indicated that industrial processes do interact with natural cycles and we have considered this problem in some detail in Chapter 14 since these interactions are of great importance for the extant network of life (see Table 15.1). Here we stress that the elements we are considering vary from those present in quantity and with fixed oxidation states, namely, Na, K, Mg, Ca, Al and Si (notice that with the other elements already described we have now mentioned all the important light elements in the first two periods of the periodic table except the halogens), to elements usually present in smaller amounts, for example (Sc), Ti, V, Cr, Mn, Fe, Co, Ni, Cu, (Zn), most of which can undergo redox changes, and with which we must include the halogens.

15.1 Annual cycling of elements through the atmosphere, hydrosphere, biosphere, and industry showing the natural transport of selected elements in volatile and soluble forms only, through the atmosphere and hydrosphere, respectively, and industrial production. There are also significant transports of dust through the atmosphere (at a level of 10^{-6} to 10^{-3} units) and of sediments through the hydrosphere (at a level up to 0.1 units)

Element	Annual flux (10^{12} kg per year) through			
	Atmosphere	Hydrosphere	Biosphere	Industry
H (as H_2O)	6×10^4	6×10^4	Very large	Large
C	200	100	150	8
N	0.25	0.1	6	0.1
O (as O_2)	300	1	300	0.1
O (as H_2O)	5×10^5	5×10^5	Very large	Large
Na	0	0.2	1*	0.001
Mg	0	0.3	1*	3×10^{-4}
Si	0	0.2	>1	0.01
P	0	0.001	1	0.15
S	0.1	0.4	0.5	0.15
Cl	0.005	0.2	1*	0.17
K	0	0.05	1*	0.05
Ca	0	0.5	0.5	0.1
As	2×10^{-5}	4×10^{-4}	—	5×10^{-5}
Se	5×10^{-6}	8×10^{-6}	—	1×10^{-6}
Hg	5×10^{-5}	5×10^{-6}	—	8×10^{-6}
Pb	0	0.01?	2×10^{-4}	0.004

* This is a rough order of magnitude based on the values for carbon and nitrogen.

As stressed in Chapter 7, the important feature of the chemistry of the first group is the easy exchange between rocks and solutions, and we note that life affects the rate of this cycling. Interaction amongst these elements is largely simple, giving compounds such as $NaCl$, $CaCO_3$ and SiO_2, whether in prebiotic or biotic chemistry (see Figs 8.19, 15.4 and 15.15). These elements are not catalytic on the whole, so they do not greatly affect the cycling of other elements in their compounds. The second group of elements is very different. There are major effects of some, especially surface iron which has had a history of change from the sulphide prebiotic and early biotic times to today's oxygen-dominated chemistry (Fig. 15.15). This change of the atmosphere has also affected the cycling of Ni, Cu and Zn, all elements that were initially present as sulphides. Now, while the quantities involved in the transfer of the heavier elements may be small, the effect of change can be dramatic since they are catalysts for the cycles of the primary elements. The change in life with the coming of dioxygen is a change in catalysis by these elements as much as anything else. Thus, the cycles of all these elements are linked to the primary six, H, C, N, O, S and P, and are extensively so linked in life. While we note that over quite long-term periods all the elements cycle, we also must see that all the cycles can drift or evolve (see Fig. 15.1 and Section 15.13) over even longer periods.

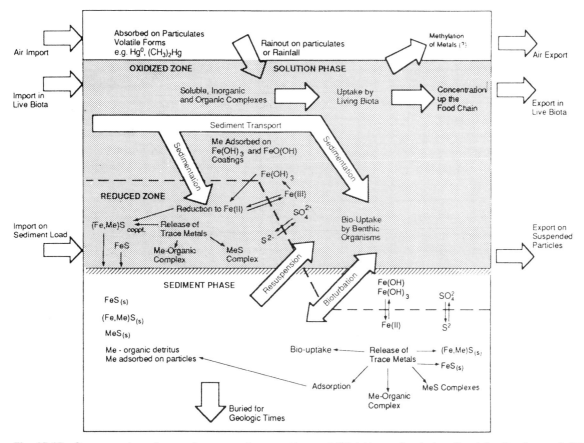

Fig. 15.15 Summary of reactions and processes important in metal (Me) biogeochemical cycling (after Butcher *et al.* (1992), reference 1 of the 'Further reading', used with permission.)

Finally, since man now uses radioactive chemicals, these cycles are carriers of the radioactivity and we must know the half-life of the activity relative to the half-life of any cycle. We could consider ^{90}Sr, for example, and ask against the background of this radioactive element how does it cycle and on what time scale. In the calculation we must consider rates of dilution, mixing, and the sizes of the different reservoirs. Do they cycle at different rates? In effect, we may well say that the cycling of *all* elements is important for life, since rare radioactive elements are carried in the cycles of other common elements.

15.13 The birth of cycles

At some time around $4–5 \times 10^9$ years ago, the Earth's surface entered into a mode resembling a chemostat which has evolved in stages to today. The centre of the Earth, the core, is a huge hot zone ($5000°C$) which is losing heat at an almost constant rate due to the insulation effect of the outer mantle. The sun, although cooling slowly, also pours out energy at a constant rate. The surface of the Earth, at an average of about $25°C$, undergoes minor fluctuations still, but its heat gain and loss is in balance from these two input sources and radiation output, and will remain so for some 5×10^9 more years (the expected life of the sun). The result is largely a structural bath of *liquid water* in a container that has temperature zones in a steady state, that is, it is a graded thermostat. This means (Chapter 7) that the physical parameters of the kinetics of chemical change on the Earth's surface are fixed in regions. In addition, there is a steady outpouring of material from just below the surface and there is an input of material to the sea and then to the surface layers by erosion, which are driven by the two energy sources, but there is also energy entering chemicals by energy capture at the surface of the Earth. In the absence of life it is probable that this physicochemical activity, slowly stirred by the energy input, became a graded chemostat with fixed values of pH and redox potential in the different parts of the flow some 4×10^9 years ago (see Fig. 15.1). There would have been some chemical losses and gains but, generally speaking, the atmosphere, the sea and the surface would have been in a chemical steady state since we have seen that the cycles of individual elements are interactive in a network. The elements fitted into a selected pattern, based on their chemical thermodynamic and kinetic properties, that would have been maintained for billions of years but for the appearance of life. This could be the situation on the surface of many other planets. There is no reason to suppose that such a surface steady state of flow would evolve since it is backed by huge reservoirs of energy and material. Of course, it is possible that zones would fluctuate locally, although these fluctuations could well have a random character about a mean. Systematic fluctuations are the changes produced by the periodic variation of the Earth and sun each day and each year, and there might be other cycles too.

Finally, on Earth and before we introduce life into the picture, we need to be constantly aware that the zones are not totally isolated. Although regions both on the surface and into the deep are graded, from north to south and from atmosphere to mantle, there is slow mixing. The oceans mix over periods of thousands of years, for example. It is this flow that ensures the steady state, much as a thermostat needs a stirrer. It is in this context that life arose, depending on a long-term thermostat control and on the presence of liquid water, which is an exacting physical demand. Although this requirement may not be satisfied uniquely on Earth, we must never forget that, even if the Earth is not unique, it is most peculiar and its development of a living system, which is still evolving after almost 4×10^9 years, is almost certain to have a unique history no matter what exists elsewhere. It is not easy to imagine the requirements for precisely such constant conditions for nearly 10^{10} years.

15.14 Life and the cycles

If we look at Fig. 15.1 we see that life is involved in chemical transformation in two ways: first, directly, due to the very nature of living systems (see Table 15.2 and this section) and, second, due to man's industry and farming (see Section 15.15). It becomes important to be aware of the effect of man's inputs since they are more strongly interactive with inorganic cycles than are the inputs from organisms. Living organism cycles are very short and are mainly a large turnover of non-metals, especially C, N, O and H. If life had remained in the simple cyclic mode of stage II in Fig. 15.1, it would have adjusted prebiotic cycles, in particular of inorganic chemicals, but slightly. As life evolved and switched to the use of light and dioxygen its effects on abiotic cycles changed. Take the case of the water cycle. Water is energised in the sense that it is moved to a higher thermal energy level as vapour and moved up in the gravitational field of the Earth. This cycle is partly controlled by life, more so on land but even in the sea by energy capture. In part, this cycle gives energy for the

Table 15.2 Chemical reactions mediated by mircoorganisms, which affect the cycles of the elements

Reactions	Microorganism	Observations
Organic compounds→methane	*Methanobacterium*	Anaerobic
	Methanococcus	Anaerobic
Oxidation of methane (for energy)	*Pseudomonas*	Chemoautotrophic
	Methanomonas	Chemoautotrophic
N_2→NH_3 (fixation of nitrogen)	*Azotobacter vinelandii*	Anaerobic
	Clostridium pasteurianum	Anaerobic
	Rhizobium	Anaerobic
NH_3→NO_2^- (for energy)	*Nitrosomonas*	Chemoautotrophic
	Nitrococcus	Chemoautotrophic
NO_2^-→NO_3^- (for energy)	*Nitrobacter*	
NO_3^-, NO_2^-→N_2O, N_2	*Thiobacillus denitrificans*	Heterotrophic
SO_4^{2-}→SO_3^{2-}→$S_2O_3^{2-}$→S^0→SH_2	*Desulphovibrio*	Anaerobic
SH_2→SO_4^{2-}	*Chromatium* (purple sulphur)	Autotrophic
(electrons to reduce CO_2)	*Chlorobium* (green sulphur)	Autotrophic
SH_2→SO_4^{2-}	*Thiobacillus denitrificans*	
Cu_2, FeS→SO_4^{2-}	*Thiobacillus thioxidans*	Chemoautotrophic
(oxidation with O_2)	*Thiobacillus ferroxidans*	
Fe(II), FeS→Fe(OH)$_3$, Fe$_2$O$_3$	Iron bacteria	Practically all facultative or
	Thiobacillus ferroxidans	anaerobic heterotrophic
Fe(III)→Fe(II), FeS	*Desulphato maculum*	

changes in weather and then for the erosion of rocks, which are first decomposed either to small, still 'energised' particles of high surface energy, or form solutions by the action of water (see Figs 15.4 and 15.5).† On entering the sea, these chemicals recombine and are deposited, sometimes again through a biological agency; for example coccoliths deposit $CaCO_3$, and, sometimes, through the effects of pressure they recrystallise and eventually return to land masses as rocks. This is why we may speak of a rock cycle (see Fig. 15.4). These biological effects on inorganic cycles increased in stage III of Fig. 15.1. They are no longer small and huge limestone deposits are often of biological origin. Within each of stages I–III there was and is a presumed balance, steady-state cycle between fast cycling of organic and slow cycling of much inorganic material. Apart from changes in the rocks, the atmosphere is now completely different from the primitive atmosphere (Table 8.7), the sea is quite different in its minor components (Tables 8.8 and 10.4), and there are deposits on land that could not have been present originally (see Figs 15.2–15.8 for examples and Table 15.3). Now, most of these products are stable with respect to the existing atmosphere but in stage IV the deposit of additional materials such as coal, oil and gas (as well as of sulphur) is unstable to oxidation by dioxygen just as the central iron core and the iron sulphides are

Table 15.3 Deposits from cycles after the advent of O_2

Deposit	Source
O_2 in atmosphere*	Biochemical action
$CaCO_3$	CO_2 of atmosphere
Fe_2O_3	Oxidation of Fe^{2+} by O_2
N_2 in atmosphere	Oxidation of NH_3
Sulphate ores	Oxidation of sulphides
Carbon oil and C_nH_{2n} gases*	Biological action

* Energised chemical system.

unstable. Thus the cycle of non-metals now has long-term (stored), slow-cycling, non-metal chemicals. We see that the evolution of life made cycling increasingly more complicated. Of course evolution is not compatible with fixed cycles and the development of man's industry further illustrates this point.

15.15 Man's input to cycles

We have described two types of cycle in Fig. 15.1 and Section 15.2. In the first there is no change since it involves the conversion of input light to output heat,

† As we have seen, atoms, molecules and crystals are formed according to well established physical and chemical laws. At the surface of the Earth, however, the crystaline structures suffer the action of the sun and other cosmic radiations, as well as that of other agents, for example, heat, water, microorganisms, etc. and energised species are formed, such as colloids and micelles. These processes are, obviously, not uniform, and depend on the region of the Earth in which they take place. The nature of the species that form the soils also varies depending on the climate, nature and activity of living species, etc. In some ways the nature of soils is as dependent on the biosphere as the biosphere is dependent on the nature of soils—there is a biunivocal relationship.

while flow of materials and conditions, temperature and pressure, are all unchanged. This cycle is as permanent as the sun and depends upon the degradation of energy, that is,

$$\text{Light energy in} \longrightarrow \text{heat energy out.}$$

It is a precise balance in this process that regulates Earth's surface temperature. It fails if the sun fails, obviously, but it also fails if either more light is absorbed rather than reflected or if more heat is retained rather than allowed to escape. Unwittingly at first, man has discovered that a further source of energy lies in the combustion of chemicals which arose as a part of stored (waste) products of earlier energised lies cycles. The major sources are oil, coal and gas. Man uses this material in a way that is part of the general C/H/O cycles, but he has catalysed one part, a slow cycle, not all of it, so that the concentration of some members with reduced carbon diminishes while those of oxidised carbon increases. (This is the effect of fire and presumably the reason why the Chinese included wood in their elements! Wood was the parent of fire but it cycles quickly.†)

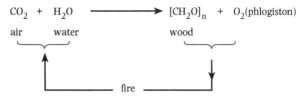

More recently man has discovered that energy (fire) can be used generally to obtain many new chemicals

$$\text{Earth} + \text{energy} \longrightarrow \text{reduced elements (metals)},$$
$$\text{Earth} + \text{energy} \longrightarrow \text{dehydrated oxides (cements)}.$$

Hence, the way energy goes into the chemical elements and their compounds has changed in the last 150 of so years (Fig. 14.7). These new materials may enter previous cycles, may become simply new stores or they may stimulate reactions and so perturb the existing cycles. Many can force materials into fast cycles, for example fluorochlorocarbons, or introduce perturbation of solutions almost accidently or at will.

Man now faces a problem since he introduces elements into new kinetic paths without any reason to suppose that these elements can cycle at a rate that does not force them to interfere with previously established cycles—Table 15.4 gives a list of abundant and available elements that have been affected.

The trace element's activities (Table 15.4 and Fig. 15.15) are most difficult to analyse, and, since the man-made products are not in equilibrium with insoluble salts or complexes of existing trace elements, there are unknowable

---

† 'The reason why *wood* gives birth to fire is that the nature of wood is mild and warmth lies concealed within it. The reason why *fire* gives birth to earth is that fire is hot and so is able to burn wood; when wood is burned it becomes ashes and these ashes are earth. The reason why *earth* gives birth to metal is that metal is found in the stones which rest on mountains, being a product of their fructification. The reason why *metal* gives birth to water is that the ether of metal moists and fructifies; furthermore, molten metal is also liquid and when mountains produce clouds they give their moisture to these clouds and precipitation follows. The reason why *water* gives birth to wood is that water moistens and is able to give life to growing plants.' Po Hu T'ung (in Proceedings of a meeting of Confucian scholars held at a place called the White Tiger Hall in AD 79, Chapter 9). Quoted in the *Wu-hsing Ta-yi* (*Great meaning of the five elements*).

Table 15.4 Man's effects on some elements

Element	Compounds and reactions	Effects (deleterious)
F	Chlorofluorocarbons (freons, halons)	Ozone hole
C	$C/H \rightarrow CO_2$, CH_4	Greenhouse effect
N	$N_2 \rightarrow NO_2^- \rightarrow NO_3^- \rightarrow N_2O$	Mutations; acid rain; ozone hole
O	$O_3 \rightarrow O_2$	Increased oxidising power; acid rain ($SO_2 \rightarrow SO_3$)
Al	Aluminosilicates $\rightarrow Al^{3+}$ (aqueous)	Acid water. Toxicity for fish and other animals
P	Fertilisers; detergents $\rightarrow H_2PO_4^-$; pesticides	Eutrophication
S	Sulphides $\rightarrow SO_2$	Acid rain
Cl	Organic chlorine compounds; insecticides	Cl free radical; deleterious for cell membranes
Trace elements	Discharge in waters; dumping of waste on soils	Pollution
Heavy elements	Radioactive isotopes Sr, Cs, Pu, etc.	Mutations
	Heavy elements—Cd, Pb, Hg	Toxicity

accumulating hazards for existing cycles. As stated in Chapter 14, many of these elements are present as traces, but cycles of major elements are based on them since trace elements are frequently catalytic. We cannot say in every case that these loadings into the environment do good or harm, but it is clear that they will change cycles, including those of life. Some of them, as we have seen in Chapter 14 and see Table 15.4, are deleterious to many forms of life, man included, particularly because they bind preferentially to thiol groups, form stable covalent bonds and are not easily removed; hence they accumulate and may reach toxic levels by blocking essential metabolic pathways. The cases of lead, cadmium and, especially, mercury are the most commonly quoted. We discuss the problem further in the last chapter where we re-examine the whole problem of the availability of elements and their natural selection.

Further reading

The subjects discussed in this chapter can be found in many excellent books and reviews. The following contain very readable accounts.

1. Butcher, S. S., Charlson, R. J., Orians, G. H., and Wolfe, G. V. (ed.) (1992). *Global biogeochemical cycles*. Academic Press, London. Chapters 9–16.
2. Press, F. and Siever, R. (1986). *Earth* (4th edn). W.H. Freeman & Co, New York. Chapters 1, 7 and 13.
3. Press, F. and Siever, R. (1994). *Understanding Earth*. W.H. Freeman & Co, New York.
4. Mason, B. and Moore, C. B. (1982). *Principles of geochemistry* (4th edn). J. Wiley & Sons, New York, Chapters 8–10 and 12.
5. Scientific Committee on Problems of the Environment (SCOPE) publications (1983–). J. Wiley, Chichester.

A recent appraisal of the chemical reactions occurring at deep sea vents, which are important from the points of view of the origin of life and composition of the sea water, is the following

6. Elderfield, H. and Rudnicki, M. (1992). Iron fountains in sea-beds. *New Sci.* **134**, 31–5.

The following are addressed to the general reader with no previous knowledge of chemistry, geochemistry and biology.

7. Gribbin, J. (1990). *Hothouse Earth—the greenhouse effect and Gaia*. Bantam Books, London.
8. Lovelock, J. E. (1979). *Gaia: a new look at life on Earth*. Oxford University Press, Oxford.
9. Lovelock, J. E. (1988). *The ages of Gaia*. Oxford University Press, Oxford.
10. Scientific American (1970). *The biosphere*. Scientific American Inc., New York.

16

The evolving natural selection of the chemical elements and the senses

Everyone gazes at the sunset
Everyone finds it awesomely beautiful
Wherein does that beauty lie
No one knows, no one will ever say.
    Anton Chekhov (1860–1904) *The beauties*, short story (1888).

We arrogate too much to ourselves
if we suppose that care of us
is the adequate work of God, an
end beyond which the divine wisdom
and power does not extend.
 Galileo Galilei (1524–1642) *Dialogues concerning two new sciences IV*.

16.1 Introduction: summary of previous chapters

In this book we have analysed the natural selection of the chemical elements in the three physical states of matter and in particular chemical combinations in an effort to understand all the material objects we see around us. By *natural* selection we mean selection based upon different driving influences, all ultimately due to atomic structure, which fall under four headings:

1. Selection through the *equilibrium distribution* of atoms between given states of matter and in chemical combinations, that is, due to the maximum affinities between elements and generated combinations of elements in molecules and in other assemblies (see Table 16.1 and Chapters 2–6 and 8).

2. Selection through *kinetic trapping* of atoms, due to an inability to reach equilibrium in states unstable in the terms of (1) above but unable to change due to physical (phase boundary) or chemical (bond) barriers at the given temperature and pressure (Table 16.2). This condition arises through the kinetic stability of energised or excited states of atoms in phases (compartments) or in chemical combinations (components) (Chapter 6–9).

These two selection modes are treated as independent of time and do not require continuous input of energy or material. They can be described by order $\rightleftharpoons$ disorder equilibria locally or generally. In Chapter 8 we analysed in these ways, and to a large degree successfully, the major selections of the elements in compounds and *condensed phases* of the Earth, formed on cooling

Table 16.1 Some chemicals in or close to phase equilibrium

Chemical	Phases
Dinitrogen	Atmosphere/ocean
Noble gases	Atmosphere/ocean
Water*	Liquid/vapour
Hydrated silica	Ocean/mineral
Calcium carbonate	Minerals and in the ocean
Many silicates	Minerals (major rock formations)
Iron hydroxide	In the ocean and precipitated
Salt (NaCl)	Rock/ocean

*See Table 16.2.

Table 16.2 Some chemicals in stationary kinetic traps*

Chemical	Phases (compartments)
Carbon (coal)	Condensed solids versus oxygen (carbon dioxide) in the atmosphere
Dioxygen	Atmosphere versus fossil fuels
Dinitrogen	Atmosphere versus dioxygen
Iron	Earth's central core versus oceans
Water (ocean)	Oceans versus iron core

* Chemicals in very slow turnover are treated as being in stationary conditions where 'slow' refers to a lifetime considerably longer than that of an organism such as man.

the original elemental gases of the solar system. We also described in Chapter 9 the nature of molecular organic compounds, as isolated from living systems or made in energised syntheses by man under selection control (2).

3. The third selection is for *functional value* in a (living) organisation. Here the principles of element selection are very different and are Darwinian to a large degree, in the sense of survival of the fittest species, but are equally applicable to a given organism. Fitness of elements is dependent only on atomic structure and is therefore natural. Each element has developed value in a flowing *organised chemical system*, but its supply must be assured, that is, the element must be always made available since an organised activity is one of constant flow with losses. The element must also be incorporated into combinations so as to secure its own functional fitness within the fitness of the whole organisation where the organisms of life provide the best examples. This ever-evolving selection, though based on chemical principles in part, that is, on (1) and (2) above, also requires a constant source of energy as well as of material. It is a co-operative steady, not a stationary, condition. Today we might think that each organism has achieved an *optimal selection* of its particular elements within an overall dynamic activity opposite the conditions which have evolved locally with time on Earth. Such a selection mode was described in Chapters 10–13 and examples with an abistic parallel are given in Table 16.3.

4. Finally, there is *man's selection of chemical elements for his own purposeful functional values* (Table 16.4). This selection is fundamentally different from that in point (3) in that it requires a large application of energy resources, which are limited, and at the present time it has no holistic control. Unlike the biological activities referred to in (3), which reached a steady state a long while ago, man's industry has developed very quickly over 5000 years but man has not come to terms with his ability to make (accidental) changes to his own environment. Again man often attempts to make stationary states of materials, not steady states, while restructuring his

Table 16.3 Some chemicals in rapid flow

Chemical	Phase (compartment)
Some water (rain)	Atmosphere (clouds)/rivers
Small amounts of many elements (examples: Na^+/K^+; Mg^{2+}/Ca^{2+}; Cu and Fe)	Aqueous and lipid phases of living organisms; nerve solutions; muscle solutions; membranes (electronic conduction)

Table 16.4 Some man-made chemicals in stationary constructs

Chemical	Compartment
Steel (Fe)	Building frameworks
Concrete (Si, Al, Ca)	Roads, buildings
Plastics (C)	Utensils, insulators
Light alloys (Mg, Al)	Building frameworks
Glass (SiO_2)	Vessels, windows

surroundings and machinery in order to develop new organisation with chemicals different from those used in (3). However, it is usually found that his materials have, in fact, relatively short lifetimes. At the moment man's management is slight and, correspondingly, risk of instability is considerable and unknowable. Even so the selections made by man are still based on the electronic structures of the elements and in this sense are a natural consequence of basic chemistry (see Chapter 14 and Section 16.14).

Given this outline of what the book has described so far we now wish to look forward to the long-term outcome of man's activities based on certain assumptions. To appreciate the future we are required to go back to basic principles since we have to understand limitations on degrees of freedom for living systems as we did for equilibrium systems in Chapters 4–6 (see Tables 16.5–16.7). To approach the problem we need to ask what are the chemical constraints that have brought about separate species, such as man, just as we asked what are the constraints that produced phases in the phase diagrams for materials given in Chapter 4. Finally, we ask what is a desirable objective for man in terms of his selection of chemical elements, which will maintain a secure long-term steady state, given his human limitations.

Table 16.5 Progression of the natural selection of the elements

Nuclear synthesis ↓	Selection on basis of nuclear forces and kinetics
Atoms ↓	Formed by electrostatic interactions (Chapter 1)
Molecular synthesis ↓	Selection on basis of abundance and electrostatics (thermodynamics) (Chapters 2 and 3)
Condensed states ↓	Selection on basis of electrostatics and barriers to reactions on cooling; *co-operative interaction* (thermodynamics largely) (Chapters 3–6)
Dissipative systems in fields of condensed states ↓	Selection on availability and kinetic trapping, gravitational separation (Chapters 6–9)
Living systems! early prokaryotes, late multicellular species ↓	Selection on basis of kinetic co-operative activity feedback relationships functional value toward *survival* (Chapters 10–13)
Man-made systems	Selection on basis of 'understanding'; deliberate and purposeful; functional value toward survival, protection, comfort and pleasure (Chapter 14)

16.2 The bases of the initial chemical selection

We take the formation of the elements in their presently observed abundances in the universe as a fixed unchanging situation existing from before the birth of Earth, even though this is a kinetic condition (see point (2) in Section 16.1), not an equilibrium. Starting from this mixture at high temperature in the solar gases there was produced on slow cooling a pattern of small molecules and then of co-operative condensates at or close to equilibrium as described in point (1) in Section 16.1. There resulted a very considerable fractionation of elements in compounds, governed by bond energy restrictions (Table 16.6). Earth formed from these condensates by accretion (Chapter 8). We could have predicted the outcome of this chemistry on Earth if cooling had produced a final equilibrium situation but it did not. We observe around us, especially in Earth's crust, a non-equilibrated set of materials that arose because the later stages of cooling were sufficiently rapid so that neither physical phase (transfer) nor chemical combination (atom exchange) equilibria could come about. In Chapter 6 we analysed such cooled non-equilibrium systems of chemicals not in terms of an overall equilibrium of phases and elements in compounds but in terms of local equilibria in compartments and of elements in components, which gave rise to a modified set of restrictions (Table 16.7). We must remember that the definitions of compartments and components are only useful in an *operational* sense since both transfer and exchange rates are adjustable by carriers and catalysts and by temperature and pressure. Again a component (a non-exchanging chemical) may or may not be in transfer equilibrium between compartments. Despite these complications we could explain broadly, given the ambient temperature, why certain elements were selected in certain physical states, solid, liquid or gas, and in certain chemical combinations, components, on Earth. The selection was seen to be a compromise between equilibrium (1) and frozen stationary systems (2) (cf. Section 16.1). While making this examination the analysis was extended in Chapters 4 and 6 to cover situations in which there were different relative amounts of two or more elements in different places. This absence of equilibrium mixing in different localities still allowed us to treat equilibrium *locally*. To do so we examined the balanced state of phases of different

Table 16.6 Restrictions and variables in an equilibrium system without boundaries at fixed temperature and pressure

Equilibrium restrictions	Chemical exchange	Physical transfer
Variables	Components $(C)^*$	Phases (P)
Example (273 K)	H_2, O_2, H_2O	Gas $\rightleftarrows$ Liquid $\rightleftarrows$ Solid

Variance (degrees of freedom) $F = C - P \ (+2)$.
* Components are those chemicals that do not exchange atoms with other chemicals at equilibrium.

Table 16.7 Restrictions and variables in a system of limited amount at fixed temperature and pressure and in local equilibrium

Equilibrium restrictions	Chemical exchange	Physical transfer	Other
Variables	Components	Phases (compartments)	Amount* (fields)
(Example	*Small* crystals* separated from solution in Earth's gravitational field)		

Variance (degrees of freedom) $F = C + P \ (+2) + 1$ (amount) $+ N$ (fields).
* At equilibrium the shape of the phase(s) is fixed.

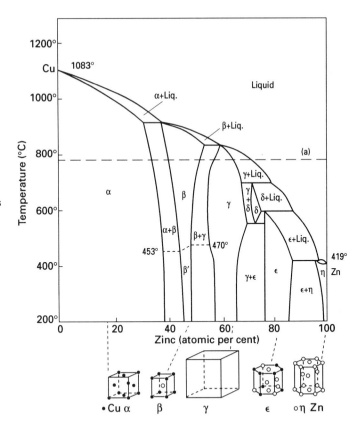

Fig. 16.1 The phase diagram for Cu/Zn alloys showing the structures of the phases (see Section 4.6). Line (a) shows the temperature at which Fig. 16.2 was derived, $T = 780$ °C.

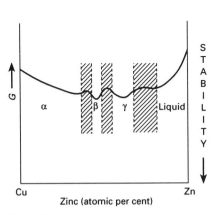

Fig. 16.2 A free energy versus composition diagram for the Cu/Zn alloys at 780°C. The shaded zones split into phases at their extremes due to lattice co-operativities.

compositions and at different temperatures using phase diagrams that revealed why the Earth contained such a complicated set of minerals (Section 4.4.2–3). We applied the (modified) phase-rule considerations also to the formation of alloys and salt solutions noting that a variety of compositions leads to a variety of differently structured phases. (In fact, because the composition across a phase boundary, say, liquid/vapour, is not identical in the two phases even at equilibrium (Fig. 4.4), cooling of a mixture from a gas to a liquid and then to a solid automatically and continuously alters the composition of the distinct phases.) While doing so we observed that in the solid state the shape of the materials (in crystals) also changed in a predetermined manner with composition. As an example, we examined changes of composition of Zn/Cu alloys at fixed temperature (Fig. 4.15), which is reproduced in Fig. 16.1. The figure shows that, at equilibrium at a given temperature, there is also a discontinuous set of phases depending on composition. We analysed the absence of continuity in terms of the relative free (binding) energies (Cu/Zn) in *co-operatively structured* lattices (Section 4.6 and Fig. 16.2). In Fig. 16.1 we show the different lattice structures across the Cu/Zn phase diagram indicating that, as we vary composition and now for *limited amounts of material*, we also observe *discontinuities of shape* enforced by the chemical and physical equilibrium conditions, giving rise to *coexisting objects*.

For the benefit of the reader who may have forgotten the discussion of equilibria given in Chapters 3–6 we next summarise generally the *numerical* constraints on degrees of freedom (variance) in phases in transfer equilibrium, which have delocalised concentrations of components in any one phase and where components (not elements) are in balanced association with one another. A reader who is familiar with the discussion should proceed directly to Section 16.3, remembering that the constraints are thermodynamic and strict.

16.2.1 Summary of variance at equilibrium

This section is added as a reminder of the equilibrium restrictions of phases and components, which we established in Chapter 4 (see Table 16.6) and which were used again in Chapter 6. The equilibrium state of phases, P, is governed by the simple algebraic (numerical) phase rule

$$P + F = C + 2$$

where boundaries of materials and fields are not considered. The equation derives from the condition that all components, C, are in equilibrium across phase boundaries. Thus one limitation on variance or the degrees of freedom (F) was the number of transfer equilibria

$$A_I \rightleftharpoons A_{II}$$

where A is a component that transfers between two phases I and II. In systems where components interact to give new phases, each new phase then gives rise to a further unit reduction in the degrees of freedom, variance. Thus variance was reduced by the presence of condensed phases from vapours or precipitates from solutions. In addition, the presence of equilibrium binding in the gas phase or the solution gave rise to new substances in a phase. However, any such *equilibrium association* represented by the equations

$$A + A \rightleftharpoons A_2$$

or

$$A + B \rightleftharpoons AB$$

does not increase variance since the concentration of the associated state was defined by the concentration of components A and B.

In Chapter 6 we expanded the discussion to limited volumes of material that had surface and therefore equilibrium shapes. In the cases of more than one separate phase, where there was no exchange, and which we described as compartments, we considered also that this separation generated fields between the compartments. The variance then increased in the description of sets of compartments with their components (Table 16.7).

Using the above approach we could understand the allowed coexistence of numbers of components in phases belonging to the stationary systems (1) and (2) of Section 16.1 and it allowed us to understand how invariant systems of phases, depicted in the phase diagrams, were restricted. This treatment does not describe why, in particular cases, some combinations of specified elements are found in a certain physical state nor why certain selected combinations of elements give a continuous or wide range of composition in any phase, for example, some alloys, while other combinations give rise to stoichiometric compounds, for example, organic molecules.

In the above we have described the case of the formation of a stoichiometric compound AB, but, and still at equilibrium, the new phase that formed could have had a variable composition. In this second case the relative supply of A and B decides the composition of AB but over a range that can be of any extent (Figs 4.6–4.18).

The nature of any phase that forms is dependent upon the co-operativity of interaction between units. In an ideal gas there is no co-operativity and order cannot be generated at any temperature. Co-operativity arises from charge–charge (nuclei/electron) interactions between large numbers of free units or between large numbers of different units of a chain polymer and it can be isotropic or anisotropic depending upon the charge distribution in the units. We have explained the formation of elementary metals, of salts and of molecular condensates in these terms in Chapter 2, and of polymer folding and polymer association in Chapter 6. The selective formation of gases, liquids or solids was shown to be related to the atomic properties of elements against the background of co-operative binding. Thus the physical state of a particular unit at ambient temperature, 300 K, is a property of the atoms in the unit. We next asked why, on mixing two types of unit, A and B, taken from any one of the three physical states, the mixtures form new phases of either stoichiometric or non-stoichiometric and usually of a wide range of continuously variable composition. When we examined condensates of A + B mixtures we observe that the greater the interaction energy between two elements in A—B as opposed to in A—A or in B—B as the ratio A:B is changed and following the free energy per mole, the less wide the A—B composition range (Figs 4.27, 4.29 and 16.2). For example NaI, formed from Na metal and molecular solid I_2, is very nearly stoichiometric, while alloys of Ni and Fe form a wide continuous range of non-stoichiometric composition. We concluded that the stronger the internal *co-operativity* of a composition A—B relative to A—A or B—B, the less extensive the regions of composition over which one phase structure can be maintained.

We see therefore how an understanding especially of specific co-operative binding energy, giving rise to condensed states, together with an understanding of the constraints on exchange and transfer equilibria can lead us to a quantitative appreciation of the way in which a system of components and phases could lead on cooling to what we observe around us in terms of specific elements in coexistant-shaped condensates.

16.3 Co-operativity and selected phase formation on Earth

We remind ourselves now of why particular condensed phases formed on Earth (Fig. 16.3) and have remained in their present stoichiometric or non-stoichiometric condition for almost 5×10^9 years. The discussion applies particularly to the Earth's crust and the first necessary point of concern is its temperature. As stated above, the initial condition of a high temperature of the solar nebula gave an atomic gas, which was followed on cooling by the formation of molecules and then co-operative condensates. Eventually some 4×10^9 years ago the energy flow on the Earth came into balance at about 300 K. The flow is of incoming high-energy quanta (light) and outgoing low-energy quanta (heat). Thus the process remains one of increasing entropy, while the retention of some of light's energy in chemicals created life within this 300 K steady state of flow. Figure 16.4, a more complete version of Fig. 1.21, shows all the processes from the big bang to today. (In this figure all processes of formation of condensates increase the entropy of radiation as energy quanta are released.) Now the rate of energy supply to Earth is fixed by the constancy of the sun as a source of energy and the rate of the energy loss has been fixed by the gases of the atmosphere, mostly by CO_2 and H_2O. Thus the liquid water/water vapour equilibrium largely governs the energy loss by feedback. Earth has been a thermostat for more than 4×10^9 years and any new energy input without changing losses will cause heating to a new

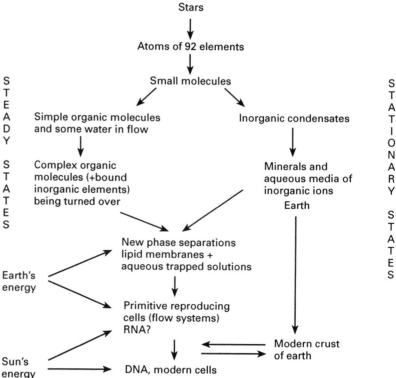

Fig. 16.3 A sequence of chemical developments leading to life in cells as outlined in Chapter 12. The designation 'steady' or 'stationary' state here distinguishes the fast cycles of life associated with mainly organic or living objects from the very slow mineral cycles which on life's time scale are stationary.

balanced temperature (see Figs 1.16 and 1.18). The Earth's temperature is an example of a non-living steady state of flow (see point (3) in Section 16.1).

In the initial cooling to this steady state, selected order⇌disorder balance in materials was created. In those materials with strong atom–atom, A—A, B—B or A—B co-operative interactions, that is, over a long range, separated phases formed. Thus we find minerals from metals with non-metals, often stoichiometric, metals alone, and alloys from metals, often non-stoichiometric. These are all combinations from atomic elements with, on average, an equivalence or an excess of the number of empty electronic orbitals over the number of binding electrons per atom, for example the Fe/Ni core of Earth and silicates. (Note that Ni—Ni and Fe—Fe are similarly cooperative to Ni—Fe, while the metal oxides, MgO, Fe_2O_3 and Al_2O_3 and so on, are very different from SiO_2 and form with it a more strongly cooperative phase, silicates.) At the same time elements and combinations of elements with excess electrons over orbitals, that is, non-metals, and their combinations remained very largely as disordered gases, N_2, Ar, CO_2 and so on. Through natural selection (as defined in Section 16.1) there was formed one molecule in this set, H_2O, with a large residual dipole. It has an intermediate co-operativity and it gave a liquid below 370 K, an intermediate order⇌disorder state. This liquid gave rise to selected solutions of variable composition in different localities. We see too that, due to the presence of fields from the condensates and due to limited amounts of phases, compartments, the particular solids that formed had shapes (at equilibrium). (The water flowed over the minerals as described in Chapter 8.)

The critical factor for the composition in the formation of Earth's coexistent phases at local equilibrium and at the temperature of 300 K was the curvature

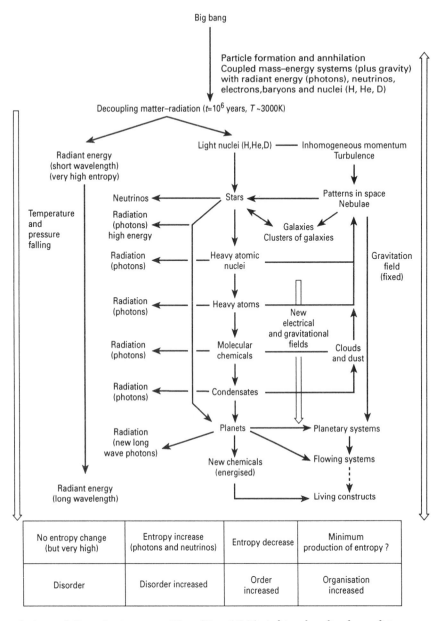

Fig. 16.4 The development of the natural selection of the chemical elements to give residual disordered, ordered and organised systems. Living constructs include objects made by man. This figure is a more complicated version of Fig. 1.21.

of plots of G against composition (Fig. 16.2), taking localised conditions as providing the elements in an equilibrated system (Fig. 16.3).

A very quick look at man's industry (Chapter 14) now shows that man's processes parallel the formation of the initial minerals of Earth except that man *selects purposefully the elements* that are to be *heated together and then cooled* in relation to the functional value of the product. Knowingly, man uses local equilibria at high temperature to make particular phases of a chosen composition which are then cooled rapidly to trap the phase structures. For example, man makes cements for concrete, iron for steel and so on through almost the entire set of the elements of the periodic table, creating *stationary* kinetically stabilised shaped structures at approximately 300 K. In the course of this construction man has introduced two novelties: (1) large (but limited) sources of energy are used outside those that were previously involved in developing the Earth's state; (2) certain new chemical elements enter into designed stationary trapped states. Of course, to some degree through waste they interact with the previously existing environment in all three possible situations described under points (1) to (3) in Section 16.1 (see Section 16.14).

We may conclude that combining the principles of thermodynamics and kinetics in a non-flowing system, that is, a stationary condition, with the principles of chemical (co-operative) bonding due to atomic properties, we can understand in a quantitative manner how the stationary features of all coexistant geological or man-made objects around us, including their shapes, have arisen. This understanding has allowed man to build his own structures in the way he wants so as to improve his lifestyle.

Before leaving the description of these stationary states we must observe that trapping in non-equilibrated compartments allows one compartment to do work on another (Section 3.6). This gives rise to the possibility of flowing systems. We now wish to go forward to an analysis of the third situation, (3), described in Section 16.1, that is, the case of steady-state flow. We picture the origin of these systems as in Fig. 16.3, which shows them to contain structure and many elements (15–20) all in a *dynamic* co-operative system. There is continuous flow because the out-of-equilibrium compartments now allow transfer of material and energy, which are constantly replenished. There are two abiotic and important groups of components in these steady states. The first is based on water in rivers and the sea, which is largely in a *physical steady state of flow between different states of matter, and, of course, it carries with it traces of many elements*. The second group of components comprises *abiotic organic molecules largely based on air and water in different phases which were and, to some degree, still are in both physical and chemical steady states since they flow and are constantly made and degraded*. We wish to consider the character of the living systems that evolved from these two flows but using selected chemicals and, in particular, we wish to discuss why coexisting inviolate biological species and not a continuous variation of living forms has resulted.

16.4 Steady states and their survival

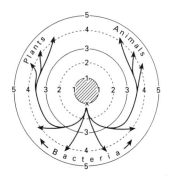

Fig. 16.5 A representation of evolution as the appearance of different shapes. The numbers 1 to 5 represent billions of years of Earth's existence. The ends of the lines shown as arrows continue to develop giving millions of species (see Fig. 10.2).

We saw in Chapters 4–6 and in the previous sections of this chapter that the appearance of phases within limited composition ranges, as in Fig. 16.2, was due to particular quantitative variations of free energy with composition. Is there an equivalence to the link between the free energy contributions to stability of certain compositions in phases and any link between the composition and emergence of living species? Emergence of a species implies survival strength as opposed to thermodynamic stability with respect to chemical composition.

First, observe that living systems are formed partly from co-operative condensed matter but are largely based on mobile and semimobile organisation, for example liquids or liquid crystal materials such as aqueous cytoplasm, lipid membranes and protein assemblies, not conventional solids. The fact that they are largely liquid makes transfer and exchange within structure possible. Structure is provided locally by somewhat elastic proteins in filamentous networks. The dynamics of the structured parts are also used to generate

Table 16.8 Connected mechanical parts of flow systems

Flow system	Mechanical parts	Engine or drive
Water mill (compare a dynamo)	Rotating wheel, rods and gears to grinding stones	River flow
Water pump (man)	Rods, valves	Petrol engine
(biology)	Proteins (helical)	Animal heart
Transporter (car)	Rods, valves, wheels	Petrol engine
(biology)	Proteins, vesicles	Muscles, filaments
Communication networks (man)	Electrons in wires	Battery
(biology)	Ions in channels	Gradients of chemicals

* Helical proteins especially are used to form rods and parts of pumps. Their value lies in their internal mechanics—that is, their ability to undergo second-order phase transitions.

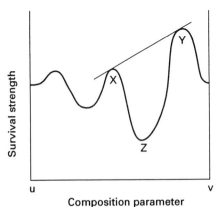

Fig. 16.6 Survival strength is plotted for systems that, because of their internal (hidden) systems, could exist at a given composition u/v. Consider a situation in which two organisms of composition X and Y grow in the same bath containing components u and v. As Y, the more favoured, grows it removes more of components v, and therefore pushes the composition of the bath so as to favour X. Organisms between X and Z and Y and Z cannot survive since the combination of X+Y always outperforms them. Such figures inevítably result if the interactions in the kinetic and thermodynamic schemes within life are differently co-operative for different ratios of components u and v. A plot of survival value against some property, here chemical composition, is often called a fitness landscape of species (see 'Further reading'). In life the number of elements is 15–20 and the resulting components, u and v, become very large indeed, so that the figure has a very large number of dimensions (see Section 4.4.4). Compare Fig. 16.2 where stability is plotted downwards.

vectorial flow in the liquids and energy must be supplied as in man's machinery (Table 16.8). The organisms are observed to maintain dynamic shape. The discussions of equilibrium systems in previous chapters showed that shape and chemical content are in fact inextricably linked (Chapter 6). Observing fixed shape in a dynamic living system we have therefore to consider its relationship with composition. Thus we must look for a parallel between the rules for the shape and phase discontinuities with composition, seen in equilibrium diagrams such as Fig. 16.1, and the shape and survival strength discontinuity of evolved biological species so familiar in diagrams of evolution (Fig. 16.5). Since the biological systems are all chemical, in effect we seek for the source of a plot of survival value against composition as shown in Fig. 16.6. How might such a plot arise?

We have seen in Chapters 2–6, using the arguments summarised above, that the materials around us were restricted by the number of chemical components, the temperature, the pressure, the fields and the total amount of material. The number of phases (compartments) and their shapes at local equilibrium were then defined by fixing the composition. These are the only possible contributors to the steady state also. Let us seek the parallel conditions for a steady state of flow of two elements in an organism,

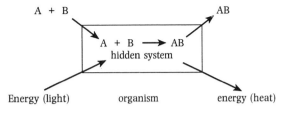

The external temperature and pressure are fixed (roughly, but see below) by the Earth. Clearly, the product and the organism are fixed only if the hidden system, the organism, controls the conversion of say A + B to AB, if the energy input (which will affect the temperature and pressure) is managed and if the relative amounts of A to B entering the organism are controlled. In the absence of any of the equilibrium constraints described in Section 16.3, the hidden system has to have a set of feedback controls on all the transfers and exchanges (Table 16.9). (Compare the thermostatic control of energy transfer giving rise to Earth's fixed temperature.) It can do this if it controls all the transfer *carriers* and exchange *catalysts* in the organism through feedback as described in

Table 16.9 Restrictions and variables for steady-state systems in flow at fixed temperature and pressure*

Variables	Restrictions
Rate of input (rate of output)	Feedbacks of all kinds on each of rate constants
Components (C)	Feedback on exchange
Compartments (P)	Feedback on transfer
Energy input	Feedback from transduced form
Energy transduction	Efficiency

* The products are the output chemicals and heat.

Sections 7.13–7.15 and in Chapters 10–13. Thus, while equilibrium conditions, represented by an equilibrium constant K, give constraints on a system of associating chemicals

$$A+B \underset{\longleftarrow}{\overset{K}{\longrightarrow}} AB$$

such that [AB] is fixed once [A] and [B] are fixed, the corresponding flow or steady-state system requires a larger set of controls, feedbacks, to fix and maintain the flow system, especially since [A] and [B] are not fixed in the environment or in food, and energy is required to be coupled to material flow in order to promote synthesis

It is necessary therefore to control also the rates of entry of A, B and energy, and the rates of exit of AB as well as the rates of conversion by specific enzymes, E, as in Fig. 16.7.

Referring to Fig. 16.7 and using the analyses in Chapters 10–13, we see that k_1, k_2 and k_5 are controlled by feedback using A, B and AB at membrane pumps, k_3 is a controlled conversion of energy, say to ATP by ATP feedback, k_4[E] is the rate of the enzyme-catalysed conversion of A+B to AB with feedback limitation due to [AB] in the hidden system and F is a feedback point. We must add a method of synthesis of the pumps, the container and the enzymes which takes us back to the relationship to DNA and for what materials DNA *must* code. Before turning to these considerations of coding, which are not absolutely necessary to sustain flow (consider a river), we wish to compare the above system with the slightly more complicated case of new co-operative phase formation in a steady state as compared with that for the equilibrium association of A+B, namely

$$A+B \underset{\longleftarrow}{\overset{}{\longrightarrow}} AB \underset{\longleftarrow}{\overset{}{\longrightarrow}} \text{precipitate (new phase)}$$

or, more generally, A+B⇋AB⇋new co-operative phase so that we can discuss the self-generation of structured compartments and transfer between compartments. The generation of a new phase limits or buffers a system at equilibrium. Similarly to the formation of such a new compartment, control by feedback of transfer of A and B, through carriers if necessary, will reduce the variance of a flow system since material in one phase is now controlled by the

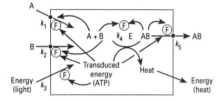

Fig. 16.7 The necessary system of feedback connections if a constant hidden system producing AB from A+B is to be maintained continuously. E, enzyme; F, feedback point. See text for explanation of k_1–k_5.

fixed flow. In effect, this is the property of a feedback pump. Biological systems, AB, can therefore give rise to a series of new compartments and, although they are not at equilibrium, the feedback, now on exchange between compartments, will have the same constraining effect as equilibrium. As we increase numbers of compartments so there are required more and more feedback connections using more and more carriers which move the components between compartments or within them (Chapter 11) so that each new compartment increases the number of systems, i.e. in composition, but the system is not increased in variance but more strongly restricted.

We begin to recognise that *a set of transfer and exchange reactions in a flow system that is controlled by feedback is reduced to a fixed system of chemicals and can generate a fixed set of compartments of given shapes, all of which are parallel to those of a chemical set of local equilibria that give rise to stationary structures.* The treatment of the degrees of freedom of such a system is not quite as restrictive as that which gave rise to the phase rule (see Tables 16.6 and 16.7), but the combination of components in flow in compartments under strict feedback controls of exchange and transfer similarly reduces the variance or degrees of freedom to certain coexisting objects.

Like a system in equilibrium, a constant flow can fail due to changes in conditions, for example of temperature, which will change all the relative k values. (Compare the change of K, equilibrium constants.) Thus maintenance of a well-controlled hidden system demands a temperature feedback control too. Now Figure 16.7 does not indicate how temperature can be controlled, or how energy can be transduced, or how the required pumps and catalysts of the pathway are made. In Chapters 11–13 we saw that, for a biological organism, these activities require some 15–20 elements so that as well as $A + B \rightarrow AB$ we need many additional reaction pathways involving all those elements.

16.5 Flow systems of many elements

Given the existing elements and given that there can be any combination of a set of available elements, that is, $A + B + C$, etc., the nature of the hidden system (organism) could be of virtually infinite variety, and continuous change in shape is then possible with change of composition. However, this is not what we observe since species are readily identified by shape, which we identify with fixed composition.

Our next step follows from Section 7.15 where we saw that the many flow paths of further elements C, D must be managed in an interactive set of controls with A, B, for example, in a coupling scheme,

In biology such sets of flows are cross-connected by feedback loops into a network (Section 7.13). This is a development of *dynamic co-operativity* between the chemical elements in different reaction paths. Just as the strength

of *static co-operativity* at equilibrium in A—B relative to A—A and B—B decides the degree to which AB is of variable composition or not, so here the strength of the feedback network system restricts the composition within a flow system. If CD is involved in the formation of new compartments and transfer is controlled by feedback as for AB, then the nature of AB relative to CD with any new compartments formed is limited to a narrow range. In a strongly co-operative network of flows of further elements and further compartments the inputs and the conversions are more strongly fixed and the hidden system and its products are then also fixed. Thus, the composition identifies the system and each species, especially an advanced species, only exists in a very limited composition range (Fig. 16.6). Again since composition and shape are connected we can deduce that species can only exist in a limited shape range. We are also led to see that the greater the feedback co-operativity in the compartments and between larger numbers of compartments the narrower the composition range over which one system can survive; the more complex the management, the less flexible the system.

This is referred to as general *homeostasis* and is seen in a full-grown organism. In conclusion, while the binding co-operativity gave rise to selected composition of components and together with the exchange and transfer equilibria gave us an understanding of stationary states, the feedback on rate constants together with the coupling of different flows gives us an understanding of speciation in flowing systems. *A code does not create this stability but it represents it and allows its reproduction.* A coded system may be flexible to some degree as in a bacteria or very constrained as in man.

16.6 Coded systems: DNA

Every reader will have noticed that we have hardly referred to DNA, the genes, in this description of the nature of the steady state of living systems. The reason is simple—DNA is no more nor less than a coded set of instructions that will see to it that a processing machinery of the above kind will *produce and reproduce efficiently* a certain desired product in given steady states. A pre-existing and desirable system of some kind of chemical turnover and feedback connections has to be available to receive the code and the production of this DNA had to be incorporated into this machinery. By studying the DNA of an organism we can discover the nature of the product species provided that we know the character of the necessary translating machine for synthesising desirable proteins in a required amount within a co-operative regulated system as described above. DNA or any other code is of no use in the absence of such a production machinery, for example in viruses. The combination of DNA and machinery will generate efficiently what is desirable in the sense that together they will survive more lastingly and can reproduce, but DNA is nothing more than a certain way of upgrading already devised machinery (see Section 16.12). We have to conclude that DNA only seemingly isolates a cell or species from DNA in another cell or species. The facts that (1) in a single-cell organism feedback is in balance with its environment, (2) in multicellular organisms differentiated cells are in balanced amounts and there are other organisms living within them in balanced amounts, that are all in balance with an internal environment and (3) today higher organisms require a balanced ecosystem, which means that the DNA of the different cells stands in a feedback relationship through metabolism with all other DNA in the ecosystem. The feedback is not unlike that of the kinetic network between compartments in cells or between organs. Thus DNA is not the arbiter of evolution but it functions to increase greatly survival of energised chemical selection in life and of course it gives a record of evolutionary changes.

The interest then lies in why a system survives since no matter what the nature of the code, DNA or any other we may find or invent, and no matter how it came to represent a survival system, it is the nature of the machinery, the survival system itself in particular forms, that is so intriguing. The exact nature of this machinery is a matter open for discussion, but as shown above it is certainly related to the elements available in a particular environment and the functions that they can be made to serve. Thus we become concerned with the environment but before we describe the connection with the environment we must summarise the best way in which to construct a system for survival in terms of the choice of chemical elements as we came to know the best ways of combining elements at equilibrium. It is the problem of fitness of elements in a self-assembling organisation that we tackle next.

16.7 Natural selection and element fitness

We have used the phrase, 'the natural selection of the chemical elements', in the sense that all selection seen in objects around us is ultimately based on atomic properties. Thus the abundance of elements is associated with the stability of their nuclei and the kinetic paths to and from their most stable nuclear conditions. Similarly, the combination of elements at thermodynamic equilibrium is based on the free energy of formation of compounds, which is itself dependent on atomic bonding capability. The properties that result from these combinations were in large part the determinants of the character of the Earth. The availability of 'inorganic' elements from the Earth is due to the ease of access, restricted by thermodynamic or kinetic considerations of atoms, from solid or especially liquid or gas phases. In the liquid phase, here largely water, the equilibrium association of the solubilised elements in complexes with one another is then a derived property of their most stable associated condition. Much of the composition of the sea is related to this condition. Most of the atmosphere is more closely a result of selection based on higher energy states of elements or their combinations, since, for example, dioxygen is in an energised trapped condition relative to the carbon compounds on the surface of the planet. The original abiotic H/C/N/O compounds of organic chemistry were equally in trapped energised conditions resulting from chemical barriers to change which are especially large for combinations of these elements. The principles underlying their behaviour were outlined in Chapters 2–7 and in our previous book where we have given reasons for the observed selection of combinations due to both the thermodynamic and kinetic properties of both organic and inorganic elements. These principles of selection refer to static concepts, whether the element is a stable condition of the lowest possible free energy or is in some permanently energised trapped condition. In total these distributions provided and provide today the components to which living steady states have access. The selections are not directly related to functional value but function is limited by them.

We saw in Section 16.6 that there is another possible natural selection which is based on the ability of an element to contribute to the survival of a whole flow system, an organisation. Here the element on its own is of little consequence since non-random flow demands at least four features: a structure to control flow; a liquid or gas for flow; a gradient or field to direct flow; and a supply of material and energy to maintain the whole system. (The reader should once again have in his mind a river.) There is a need for co-operative dynamic interaction of several elements (in compounds) in order that such a living system is made operative and, of course, it operates most effectively if each element is functionally of optimal value. We stress again that in life the functional materials are made from the co-operative combination of some 15–20 available elements which generate structure, allow flow, create fields and capture energy. In these activities different elements are selected on the basis of their atomic properties, but now all of them are not only combined suitably into stable or energised static states (see the discussion of the entatic state) but use is also made of their dynamic characteristics so that components can be constantly moved within the flowing system to occupy positions of advantage in (adjustable) space. Thus, the exchange rates as well as the binding constants are selectively used. Each element becomes associated with a 'most useful' rate of exchange together with its most useful bindings and positionings. This selection requires continuous controlled uptake of each element in appropriate proportions. Fitness is then connected to the flow of material and energy so as to maintain a specific co-operative steady state (or a developing system) opposite an environment. The evolved systems or organisms have been observed in Chapters 10–13 to optimalise the use of the elements in this way (limited by availability), for example in structure (very available H, C, N, O), in flow (very available H_2O), in messengers (less available Na^+, K^+, Cl^-, Mg^{2+}, Ca^{2+}) and in catalysts (not very available transition metals and zinc) (Table 16.10). Of course, some elements have different uses (see Chapters 10–13) in many components.

Table 16.10 The selection of elements by fitness

Element	Fitness factor	Use
Na^+, K^+, Cl^-	No binding, charge	Current carrier
Mg^{2+}, Ca^{2+}	Weak binding, charge	Fast trigger
Zn^{2+}	Moderate binding	Lewis acid catalyst, slow trigger
H, C, N, O	Covalent unstable binding, slow kinetics	Structure; code
S, P	Covalent unstable binding, faster kinetics	Transfer agents (structure); energy storage
Mn, Fe, Co, Ni, Cu, Mo	Strong binding, redox-active	Redox catalyst; slow-trigger
H	Redox-active, variable binding	Energy transduction
SiO_2, $CaCO_3$, Ca/PO_4	Insolubility	Inorganic structure

16.8 Optimal co-operativity in a steady state

As stated, co-operativity in an equilibrium chemical system is based on long-range forces. For example, a bilayer of lipid molecules in water is based on the thermodynamic co-operative interaction of H_2O molecules and of lipid molecules separately. Now, a compartmental biological structure is based on lipids, for membranes, and more generally on proteins, X, Y, etc. from which it is possible to generate co-operative structures first through folding and then of the X—X kind, for example in ferritin or in coats of viruses (Section 6.8), but it is also possible to build very complex X—Y—W—Z— and so on structures from individually different polymers. Here the thermodynamic free energy of co-operativity arises from the *matching* of many relatively short-range interactions on surfaces, say, first within X—Y and W—Z separately but then of X—Y with W—Z and so on. The building, self-assembly, of such a structure requires the *kinetically* controlled production of X, Y, W, Z, etc. in fixed ratios. An example is the combination of proteins and RNA or DNA in a virus. (Note that a virus is not living although it contains DNA and proteins since there is no co-operative flow of energy and material in it.) This is achieved through the feedback restriction of concentration levels of each X, Y, W, Z, etc. by their production unit (DNA+machinery as described in Chapter 11). It is in two senses, therefore, that we can say that self-assembly of structure is co-operative. It can be thermodynamically co-operative once X, Y, W, Z, etc. have been produced in the correct ratios but it is kinetically a co-operative system in the rate of production (and destruction) of these polymer molecules and their transfer to compartments. For this reason an organism is described as being a holistic machine. We have shown in Chapters 10–13 that such co-operativity in polymer production extends to the whole of the activity of a cell, being based also on the correct (feedback controlled) uptake and incorporation of some 15–20 chemical elements and the use of these elements in messenger systems, in control and regulation, to manage co-operative kinetics of the whole. It is the kinetics, now related to an integrated coded message, that ensures fixed flowing chemical composition, while the thermodynamics often ensures localised structure itself in self-assembly.

16.9 Summary of survival values

A living feedback flow system is now seen to be optimally operative in maintaining a steady state the better the organism matches:

(1) the supply of elements A, B, etc. due to availability, that is, by matching an environmental niche;

(2) the supply of energy, that is, again by matching an environmental niche;

and by having

(3) controlled production of four classes of polymers: lipids, proteins, nucleotides (including a code) and polysaccharides;

(4) a tightness of control over conditions of temperature (and pressure);

(5) a tightness of control over enzyme catalytic rates—not attempting to develop the fastest catalysis but the most appropriate:

(6) management of transfer, including uptake and rejection.

Elements are selected to these ends making use of their chemical bond selectivity in thermodynamic binding and in their kinetic properties, i.e. their lability. Therefore we see survival of a species as dependent on *co-operative* character related to element composition through feedback-governed flow much as co-operative contributions lead to stabilisation of phases through favourable free energy changes, ΔG. We can then plot Fig. 16.6 with confidence (compare Fig. 16.2), and we have a reason for the coexistence of species as opposed to continuous variation. We must still explain too how a series of organisms can evolve in complexity, consisting of components and compartments, to an ecosystem and how a complex organism can develop from a single cell. It is here that a code becomes an imperative necessity.

16.10 Changes in components and evolution

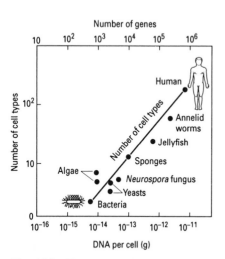

Fig. 16.8 The number of cell types in an organism seems to be related mathematically to the number of genes in the organism. In this diagram the number of genes is assumed to be proportional to the amount of DNA in a cell. The actual number of cell types in various organisms appears to rise accordingly as the amount of DNA increases. (After Kauffman, S. A. (1991). *Sci. Am.*, August 1991, pp. 64–70 (adapted).)

Once such a system settles into an *optimal* pattern of use of elements at fixed temperature and pressure, it can only develop through the involvement of new chemicals. (By definition, any mutation can only be damaging to an optimal organism.) We need then to ask how these new chemicals appear. We have first to explain diversity within evolutionary development in chemical terms. Just as in a diagram of phases versus composition the number of coexisting phases at equilibrium depends on the number of components so it is reasonable to expect that the number of organisms that survive within an availability pool of many elements (15–20) will be quite large. *They do not need to compete to elimination if they use element resources in different proportions* (see Fig. 16.6). Hence some considerable diversity will arise inevitably in any fixed environment. Let us consider that, for example, the earliest optimal pattern that evolved did not utilise some element, for example calcium, to any marked degree. Let us suppose further that calcium increased in the environment due to loss of CO_2. Amongst the organisms a branch might well develop by mutation that could make some use of calcium. In Fig. 16.6 it would occupy a new part of the composition landscape not overlapped by other organisms. Given what we know of the role of calcium, it might be that the use of this element led to development from a prokaryote ancestor (see Section 11.13) by assisting:

(1) increase in the number of compartments;

(2) increase in the internal framework;

(3) stabilisation of the membrane.

All of the above could have arisen through trial and error searching (see Section 16.6), which we suppose to be managed efficiently through chance mutations of the DNA code. These changes, together or separately, would inevitably allow the size of the organism to increase while generating eukaryote cells. By generating internal compartments, degradation (for energy and as a source of monomers) and synthesis (of polymers) could be separated using different enzymes and carriers, which has obvious advantages and, for example, could have allowed the nucleus and its activities to be well protected. Once the eukaryotes became considerably larger, they could consume or incorporate the smaller prokaryotes, later to become mitochondria and chloroplasts, within or as internal vesicles. Again the presence of internal vesicles within eukaryotes loaded with ejectable poisons or digestive proteins would not endanger the large host but, on discharge, would damage other organisms. There then arises the problem of the development of such complexity in larger eukaryotes accompanied not by the loss but by the persistence of very simple cells, bacteria. Why did both continue to exist? In fact, the growth of complexity as in Fig. 16.8 has never removed simple bacteria throughout evolution. In part such coexistence depends on living in different niches but this is far from the whole story.

16.11 Development rates and complexity

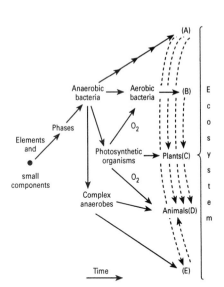

Fig. 16.9 The evolution of an ecosystem of mutually dependent species. The feedback relationship between species parallels and extends the internal chemical feedback within any one species described in this book.

Looking around ourselves at living organisms we observe that single-cell bacteria have lasted from 3.6×10^9 years ago until today and that, against the background of their simplicity, a range of complex multicompartmental organisms has appeared. What are the advantages of the simplicity of bacteria and of complexity in higher organisms? When we consider an individual bacterial cell in comparison with a larger complex anaerobe, still a single cell, then the larger cell must have a longer, that is, a disadvantageous, reproduction time, Section 11.12.2. Again, the greater the degree of complexity of the eukaryote, the more controlled the homeostasis must be so that bacteria are found to be the more adaptable in the way they manage element (or compound) supplies. As to the DNA of the eukaryotes it had to become a larger and more and more inflexible code (Fig. 16.8) the more syntheses it had to interdigitate, and once again this has a cost. Clearly, if the anaerobic eukaryote had remained vulnerable to bacteria to a greater degree than the bacteria to the anaerobes then the complex cell could not have survived. Two developments assisted the complex cell—increased protection and increased ability to poison and/or to consume bacteria. These are the advantages of complexity. We stated in Chapter 13 that, as complexity increased, it paid the complex organism to give up certain tasks, that is, to reduce complexity. The obvious chemical tasks for the eukaryote to reject are all syntheses leading to those materials fundamental to all life, so long as it could obtain their essential products from simpler organisms by digestion. The two lifestyles were bound therefore to come into balanced coexistence. It follows, as explained in Chapter 13, that ever-increasing complexity of organisms leads to ever-increasing dependence down a food chain which then becomes an ecosystem of ever-increasing diversity. Evolution had to be evolution to an ecosystem (Fig. 16.9) based on prokaryote organelles inside and on bacteria inside and outside higher cells, and it is so today.

16.12 DNA, development and evolution

Using the above reasoning we can see that different DNA came to represent different chemical systems and, given the nature of co-operative feedback, species could and did arise. (As discussed in the previous paragraphs, the need for machinery for supply of elements and for appropriate control and regulation remains essential.) Thus, different DNA at first would represent different organisms (single prokaryote cells), quite possibly in environments of different composition. Reproduction was then simple replication and division. Growth of complexity introduced a new problem in that reproduction could not be by such simple division which would involve simultaneous reproduction of all the sets of vesicles and filaments. Rather, an existing complex organism had to generate an initial simple daughter cell that developed, using an inherited timed programme, with multiple compartments and with change of shape. If we are to maintain our position that change of shape has to be due to a change in the chemical content then we should observe that continuous development is concomitant with a continuous change of chemical content, see Chapter 13. (We must insist that this timed sequence of element composition is inherited from the parent cell by the reproduction of a set of genes plus machinery in an isolated new compartment.) Thus, an egg, spore or other early form of a complex cellular system has to be chemically different from the adult cells. As far as we know this is true but it has been poorly investigated. This selection of elements, now on a continuously changing basis, is another and new example of selection for fitness of element function.

It follows from this description that it was the discovery of a combination of an inheritable *code* and *previously devised machinery* that was the major cause of initial evolutionary development. Even though improvements in basic machinery would force speciation there could have been no direct relationship between individuals or *inherited* advances without a code. It is also clear then that speciation rather than continuous variation is *stabilised* by a code although the drive to speciation exists in all *co-operative* organisation. The involvement of a code also allows an enforced development programme in complex organisms since *all activities may be based on a single set of instructions that can be activated in a timed sequence*, a property already known in the prokaryote cell cycle. It is this *timed division* programme which has been amplified to *timed production of differentiated* cells, again in controlled relative amounts, so as to make a co-operative multicellular organism. Here the self-generated environment in which series of cells are produced (within the organism) dominates organisation. This environment develops as cells surround one another (Fig. 13.31), making the code and equal not a dominant part of the organism.

In this description of development the search by the DNA for new forms is taken to be by random mutation so that the internal environment of the cell was changed in chemical content while individual vesicles each developed a specialised selected element composition. In Chapter 12 we showed that this is the case. There is no indication here that the environment had any influence on the species that evolved once the initial living cell had been created.

16.13 The environment and evolution

Now during the two billion years of anaerobic evolution the environment did in fact change (see Sections 8.5 and 8.6) in that much CO_2 was lost. Given an initial CO_2 pressure 10^3 times the present level, the slow change to today's pressure would have caused a growing presence of Ca^{2+} ions in the sea from 10^{-6} M toward 10^{-3} M. Can we be sure that it was not this increasing availability of calcium that not only allowed the development of many of the features of eukaryotes but actually stimulated it as described above? Nearly every filamentous and vesicle system of eukaryote cells is connected to calcium in one way or another, whereas the prokaryote cytoplasm does not appear to need it. It is an added-on system using added-on proteins, and it is tightly linked to the more primitive phosphate system (Section 13.3).

In the next phase of environmental change, due to the advent of dioxygen, the environment changed greatly as explained in Chapter 12 and together with these changes came multicellular organisms with a quite novel dependence upon copper and zinc chemistry especially, while the pre-existing chemistry of many other elements was somewhat modified. Again we must ask did the change in environment drive the evolution or was there just random searching?

To these questions of how organisms evolved there is a possible compromise solution to the alternatives of: (1) completely random mutational searching; and (2) environmentally driven change. Consider again the pattern in which novel elements, poisons, are introduced:

(1) a poison, of necessity, gives rise to protection from it;

(2) protection can give rise to functional use so as to increase survival.

We may imagine the process as follows. The protection against the poison involves at first greater production of the protein system most useful in handling this poison, no matter how feebly. If it is an inorganic poison, protection is obtained through sequestration and rejection (pumps), and if it is an organic poison through enzymic destruction. Thus, a poison increases the turnover of certain RNA (and proteins) and therefore of exposure of local regions of DNA. As the exposed, for some time single-stranded, DNA is more liable to mutation, then selection could drive improved protection by mutation within that short section of the code used for the pre-existing feebly protective system with little effect elsewhere on the DNA. The step from protection to functional use follows the same pattern since protection involves handling the elements and requires knowledge of poisons within the cell. The obvious step is to make the poison an extracellular messenger or part of an enzyme. The efficiency of the major evolutionary steps proposed lies in the fact that little searching occurs within the whole DNA of the centrally required primary metabolic systems. If we follow this line of discussion, we should find that there are relationships between proteins generated to handle new elements or compounds as they are introduced from the environment and those proteins previously used in binding similar elements. The parallels between zinc and copper protein chemistries are interesting in this respect. Again, the structures of pumps and channels for many elements are similar. Testing these possibilities which were developed in Chapter 13 is a task for the future. We turn to the probable effects of introducing new elements into the environment through man's industry. How do they affect the steady state?

16.14 Assumptions underlying the steady state

Now, in the above discussions there are a number of assumptions that allow the steady state to be maintained:

1. The supply of elements A and B and so on and of energy is unlimited. This is the problem of *resources* or availability. Table 16.10 indicates some biological requirements (see Chapters 10 and 12).

2. The waste products AB and so on do not accumulate so as to affect the hidden system. Sustainable systems are cyclic (Chapter 15). However, as explained in Chapters 12 and 13, biological systems themselves broke the rule when dioxygen was produced and the consequences are now clear. The evolution of living systems was generated in a search for a new cyclic steady state. Today man is breaking the rule again but the consequences are not yet clear. Waste will cause pollution and development.

3. The number of hidden systems of a given kind, that is its population, converting, for example, A + B to AB, does not increase inordinately since increase in numbers in a species will put a demand on the rate of use of A + B to make AB. This is the problem of a non-steady state of one *population* due to overprotection, which can threaten an ecosystem.

We will tackle these problems separately.

16.14.1 Resources

Resources are needed for two kinds of hidden system: (1) living organisms; and (2) man's industry. As shown throughout this book, under (1) there is a minimum demand for 15–20 elements by living systems, but under (2) a much larger range of elements is needed. The resources of living systems were the product of the original condition of the Earth and thereafter of the sequences of events from 4.5 to about 1×10^9 years ago during which life came into evolutionary balance with dioxygen (Table 16.11). Subsequently, change

Table 16.11 The evolution of chemical resources

	Years ago	Chemistry	Redox range (volts)
Sun	5.0×10^9	Nuclear + energy	
Earth	4.5×10^9	Mineral/organic?	0.0
Life	3.8×10^9	Reductive inorganic/organic	-0.4
	2.0×10^9	Oxidative (using O_2)	-0.4 to
	to today	+ reductive inorganic/organic	$+0.8$
Man	10×10^3	Mineral reductive (Cu,Sn,Zn)	
		Fire oxidative (C + O_2)	
	5×10^3	Iron reductive (Fe)	-0.5
	3×10^2	Oxidative/reductive organic	
	$<2 \times 10^2$	Mg, Al reductive	-1.0
	Today	Oxidative (Cl_2, NO_3^-)	$+1.2$
		Nuclear for energy	

appears to have slowed. The changes are described in Chapters 8 and 12. Most recently, the resources, available elements, are being changed again by man's industry (Tables 16.12–16.15).

In principle, the ultimate resources for man are the same as those of living systems, that is, the 90 elements of Earth, but man's processing, even of the 15–20 elements essential for life, is quite different. Living organisms are restricted to temperatures ranging from 0 to 100°C, and this low temperature and the manner of the use of light energy did not enable organisms to mine and then exploit very stable materials from the Earth. As explained in Chapter 14, man has achieved a much greater ability to exploit these materials by reverting to extremely high temperatures and to reducing atmospheres. At tempera-

Table 16.12 Global problems

Problem	Source (agent)	Effects
Greenhouse effect	Carbon dioxide, carbon monoxide, methane, water vapour, clorofluoro-carbons, etc.	Global warming and associated effects
Acid rain	Sulphur dioxide, nitrogen oxide, carbon dioxide, dimethyl sulphide (after oxidation)	Acidification of rivers and lakes; solubilisation of rocks and minerals (e.g. with aluminium); destruction of vegetation
Ozone hole (in the stratosphere)	Atomic chlorine derived from chlorofluorocarbons (freons), methyl chloride and other products	Skin diseases; possibility of increased incidence of skin cancer
Photochemical smog	Carbon monoxide and nitrogen oxides (from exhausts), and hydrocarbons (terpenes)	Respiratory affections and diseases
Radioactivity	Nuclear power stations (release of radioactive isotopes)	Mutations (cancer)

Table 16.13 Atmospheric pollution (local and regional)

Pollutant	Source	Effect
Sulphur dioxide (SO_2)	Burning of fossil fuels	Respiratory diseases; acid rain, release of aluminium
Hydrogen fluoride (HF)	Glass factories	Destruction of grazing fields and pinewoods
Hydrogen sulphide (H_2S); ammonia (NH_3); methane (CH_4)	Bacterial activity	Poisons
Nitrogen oxides (NO_x)	Burning of fuels and of vegetation	Photochemical smog
Lead tetraethyl ($Pb(CH_3)_4$)	Combustion of petrol (exhaust)	Destruction of vegetation (in zones close to transport systems)
Dust and smoke (with heavy metals)	Industry: traffic	Respiratory diseases; toxicity of heavy metals

Table 16.14 Aquatic pollution

Source	Nature	Effects
Domestic sewage	Excrements; detergents; soap; paper	Pathogenic micro-organisms, skin and intestinal infections
Industrial effluents	Oxidisable organic materials; foams; heavy metals	Toxicity for plants and animals; deoxygenation of rivers, lakes and estuaries; death of fish and vegetation
Agriculture and cattle breeding effluents (rain and irrigation water)	Manure; fertilisers (nitrate and phosphate); insecticides, pesticides, herbicides, etc.	Toxicity for fish and vegetation; eutrophication
Transport of crude oil (in the sea)	Heavy hydrocarbons	'Black tides'; death of birds and sea animals
Storage of toxic and radioactive wastes	Variable	Risks of toxicity and radioactivity at dangerous levels

Table 16.15 Soil pollution

Source	Nature	Effects
Fungicides, herbicides, pesticides and insecticides	Especially chlorinated and phosphorylated compounds	Poisoning of animals; metabolic deficiencies (e.g. absence of eggs, too thin egg shells, dead embryos, etc.)
General household and industrial rubbish	Variable including trace metals	Crop limitations

tures in the range 1000–2000°C man is able to activate the reactions of many minerals including silicates and metal ores. For example the metal ores can then be reduced to metals. As a consequence, living organisms have an available resource of some 20–25 elements in limited components, while man can obtain all 90 in an almost limitless variety of components. (Note that in evolution biological systems switched from reducing to oxidising conditions, while much of man's activities reverts to reduction stronger even than that available at the earliest times (Table 16.11).)

16.14.2 General considerations concerning waste products

We have already mentioned the first major occurrence of pollution—the emergence of an oxygenated atmosphere due to the abnormal (for the time) production of increasing amounts of dioxygen by cyanobacteria which resorted to H_2O instead of H_2S as a terminal source of reducing equivalents (hydrogen atoms). Dioxygen was then a waste product and became a poison to the degree that many living species were excluded from large regions of Earth. The ones that survived were those that could hide in anaerobic niches—today the bottom of lakes, polluted waters or the digestive system of ruminants, for example—or learned how to use oxygen in energy-producing oxidation reactions. We feel that the advent of dioxygen was a major driving force for natural evolution through the liberation by dioxygen of many other poisonous inorganic chemicals (Chapter 12), and the subsequent synthesis of organic poisons. The utilisation of energy was changed too through the storage of O_2 in the atmosphere and reduced carbon dioxide in plants.

Another ancient (and also recent) form of natural pollution is that due to the volcanic eruptions which contribute high amounts of, for example, SO_2, as well as ashes and dust in such a quantity that they may well hinder the passage of sunlight and so decrease the Earth's surface temperature to a considerable extent even in regions far away from the active volcano.

Man's activities in industry and farming are, however, those that are now giving more reasons for concern (Tables 16.12–16.15). Generally, we may say that the problems can be examined at two levels—local and global (some being of intermediate character, that is, regional (see Tables 16.12–16.15)). Naturally, the global problems (Table 16.12) receive wider attention and in most respects are more acute since they carry higher and more extensive risks for the extant ecosystems.

We comment briefly on the global problems—greenhouse effect, acid rain, ozone hole in the stratosphere, photochemical smog, halogen compounds, radioactivity, and trace elements—to show that some will cause considerable impact on the cycles of the elements in the next 100 years.

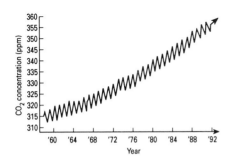

Fig. 16.10 The carbon-dioxide content of the atmosphere. Measurements of carbon dioxide (in parts per million, ppm) taken at the National Oceanic and Atmospheric Administration's station on Hawaii show that concentrations of this gas have increased by 10 per cent since 1958, when the observations started. The saw-toothed pattern is probably due to the absorption of carbon dioxide by plants during the spring and summer.

Greenhouse effect : loss of temperature homeostasis

The so-called greenhouse gases (which absorb part of the sun's radiation re-emitted by the Earth and cause an increase in the average atmospheric temperature (greenhouse effect see Fig. 15.11)) are carbon dioxide, CO_2, and carbon monoxide, CO, but also methane, CH_4, water vapour, H_2O, and others such as chlorofluorocarbons, CFCs (see below 'The ozone content of the atmosphere'). Usually, when this effect is mentioned one thinks of the increase of the contents of CO_2 in the atmosphere due to the burning of carbon-containing fuels, for example, wood, oil and gasoline, natural gas and coal, which, as we have seen, are in an out-of-equilibrium state. There are, in fact, many other contributing factors and here we refer to the texts given in the 'Further reading'.

The concentration of CO_2 in the atmosphere (Fig. 16.10) has increased from 295 p.p.m. in 1940 to about 355 p.p.m. in 1994 (the excess of production over consumption of CO_2 by green plants and algae, through photosynthesis, is of the order of 5×10^9 tonnes per year, the dissolution in the oceans being taken into account). At present, we cannot say that this increase in the concentration of CO_2 represents a *real* risk for animal or plant life (it may even have increased photosynthesis, especially in the tropical regions) but the *rate* of the process gives reasons for concern since estimates point to an increase of $0.5°C$ by the year 2005 and of $1°C$ by the year 2030 in the average temperature of the atmosphere. The effects to be expected from such a global warming are well known—warming of the sea, melting of polar ice caps, increase of the volume of the oceans, submersion of lower lands, etc.—but so far the results are controversial and no rise of average temperature is definitely proved, that is, there is not yet a proven simple relationship yet between concentration of CO_2 and the average temperature of the atmosphere.

Acid rain

A second major problem is the production of oxides which are acidic by the burning of organic fuels such as coals and oils. The major waste product is sulphur dioxide (but see Table 16.13). The effect of the acidity generated by this gas and its oxidation product is not only direct but it also liberates a toxic metal, aluminium, from soils into ground water (see Table 16.12). There could hardly be a closer parallel with the advent of dioxygen some 3 to 1×10^9 years ago. If acidity were increased to, say, pH $= 5$ over a period of 1×10^6 years, there could undoubtedly be evolution taking into account the fact that aluminium would become *available* and more *exchangeable*. First, it will act as a poison, as now, and then it will be converted to a useful element. Do not forget that man or gigantic volcanic activity can cause parallel insults acidity.

The ozone content of the atmosphere

There are two apparent threats related to ozone in the atmosphere: (1) the loss of ozone from the stratosphere, which is largely due to the reactions of chlorine radicals from escaped refrigerant gases, see Table 16.12; (2) the increase in ozone at a low (troposphere) level, due largely to oxidation reactions of some natural products and during the sparking of fuels.

Ozone is an oxidant, stronger than oxygen, and, when subject to the UV radiation of the sun in the presence of water vapour, it produces the hydroxyl radical $OH^{\cdot}$ which is an even stronger oxidant and very fast in its reactions, hence highly dangerous for organic materials (it may even be a mutagen).

The reactions of ozone in the troposphere, eventually through $OH^{\cdot}$, help to generate carbon dioxide (from carbon monoxide and methane), sulphur trioxide (from sulphur dioxide) and nitrogen oxides. Ozone is, therefore, linked to both the greenhouse effect and to the occurrence of acid rain. Furthermore, it oxidises connective tissue, for example, the skin (thus causing 'ageing'), it is a cause of bronchitis and asthma, and may destroy trees and other vegetation.

In contrast, in the stratosphere we are particularly worried with the lack of ozone, that is, with the ozone hole due to the diminishing thickness of the ozone layer, especially, but not only, over the Antarctic and Arctic regions. The reason for concern is obviously different from those above: the ozone layer absorbs in part the strong UV radiation of the sun and so protects the organisms on the Earth from its deleterious effects which may cause, in our case, skin diseases and even cancer. Note that life only expanded into the land from the sea after the ozone layer was formed.

Oxides of nitrogen

Oxides of nitrogen are generated from N_2 during oxidative reactions especially in car engines. Oxidation state diagrams (Fig. 5.14) indicate that dioxygen is capable of oxidising very large amounts of nitrogen or ammonia even to nitrate, but there are kinetic restrictions. Thus, these reactions together with the use of ammonium and nitrate fertilisers increase nitrate in ground water. Nitrate, through the lower oxidation states of nitrogen, is dangerous especially during both embryonic and fetal life stages.

Table 16.16 Production of halogenated compounds*

Product	Production (1000 tonnes.year^{-1})	Source
Methyl chloride (CH_3Cl)	5000–10 000	Seaweeds; burning of vegetation; reaction of methyl iodide with seawater
	26	Industry
	10	Seaweeds
Brominated methanes (CH_3Br; CH_2Br_2; $CHBr_3$)	150	Industry
Methyl iodide (CH_3I)	1000†	Seaweeds
Chlorofluorocarbons (CFCs, freons)	1500	Industry (refrigerators, sprays, solvents, etc.)

* Source: Rogers, J. E. and Whitman, W. B. (ed.) (1991). *Microbial production and consumption of greenhouse gases: methane, nitrogen oxides and halomethanes.* American Society for Microbiology, Washington, DC. See also: *J. geophys. Res.* **85**, 7350 (1980); *J. geophys. Res.* **86**, 7210 (1981); *J. geophys. Res.* **88**, 3684 (1983); *Science* **227**, 1035 (1985).
†Probably higher, but it reacts in part with seawater.

Halogenated compounds (F, Cl and Br)

Man has discovered that halogenated compounds especially chlorinated carbon compounds, are very valuable in sanitation and medicine. Note in the oxidation state diagram (Fig. 5.14) that the halogens and hence C—halogen bonds are only just possible to prepare using dioxygen as an oxidant. There are halogenated compounds made by living systems, mainly in lowly sea plants (see Table 16.16 and references), but much of life and higher animals in particular use very little chlorinated compounds. Today, however, man's production and use of chlorine-containing insecticides, disinfectants, and antibiotics is considerable and fluorine compounds are also used as fungal poisons in agriculture. There are considerable problems with many of these compounds apart from their damaging effect on the ozone layer. Many such compounds may attack DNA so that they can cause cancer as well as kill bacteria. But biological systems can evolve to use organic F, Cl and Br just as they can use organic I. Is 'drug' medicine safe?

Energy and radioactivity

Like any other living system, man is internally a chemical factory no matter what other attributes may be claimed for him. In principle, he could be fed chemically and independently from any other organism since in principle we know his requirements for component chemicals. Generally, however, man uses animals and plants today as the supply of chemicals and treats them as his source of food. To live he also needs energy, which he can get in limited amounts from the same source or from the degradation products of plants, that is, coal, oil and gas. However, if he continues to demand such a high level of protection and also of comfort and pleasure as he does today, he will require much more energy than he can get from such sources. His problem is immense when viewed on a long time scale if he is to maintain present-day attitudes to material wealth. Unfortunately, most organically based fuels will last but a few generations as a source of energy and man must then switch to other sources of which only nuclear power seems viable at present. If there is no choice and nuclear power is used, there are three further consequences:

(1) increase in heat generation;

(2) increase in Earth's radioactivity;

(3) increase in the turnover of elements.

The risks involved with all three are clear. An increase in radioactivity will increase mutation rates. It could be that this assists evolution but not particularly in such species as man; heat will not in itself cause evolutionary problems but will adjust many zones in which living forms are in balance; an increase in the turnover of elements has already been analysed and was shown to affect evolution. We will look at this point again in the next paragraph.

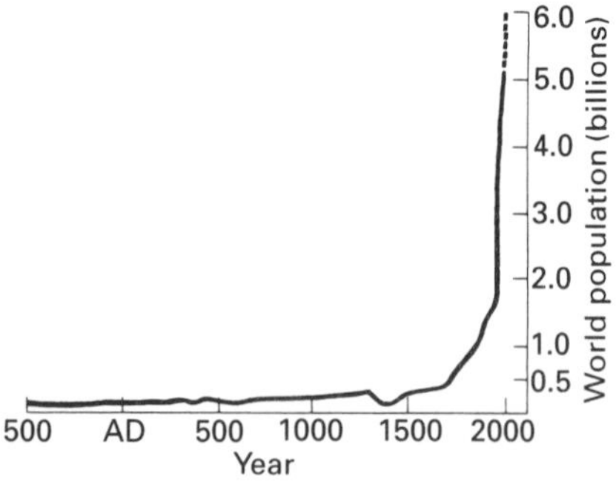

Trace elements and catalysts

Man produces a huge range of elements new to biological systems and some in relatively stable compounds. Many of these elements and compounds are catalysts of chemical change and could then be incorporated (as trace elements) into living systems. Their impact on evolution is unknowable but the release of Al, Cr, Ni and numerous heavy metals such as Cd, Hg and Pb is bound to have some effect, Table 16.17. Some organisms will evolve beneficially while others will suffer. The greatest benefit will come first to the most adaptable—bacteria—and the greatest suffering to the most developed—

Fig. 16.11 The human population. Data and estimates from the US Bureau of Census, the Population Reference Bureau and the United Nations Population Fund were combined to produce this graph showing that the human population began to rise in the eighteenth century and the astonishing population increase in the present century.

Table 16.17 Effects of excesses of non-biological metals on humans and animals

Metal	Effect of excess	Biochemical observations
Al	Implicated in dialysis encephalopathies and in Alzheimer's disease?	Interaction with phosphate? Cross-linking of proteins
Cd	Reduction of effective filtration capacity of glomerulus (renal toxicity)	Blocks sulphydryl groups in enzymes and competes with zinc for its sites. Stimulates metallothionein synthesis and interferes with Cu^{2+}, Zn^{2+} metabolism
Hg	Damage to the central nervous system; neuropsychiatric disorders	CH_3Hg^+ compounds are lipid-soluble
Pb	Saturnism; injuries to the peripheral nervous system; disturbs haem synthesis and affects the kidneys	Pb^{2+} may replace Ca^{2+} with loss of functional (and structural) integrity. Replaces Zn^{2+} in δ-aminolaevulinic acid dehydratase; reacts with sulphydryl groups
Tl^+	Poisonous to nervous systems; enters cells via K^+ channels	Tl^+ is similar to K^+ and binds more tightly, including to N and S ligands

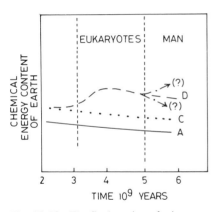

Fig. 16.12 The final version of what can happen due to chemical element evolution. The question marks indicate the possible results of the increase in chemical energy on the Earth's surface no matter what source man develops. A, without life; C, with prokaryote life; D, with eukaryote life. B, absotic organic systems are not included. See Figs 7.47, 10.29 and 14.13.

man. We stress none of the above activities is of necessity deleterious if it is managed properly.

16.14.3 Selfish protection by individual species: population increase

We have constantly stated that a considerable part of evolution concerns the reaction of living systems to the development of poisons and of protection. Man externalises selective poisoning through bactericides, insecticides, pesticides and so on. By removing organisms deleterious to man's purposes he biases the distribution of an ecosystem in three ways not necessarily linked to his long-term survival: (1) the increase of his own population (Fig. 16.11); (2) the decrease of the population of every other organism which is looked upon as a nuisance; (3) the changes of evolution forced on other organisms which adapt. The developing problems under (1) are now much discussed but those under (2) and (3) are also considerable. The deliberate release of new components as poisons could then be a risky development within a stable ecosystem since it forces a greater and greater reliance upon man's own ability to protect himself and even to produce organisms essential for his own lifestyle, namely, domestic species. There must be a recognition that the ecosystem is dependent on very lowly life—bacteria—many of which are essential to all higher life. Man must change his demands or knowingly risk the consequences. He has to become aware of facts such as that 90 per cent of the cells in his own body are from other, mainly unicellular organisms, usually bacteria. It is these organisms that help to ensure an appropriate supply of vital components (Section 13). The spiral of increase in population with increasing control of the whole organisation of an ecosystem of other species is doomed in another way. It is forever increasing the risks of collapse due to an over-complex system (see Fig. 16.9) which is the final version of a figure we have used in several previous chapters (see Figs 7.47, 10.29 and 14.13). Figure 16.12 indicates that man has a choice between extremes of energy use in the manipulation of elements. The choice is open to man since he is a self-conscious animal. The nature of man in this respect brings us to our final section of this book.

16.15 The natural selection of elements and the nature of man

In finishing this book we return to the first quotation of Chapter 1 from Democritus in 420 BC.

The Intellect: Apparently there is colour, apparently sweetness, apparently bitterness, actually there are only atoms and the void.
The Senses: Poor intellect, do you hope to defeat us while from us you borrow your very evidence.

In this dialogue between 'The intellect' and 'The senses' the Greek philosopher touches with far reaching insight a major problem of contemporary neurosciences—the wholeness (oneness) of the mind and the body, which implies an indissoluble link between consciousness, reason, emotions, perceptions and sensations, all related to a single 'reality' for a person as opposed to that 'reality' which science reduces to the properties of the atoms of the elements (and the void. . .).

Democritus' dialogue could have been written today, even if 'the intellect's' argument is now more firmly supported by sophisticated instrument measurements, beyond those possible using the senses directly, and on elaborated mathematical concepts and descriptions. Nevertheless, and although 'the senses' cannot now claim to be the sole source of 'the intellect's' evidence, this does not mean that they can be left to one side; it is now known that the whole of an organism cannot function properly if it is amputated in one of its parts, so that the philosophical mind–body dicotomy is today hardly more than just an academic question.

Today 'the intellect' is able to give a general description of all that is around us (and of ourselves) in an analytical form based on the properties of the chemical elements of the Periodic Table (which have been selected *naturally*). Each object is then a special selected combination of elements where selection has been made at three levels all based fundamentally on atomic properties and hence called 'natural' selection by us. As repeatedly stated the three levels of selection lead to the following types of system:

1. Effectively stationary conditions, of inanimate materials at equilibrium or in kinetic traps, observed largely in Earth's materials but on a much smaller scale in many of man's constructs. They are selected by thermodynamic or kinetic stability and may last for long periods of time though nothing is permanent, of course (Second Law of Thermodynamics).

2. Cycling systems of inanimate materials which are not self-assembling such as the patterns of weather, waterways and drifting continents. Some of man's products fall in this class too and again are on a relatively small scale. They are selected by their ability to absorb energy and remain stable for relatively short periods of time, but the cycling time is very variable.

3. Animate flowing organisations that self-assemble and self-reproduce. The scale of activity in any cycle is today close to that of man's chemical industry. They are selected for their ability to absorb energy and become self-structured systems in which there is flow. The cycling time is very short compared with the change in the systems described under (1) but comparable with some systems under (2).

It is our belief that further intellectual understanding is mainly an unravelling of the complexity of these systems.

A major step in this recognition of what has been uncovered is to see that the division into (1), (2) and (3) above is largely a matter of the time scale of change. For living systems this is a matter of from an hour to a few hundred years. For systems under (2) we might consider that the weather and water flow are on the scale of days and continental flow patterns are on the scale of millions of years. Major changes under (1) as far as Earth is concerned are on the scale of hundreds of millions of years but there is the exception of a new factor—man's industry—which could change chemical selectivity in periods of less than 100 years. The scale is then comparable to that of (3). While living organisms belong in an ecosystem they also belong in an environment. With this knowledge how should man respond? Clearly within the next 100 years he must come to terms with his position in a chemical ecosystem linked to a chemical environment that requires him to look after rather than exploit not just the animate but also the inanimate world for reasons explained in Section 16.13. Man introduces new chemicals and therefore can affect parts of (2) and all of (3) if he is not careful. This requires him to have an understanding of everything around him where the word 'understanding' introduces not just semantic or even philosophical difficulties but chemical problems for us all.

16.15.1 Understanding

We consider that stationary states of materials, whether energised or equilibrated, can be understood supposing that we have a full knowledge of the nature of atoms, although the conceptual problems discussed in Chapter 1 remain. We cannot grasp the significance of the requirement to use wave/particle duality in the mathematical modelling of the atom even though we need to use such an approach. A similar problem arises with the separation of phases in a gravitational field. What is gravity? Notwithstanding these problems, knowledge has given mankind much power over the existing stationary state systems, for example the minerals of the Earth, and has allowed us to construct many new systems—buildings, roads and so on.

There is a different difficulty in the understanding of a flowing system. While we can break down the components of a flowing system into parts, revealing to some degree the functional significance of the natural selection of the elements in such systems, for example in life, these explanations of parts of a flow are intrinsically unsatisfactory even if we accept that all can be reduced analytically to 'understandable' atoms. To explain continuous flow we need to uncover the origin of its structure, the connection to fields causing flow, which are constantly re-energised, and the sources and fates of flowing material. The simple example of a river reveals the difficulty. To explain a river we must explain the structure of the land and how it arose relative to the sea; the nature and source of energy, the sun, and how it interacts with water in the sea to create flow of the atmosphere; and then the nature of clouds, rain and flowing water. Today we see little chance of predicting the weather with any accuracy over long periods and similarly a river is without full explanation, not because its parts cannot be described by atoms but because it demands a complex holistic dynamic description as well as a reductive analytical one.

In a living cellular system this problem is compounded by a vast increase in the number of components (linked to a genetic plan), the forms of energy and materials, the self-synthesis of a developing structure, uptake and rejection of chemicals in a controlled manner and, in the human at least, by self-conscious response to many activities. All are a product of flows of materials and, in a deep sense, of the sources of energy—largely the sun again. Undoubtedly all the parts will be described eventually in terms of natural selection of chemical elements and we have gone some way along this trail in Chapters 10–14. However, we do not consider that this analysis of parts can lead to the understanding of the whole. The whole is a co-operative network of flows of immense complexity.

Legitimately, however, we can claim that some parts of the problems of flow can now be appreciated in far greater detail than 2000 years ago (see Chapter 1 and the revised version of Fig. 1.1, namely, Fig. 16.13). We have been able to disentangle the nature of complexity at the level of the selection of the

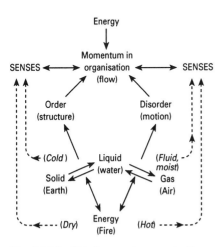

Fig. 16.13 This figure returns us to Fig. 1.1 of the book. It still represents the problem of the interaction of energy and material that is inherent in any evolving system. Our development must not hide the underlying nature of the whole.

chemical elements from that of the perception of what is around us based just on our senses. It can readily be seen that, in many respects, classical thought failed since it relied on our senses only. Thus, in the above quotation, 'the senses' can no longer claim to be the source of the 'intellect's' evidence. However, the case made by 'the intellect' above and sometimes today, is that it can explain sense experience based on 'objective' data provided by machines. Let us turn aside and look at sense experience, since it is sense experience as well as scientific investigation that guides the latest form of flow—man's development—and with it intellectual activity connected to 'understanding'.

16.15.2 The senses

The senses are limited as instruments, which means that they cannot know about very small objects such as atoms nor about very big objects such as nebulae (Fig. 16.14). Our ears and our noses are limited as detectors even as our eyes are. Different animals even have different detection ranges and their senses undoubtedly give a different 'picture' of the world around them. Moreover, the picture for an animal such as man is co-operatively summed from these different sense organs in the brain.

When scientists develop instruments that multiply or divide sizes so as to bring pictures into the range of man's perceptions the images do not bewilder. Thus the microscopes and telescopes used until 1900 to describe the very small and the very large gave pictures that could be called conceivable in that the linguistic constructions used did not change from those used to describe what is seen by the eye. They represented extension only. The description of atoms as very small balls and of chemicals as the coming together of balls sticking together in particular ways is just within the limits of the imaginable in these terms. We have based this book on the need to remain within this conceptual level. Again the forces involved, gravitational and electrical, have remained a mystery from the earliest of times and still are, but everyday experience tells us that they exist. Though baffling, they are accepted. Finally, the general reasoning behind the discussion of gases, liquids and solids and the effect of temperature and pressure upon them is readily acceptable from sense

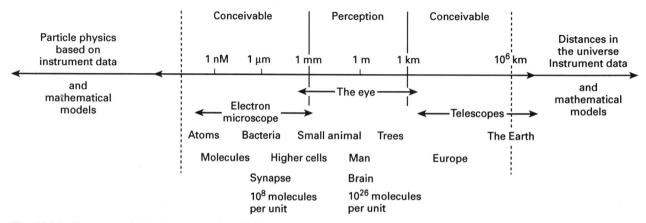

Fig. 16.14 The extent of the electromagnetic radiation spectrum open to man and two extensions of it to the very, very large and very, very small. The regions beyond the dashed lines are not open to conception by the senses. The figure is not to scale.

impressions. The development of science until about 1900, therefore, did not give views of nature that clashed with conceivable (by senses) objects. However, scientific investigation after 1900 has changed science from being based on conceivable objects within the conceptual bounds of the human senses to being based on a set of mathematical equations often with language stolen from perception but misapplied. If there is an observable that scientists state behaves as if it were a wave and a particle at the same time, because an equation including both fits its behaviour, then it is neither a wave nor a particle in common language, that is, for our senses. It is just a mystery. Man's brain is really a device for interpreting and interacting with the *perceived* character of his surrounds not with this mysterious '*scientific*' character. If we are to come to grips with our understanding we must attempt to clarify the functioning of the brain, seeing its relationship to every kind of 'reality'.

16.15.3 The brain

It is through the brain's functions that man has come to interpret the material world in terms of images of atoms. It is only through these images that chemistry can be given an intellectual description. The brain itself can be described in the same chemical language as all other objects. We say that anatomically the brain is a vast series of chemical connections that form a three-dimensional network of current-carrying tubes, neurons. The carriers of current are stored inorganic ions and organic transmitters. Before birth the tubes have only a low level of connectivity automatically formed. They respond to sense experience after birth by growth and division while storing more and more ions and transmitters at junctions (Section 13.8). Thus, *the growth pattern is a memory of an individual's life superimposed on inherited skills*, which are integrated (Fig. 16.15). Thus, the brain responds in no simple way related to sense experience since it conjures up connections confused by primitive animal reactions and rational constructs. One of these rational activities is the ability to formulate from observations the abstractions that we have used throughout the book on the underlying nature of materials in terms of atoms and energy. The value of this description has limitations, which can be appreciated if we now return to Fig. 1.1 of this book and compare it with Fig. 16.13. While we have explained the classical 'elements', earth, water, air and even fire, in the language of atoms and photons, we have not tackled the words dry and moist, hot and cold, sweet and bitter or even words describing colour which the classical world mistakenly used to provide an intellectual connection between material concepts. The reason is simple. This last group of words belongs to our primitive senses and for several reasons cannot be given atomic chemical description. This is not because we believe them to be based on anything other than the natural selection and behaviour of chemical elements, but because the response engendered by the brain is not in the category of intellectual response that can be reductively analysed and is of too great a co-operative complexity (Fig. 16.15). It is a holistic response belonging to the evolution of life and hence is a mixture of the individual genetic make-up and of individual phenotypic construction of the brain after birth due to experience. Thus, response by the human brain cannot be introduced into this book, and as chemists (scientists) we cannot expect to offer a new description of individual experience of dryness, warmth and so on even though we can

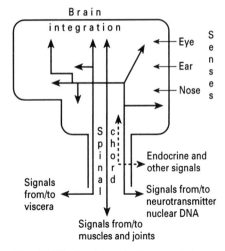

Fig. 16.15 A schematic diagram of the brain and spinal chord in constant exchange condition with the body and the environment (through the senses). The mental and physical circumstances are interpreted by inherited and experience-based connectivities in the brain such that every individual is different. The scientific method used throughout this book depends upon the observing machines all 'seeing' in the same way. (After A. R. Damasio.)

provide a vastly improved and quantitative appreciation of external events concerning fire, air, earth and water.

Insofar as the progress of man is guided by the senses, the scientist must beware of forcing his views, an intellectual conceit, just as arguments concerning science cannot be based on sense conceits. The use of the parallel language of computers, man's machines, for the analysis of brain functions is a mistake and not just one of scale. The stress on 'understanding' the brain in materialistic terms, which characterises such studies, is surely misplaced. A brain's activity is a confusion based on evaluation of impressions which were helpful to survival of an organism long before self-conscious rational thought was added to the organism's armoury. Scientists must not claim that mathematical abstractions, model constructs, can describe 'reality' in an objective sense or can evaluate it using such words as 'beautiful'. It is for this reason that we have opened the chapter with a quotation from Chekhov.

16.16 Future selection of the chemical elements

The conclusion from the analysis in this book is that man has to appreciate the nature of the interdependence of the Earth and all materials in it upon three bases. The first is the natural selection of the chemical elements in all non-living materials. Here relative chemical analysis is supreme and the possibilities of the use of this knowledge for progressive improvement in lifestyle are clear, provided that the risks are avoided. The second derives from the fact that natural selection of the chemical elements has generated a co-operative or holistic living ecosystem which is not open to reductive analysis and certainly not in terms of DNA or any other code alone. It is clear from our discussion that the code either evolved after or with the machinery of the cell and within the context of a chemically specific initial environment. It was forced to write new programmes by environmental change. Man must now take care of both living and non-living ('dead') materials for his own sake (see the Galileo quotation in the opening of this chapter). Here the knowledge and application of chemistry is absolutely essential and must be included within the desire to achieve sustainable development.

The third basis of appreciation, by the senses, must also bias the choice of direction of man's evolution, which raises the problem of values and codes of behaviour that concerns not only the isolated individual but the individual in a society. How are an individual's intelligence and knowledge to be used in an interactive society? Moreover how shall we use our new skills based on new selections of chemical materials in making science-based information systems, equivalent in some ways to a brain but where judgements are on science-based knowledge not on senses? To what purposes are these changes designed? The answer must be, as always, to ensure survival, more comfort, more pleasure, but these words must be within the context of the above two bases and senses of man including the features which are seemingly irrational but have evolved over billions of years.

Putting to one side any new social order or constraint we again refer to the possibility (now already a reality) of a new natural selection of chemical elements, for example based on genetic manipulation as well as on industry but there is the risk of man mistaking his original chemical nature and hence of starting a conflict in which his intellectual designing (including that of DNA) and his innate instincts are on opposite sides. The outcome is hard to predict. Will it be considered a victory or a defeat of the intellect? It is the task of the chemist to supply his background principles and experimental findings for all to see so that a conflict of objectives is avoided. We hope that this book, which is an attempt to set the holistic nature of the world in a materialist manner, may contribute toward the achievement of a sustainable cyclic flow of material and energy in all dead and living systems. In effect this idea of an all-embracing and humanly satisfying cyclic flow, so often expressed as desirable in classical literature, must be the ultimate refinement of the natural selection of the chemical elements.

Further reading

This chapter is in large part a summary of the main conclusions of the discussions in this book, so that specific references to the particular topics are to be found in the appropriate chapters. Some new aspects, however, have been brought in, and the following references provide recommended complementary reading:

1. Blumenfeld, L. A. and Tikhonor, A. N. (1994). *Biophysical thermodynamics of intracellular processes*. Springer-Verlag, New York.
2. Depew, D. J. and Weber, B. H. (1995). *Darwinism evolving systems dynamics and the genealogy of natural selection*. The MIT Press, Cambridge, MA. See references therein.
3. Symonds, N. (1994). Direct mutation—a current perspective. *Journal of Theoretical Biology* **169**, 317–22.

On environmental problems:

4. O'Neill, P. (1993). *Environmental chemistry*, (2nd edn). Chapman and Hall, London.
5. Baird, C. (1994). *Environmental chemistry*. W.H. Freeman, New York.
6. Wayne, R. P. (1993). *Chemistry of atmospheres*, (2nd edn). Oxford University Press, Oxford. Chapters 1, 2, 4, 5.
7. Paul, J. and Pradier, C. M. (ed.) (1994). *Carbon dioxide chemistry : environmental issues*. Royal Society of Chemistry Special Publication no. 153, Cambridge.

Still on environmental problems, at the layman level:

8. Lovelock, J. E.: (a) *Gaia—a new look at life on Earth*. Oxford University Press, Oxford (1979). (b) *The ages of Gaia*. Oxford University Press, Oxford (1988). (c) *Gaia—the practical science of planetary medicine*. Gaia Books Ed., London (1991).
9. Gribbin, J.: (a) *The hole in the sky*. Bantam Press, London (1988). (b) *Hothouse Earth—the greenhouse effect and Gaia*. Bantam Press, London (1990).
10. Gribble, G. W. (1994). Natural organohalogens—many more than you think. *Journal of Chemical Education* **71**, 907–11.

On the relations between the mind, the brain and the understanding of the external world:

11. Damasio, A. R. (1994). *Descartes' error—emotion, reason and the human brain*. G. P. Putnam and Sons, New York.

12. *Journal of Theoretical Biology* (1994). Special issue on mind and matter, **171**, 1–122.
13. *Scientific American* (1992). Special issue devoted to mind and brain (September issue).

Some more general references are included which give overall views of evolution in the universe or on Earth, most at the layman level but some (references 17, 18, 23, 24, 25) at more advanced level

14. Allègre, C. (1993). *Introduction à une Histoire Naturelle—du Big-Bang à la Disparition de l'Homme*. Librairie Arthème-Fayard, Paris.
15. Gribbin, J. (1994). *In the beginning—the birth of the living universe*. Penguin Books, London.
16. Layzer, D. (1990). *Cosmogenesis—the growth of order in the universe*. Oxford University Press, Oxford.
17. Mason, S. F. (1992). *Chemical evolution*. Oxford University Press, Oxford.
18. Barrow, J. and Tipler, F. (1986). *The anthropic cosmological principle*. Oxford University Press, Oxford.
19. *Scientific American* (1994). Special issue devoted to life in the universe (October issue).
20. Day, W. (1981). *Genesis on Planet Earth*. Shiva Publishing Limited, Nantwich (Ches.).
21. Dawkins, R. (1986). *The blind watchmaker*. Longman, London.
22. Maynard Smith, J. and Szathmáry, E. (1995). *The major transitions in evolution*. W.H. Freeman/Spektrum, New York.
23. Burns, T. T. (1994). Fittedness of organisms. *Journal of Theoretical Biology* **170**, 115–27. See references therein.
24. Weber, B. H., Depew, D. J. and Smith, J. D. (1988). *Entropy, information and evolution*. MIT Press, Cambridge, MA. A series of interesting papers on the dynamics of biological order and evolution.
25. Kauffman, S. (1993). *The origins of order*. Oxford University Press, New York. An original view, but a difficult book for non-mathematically inclined readers.
26. Blum, H. F. (1968). *Time's arrow and evolution* (3rd edn). Princeton University Press, Princeton, New Jersey. An approach to the problems of order, entropy and evolution.

Index

In this index detailed headings are given only for the twenty or so chemical elements which have made a particularly strong impact on the development of geological and/or biological systems. Examples are iron, copper, phosphorus and carbon. To find the involvement of other chemical elements in the index the reader should look under general headings of properties and uses such as abundance, availability, catalysts, and element functions.